# Introduction to Remote Sensing

**SECOND EDITION**

# Introduction to Remote Sensing

## SECOND EDITION

JAMES B. CAMPBELL
Virginia Polytechnic Institute
and State University

THE GUILFORD PRESS
New York · London

A Division of Guilford Publications, Inc.
72 Spring Street, New York, NY 10012



Printed in the United States of America

This book is printed on acid-free paper.

Last digit is print number: 9 8 7 6 5 4 3 2

**Library of Congress Cataloging-in-Publication Data**

Campbell, James B., 1944–
Introduction to remote sensing / James B. Campbell. — 2nd ed.
p. cm.
Incluces bibliographical references and index.
ISBN 1-57230-041-8.
1. Remote sensing. I. Title.
G70.4.C23 1996
621.36′78—dc20 96-15196
CIP

Cover photo courtesy of NASA (STS-40-614-047).

*To the memory of my parents,*

**JAMES BATCHELDER CAMPBELL**
**1911–1982**

**and**

**DOROTHY ANN TISON CAMPBELL**
**1917–1987**

# Preface

This volume presents an introductory course of study in remote sensing, designed to be comprehensive within constraints imposed by the length of the volume and the backgrounds of students.

Readers are already aware, if only in a general way, of the dramatic advances in instrumentation and analysis that our field has experienced since the first edition of *Introduction to Remote Sensing* was published. These technical achievements have expanded both commercial and scientific applications of remote sensing and geographic information systems (GISs). Both kinds of developments elevate the need for increased spatial literacy; this book attempts to meet this need for both specialists and generalists alike.

Compared to the first edition, this edition introduces digital imagery earlier in the text and assigns it greater significance by devoting a separate chapter to this topic. New material on RADARSAT, CORONA, the Internet, the World Wide Web, and other subjects has been included in appropriate chapters. New material on hyperspectral remote sensing and global remote sensing is presented in separate chapters, in part to emphasize the growing significance of these subjects.

Because this textbook is intended for use in a variety of curricula, it is not tailored to any single academic discipline. The field of remote sensing itself illustrates the difficulty of defining clear boundaries between disciplines, so this volume seeks to encourage students to explore subjects outside their usual curricula. To this end, the text attempts to integrate knowledge from the many fields that contribute to remote sensing and to avoid preoccupation with any single perspective.

Instructors will want to supplement the content of this volume with material of special significance in their own programs. Supplementary materials will of course vary greatly from one institution to the next, depending on access to facilities and equipment as well as the varying expectations and interests of instructors, students, and curricula.

It is assumed that the text will be used as the basis for readings and lectures, and that most courses will include at least brief laboratory exercises that permit students to examine many more images than can be presented here. Because access to specific equipment and software varies so greatly, and because of the great variation in emphasis noted above, this book does not include laboratory exercises. Each chapter concludes with a set of review questions and problems that can assist in review and assessment of concepts and material.

For students who intend to specialize in remote sensing, this text forms not only an introduction but also a framework for subjects to be studied in greater detail. Students who do intend such specialization should of course consult their instructors to plan a comprehensive course of study based on work in several disciplines as discussed on pages 16–19. This philosophy is reflected in the text itself—it introduces the student to each of the main topics of significance for remote sensing but recognizes that students will require additional depth in their chosen fields of study.

For those students who do not intend to pursue remote sensing beyond the introductory level, this book serves as an overview and introduction, so that they can understand remote sensing, its applications in varied disciplines, and its significance in today's world. For many, the primary emphasis may focus on study of those chapters and methods of greatest significance in the student's major field of study. Thus, a student who will not specialize in remote sensing should be well informed of the significance of remote sensing and its applications specific to his or her major field of study.

Brief essays included at the end of some chapters form specific examples of principles discussed in the main body of the text. These capsule summaries often pertain to several chapters but follow the chapters that seem most directly related.

To permit instructors to alter assignments to meet requirements of a specific course, material is organized at several levels. At the broadest level, rough division into four units offers a progression in the kinds of knowledge presented, with only a few concessions to practicality (such as placing the "Image Interpretation" chapter under "Image Acquisition" rather than in its logical place in "Analysis"). Each division consists of two or more chapters:

*Foundations*

1. History and Scope of Remote Sensing
2. Electromagnetic Radiation

*Image Acquisition*

3. Photographic Sensors
4. Digital Data
5. Image Interpretation
6. Land Observation Satellites
7. Active Microwave
8. Thermal Radiation
9. Image Resolution

*Analysis*

10. Preprocessing
11. Image Classification
12. Field Data
13. Accuracy Assessment
14. Hyperspectral Remote Sensing

*Applications*

15. Geographic Information Systems
16. Plant Sciences
17. Earth Sciences
18. Hydrospheric Sciences
19. Land Use and Land Cover
20. Global Remote Sensing

These 20 chapters each constitute more or less independent units that can be selected as necessary to meet specific needs of each instructor. Numbered sections within chapters form even smaller units that can be selected and combined with other material as desired by the instructor.

In a field in which important changes occur so rapidly, it is impossible to provide a text that can be current for more than a short time. Nonetheless, current innovations are likely to lead to broader developments that have been outlined, at least in a rudimentary fashion, in this edition. Therefore, this text can provide a base from which students can assess future developments that may not be explicitly covered here.

I gratefully acknowledge the contributions of those who assisted in identifying and acquiring images used in this book. Individuals and organizations in both private industry and governmental agencies have been generous with advice and support. Daedalus Enterprises Incorporated, EROS Data Center, Environmental Research Institute of Michigan, EOSAT, GeoSpectra Corporation, North Pacific Aerial Surveys, and SPOT Image Corporation are among the organizations that assisted with the search for suitable images. For the second edition, I am grateful for the continued support of these organizations and for the assistance of the U.S. Geological Survey, RADARSAT International, Earth Satellite Corporation, and the Jet Propulsion Laboratory.

Much of what is good about this book is the result of the assistance of colleagues in many disciplines in universities, corporations, and research institutions who have contributed through their correspondence, criticisms, explanations, and discussions. Students in my classes have, through their questions, mistakes, and discussions, contributed greatly to my own learning and, therefore, to this volume. Faculty who use this text at other universities have provided suggestions and responded to questionnaires designed by The Guilford Press.

At Guilford, Janet Crane nurtured my work on the first edition, then Peter Wissoker guided preparation of this second edition; I am especially grateful for the efforts of Judith Grauman, Editorial Supervisor, who magically transformed, under deadline, manuscripts for both editions into page proofs and then bound volumes. At Virginia Tech, Jane Price and Vanessa Scott typed the many drafts of the manuscript for the first edition; Sharon C. Chiang, Caren Ertmann Gallimore, and Gerrit Mellen assisted with drafting of illustrations for the first edition, and Buela Prestrude assisted with preparation of illustrations for the second edition. Many others, including Russ Congalton, Mike Story, Gordon Grender, L. W. Carstensen, F. M. Henderson, L. Grossman, R. W. Ehrich, Liz Garland, and Margaret Mayers, were generous in their help in reading, checking, and editing the manuscript, and in providing comments and suggestions.

In addition, for the second edition, I acknowledge suggestions and criticisms of Laurence W. Carstensen, Don Light, David Pitt, Jim Merchant, Margaret Mayers, and the detailed comments of two anonymous reviewers. All too often, constraints of time or length prevented implementation of their thoughtful and well-informed suggestions.

Users of this text can inform the author of errors, suggestions, and other comments at

jayhawk@vt.edu

Corrections, updates, revisions, and supplementary materials will be available at

http://ptolemy.geog.vt.edu/info/

This will bring you to my department's home page; look for my name under the faculty listing, then for the title of this book.

JAMES B. CAMPBELL
*Blacksburg, Virginia*

# List of Tables

# List of Figures and Plates

# Contents

# Introduction to Remote Sensing

**SECOND EDITION**

CHAPTER ONE

# History and Scope of Remote Sensing

## 1.1. Introduction

A picture is worth a thousand words. Have you ever wondered why?

Pictures concisely convey information about positions, sizes, and interrelationships between objects—by their nature, they portray spatial information that we can recognize as objects. These objects in turn tell a story that can convey a different kind of meaning (Figure 1.1). Human beings are good at deriving information from such images, so we experience little difficulty in interpreting even scenes that are visually complex because of our innate visual and mental abilities. Humans are so competent in such tasks that it is only when we attempt to replicate these capabilities using computer programs that we realize how powerful are our abilities to derive information from visually complex scenes. Each picture therefore can truthfully be said to distill the meaning of thousands of words.

This textbook is devoted to the analysis of pictures that employ an overhead perspective (e.g., maps, aerial photographs, and similar images), including many that are based on radiation not visible to the human eye. These images form especially effective pictures that permit us to apply their power to the study of the earth's surface—to see patterns instead of isolated points and relationships between different distributions. Such pictures are especially powerful because they permit us to see differences over time; to measure sizes, areas, depths, and heights; and, in general, to acquire information that is difficult to acquire by other means. However, our ability to use these images is not innate—we must work hard to develop the knowledge and skills that allow us to use such images as those shown in Figure 1.1.

Specialized knowledge is important because remotely sensed images have qualities that differ from those we encounter in everyday experience:

- Image presentation
- Unfamiliar scales and resolutions
- Overhead views from aircraft or satellites
- Use of several regions of the electromagnetic spectrum

**FIGURE 1.1.** Two examples of visual interpretation of images. Humans have innate ability to derive meaning from the complex patterns of light and dark that form this image—we can interpret patterns of light and dark as people, and objects. At another, higher level of understanding, we learn to derive meaning beyond mere recognition of objects—to interpret the arrangement of figures and subtle differences in posture, and to assign meaning not present in the arbitrary pattern of light and dark. Thus, this picture tells a story—it conveys a meaning that can be received only by observers who can understand the significance of the figures, the statue, and their relationship.

**FIGURE 1.1 (cont.).** So it is also with this second image, a satellite image of southwestern Virginia. With only modest effort and experience, we can interpret these patterns of light and dark, to recognize topography, drainage, rivers, and vegetation. There is a deeper meaning here as well, as the pattern of white tones tells a story about the interrelated human and natural patterns within this landscape—a story that can be understood by those prepared with the necessary knowledge and perspective. Because this image employs an unfamiliar perspective, and is derived from radiation outside the visible portion of the electromagnetic spectrum, our everyday experience and intuition are not adequate to interpret the meaning of the patterns recorded here, and it is necessary to consciously learn and apply acquired knowledge to understand the meaning of this pattern.

This text explores other elements of remote sensing and some of its many applications. Our purpose in Chapter 1 is to briefly outline its content, origins, and scope as a foundation for the more specific chapters that follow.

## 1.2. Definitions

The field of remote sensing has been defined many times (Table 1.1). By examining the common elements of these varied definitions, it may be possible to identify some of its most important themes. It is easy to identify a central concept—the gathering of information at a distance. This broad definition, however, must be refined. The kind of remote sensing discussed here is devoted to observation of the earth's land and water surfaces by

**TABLE 1.1. Remote Sensing: Some Definitions**

Remote sensing has been variously defined but basically it is the art or science of telling something about an object without touching it. (Fischer et al., 1976, p. 34)

Remote sensing is the acquisition of physical data of an object without touch or contact. (Lintz & Simonett, 1976, p. 1)

. . . imagery is acquired with a sensor other than (or in addition to) a conventional camera through which a scene is recorded, such as by electronic scanning, using radiations outside the normal visual range of the film and camera—microwave, radar, thermal, infrared, ultraviolet, as well as multispectral, special techniques are applied to process and interpret remote sensing imagery for the purpose of producing conventional maps, thematic maps, resources surveys, etc., in the fields of agriculture, archaeology, forestry, geography, geology, and others. (American Society of Photogrammetry)

Remote sensing is the observation of a target by a device separated from it by some distance. (Barrett & Curtis, 1976, p. 3)

The term "remote sensing" in its broadest sense merely means "reconnaissance at a distance." (Colwell, 1966, p. 71)

Remote sensing, though not precisely defined, includes all methods of obtaining pictures or other forms of electromagnetic records of the Earth's surface from a distance, and the treatment and processing of the picture data. . . . Remote sensing then in the widest sense is concerned with detecting and recording electromagnetic radiation from the target areas in the field of view of the sensor instrument. This radiation may have originated directly from separate components of the target area; it may be solar energy reflected from them; or it may be reflections of energy transmitted to the target area from the sensor itself. (White, 1977, pp. 1–2)

"Remote sensing" is the term currently used by a number of scientists for the study of remote objects (earth, lunar, and planetary surfaces and atmospheres, stellar and galactic phenomena, etc.) from great distances. Broadly defined . . . , remote sensing denotes the joint effects of employing modern sensors, data-processing equipment, information theory and processing methodology, communications theory and devices, space and airborne vehicles, and large-systems theory and practice for the purposes of carrying out aerial or space surveys of the earth's surface. (National Academy of Sciences, 1970, p. 1)

Remote sensing is the science of deriving information about an object from measurements made at a distance from the object, i.e., without actually coming in contact with it. The quantity most frequently measured in present-day remote sensing systems is the electromagnetic energy emanating from objects of interest, and although there are other possibilities (e.g., seismic waves, sonic waves, and gravitational force), our attention . . . is focused upon systems which measure electromagnetic energy. (D. A. Landgrebe, in Swain & Davis, 1978, p. 1) means of reflected or emitted electromagnetic energy. This refined definition excludes applications that could be reasonably included in such broader definitions as sensing the earth's magnetic field, cloud patterns, or the temperature of the human body. Also, this text discusses remote sensing instruments that present image information in an image format (similar to Figure 1.1), so we must exclude instruments (e.g., certain lasers) that collect data at a distance but do not form images. Such excluded applications can, of course, be considered "remote sensing" in its broad meaning, but they are omitted here as a matter of convenience. For our purposes, the definition can be based on modification of concepts given in Table 1.1:

> *Remote sensing is the practice of deriving information about the earth's land and water surfaces using images acquired from an overhead perspective, using electromagnetic radiation in one or more regions of the electromagnetic spectrum, reflected or emitted from the earth's surface.*

This definition serves as a concise expression of the scope of this volume. It is not, however, universally applicable, and is not intended to be so, as practical constraints limit the scope of this volume. So, although this text must omit many interesting topics (e.g., meteorological or extraterrestrial remote sensing), it can preview knowledge and perspectives necessary for pursuit of topics that cannot be covered here.

## 1.3. Milestones in the History of Remote Sensing

The scope of the field of remote sensing can be elaborated by examining its history to trace the development of some of its central concepts. A few key events trace the evolution of the field (Table 1.2). More complete accounts are given by Fisher (1975), Stone (1974), Simonett (1983), and others.

An essential element in remote sensing is the ability to record an image of the earth's surface, so the starting point in its history is the beginning of the practice of photography. The first attempts to form images by photography date from the early 1800s, when a number of scientists, now largely forgotten, conducted experiments with photographic chemicals. In 1839, Louis Daguerre (1789–1851) publicly reported results of his experiments with photographic chemicals; this date forms a convenient, although arbitrary, milestone for the beginning of photography.

The use of photography to record an aerial view of the earth's surface (from a captive balloon) dates from 1858. In succeeding years, numerous improvements were made in photographic technology and in methods of acquiring photographs of the earth from balloons and kites. These aerial images of the earth are among the first to fit the definition of remote sensing given previously, but most must be regarded as curiosities rather than the basis for a systematic field of study.

The next milestone is the use of powered airplanes as platforms for aerial photography. In 1910 Wilbur Wright piloted the plane that acquired motion pictures of the Italian landscape near Centocelli, said to be the first aerial photographs taken from an airplane. The maneuverability of the airplane provided the capability for the control of speed, altitude, and direction required for systematic use of the airborne camera.

World War I (1914–1918) marked the beginning of the acquisition of aerial photogra-

**TABLE 1.2. Milestones in the History of Remote Sensing**

| | |
|---|---|
| 1800 | Discovery of infrared by Sir William Herschel |
| 1839 | Beginning of practice of photography |
| 1847 | Infrared spectrum shown by A. H. L. Fizeau and J. B. L. Foucault to share properties with visible light |
| 1850–1860 | Photography from balloons |
| 1873 | Theory of electromagnetic energy developed by James Clerk Maxwell |
| 1909 | Photography from airplanes |
| 1910–1920 | World War I: aerial reconnaissance |
| 1920–1930 | Developmental and initial applications of aerial photography and photogrammetry |
| 1930–1940 | Development of radar in Germany, United States, and United Kingdom |
| 1940–1950 | World War II: applications of nonvisible portions of electromagnetic spectrum; training of persons in acquisition and interpretation of airphotos |
| 1950–1960 | Military research and development |
| 1956 | Colwell's research on disease detection with infrared photography |
| 1960–1970 | First use of term "remote sensing" |
| | TIROS weather satellite |
| | Skylab remote sensing observations from space |
| 1972 | Launch of Landsat 1 |
| 1970–1980 | Rapid advances in digital image processing |
| 1980–1990 | Landsat 4: new generation of Landsat sensors |
| 1986 | (SPOT) French Earth Observation Satellite |
| 1980s | Development of hyperspectral sensors |

phy on a routine basis. During the war, equipment was designed specifically for aerial photography, and many people were trained in the process of acquiring, processing, and interpreting aerial photos. Although the practice of aerial photography then was primitive by later standards, many of these individuals became pioneers in the application of aerial photography in civilian endeavors.

Numerous improvements followed from these beginnings. Camera designs were improved and tailored specifically for use in aircraft. The science of *photogrammetry*—the practice of making accurate measurements from photographs—was applied to aerial photography, with the development of instruments specifically designed for analysis of aerial photos. Although the fundamentals of photogrammetry were defined much earlier, the field developed toward its modern form in the 1920s, with the application of accurate photogrammetric instruments. From these origins, another landmark was established—the more or less routine application of aerial photography in government programs, initially for topographic mapping but later for soil survey, geologic mapping, forest surveys, and agricultural statistics.

During this period, the well-illustrated volume by Lee (1922), *The Face of the Earth as Seen from the Air,* surveyed the range of possible applications of aerial photography in a variety of disciplines from the perspective of those early days. Although the applications that Lee envisioned were achieved at a slow pace, the expression of governmental interest ensured a continuity in the scientific development of the acquisition and analysis of aerial photography, increased the number of photographs available, and trained many people in uses of aerial photography. The acceptance of the use of aerial photography in many scientific endeavors occurred quite slowly because of resistance among traditionalists, imperfections in equipment and technique, and genuine uncertainties regarding the proper role of aerial photography in scientific inquiry and practical applications.

These developments led to the eve of World War II (1939–1945), which forms the next milestone in our history. During the war years, use of the electromagnetic spectrum was extended from almost exclusive emphasis on the visible spectrum to other regions, most notably the infrared and microwave regions (far beyond the range of human vision). Knowledge of these regions of the spectrum had been developed in both basic and applied sciences during the preceding 150 years (Table 1.2). However, during the war years, application and further development of this knowledge accelerated, as did dissemination of the means to apply it. The potential of the nonvisible spectrum had been previously known to research scientists, although the equipment, materials, and experience necessary to apply it to practical problems were not at hand. Wartime research and operational experience provided the theoretical and practical knowledge required for everyday use of the nonvisible spectrum in remote sensing.

Furthermore, the training and experience of the large numbers of pilots, camera operators, and photointerpreters provided a large pool of experienced personnel who were able to transfer their skills and experience to civilian occupations after the war. Many of these people assumed leadership positions in the efforts of business, scientific, and governmental programs to apply aerial photography and remote sensing to a broad range of problems.

The postwar era saw the continuation of trends set in motion by wartime research. On one hand, established capabilities found their way into civilian applications. At the same time, the beginnings of the Cold War between the Western democracies and the Soviet Union created the environment for further development of reconnaissance techniques, which were often held as defense secrets for many years until they were replaced by more sophisticated methods.

Among the most significant developments in the civilian sphere was the work of Robert Colwell, published in 1956, applying the color infrared film (popularly known as camouflage detection film) to problems of identifying small-grain cereal crops and their diseases. His work applied color infrared aerial photography, first used widely during World War II, to important problems in the plant sciences. Although many of the basic principles of his research were established earlier, his systematic investigation of their practical dimensions forms a clear milestone in the development of the field of remote sensing. Even at this early date, Colwell delineated the outlines of modern remote sensing and anticipated many of the opportunities and difficulties of this field of inquiry.

The 1960s saw a series of important developments occur in rapid sequence. The first meteorological satellite (TIROS-1) was launched in April 1960. This satellite was designed for climatological and meteorological observations but provided the basis for later development of land observation satellites. During this period, some of the remote sensing instruments originally developed for military reconnaissance, and classified as defense secrets, were released for civilian use as more advanced designs became available for military application. These instruments extended the reach of aerial observation outside the visible spectrum into the infrared and microwave regions.

It was in this context that the name "remote sensing" was first used. Evelyn Pruit, a scientist working for the U.S. Navy's Office of Naval Research, coined this term when she recognized that the term "aerial photography" no longer accurately described the many forms of imagery collected using radiation outside the visible region of the spectrum. Early in the 1960s, the U.S. National Aeronautics and Space Administration (NASA) established a research program in remote sensing—a program that, during the next decade, was to support remote sensing research at institutions through the United States. During this same period, a committee of the United States National Academy of Sciences (NAS) studied opportunities for application of remote sensing in the field of agriculture and forestry. In 1970, the NAS reported the results of its work in a document that outlined many of the opportunities offered by this emerging field of inquiry.

In 1972, the launch of Landsat 1, the first of many earth-orbiting satellites designed for observation of the earth's land areas, formed another milestone. Landsat provided, for the first time, systematic, repetitive observation of the earth's land areas. Each Landsat image depicted large areas of the earth's surface in several regions of the electromagnetic spectrum yet provided modest levels of detail sufficient for practical applications in many fields. Landsat's full significance may not yet be fully appreciated, but it is possible to recognize two of its most important contributions. First, the routine availability of multispectral data of large regions of the earth's surface greatly expanded the number of people who acquired experience and interest in analysis of multispectral data. Multispectral data had been acquired previously but were largely confined to specialized research laboratories. Landsat's data greatly expanded the population of scientists with interest in multispectral analysis.

Landsat's second contribution was the rapid and broad expansion of uses of digital analyses. Before Landsat, most analyses were completed visually by examining prints and transparencies of aerial images. Analyses of digital images by computer were possible mainly in specialized research institutions; personal computers, and the variety of image analysis programs that we now regard as commonplace, did not exist. Although Landsat data were primarily available as prints or transparencies, they were also provided in digital form. The routine availability of digital data in a standard format created the environment that increased the popularity of digital analysis and set the stage for the applications of image analysis software for small computers that are now commonplace. Landsat also formed a model for other land observation satellites designed and operated by organizations in many nations.

In the 1980s, scientists at the Jet Propulsion Laboratory (Pasadena, California) began, with NASA support, to develop instruments that could create images of the earth at unprecedented levels of detail. These instruments created the field of *hyperspectral remote sensing,* which, in the 1990s, is still developing as a field of inquiry. Hyperspectral remote sensing will form the basis for a more thorough understanding of how to best apply more conventional remote sensing capabilities.

## 1.4. Overview of the Remote Sensing Process

Because remotely sensed images are formed by many interrelated processes, an isolated focus on any single component produces a fragmented picture. Therefore, our initial view of the field can benefit from a broad perspective that identifies the kinds of knowledge required for the practice of remote sensing (Figure 1.2).

Consider first the *physical features,* including buildings, vegetation, soil, water, and the like. These are the features that applications scientists wish to continue. Knowledge of the physical features resides within such specific disciplines as geology, forestry, soil science, geography, and urban planning.

*Sensor data* are formed as an instrument (e.g., a camera or radar) views the physical features by recording electromagnetic radiation emitted or reflected from the landscape. Chapters 3–9 are devoted to a description of the acquisition of image data using cameras and other sensors. Although the image domain can consist of pictorial images familiar to us all (Figure 1.1), often images are most useful in their digital forms, which present information as numerical arrays that can be displayed and analyzed by computers. For many of us, sensor data often seem to be abstract and foreign because of their unfamiliar overhead perspective, unusual resolutions, and the use of spectral regions outside the visible spectrum. As a result, the effective use of sensor data requires analysis and interpretation to convert data to information for addressing such practical problems as siting landfills or searching for mineral deposits. Chapters 10–14 describe methods used to interpret remotely sensed data.

These interpretations create *extracted information,* which consists of transformations of sensor data designed to reveal specific kinds of information (Figure 1.2). Actually, a more realistic view (Figure 1.3) demonstrates that the same sensor data can be examined from alternative perspectives to yield different interpretations. Therefore, a single image can be interpreted to provide information about, for example, soils, land use, or hydrography, depending on the specific image and the purpose of the analysis. In this text, Chapters 10–14 are devoted to image analysis broadly defined—the extraction of information, specific to particular disciplines, from raw remotely sensed data.

Finally, we can proceed to the *applications,* in which the analyzed remote sensing data can be combined with other data to address a specific practical problem, such as land use planning, mineral exploration, or water quality mapping (Chapters 15–20). When digital remote sensing data can be combined with other digital data, applications can be implemented in the context of a *geographic information system* (GIS), designed to bring varied

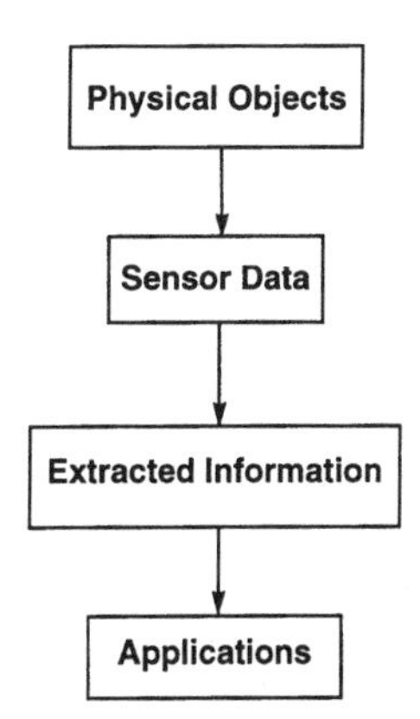


**FIGURE 1.2.** Schematic overview of knowledge used in remote sensing.

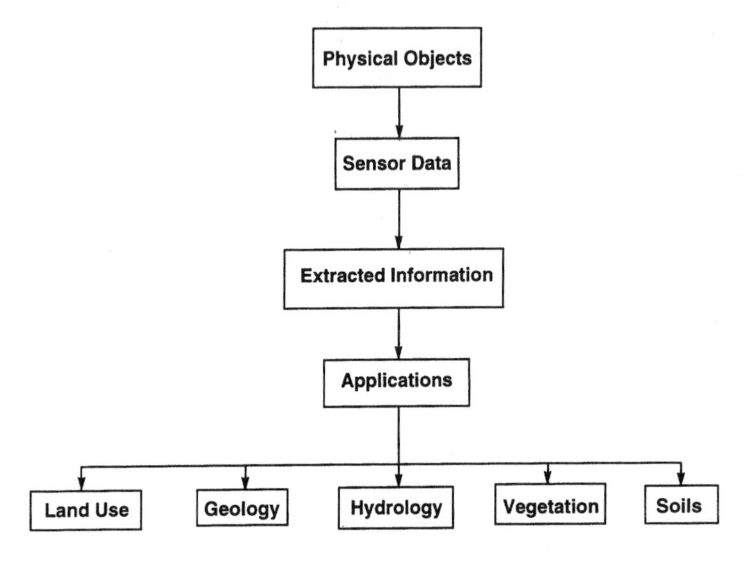


**FIGURE 1.3.** Expanded view of the process outlined in Figure 1.2.

data together in a format that allows efficient statistical and geographic analysis. For example, remote sensing data may provide accurate land use information that can be combined with soils, geologic, transportation, and other information to guide the siting of a new landfill. This process is largely beyond the scope of this text, although Chapter 15 introduces readers to GIS, outlining some of the ways that remotely sensed data can be used in GIS.

## 1.5. A Specific Example

Consider how a remotely sensed image is acquired, by tracing the path of energy used to make an image. A cursory examination of an aerial image of a deciduous forest (Figure 1.4) introduces some dimensions of the field of remote sensing. (All the concepts mentioned here are discussed in detail in subsequent chapters.)

### *Physical Features*

Energy that reaches the earth from the sun is composed of many kinds of radiation, including that in the visible region of the spectrum (including blue, green, and red light), as well as other radiation (such as the infrared) outside the range of human vision. The solar beam must pass through the earth's atmosphere to reach the forest canopy; some of this energy is absorbed and scattered before it reaches the forest. Much scattering is selective by wavelength—for example, the atmosphere scatters blue light much more than it scatters green or red light. Of the remaining energy, some will reach the leaves of the tree

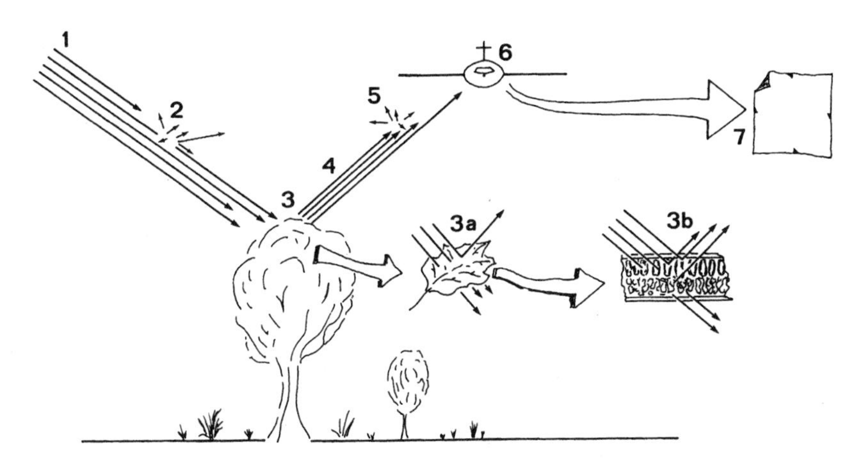


**FIGURE 1.4.** Idealized overview of acquisition of a remotely sensed image. This diagram depicts a schematic view of some of the elements that must be considered in the acquisition and interpretation of a remote sensing image, as a means of illustrating the kinds of knowledge that constitute the field of remote sensing. In this example, the image is derived from reflected solar radiation. Incoming solar radiation (1) is in part scattered or attenuated by the earth's atmosphere (2). The remaining energy reaches the earth's surface, to interact with the features present on the landscape—in this example, a living tree (3). Much of the reflection of energy from the tree canopy is controlled by interactions with individual leaves (3a), which selectively absorb, transmit, and reflect energy depending on its wavelength (3b), as discussed in the text, and in later chapters. The reflected energy (4) is again subject to atmospheric attenuation (5) before it is recorded by an airborne (6) or satellite sensor (6). The results of a multitude of such interactions are recorded as a photograph-like image (7), or as an array of quantitative values. The image is then available for interpretation and analysis to derive information concerning the landscape, based on the interpreter's examination of the image and knowledge of items 1 through 7.

canopy; of energy reaching the leaves, much of the infrared radiation (wavelengths just longer than those of visible radiation) will be reflected. The green region of the spectrum will be reflected by a different portion of the leaf, whereas blue and red radiation will be absorbed for use in photosynthesis. Energy that reaches the camera lens must *again* pass through the atmosphere, where it is again subject to attenuation.

### *Sensor Data*

Thus, energy recorded by the camera lens is much different from the sunlight that entered the atmosphere. Some of the blue light was scattered by the atmosphere and reaches the camera without being reflected from the earth's surface. Blue, red, green, and infrared have been reflected from the canopy but reach the camera in proportions that differ from the proportions that were intercepted by the canopy. The film portrays these different kinds of radiation in different colors, which may not (as explained in Chapter 3) match those of the radiation they represent. This image represents a portion of the earth's surface, but its meaning is hidden from casual observation.

### *Extracted Information*

Data within the image can be translated into information only through the process of *image interpretation,* or *image analysis.* An *image analyst,* skilled in the examination of images, studies the image to identify, for example, timber types, and areas damaged by disease and insect activity. For some, much of this knowledge is based on everyday experience rather than formally structured learning. We do not need to "learn" that deciduous trees in temperate climates lose their leaves in the winter, or that wheat tends to be planted in fields of regular size and shape. This store of *implicit* knowledge is important because it forms such a large portion of what we know about a scene, and because we often are unaware of its role in image analysis. More *formal, explicit,* knowledge of the landscape is especially significant in two contexts. First, we have no means of acquiring an implicit understanding of how radiation behaves outside the visible spectrum. So, knowledge that living vegetation is highly reflective in the near infrared portion of the spectrum must be acquired in a formalized learning situation. Second, our implicit knowledge has its geographic limits; when we travel outside the range of our experience, it is no longer valid in our assessment of the landscape. For example, there are large regions of the earth where crops are not customarily planted in regular fields and where trees do not all shed their leaves during the same season. Therefore, analysts often must consciously acquire information necessary to analyze images acquired in areas outside the range of their direct experience.

### *Applications*

Applications of remotely sensed data to practical problems usually require links to other kinds of information, including for example, topography, political boundaries, pedologic, geologic, or hydrologic data. In recent years, such links have increasingly been made within the framework of geographic information systems. GISs are devoted to the analysis of spatially distributed data, and especially to analysis of the interrelationships among different kinds of data, matched to each other within a specific geographic region. Although there is no sharp line between remote sensing and GISs, it is only partially incorrect to state that remote sensing is primarily a means of collecting data, and that GISs are primarily a means of analyzing and storing data.

At local levels of government, remote sensing imagery records large-scale representations of topography and drainage and the basic infrastructure of highways, buildings, and utilities. At county and regional levels of detail, remotely sensed data provide a basis for outlining broad-scale patterns of development, to coordinate the relationships among transportation; residential, industrial, and recreational land uses; siting of landfills; and planning of the future development. State governments require information from remotely sensed data for broad-scale inventories of natural resources and monitoring environmental issues, including land reclamation, water quality, and planning economic development.

At broader levels of examination, national governments apply the methods of remote sensing and image analysis for environmental monitoring (both domestically and internationally); for management of federal lands, crop forecasting, and disaster relief; and to support activities to further national security and international relations. Remote sensing is also used to support activities of international scope, including analysis of broad-scale environmental issues, international development, disaster relief, aid for refugees, and investigation of environmental issues of global scope.

These activities are conducted or supported by industrial and commercial enterprises that design and manufacture equipment, instruments, and materials. Others operate systems to collect and distribute data to customers. And still others provide services to plan for effective acquisition of remotely sensed data and to analyze and interpret data to provide information for customers. Other industrial and commercial organizations conduct basic research to develop and refine instruments and analyses.

Within subject area disciplines, remote sensing imagery is almost always combined with other kinds of data. Geologists and geophysicists use remotely sensed images to study lithologies, structures, surface processes, and geologic hazards. Hydrologists examine images that show land cover patterns, soil moisture status, drainage systems, sediment content of lake and rivers, ocean currents, and other characteristics of water bodies. Geographers and planners examine imagery to study settlement patterns and inventory land resources and track changes in human uses of the landscape. Foresters use remote sensing and GIS to map timber stands, estimate timber volume, monitor insect infestations, fight forest fires, and plan harvesting of timber. Agricultural scientists can examine the growth, maturing, and harvesting of crops and monitor the progress of diseases, infestations, and droughts to forecast their impact on crop yields. Soil scientists use remotely sensed imagery to plot boundaries of soil units and to examine relationships between soil patterns and those of land use and vegetation. In brief, remotely sensed imagery has found applications in virtually all fields that require analysis of distributions of natural or human resources on the earth's surface.

## 1.6. Key Concepts of Remote Sensing

Scientists usually attempt to avoid the pitfalls of relying on improvised approaches for conducting research. Improvised, or ad hoc methods may repeat previously discovered methodological errors, lead to trivial results, promote inefficiency, or simply create inaccuracies. Instead, scientists prefer to base their methods on systems of facts and methods found to yield consistent, accurate results.

The practice of remote sensing is young enough that the basic facts and methods are not known completely. Scientists are still investigating approaches to define many of the fundamental methods and concepts central to remote sensing. Nonetheless, it is useful to base our study of remote sensing on a set of principles that seem to convey the essential dimensions of the practice of remote sensing. The principles outlined below are tentatively proposed as concepts that address ideas central to the practice of remote sensing, regardless of specific disciplinary applications.

### *Spectral Differentiation*

Remote sensing depends on observed spectral differences in the energy reflected or emitted from features of interest. Expressed in everyday terms, one might say that we look for differences in "colors" of objects. (This analogy may be useful even though remote sensing is often conducted outside the visible spectrum, where "colors," in the usual meaning of the word, do not exist.) This principle is the basis of *multispectral remote sensing,* the science of observing features at varied wavelengths in an effort to derive information about the features and their distributions. The term "spectral signature" has been used to refer to the spectral response of a feature, as observed over a range of wavelengths (Parker & Wolff, 1965). For the beginning student this term can be misleading because it implies a distinctiveness and a consistency that seldom can be observed in nature. For example, consider the notion of a spectral signature for "corn." A cornfield has one spectral response when it is planted, another as the plants emerge, another as the plants mature, and yet another after harvest. Selection of a single time and place to designate a single spectral signature for corn is futile. Yet the term conveys a more general concept that is useful because at specific times and places the spectral response of cornfields may form the basis for reliable separation of corn from what, for example, even though neither crop forms universally applicable spectral signatures.

### *Radiometric Differentiation*

Examination of any image acquired by remote sensing ultimately depends on detection of differences in brightness of objects and features. The scene itself must have sufficient contrast (at a specific spectral region) in brightness, and the remote sensing instrument must be capable of recording this contrast, before information can be derived from the image. As a result, the sensitivity of the instrument and the existing contrast in the scene between objects and their backgrounds are always issues of significance in remote sensing investigations.

### *Spatial Differentiation*

Every sensor is limited in respect to the size of the smallest area that can be separately recorded as an entity on an image. This minimum area determines the spatial detail—the fineness of the patterns—on the image. For some remote sensing systems these smallest areal units—picture elements ("pixels")—are in fact discrete, distinct units, identifiable on the image (Figure 1.5). In other instances, the spatial detail is less obvious (determined by the quality of a camera lens, or the film that has been used) but is nonetheless present. Our ability to record spatial detail is influenced primarily by the choice of sensor and the altitude at which it is used to record images of the earth. Note also that some landscapes vary greatly in their spatial complexity—some may be represented clearly at coarse levels of detail, whereas others are so complex that the finest level of detail is required to record their essential characteristics.

### *Geometric Transformation*

Every remotely sensed image represents a landscape in a specific geometric relationship determined by the design of the remote sensing instrument, specific operating conditions, terrain relief, and other factors. The ideal remote sensing instrument hypothetically would be able to create an image with accurate, consistent geometric relationships between

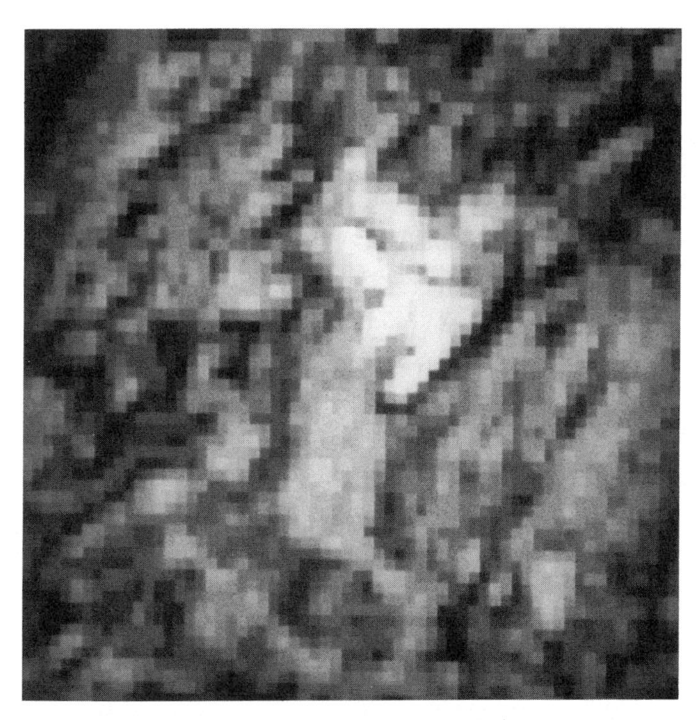

**FIGURE 1.5.** Digital remote sensing image. This image is a small section of Figure 1.1 shown in much greater detail, so that individual picture elements ("pixels") are visible now as separate square areas of varied brightness. An alternative form of this image represents each of these areas as integer values between 0 (black) and 127 (bright white) in proportion to the brightness of the ground area at a specific time and location. (Most of the areas have intermediate brightnesses and appear here as grays of varying intensities.) In its pictorial form (Figure 1.1), the image is more easily examined by image interpreters, but in its digital form (an array of integer values) the image is easily examined by quantitative methods of analysis to enhance image features, or to derive information from the original image. Any pictorial image can be represented in digital form (this image), and any digital image can be portrayed in pictorial form (Figure 1.1).

points on the ground and their corresponding representations on the image. Such an image could form the basis for accurate measurements of areas and distances. In reality, of course, each image includes positional errors caused by the perspective of the sensor optics, motion of scanning optics, terrain relief, and earth curvature. Each source of error can vary in significance in specific instances, but the result is that geometric errors are inherent, not accidental, characteristics of remotely sensed images. In some instances we may be able to remove or reduce locational error, but it must *always* be taken into account before images are used as the basis for measurements of areas and distances.

### *Interchangeability of Pictorial and Digital Formats*

Many remote sensing systems generate, as their primary output, digital arrays that represent brightnesses of areas of the earth's surface. These arrays can be presented in pictorial form by portraying each digital value as a photographic brightness level scaled to the magnitude of the value (Figure 1.1). Or, pictorial images can be represented in digital form by systematically subdividing the image into tiny areas of equal size and shape, then representing the brightness of these areas by discrete values (Figure 1.5).

The two forms for remote sensing data represent different methods of display and representation, but there is no real difference in the information conveyed by the two forms. Any image can be portrayed in either form (sometimes with a loss of detail in converting from one form to another) according to the purposes of our investigation.

### *Remote Sensing Instrumentation Acts as a System*

The image analyst must always be conscious of the fact that the many components of the remote sensing process *act as a system* and cannot be isolated from one another. For example, upgrading the quality of a camera lens makes little sense unless we also use a film of sufficient quality to record improvements produced by the superior lens which will give analyst the ability to derive better information from the image.

Components of the system must be appropriate for the task at hand. This means that the interpreter must know intimately not only the remote sensing system but also the subject of the interpretation, to include the amount of detail required, appropriate time of year to acquire the data, best spectral regions to use, and so on. Like the physical components of the system, the interpreter's knowledge and experience also interact to form a whole.

### *Role of the Atmosphere*

All energy reaching the remote sensing instrument must pass through a portion of the earth's atmosphere. For satellite remote sensing in the visible and near infrared, energy received by the sensor must pass through a considerable depth of the earth's atmosphere. In doing so, the sun's energy is altered in intensity and wavelength by particles and gases in the earth's atmosphere. These changes appear on the image in ways that degrade image quality or influence the accuracy of interpretations.

## 1.7. Career Preparation and Professional Development

For the student, a course of study in remote sensing offers opportunities to enter a field of knowledge that can contribute to several dimensions of a university education and subsequent personal and professional development. Students enrolled in introductory remote sensing courses often view the topic as an important part of their occupational and professional preparation. It is certainly true that skills in remote sensing are valuable in the initial search for employment. But it is equally important to acknowledge that this topic should form part of a comprehensive program of study that includes work in GISs and in-depth study of a specific discipline. A well-thought-out program appropriate for a student's specific interests and strengths should combine studies in such interrelated topics as:

- Geology, hydrology, geomorphology, soils
- Urban planning, transportation, urban geography
- Forestry, ecology, soils

Such programs are based on a foundation of supporting courses including statistics, computer science, and the physical sciences.

Students should avoid studies that provide only narrowly based or technique-oriented content. Such highly focused studies, perhaps with specific equipment or software, may provide immediate skills for entry-level positions but leave the student unprepared to participate in the broader assignments required for effective performance and professional advancement. Employers report that they seek employees who:

- Have a good background in at least one traditional discipline
- Are reliable, and able to follow instructions without detailed supervision
- Can write and speak effectively
- Work effectively in teams with others in other disciplines
- Are familiar with common business practices

Because this kind of preparation is seldom encompassed in a single academic unit within a university, students often have to apply their own initiative to identify the specific courses they will need to best develop these qualities.

Possibly the most important but least visible contributions are to the development of conceptual thinking concerning the role of basic theory and method, integration of knowledge from several disciplines, and proficiency in identifying practical problems in a spatial context. Although skills and knowledge of remote sensing are very important, it is usually a mistake to focus exclusively on methodology and technique. At least two pitfalls are obvious. First, emphasis on fact and technique without consideration of basic principles and theory provides a narrow, empirical foundation in a field that is characterized by diversity and rapid change. A student equipped with a narrow background is ill-prepared to compete with those trained in other disciplines or to adjust to unexpected developments in science and technology. Thus, any educational experience is best perceived not as a catalogue of facts to be memorized but as an experience in *how to learn* to equip oneself for independent learning later, outside the classroom. This task requires a familiarity with basic references, fundamental principles, and the content of related disciplines, as well as the core of facts that form the substance of a field of knowledge.

Second, many employers have little interest in hiring employees with shallow preparation in either their major discipline or remote sensing. Lillesand (1982) reports that a panel of managers from diverse industries concerned with remote sensing recommended that prospective employees develop "an ability and desire to interact at a conceptual level with other specialists" (p. 290). Campbell (1978) quotes other supervisors who are also concerned that students receive a broad preparation in remote sensing and in their primary field of study:

> It is essential that the interpreter have a good general education in an area of expertise. For example, you can make a geologist into a good photo geologist, but you cannot make an image interpreter into a geologist.
>
> Often people lack any real philosophical understanding of why they are doing remote sensing, and lack the broad overview of the interrelationships of all earth science and earth-oriented disciplines (geography, geology, biology, hydrology, meteorology, etc.) This often creates delays in our work as people continue to work in small segments of the (real) world and don't see the interconnections with another's research. (p. 35)

These same individuals have recommended that students interested in remote sensing complete courses in computer science, physics, geology, geography, biology, engineering, mathematics, hydrology, business, statistics, and a wide variety of other disciplines. No student could possibly take all recommended courses during a normal program of study, but it is clear that a haphazard selection of university courses, or one that focused exclusively on remote sensing courses, would not form a substantive background in remote sensing. In addition, many organizations have been forceful in stating that they desire employees who can write well, and several expressed an interest in persons with expertise in remote sensing who have knowledge of a foreign language. The point to be emphasized is that education in remote sensing should be closely coordinated with study in traditional academic disciplines and supported by a program of courses (perhaps not so traditional) carefully selected from offerings in related disciplines.

Students should consider joining a professional society devoted to the field of remote sensing. In the United States and Canada, the American Society for Photogrammetry and Remote Sensing (ASPRS) (5410 Grosvenor Lane, Suite 210, Bethesda, MD 20814-2160; 301-493-0290) is the principal professional organization in this field. ASPRS offers student membership rates and discounts on publications and meeting registration and conducts job fairs at annual meetings. ASPRS is organized on a regional basis, so local chapters conduct their own activities, which are open to student participation. Other professional organizations often have interest groups devoted to applications of remote sensing within specific disciplines, often with similar benefits for student members.

Students should also investigate local libraries to become familiar with professional journals in the field. Among principal journals are:

*Photogrammetric Engineering and Remote Sensing*
*Remote Sensing of Environment*
*International Journal of Remote Sensing*
*IEEE Transactions on Geoscience and Remote Sensing*
*Computers and Geosciences*

Although beginning students may not yet be prepared to read research articles in detail, those who make the effort to familiarize themselves with these journals will have prepared the way take advantage of their content later. In particular, students may find *Photogrammetric Engineering and Remote Sensing* useful because of its listing of job opportunities, scheduled meetings, and new products.

Even though many students who study remote sensing do not plan to specialize in the field, there are sound reasons to acquire a good background in the methods and principles of remote sensing. For many years strong forces in commerce and industry have been expanding the demand for a spatially literate workforce, capable of understanding spatial processes and using the tools of spatial analysis. The fields of GIS and remote sensing, once seen as specialized techniques, are now shaped by technological and market forces into methods that are becoming everyday tools for analysts in a broad cross-section of enterprises. These developments have increased the number of employers who seek to hire people prepared with the knowledge and skills acquired in the study of remote sensing, and have also increased the level of expectation that more of their employees will be proficient in these areas. Thus, for university students these developments have increased opportunities, but also the requirement to acquire additional skills and knowledge.

## Review Questions

1. Aerial photography and other remotely sensed images have found rather slow acceptance into many, if not most, fields of study. Imagine that you are director of a unit engaged in geological mapping in the early days of aerial photography (e.g., in the 1930s). Can you suggest reasons why you might be reluctant to devote your efforts and resources to use of aerial photography rather than to continued use of your usual procedures?
2. Satellite observation of the earth provides many advantages over aircraft-borne sensors. Consider fields such as agronomy, forestry, or hydrology. For one such field of study, list as many of the advantages as you can. Can you suggest some disadvantages?
3. Much (but not all) information derived from remotely sensed data is derived from spectral information. To understand how spectral data may not always be as reliable as one might first think, briefly describe the spectral properties of a maple tree and of a corn field. How might these properties change over the period of a year? Or a day?
4. All remotely sensed images observe the earth from above. Can you list some advantages to the overhead view (as opposed to ground-level views) that make remote sensing images inherently advantageous for many purposes? List some *disadvantages* to the overhead view.
5. Remotely sensed images show the combined effects of many landscape elements, including vegetation, topography, illumination, soil, drainage, and others. Is this diverse combination in your view an advantage or a disadvantage? Explain.
6. List ways in which remotely sensed images differ from maps. Also list advantages and disadvantages of each. List some of the tasks for which each might be more useful.
7. Chapter 1 emphasizes how the field of remote sensing is formed by knowledge and perspective from many different disciplines. Examine the undergraduate catalogue for your college or university and prepare a comprehensive program of study in remote sensing from courses listed. Identify gaps—courses or subjects that would be desirable but are not offered.
8. In your university library, find copies of *Photogrammetric Engineering and Remote Sens-*

*ing, International Journal of Remote Sensing,* and *Remote Sensing of Environment,* some of the most important English-language journals reporting remote sensing research. Examine some of the articles in several issues of each journal. Although titles of some of these articles may now seem rather strange, as you progress through this course, you will be able to judge the significance of most. Refer to these journals again as you complete the course.

9. Inspect library copies of some of the remote sensing texts listed in the references for Chapter 1. Examine the tables of contents, selected chapters, and lists of references. Many of these volumes may form useful references for future study or research in the field of remote sensing.

10. Examine some of the journals mentioned in question 9, noting the affiliations and institutions of authors of articles. Be sure to look at issues that date back for several years, so you can identify some of the institutions and agencies that have been making a continuing contribution to remote sensing research.

## References

Alföldi, T., P. Catt, and P. Stephens. 1993. Definitions of Remote Sensing. *Photogrammetric Engineering and Remote Sensing,* Vol. 59, pp. 611–613.

Avery, Thomas E., and G. L. Berlin. 1992. *Fundamentals of Remote Sensing and Airphoto Interpretation.* New York: Macmillan, 472 pp.

Barrett, E. C., and L. F. Curtis. 1976. *Introduction to Environmental Remote Sensing.* New York: Wiley, 336 pp.

Campbell, J. B. 1978. Employer Needs in Remote Sensing in Geography. *Remote Sensing of the Electromagnetic Spectrum,* Vol. 5, No. 4, pp. 29–36.

Colwell, Robert N. 1956. Determining the Prevalence of Certain Cereal Crop Diseases by Means of Aerial Photography. *Hilgardia,* Vol. 26, No. 5, pp. 223–286.

Colwell, Robert N. 1966. Uses and Limitations of Multispectral Remote Sensing. In *Proceedings of the Fourth Symposium on Remote Sensing of Environment.* Ann Arbor: University of Michigan, pp. 71–100.

Colwell, Robert N. (ed.). 1983. *Manual of Remote Sensing* (2nd edition). Falls Church, VA: American Society of Photogrammetry. 2 vols., 2240 pp.

Curran, P. 1985. *Principles of Remote Sensing.* New York: Longman, 282 pp.

Curran, P. 1987. Commentary: On Defining Remote Sensing. *Photogrammetric Engineering and Remote Sensing,* Vol. 53, pp. 305–306.

Estes, John E., John R. Jensen, and David S. Simonett. 1977. The Impact of Remote Sensing on United States Geography: The Past in Perspective, Present Realities, Future Potentials. In *Proceedings of the Eleventh International Symposium on Remote Sensing of Environment.* Ann Arbor: University of Michigan, Institute of Science and Technology, pp. 101–121.

Estes, J. E., et al. 1993. The NCGIA Core Curriculum in Remote Sensing. *Photogrammetric Engineering and Remote Sensing,* Vol. 59, pp. 945–948.

Fischer, William A. (ed.). 1975. History of Remote Sensing. Chapter 2 in *Manual of Remote Sensing* (R. G. Reeves, ed.). Falls Church, VA: American Society of Photogrammetry, pp. 27–50.

Fischer, William A., W. R. Hemphill, and A. Kover. 1976. Progress in Remote Sensing. *Photogrammetria,* Vol. 32, pp. 33–72.

Fussell, Jay, D. Rundquist, and J. A. Harrington. 1986. On Defining Remote Sensing. *Photogrammetric Engineering and Remote Sensing,* Vol. 52, pp. 1507–1511.

Hall, Stephen S. 1992. *Mapping the Next Millennium: The Discovery of New Geographies.* New York: Random House, 384 pp.

Hall, Stephen S. 1993. *Mapping the Next Millennium: How Computer Driven Cartography Is Revolutionizing the Face of Science.* New York: Random House, 360 pp.
Landgrebe, David. 1976. Computer-Based Remote Sensing Technology: A Look to the Future. *Remote Sensing of Environment,* Vol. 5, pp. 229–246.
Lillesand, T. M. 1982. Trends and Issues in Remote Sensing Education. *Photogrammetric Engineering and Remote Sensing,* Vol. 48, pp. 287–293.
Lillesand, T. M., and R. W. Kiefer. 1994. *Remote Sensing and Image Interpretation.* New York: Wiley, 750 pp.
Lintz, Joseph, and D. S. Simonett. 1976. *Remote Sensing of Environment.* Reading, MA: Addison-Wesley, 694 pp.
Lee, William T. 1922. *The Face of the Earth as Seen from the Air.* American Geographical Society Special Publication No. 4. New York: American Geographical Society.
National Academy of Sciences. 1970. *Remote Sensing with Special Reference to Agriculture and Forestry.* Washington, DC: National Academy of Sciences, 424 pp.
Parker, Dana C., and Michael F. Wolff. 1965. Remote Sensing. *International Science and Technology,* Vol. 43, pp. 20–31.
Philipson, Warren R. 1980. Problem-Solving with Remote Sensing. *Photogrammetric Engineering and Remote Sensing,* Vol. 46, pp. 1335–1338.
Ray, R. G. 1960. *Aerial Photographs in Geological Interpretation and Mapping.* U.S. Geological Survey Professional Paper 373. Washington, DC: U.S. Geological Survey, 230 pp.
Reeves, Robert G. (ed.). 1975. *Manual of Remote Sensing.* Falls Church, VA: American Society of Photogrammetry, 2 vols., 2144 pp.
Simonett, David S. 1966. Present and Future Needs of Remote Sensing in Geography. In *Proceedings of the Fourth International Symposium on Remote Sensing of the Environment.* Ann Arbor: University of Michigan, Institute of Science and Technology, pp. 37–47.
Simonett, David S. (ed.). 1983. Development and Principles of Remote Sensing. Chapter 1 in *Manual of Remote Sensing* (R. N. Colwell, ed.). Falls Church, VA: American Society of Photogrammetry, pp. 1–35.
Stone, K. 1974. Developing Geographical Remote Sensing. Chapter 1 in *Remote Sensing: Techniques for Environmental Analysis* (J. E. Estes and L. W. Senger, eds.). Santa Barbara, CA: Hamilton, pp. 1–13.
Swain, P. H., and S. M. Davis. 1978. *Remote Sensing: The Quantitative Approach.* New York: McGraw-Hill, 396 pp.
White, L. P. 1977. *Aerial Photography and Remote Sensing for Soil Survey.* Oxford: Clarendon Press, 104 pp.

CHAPTER TWO

# Electromagnetic Radiation

## 2.1. Introduction

With the exception of objects that might be at absolute zero, all objects emit electromagnetic radiation, and some objects also reflect radiation that has been emitted by other objects. By recording this emitted or reflected radiation and applying a knowledge of its behavior as it passes through the earth's atmosphere and interacts with objects, remote sensing analysts develop a knowledge of the character of features (vegetation, structures, soils, rock, water bodies, etc.) on the earth's surface. Virtually all practical questions of applications of remote sensing imagery hinge on a sound understanding of electromagnetic radiation and its interaction with surfaces, the atmosphere, and our instruments. Therefore, the discussion of electromagnetic radiation in this chapter builds a foundation to permit development in subsequent chapters of the many other important topics within the field of remote sensing.

The most familiar form of electromagnetic radiation is visible light, which forms only a small (but important) portion of the full electromagnetic spectrum. The large segments of this spectrum that lie outside the range of human vision require our special attention because they may behave in ways that are quite foreign to our everyday experience with visible radiation.

## 2.2. The Electromagnetic Spectrum

Electromagnetic energy is generated by several mechanisms, including changes in the energy levels of electrons, acceleration of electrical charges, decay of radioactive substances, and the thermal motion of atoms and molecules. Nuclear reactions within the sun produce a full spectrum of electromagnetic radiation, which is transmitted through space without experiencing major changes. As this radiation approaches the earth, it passes through the atmosphere before reaching the earth's surface. Some is reflected upwards from the earth's surface; it is this radiation that forms the basis for photographs and similar images. In addition, some solar radiation is absorbed at the surface of the earth and is re-radiated as thermal energy. This thermal energy can also be used to form remotely sensed images, although they differ greatly from the aerial photographs formed from re-

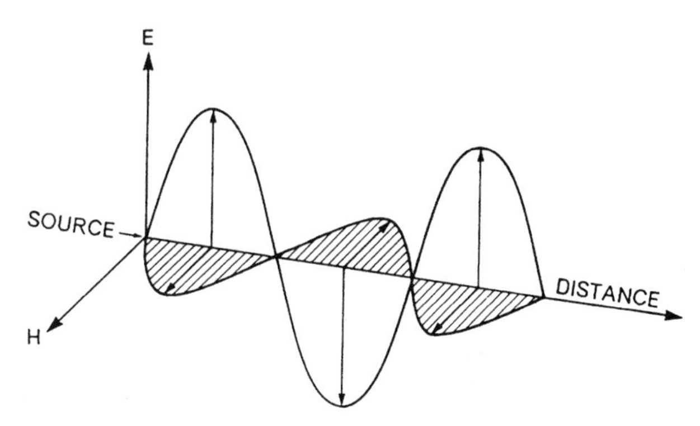


**FIGURE 2.1.** Electric ($E$) and magnetic ($H$) components of electromagnetic radiation. The electric and magnetic components are oriented at right angles to one another and vary along an axis perpendicular to the axis of propagation.

flected energy. Finally, man-made radiation, such as that generated by imaging radars, is also used for remote sensing.

Electromagnetic radiation consists of an electrical field ($E$) that varies in magnitude in a direction perpendicular to the direction of propagation (Figure 2.1). In addition, a magnetic field ($H$), oriented at right angles to the electrical field, is propagated in phase with the electrical field.

Electromagnetic energy displays three properties (Figure 2.2):

1. *Wavelength* is the distance from one wave crest to the next. Wavelength can be measured in everyday units of length, although very short wavelengths have such

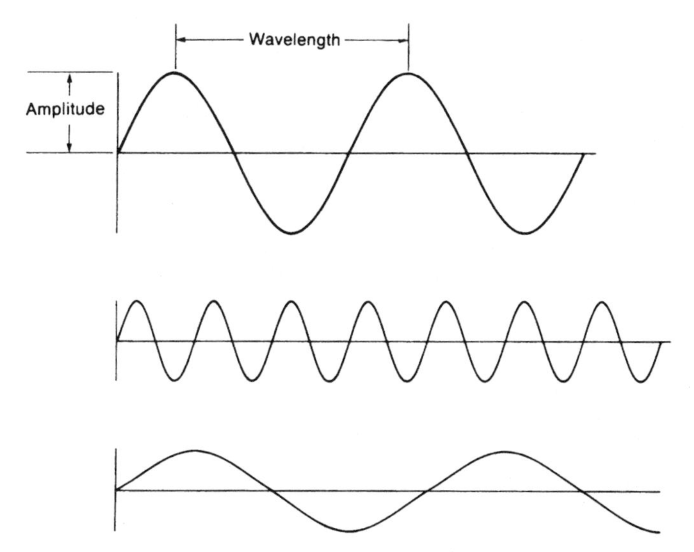


**FIGURE 2.2.** Amplitude, frequency, and wavelength. The center diagram represents high frequency, short wavelength; the bottom diagram shows low frequency, long wavelength.

**TABLE 2.1. Units of Length Used in Remote Sensing**

| Unit | Distance | |
|---|---|---|
| Kilometer (km) | 1,000 m | |
| Meter (m) | 1.0 m | |
| Centimeter (cm) | 0.01 m | $= 10^{-2}$ m |
| Millimeter (mm) | 0.001 m | $= 10^{-3}$ m |
| Micrometer (μm)[a] | 0.000001 m | $= 10^{-6}$ m |
| Nanometer (nm) | | $10^{-9}$ m |
| Ångstrom unit (Å) | | $10^{-10}$ m |

[a]Formerly called the "micron" (μ); the term "micrometer" is now used by agreement of the General Conference on Weights and Measures.

short distances between wave crests that extremely short (and therefore less familiar) measurement units are required (Table 2.1).

2. *Frequency* is measured as the number of crests passing a fixed point in a given period. Frequency is often measured in *hertz,* units each equivalent to one cycle per second (Table 2.2), and multiples of the hertz.
3. *Amplitude* is equivalent in Figure 2.2 to the height of each peak. Amplitude is often measured as energy levels (formally known as *spectral irradiance*), expressed as watts per square meter per micrometer (i.e., as energy level per wavelength interval).

The speed of electromagnetic energy ($c$) is constant at 299,893 kilometers (km) per second. Frequency ($v$) and wavelength ($\lambda$) are related:

$$c = \lambda v \qquad \text{(Eq. 2.1)}$$

Therefore, characteristics of electromagnetic energy can be specified using either frequency or wavelength. Varied disciplines, and varied applications, follow different conventions for describing electromagnetic radiation, using either wavelength (often measured in Angstrom units [Å], microns, micrometers, nanometers, millimeters, etc.) or frequency (using hertz, kilohertz, megahertz, and so on, as appropriate). Although there is no authoritative standard, a common practice in the field of remote sensing is to define regions of the spectrum on the basis of wavelength, often using micrometers (each equal to one one-millionth of a meter, symbolized as μm), millimeters (mm), and meters (m) as units of length. Departures from this practice are common; for example, electrical engi-

**TABLE 2.2. Frequencies Used in Remote Sensing**

| Unit | Frequency (cycles per second) |
|---|---|
| Hertz (Hz) | 1 |
| Kilohertz (kHz) | $10^3$ (= 1.000) |
| Megahertz (MHz) | $10^6$ (= 1,000,000) |
| Gigahertz (GHz) | $10^9$ (= 1,000,000,000) |

neers who work with microwave radiation traditionally use frequency to designate subdivisions of the spectrum. This volume usually employs wavelength designations. The student should, however, be prepared to encounter different usage in scientific journals and in references.

## 2.3. Major Divisions of the Electromagnetic Spectrum

Major divisions of the electromagnetic spectrum (Table 2.3) are, in essence, arbitrarily defined. In a full spectrum of solar energy there are no sharp breaks at the divisions indicated in Figure 2.3. Subdivisions are established for convenience and by traditions within the different disciplines, so do not be surprised to find different definitions in other sources or in references pertaining to other disciplines.

Two important categories are not shown in Table 2.3. The *optical spectrum*, from 0.30 to 15 µm, defines those wavelengths that can be reflected and refracted with lenses and mirrors. The *reflective spectrum* extends from about 0.38 to 3.0 µm; it defines that portion of the solar spectrum used directly for remote sensing.

### *The Ultraviolet Spectrum*

For practical purposes, radiation of significance for remote sensing can be said to begin with the ultraviolet region, a zone of short-wavelength radiation that lies between the X-ray region and the limit of human vision. Often the ultraviolet region is subdivided into the *near ultraviolet* (sometimes known as *UV-A*) (0.32 to 0.40 µm), the *far ultraviolet* (*UV-B*) (0.32 to 0.28 µm), and the extreme ultraviolet (*UV-C*) (below 0.28 µm). The ultraviolet region was discovered in 1801 by the German scientist Johann Wilhelm Ritter (1776–1810). Literally, ultraviolet means "beyond the violet," designating it as the region just outside the violet region, the shortest wavelengths visible to humans. Near ultraviolet radiation is known for its ability to induce *fluorescence* (emission of visible radiation) in some materials, so it has significance for a specialized form of remote sensing. However,

**TABLE 2.3. Principal Divisions of the Electromagnetic Spectrum**

| Division | Limits |
|---|---|
| Gamma rays | <0.03 nm |
| X rays | 0.03–300 nm |
| Ultraviolet radiation | 0.30–0.38 µm |
| Visible | 0.38–0.72 µm |
| Infrared radiation | |
| Near infrared | 0.72–1.30 µm |
| Mid infrared | 1.30–3.00 µm |
| Far infrared | 7.0–1,000 µm (1 mm) |
| Microwave radiation | 1 mm–30 cm |
| Radio | ≥ 30 cm |

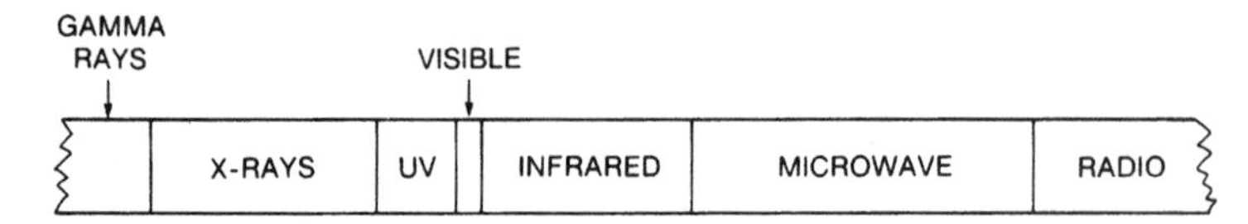


**FIGURE 2.3.** Major divisions of the electromagnetic spectrum. This diagram gives only a schematic representation—sizes of divisions are not shown in correct proportions. (See Table 2.3.)

ultraviolet radiation is largely scattered by the earth's atmosphere, so it is not generally used in the field of remote sensing.

### *The Visible Spectrum*

Although the visible spectrum constitutes a small portion of the full electromagnetic spectrum, it has obvious significance in remote sensing. Limits of the visible spectrum are defined by the sensitivity of the human visual system. Optical properties of visible radiation were first investigated by Isaac Newton (1641–1727), who, during 1665 and 1666, conducted experiments that revealed that visible light can be divided (using prisms, or, in our time, diffraction gratings) into three segments. Today we know these segments as the *additive primaries,* defined approximately from 0.4 to 0.5 μm (blue), 0.5 to 0.6 μm (green), and 0.6 to 0.7 μm (red) (Figure 2.4). Primary colors are defined such that no single primary can be formed from a mixture of the other two, and that all other colors can be formed by mixing the three primaries in appropriate proportions. Equal proportions of the three additive primaries combine to form white light.

The color of an object is defined by the color of the light it reflects (Figure 2.4). Thus, a "blue" object is blue because it reflects blue light. Intermediate colors are formed

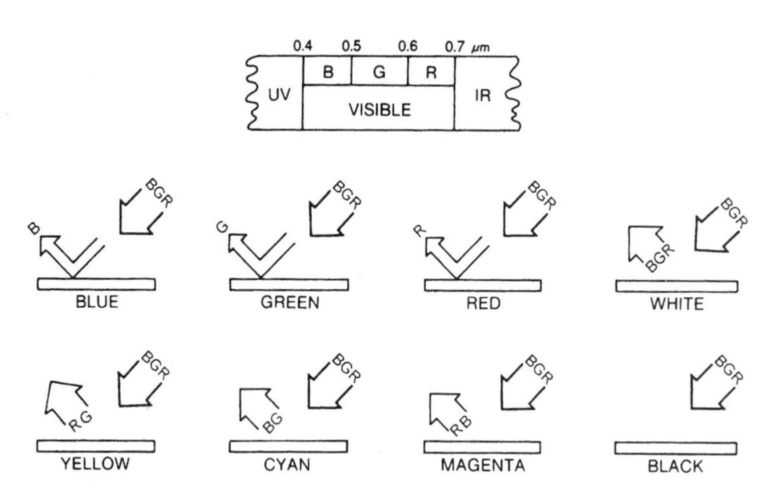


**FIGURE 2.4.** Colors.

when an object reflects two or more of the additive primaries, which combine to create the sensation of "yellow" (red and green), "purple" (red and blue), or other colors. The additive primaries are significant whenever we consider the colors of light, as, for example, in the exposure of photographic films.

In contrast, *representations* of colors in films, paintings, and similar images are formed by combinations of the three *subtractive primaries* that define colors of pigments and dyes. Each of the three subtractive primaries absorbs a third of the visible spectrum (Figure 2.4). *Yellow* absorbs blue light (and reflects red and green), *cyan* (a greenish-blue) absorbs red light (and reflects blue and green), and *magenta* (a bluish red) absorbs green light (while reflecting red and blue light). A mixture of equal proportions of pigments of the three subtractive primaries yields black (complete absorption of the visible spectrum). Additive primaries are of interest in matters concerning radiant energy, whereas the subtractive primaries specify colors of the pigments and dyes used in reproducing colors on films, photographic prints, and other images.

### *The Infrared Spectrum*

Wavelengths longer than the red portion of the visible spectrum are designated as the *infrared* region, discovered in 1800 by the British astronomer William Herschel (1738–1822). This segment of the spectrum is large relative to the visible region, as it extends from 0.72 to 15 µm—more than 40 times as wide as the visible. Because of its broad range, the infrared region encompasses radiation with varied properties. Two important categories can be recognized. The first consists of *near infrared* and *mid infrared* radiation—defined as those regions of the infrared spectrum closest to the visible. Radiation in the near infrared region behaves, with respect to optical systems, in a manner analogous to radiation in the visible spectrum. Therefore, remote sensing in the near infrared can use films, filters, and cameras with designs similar to those intended for use with visible light.

The second category of infrared radiation consists of the *far infrared* region, consisting of wavelengths well beyond the visible, extending into regions that border the microwave region (Table 2.3). This radiation is fundamentally different from that in the visible and the near infrared. Whereas near infrared radiation is essentially solar radiation reflected from the earth's surface, far infrared radiation is emitted by the earth. In everyday language, the far infrared consists of "heat," or "thermal energy." Sometimes this portion of the spectrum is referred to as the *emitted infrared.*

### *Microwave Energy*

The longest wavelengths commonly used in remote sensing are those from about 1 mm to 1 m in wavelength. The shortest wavelengths in this range have much in common with the thermal energy of the far infrared. The longer wavelengths merge into the radio wavelengths used for commercial broadcasts. Our knowledge of the microwave region originates from the work of the Scottish physicist James Clerk Maxwell (1831–1879) and the German physicist Heinrich Hertz (1857–1894).

## 2.4. Radiation Laws

The propagation of electromagnetic energy follows certain physical laws. In the interests of conciseness, some of these laws are outlined in abbreviated form because our interest here is the basic relationship they express rather than the formal derivations that are available to the student in more comprehensive sources.

Isaac Newton was among the first to examine the dual nature of light (and, by extension, all forms of electromagnetic radiation) as displaying behavior simultaneously associated with both discrete and continuous phenomena. Newton maintained that light is a stream of minuscule particles ("corpuscles") that travel in straight lines. This notion is consistent with modern theories of Max Planck (1858–1947) and Albert Einstein (1879–1955). Planck discovered that electromagnetic energy is absorbed and emitted in discrete units called *quanta*, or *photons*. The size of each unit is directly proportional to the frequency of the energy's radiation. Planck defined a constant ($h$) to relate frequency ($v$) to radiant energy ($Q$):

$$Q = hv \quad \text{(Eq. 2.2)}$$

His model explains the photoelectric effect (the generation of electric currents by the exposure of certain substances to light) as the effect of the impact of these discrete units of energy (quanta) upon surfaces of certain metals, causing the emission of electrons.

Newton knew of other phenomena, such as the refraction of light by prisms, that are best explained by assuming that electromagnetic energy travels in a wave-like manner. James Clerk Maxwell was the first to formally define the wave model of electromagnetic radiation. His mathematical definitions of the behavior of electromagnetic energy are based on the assumption from classical (mechanical) physics that light and other electromagnetic energy propagate as a series of waves. The wave model best explains some aspects of the observed behavior of electromagnetic energy (e.g., refraction by lenses and prisms and diffraction), whereas quantum theory provides explanations of other phenomena (notably the photoelectric effect).

The rate at which photons (quanta) strike a surface is the *radiant flux* ($\Phi_e$), measured in watts (W); this measure specifies energy delivered to a surface in a unit of time. We also need to specify a unit of area; the *irradiance* ($E_e$) is defined as radiant flux per unit area (usually measured as watts per square meter). Irradiance measures radiation that strikes a surface, whereas the term *radiant exitance* ($M_e$) defines the rate at which radiation is emitted from a unit area (also measured in watts per square meter).

All objects with temperatures above absolute zero have temperature and emit energy. The amount of energy and the wavelengths at which it is emitted depend on the temperature of the object. As the temperature of an object increases, the total amount of energy emitted also increases, and the wavelength of maximum (peak) emission becomes shorter. These relationships can be expressed formally using the concept of the *blackbody*. A blackbody is a hypothetical source of energy that behaves in an idealized manner. It absorbs all incident radiation; none is reflected. A blackbody emits energy with perfect efficiency; its effectiveness as a radiator of energy varies only as temperature varies.

The blackbody is a hypothetical entity because in nature all objects reflect at least a small proportion of the radiation that strikes them and do not act as perfect re-radiators of absorbed energy. Although truly perfect blackbodies cannot exist, their behavior can be approximated by laboratory instruments. Such instruments have formed the basis for the scientific research that has defined relationships between the temperatures of objects and the radiation they emit. *Kirchhoff's law* states that the ratio of emitted radiation to absorbed radiation flux is the same for all blackbodies at the same temperature. This law forms the basis for the definition of *emissivity* ($\epsilon$), the ratio between the emittance of a given object ($M$) and that of blackbody at the same temperature ($M_b$):

$$\epsilon = M/M_b \tag{Eq. 2.3}$$

The emissivity of a true blackbody is 1, and that of a perfect reflector (a *whitebody*) would be 0. Blackbodies and whitebodies are concepts, approximated in the laboratory under contrived conditions. In nature, all objects have emissivities that fall between these extremes (*graybodies*). For these objects, emissivity is a useful measure of their effectiveness as radiators of electromagnetic energy. Those objects that tend to absorb high proportions of incident radiation and to re-radiate this energy will have high emissivities. Those that are less effective as absorbers and radiators of energy have low emissivities (i.e., they return much more of the energy that reaches them). (In Chapter 8, a further discussion of emissivity explains that emissivity of an object can vary with its temperature.)

The *Stefan–Boltzmann law* defines the relationship between the total emitted radiation ($W$) (often expressed in watts · $cm^{-2}$) and temperature ($T$) (absolute temperature, K):

$$W = \sigma T^4 \tag{Eq. 2.4}$$

Total radiation emitted from a blackbody is proportional to the fourth power of its absolute temperature. The constant ($\sigma$) is the Stefan–Boltzmann constant ($5.6697 \times 10^{-8}$) (watts · $m^{-2}$ · $K^{-4}$). In essence, the Stefan–Boltzmann law states that hot blackbodies emit more energy per unit area than do cool blackbodies.

*Wien's displacement law* specifies the relationship between the wavelength of radiation emitted and the temperature of the object:

$$\lambda = 2{,}897.8/T \tag{Eq. 2.5}$$

where $\lambda$ is the wavelength at which radiance is at a maximum, and $T$ is the absolute temperature (K). As objects become hotter, the wavelength of maximum emittance shifts to shorter wavelengths (Figure 2.5).

All three of these radiation laws are important for understanding electromagnetic radiation and have special significance later in discussions of detection of radiation in the far infrared spectrum (Chapter 8).

## 2.5. Interactions with the Atmosphere

All radiation used for remote sensing must pass through the earth's atmosphere. If the sensor is carried by a low-flying aircraft, effects of the atmosphere on image quality may be negligible. In contrast, energy that reaches sensors carried by earth satellites (Chapter 6)

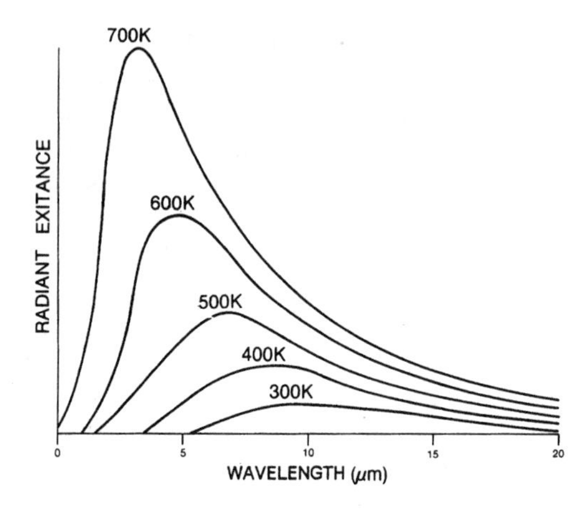


**FIGURE 2.5.** Wien's displacement law. For blackbodies at high temperatures, maximum emission occurs at short wavelengths. Blackbodies at low temperatures emit maximum of radiation at longer wavelengths.

must pass through the *entire depth* of the earth's atmosphere. Under these conditions, atmospheric effects may have substantial impact upon the quality of images and data that the sensors generate. Therefore, the practice of remote sensing requires knowledge of interactions of electromagnetic energy with the atmosphere.

In cities we are often acutely aware of the visual effects of dust, smoke, haze, and other atmospheric impurities due to their concentrations. We easily appreciate their effects on brightnesses and colors we see. But even in clear air, visual effects of the atmosphere are numerous, although so commonplace that we may not recognize their significance. In both settings, as solar energy passes through the earth's atmosphere, it is subject to modification by several physical processes, including: (1) scattering, (2) absorption, and (3) refraction.

### *Scattering*

Scattering is the redirection of electromagnetic energy by particles suspended in the atmosphere or by large molecules of atmospheric gases (Figure 2.6). The amount of scatter-

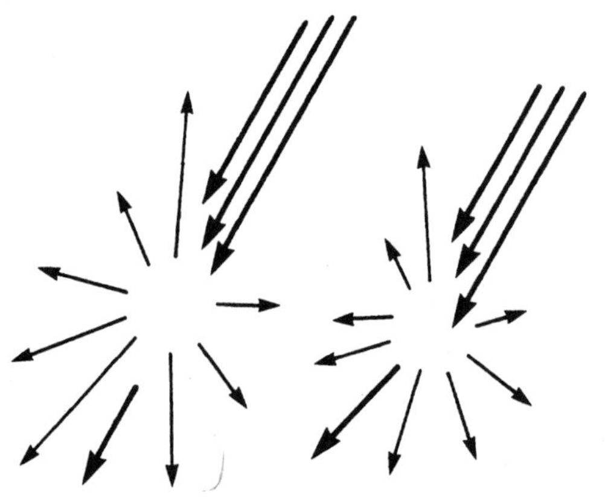

**FIGURE 2.6.** Scattering.

ing that occurs depends on sizes of these particles, their abundance, the wavelength of the radiation, and the depth of the atmosphere through which the energy is traveling. The effect of scattering is to redirect radiation so that a portion of the incoming solar beam is directed back toward space, as well as toward the earth's surface.

A common form of scattering was discovered by the British scientist, Lord J. W. S. Rayleigh (1824–1919) in the late 1890s. He demonstrated that a perfectly clean atmosphere, consisting only of atmospheric gases, causes scattering of light in a manner such that the amount of scattering increases greatly as wavelength becomes shorter. *Rayleigh scattering* occurs when atmospheric particles have diameters that are small relative to the wavelength of the radiation. Typically such particles could be very small specks of dust, or some of the larger molecules of atmosphere gases, such as $N_2$ and $O_2$. These particles have diameters that are much smaller than the wavelength of visible and near infrared radiation (on the order of diameters less than λ). Because Rayleigh scattering can occur in the absence of atmospheric impurities, it is known sometimes as "clear atmosphere" scattering. It is the dominant scattering process high in the atmosphere, up to altitudes of 9 to 10 km, the upper limit for atmospheric scattering. Rayleigh scattering is *wavelength dependent,* meaning that the amount of scattering changes greatly as one examines different regions of the spectrum (Figure 2.7). Blue light is scattered about 4 times as much as is red light, and ultraviolet light is scattered almost 16 times as much as is red light. *Rayleigh's law* states that this form of scattering is in proportion to the inverse of the fourth power of the wavelength.

Rayleigh scattering is the cause both for the blue color of the sky and for the brilliant red and orange colors often seen at sunset. At midday, when the sun is high in the sky, the atmospheric path of the solar beam is relatively short and direct, so an observer at the earth's surface sees mainly the blue light preferentially redirected by Rayleigh scattering. At sunset, observers on the earth's surface see only those wavelengths that pass through the longer atmospheric path caused by the low solar elevation; because only the longer wavelengths penetrate this distance without attenuation by scattering, we see only the reddish component of the solar beam. Variations of concentrations of fine atmospheric dust or of tiny water droplets in the atmosphere may contribute to variations in atmospheric clarity and therefore to variations in colors of sunsets.

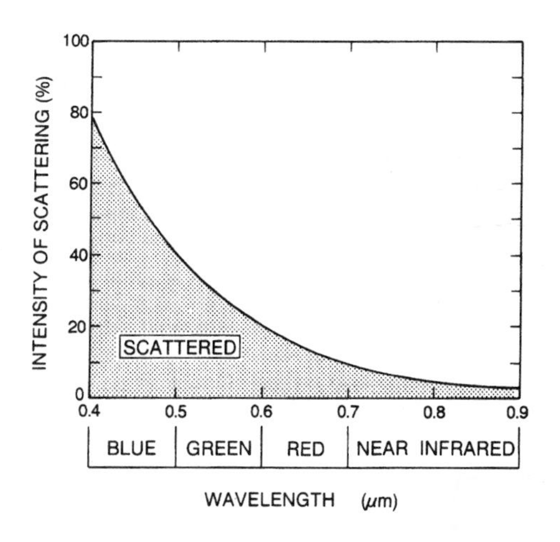


**FIGURE 2.7.** Rayleigh scattering. Scattering is much higher at shorter wavelengths.

*Mie scattering* is caused by larger particles present in the atmosphere, including dust, pollen, smoke, and water droplets. Such particles may seem to be small by standards of everyday experience, but they are many times larger than those responsible for Rayleigh scattering. Those particles that cause Mie scattering have diameters that are roughly equivalent to the wavelength of the scattered radiation. Mie scattering can influence a broad range of wavelengths in and near the visible spectrum; its effects are wavelength dependent, but not in the simple manner of Rayleigh scattering. Mie scattering tends to be greatest in the lower atmosphere (0 to 5 km), where larger particles are more abundant.

*Nonselective scattering* is caused by particles that are much larger than the wavelength of the scattered radiation. For radiation in and near the visible spectrum, such particles might be larger water droplets, or large particles of airborne dust. Nonselective means that scattering is *not* wavelength dependent, so we observe it as a whitish or grayish haze—all visible wavelengths are scattered equally.

### EFFECTS OF SCATTERING

Scattering causes the atmosphere to have a brightness of its own. In the visible portion of the spectrum, shadows are not jet black (as they would be in the absence of scattering) but merely dark; we can see objects in shadows because of light re-directed by particles in the path of the solar beam. The effects of scattering are also easily observed in vistas of landscapes—the colors and brightness of objects are altered as they are positioned at locations more distant from the observer. Landscape artists take advantage of this effect (called *atmospheric perspective*) to create the illusion of depth by painting more distant features in subdued colors and those in the foreground in brighter, more vivid colors.

For remote sensing, scattering has several important consequences. Because of the wavelength dependency of Rayleigh scattering, radiation in the blue and ultraviolet regions of the spectrum (most strongly affected by scattering) is usually not considered useful for remote sensing. Images that record these portions of the spectrum tend to record the brightness of the atmosphere rather than the brightness of the scene itself. For this reason, remote sensing instruments often exclude short wave radiation (blue and ultraviolet wavelengths) by use of filters, or by decreasing sensitivities of films to these wavelengths. (However, some specialized applications of remote sensing, not discussed here, do use ultraviolet radiation.) Scattering also directs energy from outside the sensor's field of view toward the sensor's aperture, thereby decreasing the spatial detail recorded by the sensor. Furthermore, scattering tends to make dark objects appear brighter than they would otherwise be and bright objects appear darker, thereby decreasing the *contrast* recorded by a sensor (Chapter 3). Because "good" images preserve the range of brightnesses present in a scene, scattering degrades the quality of an image.

Effects of scattering

Some of these effects are illustrated in Figure 2.8. Observed radiance at the sensor, $I$, is the sum of $I_S$, radiance reflected from the earth's surface, conveying information about surface reflectance, $I_O$, radiation scattered from the solar beam directly to the sensor without reaching the earth's surface, and $I_D$, diffuse radiation, directed to the ground, then to the atmosphere before reaching the sensor. Effects of these components are additive within a given spectral band (Kaufman, 1984):

$$I = I_S + I_O + I_D \quad \text{(Eq. 2.6)}$$

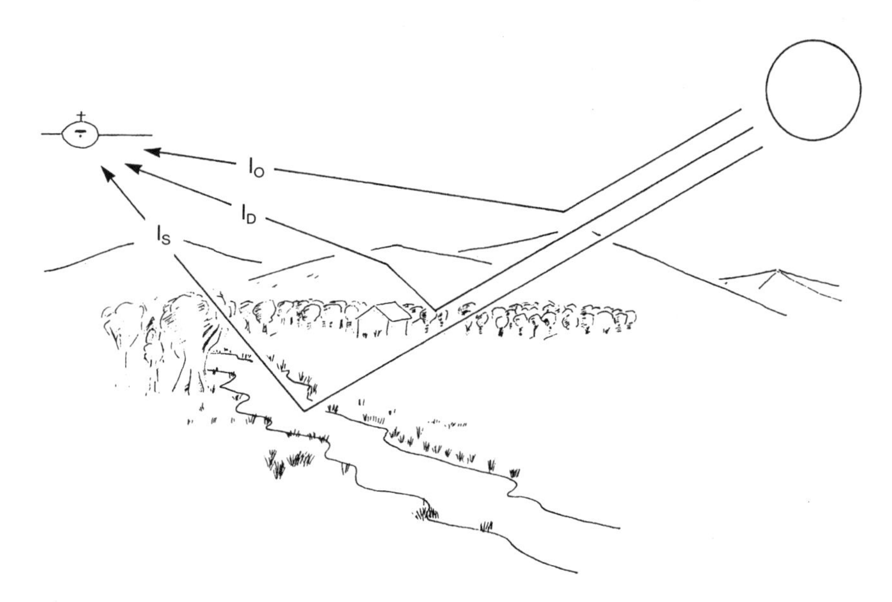


**FIGURE 2.8.** Principal components of observed brightness. $I_S$ represents radiation reflected from the ground surface, $I_O$ is energy scattered by the atmosphere directly to the sensor, and $I_D$ represents diffuse light, directed to the ground, then to the atmosphere before reaching the sensor. This diagram describes behavior of radiation in and near the visible region of the spectrum. From Campbell and Ran (1993). Copyright 1993 by Elsevier Science Ltd. Reproduced by permission.

$I_S$ varies with differing surface materials, topographic slope and orientation, and angles of illumination and observation. $I_O$ is often assumed to be more or less constant over large areas, although most satellite images represent areas large enough to encompass atmospheric differences sufficient to create variations in $I_O$. Diffuse radiation, $I_D$, is expected to be small relative to other factors, but varies from one land surface type to another, so in practice would be difficult to estimate. We should note the special case presented by shadows, in which $I_S = 0$, because the surface receives no direct solar radiation. However, shadows have their own brightness, derived from $I_D$, and their own spectral patterns, derived from the influence of local land cover upon diffuse radiation. Remote sensing is devoted to the examination of $I_S$ at different wavelengths to derive information about the earth's surface. Figure 2.9 illustrates how $I_D$, $I_S$, and $I_O$ vary with wavelength for surfaces of differing brightness.

### *Refraction*

*Refraction* is the bending of light rays at the contact between two media that transmit light. Familiar examples of refraction are the lenses of cameras or magnifying glasses (Chapter 3), which bend light rays to project or enlarge images, and the apparent displacement of objects submerged in clear water. Refraction also occurs in the atmosphere as light passes through atmospheric layers of varied clarity, humidity, and temperature.

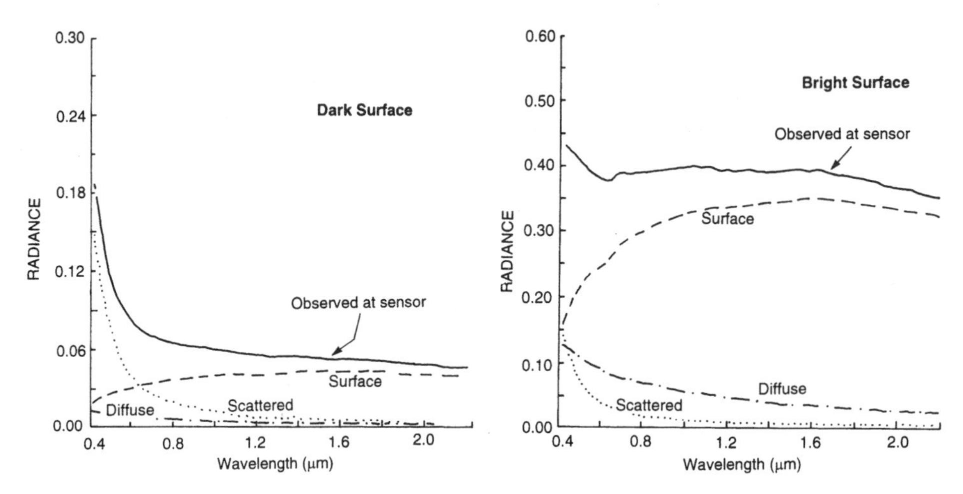


**FIGURE 2.9.** Changes in reflected, diffuse, scattered, and observed radiation over wavelength for dark (left) and bright (right) surfaces. This diagram shows the magnitude of the components illustrated in Figure 2.8. Atmospheric effects constitute a larger proportion of observed brightness for dark objects than for bright objects, especially at short wavelengths. Radiance has been normalized; note also differences in the scaling of the vertical axes for the two diagrams. Redrawn from Kaufman (1984). Reproduced by permission of the author and the Society of Photo-Optical Instrumentation Engineers.

These variations influence the density of atmospheric layers, which in turn causes a bending of light rays as they pass from one layer to another. Everyday examples are the shimmering appearances on hot summer days of objects viewed in the distance as light passes through hot air near the surface of heated highways, runways, and parking lots. Under usual conditions the maximum refraction of light entering or exiting the atmosphere is about 0.5°, such as when an observer at the surface of the earth watches the rising or setting sun.

The *index of refraction* ($n$) is defined as the ratio between the velocity of light in a vacuum ($c$) to its velocity in the medium ($c_n$):

$$n = c/c_n \qquad \text{(Eq. 2.7)}$$

Assuming uniform media, as the light passes into a denser medium it is deflected *toward* the surface normal (a line perpendicular to the surface at the point when the light ray enters the denser medium, as represented by the solid line in Figure 2.10). The angle $\theta'$ (that defines the path of the refracted ray) is given by *Snell's law*:

$$n \sin \theta' = n' \sin \theta' \qquad \text{(Eq. 2.8)}$$

where $n$ and $n'$ are the indices of refraction of the first and second media, respectively, and $\theta$ and $\theta'$ are angles measured with respect to the surface normal, as defined in Figure 2.10.

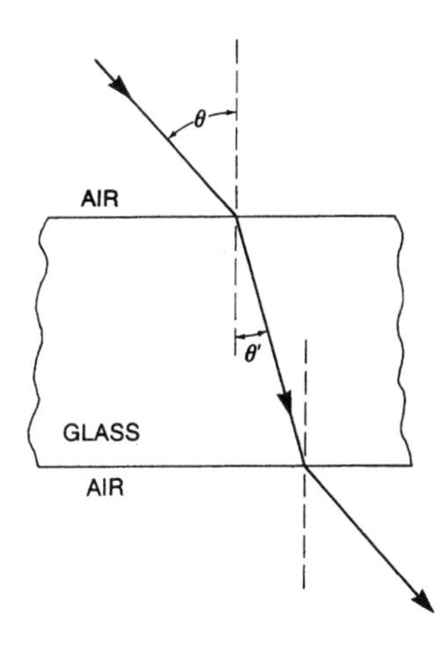


**FIGURE 2.10.** Refraction. The path of a ray of light as it passes from one medium (air) to another (glass), and then again as it passes back to the first.

### *Absorption*

Absorption of radiation occurs when the atmosphere prevents, or strongly attenuates, transmission of radiation or its energy through the atmosphere. (Energy acquired by the atmosphere is subsequently re-radiated at longer wavelengths.) Three gases are responsible for most absorption of solar radiation. Ozone ($O_3$) is formed by the interaction of high-energy ultraviolet radiation with oxygen molecules ($O_2$) high in the atmosphere (maximum concentrations of ozone are found at altitudes of about 20 to 30 km in the stratosphere). Although naturally occurring concentrations of ozone are quite low (perhaps 0.07 parts per million at ground level, 0.1 to 0.2 parts per million in the stratosphere), ozone plays an important role in the earth's energy balance. Absorption of the high-energy, short wavelength, portions of the ultraviolet spectrum (mainly $\lambda$ less than 0.24 $\mu$m) prevents transmission of this radiation to the lower atmosphere.

Carbon dioxide ($CO_2$) also occurs in low concentrations (about 0.03% by volume of a dry atmosphere), mainly in the lower atmosphere. Aside from local variations caused by volcanic eruptions and mankind's activities, the distribution of $CO_2$ in the lower atmosphere is probably relatively uniform (although mankind's burning of fossil fuels is apparently contributing to increases during the past 100 years or so). Carbon dioxide is important in remote sensing because it effectively absorbs radiation in the mid and far infrared regions of the spectrum. Its strongest absorption occurs in the region from about 13 to 17.5 $\mu$m.

Finally, water vapor ($H_2O$) is commonly present in the lower atmosphere (below 100 km) in amounts that vary from 0 to about 3% by volume. (Note the distinction between *water vapor*, discussed here, and droplets of *liquid* water, mentioned previously.) From everyday experience we know that the abundance of water vapor varies greatly from time to time and from place to place. Consequently, the role of atmospheric water vapor, unlike the roles of ozone and carbon dioxide, changes greatly with time and location. It may be almost insignificant in a desert setting or in a dry airmass but highly significant in humid climates and in moist airmasses. Furthermore, water vapor is several times as effective in absorbing radiation as are all other atmospheric gases combined. Two of the most important regions of absorption are in several bands between 5.5 and 7.0 μm, and above 27.0 μm; absorption in these regions can exceed 80% if the atmosphere contains appreciable amounts of water vapor.

### *Atmospheric Windows*

Thus, the earth's atmosphere is by no means completely transparent to electromagnetic radiation because these gases together form important barriers to transmission of electromagnetic radiation through the atmosphere. It selectively transmits energy of certain wavelengths; those wavelengths that are relatively easily transmitted through the atmosphere are referred to as *atmospheric windows* (Figure 2.11). Positions, extents, and effectiveness of atmospheric windows are determined by the absorption spectra of atmospheric gases. Atmospheric windows are of obvious significance for remote sensing—they define those wavelengths that can be used for forming images. Energy at other wavelengths, not within the windows, is severely attenuated by the atmosphere and therefore cannot be effective for remote sensing. In the far infrared region, the two most important windows extend from 3.5 to 4.1 μm, and from 10.5 to 12.5 μm. The latter is especially important because it corresponds approximately to wavelengths of peak emission from the earth's surface. A few of the most important atmospheric windows are tabulated in Table 2.4; other, smaller windows are not given here but are listed in reference books.

### *Overview of Energy Interactions in the Atmosphere*

Remote sensing is conducted in the context of all the atmospheric processes discussed thus far, so it is useful to summarize some of the most important points by outlining a perspective that integrates much of the preceding material. Figure 2.12 is an idealized diagram of interactions of shortwave solar radiation with the atmosphere; values are based on typical, or average, values derived from many places and many seasons, so they are by no

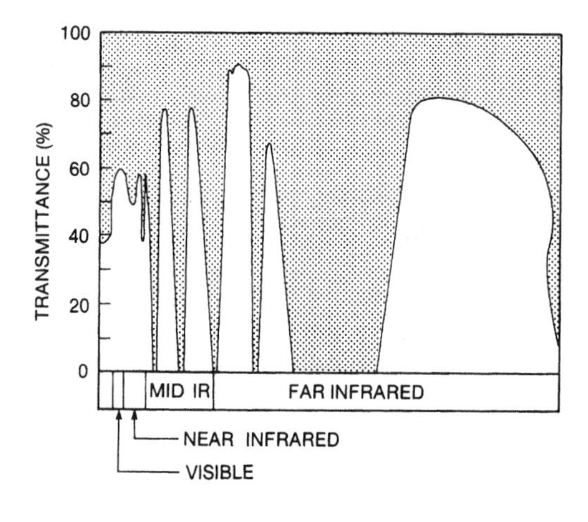


**FIGURE 2.11.** Atmospheric windows. This is a schematic representation that can depict only a few of the most important windows. The shaded area represents absorption of electromagnetic radiation.

**TABLE 2.4. Major Atmospheric Windows**

| | |
|---|---|
| Ultraviolet and visible | 0.30–0.75 μm |
| | 0.77–0.91 μm |
| Near infrared | 1.55–1.75 μm |
| | 2.05–2.4 μm |
| Thermal infrared | 8.0–9.2 μm |
| | 10.2–12.4 μm |
| Microwave | 7.5–11.5 mm |
| | 20.0+ mm |

*Note.* Data selected from Fraser and Curran (1976, p. 35). Reproduced by permission of Addison-Wesley Publishing Co., Inc.

means representative of values that might be observed at a particular time and place. This diagram represents only the behavior of "shortwave" radiation (defined loosely here to include radiation with wavelengths less than 4.0 μm). It is true that the sun emits a broad spectrum of radiation, but the maximum intensity is emitted at approximately 0.5 μm within this region, and little solar radiation at wavelengths greater than 4.0 μm reaches the ground surface.

Of 100 units of shortwave radiation that reach the outer edge of the earth's atmosphere, about 3 are absorbed in the stratosphere as ultraviolet radiation interacts with oxygen to form ozone. Of the remaining 97 units, about 25 are reflected from clouds and about 19 are absorbed by dust and gases in the lower atmosphere. About 8 units are re-

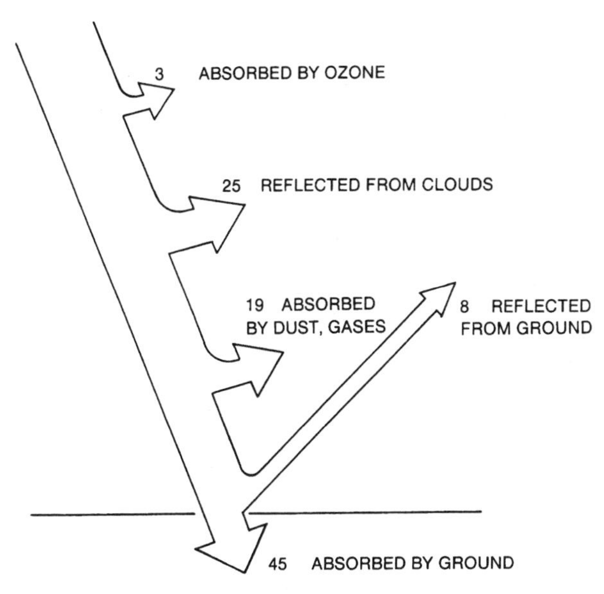


**FIGURE 2.12.** Incoming solar radiation. This diagram represents radiation at relatively short wavelengths, in and near the visible region. Values represent approximate magnitudes for the earth as a whole—conditions at any specific place and time would differ from those given here.

flected from the ground surface (this value varies greatly with different surface materials), and about 45 units ("about 50%") are ultimately absorbed at the earth's surface. For remote sensing in the visible spectrum, it is the portion reflected from the earth's surface that is of primary interest (Figure 1.4, Chapter 1), although knowledge of the quantity scattered is also important.

The 45 units that are absorbed are then re-radiated by the earth's surface. From Wien's displacement law (equation 2.5), we know that the earth, being much cooler than the sun, must emit radiation at much longer wavelengths than does the sun. The sun at 6,000 K, has its maximum intensity at 0.5 μm (in the green portion of the visible spectrum); the earth, at 300 K, emits with maximum intensity near 10 μm, in the far infrared spectrum.

Terrestrial radiation, with wavelengths longer than 10 μm, is represented in Figure 2.13. There is little, if any, overlap between wavelengths of solar radiation depicted in Figure 2.12, and terrestrial radiation shown in Figure 2.13. About 8 units are transferred from the ground surface to the atmosphere by "turbulent transfer" (heating of the lower atmosphere by the ground surface, which causes upward movement of air, then movement of cooler air to replace the original air). About 22 units are lost to the atmosphere by evaporation of moisture in the soil, water bodies, and vegetation (this energy is transferred as the latent heat of evaporation).

About 113 units are radiated directly to the atmosphere. Because of the effectiveness of atmospheric gases in absorbing infrared radiation, most of this energy (98 units) is retained by the atmosphere. About 15 units pass directly through the atmosphere to space; this is energy emitted at wavelengths that correspond to atmospheric windows (chiefly 8 to 13 μm). Energy absorbed by the atmosphere is re-radiated to space (49 units) and back to the earth (98 units). In meteorology it is these re-radiated units that are of interest because they form the source of energy for heating of the earth's atmosphere. For remote sensing it is the 15 units that pass through the atmospheric windows

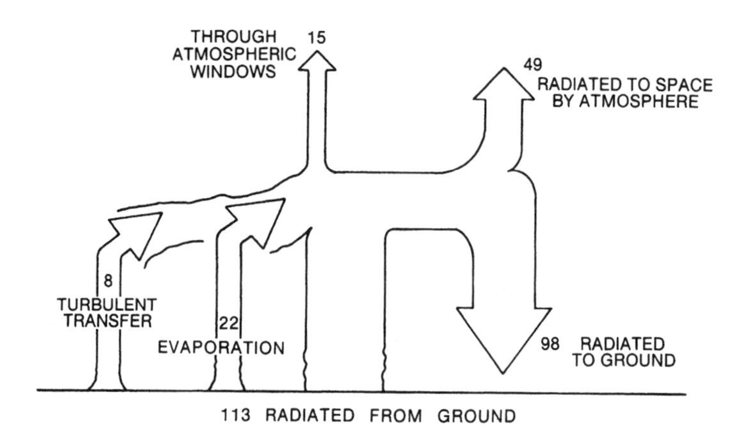


**FIGURE 2.13.** Outgoing terrestrial radiation. This diagram represents radiation at relatively long wavelengths—what we think of as sensible heat, or thermal radiation. Because the earth's atmosphere absorbs much of the radiation emitted by the earth, only those wavelengths that can pass through the atmospheric windows can be used for remote sensing.

that are of significance, as it is this radiation that presents information concerning the relative radiometric properties of earth features that can pass through the lower atmosphere without significant attenuation.

## 2.6. Interactions with Surfaces

As electromagnetic energy reaches the earth's surface, it must be reflected, absorbed, or transmitted. The proportions accounted for by each process depend on the nature of the surface, the wavelength of the energy, and the angle of illumination.

### *Reflection*

*Reflection* occurs when a ray of light is re-directed as it strikes a nontransparent surface. The nature of the reflection depends on sizes of surface irregularities (roughness or smoothness) in relation to the wavelength of the radiation considered. If the surface is smooth relative to wavelength, *specular* reflection occurs (Figure 2.14a). Specular reflection redirects all, or almost all, incident radiation in a single direction. For such surfaces, the angle of incidence is equal to the angle of reflection (i.e., in equation 2.8, the two media are identical, so $n = n'$, and therefore $\theta = \theta'$). For visible radiation, specular reflection can occur with surfaces such as a mirror, smooth metal, or a calm water body.

If a surface is rough relative to wavelength, it acts as a *diffuse,* or *isotropic* reflector. Energy is scattered more or less equally in all directions. For visible radiation, many natural surfaces might behave as diffuse reflectors, including, for example, uniform grassy surfaces. A perfectly diffuse reflector (often designated as a *Lambertian surface*) would have equal brightnesses when observed from any angle (Figure 2.14b).

The idealized concept of a perfectly diffuse reflecting surface is derived from the work of Johann H. Lambert (1728–1777), who conducted many experiments designed to describe the behavior of light. One of Lambert's laws of illumination states that the perceived brightness (radiance) of a perfectly diffuse surface does not change with the angle of view. This is Lambert's cosine law, which states that that the observed brightness ($I'$) of

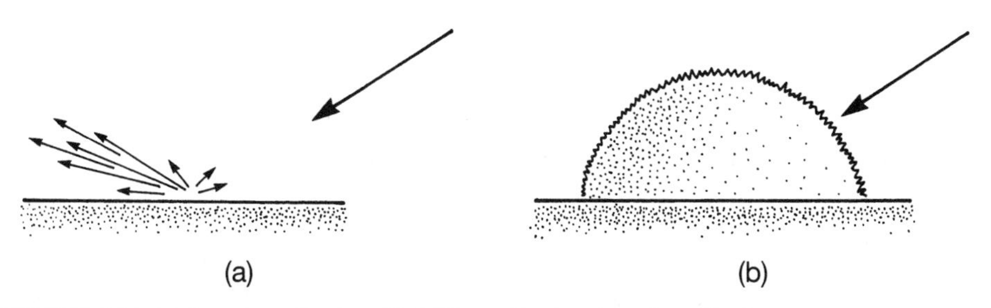


**FIGURE 2.14.** (*a*) Specular reflection. (*b*) Diffuse reflection. Specular reflection occurs when a smooth surface tends to direct incident radiation in a single direction. Diffuse reflection occurs when a rough surface tends to scatter energy more or less equally in all directions.

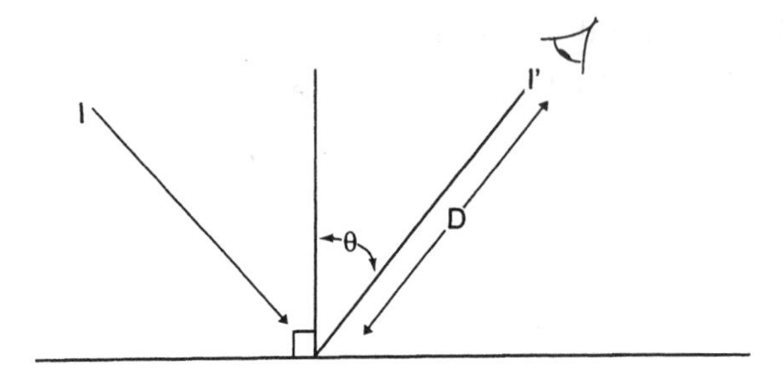


**FIGURE 2.15.** Inverse square law and Lambert's cosine law.

such a surface is proportional to the cosine of the incidence angle $\theta$, where $I$ is the brightness of the incident radiation as observed at zero incidence:

$$I' = I/\cos\theta \tag{Eq. 2.9}$$

This relationship is often combined with the equally important inverse-square law, which states that observed brightness decreases according to the square of the distance from the observer to the source:

$$I' = \frac{I}{D^2}(\cos\theta) \tag{Eq. 2.10}$$

(Both the cosine law and the inverse square law are depicted in Figure 2.15.)

### *Bidirectional Reflectance Distribution Function*

Because of its simplicity and directness, the concept of a Lambertian surface is frequently used as an approximation of the optical behavior of objects observed in remote sensing. However, the Lambertian model does not hold precisely for many, if not most, natural surfaces. Actual surfaces exhibit complex patterns of reflection determined by details of surface geometry (e.g., the sizes, shapes, and orientations of plant leaves). Some surfaces may approximate Lambertian behavior at some incidence angles but are clearly non-Lambertian properties at other angles.

Reflection characteristics of a surface are described by the *bidirectional reflectance distribution function* (BRDF). The BRDF is a mathematical description of the optical behavior of a surface with respect to angles of illumination and observation, given that it has been illuminated with a parallel beam of light at a specified azimuth and elevation. (The function is bidirectional in the sense that it accounts both for the angle of illumination and the angle of observation.) The BRDF for a Lambertian surface has the shape depicted in Figure 2.14b, with even brightnesses as the surface is observed from any angle. Actual surfaces have more complex behavior. The description of BRDFs for actual, rather than idealized, surfaces permits assessment of the degrees to which they approach the ideals of specular and diffuse surfaces (Figure 2.16).

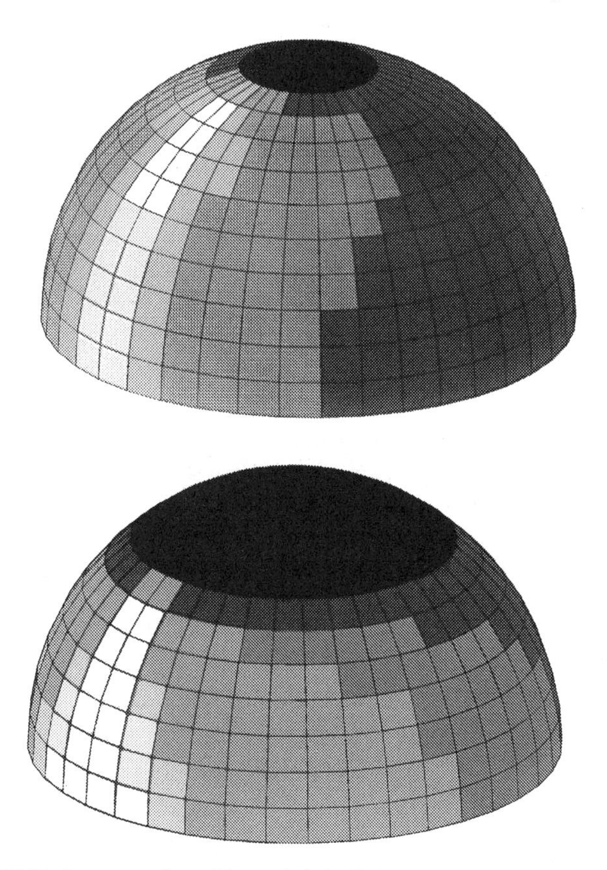

**FIGURE 2.16.** BRDFs for two surfaces. The varied shading represents differing intensities of observed radiation. (Calculated by Pierre Villeneuve.)

### *Transmission*

*Transmission* of radiation occurs when radiation passes through a substance without significant attenuation (Figure 2.17). From a given thickness, or depth, of a substance, the ability of a medium to transmit energy is measured as the transmittance ($t$):

$$t = \frac{\text{Transmitted radiation}}{\text{Incident radiation}} \qquad \text{(Eq. 2.11)}$$

In the field of remote sensing, the transmittance of films and filters is often important. With respect to naturally occurring materials we often think only of water bodies as capable of transmitting significant amounts of radiation. However, the transmittance of many materials varies greatly with wavelengths so our direct observations in the visible spectrum do not transfer to other parts of the spectrum. For example, leaves are generally opaque to visible radiation but transmit significant amounts of radiation in the infrared.

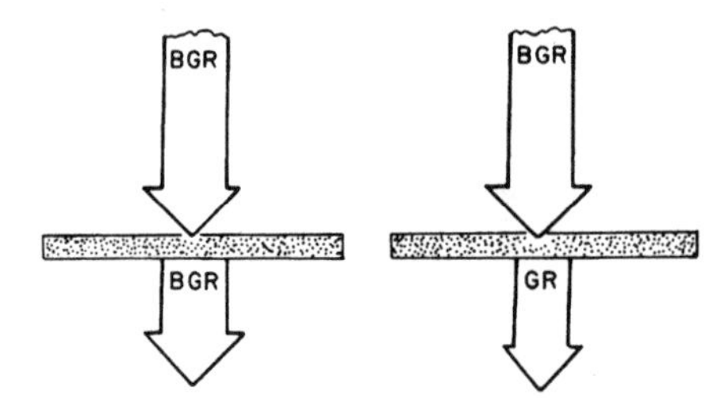


**FIGURE 2.17.** Transmission. Incident radiation passes through an object without significant attenuation (left), or may be selectively transmitted (right). The object on the right would act as a yellow ("minus blue") filter, as it would transmit all visible radiation except for blue light.

### *Fluorescence*

Fluorescence occurs when an object illuminated with radiation of one wavelength emits radiation at a different wavelength. The most familiar examples are some sulfide minerals, which emit visible radiation when illuminated with ultraviolet radiation. Other objects also fluoresce, although observation of fluorescence requires accurate and detailed measurements not now routinely available for most applications. Figure 2.18 illustrates the fluorescence of healthy and senescent leaves, using one axis to describe the spectral distribution of the illumination and the other to show the spectra of the emitted energy. These contrasting surfaces illustrate the effectiveness of fluorescence in revealing differences between healthy and stressed leaves.

### *Spectral Properties of Objects*

Remote sensing consists of the study of radiation emitted and reflected from features at the earth's surface. In the instance of emitted (far infrared) radiation, the object itself is the immediate source of radiation. For reflected radiation, the source may be the sun, the atmosphere (by means of scattering of solar radiation), or man-made radiation (chiefly imaging radars).

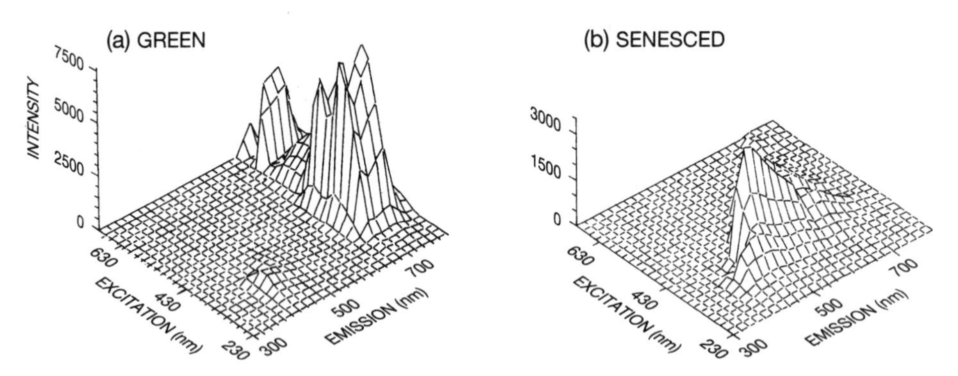


**FIGURE 2.18.** Fluorescence. Excitation and emission are shown along the two horizontal axes (with wavelengths given in nanometers). The vertical axes show strength of fluoresence, with the two examples illustrating the contrast in fluoresence between (*a*) healthy and (*b*) senesced leaves. From Rinker (1994).

A fundamental premise in remote sensing is that we can learn about objects and features on the earth's surface by studying the radiation reflected and/or emitted by these features. Using cameras and other remote sensing instruments, the brightnesses of objects can be observed over a range of wavelengths, so that there are numerous points of comparison between brightnesses of separate objects. A set of such observations or measurements constitute a spectral response pattern, sometimes called the *spectral signature* of an object (Figure 2.19). Ideally, detailed knowledge of a spectral response pattern might permit identification of such features of interest as separate kinds of crops, forests, or minerals. This idea has been expressed as follows: "Everything in nature has its own unique distribution of reflected, emitted, and absorbed radiation. These spectral characteristics can—if ingeniously exploited—be used to distinguish one thing from another or to obtain information about shape, size, and other physical and chemical properties" (Parker & Wolff, 1965, p. 21).

This statement expresses the fundamental concept of the spectral signature—the notion that features display unique spectral responses that would permit clear identification—from spectral information alone—of individual crops, soils, and so on, from remotely sensed images. In practice, it is now recognized that spectra of features change both over time (e.g., as a cornfield grows during a season) and over distance (e.g., as pro-

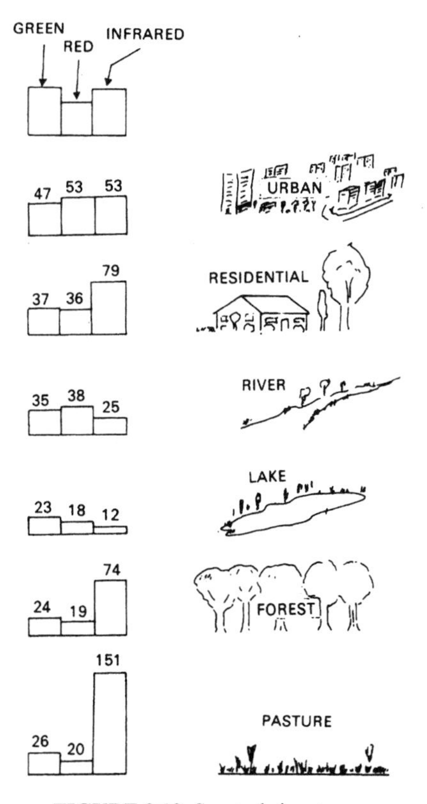


**FIGURE 2.19.** Spectral signatures.

portions of specific tree species in a forest change from place to place). Furthermore, it is also clear that few, if any, objects in nature can be uniquely identified at the desired levels of detail using the spectral signature concept.

Nonetheless, the study of spectral properties of objects forms an important part of remote sensing. Some research has been focused on examination of spectral properties of different classes of features. Thus, although it may not be feasible to define unique signatures for specific kinds of vegetation, we can recognize distinctive spectral patterns for vegetated and nonvegetated areas, for certain classes of vegetation, and can sometimes detect the existence of diseased or stressed vegetation. In other instances, we may be able to define spectral patterns that are useful within restricted geographic and temporal limits as a means of studying the distributions of certain plant and soil characteristics. Chapter 14 describes how very detailed spectral measurements permit application of some aspects of the concept of the spectral signature.

## 2.7. Summary: Three Models for Remote Sensing

Remote sensing typically takes one of three basic forms depending on the wavelengths of energy detected and on the purposes of the study. The simplest is to record the reflection of solar radiation from the earth's surface (Figure 2.20). This is the kind of remote sensing most nearly similar to everyday experience. For example, film in a camera records radiation from the sun after it is reflected from the objects of interest, regardless of whether one uses a simple hand-held camera to photograph a family scene or an aerial camera to photograph a large area of the earth's surface. This form of remote sensing mainly uses energy in the visible and near infrared portions of the spectrum. Key variables include atmospheric clarity, spectral properties of objects, angle and intensity of the solar beam, choices of films, filters, and other variables as explained in Chapter 3.

A second strategy for remote sensing is to record radiation *emitted* from (rather than reflected from) the earth's surface. Because the emitted energy is strongest in the far infrared spectrum, this kind of remote sensing requires special instruments designed to

**FIGURE 2.20.** Remote sensing using reflected solar radiation. The sensor detects solar radiation that has been reflected from features at the earth's surface. (See Figure 2.12.)

**FIGURE 2.21.** Remote sensing using emitted terrestrial radiation. The sensor records solar radiation that has been absorbed by the earth, then re-emitted as thermal infrared radiation. (See Figures 2.12 and 2.13.)

record these wavelengths. (There is no direct analogue to everyday experience for this kind of remote sensing.) Emitted energy from the earth's surface is mainly derived from shortwave energy from the sun that has been absorbed, then re-radiated at longer wavelengths (Figure 2.21). Emitted radiation from the earth's surface reveals information concerning thermal properties of materials, which can be interpreted to suggest patterns of moisture, vegetation, surface materials, and man-made structures. Other sources of emitted radiation (of secondary significance here, but often of primary significance elsewhere) include geothermal energy, heat from steam pipes, power plants, buildings, and forest fires. This example also represents "passive" remote sensing because the energy we sense is emitted from the earth, not the sensor itself.

Finally, a third class of remote sensing instruments generates their own energy, then records the reflection of that energy from the earth's surface (Figure 2.22). These are "active" sensors—"active" in the sense that they provide their own source of energy and are independent of solar and terrestrial radiation. As an everyday analogy, a camera with a flash attachment can be considered an active sensor. In practice, active sensors are best represented by imaging radars (Chapter 7) which transmit a microwave signal toward the earth's surface from an aircraft or satellite, then use the reflected energy to form an image.

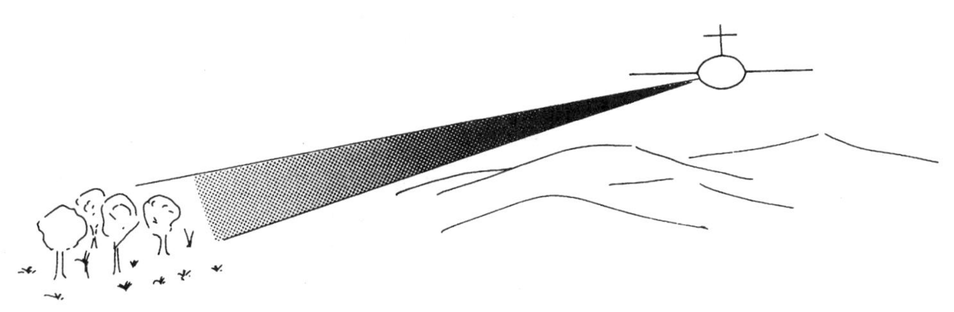

**FIGURE 2.22.** Active remote sensing. The sensor illuminates the terrain with its own energy, then records the reflected energy as it has been altered by the earth's surface.

Because they sense energy provided directly by the sensor itself, such instruments have the capability to operate at night and during cloudy weather (but not through intense thunderstorms).

## Review Questions

1. Using books provided by your instructor or available in your library, examine reproductions of landscape paintings to identify artistic use of atmospheric perspective. Perhaps some of your own photographs of landscapes illustrate the optical effects of atmospheric haze.

2. Some streetlights are deliberately manufactured to provide illumination with a reddish color. From material presented in this chapter, can you suggest why?

3. Although this chapter has largely dismissed ultraviolet radiation as an important aspect of remote sensing, there may well be instances where it might be effective, despite the problems associated with its use. Under what conditions might it prove practice to use ultraviolet radiation for remote sensing?

4. The human visual system is most nearly similar to which model for remote sensing as described in the last sections of this chapter?

5. Can you identify analogues from the animal kingdom for each of the models for remote sensing discussed in Section 2.7?

6. Examine Figures 2.12 and 2.13, which show the radiation balance of the earth's atmosphere. Explain how it can be that 100 units of solar radiation enter at the outer edge of the earth's atmosphere, yet 113 units are emitted by the atmosphere.

7. Examine Figures 2.12 and 2.13 again. Discuss how the values in this figure might change in different environments, including (a) a desert, (b) the arctic, (c) an equatorial climate. How might these differences influence our ability to conduct remote sensing in each region?

8. Spectral signatures can be illustrated using values indicating the brightness in several spectral regions.

| | UV | Blue | Green | Red | IR |
|---|---|---|---|---|---|
| Forest | 28 | 29 | 36 | 27 | 56 |
| Water | 22 | 23 | 19 | 13 | 8 |
| Corn | 53 | 58 | 59 | 60 | 71 |
| Pasture | 40 | 39 | 42 | 32 | 62 |

   Assume for now that these are "pure" signatures, not influenced by effects of the atmosphere. Can all categories be reliably separated, based upon these spectral values? Which bands are most useful for distinguishing between these classes?

9. Describe ideal atmospheric conditions for remote sensing.

## References

Bohren, Craig F. 1987. *Clouds in a Glass of Beer: Simple Experiments in Atmospheric Physics*. New York: Wiley, 195 pp.

Campbell, J. B., and L. Ran. 1993. CHROM: A C Program to Evaluate the Application of the Dark Object Subtraction Technique to Digital Remote Sensing Data. *Computers and Geosciences,* Vol. 19, pp. 1475–1499.

Chahine, Moustafa T. 1983. Interaction Mechanisms within the Atmosphere. Chapter 5 in *Manual of Remote Sensing* (R. N. Colwell, ed.). Falls Church, VA: American Society of Photogrammetry, pp. 165–230.

Chameides, William L., and Douglas D. Davis. 1982. Chemistry in the Troposphere. *Chemical and Engineering News*, Vol. 60, pp. 39–52.

Estes, John E. 1978. The Electromagnetic Spectrum and Its Use in Remote Sensing. Chapter 2 in *Introduction to Remote Sensing of Environment* (B. F. Richason, ed.). Dubuque, IA: Kendall-Hunt, pp. 15–39.

Fraser, R. S., and R. J. Curran. 1976. Effects of the Atmosphere on Remote Sensing. Chapter 2 in *Remote Sensing of Environment* (C. C.Lintz and D. S. Simonett, eds.). Reading, MA: Addison-Wesley, pp. 34–84.

Goetz, A. F. H., J. B. Wellman, and W. L. Barnes. 1985. Optical Remote Sensing of the Earth. *Proceedings of the IEEE,* Vol. 73, pp. 950–969.

Kaufman, Yoram. J. 1984. Atmospheric Effects on Remote Sensing of Surface Reflectance. *Remote Sensing* (P. N. Slater, Ed.). *Proceedings, SPIE,* Vol. 475, pp. 20–33.

Kaufman, Y. J. 1989: The Atmospheric Effect on Remote Sensing and Its Correction. Chapter 9 in *Theory and Applications of Optical Remote Sensing* (Ghassam Asrar, ed.). New York: Wiley, pp. 336–428.

Nunnally, Nelson R. 1973. Introduction to Remote Sensing: The Physics of Electromagnetic Radiation. Chapter 3 in *The Surveillant Science: Remote Sensing of the Environment* (R. K. Holz, ed.). New York: Houghton Mifflin, pp. 18–27.

Minnaert, M. 1954. *The Nature of Light and Color*. New York: Dover. 362 pp. (Revision by H. M. Kremer-Priest; translation by K. E. Brian Jay.)

Parker, D. C., and M. F. Wolff. 1965. Remote Sensing. *International Science and Technology,* Vol. 43, pp. 20–31.

Rinker, J. N. 1994. ISSSR Tutorial I: Introduction to Remote Sensing. In *Proceedings of the International Symposium on Spectral Sensing Research '94*. Alexandria, VA: U.S. Army Topographic Engineering Center, pp. 5–43.

Rees, W. G. 1990. *Physical Principles of Remote Sensing.* New York: Cambridge University Press, 247 pp.

Slater, Philip N. 1980. *Remote Sensing: Optics and Optical Systems*. Reading, MA: Addison-Wesley, 575 pp.

Stimson, Allen. 1974. *Photometry and Radiometry for Engineers*. New York: Wiley, 446 pp.

Suits, Gwynn H. 1983. The Nature of Electromagnetic Radiation. Chapter 2 in *Manual of Remote Sensing* (R. N. Colwell, ed.). Bethesda, MD: American Society of Photogrammetry, pp. 37–60.

Swain, Philip H., and Shirley M. Davis. (eds.). 1978. *Remote Sensing: The Quantitative Approach*. New York: McGraw-Hill, 396 pp.

Turner, R. E., Malila, W. A., and Nalepka, R. F. 1971. Importance of Atmospheric Scattering in Remote Sensing. In *Proceedings of the 7th International Symposium on Remote Sensing of Environment,* pp. 1651–1697.

PART ONE

# IMAGE ACQUISITION

CHAPTER THREE

# Photographic Sensors

## 3.1. Introduction

The word *photography* was coined in France in the mid-1800s, when it was fashionable to seek Greek and Latin words to name new scientific discoveries. The word *photography* means "to write with light"—a literal description of the use of the then newly invented camera. Today our meaning is expanded to include radiation outside the visible spectrum, in the ultraviolet and near infrared regions.

Despite the many recent innovations in imaging technology, photography remains the most practical, inexpensive, and widely used means of remote sensing. Further, the basic optical principles used for photography are also employed in optical systems of nonphotographic sensors, and we often use photographic film to record images generated by nonphotographic sensors. Therefore, knowledge of photography forms a central core for understanding the field of remote sensing.

The formation of images by refraction of light is a surprisingly old practice. In antiquity, Greek and Arab scholars knew that images could be formed as light passed through a pinhole opening in a dark enclosure. Refraction of light at the tiny opening bends light rays to form an inverted image in a manner analogous to the effect of a simple lens. In medieval Europe, a device known as the *camera obscura* ("dark chamber") employed this principle to project an image onto a screen as an aid for artists, who could then trace outlines of an image as the basis for more elaborate drawings. During the Renaissance, the addition of a simplex convex lens improved the *camera obscura,* although there was still no convenient means of recording the image formed on the screen. Later, with the development of photographic emulsions (described below) as a means of making a detailed record of the image, the *camera obscura* began its evolution toward the popular cameras that we know today, which in turn are models for the more complex cameras used for aerial survey.

Despite the current use of much more sophisticated imaging systems, aerial photography remains one of the most reliable and most widely used forms of remotely sensed imagery. Routine use of aerial photography has incalculable value throughout the world as a source of information concerning the landscape and as the primary means of producing modern topographic maps. Its economic contributions to effective planning and surveying of the earth are considerable, and it is clear that aerial photography will remain as an important source of remote sensing imagery for many years to come.

## 3.2. The Aerial Camera

In their most basic elements, aerial cameras are similar to the simple hand-held cameras we all have used. Both share the four main components of all cameras: (1) a lens to focus light on the film, (2) a light-sensitive film to record the image, (3) a shutter that controls entry of light into the camera, and (4) the camera body, a light-tight enclosure that holds the film, lens, and shutter in their correct positions.

In addition, aerial cameras include three other elements not usually encountered in our personal experiences with photography: the film magazine, the drive mechanism, and the lens cone (Figure 3.1).

### *The Lens*

The lens gathers reflected light and focuses it on the film. In its simplest form, a lens is formed from a glass disk carefully ground into a shape with nonparallel curved surfaces (Figure 3.2). The change in optical densities as light rays pass from the atmosphere to the lens and back to the atmosphere causes refraction of light rays; the sizes, shapes, arrangements, and compositions of lenses are carefully designed to control bending of light rays, to maintain color balance, and to minimize optical distortions. Optical characteristics of lenses are determined largely by the refractive index of the glass (Chapter 2) and the degree of curvature present in the lens surface. The quality of a lens is determined by the

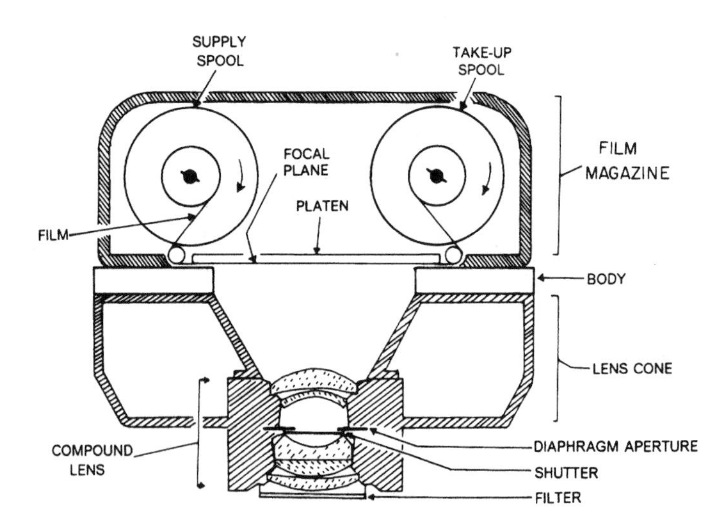


**FIGURE 3.1.** Schematic diagram of an aerial camera. Cross-sectional view. Labeled items are discussed in the text.

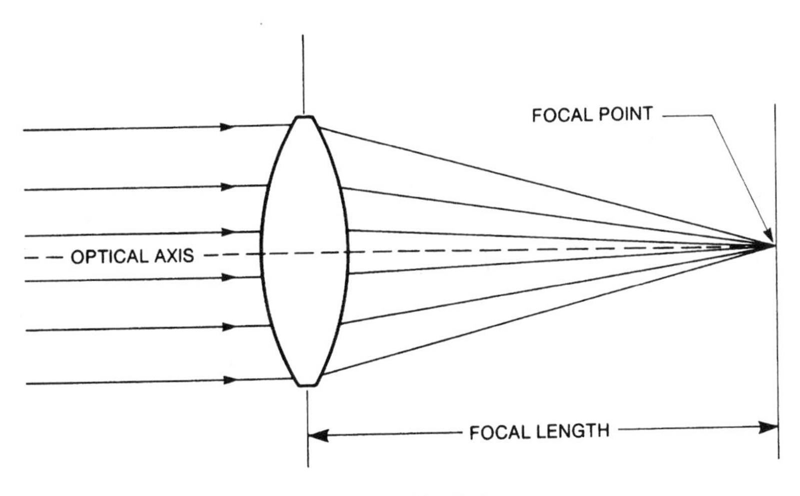


**FIGURE 3.2.** Simple lens.

quality of glass, the precision with which the glass is shaped, and the accuracy with which it is positioned within a camera. Imperfections in lens shape contribute to spherical aberration, a source of error that contributes to loss of clarity and distortion of images. For modern aerial photography, spherical aberration is usually not a severe problem because most modern aerial cameras use lenses of very high quality.

Figure 3.2 shows the simplest of all lenses—a simple positive lens. Such a lens is formed from a glass disk with equal curvature at both sides; light rays are refracted at both edges to form an image. Most aerial cameras use compound lenses, formed from many separate lenses of varied sizes, shapes, and properties. These many components are designed to correct for the errors that may be present in any single component, so the whole unit is much more accurate than any single element. For our purposes, consideration of a simple lens will be sufficient to define the most important feature of lenses, even though they differ from those actually used in modern aerial cameras.

The *optical axis* joins the centers of curvature of the two sides of the lens. Although refraction occurs throughout a lens, a plane passing through the center of the lens, known as the image principal plane, is considered to be the center of refraction within the lens (Figure 3.3). The image principal plane intersects the optical axis at the nodal point.

Parallel light rays reflected from an object at a great distance (at an "infinite" distance) pass through the lens and are brought to focus at the principal focal point, the point at which the lens forms an image of the distant object. The chief ray passes through the nodal point without changing direction; all other rays are bent by the lens. A plane passing through the focal point parallel to the image principal plane is defined as the *focal plane.* For hand-held cameras, the distance from the lens to the object is important because the image is brought into focus at distances that increase as the object is positioned closer to the lens. For such cameras, it is important to use lenses that can be adjusted to bring each object to a correct focus as the distance from the camera to the object changes. For aerial cameras, the scene to be photographed is always at such large distances that the focus can be fixed at infinity, with no need to change the focus of the lens.

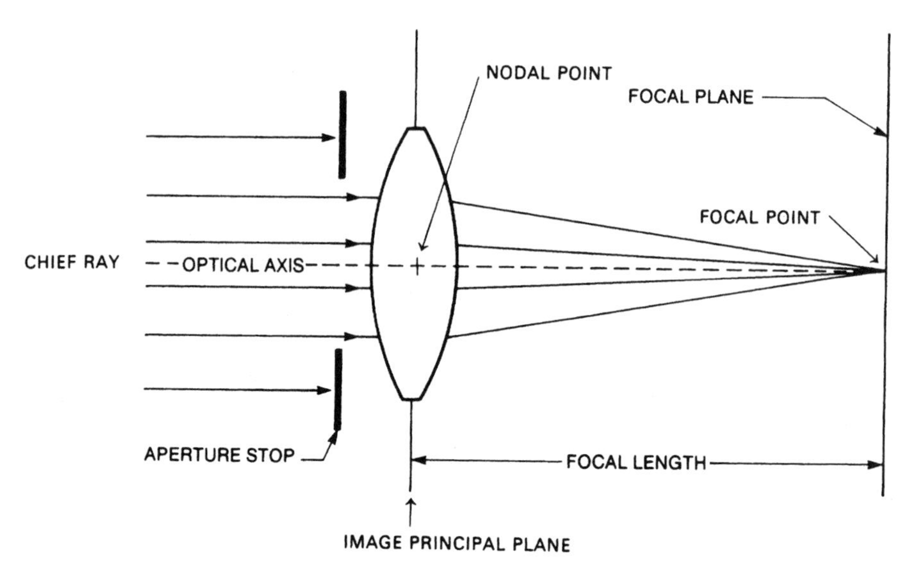


**FIGURE 3.3.** Cross-sectional view of an image formed by a simple lens.

For a simple positive lens, the *focal length* is defined as the distance from the center of the lens to the focal point, usually measured in inches or millimeters. (For a compound lens, the definition is more complex.) For a given lens, the focal length is not identical for all wavelengths. Blue light is brought to a focal point at a shorter distance than are red or infrared wavelengths (Figure 3.4). This effect is the source of *chromatic aberration.* Unless corrected by lens design, chromatic aberration would cause individual colors of an image to be out of focus in the photograph. Chromatic aberration is corrected in high-quality aerial cameras to ensure that the radiation used to form the image is brought to a common focal point.

The field of view of a lens can be controlled by a *field stop*—a mask positioned just

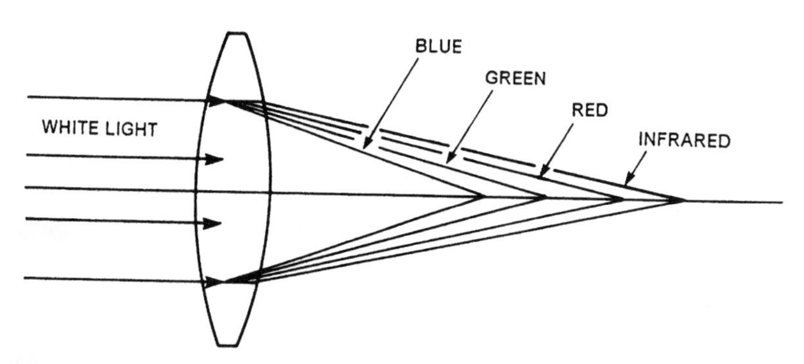


**FIGURE 3.4.** Chromatic aberration. Energy of differing wavelengths is brought to a focus at varying distances from the lens. More complex lenses are corrected to bring all wavelengths to a common focal point.

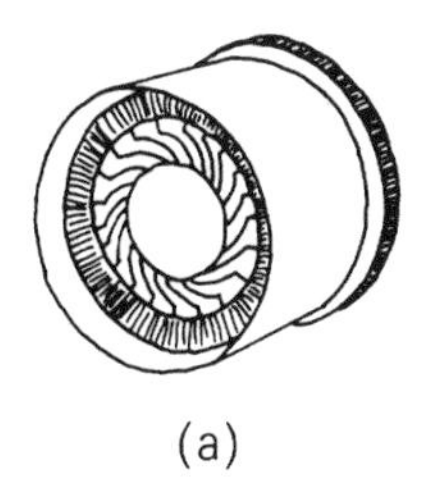
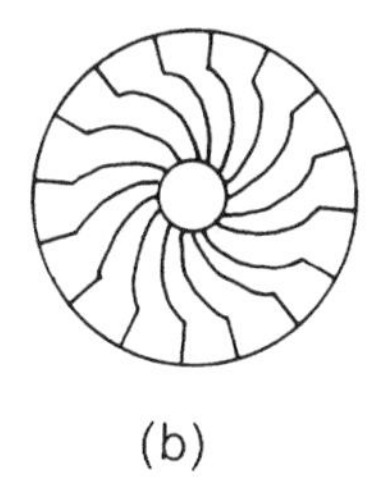
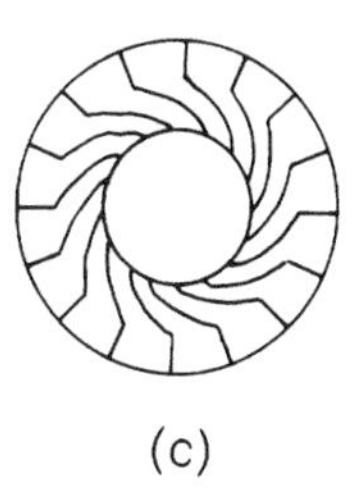

(a) (b) (c)

**FIGURE 3.5.** Diaphragm aperture stop. (*a*) Perspective view. (*b*) Narrow aperture. (*c*) Wide aperture.

in front of the focal plane. An aperture stop is usually positioned near the center of a compound lens; it consists of a mask with a circular opening of adjustable diameter (Figure 3.5). An aperture stop can control the intensity of light as the focal plane but does not influence the field of view or the size of the image. Manipulation of the aperture stop controls only the brightness of the image without changing its size. Usually aperture size is measured as the diameter of the adjustable opening that admits light to the camera.

Relative aperture is defined as

$$f = \frac{\text{Focal length}}{\text{Aperture size}} \qquad \text{(Eq. 3.1)}$$

where focal length and aperture size are measured in the same units of length, and $f$ is the *f number*—the relative aperture. A large $f$ number means that the aperture opening is small relative to focal length; a small $f$ number means that the opening is large relative to focal length.

Why use $f$ numbers rather than direct measurements of aperture? One reason is that standardization of aperture with respect to focal length permits specification of aperture sizes using a value that is independent of camera size. Specification of an aperture as "23 mm" has no practical meaning unless we also know the size (focal length) of the camera. Specification of aperture as "$f$ 4" has meaning for cameras of all sizes; we know that it is one fourth of the focal length, for any size camera.

The standard sequence of apertures is: $f1, f1.4, f2, f2.8, f4, f5.6, f8, f11, f16, f22, f32, f64, \ldots$. This sequence is designed to change the amount of light by a factor of 2 as the $f$ stop is changed by one position. For example, a change from $f2$ to $f2.8$ halves the amount of light entering the camera; a change from $f11$ to $f8$ doubles the amount of light. A given lens, of course, is capable of using only a portion of the range of apertures mentioned above.

### *The Shutter*

The shutter controls the length of time the film is exposed to light. The simplest shutters are often metal blades positioned between elements of the lens, forming "intralens," or "between the lens," shutters. An alternative form of shutter is the focal plane shutter, consisting of a metal or fabric curtain positioned just in front of the film, near the focal plane. The curtain is constructed with a number of slits; the choice of shutter speed by the operator selects the opening that produces the desired exposure. Although some aerial cameras use focal plane shutters, the between-the-lens shutter is preferred for use in most aerial cameras. The between-the-lens shutter subjects the entire negative to illumination simultaneously and presents a clearly defined perspective that permits use of the image negative as the basis for precise measurements.

### *The Film Magazine*

The film magazine (Figure 3.1) is a light-tight container that holds the supply of film. The magazine usually includes a supply spool, holding perhaps several hundred feet of unexposed aerial film, and a take-up spool to accept exposed film.

### *The Lens Cone*

The lens cone (Figure 3.1) supports the lens and filters and holds them in their correct positions in relation to the film. The lens cone is usually detachable to permit use of different lenses with the same camera body. The camera manufacturer carefully aligns the lens with the other components of the camera to ensure geometric accuracy of photographs. Common focal lengths for typical aerial cameras are 150 mm (about 6 in.), 300 mm (about 12 in.), and 450 mm (about 18 in.). Slater (1975) lists characteristics (including focal lengths and apertures) for a number of specific models of aerial cameras.

### *The Drive Mechanism*

The drive mechanism advances the film after each exposure. At the time of exposure, many cameras use a vacuum platen to hold the film during exposure. The hand-held cameras that most of us use have very simple platens—small spring-mounted metal plates designed to hold the film flat at the focal plane. Because of the difficulty of holding large sheets of film flat, aerial cameras use special platens. A *vacuum platen* consists of a flat plate positioned at the focal plane; a vacuum pump draws air through small holes in the plate to hold the film flat and stationary during exposure. The vacuum is released after exposure to allow the film to advance for the next exposure, then is applied again as the next frame is ready for exposure. The vacuum sucks the film flat against the platen to prevent bending of the film or formation of bubbles of air as the film is positioned in the focal plane.

## 3.3. Kinds of Aerial Cameras

Most civilian aerial photography has been acquired using *metric cameras* (sometimes called *cartographic cameras*) (Figure 3.1). These are aerial cameras designed to provide high-quality images with a minimum of optical and geometric error. Metric cameras used for professional work have been calibrated at special laboratories operated by the manufacturer or by governmental agencies. Each camera is used to photograph a target image having features positioned with great accuracy. Then precise measurements are made of focal length, flatness of the focal plane, and other variables. Such precise knowledge of the internal geometry of a camera permits photogrammetrists to make accurate measurements from photographs.

Other kinds of aerial cameras are less frequently used for routine photography but may have uses for special applications. Reconnaissance cameras have been chiefly designed for military use. For such applications, geometric accuracy may be less important than the ability to take photographs at high air speed, at low altitude, or under unfavorable illumination. As a result, photographs from reconnaissance cameras do not have the geometric accuracy expected from those taken by metric cameras.

Strip cameras acquire images by moving film in front of a fixed slit that serves as a form of shutter (Figure 3.6). The speed of film movement as it passes the slit is coordinated with the speed and altitude of the aircraft to provide proper exposure. The image is a long, continuous strip of imagery without the individual frames formed by conventional cameras. Strip cameras are capable of acquiring high-quality images from planes flying at high speed and low altitudes—optical conditions that are so extreme that conventional cameras often cannot provide the fast shutter speeds necessary to acquire sharp images.

Panoramic cameras (Figure 3.6) are designed to record a very wide field of view. Usually, a lens with a narrow field of view scans across a wide strip of land, forming an image by the side-to-side motion of the lens, as the aircraft moves forward. Photographs from panoramic cameras show a long, narrow strip of terrain that extends perpendicular to the flight track from horizon to horizon. Because of the forward motion of the aircraft during the side-to-side scan of the lens, panoramic photographs have serious geometric distortions that require correction before they can be used as the basis for measurements.

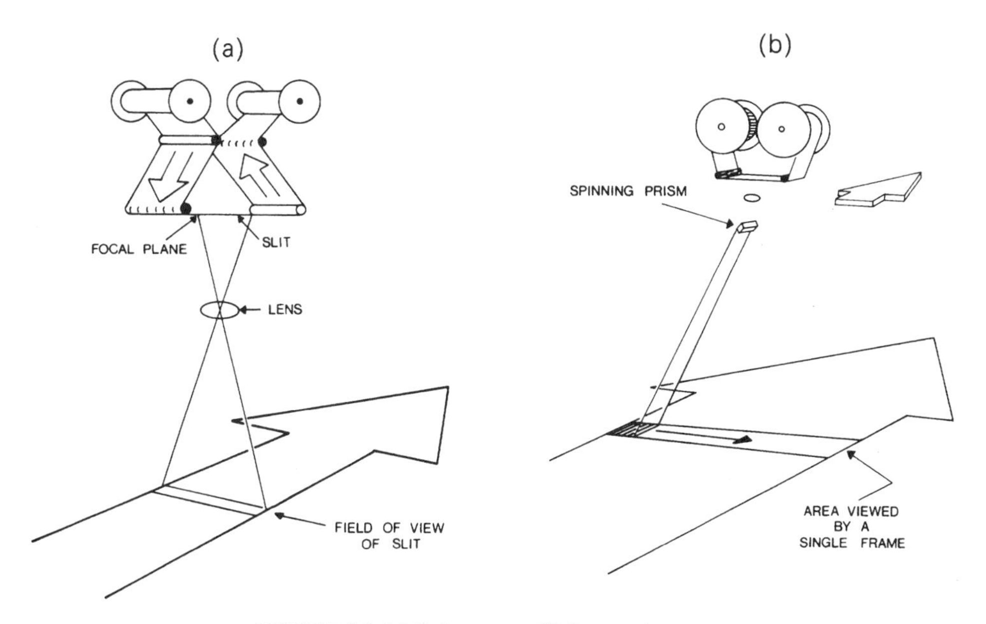


**FIGURE 3.6.** (*a*) Strip camera. (*b*) Panoramic camera.

Panoramic aerial photographs are useful because of the large areas they represent, but only the central portions are suitable for detailed interpretation because of the large variations in scale and detail present near the outside edges of the images.

## 3.4. Black-and-White Aerial Films

Today the field of remote sensing encompasses a wide variety of sensors, both photographic and nonphotographic. Yet we have only one practical medium for recording images on paper or film—the photographic emulsion. Therefore, even if we do not use a camera to record an image, we must still use photographic film to prepare film or paper copies. Knowledge of the qualities and limitations of photographic film is central to understanding the field of remote sensing.

In the late 1700s and early 1800s, a number of amateur scientists experimented with light-sensitive (*photosensitive*) chemicals. For example, silver nitrate ($AgNO_3$), familiar to high school chemistry students, darkens when exposed to sunlight; therefore, a glass or metal plate coated with silver nitrate formed the basis for recording a crude image. Those areas on the plate where the light is brightest became dark; those where the light is dim remained light in tone.

Joseph Nicephone Niepce (1765–1833), a French chemist, is one of the many who experimented with such chemicals. He is often assigned credit for devising the first negative image (1826). Niepce worked with Louis Daguerre (1789–1851), a French scientist and artist, to design a silver-coated metal plate treated with iodine vapor. Their invention formed the first practical means of recording projected images. Their experiments were conducted over many years; by tradition, the year that Daguerre ceded their invention to the French Academy of Sciences (1839) is given as the date for the beginning of photography.

Daguerrotypes (an early name for photographic images made using Daguerre's method) were used for many years, with many modifications. Many features of early photography differ greatly from modern equipment and practice and were clearly impractical for routine aerial photography. Then, equipment was large, heavy, and fragile. Exposures times were long, required bright light, and were recorded on metal or glass plates, which themselves were heavy, fragile, and awkward. Nonetheless, many aerial photographs were taken, mainly by using balloons or large kites as a means of elevating the camera. Of course, such photographs were primarily curiosities rather than scientific tools because of the difficulty of controlling the orientation of the camera. Furthermore, each photographer tended to have tailor-made equipment; photographers often prepared their own chemicals and used individually formulated emulsions. The lack of standardization of equipment, materials, and practice meant that even the fundamentals of photography were as much an art as a science.

The more compact lightweight photographic equipment required for modern aerial photography was made possible by developments started by George Eastman (1854–1932), who invented roll film and improved and standardized methods of photographic processing. His invention of the Kodak camera in 1888, and formation of the Eastman Kodak Company (1892), popularized the practice of photography by mass production of standardized photographic products. Widening the scope of photography increased the number of people knowledgeable about photographic practice, standardized photographic practice, and decreased costs of photographic materials. In brief, his work created the environment in which modern aerial survey could develop and grow into its present form.

Initially, photographic films were designed to be sensitive primarily to portions of the visible spectrum, so any single image could portray only those brightnesses in a single broad region of the spectrum. Modern films are sensitive to nonvisible portions of the spectrum and can represent reflectances in much more specific spectral regions.

### *Major Components*

Aerial films have essentially the same structure as photographic films used in hand-held cameras. The film *base,* or *support,* is usually a thin (40 to 100 μm), flexible, transparent material that holds a light-sensitive coating. In the early days of photography, the support was often formed from metal or glass plates, but today such materials are inconvenient for everyday use. Modern films have bases of polyester film. These materials are useful because they can be fabricated into thin, lightweight, flexible strips that are strong enough to withstand the forceful motions of winding and unwinding as film is moved within the camera.

The base must be able to resist changes in size as temperature and humidity vary. Photogrammetrists measure distances on images so precisely that even small differences in image size due to shrinking or expansion of the base can introduce significant errors. Therefore, glass plates are still used for images that are to be used with some photogrammetric instruments because they are insensitive to variations in temperature and humidity.

The base is coated with a light-sensitive coating—the photographic emulsion (Figure 3.7). Photosensitive coatings used at the beginnings of photography were formed from silver nitrate (metallic silver dissolved in nitric acid). When a surface coated with silver nitrate is exposed to light, the silver nitrate darkens as the action of light changes it to metallic silver. The darkening effect increases as the light becomes more intense or as the length of exposure is increased. This effect provided a crude means of recording the image of a scene, but a number of practical problems (including the long exposures required to darken the coating) provided the incentive to develop improved photosensitive coatings.

Modern emulsions consist of extremely small crystals of silver halide (typically, silver bromide [95%] and silver iodide [5%]) suspended in a gelatin matrix (possibly 5 μm thick). These crystals form the light-sensitive portion of modern films, just as silver ni-

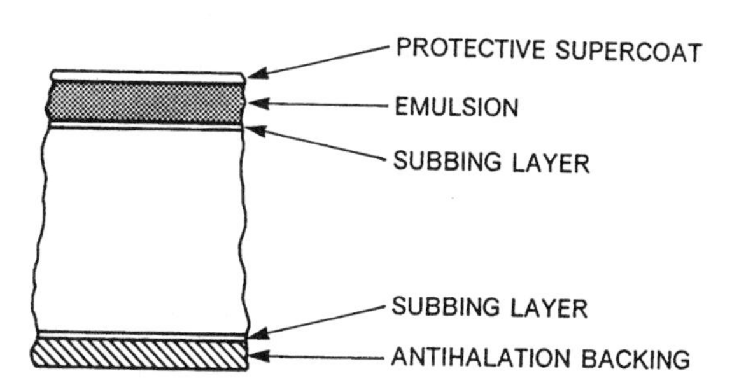


**FIGURE 3.7.** Schematic cross-sectional view of black-and-white photographic film.

trate was the light-sensitive agent in the early days of photography. Although the gelatin that holds the grains is ostensibly a mundane substance, it possesses several important characteristics. Silver halide crystals are insoluble and possess other physical characteristics that prevent them from adhering directly to the base. Gelatin holds the crystals in suspension, permitting the manufacturer to spread them evenly on the base. Furthermore, gelatin is transparent and porous (to allow photographic chemicals to contact crystals) and absorbs halogen gases released when light strikes the emulsion.

Physical characteristics of the silver halide crystals assume some importance. They are extremely small and irregular in shape, with many sharp edges—shapes that favor interception of photons that pass into the emulsion. The finer the size of the grains, the finer the detail that can be recorded. Coarser grains can record less detail but produce a film with greater sensitivity to light. Thus, the spatial resolution of a film is inversely related to its speed; as we increase the sizes of crystals to improve sensitivity to light, the finest level of detail that the emulsion can record becomes coarser. Alternatively, if we design a film with very fine grains to record fine detail, the emulsion exhibits decreased sensitivity—we must have brighter light or use longer exposures.

Recent research by Eastman Kodak Company has produced film emulsions with grains that are flat in shape. These new grains have essentially the same volume as those in older emulsions, but their surface area is greatly increased. The flat grains, oriented parallel to the film surface, expose large surface areas to the light, thereby increasing the speed of the film without decreasing the resolution of the film. At present, such films are marketed for popular photography (color prints only) rather than for aerial survey.

The film emulsion is coated with a thin layer of clear gelatin—the *protective supercoat*—designed to shield the emulsion from scratches during handling. Despite the presence of the supercoat, the emulsion is still vulnerable to damage from dust and from moisture and oil from handling with bare hands. As a result, analysts and interpreters should always wear cotton gloves when handling film or should protect film with transparent plastic sleeves.

Below the emulsion is a *subbing layer,* designed to ensure that the emulsion adheres to the base. The back side of the base has an *antihalation* backing. This backing absorbs light that passes through the emulsion and the base, to prevent reflection back to the emulsion. In the absence of such a backing, the images of bright objects will be surrounded by halos caused by these reflections. The backing also acts as an anticurl agent, to counteract the curling effect of the emulsion that coats the upper side of the film.

When the shutter opens, it allows light to enter and strike the emulsion. The silver halide crystals are so small that even a small area of the film contains many thousands of crystals. When light strikes a crystal, it changes a small portion of the crystal (perhaps only a single molecule) to metallic silver. The more intense the light striking a portion of the film, the greater the number of crystals affected. Thus the pattern of crystals influenced by light forms a record of patterns of light reflected from the scene. If it were possible to examine the exposed film without again subjecting it to the effects of light, it would appear no different than before exposure because of the extremely subtle effect of light on the emulsion. At this point the image is recorded only as a latent image; processing is required to reveal this image.

Development is the process of bathing the exposed film in an alkaline chemical that reduces the silver halide grains that have been exposed to light (Figure 3.8). Crystals in the latent image that were altered minimally are now completely changed to metallic silver in a process which, in effect, amplifies the pattern recorded by the latent image. In the latent image only a tiny portion of each grain has been altered by the effect of light, whereas after development each grain exposed to light is changed entirely to metallic silver. The developer acts most rapidly on those grains that have been exposed to light, so those areas that were exposed to the most intense light have the greatest density of metallic silver in the final image. Application of an acid stop bath allows exact control of the time the film is in contact with the developer by counteracting the chemical effect of the alkaline developer. Next a *fixer* is applied to dissolve, then remove, unexposed silver halide grains. If the fixer were not used, these unexposed grains would darken when the film was next exposed to daylight.

After development and fixing, the resulting image is a negative representation of the scene, because those areas that were brightest in the scene are represented by the greatest

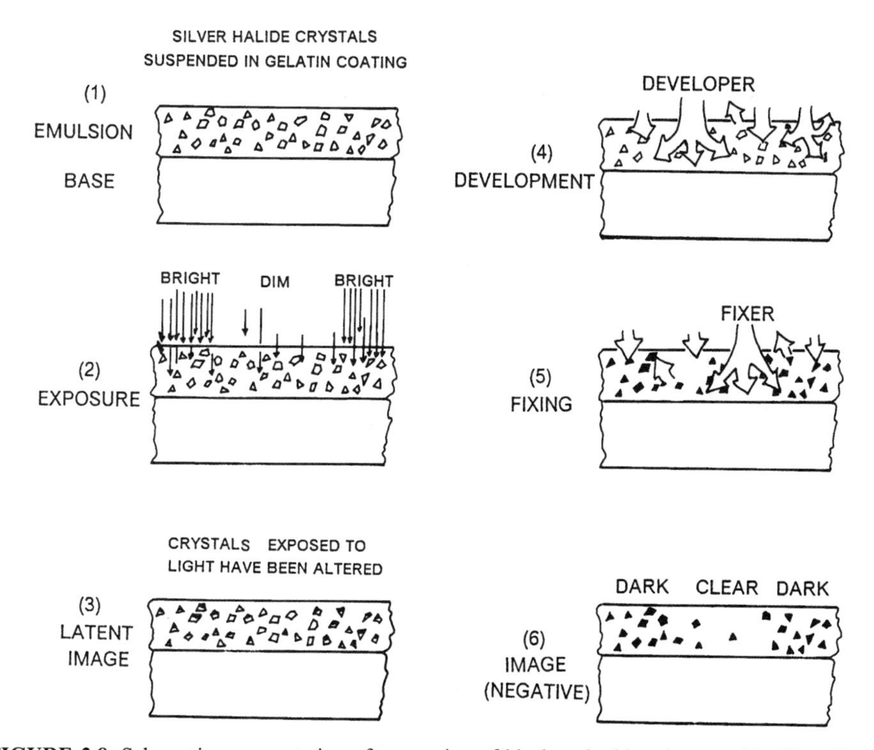


**FIGURE 3.8.** Schematic representation of processing of black-and-white photographic film. (1) Basic structure of the film as seen in cross section; photosensitive chemicals are suspended in a gelatin coating on the film base. (2) During exposure, light strikes the emulsion at varied intensities, depending on the brightness levels in the scene. (3) Light creates a chemical reaction in the photosensitive chemicals that changes only a few molecules of each grain, creating the latent image. (4) During development, the emulsion is bathed in an alkaline chemical that changes to metallic silver all grains modified in step (3); not shown here is the addition of an acidic chemical, the stop bath, that stops the action of the developer. (5) During fixing, unexposed grains are removed from the emulsion, leaving only those that had been exposed to light in (2). (6) The final image is a negative; those areas exposed to the most intense light in (2) are darkest; those exposed to dim light are clear.

concentrations of metallic silver, which appears dark on the processed image (Figure 3.9). Thus, in the negative, brightnesses are reversed from their original values in the scene.

*Film speed* is a measure of the sensitivity of an emulsion to light. A fast film requires relatively low intensity of light for proper exposure; a slow film requires more light, meaning that the aperture must be opened wider, or that a longer exposure time must be used. As mentioned previously, film speed is directly related to grain size and is inversely related to the ability of the film to record fine detail. Amateur photographers are familiar with the DIN and ASA ratings for assessing the speeds of films for hand-held cameras. The analogous scales for aerial films include the aerial film speed (AFS) and the aerial exposure index (AEI).

*Contrast* indicates the range of gray tones recorded by a film. High contrast means that the film records the scene largely in blacks and whites, with few intermediate gray tones. Low contrast indicates a representation largely in grays, with few really dark or really bright tones. Often the interpreter needs information about the intermediate brightnesses in a film, so for aerial photography, low contrast representation may be desirable. Fine-grained emulsions tend to have low contrast, so slower films tend to have higher spatial resolution and lower contrast than do the coarser-grained fast films. Emulsions on photographic papers typically have higher contrast than emulsions on films, so interpreters often prefer to use film transparencies if they are available.

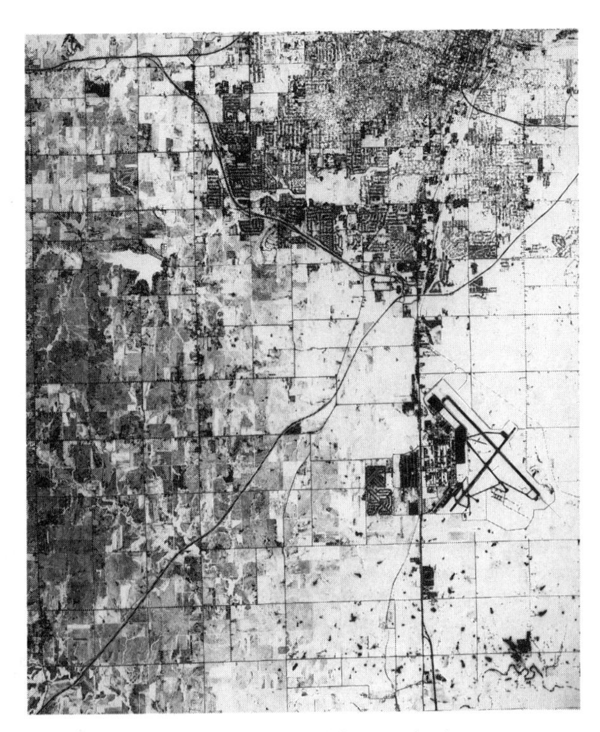

**FIGURE 3.9.** Black-and-white negative image

*Spectral sensitivity* records the spectral region to which a film is sensitive (Figure 3.10). The spectral sensitivity curve for Kodak Tri-X Aerographic Film 2403 shows typical features of black-and-white films (Figure 3.10a). It is sensitive throughout the visible spectrum, but it is also sensitive to ultraviolet radiation. Because of the scattering of these shorter (ultraviolet and blue) wavelengths, filters are often used with black-and-white aerial films to screen out blue light (Figure 3.11). This film presents a black-and-white representation of a scene (Figure 3.12a) that is essentially in accord with our view of the scene as we see it directly with our own eyes. This is because the Tri-X film is an example of a *panchromatic* film—an emulsion that is sensitive to radiation throughout the visible spectrum, much the same as the human visual system is sensitive throughout the visible spectrum. The term *orthochromatic* designates films with preferred sensitivity in the blue and green, usually with peak sensitivity in the green.

Figure 3.10 also shows the spectral sensitivity curve for Kodak Infrared Aerographic Film 2424, a black-and-white infrared film. Note that its sensitivity extends well beyond the visible into the infrared portion of the spectrum. Usually it is desirable to exclude visible radiation, so this film is often used with a deep red filter that blocks visible radiation but allows infrared radiation to pass (Figure 3.11). An image recorded by black-and-white infrared film (Figure 3.12b) is quite different from its representation in the visible spectrum. For example, living vegetation is many times brighter in the near infrared portion of the spectrum than it is in the visible, so vegetated areas appear bright white on the black-and-white infrared image.

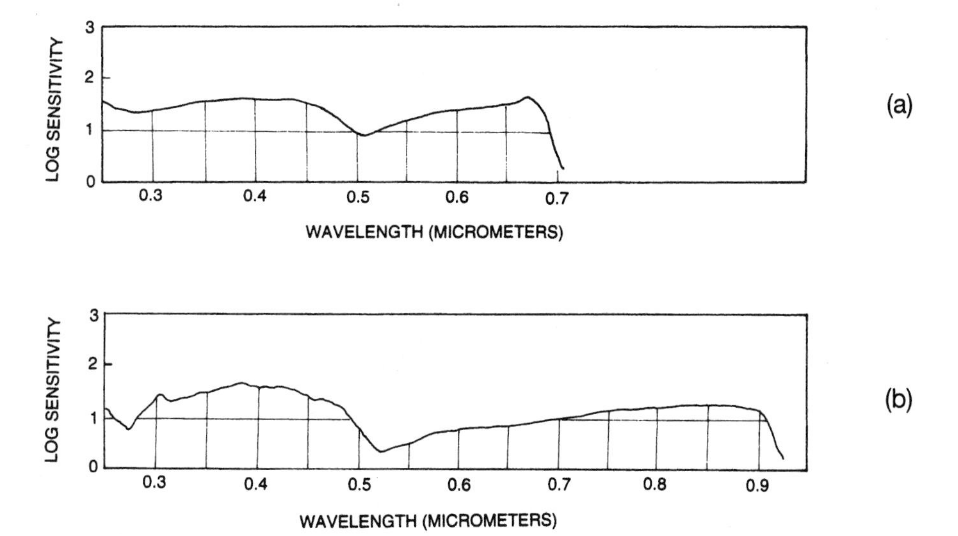


**FIGURE 3.10.** Spectral sensitivities of two photographic films. (*a*) Black-and-white panchromatic film (Kodak TRI-X Aerographic Film 2403). (*b*) Black-and-white infrared film (Kodak Infrared Aerographic Film 2424). Copyright Eastman Kodak Company. Permission has been granted to reproduce this material from *KODAK Data for Aerial Photography* (Code: M-29), courtsey of Silver Pixel Press, official licensee and publisher of Kodak books.

(a)

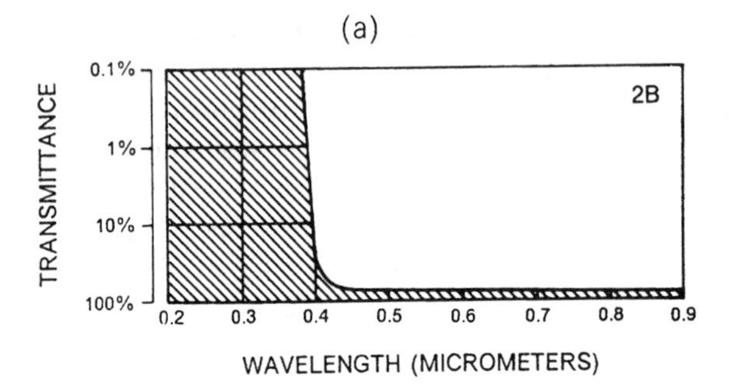


(b)

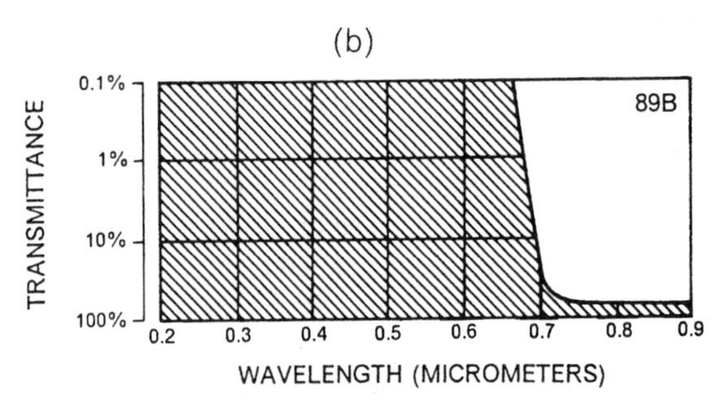


**FIGURE 3.11.** Transmission curves for two filters. (*a*) Pale yellow filter (Kodak filter 2B) to prevent ultraviolet radiation from reaching the film; it is frequently used with panchromatic film. (*b*) Kodak 89B filter used to exclude visible light, used with black-and-white infrared film. Copyright Eastman Kodak Company. Permission has been granted to reproduce this material from *KODAK Photographic Filters Handbook* (Code: B-3), courtesy of Silver Pixel Press, official licensee and publisher of Kodak books.

### *The Characteristic Curve*

If we examine a negative after development and fixing, we find a pattern of dark and light related to the patterns of metallic silver formed in the processed film. Where the original scene was bright, the negative now has large amounts of silver, which create the dark areas. Where the original scene was dark, the film is clear due to the absence of metallic silver.

(a)

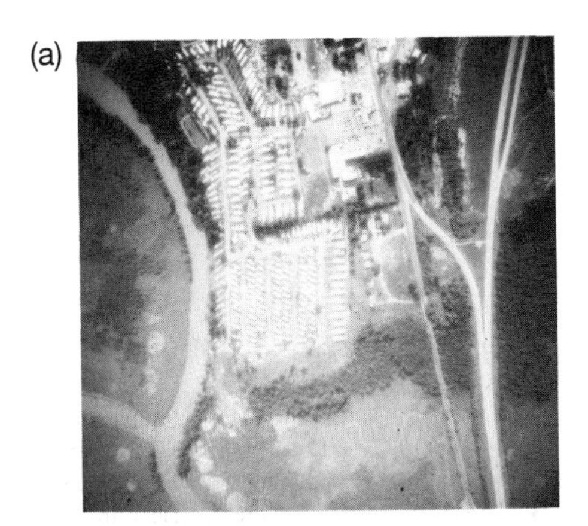

(b)

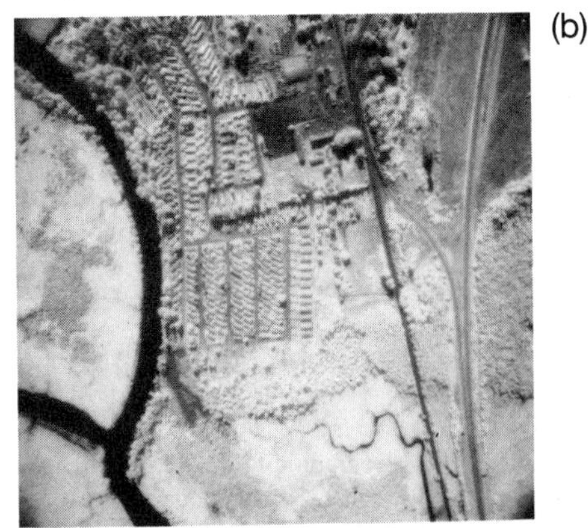

**FIGURE 3.12.** Aerial photographs. (*a*) Panchromatic film; (*b*) Black-and-white infrared film. Both show Wayson's Corner, Maryland. Photographs courtesy of NASA.

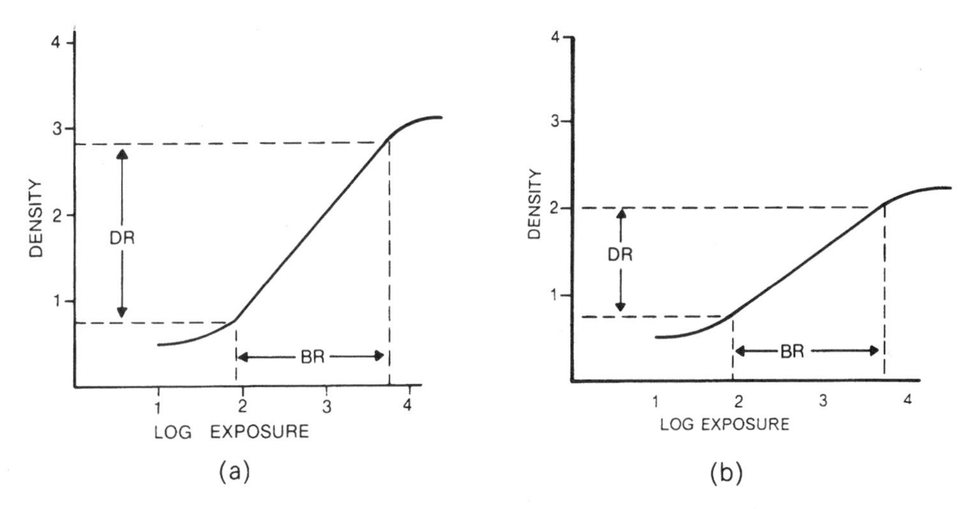


**FIGURE 3.15.** Examples of characteristic curves.

between exposure and density. This means that for very high and very low exposures, the film will not produce the predictable densities that it will for exposures in the straight-line segment. That is, the film is no longer a scientific tool for portraying scene brightness because we cannot measure a density and then relate the density to the brightness of the original scene. Often, professional photographers will use the toe or the shoulder to create artistic effects in photographs, but scientists always avoid use of those image measurements that may be based on very high or very low image densities—they have an unknown relationship to brightnesses in the scene they portray.

Because photographic films are used to record images acquired by nonphotographic sensors (as described in subsequent chapters), knowledge of the characteristic curve is especially important in the field of remote sensing. Such sensors often record a large range of brightnesses—a range so large that it may exceed the capability of the film to record it. If so, the photographic record of the image will inevitably be inaccurate as a record of scene brightness. Either the very dark areas, the very bright areas, or perhaps both will be represented in the nonlinear part of the characteristic curve, and the image will show only a small portion of the brightness information actually present. It is partially for this reason that digital image analyses, which do not rely on film images, are often advantageous.

## 3.5. Color Reversal Films

Many of the color films used in remote sensing are reversal films, similar to those often used in hand-held cameras for color slides. Their basic elements are similar to those of black-and-white photographic film except that there are three separate emulsions, each sensitive to one of the three additive primaries (Figure 3.16). The protective supercoat, the backing, and the subbing layer are present, and between the several emulsions are spacer layers of gelatin to prevent mixing of adjacent emulsions. The layer between the uppermost blue-sensitive emulsion and the middle (green-sensitive) emulsion is treated to act as

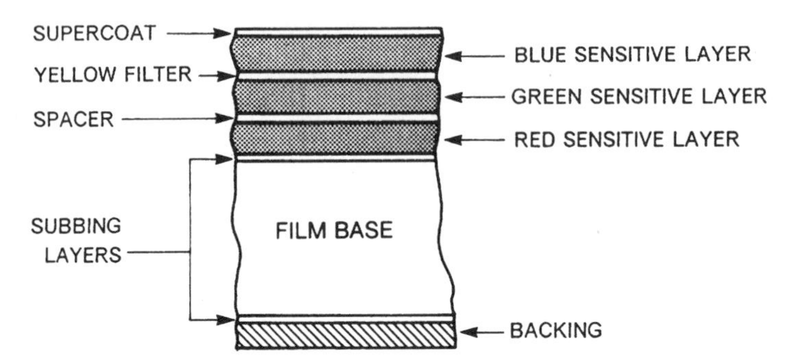


**FIGURE 3.16.** Idealized cross-sectional diagram of color reversal film.

a yellow filter to prevent blue light from passing through the upper layers to expose the lower emulsions. This filter is necessary because of the difficulty of manufacturing emulsions sensitive to red and green light without also sensitizing them to blue light.

Upon exposure, blue light exposes the blue layer, passes through the blue layer, but is prevented from exposing the other two layers by the yellow filter (Figure 3.17). Green light passes through the blue layer and exposes the green-sensitive emulsion. Red light passes through the upper emulsions to expose only the lower, red-sensitive layer.

After processing, all areas *not* exposed to blue light are represented by a yellow dye, while those areas exposed to blue are left clear on the blue-sensitive emulsion. Areas exposed by green light on the green-sensitive emulsion are left clear; other areas are shown in magenta dye. Areas on the red-sensitive layer not exposed to red light are represented

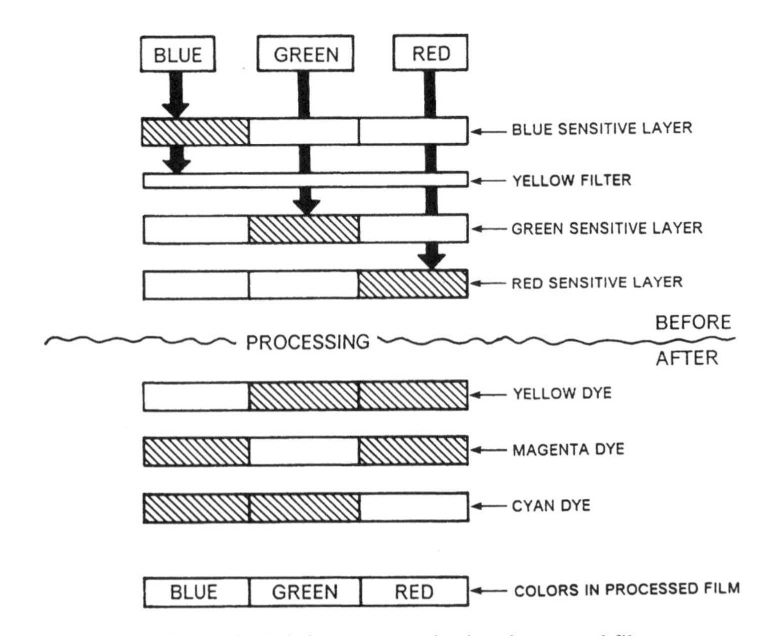


**FIGURE 3.17.** Color representation in color reversal film.

by cyan dye; images of red objects are clear on this emulsion. Thus, each emulsion is sensitive to one of the additive primaries; after processing, each emulsion contains one of the subtractive primaries. The strategy of dying all areas not exposed in each separate emulsion differs fundamentally from the process used for the black-and-white films described earlier. Here there is no negative image; the dyes in the film processed combine to form a positive image in which brightness in the image correspond (approximately) to brightness in the original scene.

When the processed film is viewed as a transparency against a light source, the magenta and cyan dyes present in those areas exposed to blue light combine to form a blue color. Likewise yellow and cyan combine to represent green, and yellow and magenta combine to form red (Figure 3.17 and Plate 1). This process is the same basis used for production of the 35 mm color slides that are familiar to many readers, so it should be easy to visualize the result of this process, even though the explanation shown in Figure 3.17 may seem rather abstract.

## 3.6. Color Infrared Films

Color infrared (CIR) films are based on the same principles as are color reversal films except for differences in the sensitivity of the emulsions and conventions in representation of colors. The blue-sensitive layer is replaced by an emulsion sensitive to a portion of the near infrared (Figure 3.18). After developing, representation of colors in the scene is shifted one position in the spectrum, so that green in the scene appears as blue on the image, red appears as green, and objects reflecting strongly in the near infrared are depicted in red. The comparison with normal color films can be represented schematically as follows:

| | | | | |
|---|---|---|---|---|
| Object in the scene reflects: | Blue | Green | Red | Infrared |
| Color reversal film represents the object as: | Blue | Green | Red | ***** |
| Color infrared film represents the object as: | ** | Blue | Green | Red |

Most objects, of course, reflect in several portions of the spectrum, so the CIR image shows a variety of colors, derived from the varied reflectances in the scene. CIR film is designed in a manner analogous but not identical, to that of color reversal film (Figure 3.19). A yellow filter over the camera lens excludes all blue light. Green light exposes the

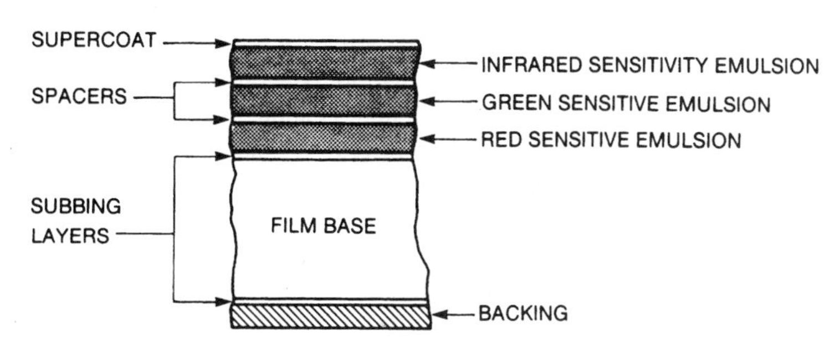


**FIGURE 3.18.** Idealized cross-sectional diagram of color infrared film.

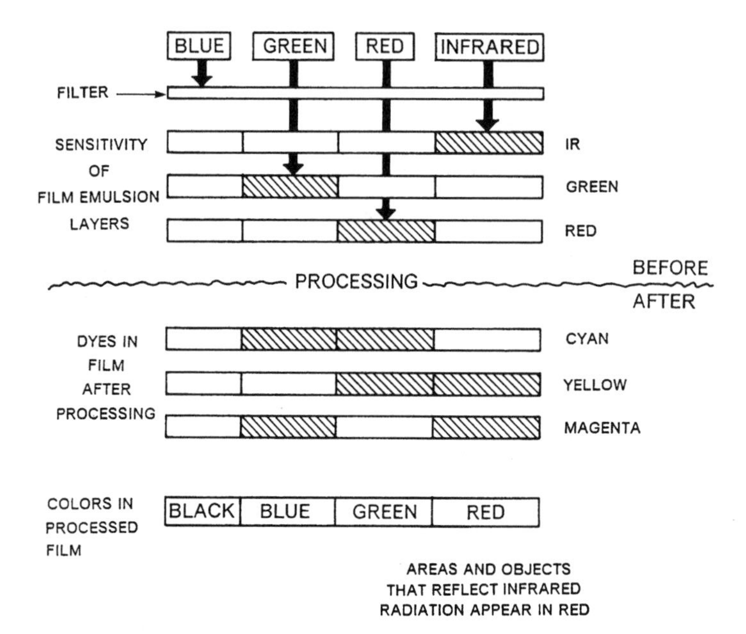


**FIGURE 3.19.** Color representation in color infrared film.

green-sensitive layer; red light exposes the red-sensitive layer, and infrared radiation exposes the infrared-sensitive layer. After processing, areas not exposed to the green light are colored in yellow dye. Areas not exposed to red are colored magenta. And areas not exposed to IR radiation are represented as cyan. In the final transparency, cyan and magenta combine to represent green areas as blue, cyan and yellow combine to represent red areas as green, and yellow and magenta combine to form the red color that represents objects that reflect strongly in the near infrared (Plate 1).

## 3.7. Film Format and Annotation

The term *format* designates the size of the image acquired by a camera. Mapping cameras generally produce a square image, with a size controlled by the width of the film. Common formats are 23 cm × 23 cm (approximately 9 in. × 9 in.), and 5.7 cm × 5.7 cm (approximately 2.5 in. × 2.5 in.). Film 35 mm in width (24 mm × 36 mm image format) has also been used for specialized purposes, although generally it is not practical for routine applications of remote sensing. The 23-cm format is a standard for most cartographic cameras, and paper prints made directly from the negative (*contact prints*) are among the most frequently used form of aerial photography. Although enlargements of the original negative can be made to any size desired, the 23-cm format is convenient for storage and handling, and because the prints are made directly from the negative without enlargement, the image retains maximum detail and sharpness.

In some instances, paper prints are not made, and the film is examined simply as a

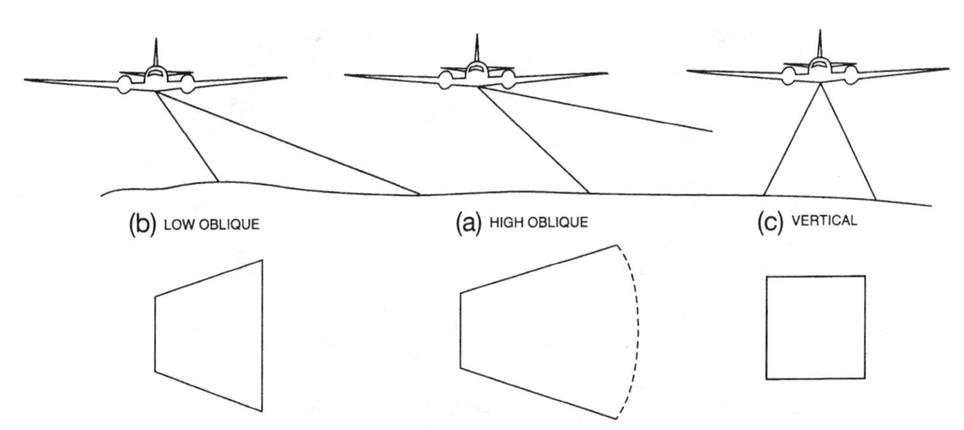


**FIGURE 3.20.** Oblique and vertical aerial photographs.

strip of film—a *positive transparency*—wound on large spools, and viewed against an illuminated background (Chapter 5). Because emulsions of transparencies typically represent a greater range of image tones than do paper prints, positive transparencies are often preferred for detailed interpretations, and especially for color and color infrared films.

Most aerial photographs carry some form of annotation—markings that identify the photographs and details of their acquisition. Typically aerial photographs are annotated in the forward edge of the photograph (Figure 3.20). Annotation consists of a series of letters and numerals that can vary in meaning from one aerial survey firm to the next, but usually they consist of the date of the photography, a series of letters and numbers that identify each project, and the film roll number. Usually the last three digits specify the frame number, which shows the sequence in which photographs were taken. Other annotations, such as image scale, may also be shown. In some instances the information may be recorded directly by the camera itself as the image is acquired, or in other instances it may be added later as the film is processed and prepared for dissemination. The most sophisticated cameras may record on each frame information such as date, project identifier, focal length, time, image of a bubble level to indicate degree of tilt, and locational coordinates provided by on-board global positioning systems.

## 3.8. Geometry of the Vertical Aerial Photograph

Aerial photographs can be classified according to the orientation of the camera in relation to the ground at the time of exposure (Figure 3.21). Oblique aerial photographs have been acquired by cameras oriented toward the side of the aircraft. *High oblique* photographs (Figure 3.21b and Plate 2) show the horizon; *low oblique* photographs (Figures 3.21a and Plate 3), taken with the camera aimed more directly toward the ground surface, do not show the horizon. Oblique photographs provide the advantage of showing large areas in a single image. Often those features in the foreground are easily recognized, as the view in an oblique photograph may resemble that from a tall building or mountain top. However,

**FIGURE 3.21.** Vertical aerial photograph.

oblique photographs are not widely used, primarily because the drastic changes in scale that occur from foreground to background prevent convenient measurement of distances, areas, and elevations.

Vertical photographs are acquired by a camera aimed directly at the ground surface from above (Figure 3.21c and Figure 3.20). Although objects and features are often difficult to recognize from their representations on vertical photographs, the map-like view of the earth, and the predictable geometric properties of vertical photographs provide practical advantages. It should be noted that few, if any, aerial photographs are truly vertical; most have some small degree of tilt due to aircraft motion and other factors. The term *vertical photograph* is commonly used to designate aerial photographs that are within a few degrees of a corresponding (hypothetical) truly vertical photograph.

Because the geometric properties of vertical and nearly vertical aerial photographs are well understood and can be applied to many practical problems, they form the basis for making accurate measurements using aerial photographs. The science of making accurate measurements from aerial photographs (or from any photograph) is known as *photogrammetry*. The following paragraphs outline some of the most basic elements of introductory photogrammetry; the reader should consult a photogrammetry text (e.g., Wolf,

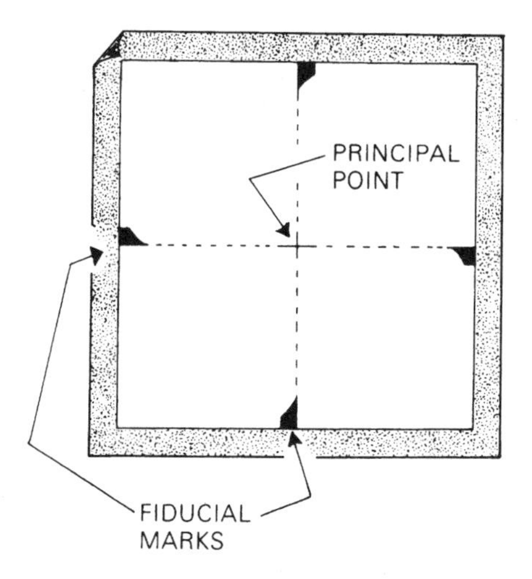


**FIGURE 3.22.** Fiducial marks and principal point.

1983) for a complete discussion of this subject. Aerial cameras are manufactured to include adjustable index marks attached rigidly to the camera so that the positions of the index marks are recorded on the photograph during exposure. These *fiducial marks* (usually four or eight in number) appear as silhouettes at the edges and/or corners of the photograph (Figures 3.20 and 3.22). Lines that connect opposite pairs of fiducial marks intersect to identify the principal point, the optical center of the image. The ground nadir is defined as the point on the ground vertically beneath the center of the camera lens at the time the photograph was taken (Figure 3.23). The *photographic nadir* is defined by the intersection with the photograph of the vertical line that intersects the ground nadir and the center of the lens (i.e., the image of the ground nadir).

The isocenter can be defined informally as the focus of tilt. Imagine a truly vertical photograph that was taken at the same instant as the real, almost-vertical, image. The almost-vertical image would intersect with the (hypothetical) perfect image along a line that would form a "hinge"; the isocenter is a point on this hinge. On a truly vertical photograph, the *isocenter*, the principal point, and the photographic nadir coincide. Positional, or geometric, errors in the vertical aerial photograph can be outlined as follows.

1. *Optical distortions* are perhaps caused by an inferior camera lens, camera malfunction, or similar problems. These distortions are probably of minor significance in most modern photography flown by professional aerial survey firms.

2. *Tilt* is caused by displacement of the focal plane from a truly horizontal position by aircraft motion (Figure 3.23). The focus of tilt, the *isocenter,* is located at or near the principal point. Image areas on the upper side of the tilt are displaced further away from the ground than is the isocenter; these areas are therefore depicted at scales smaller than the nominal scale. Image areas on the lower side of the tilt are displaced down; these areas are depicted at scales larger than the nominal scale. Therefore, because all photographs have some degree of tilt, measurements confined to one portion of the image run the risk of including systematic error caused by tilt (i.e., measurements may be consistently too

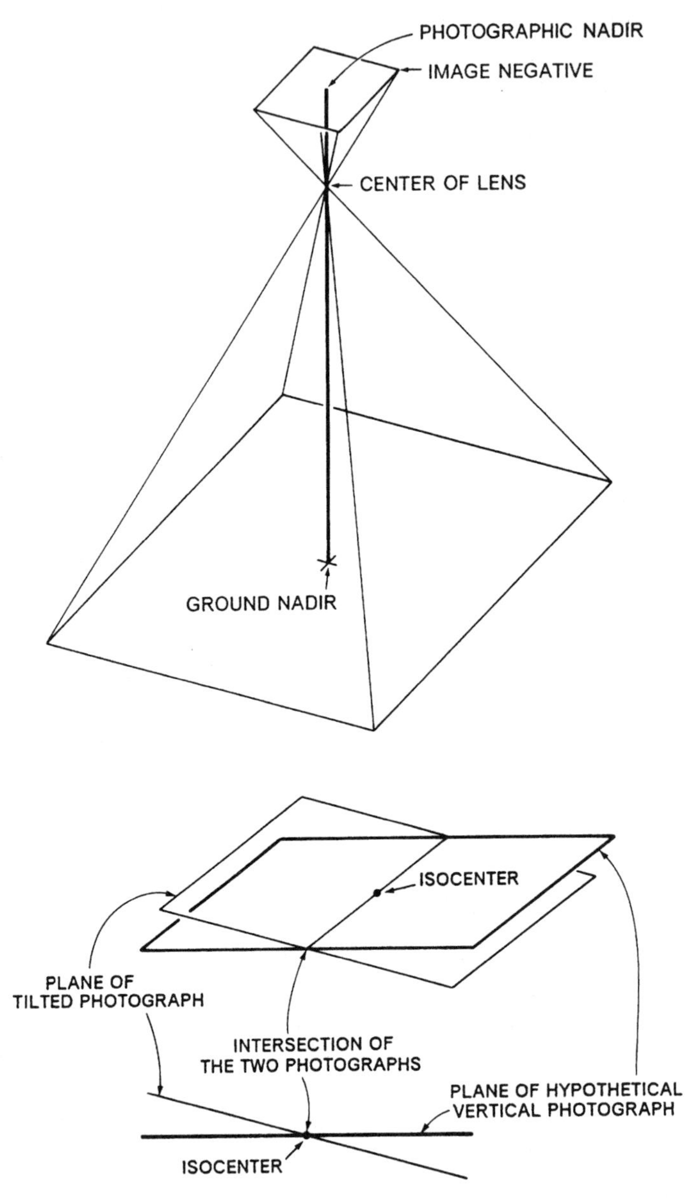


**FIGURE 3.23.** Schematic representation of terms to describe geometry of vertical aerial photographs.

large or too small). To avoid this effect, it is a good practice to select distances used for scale measurements (Chapter 5) as lines that pass close to the principal point; errors caused by the upward tilt compensate for errors caused by the downward tilt. The resulting value for image scale is not, of course, precisely accurate for either portion of the image, but it will not include the large errors that can arise in areas located further from the principal point.

Because of routine use of high-quality cameras and careful inspection of photography to monitor photo quality, today the most important source of positional error in vertical aerial photography is probably *relief displacement* (Figure 3.24). Objects positioned directly beneath the center of the camera lens will be photographed so that only the top of the object is visible (e.g., object *A* in Figure 3.24). All other objects are positioned such that both their tops and their sides are visible from the position of the lens. That is, these objects appear to lean outward from the central perspective of the camera lens (e.g., see objects in Figure 3.24). Correct planimetric positioning of these features would represent only the top view, yet the photograph shows both the top and sides of the object. For tall features, it is intuitively clear that the base and the top cannot both be in their correct planimetric positions.

This difference in apparent location is due to the height (*relief*) of the object and forms an important source of positional error in vertical aerial photographs. The direction of relief displacement is always radial from the nadir; the amount of displacement depends on (1) the height of the object and (2) the distance of the object from the nadir. Relief displacement increases with increasing heights of features and with increasing distances from the nadir. (It also depends on focal length and flight altitude; these may be regarded as constant for a few sequential photographs.)

Relief displacement forms the basis of measurements of heights of objects, but its

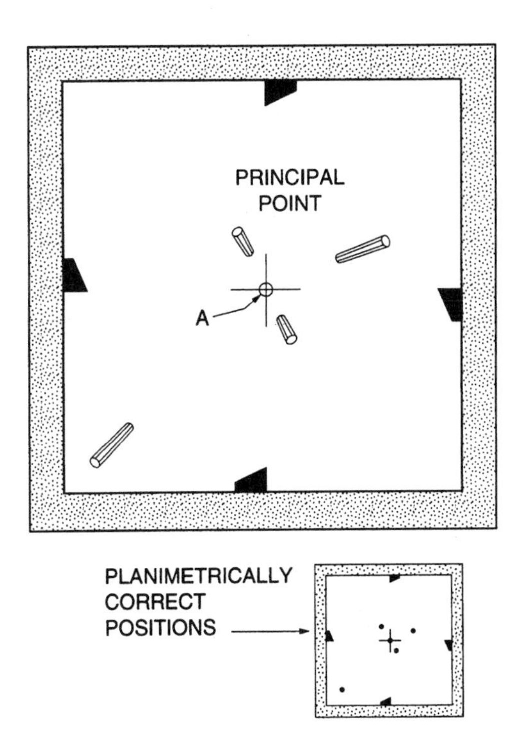


**FIGURE 3.24.** Relief displacement. The diagram depicts a vertical aerial photograph of a level terrain with five towers of equal height located at different positions with respect to the principal point. Images of the tops of towers are displaced away from the principal point along lines that radiate from the nadir, as discussed in the text.

greatest significance is its role as a source of positional error. Uneven terrain can create significant relief displacement, so all measurements made directly from uncorrected aerial photographs are suspect.

## 3.9. Coverage by Multiple Photographs

Pilots normally acquire vertical aerial photographs by flying a series of parallel flight lines that together build up complete coverage of a specific region. Each flight line consists of individual frames, usually numbered in sequence (Figure 3.25). Often the camera operator can view the area to be photographed through a viewfinder attached to the camera, or through a telescope-like instrument that observes the ground area below the plane. With the aid of these devices, the operator can manually trigger the shutter as aircraft motion brings predesignated landmarks into the field of view or can set controls to automatically acquire photographs at intervals tailored to provide the desired coverage.

Individual frames form ordered strips, as shown in Figure 3.25a. If the plane's course is deflected by a crosswind, the positions of ground areas shown by successive photographs form the pattern shown in Figure 3.25b, known as *drift*. *Crab* (Figure 3.25c) is caused by correction of the flight path to compensate for drift without a change in the orientation of the camera.

Usually flight plans call for a certain amount of forward overlap (Figure 3.26), duplicate coverage by successive frames in a flight line, usually by 50% to 60% of each frame. If forward overlap is 50% or more, the image of the principal point of one photograph is visible on the next photograph in the flight line. These are known as *conjugate principal points* (Figure 3.26). When it is necessary to photograph large areas, coverage is built up by means of several parallel strips of photography; each strip is called a *flight line*. Sidelap between adjacent flight lines may be from about 5% to 15%, in an effort to prevent gaps in coverage of adjacent flight lines.

However, as pilots and other crew members collect complete photographic coverage of a region, there may still be gaps (known as *holidays*) in coverage due to equipment malfunction, navigation errors, cloud cover, or other problems. Sometimes photography flown later to cover holidays differs noticeably from adjacent images with respect to sun angle, vegetative cover, and other qualities. For planning flight lines, the number of photographs required for each line can be estimated using the relationship

$$\text{Number of photos} = \frac{\text{Length of flight line}}{(gd \text{ of photo}) \times (1 - \text{overlap})} \qquad \text{(Eq. 3.4)}$$

where *gd* represents the ground distance represented on a single frame, measured in the same units as the length of the planned flight line. For example, if a flight line is planned to be 33 mi. in length, each photograph is planned to represent 3.4 mi. on a side, and forward overlap is to be 0.60, then $33/[3.4 \times (1 - 0.60)] = 33/(1.36) = 24.26$, or about 25 photographs are required. (Chapter 5 shows how to calculate the coverage of a photograph for a given negative size, focal length, and flying altitude.)

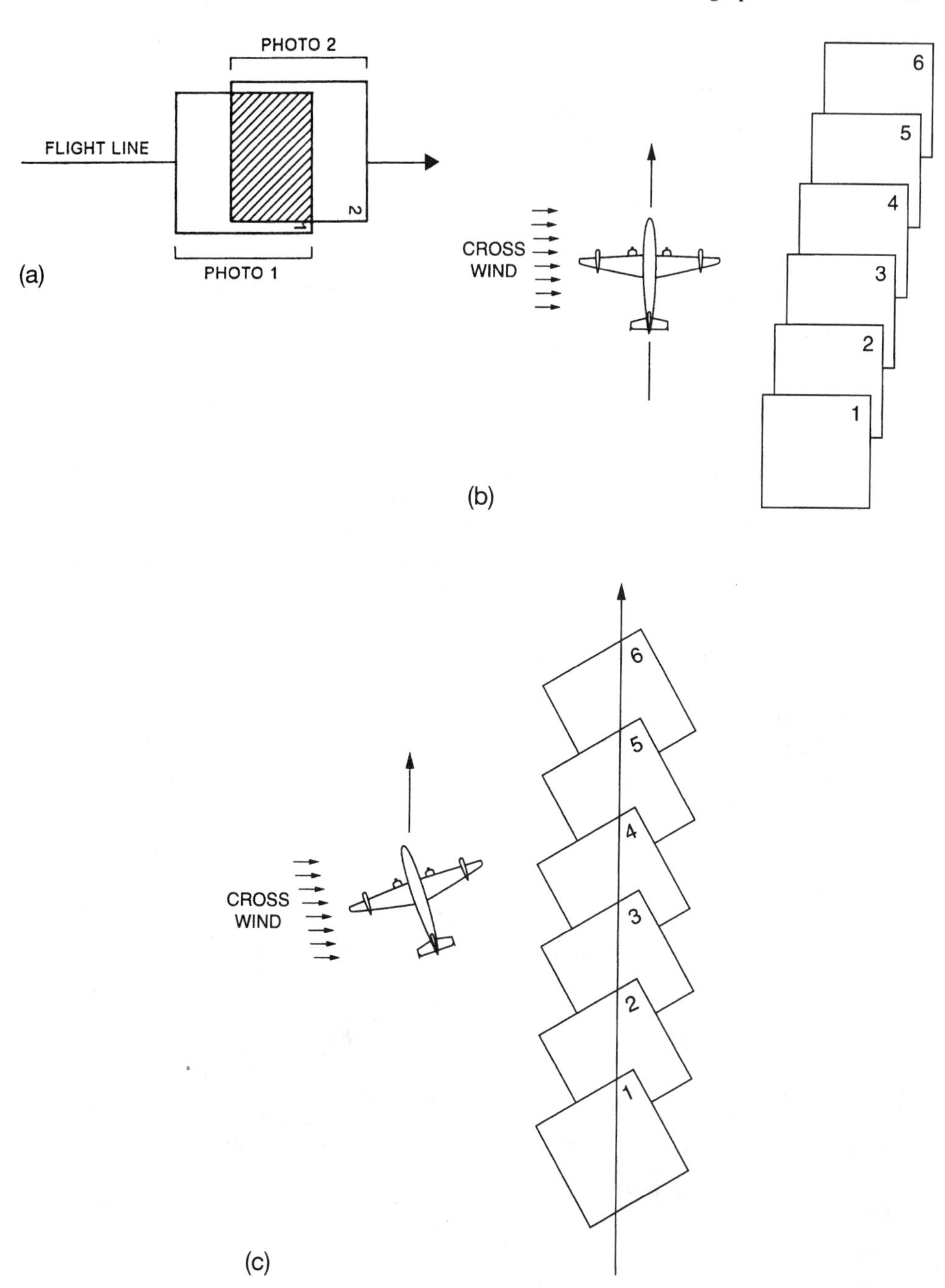


**FIGURE 3.25.** Aerial photographic coverage. (*a*) Forward overlap. (*b*) Drift. (*c*) Crab.

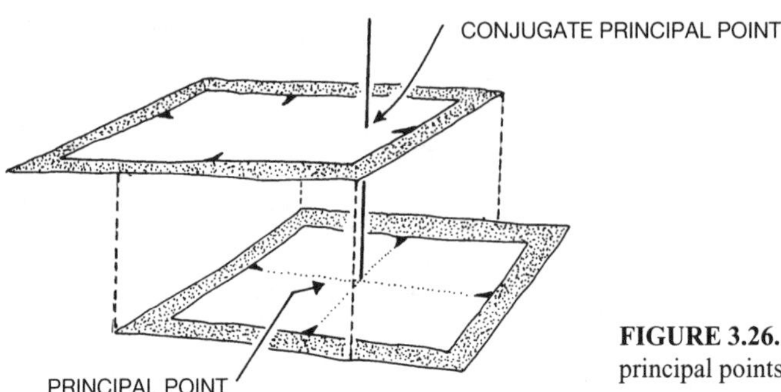


**FIGURE 3.26.** Forward overlap and conjugate principal points.

### *Stereoscopic Parallax*

If we have two photographs of the same area taken from different perspectives (i.e., from different camera positions), we observe a displacement of images of objects from one image to the other. You can observe this effect now by simple observation of nearby objects. Look up from the book at the objects near you. Close one eye, then open it and close the other. As you do this you observe a change in the appearance of objects from one eye to the next. Nearby objects are slightly different in appearance because one eye tends to see, for example, only the front of an object, whereas the other, because of its position (about 2.5 in.), from the other, sees the front and some of the side of the same object. This difference in appearances of objects due to change in perspective is known as *stereoscopic parallax*. The amount of parallax decreases as objects increase in distance from the observer (Figure 3.27). If you repeat the experiment looking out the window at a landscape you can confirm this effect by noting that distant objects display little or no observable parallax.

**FIGURE 3.27.** Stereoscopic parallax. These two photographs of the same scene were taken from slightly different positions. Note the differences in the appearances of objects due to the difference in perspective; note also that the differences are greatest for objects nearest the camera and least for objects in the distance.

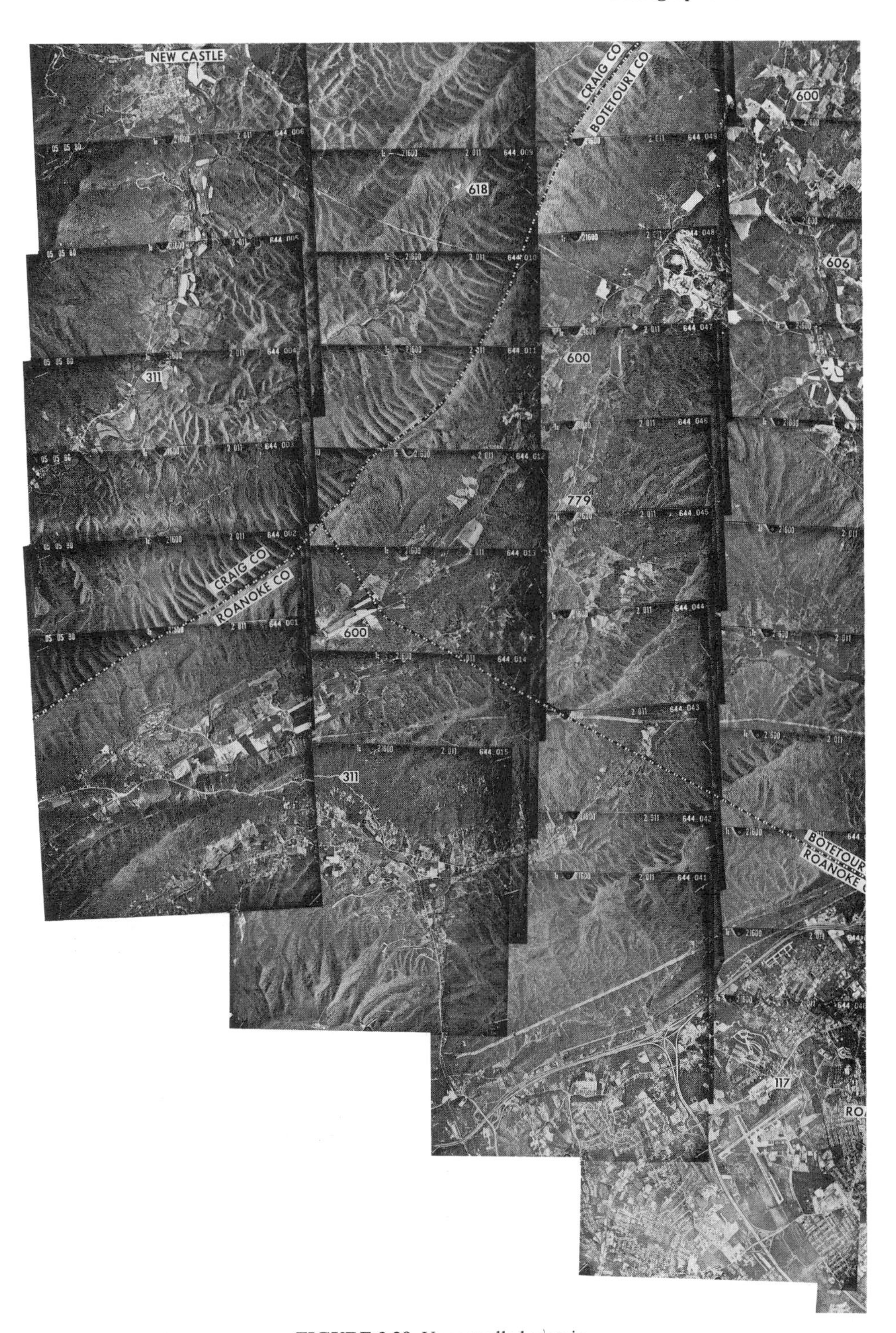


FIGURE 3.29. Uncontrolled mosaic.

en instant, project a corrected version of a small portion of an image. An orthophotoscope is an optical–mechanical instrument that, instead of exposing an entire image from a central perspective (i.e., through a single lens), exposes each small section individually in a manner that corrects for the elevation of that small section. The result is an image that has orthographic properties rather than those of the central perspective of the original aerial photograph. The orthophotoscope is capable of scanning an entire image piece by piece to generate a corrected version of the entire image. The projection orientation is adjusted to correct for tilt, and the instrument continuously varies the projection distance to correct for relief displacement. Thus, as the instrument scans an image the operator views the ground surface in stereo and the new image is formed as a geometrically correct version of the original image. The result is a photo image that shows the same detail as the original aerial photograph but without the geometric errors introduced by tilt and relief displacement. *Orthophotomaps* therefore can be used for most purposes as maps, because they show correct planimetric position and preserve consistent scale throughout the image. Orthophotographs form the basis for orthophotomaps, which are orthophotographs presented in map format, with annotations, scale, and geographic coordinates.

Orthophotomaps (Figure 3.30) are valuable because they show the fine detail of an aerial photography without the geometric errors that are normally present, and they can be compiled much more quickly and cheaply than the usual topographic maps. Therefore, they can be very useful as map substitutes when topographic maps are not available, or as map supplements when maps are available, but the analyst requires the finer detail and more recent information provided by an image.

**FIGURE 3.30.** Orthophoto.

## 3.10. Photogrammetry

Photogrammetry is the science of making accurate measurements from photographs. Photogrammetry applies principles of optics and knowledge of the interior geometry of the camera and its orientation to reconstruct dimensions and positions of objects represented within photographs. Therefore, its practice requires detailed knowledge of specific cameras and the circumstances under which they were used, and accurate measurements of features within photographs. Thus, photographs used for photogrammetry have traditionally been prepared on glass plates or other dimensionally stable materials (i.e., materials that do not change in size as temperature and humidity change).

Photogrammetry can be applied to any photograph, provided the necessary information is at hand. However, by far the most frequent application of photogrammetry is the analysis of stereo aerial photography to derive estimates of topographic elevation for topographic mapping. With the aid of accurate locational information of key features within a scene ("ground control"), photogrammetry estimates topographic relief by estimating stereo parallax for any array of points within a region. Although stereo parallax can be measured manually, it is far more practical to employ specialized instruments designed for stereoscopic analysis.

Such instruments (*analytical stereoplotters*), first designed in the 1920s, reconstruct the orientations of photographs at the time they were taken. Operators then can view the image in stereo and by maintaining constant parallax (visually), trace lines of uniform elevation. The quality of information derived from such instruments depends on the quality of the photography, the accuracy of the data, and the operator's skill in setting up the stereo model, as well as tracing lines of uniform parallax. As the design of instruments improved, it eventually became possible to automatically match corresponding points on stereo pairs and thereby identify lines of uniform parallax with limited assistance from the operator.

## 3.11. Digital Photography

Photographs can be electronically scanned to record the patterns of blacks, grays, and whites as digital values, each representing the brightness of a specific point within the image. Although these values can be displayed in the form of a conventional photograph, the digital format offers advantages of compact storage and the power of numerical representation (Chapters 4 and 11). Further, it is possible to manufacture cameras that replace the film in the focal plane with an array of light-sensitive detectors (Chapter 4) that directly record images in digital form, thereby bypassing the digitization step.

## 3.12. Softcopy Photogrammetry

With further advances in instrumentation, it became possible to extend automation of the photogrammetric process to conduct the analysis completely within the digital domain. Satellite images or aerial photographs can be acquired digitally (or conventional photographs can be scanned to create digital products). With the use of global positioning systems (GPSs) (Chapter 12) to acquire accurate positional information and the use of data recorded from the aircraft's navigational system to record the orientations of photographs, it then became feasible to reconstruct the geometry of the image using those data gathered as the image was acquired.

This process forms the basis for *softcopy photogrammetry*, so named because it does not require the physical (*hardcopy)* form of the photograph necessary for traditional photogrammetry. Instead, the digital ("soft") version of the image is used as input for a series of mathematical models that reconstruct the orientation of each image to create planimetrically correct representations. This process requires specialized computer software, installed in workstations, which analyzes digital data specifically acquired for the purpose of photogrammetric analysis. Softcopy photogrammetry offers advantages of speed and accuracy and also creates output data that are easily integrated into other production and analytical systems, including GISs (Chapter 15).

## 3.13. Sources of Aerial Photography

Aerial photography can be acquired by (1) the user, or (2) purchased from organizations that serve as repositories for imagery flown by others (*archival imagery*). In the first instance, aerial photography is produced upon request by firms that specialize in taking high-quality aerial photography. Such firms are listed in the business directories of most metropolitan phone directories. Customers may be individuals, governmental agencies, or other businesses that use aerial photography. Such photography is, of course, customized to meet specific needs of customers with respect to date, scale, film, and coverage. As a result, costs may be prohibitive for many noncommercial uses.

Thus, for financial reasons, many users of aerial photography turn to archival photography as a means of acquiring the images they need. Although such photographs may not exactly meet users' requirements with respect to scale or date, low costs and ease of access may compensate for any shortcomings. For some tasks that require reconstruction of conditions at earlier dates (such as the Environmental Protection Agency's search for abandoned toxic waste dumps), the archival images may form the only source of information.

It is feasible to take "do-it-yourself" aerial photography. Many small cameras are suitable for aerial photography, and often costs of local air charter services for an hour or so of flight time are relatively low. Small-format cameras, such as the usual 35-mm cameras, can be used for aerial photography if the photographer avoids effects of aircraft vibration (do not rest the camera against the aircraft). If the altitude is low and the atmosphere is clear, ordinary films can produce satisfactory results. Be sure to use a high-wing aircraft to ensure that the photographer will have a clear view of the landscape. If the camera can accommodate filters, it is possible to use other films (such as infrared or color infrared) similar to those described above. Some experimentation may be necessary for the first-time user to obtain proper exposures, but most people can learn rather quickly to take satisfactory photographs. Usually the best lighting is obtained when the camera is aimed away from the sun. Photographs acquired in this manner (Figure 3.31) may be useful for illustrative purposes, although for scientific or professional work, the large-format, high-quality work of the fully equipped air survey firm is probably necessary.

**FIGURE 3.31.** Aerial photograph taken with hand-held camera.

### *EROS Data Center*

The EROS Data Center (EDC), Sioux Falls, South Dakota, is operated by the U.S. Geological Survey (USGS) as a repository for aerial photographs and satellite images acquired by NASA, the USGS, and other agencies. A computerized data base at EDC provides an indexing system for information pertaining to aerial photographs and satellite images.

### *Earth Science Information Centers*

The USGS operates a network of Earth Science Information Centers (ESICs) as sources for information pertaining to maps and aerial photographs. ESICs have a special interest in information pertaining to federal programs and agencies but also collect data pertaining to maps and photographs held by state and local governments. ESIC headquarters is located at USGS headquarters, Reston, Virginia (703-648-6045), but maintains other offices throughout the United States, and other federal agencies have affiliated offices. ESICs provide information to the public concerning the availability of maps and remotely sensed images. The following sections describe two programs administered by ESIC that can provide access to archival aerial photography.

### *Aerial Photography Summary Record System*

The Aerial Photography Summary Record System (APSRS) is maintained by ESIC as a computer-based information system for recording detailed information pertaining to aerial photography held by numerous federal, state, and private organizations. Prior to the establishment of ESIC in 1975, citizens desiring comprehensive information on coverage by aerial photographs were required to query holdings by numerous agencies, which each followed different conventions in reporting coverage. After 1975, users could obtain information on the integrated holdings of numerous agencies, reported in a standard form.

Those who request information from APSRS receive information organized on a state-by-state basis, or by segments of states, in the form of line-printer computer listings, outline maps that depict coverage of various photographic missions, or microfiche. Coverage is indexed by USGS 7.5-minute quadrangles, which are listed by the latitude and longitude of the southeastern corner of each quadrangle. Users first identify the quadrangle that covers their area of interest, then use the latitude and longitude to search for coverage of the region. For each photographic mission, APSRS provides a listing giving the date of coverage, amount of cloud coverage, scale, film type and format, focal length, and other qualities. Also listed is the agency that holds the photography; these listings are keyed to a directory that lists addresses so that the user can order copies of the photographs. Listings also include the Federal Information Processing Standards (FIPS) code for each area, which permits cross-reference to political and census units. APSRS data are provided on CD-ROMs, which are reissued with new and revised data at about 9-month intervals.

### *National Aerial Photography Program*

The National Aerial Photography Program (NAPP) is designed to acquire aerial photography for the coterminous United States, according to a systematic plan that ensures uniform standards. NAPP was initiated in 1987 as a replacement for the National High Altitude Aerial Photography Program (NHAP), begun in 1980 to consolidate the many federal programs for aerial photography. The USGS manages the NAPP, but it is funded by the federal agencies that are primary users of its photography. Program oversight is proved by a committee of representatives from several federal agencies, including the USGS, the Soil Conservation Service, the Agricultural Stabilization and Conservation Service, the U.S. Forest Service, the Bureau of Land Management, and the Tennessee Valley Authority. Light (1993) and Plasker and TeSelle (1988) provide further details of NAPP.

Under NHAP, photography was first acquired in 1978 under a plan to obtain complete coverage of the 48 coterminous states and update coverage as necessary to keep pace with requirements for current photography. Current plans call for updates at intervals of 5 years.

Flight lines are oriented in a north–south direction centered on each of four quadrants systematically positioned within USGS 7.5-minute quadrangles, with full stereoscopic coverage at 60% forward overlap and sidelap of at least 27%. Ten frames, acquired in two flight lines, provide full stereoscopic coverage of each quadrangle. Photography is acquired using cameras with focal lengths of 6 in. at a flying altitude of 20,000 ft. above the terrain, to provide coverage at 1:40,000. Either black-and-white or color infrared film is used (Plate 4).

Dates of photography vary according to geographic region; flights are timed to provide optimum atmospheric conditions for photography and to meet specifications for sun angle, snow cover, and shadowing. Areas are usually photographed during autumn or winter to provide images that show the landscape without the cover of deciduous vegetation.

Photographs are available to all who may have an interest in their use. Their detail and quality permit use for land cover surveys, assessment of agricultural, mineral, and forest resources, as well as examination of patterns of soil erosion and water quality. Light

(1993) assess uses of NAPP photographs for use in GISs. Photography is archived and available from the USGS EROS Data Center (Sioux Falls, SD 57198) or the Aerial Photography Field Office (P.O. Box 30010, 2222 West 2300 South, Salt Lake City, UT 84130-0010).

### *Sources of Old Aerial Photographs*

Many aerial photographs taken prior to 1940 have been collected at the Cartographic and Architectural Branch of the U.S. National Archives (8th and Pennsylvania Avenue NW, Washington, DC 20408; 703-756-6700). Included are a wide selection of photographs from throughout the 48 coterminous U.S. states, and collections of U.S. military photographs (1940–1960) and photographs acquired by German military forces in Europe during World War II.

The Library of Congress (Prints and Photographs Division, James Madison Memorial Building, 101 Independence Avenue SE, Washington, DC 20540; 202-707-6277) also maintains a large collection of historical photographs encompassing the period 1900–1940, as well as examples of early balloon photographs from France, panoramic views of U.S. cities from the early 1900s, and similar photographs.

## 3.14. Summary

Aerial photography is a simple, reliable, and inexpensive means of acquiring remotely sensed images. It has been used to make images from low altitudes and from earth satellites, so it can be said to be one of the most flexible strategies for remote sensing. Aerial photography is useful mainly in the visible and near infrared portions of the spectrum, but its principles are important throughout the field of remote sensing. For example, lenses are used in many nonphotographic sensors, and photographic films are used to record images acquired by a variety of instruments.

Aerial photographs form the primary source of information for compilation of many maps, especially large-scale topographic maps. Vertical aerial photographs are valuable as map substitutes or map supplements. Geometric errors in the representation of location prevent direct use of aerial photographs as the basis for measurement of distance or area. But, as these errors are known and are well understood, it is possible for photogrammetrists to use photographs as the basis for reconstruction of correct positional relationships and the derivation of accurate measurements. Aerial photographs record complex detail of the varied patterns that constitute any landscape. Each image interpreter must develop the skills and knowledge necessary to resolve these patterns by disciplined examination of aerial images.

## Review Questions

1. List several reasons why time of day might be very important in flight planning for aerial photography.

2. Outline advantages and disadvantages of high-altitude photography. Explain why routine high-altitude aerial photography was not practical until infrared films were available.

3. List several problems that you would encounter in acquiring and interpreting large-scale aerial photography of a mountainous region.

4. Speculate upon the likely progress of aerial photography since 1890 if George Eastman had not been successful in popularizing the practice of photography to the general public.

5. Is an aerial photograph a "map"? Explain.

6. Assume you have recently accepted a position as an employee of an aerial survey company; your responsibilities include preparation of flight plans for the company's customers. What are the factors that you must consider as you plan each mission?

7. If color films and color infrared films are now available, why are black-and-white films still widely used?

8. Suggest circumstances in which oblique aerial photography might be more useful than vertical aerial photography. Identify situations in which oblique aerial photography would clearly *not* be suitable.

9. It might seem that large-scale aerial photographs might always be more useful than small-scale aerial photographs; yet larger scale images are not always the most useful. What are disadvantages to the use of large-scale images?

10. A particular object will not always appear the same when photographed by an aerial camera. List some of the factors that can cause the appearance of an object to change from one photograph to the next.

## References

American Society of Photogrammetry and Remote Sensing. 1985. Interview: Frederick J. Doyle and Gottfried Konecny. *Photogrammetric Engineering and Remote Sensing,* Vol. 5, pp. 1160–1169.

Anon. 1968. *Applied Infrared Photography*. Kodak Technical Publication M-28. Rochester, NY: Eastman Kodak Co., 88 pp.

Anon. 1976. *Kodak Data for Aerial Photography*. Kodak Publication M-29. Rochester, NY: Eastman Kodak Co., 92 pp.

Anon. 1970. *Kodak Filters for Scientific and Technical Uses*. Kodak Publication B-3. Rochester, NY: Eastman Kodak Co., 90 pp.

Anon. 1974. *Kodak Infrared Films*. Kodak Publication N-17. Rochester, NY: Eastman Kodak Co., 16 pp.

Doyle, Frederick J. 1985. The Large Format Camera on Shuttle Mission 41-G. *Photogrammetric Engineering and Remote Sensing,* Vol. 51, pp. 200–201.

Estes, John E. 1974. Imaging with Photographic and Non-Photographic Sensor Systems. Chapter 2 in *Remote Sensing: Techniques for Environmental Analysis* (J. E. Estes and L. Senger, eds.). Santa Barbara, CA: Hamilton, pp. 15–50.

Langford, Michael J. 1965. *Basic Photography: A Primer for Professionals*. New York: Focal Press, 376 pp.

Light, Donald L. 1993. The National Aerial Photography Program as a Geographic Information System Resource. *Photogrammetric Engineering and Remote Sensing,* Vol. 59, pp. 61–65.

Miller, S. B., U. V. Helavea, and K. D. Helavea. 1992. Softcopy Photogrammetric Workstations. *Photogrammetric Engineering and Remote Sensing,* Vol. 58, pp. 77–83.

Plasker, J. R., and G. W. TeSelle. 1988. Present Status and Future Applications of the National Aerial Photography Program. In *Proceedings of the ACSM/ASPRS Convention.* Bethesda, MD: American Society for Photogrammetry and Remote Sensing, pp. 86–92.

Silva, LeRoy F. 1978. Radiation and Instrumentation in Remote Sensing. Chapter 2 in *Remote Sensing: The Quantitative Approach* (P. H. Swain and S. M. Davis, eds.). New York: McGraw-Hill, pp. 21–135.

Skalet, C. D., G. Y. G. Lee, and L. J. Ladner. 1992. Implementation of Softcopy Photogrammetric Workstations at the U.S. Geological Survey. *Photogrammetric Engineering and Remote Sensing,* Vol. 58, pp. 57–63.

Slater, Phillip N. 1975. Photographic Systems for Remote Sensing. Chapter 6 in *Manual of Remote Sensing* (R. G. Reeves, ed.). Bethesda, MD: American Society for Photogrammetry and Remote Sensing, pp. 235–323.

Stimson, Alan. 1974. *Photometry and Radiometry for Engineers*. New York: Wiley, 446 pp.

Tani, Tadaki. 1989. Physics of the Photographic Latent Image. *Physics Today,* Vol. 42, No. 9, pp. 36–41.

Wolf, Paul R. 1983. *Elements of Photogrammetry*. New York: McGraw-Hill, 628 pp.

## YOUR OWN INFRARED PHOTOGRAPHS

Anyone with even modest experience with amateur photography can take infrared photographs using commonly available materials. A 35-mm camera, with some of the usual filters, will be satisfactory. Infrared films can be purchased at camera stores (but are unlikely to be available at stores that do not specialize in photographic supplies). Infrared films are essentially similar to the usual films but should be used promptly as the emulsions deteriorate much more rapidly than do those of normal films. To maximize life of the film, it should be stored under refrigeration according to the manufacturer's instructions.

Black-and-white infrared films should be used with a deep red filter to exclude most of the visible spectrum. Black-and-white infrared film can be developed using normal processing for black-and-white emulsions, as specified by the manufacturer.

Color infrared films are also available in 35-mm format. They should be used with a yellow filter, as specified by the manufacturer. Processing of CIR film will require the services of a photographic laboratory that specializes in customized work rather than the laboratories that handle only the more popular films. Before purchasing the film, it is best to inquire concerning the availability and costs of processing.

Results are best with bright illumination. The photographer should take special care to face away from the sun while taking photographs. Because of differences in the reflectances of objects in the visible and the near infrared, the photographer should anticipate the nature of the scene as it will appear on the infrared film. Artistic photographers have sometimes used these differences to create special effects.

The camera lens will bring infrared radiation to a focal point that differs from that for visible radiation, so infrared images may be slightly out of focus if the normal focus is used. Some lenses have special markings to show the correct focus for infrared films.

Figure 3.32 illustrates the contrast between the usual black-and-white films and a black-and-white infrared film. Open water absorbs infrared radiation, so it appears unusually dark relative to the visible spectrum; vegetation appears very bright in the near infrared (Chapter 16). In the infrared, atmospheric scattering is minimal (Figure 2.7) so the sky is dark and shadows dark and sharply defined.

**FIGURE 3.32.** Black-and-white infrared photograph (top), with normal black-and-white photograph of the same scene (bottom) shown for comparison.

## YOUR OWN 3D PHOTOGRAPHS

You can take your own stereo photographs using a hand-held camera simply by taking a pair of overlapping photographs (Figure 3.33). The two photographs of the same scene, taken from slightly different positions, create a stereo effect in the same manner that overlapping aerial photographs provide a three-dimensional view of the terrain. This effect can be accomplished by aiming the camera to frame the desired scene, taking the first photograph, moving the camera laterally a short distance, then taking a second photograph that overlaps the field of view of

**FIGURE 3.33.** Stereo photographs.

the first. The lateral displacement need only be a few inches (equivalent to the distance between the pupils of a person's eyes), but a displacement of a few feet will often provide a modest exaggeration of depth that can be useful in distinguishing depth. However, if the displacement is too great, the eye cannot fuse the two images to simulate the effect of depth.

Prints of the two photographs can then be mounted side by side to form a stereo pair that can be viewed with a stereoscope, just as a pair of aerial photos can be viewed in stereo. Stereo images can provide three-dimensional ground views that illustrate conditions encountered within different regions delineated on aerial photographs. Section 5.3 provides more information on viewing of stereo photographs.

CHAPTER FOUR

# Digital Data

## 4.1. Introduction

Thus far, our discussion has focused on examination of remotely sensed images as photographs or photograph-like images. Such images can also be represented in digital form, so that the pattern of image brightness (often in several spectral channels) forms an array of numbers recorded in digital form. When represented as numbers, brightnesses can be added, subtracted, multiplied, divided, and, in general, subjected to statistical manipulations that are not possible if an image is presented only as a photograph. Thus, the digital representation greatly increases our ability to examine, display, and analyze remotely sensed data.

Although digital analysis of remotely sensed data dates from the early days of remote sensing, the launch of the first Landsat earth observation satellite in 1972 began an era of increasing interest in machine processing. Previously, digital remote sensing data could be analyzed only at specialized remote sensing laboratories. Specialized equipment and trained personnel necessary to conduct routine machine analysis of data were not widely available, in part because of limited availability of digital remote sensing data and a lack of appreciation of their qualities.

After Landsat 1 began to generate a stream of digital data, analysts realized that the usual visual examination of Landsat images would not permit full exploitation of the information they conveyed. In time, routine availability of digital data increased interest among businesses and institutions, computers and peripheral equipment became less expensive, more personnel acquired training and experience, software became more widely available, and managers developed knowledge of the capabilities of digital analysis for remote sensing applications. Today, digital analysis has a significance that probably exceeds that of purely visual interpretation of remotely sensed data.

## 4.2. Electronic Imagery

Digital data can be created by a family of instruments that can systematically scan portions of the earth's surface, recording photons reflected or emitted from individual patches of ground, known as *pixels* ("picture elements"). A digital image is composed of many thousands of pixels, each usually too small to be individually resolved by the human eye. Each pixel represents the brightness of a small region on the earth's surface, recorded digitally as a numeric value, usually with separate values for each of several regions of the electromagnetic spectrum (Figure 4.1).

Digital images can be generated by either of two kinds of instruments, each described in later chapters. *Mechanical scanners* physically move mirrors or lenses to systematically aim the field of view over the earth's surface (Figure 4.2). As the instrument scans the earth's surface, it generates an electrical current that varies in intensity as the land surface varies in brightness. Sensors sensitive in several regions of the spectrum use filters to separate energy into several spectral regions, each represented by a separate electrical current. Each electrical signal must be subdivided into distinct units to create the discrete values necessary for digital analysis. This conversion from the continuously varying analog signal to the discrete digital values is accomplished by sampling the current at a uniform interval (analog-to-digital, or *A*-to-*D*, *conversion*) (Figure 4.3). Because all signal values within this interval are represented as a single average, all variation within this interval is lost. Thus, the choice of sampling interval forms one dimension to the resolution of the sensor (Chapter 9).

A second basic design for sensors uses a *charge-coupled device* (CCD) (Figure 4.4). A CCD is formed from light-sensitive material embedded in a silicon chip. The *potential well* receives photons from the scene, usually through an optical system to collect, filter, and focus radiation. The sensitive components of CCDs can be manufactured to be very small, perhaps as small as 1 μm in diameter, sensitive to visible and near infrared radiation. These elements can be connected using microcircuitry to form *arrays*—detectors arranged in a single line form a *linear array*; several rows and columns form *two-dimensional arrays*. Individual detectors are so small that a linear array smaller than 2 cm in

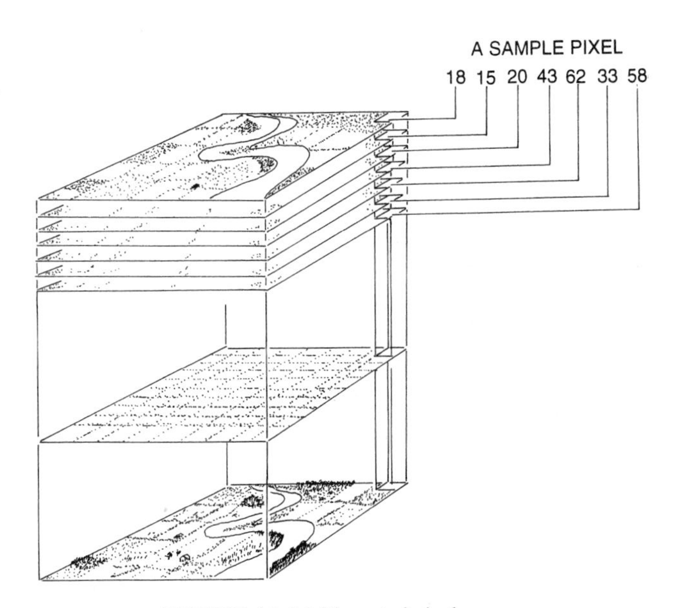


**FIGURE 4.1.** Multispectral pixels.

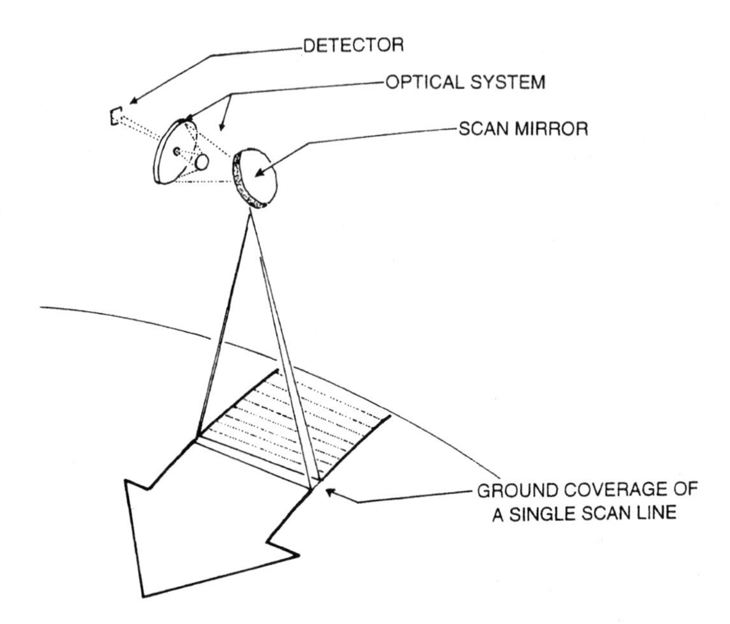


**FIGURE 4.2.** Optical–mechanical scanner.

length might have as many as 1,000 separate detectors. Two-dimensional arrays of 800 × 800 have been used, and larger arrays can be formed by mosaicking several small arrays together.

Each detector collects photons that strike its surface and accumulates a charge proportional to the intensity of the radiation. At a specified interval, charges accumulated at each detector pass through a *transfer gate,* which controls the flow of information from the detectors. Microcircuits connect detectors within an array to form *shift registers.* Shift registers permit charges received at each detector to be passed to adjacent elements, temporarily recording the information until it is convenient to transfer it to another portion of the instrument. Through this process, information read from the shift register is read se-

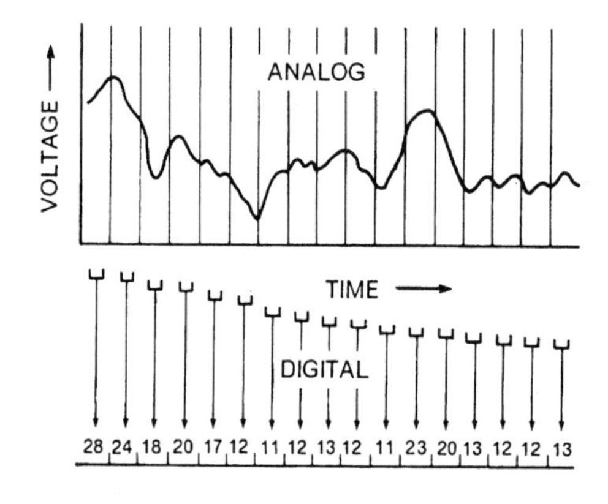


**FIGURE 4.3.** Analog-to-digital conversion.

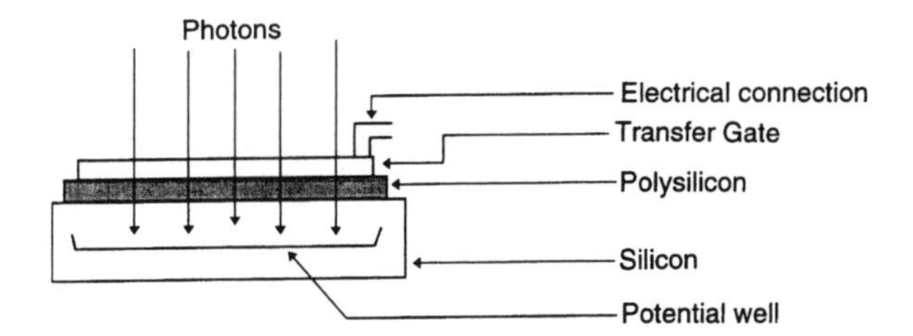


**FIGURE 4.4.** Charge-coupled device.

quentially, much in the same manner that a mechanical scanner collects a line of data through its side-to-side motion.

A CCD therefore scans electronically, without the mechanical motion necessary for the optical scanners described above. CCDs are compact; relative to other sensors they are efficient in detecting photons, so CCDs are especially effective when intensities are dim. Further, they tend to respond linearly to brightness, so they produce images that have more consistent relationship to scene brightness than is possible using photographic processes (Chapter 3).

Optical sensors often use prisms and filters to separate light into separate spectral regions. Electronic sensors usually use *diffraction gratings,* considered more efficient because of their effectiveness, small size, and light weight. Diffraction gratings are closely spaced, transmitting slits cut into a flat surface (a *transmission grating*) or grooves cut into a polished surface (*reflection grating*). Effective transmittion gratings must be very accurately, consistently spaced, and have very sharp edges. Light from a scene is passed through a *colluminating lens,* designed to produce a beam of parallel rays of light, which is oriented to strike the diffraction grating at an angle (Figure 4.5).

Light striking a diffraction grating experiences both destructive and constructive interference as wavefronts interact with the grating. Destructive interference causes some wavelengths to be suppressed, whereas constructive interference causes others to be reinforced. Because the grating is oriented at an angle with respect to the beam of light, different wavelengths are diffracted at different angles and the radiation can be separated spectrally. This light can then illuminate detectors to achieve the desired spectral sensitivity.

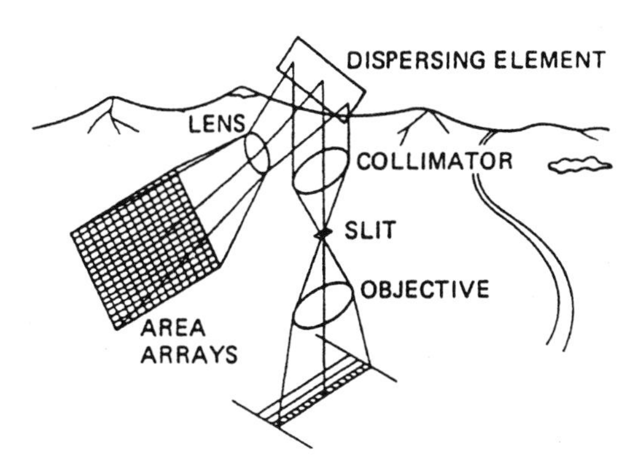


**FIGURE 4.5.** Diffraction grating and colluminating lens.

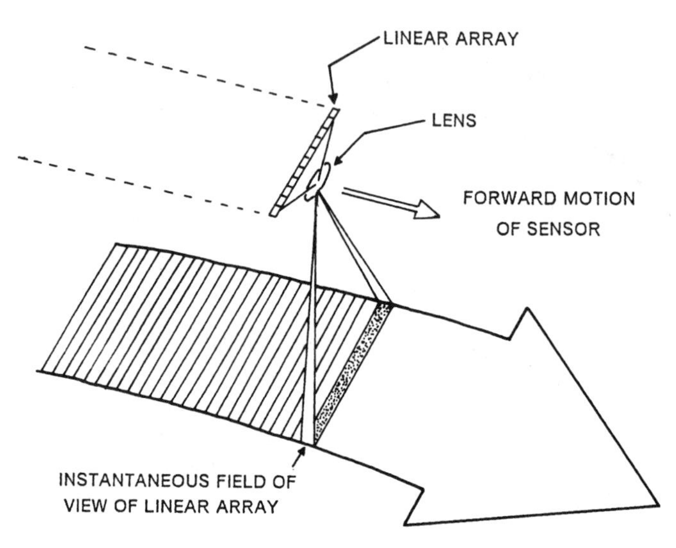


**FIGURE 4.6.** Linear array used in pushbroom scanning.

CCDs can be positioned in the focal plane of a sensor, such that they view a thin rectangular strip oriented at right angles to the flight path (Figure 4.6), and the forward motion of the aircraft or satellite moves the field of view forward along the flight path, building up coverage. By analogy, this means of generating an image is known as *pushbroom scanning.* (In contrast, mechanical scanning can be visualized by analogy to a whisk broom, which creates an image using the back-and-forth motion of the scanner.)

The *instantaneous field of view* (IFOV) represents the area viewed by the instrument if it were possible to suspend the motion of the aircraft and the scanning of the sensor for an instant (Figure 4.7). The IFOV therefore defines the smallest area viewed by the sensor, and a lower limit for the level of spatial detail that can be represented in a digital image.

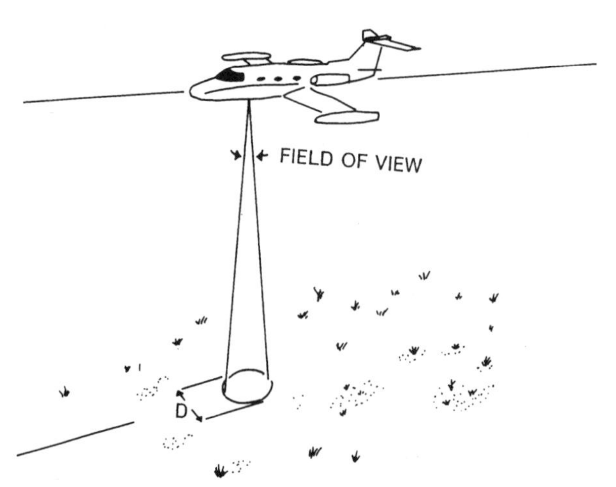


**FIGURE 4.7.** Instantaneous field of view.

Although data in the final image can be aggregated so that an image pixel represents a ground area larger than the IFOV, it is not possible for pixels to carry information about ground areas smaller than the IFOV.

Electronic sensors must be operated within the limits of their design capabilities. Altitudes and speeds of aircraft and satellites must be designed to match sensitivities of the sensors so that detectors view a given ground area long enough to accumulate enough photons to generate strong signals (this interval is known as *dwell time*). At the lower end of an instrument's sensitivity is the *dark current* signal (Figure 4.8). At low levels of brightness, a sensor can record a small level of brightness even when there is none in the scene. At an instrument's upper threshold, bright targets *saturate* the sensor's response—the instrument fails to record further increases in target brightness. Between these limits, sensors are designed to generate signals that have predictable relationships with scene brightness—relationships established by careful calibration of each individual sensor. These characteristics of electronic sensors are analogous to the characteristic curve previously described for photographic films (Chapter 3). In other words, they define the upper and lower limits of the system's sensitivity to brightness and the range of brightnesses over which a system can generate measurements with consistent relationships to scene brightnesses.

The range of brightnesses that can be accurately recorded is known as the sensor's *dynamic range*. Most electronic sensors have rather large dynamic ranges compared to those of photographic films, computer displays, or the human visual system. Therefore, photographic representations of electronic imagery tend to lose information at the upper and/or lower ranges of brightness. Because visual interpretation forms such an important dimension of our understanding of images, the way that photographic films (Chapter 3), image displays (discussed below), and image enhancement methods (also discussed below) handle this problem forms an important dimension of the field of image analysis.

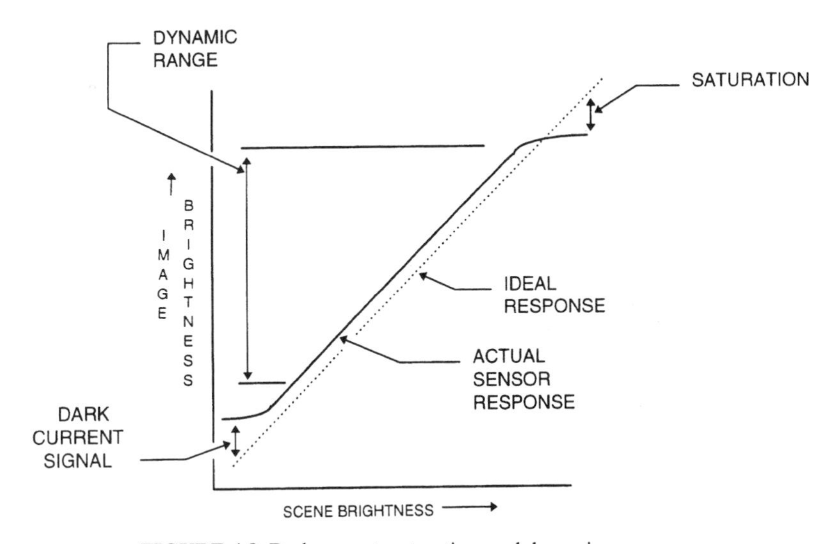


**FIGURE 4.8.** Dark current, saturation, and dynamic range.

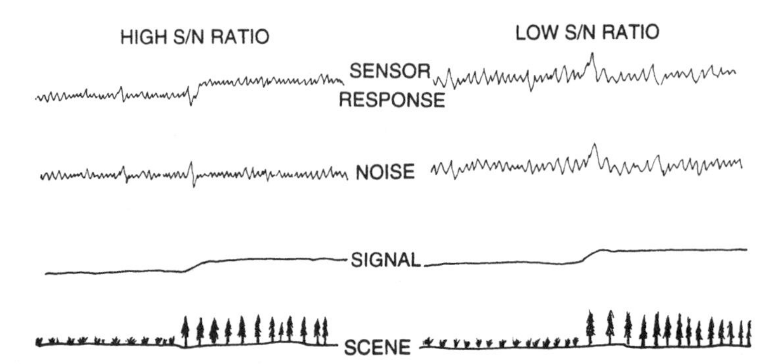


**FIGURE 4.9.** Signal-to-noise ratio. At the bottom, a hypothetical scene is composed of two cover types. The signal records this region, with only a small difference in brightness between the two classes. Atmospheric effects, sensor error, and other factors contribute to noise, which is added to the signal. The sensor then records a combination of signal and noise. When noise is small relative to the signal (left: high S/N ratio), the sensor conveys the difference between the two regions. When the signal is small relative to noise (right, low S/N ratio) the sensor cannot portray the difference in brightness between the two regions.

Each sensor creates responses unrelated to target brightness (i.e., *noise*), created in part by accumulated electronic errors from various components of the sensor. (In this context noise refers specifically to noise generated by the sensor, although the noise that the analyst receives originates not only in the sensor but also in the atmosphere, the interpretation process, etc.) For effective use, instruments must be designed so that the noise levels are small relative to the signal (brightness of the target). This is measured as the *signal to noise ratio* (S/N or SNR) (Figure 4.9), the ratio of the measured brightness to the variation of the noise. Analysts desire signals to be large relative to noise, so the SNR should be large not only for bright targets, when the signal is large, but over the entire dynamic range of the instrument, especially at the lower levels of sensitivity where the signal is small relative to noise. Therefore, the lowest brightness that can be reliably measured is an indication of the sensitivity of a sensor; this quality is measured as the *noise-equivalent radiance*, the radiance when the SNR = 1. A low noise-equivalent radiance indicates a sensor with high sensitivity. Engineers who design sensors must balance the radiometric sensitivity of the instrument with pixel size, dynamic range, operational altitude, and other factors to maintain acceptable SNRs.

## 4.3. Digital Data

Output from electronic sensors reach the analyst as a set of numeric values. Each digital value is recorded as a series of binary digits known as *bits*. Each bit records an exponent of a power of 2, with the value of the exponent determined by the position of the bit in the sequence. As an example, consider a system designed to record 7 bits for each digital value. This means that seven binary places are available to record the brightness sensed for each band of the sensor. The seven values record, in sequence, successive powers of 2. A "1" signifies that a specific power of 2 (determined by its position within the sequence) is

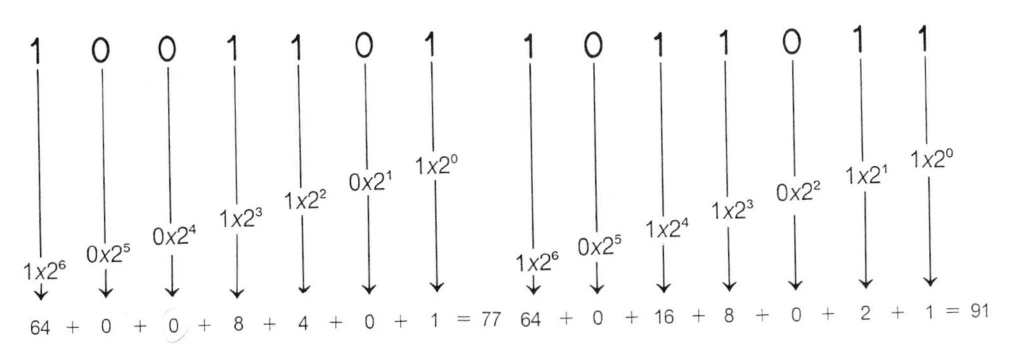


**FIGURE 4.10.** Digital representation of values in seven bits.

to be evoked; a "0" indicates a value of zero for that position. Thus, the 7-bit binary number "1111111" signifies $2^6 + 2^5 + 2^4 + 2^3 + 2^2 + 2^1 + 2^0 = 64 + 32 + 16 + 8 + 4 + 2 + 1 = 127$. And "1001011" records $2^6 + 0^5 + 0^4 + 2^3 + 0^2 + 2^1 + 2^0 = 64 + 0 + 0 + 8 + 0 + 2 + 1 = 75$. Figure 4.10 shows different examples, and Table 4.1 provides designations for digital data storage capacities.

In this manner, discrete digital values for each pixel are recorded in a form suitable for storage on tapes or disks and for analysis by digital computer. Often these values, as read from tape or disk, are popularly known as *digital numbers* (DN), *brightness values* (BV), or *digital counts,* in part as a means of signifying that these values do not record true radiances from the scene.

The number of brightness values within a digital image is determined by the number of bits available. The 7-bit example given above permits a maximum range of 128 possible values (0 to 127) for each pixel. A decrease to 6 bits would decrease the range of brightness values to 64 (0 to 63); an increase to 8 bits would extend the range to 256 (0 to 255). Thus, the number of bits determines the radiometric resolution (Chapter 9) of a digital image. The number of bits available is determined by the design of the system, especially the sensitivity of the sensor (adding too many extra bits would simply record system noise rather than provide additional information about scene brightness), and the capabilities for recording and transmitting data (each added bit increases transmission requirements). If we assume that transmission and storage resources are fixed, increasing the number of bits for each pixel means that we will have fewer pixels per image and that pixel sizes must be larger. Thus, design of remote sensing systems requires trade-offs between image coverage and radiometric, spectral, and spatial resolutions.

**TABLE 4.1. Terminology for Computer Storage**

| | |
|---|---|
| Bit | A binary digit (0 or 1) |
| Byte | 8 bits, 1 character |
| Kilobyte (K or KB) | 1,024 bytes ($2^{10}$) |
| Megabyte (MB) | $2^{20}$ bytes |
| Gigabyte (GB) | $2^{30}$ bytes |
| Terabyte (TB) | $2^{40}$ bytes |

### *Radiances*

Brightness of radiation reflected from the earth's surface is measured as brightness (watts) per wavelength interval (micrometer) per angular unit (steradian); thus, the measured brightness is defined with respect to wavelength (i.e., "color"), spatial area (angle), and intensity (brightness). Radiances record actual brightnesses, measured in physical units, given as real values (to include decimal fractions).

Use of digital counts facilitates the design of instruments and data communications and the visual display of image data. For visual comparison of different scenes, or analyses that examine relative brightnesses, use of digital counts is satisfactory. However, because a digital count from one scene does not represent the same brightness as the same digital count from another scene, digital counts are not comparable from scene to scene if an analysis must examine actual scene brightnesses for purposes that require use of original physical units.

For such purposes, it is necessary to convert the DNs to the original radiances. This process requires knowledge of calibration data specific to each instrument, as outlined in the Appendix, which also provides calibration data for some of the most frequently used sensors described in later chapters. To ensure that a given sensor provides an accurate measure of brightness it must be calibrated against targets of known brightness. The sensitivity of electronic sensors tends to drift over time, so to maintain accuracy it is necessary to recalibrate the instrument. Although those sensors used in aircraft can be recalibrated periodically, those used in satellites are not available after launch for the same kind of recalibration. Typically, such sensors are designed so that they can observe calibration targets on board the satellite or are calibrated using landscapes of uniform brightness (in desert regions for example). Nonetheless, such calibration errors as those described in Chapter 10 sometimes remain.

## 4.4. Data Formats

Digital image analysis is usually conducted using *raster* data structures—each image is treated as an array of values (Figure 4.11). Additional spectral channels form additional arrays that register to one another. Each pixel is treated as a separate unit, which can always be located within the image by its row and column coordinates. In most remote sensing analyses, coordinates originate in the upper left-hand corner of an image, and are referred to as *rows* and *columns,* or as *lines* and *pixels,* to measure position down and to the right, respectively.

Raster data structures offer advantages for manipulation of pixel values by image processing systems, as it is easy to find and locate pixels and their values. The disadvantages are usually apparent only when we need to represent not the individual pixels but areas of pixels as discrete patches or regions. Then the alternative structure—*vector format*—becomes more attractive. Vector format (discussed in Chapter 17) uses polygonal patches and their boundaries as fundamental units for analysis and manipulation. The vector format is not appropriate for digital analysis of remotely sensed data, although sometimes we may wish to display the results of our analysis using a vector format. Equipment

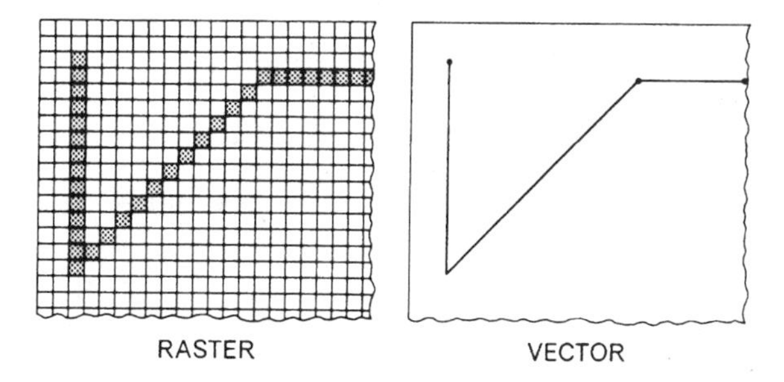


**FIGURE 4.11.** Raster and vector formats.

and software for digital processing of remotely sensed data almost always must be tailored for raster format.

Digital remote sensing data are often organized using one of the three common formats used to organize image data. Consider an image consisting of four spectral channels, which can be visualized as four superimposed images, with corresponding pixels in one band registering exactly to those in the other bands.

One of the earliest formats for digital remote sensing data was *band interleaved by pixel* (BIP). Data are organized in sequence values for line 1, pixel 1, band 1; then line 1, pixel 1, band 2; then line 1, pixel 1, band 3; and line 1, pixel 1, band 4. Next are the four bands for line 1, pixel 2, and so on (Figure 4.12). Thus, values for all four bands are written before values for the next pixel are represented. Any given pixel, once located within the data, is found with values for all four bands written in sequence one directly after the other. This arrangement may be advantageous in some situations, but for most applications it is awkward to sort through the entire sequence of data (which typically are very large in number) in order to sort the four bands into their respective images.

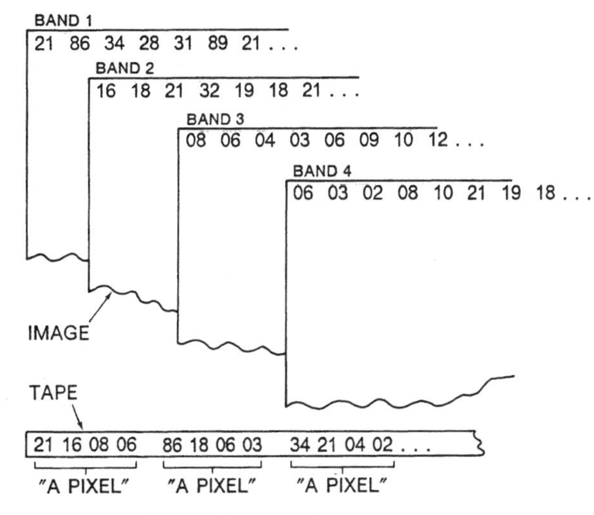


**FIGURE 4.12.** BIP format.

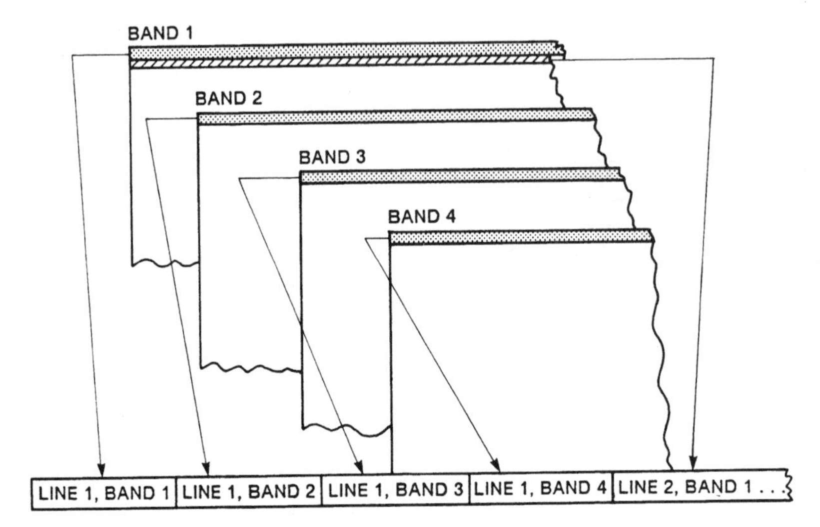


**FIGURE 4.13.** BIL format.

The *band interleaved by line* (BIL) format treats each line of data as a separate unit (Figure 4.13). In sequence, the analyst finds line 1 for band 1, line 1 for band 2, line 1 for band 3, line 1 for band 4, line 2 for band 1, line 2 for band 2, and so on. Each line is represented in all four bands before the next line is encountered.

A third convention for recording remotely sensed data is the *band sequential* (BSQ) format (Figure 4.14). All data for band 1 are written in sequence, followed by all data for band 2, then bands 3 and 4 in sequence. Each band is treated as a separate unit. For many applications, this format is among the most practical as it presents data in the format that most closely resembles the data structure used for the display and analysis. However, if areas smaller than the entire scene are to be examined, the analyst must read all four images before the subarea can be identified and extracted.

Actual data formats used to distribute digital remote sensing data are usually varia-

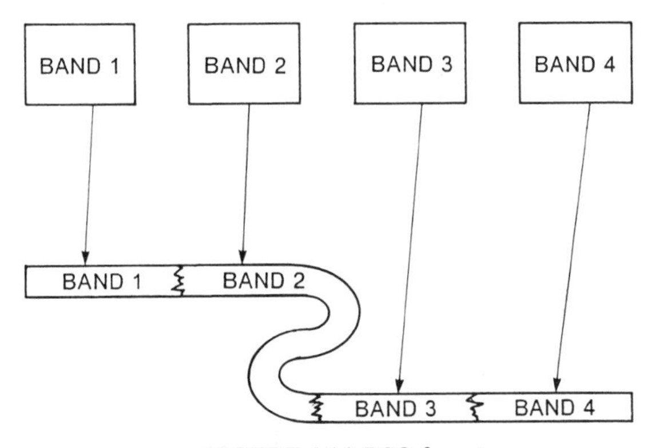


**FIGURE 4.14.** BSQ format.

tions on these basic alternatives. For example, the fast format used by Earth Observation Satellite Company (EOSAT) to distribute some of its Landsat data (introduced in Chapter 6) is a variation on the BSQ format. Likewise, SPOT Image Corporation (also discussed in Chapter 6) uses a form of the BIL format for much of its image data. Exact details of data formats are specific to particular organizations, and to particular forms of data, so whenever an analyst acquires data, it is important also to acquire detailed information regarding the data format. Although organizations attempt to standardize formats for specific kinds of data, it is also true that data formats change as new mass storage media come into widespread use, and as user communities employ new kinds of hardware or software.

The "best" data format depends on immediate context, and often the specific software and equipment available. If all bands for an entire image must be used, the BSQ and BIL formats are useful because they are convenient for reconstructing the entire scene in all four bands. If the analyst knows beforehand the exact position on the image of the subarea that is to be studied, the BIP format is useful because values for all bands are found together and it is not necessary to read through the entire data set to find a specific region. In general, however, the analyst must be prepared to read the data in the format that they are received, and to convert them into the format most convenient for use at a specific laboratory.

*Data compression* reduces the volume of data required to store or transmit information by exploiting redundancies within the data set. Because remotely sensed images require large amounts of storage, and usually contain modest redundancies, data compression is of significance for storage, transmission, and handling of digital remote sensing data. Compression and decompression are accomplished by running computer programs that receive, for example, compressed data as input and produce a decompressed version of the same data as output.

The *compression ratio* compares the size of the original image to the size of the compressed image. A ratio of 2:1 indicates that the compressed image is one half the size of the original. *Lossless compression* techniques restore compressed data to their exact original form; *lossy* techniques degrade the reconstructed image, although in some applications the visual impact of a lossy technique may be imperceptible. For digital satellite data, lossless compression techniques can achieve ratios of 1.04:1 to 1.9:1. For digitalized cartographic data, ratios of 24:1 using lossy techniques have been reported to exhibit good quality. It is beyond the scope of this discussion to describe the numerous techniques and algorithms for image compression. Probably the most well-known compression technique is the JPEG (Joint Photographics Experts Group), a lossy technique that applies the discrete cosine transform (DCT) as a compression–decompression algorithm. Although JPEG has been accepted as a useful technique for compression of continuous-tone photographs, it is not likely to be ideal for remote sensing and GIS data in general.

Generally stated, lossy techniques should not be applied to archival or original data to be analyzed. Lossy compression may be appropriate for images that present visual results of analytical processes, provided they do not form input for other analyses.

## 4.5. Equipment for Digital Analysis

Digital analysis requires specialized equipment tailored for the storage, display, and analysis of the large amounts of data that are necessary to record remotely sensed images.

Remote sensing analysis has specific needs for input, storage, analysis, and display of data that are often not fully met by standard equipment and programs.

## *Computers*

Although there was a time when some computers were specifically designed to analyze and display remotely sensed data, these computers were found to be too expensive and inflexible for widespread use. Therefore, although remote sensing analysis may sometimes require extra memory, disk storage, or peripheral devices, today the vast majority of remote sensing analysis uses general purpose computers using specialized remote sensing software.

The heart of any digital computer is its *central processing unit* (CPU), which consists of three components (Figure 4.15). The control unit manages the flow of data and instructions to and from the CPU and to and from input and output devices. Once CPUs were massive networks of vacuum tubes, wires, and switches that filled several rooms. Today the electrical pathways for a CPU are tiny circuits printed on silicon chips perhaps half the size of a credit card. Successive generations of chips have become increasingly more powerful and efficient, so the once arcane designations assigned by manufacturers to chips have become common shorthand for designating the power of a computer. Intel's 80286, 80386, 80486, and Pentium designations and Motorola's 68000, 68020, 68030, and 68040 designations identify increasingly more sophisticated chips.

Within the CPU, the *arithmetic/logic unit* (ALU) manipulates input data to produce new (output) values that form results of a particular operation. Any computer program, when resolved into its simplest components, consists of a sequence of such fundamental operations as addition, multiplication, and comparisons of values. These operations are often performed by specially designed silicon chips that form the ALU. Whereas most chips simply store bits that record data without altering the values, those in the ALU are designed to perform arithmetic or logical operations, so that the data they return to the CPU are altered from those that they receive. The ALU receives data from the control unit, performs the required operations, then returns the results to the control unit. Although each operation of the ALU is itself very simple, a computer's power is achieved by

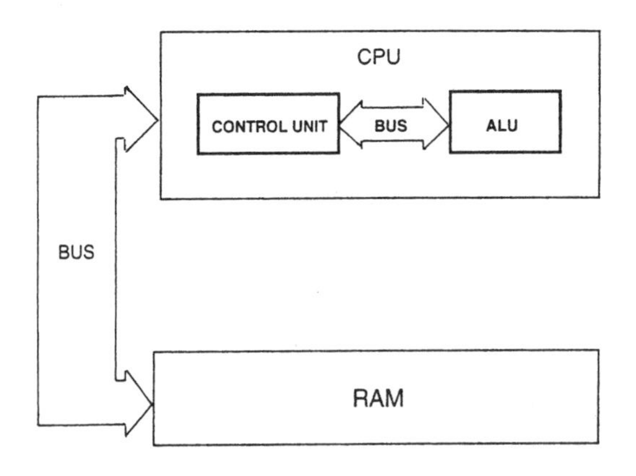


**FIGURE 4.15**. Schematic representation of the CPU.

its ability to perform thousands of such simple operations quickly, with accuracy, and with precision.

The third component of the CPU is its *primary storage*—the CPU's internal memory. Primary storage is *random access memory* (RAM), meaning that the CPU has direct access to each location in primary memory without sequential searching through preceding addresses and without mechanical movement (as are required with other forms of memory). Every value is stored in binary form, as described previously. The capabilities of main memory are a function of the number of bits used to record each value in memory, which is a function of the design of the chips that form the computer's main memory. Computers used for image processing now generally have 16- or 32-bit processors, meaning that each value stored in main memory is recorded with precision of 16 or 32 bits, as explained previously.

Because the CPU must load a program and data into primary storage, the amount of primary storage determines many of the capabilities of the computer. Small primary storage means that the computer can run rather short programs on small data sets; large primary storage means that the computer can run longer programs on larger data sets. For image processing work, 2.5 megabytes (2,500 K) of memory forms a rather respectable primary storage—sufficient to permit rather sophisticated programs to operate upon large images.

The CPU, RAM, and ALU are connected by *data buses*, electrical pathways that transfer data from one portion of the computer to another. A bus consists of wires, connectors, and controllers; a bus must provide one wire for each bit required to specify the address of a location in memory plus additional pathways to distinguish among the alternative data transfer options possible. For PCs, the design of the bus, often known as the *expansion bus*, determines the extent to which a computer's memory, speed, and peripheral support can be expanded by adding or upgrading components. The capacity of the bus to transmit data determines the speed at which it can operate. Industry standards specify capacities of 8 or 16 bits for data buses, although 32-bit buses have been developed to match to the higher capacities of some microprocessors.

### *Mass Storage*

Because remotely sensed data typically require large storage capacities, image processing systems use substantial amounts of disk storage. Disks can be manufactured from either rigid or flexible materials that are coated with a magnetized substance sensitized to record the bits that represent each digital value. The computer's disk drives rapidly spin the disk, allowing the computer to read data from both sides. Rigid (*hard*) disks provide rapid access to large amounts of storage required for image processing. The flexible (*floppy*) disks have smaller storage capacities, and provide slower access to the data; they are used primarily for smaller computers, whose demands for storage and access are not as critical. Also available are a variety of removable disk drives that can store from 20 to 230 MB of data in a compact, transportable form.

*Computer tapes* are strips of coated plastic film that have been sensitized to record digital data when the write heads of tape drives encode them with magnetic marks that record bits and bytes. Computer tapes are now available in a range of sizes and capacities, some as small as the audio cassettes familiar to most readers.

Computer tapes permit compact storage and convenient transportation of the large amounts of digital data required for remote sensing. Data storage is a concern because remotely sensed images are so large that usually it is possible to store only a few on hard disk at a given time. Those images not in immediate use are usually written to a magnetic tape, which is then stored until the images are required once again. Furthermore, magnetic tapes usually form the primary means of providing archive (*backup*) copies for use in the event a primary copy is accidentally destroyed. Although magnetic tapes must be carefully maintained under proper temperature and humidity, tape storage is relatively convenient and inexpensive. Computers read from and write to magnetic tape by means of *tape drives*, which can rapidly wind and unwind spools of tape. Sensitized read and write heads sense the magnetic record of bits written on the tape or can sensitize the tape's coating to record data.

Older computer systems once used long tapes stored on large (10-in. diameter) reels. Each tape recorded data in nine tracks that ran parallel to each other for the entire length of the tape. Each track recorded one bit, so an eight-bit word could be written across the width of the tape. (The ninth track recorded the *parity bit*, used to detect errors in reading and writing data to and from tape.) The amount of data recorded on a tape was determined in part by the density of data as written to the tape; usually data were written at either 1,600 bits per inch (BPI) or 6,250 BPI (although even older tapes may be written at 800 BPI). Today, use of these tapes has declined rapidly as they have been replaced by other mass storage media (discussed below). Nonetheless, reel tapes still hold older data in archives, so they will be used in some institutions until the data can be transferred to more compact media.

*Data cartridges* and *cartridge tape units* are more compact versions of the tapes just described, usually using high-density tapes stored in small cassettes. Exact sizes, capacities, and data formats vary with manufactuters; cassettes vary in size from about 4 in. × 6 in. to about 2.5 in. × 3.5 in. Some of these smaller high-density tapes, although as small as a common audiocassette, can store more data than can the older reel tapes. Because the drives for these smaller tapes are also compact and inexpensive, they greatly expand the capabilities of desktop computer systems and the ability of individuals and organizations to store and use remote sensing data.

CD-ROM (*Compact Disk, Read-Only Memory*) storage is familiar to most readers as the prevailing means of recording audio information for mass distribution. CD-ROMs are the oldest and most reliable form of *optical storage*: Each CD-ROM is a plastic disk, encoded by irregularities in very fine, textured grooves that can be read by a laser within a CD-ROM drive. Because each disk can hold 680 MB, they form a compact, durable means of storing remote sensing data.

Once encoded, CD-ROMs cannot be erased or rewritten, so they are suitable primarily, in the context of remote sensing, as a means of transmitting raw image data that will be read many times. Although the cost of preparing CD-ROMs was once so expensive that they were used only for data that were needed in many copies, costs are now low enough that they can be efficient for many fewer copies, and image data are now often distributed on CD-ROMs. CD-ROMs must be read using special drives available as peripheral units but now also increasingly offered as standard equipment on personal computers.

WORM (*Write Once, Read Many*) disks are another form of optical storage that permits the user, rather than the manufacturer, to write data to the disk. Once written, however, the information can be read but not altered. WORM disks vary in size from about 5.25 in. in diameter to 14 in., with larger disks holding as much as 3 GB on each side. Although WORM disks offer important advantages for some users, manufacturers have yet to agree on common standards for data formats, so they are not yet practical for uses that require many transfers between different systems. *Erasable optical disks* are yet another form of optical storage, with the ability to store large amounts of data cheaply. Rewritable optical disk cartridges are 5.5 in. in diameter; they offer improved flexibility over WORMs and CD-ROMs, although current drives are slow compared to magnetic disks.

### *Video Display Terminal and Keyboard*

The analyst communicates with the computer by means of a terminal consisting of a display screen and keyboard. The screen is used only for displaying commands entered from the keyboard or responses and data provided by the computer. Usually there is a special display (described below) tailored for the specific purpose of displaying images. The computer terminal, together with the image display unit, forms a single workstation at which the analyst can both communicate with the computer and examine the images to be analyzed. Often a large computer is configured to service several workstations by means of a time-sharing system, so several analysts can use the computing facilities at the same time.

### *Image Display*

For remote sensing computing, the image display is especially important because the analyst must be able to examine images and inspect results of analyses, which often are themselves images. At the simplest level, an image display can be thought of as a high-quality television screen, although those tailored specifically for image processing have image display processors, which are special computers designed to receive rapidly digital data from the main computer and display them as brightnesses on the screen. The capabilities of an image display are determined by several factors . First is the size of the image it can display, usually specified by the number of rows and columns it can show at any one time. A large display might show 1,024 rows and 1,280 columns. A smaller display with respectable capabilities could show a 1,024 × 708 image; others of much more limited capabilities could show only smaller sizes, perhaps 640 × 480 (typical of displays for the IBM PC).

Second, a display has a given radiometric resolution (Chapter 9). That is, for each pixel, it has a capability to show a range of brightnesses. One-bit resolution would give the capability to show either black or white—certainly not enough detail to be useful for most purposes. In practice, six bits (64 brightness levels) are probably necessary for images to appear "natural," and high-quality displays typically display eight bits (256 brightness levels) or more.

A third factor controls the rendition of color in the displayed image. The method of depicting color is closely related to the design of the image display and the display processor. Image display data are held in the frame buffer—a large segment of computer memory dedicated to handling data for display. The frame buffer provides one or more bits to record the brightness of each pixel to be shown on the screen (the *bit plane*); thus, the displayed image is generated, bit by bit, in the frame buffer. The more bits that have been designed in the frame buffer for each pixel, the greater the range of brightnesses that can be shown for that pixel, as explained above. For actual display on the screen, the digital value for each pixel is converted into an electrical signal that controls the brightness of the pixel on the screen. This requires a digital-to-analog (D-to-A) converter that translates discrete digital values into continuous electrical signals (the opposite function of the A-to-D converter mentioned previously).

Display of color images requires a more complex process. For the simplest and cheapest color displays (sometimes known as pseudo-color displays) the image is stored in a single memory plane (Figure 4.16). The color for each pixel is determined by matching each pixel to a set of colors (a "palette") stored in a segment of memory known as a look-up table. Values in the look-up table control the intensities of the three A-to-D converters (one for each primary) that create colors on the display screen. Therefore, the number of bits in the look-up table forms one limit on the number of colors that can be shown by a given display. The look-up table stores the current selection of colors; the number of colors that can be shown at a given time is controlled by the number of bits assigned to each pixel in the screen's RAM. The values in the look-up table control the brightness on the screen in the manner illustrated by Table 4.2. In this simple example, the intensity of each additive primary is controlled by only one bit, so a maximum of only eight colors can be displayed at a given time. Each primary is either "on" (1) or "off" (0), resulting in the eight combinations shown in Table 4.2. If the look-up table has more bits, the primaries can be represented more subtly (rather than simply as on or off), so, for example, 4 bits give intensities from 0 to 15, 5 bits intensities from 0 to 31, and 7 bits intensities from 0 to 127. With more subtle variations of the brightnesses of the intensities, the colors on the screen assume a wider range of variations, more nearly approaching those we observe in nature. For example, with 7 bits the display could show red = 127, green = 0, and blue = 0 to produce a pure red on the screen, as is possible with the simple example in Table 4.2. But it could also show a more subtle shade of color by setting red = 127, green = 37, and blue = 65, a choice that would not be possible with the simpler system.

Thus, the look-up table translates pixel values into intensities that can be displayed

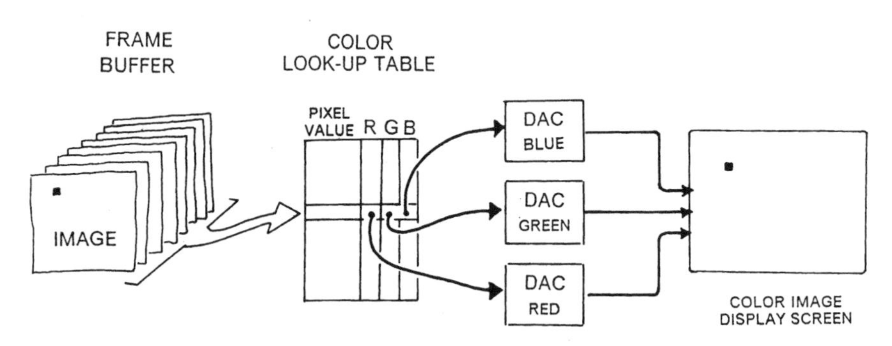


**FIGURE 4.16.** Schematic diagram of a pseudo-color display.

**TABLE 4.2. Example of Look-Up Table for Three-Bit Frame Buffer**

| | Values from bit plane | | |
|---|---|---|---|
| Color on screen | Red | Green | Blue |
| Black | 0 | 0 | 0 |
| Red | 1 | 0 | 0 |
| Green | 0 | 1 | 0 |
| Blue | 0 | 0 | 1 |
| Yellow | 1 | 1 | 0 |
| Cyan | 0 | 1 | 1 |
| Magenta | 1 | 0 | 1 |
| White | 1 | 1 | 1 |

on the screen. Note that the look-up table may not have sufficient precision to accurately represent the varied detail within the data. (Certainly the example in Table 4.2 would not provide very much detail.) The look-up table limits the number of colors that can be shown at a given time, but the analyst can reconfigure (or "reload") the look-up table to select another set of colors. The limitation of pseudo-color technology is both the restriction in the range of colors that can be shown at a given time and the time and inconvenience entailed in reloading and displaying colors.

The alternative form of representing colors is known as true, full, or analog color display (Figure 4.17). A separate memory plane is required for each primary color, each with its own look-up table and D-to-A converter. Because each of the three primary colors resides in independent memory planes, manipulation of colors is much faster and easier

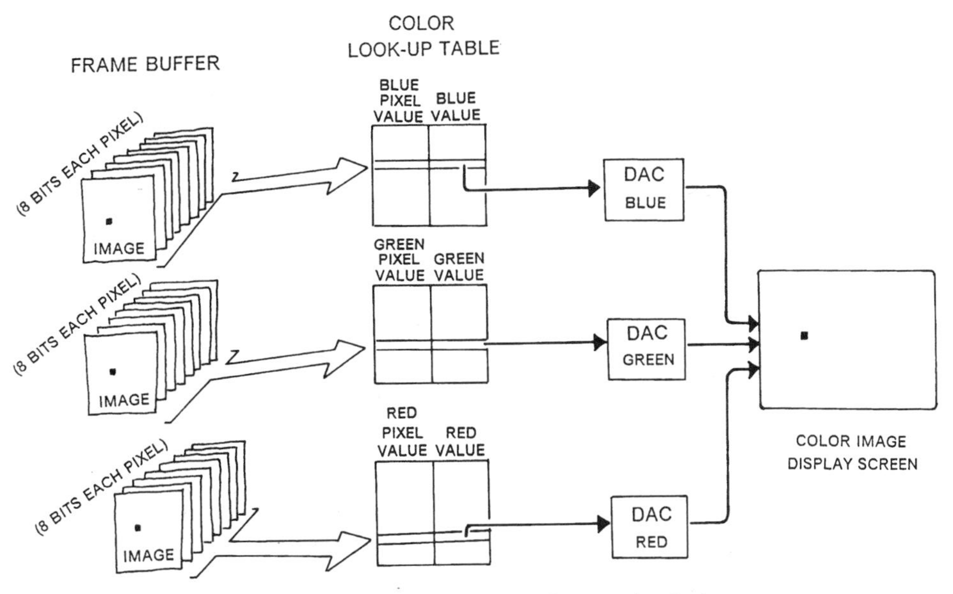


**FIGURE 4.17.** Schematic diagram of a true color display.

**TABLE 4.3. Graphic Standards for PCs**

| Name and designation | Colors | Screen size |
|---|---|---|
| Color graphics adapter (CGA) | 4 | 320 × 200 |
| Extended graphics adapter (EGA) | 16 | 640 × 350 |
| Video graphics array (VGA) | 16 | 640 × 450 |
| Super VGA (S-VGA) | 256 | 800 × 600 |
| Extended graphics array (XGA)[a] | 256 | 1,024 × 768 |

[a]Some PC displays, not now widely used, employ screen sizes of 1,280 × 1,024.

than with pseudo-color systems. In addition, each primary color has its own look-up table, which permits a much wider range of intensities for each primary, and therefore a wide range of combinations on the screen, which in practice approximates the range of colors that might be observed in nature—hence, the term *true color.* True color display shows much more realistic and attractive images than do pseudo-color displays but are also more expensive.

Table 4.3 lists characteristics of current graphics standards. Image display is important because remote sensing instruments typically record a much wider range of values than can be accurately displayed by film products (Chapter 3) or by any single representation on an image display device. As a result, displays that provide the analyst with the capability to conveniently change image scale, color assignments, viewing area, and so forth, provide a capability to explore visually the varied dimensions of an image. Such capabilities are not available with film images, which require considerable effort to study varied representations of a digital image

### *Film Recorders*

Film recorders are instruments designed to write image information directly to photographic paper or film. Usually the film is mounted on a cylindrical drum. During operation, the drum spins rapidly. A beam of light is varied in intensity in proportion to image brightness. As the drum spins, the light slowly moves the length of the image, exposing the entire image line by line. If a color image is to be made, usually it is necessary to make three separate exposures—one for each primary color.

## 4.6. Image Processing Software

Digital remote sensing data can be interpreted by computer programs that manipulate the data recorded in pixels to yield information about specific subjects, as described in subsequent chapters. This kind of analysis is known as *image processing,* a term that encompasses a wide range of techniques. Image processing requires a system of specialized computer programs tailored to manipulate digital image data. Although such programs vary greatly in purpose and detail, it is possible to identify the major components likely to be found in most image processing systems.

A separate specific portion of the system is designed to read image data, usually from tape, CD-ROM, or other storage media, and reorganize them into the form to be used by the program. For example, many image processing programs manipulate the data in BSQ format. Thus, the first step may be to read BIL or BIP data and reformat the data into the BSQ format required for the analytical components of the system. Another portion of the system may permit the analyst to subdivide the image into subimages; to merge, superimpose, or mosaic separate images; and in general to prepare the data for analysis, as described later in Chapter 10. The heart of the program is formed by a suit of programs that analyze, classify (Chapter 11), and manipulate data to produce output images and the statistics and data that may accompany them. Finally, a section of the image processing system must prepare data for display and output, either to the display processor or to the line printer. In addition, the program requires "housekeeping" subprograms that monitor movement and labeling of files from one portion of the program to another, generate error messages, and provide on-line documentation and assistance to the analyst.

All the popular image processing systems run on PCs or workstations that can reside in an office or a small laboratory. More elaborate systems can be supported by peripheral equipment, including extra mass storage, digitizers, scanners, color printers, disk drives, tape units, and related equipment. Almost all such systems are directed by menus and graphic user interfaces that permit the analyst to select options from a list of options on the screen.

Although there are many good image processing systems available, some of the most commonly used are:

- **ER Mapper** (Earth Resources Mapping, 4370 La Jolla Village Drive, Suite 900, San Diego, CA 92122-1253; 619-558-4709)
- **EASIPACE** (PCI Enterprises, 2 Guidwara Road, Suite 220, Nepean, Ontario K2E 1A2, Canada; 613-226-8588)
- **ERDAS** (ERDAS, Inc., 2801 Buford Highway, Suite 300, Atlanta, GA 30329; 404-248-9000)
- **IDRISI** (The IDRISI Project, The Clark Labs for Cartographic Technology and Geographic Analysis, Clark University, 950 Main Street, Worcester, MA 01610-1477; 508-793-7526)

IDRISI is designed for use on PCs; the others listed here are designed for use on workstations but may have PC versions available.

The specific systems listed here can be considered general-purpose image processing systems; others have been designed specifically to address requirements for specific kinds of analysis (geology, hydrology, etc.), and some of the general-purpose systems have developed optional modules that focus on more specific topics. Further details of image analysis systems are given by user manuals for specific systems. Authors of image processing systems usually prepare new versions to add new or improved capabilities, accommodate new equipment, or address additional application areas.

## 4.7. The Internet

Digital data once were available only through physical transfer of disks or tapes, usually an inconvenient, awkward process. Now the Internet has facilitated not only the acquisition of digital data but also the equally troublesome task of searching indexes and archives to identify the appropriate coverage. The *Internet* is a network of computers connected by the world's telecommunications infrastructure of phone lines, microwave relays, and satellite links. This hardware infrastructure is supported by software that employs common conventions for electronic transmission of data. Thus, the Internet consists of *servers* (computers that run programs designed to share their data with others by permitting remote access) and *clients* (computers that run programs designed to access servers).

A large part of the power of the Internet derives from the simultaneous use of some computers as clients and as servers by establishing temporary links between computers at distant locations holding related data. Internet users, therefore, access vast amounts of data residing at varied locations as if the data were immediately at hand. Further, because different users can define their own pathways through the network, it can be tailored with specific requirements of users. Because servers are maintained by a vast and growing number of individuals, businesses, universities, and other insitituions, the amount and kinds of information provide an unprecedented resource for students and researchers in the field of remote sensing. Although such commercial networks (*on-line services*) as Prodigy, CompuServe, and America Online, provide access to a wide range of information, Internet's scope and ease of access have made it the primary network for scientific and research applications.

To utilize these networks, users require access to a computer (IBM-compatible PC, Macintosh, or UNIX) equipped with communications software, a modem or ethernet card, and such communications as a phone line or fiber-optic cable. Finally, Internet users require accounts or user identification, such as those issued by university computing centers or commercial networks, known as *Internet providers*.

The only real way for students, to learn about the Internet is to actually use it. For university students this is often relatively easy, as access to the Internet is without charge and most universities offer access for students. Local resources vary greatly, so students need to consult local service guides. Further, Internet resources are changed frequently, so beginners should verify that their information is current.

### *FTP and TELNET*

File Transfer Protocol (FTP) is a program for transferring computer files between two computers connected by the Internet. FTP permits transfer of either ASCII text or binary files. Remotely sensed images are almost always stored as binary files, although they may be accompanied by text files that convey important information about the binary images.

FTP usually requires that a user know the user ID and password on the other ("target") computer as well as those of the user's own computer. FTP is a powerful tool for practitioners of remote sensing because of the convenience it offers in transferring large images from one computer to another, even those separated by large distances. Equally valuable, however, is the access that FTP provides to archives of free, public domain software and imagery. Some computers permit outside users to log on without a user ID or password using a convention called *anonymous FTP*. When prompted for a user ID, the analyst enters "anonymous," and his or her own Internet address when prompted for the password.

Whereas FTP permits users to transfer data to or from a remote computer, Telnet permits users to log on to other computers as if they were owners. Telnet therefore requires knowledge of the user ID and password and permission of the owner. Exceptions might be public computers (e.g., those of some libraries or governmental agencies, which may publish the user ID and password or may configure the account so that a password is not required).

### *Gopher*

Gopher, developed at the University of Minnesota, is a means of interconnecting files installed on a wide variety of computers, including many popular PCs. A user must run a gopher program (a form of *client application)* on a local computer. This program can connect to gopher servers—programs installed on other computers that provide access to files, data, and other servers. Usually gopher menus appear as a series of nested lists that permit the user to search for information and data. An important characteristic of gopher files is that users can both read information in the files and also transfer data to their own computers. Therefore, gopher systems can include archives of remote sensing data that are available for public access. Gopher servers are supported by search programs that can search the Internet for files and data by accessing an index of files names that is updated on a regular basis. One of the most widely known search programs is *archie* (a name derived from the word *archive*), although many others are also available.

### *World Wide Web*

*World Wide Web* (W3, or WWW) is a system for organizing information on the Internet. Mosaic and Netscape Navigator are widely used programs that permits users to access information on the WWW using a computer mouse to point and click to select topics or icons presented on the screen.

World Wide Web was proposed in 1989 at CERN (the European Laboratory for Particle Physics in Geneva, Switzerland), the home of the World Wide Web. Key features of the WWW are rules that standardize communications between computers so that users can easily access information at varied locations around the world without changing from one computer to the other. Users access the WWW through specialized programs called *browsers,* which permit convenient access to remote sites and rapid transfer of text and graphics over the Internet. Two of the most widely used browsers are *Mosaic* (National Center for Supercomputing Applications, University of Illinois) and *Netscape Navigator* (Netscape Communications Corporation, Mountain View, CA).

Users of these programs view a screen, transmitted from the remote computer, that resembles the cover of a magazine, with images and lists of topics. This image forms the *home page* for a specific computer or a set of information on a specific computer (Figure 4.18). The home page is both an index and an advertisement for the information at that site. This information is a form of *hypertext* in which highlighted words or phrases are linked to additional text that explains or elaborates on the topic identified by the highlighted expression. (Data are prepared for access in this form using *hypertext mark-up language* [HTML].) By employing the mouse to select highlighted items, the user can access the more detailed information. These cross-references may identify information resi-

**FIGURE 4.18**. Sample home page. Underlined items, shown in blue on the computer screen, link to more detailed information, which can be viewed by placing the pointer on the underlined item and clicking the computer mouse.

dent on computers at completely different locations, perhaps separated by thousands of miles (although necessarily linked to the Internet). The browser enables the user to access information at diverse sites with a speed and convenience that present the impression of having the information at hand locally. Browsers extend the reach of gopher and related systems by permitting users to receive graphic information (maps, movies, images, photographs, sound, and diagrams) in addition to text and data.

A large number of public domain data sets are available over the WWW. WWW addresses are known as *uniform resource locators* (URLs) and have the distinctive format illustrated in Table 4.4, which is a sample of URLs of interest to students of remote sensing. Any list of Internet addresses will become quickly outdated as new sites are added and older addresses are changed or removed altogether. Most users prepare their own lists of URLs (bookmarks, in WWW terminology), to be updated as they obtain new addresses from the Internet itself.

New users of the Internet should note some of its unwritten practices. Not all sites are open to remote users throughout the day, and some restrict access during peak demand times, so it may be necessary to access some sites during evening or weekend hours. Before downloading data, be sure to verify that you have sufficient disk space to receive the

**TABLE 4.4. Some WWW Addresses of Interest to Remote Sensing Students**

| | |
|---|---|
| http://www.ncsa.uiuc.edu/SDG/Software/Mosaic/NCSAMosaicHome.html | Mosaic home page |
| http://home.netscape.com/ndx.html | Netscape home page |
| http://www.sun.com | SUN Computers |
| http://www.erdas.com | ERDAS, Inc. |
| http://sun1.cr.usgs.gov/eros-home.html | USGS EROS DC |
| http://eosdis.larc.nasa.gov | NASA Langley |
| http://info.er.usgs.gov/ | USGS Home Page |
| http://www.mtpe.hq.nasa.gov/ | NASA HQ |
| http://www.jpl.nasa.gov/ | JPL Home Page |
| http://www.noaa.gov/ | NOAA Home Page |
| http://www.ccrs.nrcan.gc.ca/gcnet/ | Geomatics Canada |
| http://daac.gsfc.nasa.gov/ | NASA DAAC |
| http://www.esrin.esa.it/ | Eur. Space Agency |
| http://services.esrin.esa.it/ | Earth Observ. RS directory |
| http://web.ngdc.noaa.gov/dsmp/dsmp.html | Defense Met. Satellite |
| http://sun1.cr.usgs.gov/glis/glis.hmtl | Global Land Info |
| http://www.ngdc.noaa.gov/mgg/mggd.hmtl | Nat. Geop Data Cen |
| http://sun1.cr.usgs.gov/landdaac/landdaac.html | EDC Land Processes |
| http://www.hq.nasa.gov/office/mtpe | Mission to Planet Earth |
| http://www.calmit.unl.edu/calmit.html | CALMIT, U of Neb-L. |
| http://ltpwww.gsfc.nasa.gov/MODIS/MODIS.html | MODIS Home Page |
| http://psbsgil.nesdis.noaa.gov:8088/OSDPD/OSDPD.html | NOAA Sat. Archive |
| http://eos.nasa.gov/ | NASA EOS |
| http://adro.radar1.sp-agency.ca/adrohomepage.html | RadarSat Home Page |
| http://daac.gsfc.nasa.gov./DATASET_DOCS/avhrr_dataset.html | AVHRR Land Data Sets |
| http://daac.gsfc.nasa.gov/DAAC_DOCS/gdaac_home.html | GSFC DAAC |
| http://xtreme.gsfc.nasa.gov/ | AVHRR Pathfinder |
| http://cen.cenet.com/htmls/Services.html | Russian Satellite Data |
| http://www.erim.org | ERIM |
| http://www.ieee.org/grs/index.html | IEEE Geosciences & RS |
| http://edcwww.cr.usgs.gov/dclass/dclass.html | CORONA imagery |
| http://nssdc.gsfc.nasa.gov/ | Nat. Space Sci. Data Center |
| http://www.asprs.org/asprs | Am. Society for Photog & RS |
| http://www.odyssey.maine.edu/gisweb/ | UM/NGIS archive of papers |
| http://www.lib.berkeley.edu/ | Dictionary of RS abbrev |
| http://www.cla.sc.edu/geog/rslab/rslab.html | Index for RS archives |

*Note.* Check with current sources for up-to-date addresses, as entries on this list are likely to change over time. Some of these addresses are given by Merchant (1995), who also lists many other useful Internet addresses. This list can only be a selection of the many possible entries, so omission of other addresses does not imply they are not of interest. Abbreviations and acronyms used here for conciseness are explained in later chapters or are evident upon accessing the address.

requested data, and remember that image files are often very large and require long intervals to download (perhaps hours, if you have access by modem). Executable files can transmit computer viruses to your computers, so they should be screened by a reliable virus-checking program. Further, all users of the Internet should remember to exercise common courtesy. The WWW is feasible only because some people and institutions are willing to provide access to their computers as an act of generosity. Users therefore are guests who benefit from the goodwill of others. Do not use the connection longer than necessary as your use denies access to others who may be attempting to reach the same address.

Once data are in hand, either from the Internet or from more traditional sources, analysts must bring them into a format that can be used by the specific image processing system at a particular laboratory. This step requires knowledge of the file structure (number and sequence of files and header records), organization of image data (BIP, BIL, BSQ, etc.), dimensions of images, number of bands, and number of bits assigned to each pixel. Such information is usually provided in the text files that accompany each image, or in paper printouts that are provided with tapes or CD-ROMs. Given these data, utility programs within image processing systems can reorganize data for use by other components of the image processing system. Once data have entered the system, users typically are unaware of the internal file structure employed by a given system to organize the bits on the tape or the CD-ROM into an image that can be displayed and analyzed.

## 4.8. An Overview of the Image Analysis Process

Once data are available, in either numeric or image form, the analyst is prepared to conduct an interpretation using some combination of manual or digital procedures.

Figure 4.19 suggests an idealized sequence for a typical digital analysis, which is proposed as a learning device for the student rather than a detailed guide for actual analysis. Each step is briefly outlined here and discussed in greater detail in subsequent chapters.

Preprocessing consists of those operations that prepare data for subsequent analysis, usually by attempts to correct or compensate for systematic errors (Chapter 10). Three classes of preprocessing operations can be defined. The first are those that simply display, or summarize, the data as a means of inspecting characteristics and quality. These operations present histograms, scattergrams, or statistical summaries that permit the operator to assess image quality and thereby determine subsequent preprocessing steps (if any) that may be necessary.

A second group of preprocessing operations compensate for radiometric errors. These errors result from defects in sensor operation, atmospheric absorption and scattering, variations in scan angle, variations in illumination of the scene, and system noise.

A third class of preprocessing operations corrects positional errors in relationships between image representations of ground features and their actual geographic relationships as they occur on the earth's surface. Geometric corrections are necessary because of systematic variations in the orientation of the vehicle that carries the sensor. Some are *intrinsic* to the sensor, caused, for example, by variations in the motion of the instrument; others, referred to as *extrinsic errors,* are the result of variations in orientation of the

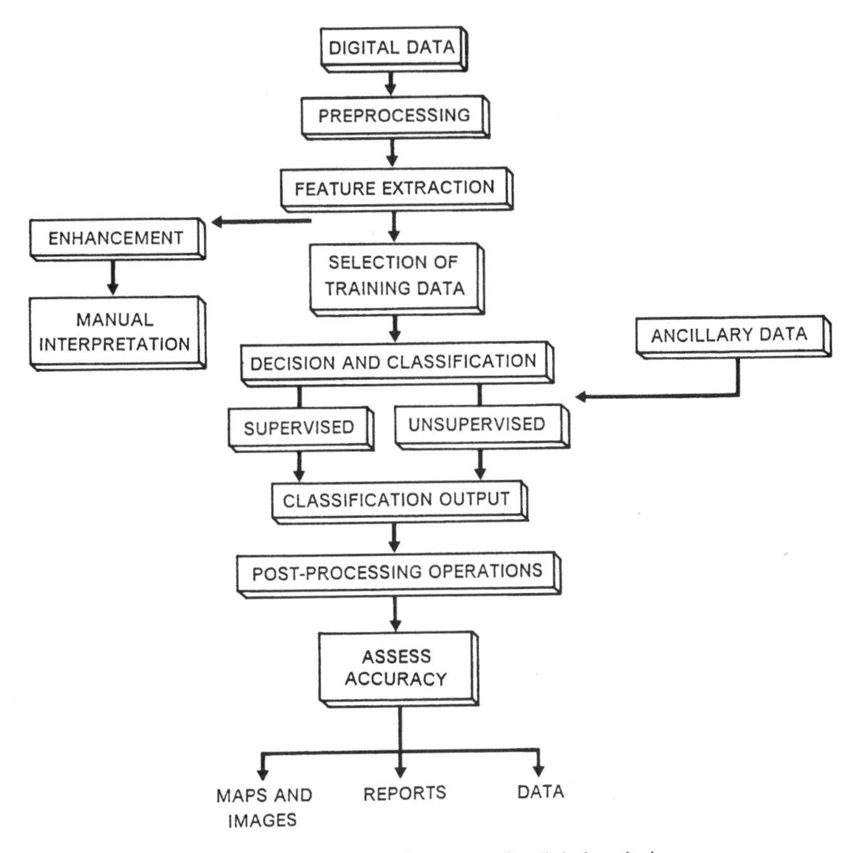


**FIGURE 4.19.** Idealized sequence for digital analysis.

platform, or errors in recording platform position. Many of these errors are systematic and can be removed before they reach the user. Other geometric transformations are not related to errors—changing of scales and projection conversion are operations that simply alter sizes or locational reference systems without consideration of the presence of errors.

After preprocessing is complete, the analyst may use *feature extraction* to reduce the dimensionality of the data. In this context, "a feature" should be thought of as useful information in the data rather than a physical feature on the earth's surface. Thus, feature extraction is the process of isolating the most useful components of the data for further study while discarding the less useful aspects (errors, noise, etc.). (For many, this term is misleading but is established by long usage.) Feature extraction reduces the number of variables that must be examined, thereby saving time and resources.

At this point the analyst may choose to apply techniques to improve the visual appearance of the image, using image enhancement to prepare the image for subsequent manual interpretation. Because image enhancement often drastically alters the original numeric data, it is normally used only for visual (manual) interpretation and not for further numeric analysis.

There are several alternative approaches to decision and classification. In this step, data that have been subjected to preprocessing and feature extraction are subjected to quantitative analysis to assign individual pixels to specific classes. That is, individual pixels are grouped together with other pixels that have similar brightnesses. These groups may correspond to regions on the ground that have common geologic or biologic properties and therefore may be useful to the analyst as a means of deriving information concerning subjects such as land use, water quality, geology, and plant vigor.

Chapter 11 discusses image classification in detail, but Figure 4.19 indicates the choice between the two alternative approaches to image classification. *Supervised classification* uses information derived from a few areas of known identity to classify the remainder of the image, consisting of those pixels of unknown identity. The analyst must carefully locate, both on the ground and on the image, areas that represent the classes of interest on the final product. These areas must be carefully selected as accurate samples of the broader categories they are to represent, and their location must be carefully recorded to assist in delineation of these corresponding regions on the image. Once these "training areas," or "training fields," have been accurately located on the image, the classification procedure uses characteristics of these areas to assign pixels of unknown identity to each of the categories specified by the operator.

*Unsupervised classification*, the alternative to supervised classification, is the search for "natural" groupings of pixels based on their brightnesses in several spectral channels. The analyst then attempts to assign these natural classes to the user-defined categories (e.g., "forest," or "water") that are of value to the scientists and planners who will use the results.

After classification is complete, it is necessary to evaluate its accuracy by comparing the categories on the classified images with areas of known identity on the ground. If such comparisons can be conducted for each category, the analyst can develop information that permits assessment of the categories that are most accurately classified and can document the quality of the results (Chapter 12).

The final results of the analysis consist of maps (or images), data, and a report. The maps and images provide the user with knowledge of spatial distributions of the categories of interest and their patterns on the landscape. Data are provided by tabulations of classes and their areas. Finally, the report provides a clear account of the images examined, the procedures that were used, and the results of the accuracy assessment. These three components of the results provide the user with full information concerning the source data, the method of analysis, and the outcome and its reliability.

## 4.9. Summary

Image processing hardware is a means to accomplish an end—the display and analysis of remotely sensed data. In this context, the details of the hardware may seem rather mundane and peripheral to our primary concerns. Yet it is important to recognize that the role of the equipment is much more important, and much more subtle, than we may first appreciate. Display and analysis hardware determine in part how we perceive the data, and therefore how we use them. The equipment is, in effect, a filter through which we visualize data. We can never see them without this filter because there are so many data, and so many details, that we can never see them directly. Thus we depend on image analysis hardware and software to assist us in viewing and understanding the data. The best equipment permits us flexibility as we choose between different levels of radiometric, spatial, and spectral detail and allows us convenience as we select alternative ways of examining images.

## Review Questions

1. It may be useful to practice conversion of some values from digital to binary form, as confirmation that the student understands the concepts. Convert the following digital numbers to eight-bit binary values:

   a. 100 c. 24 e. 2 g. 256
   b. 15 d. 31 f. 111 h. 123

2. Convert the following values from binary to digital form:

   a. 10110 c. 10111 e. 0011011
   b. 11100 d. 1110111 f. 1101101

3. Consider the implications of selecting the appropriate number of bits for recording remotely sensed data. One might be tempted to say, "Use a large number of bits to be sure that all values are recorded precisely." What would be the disadvantage of using, for example, seven bits to record data that are accurate to only five bits?

4. Describe in a flow chart or diagram the steps required to read BIP data, and place the data in BSQ format.

5. What are the minimum number of bits required to represent the following values precisely?

   a. 1,786 d. 32,000
   b. 32 e. 17
   c. 689 f. 3
   g. 29

6. Assume that radiometric detail for a sensor is increased from 8 to 10 bits; how many bits would then be required to record the data for an entire scene? What would be advantages and disadvantages of such an increase?

7. One of the primary effects of the routine availability of Landsat digital data in the 1970s was the increased availability of image processing systems, and especially the development of image processing systems for microcomputers. One limitation of such systems is that they typically work with subsets of satellite images. Examine a small-scale map of your city or county to develop a plan for selecting the smallest feasible subset that will be satisfactory for studying the region.

8. Using your own computer, or those in your classroom, visit several of the WWW sites listed in Table 4.4. From information at these sites, add to the list.

9. Would it be possible to estimate a sensor's SNR by visual examination of an image? How?

## References

Bracken, P. A. 1983. Remote Sensing Software Systems. Chapter 19 in *Manual of Remote Sensing* (R. N. Colwell, ed.). Falls Church, VA: American Society of Photogrammetry, pp. 807–839.

Davidson, D. B., and E. Chen. 1995. A Brief Introduction to the Internet. *Computers and Geosciences,* Vol. 21, pp. 731–735.

Holkenbrink, Patrick F. 1978. *Manual on Characteristics of Landsat Computer-Compatible Tapes,* Produced by the EROS Data Center Digital Image Processing System, U.S. Geological Survey. Washington, DC: U.S. Government Printing Office. (Stock No 024-001-03116-7) (with change 1 August 1979)

Hopper, G. M., and S. L. Mandell. 1984. *Understanding Computers.* New York: West Publishing Company, 490 pp.

Hutchinson, Sarah, and S. C. Sawyer. 1992. *Microcomputers: The User Perspective.* Boston: Irwin, 769 pp.

Jensen, John R. 1996. *Introductory Digital Image Processing: A Remote Sensing Perspective.* Englewood Cliffs, NJ: Prentice Hall, 316 pp.

Kidrer, David B., and Derek H. Smith. 1992. Compression of Digital Elevation Models by Huffman Coding. *Computers and Geosciences,* Vol. 18, pp. 1013–1034.

Merchant, James W. 1995. A Guide to GIS, Remote Sensing and Other Useful Internet Addresses. *RSSG Newsletter* (Association of American Geographers, Remote Sensing Specialty Group), Vol. 16, No. 2, pp. 13–19.

Nichols, D. 1983. Digital Hardware. Chapter 20 in *Manual of Remote Sensing* (R. N. Colwell, ed.). Falls Church, VA: American Society of Photogrammetry, pp. 841–871.

Rogers, David F. 1985. *Procedural Elements for Computer Graphics* New York: McGraw-Hill, 433 pp.

Root, Ralph R. 1995. Introduction to E-mail and Other Internet Services. *Photogrammetric Engineering and Remote Sensing*, Vol. 61, pp. 875–880.

Rose, Albert, and P. K. Weimer. 1989. Physical Limits to Performance of Imaging Systems. *Physics Today,* Vol. 42, No. 9, pp. 24–32.

Southard, David A. 1992. Compression of Digitized Map Images. *Computers and Geosciences,* Vol. 18, pp. 1213–1253.

Thomas, Brian J. 1995. *The Internet for Scientists and Engineers: Online Tools and Resources*. Bellingham, WA: SPIE—The International Society for Optical Engineering, 450 pp.

Thomas, Valerie L. 1975. *Generation and Physical Characteristics of the LANDSAT 1 AND 2 MSS Computer Compatible Tapes* (X-563-75-223). Greenbelt, MD: Goddard Space Flight Center.

Woronow, Alex, and Scott Dare. 1995. On the Internet with a PC. *Computers and Geosciences,* Vol. 21, pp. 753–757.

In order to address this subject at an early point in the text, we must confine the discussion to interpretation of aerial photography—the only form of remote sensing imagery we have looked at thus far. But, the principles, procedures and equipment described here are equally applicable to other kinds of imagery acquired by the sensors described in later chapters.

Manual image interpretation is discussed in greater detail by Avery and Berlin (1992); other references that may also be useful are the text by Lueder (1959), the *Manual of Photographic Interpretation* (Colwell, 1960), and *Photointerpretation* (Philipson, 1996), published by the American Society for Photogrammetry and Remote Sensing.

## 5.3. Image Interpretation Tasks

The image interpreter must routinely conduct several kinds of tasks, many of which may be completed together in an integrated process. Nonetheless, for purposes of clarification it is important to distinguish between these separate functions (Figure 5.1).

### *Classification*

Classification is the assignment of objects, features, or areas to classes based on their appearance on the imagery. Often the distinction is made between three levels of confidence and precision. *Detection* is the determination of the presence or absence of a feature. *Recognition* implies a higher level of knowledge about a feature or object such that the object can be assigned an identity in a general class or category. Finally, *identification*

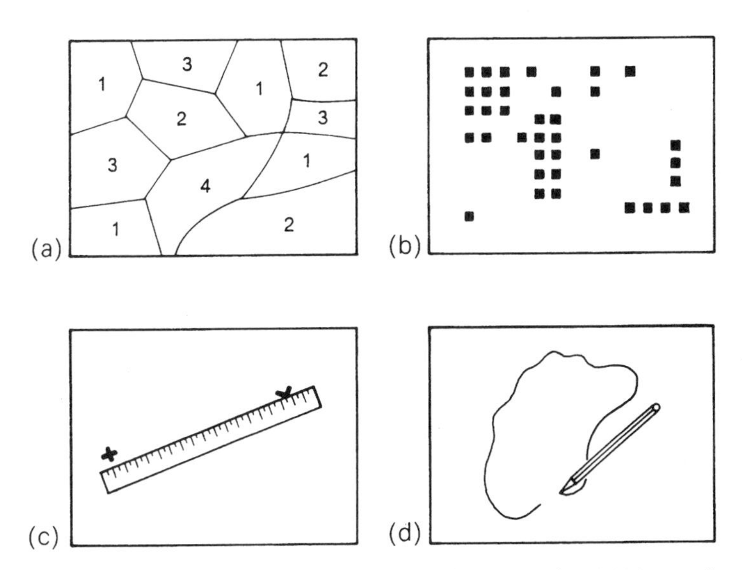


**FIGURE 5.1.** Image interpretation skills. (*a*) Classification. (*b*) Enumeration. (*c*) Mensuration. (*d*) Delineation.

means that the identity of an object or feature can be specified with enough confidence and detail to place it in a specific class. Often an interpreter may quality his or her confidence in an interpretation by specifying the identification as "possible" or "probable."

### *Enumeration*

Enumeration is the task of listing or counting discrete items visible on an image. For example, housing units can be classified as "detached single family," "multifamily complex," "mobile home," and "multistory residential" and then reported as numbers present within a defined area. Clearly, the ability to conduct such an enumeration depends on an ability to accurately identify and classify items as discussed above.

### *Measurement*

Measurement, or *mensuration,* is an important function in many image interpretation problems. Two kinds of measurement are important. First is the measurement of distance and height, and by extension, volumes and areas as well. The practice of making such measurements forms the subject of *photogrammetry* (Chapter 3), which applies knowledge of image geometry to the derivation of accurate distances. Although strictly speaking photogrammetry applies only to measurements from photographs, by extension it has analogs for derivation of measurements from other kinds of remotely sensed images.

A second form of measurement is quantitative assessment of image brightness. The science of *photometry* is devoted to measurement of the intensity of light, to include estimation of scene brightness by examination of image tone, using special instruments known as *densitometers,* described later. If the measured radiation extends outside the visible spectrum, the term *radiometry* applies. Both photometry and radiometry apply similar instruments and principles, so they are closely related.

### *Delineation*

Finally, the interpreter must often delineate, or outline, regions as they are observed on remotely sensed images. The interpreter must be able to separate distinct aerial units that are characterized by specific tones and textures and to identify edges or boundaries between separate areas. Typical examples include delineation of separate classes of forest, or of land use—both occur only as areal entities (rather than discrete objects). Typical problems include (1) selection of appropriate levels of generalization (when boundaries are intricate, or when many tiny but distinct parcels are present), and (2) placement of boundaries when there is a gradation (rather than a sharp edge) between two units.

The image analyst may simultaneously apply several of these skills in examining an image. Recognition, delineation, and mensuration may all be required as the interpreter examines an image. Yet specific interpretation problems may emphasize specific skills. Military photo interpretation often depends on accurate recognition and enumeration of specific items of equipment, whereas land-use inventory emphasizes delineation, although other skills are obviously important. Image analysts therefore need to develop proficiency in all these skills.

## 5.4. Elements of Image Interpretation

By tradition, image interpreters are said to employ some combination of the eight *elements of image interpretation,* which describe characteristics of objects and features as they appear on remotely sensed images. Image interpreters quite clearly use these characteristics together in complex, poorly understood processes as they examine images. Nonetheless, it is convenient to list them separately as a way of emphasizing their significance.

### *Image Tone*

Image tone denotes the lightness or darkness of a region within an image (Figure 5.2). For black-and-white images, tone may be characterized as *light, medium gray, dark gray, dark,* and so on, as the image assumes varied shades of white, gray, or black. For color or CIR imagery, image tone refers simply to color, described informally perhaps in such terms as *dark green, light blue,* or *pale pink.* Image tone refers ultimately to the brightness of an area of ground as portrayed by the film in a given spectral region (or in *three* spectral regions, for color or CIR film).

Image tone can be influenced also by the intensity and angle of illumination, and by processing of the film. Within a single aerial photography, vignetting (Chapter 3) may create noticeable differences in image tone due solely to the position of an area within a frame of photography—the image becomes darker near the edges. Thus, the interpreter

**FIGURE 5.2.** Image tone.

must employ caution in relying solely on image tone for an interpretation, as it can be influenced by factors other than the absolute brightness of the earth's surface. Analysts should also remember that very dark or very bright regions on an image may be exposed in the nonlinear portion of the characteristic curve (Chapter 3) so they may not be represented in their correct relative brightness. Also, nonphotographic sensors may record such a wide range of brightness values that they cannot all be accurately represented on photographic film—in such instances digital analyses (Chapter 4) may be more accurate.

Experiments have shown that interpreters tend to be consistent in interpretation of tones on black-and-white imagery but less so in interpretation of color imagery (Cihlar & Protz, 1972). Interpreters' assessment of image tone is much less sensitive to subtle differences in tone than are measurements by instruments (as might be expected). For the range of tones used in the experiments, human interpreters' assessment of tone expressed a linear relationship with corresponding measurements made by instruments. Their results imply that a human interpreter can provide reliable estimates of relative differences in tone, although they may not be capable of accurate description of absolute image brightness.

### *Image Texture*

Image texture refers to the apparent roughness or smoothness of an image region. Usually texture is caused by the pattern of highlighted and shadowed areas as an irregular surface is illuminated from an oblique angle. Contrasting examples (Figure 5.3) include the rough textures of a mature forest and the smooth textures of a mature wheat field. The human interpreter is good at distinguishing subtle differences in image texture, so it is a valuable aid to interpretation—certainly equal in importance to image tone in many circumstances.

Image texture depends not only on the surface itself but also on the angle of illumination, so it can vary as lighting varies. Also, good rendition of texture depends on favorable image contrast, so images of poor or marginal quality may lack the distinct textural differences so valuable to the interpreter.

### *Shadow*

Shadow is an especially important clue in the interpretation of objects. A building or vehicle, illuminated at an angle, casts a shadow that may reveal characteristics of its size or shape that would not be obvious from the overhead view alone. Because military photointerpreters often are primarily interested in identification of individual items of equipment, shadow has been of great significance in distinguishing subtle differences that might not otherwise be visible. By extension, we can emphasize this role of shadow in interpretation of any man-made landscape in which identification of separate kinds of structures or objects is significant.

Shadow is of great significance also in the interpretation of natural phenomena, even though its role may not be as obvious. For example, Figure 5.4a depicts an open field occupied by scattered shrubs and bushes separated by areas of open land. Without shadows, the individual plants might be too small (as seen from above), and so nearly similar in tone to their background to be visible. Yet their shadows are large enough, and dark

**FIGURE 5.3.** Image texture.

enough, to create the streaked pattern on the imagery typical of this kind of land. A second example is shown in Figure 5.4b. At the edge between mature forest and open land, the forest often casts a shadow that, at small scale, appears as a dark strip that enhances the boundary between the two zones on the imagery.

### *Pattern*

*Pattern* refers to the arrangement of individual objects into distinctive, recurring forms that permit recognition on aerial imagery (Figure 5.5). Pattern on an image usually follows from a functional relationship between the individual features that compose the pattern. Thus, the buildings in an industrial plant may have a distinctive pattern due to their

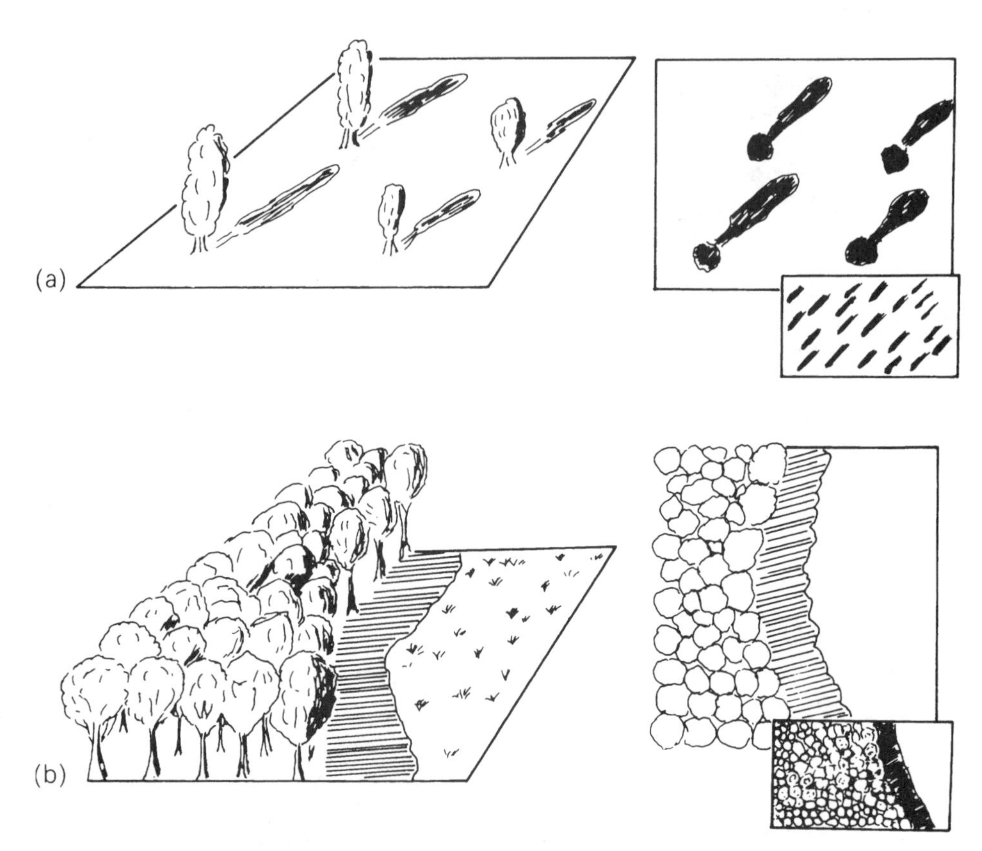


**FIGURE 5.4.** Significance of shadow. (*a*) Characteristic pattern caused by shadows of shrubs cast on open field. (*b*) Shadow at the edge of a forest enhances the boundary between two different land types. From Campbell (1983). Copyright 1983 by the Association of American Geographers. Reproduced by permission.

**FIGURE 5.5.** Pattern.

organization to permit economical flow of materials through the plant from receiving raw material to shipping of the finished product. The distinctive spacing of trees in an orchard arises from careful planting of trees at intervals, which prevents competition between individual trees and permits convenient movement of equipment through the orchard.

### *Association*

*Association* specifies the occurrence of certain objects or features usually without the strict spatial arrangement implied by pattern. In the context of military photointerpretation, association of specific items has great significance, as, for example, when the identification of a specific class of equipment implies that other, more important items are likely to be found nearby.

### *Shape*

*Shapes* of features are obvious clues to their identity. For example, individual structures and vehicles have characteristics shapes, which, if visible in sufficient detail, provide the basis for identification. Features in nature often have such distinctive shapes that shape alone might be sufficient to provide clear identification. For example, ponds, lakes, and rivers occur in specific shapes unlike others found in nature. Often, specific agricultural crops tend to be planted in fields that have characteristic shapes (perhaps related to constraints of equipment used, or the kind of irrigation that the farmer uses).

### *Size*

*Size* is important in two ways. First, the size of an object or feature is relative in relation to other objects on the image. This is probably the most direct and important function of size, as it provides the interpreter with an intuitive notion of the scale and resolution of an image even though no measurements or calculations may have been made. This role is achieved by recognition of familiar objects (dwellings, highways, rivers, etc.) and extrapolation of these known features to estimate sizes and identities of those that might not be easily identified.

Second, absolute measurement can be equally valuable as interpretation aids. Measurements of the size of an object can confirm its identification based on other factors, especially if its dimensions are so distinctive that they form definitive criteria for specific items or classes of items. Furthermore, absolute measurements permit derivation of quantitative information, including lengths, volumes, or (sometimes) even rates of movement (of vehicles or ocean waves, for example, as they are shown in successive photographs).

### *Site*

*Site* refers to topographic position. For example, sewage treatment facilities are positioned at low topographic sites near streams or rivers to collect waste flowing through the system from higher locations. Orchards may be positioned at characteristic topographic sites—often on hillsides (to avoid cold air drainage to low-lying areas) or near large water bodies (to exploit cooler spring temperatures near large lakes to prevent early blossoming).

## 5.5. Image Interpretation Strategies

An image interpretation strategy can be defined as a disciplined procedure that enables the interpreter to relate geographic patterns on the ground to their appearance on the image. Campbell (1978) defined five categories of image interpretation strategies.

### *Field Observations*

Field observations, as an approach to image interpretation, are required when the image and its relationship to ground conditions are so imperfectly understood that the interpreter is forced to go to the field to make an identification. In effect, the analyst is unable to interpret the image from knowledge and experience at hand and must gather field observations to ascertain the relationship between the landscape and its appearance on the image. Field observations are, of course, a routine dimension to any interpretation as a check on accuracy or a means of familiarization with a specific region. When they are required for the interpretation, their use reflects a rudimentary understanding of the manner in which a landscape is depicted on a specific image.

### *Direct Recognition*

Direct recognition is the application of an interpreter's experience, skill, and judgment to associate the image patterns with informational classes. The process is essentially a qualitative, subjective analysis of the image using the elements of image interpretation as visual and logical clues. In everyday experience, direct recognition is applied in an intuitive manner; for image analysis, it must be a disciplined process, with a careful, systematic examination of the image.

### *Interpretation by Inference*

Interpretation by inference is the use of a visible distribution to map one that is not itself visible on the image. The visible distribution acts as a surrogate, or proxy (i.e., a substitute) for the mapped distribution. For example, soils are defined by vertical profiles that cannot be directly observed by remotely sensed imagery. But soil distributions are sometimes closely related to patterns of landforms and vegetation that are recorded on the image. Thus, they can form surrogates for the soil pattern; the interpreter infers the invisible soil distribution from those that are visible. Application of this strategy requires a complete knowledge of the link between the proxy and the mapped distribution; attempts to apply imperfectly defined proxies produce inaccurate interpretations.

### *Probabilistic Interpretation*

Probabilistic interpretations are efforts to narrow the range of possible interpretations by formally integrating nonimage information into the classification process, often by means of quantitative classification algorithms. For example, knowledge of the crop calendar can restrict the likely choices for identifying crops of a specific region. If it is known that winter wheat is harvested in June, the choice of crops for interpretation of an August image can be restricted to eliminate wheat as a likely choice and thereby avoid a potential classification error. Often such knowledge can be expressed as a statement of probability. Possibly, certain classes might favor specific topographic sites but occur over a range of sites, so a decision rule might express this knowledge as a 0.90 probability of finding the class on a well-drained site, but only a 0.05 probability of finding it on a poorly drained site. Several such statements systematically incorporated into the decision-making process can improve classification accuracy.

### *Deterministic Interpretation*

A fifth strategy for image interpretation is deterministic interpretation, the most rigorous and precise approach to image interpretation. Deterministic interpretations are based on quantitatively expressed relationships that tie image characteristics to ground conditions. In contrast with the other methods, most information is derived from the image itself. Photogrammetric analysis of stereo pairs for terrain information is a good example. A scene is imaged from two separate positions along a flight path and the photogrammetrist measures the apparent displacement. Based on his knowledge of the geometry of the photographic system, a topographic model of the landscape can be reconstructed. The result is therefore the derivation of precise information about the landscape using only the image itself and a knowledge of its geometric relationships with the landscape. Relative to the other methods, very little nonimage information is required.

Image interpreters, of course, may apply a mixture of several strategies in a given situation. For example, interpretation of soil patterns may require direct recognition to identify specific classes of vegetation, then application of interpretation by proxy to relate the vegetation pattern to the underlying soil pattern.

## 5.6. Collateral Information

Collateral, or ancillary, information refers to nonimage information used to assist in the interpretation of an image. Actually, all image interpretations use collateral information in the form of the implicit, often intuitive knowledge that every interpreter brings to an interpretation in the form of everyday experience and also formal training. In its narrower meaning, it refers instead to the explicit, conscious effort to employ maps, statistics, and similar material to aid in analysis of an image. In the context of image interpretation, use of collateral information is permissible, and certainly desirable, provided two conditions are satisfied. First, the use of such information is to be explicitly acknowledged in the written report; second, the information must not focus on a single portion of the image or map to the extent that it produces uneven detail or accuracy in the final map. For example, it would be inappropriate for an interpreter to focus on acquiring detailed knowledge of tobacco farming in an area of mixed agriculture if he or she then produced highly detailed, accurate delineations of tobacco fields but mapped other fields at lesser detail or accuracy.

Collateral information can consist of information from books, maps, statistical tables, field observations, or other sources. Written material may pertain to the specific geographic area under examination, or, if such material is unavailable, it may be appropriate to search for information pertaining to analogous areas—similar geographic regions (possibly quite distant from the area of interest) characterized by comparable ecology, soils, landforms, climate, or vegetation.

## 5.7. Imagery Interpretability Rating Scales

Remote sensing imagery can vary greatly in quality, due to both environmental and technical conditions influencing acquisition of the data. In the United States, some governmental agencies use rating scales to evaluate the suitability of imagery for specific purposes. The National Imagery Interpretability Rating Scale (NIIRS) has been developed for single channel and panchromatic imagery, and the Multispectral Imagery Interpretability Rating Scale (MS IIRS) (Erdman et al., 1994) for multispectral imagery. Such scales are based on evaluations using a large number of experienced interpreters to independently evaluate images of varied natural and man-made features, as recorded by images of varying characteristics. They provide a guide for evaluation if a specific form of imagery is likely to be satisfactory for specific purposes.

## 5.8. Image Interpretation Keys

Image interpretation keys are valuable aids for summarizing complex information portrayed as images and have been widely used for image interpretation (Coiner, 1971). Such keys serve either or both of two purposes: (1) a means of training inexperienced personnel in the interpretation of complex or unfamiliar topics, and (2) a reference aid for experienced interpreters to organize information and examples pertaining to specific topics.

An image interpretation key is simply reference material designed to permit rapid and accurate identification of objects or features represented on aerial images. A key usually consists of two parts: (1) a collection of annotated or captioned images or stereograms, and (2) a graphic or word description, possibly including sketches or diagrams. These materials are organized in a systematic manner that permits retrieval of desired images by, for example, data, season, region, or subject.

Keys of various forms have been used for many years in the biologic sciences, especially botany and zoology. These disciplines rely on complex taxonomic systems that are so extensive that even experts cannot master the entire body of knowledge. The key therefore is a means of organizing the essential characteristics of a topic in an orderly manner. It must be noted that scientific keys of all forms require a basic familiarity with the subject matter. A key is not a substitute for experience and knowledge but a means of systematically ordering information so that an informed user can learn quickly.

Keys were first routinely applied to aerial images during World War II, when it was necessary to train large numbers of inexperienced photointerpreters in the identification of equipment of foreign manufacture and in the analysis of regions far removed from the experience of most interpreters. The interpretation key formed an effective way of organizing and presenting the expert knowledge of a few individuals. After the end of the war, interpretation keys were applied to many other subjects, including agriculture, forestry, soils, and landforms. Their use has been extended from aerial photography to other forms of remotely sensed imagery. Today interpretation keys may be used for instruction and training, but they may have somewhat wider use as reference aids. Also, it is true that construction of a key tends to sharpen one's interpretation skills and encourages the interpreter to think more clearly about the interpretation process.

Keys designed solely for use by experts are referred to as technical keys. Nontechnical keys are those designed for use by those with a lower level of expertise. Often it is more useful to classify keys by their formats and organizations. *Essay keys* consist of extensive written descriptions, usually with annotated images as illustrations. A *file key* is essentially a personal image file with notes; its completeness reflects the interests and knowledge of the compiler. Its content and organization suit the needs of the compiler, so it may not be organized in a manner suitable for use by others.

## 5.9. Interpretive Overlays

Often in resource-oriented interpretations it is necessary to search for complex associations of several related factors that together define the distribution or pattern of interest. For example, soil patterns may often be revealed by distinctive relationships between separate patterns of vegetation, slope, and drainage. The *interpretive overlays* approach to image interpretation is a way of deriving information from complex interrelationships between separate distributions recorded on remotely sensed images. The correspondence between several separate patterns may reveal other patterns not directly visible on the image (Figure 5.6).

The method is applied by means of a series of individual overlays for each image to be examined. The first overlay might show the major classes of vegetation, perhaps consisting of dense forest, open forest, grassland, and wetlands. A second overlay maps slope

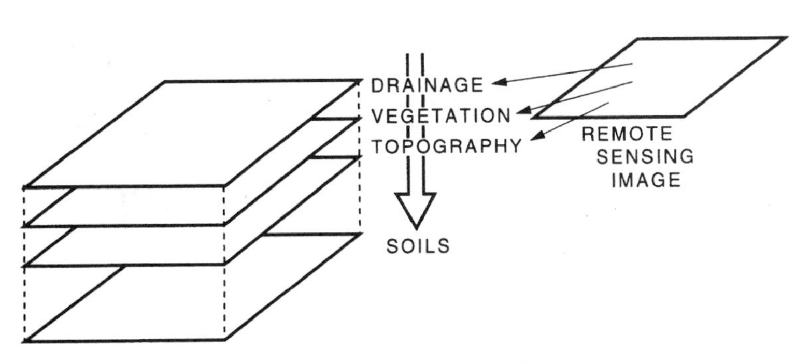


**FIGURE 5.6.** Interpretive overlays.

classes, including, perhaps, level, gently sloping, and steep slopes. Another shows the drainage pattern, and still others might show land use and geology. Thus, for each image, the interpreter may have as many as five or six overlays, each depicting a separate pattern. By superimposing these overlays, the interpreter can derive information presented by the coincidence of several patterns. From his or her knowledge of the local terrain, the interpreter may know that certain soil conditions can be expected where the steep slopes and dense forest are found together, and that others are expected where the dense forest matches to the gentle slopes. From the information presented by several patterns, the interpreter can resolve information not conveyed by any single pattern.

## 5.10. Photomorphic Regions

Another approach to interpretation of complex patterns is the search for photomorphic regions—regions of uniform appearance on the image. The interpreter does not attempt to resolve the individual components within the landscape (as he or she does when using interpretive overlays) but instead looks for their combined influence on image pattern (Figure 5.7). For this reason, application of photomorphic regions may be most applicable to small-scale imagery where the coarse resolution tends to average together the separate components of the landscape. *Photomorphic regions,* then, are simply image regions of relatively uniform tone and texture.

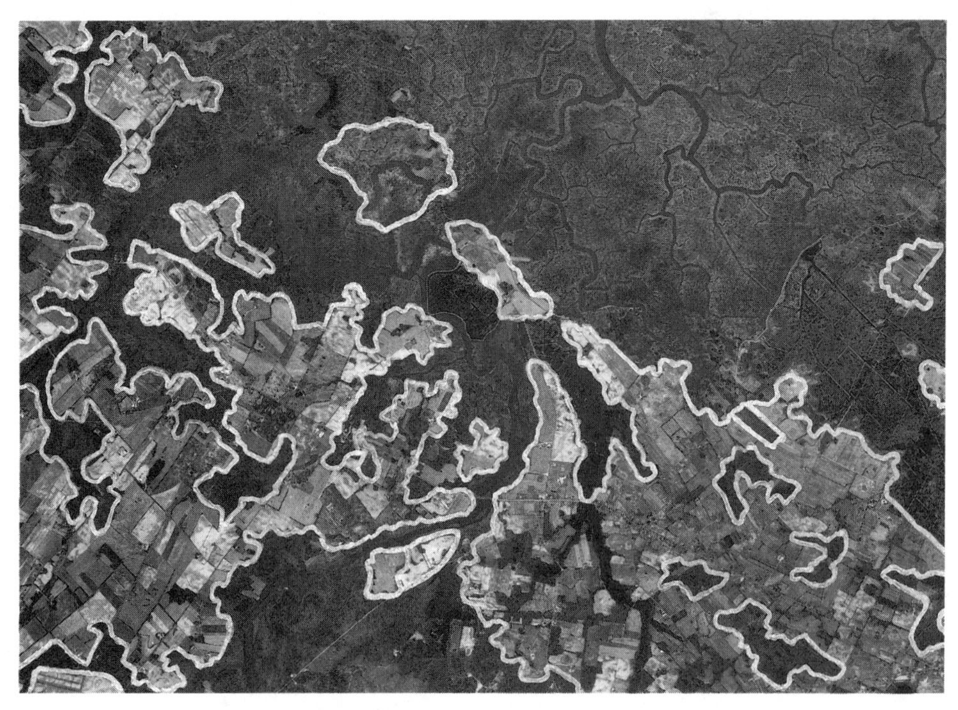

**FIGURE 5.7.** Photomorphic regions.

In the first step the interpreter delineates regions of uniform image appearance, using tone, texture, shadow, and the other elements of image interpretation as a means of separating regions. In some instances, interpreters have used densitometers to quantitatively measure image tone, and variation in image tone, as an aid to more subjective interpretation techniques (Nunnally, 1969).

In the second step the interpreter must be able to match photomorphic regions to useful classes of interest to the interpreter. For example, the interpreter must determine whether specific photomorphic regions match to vegetation classes. This step obviously requires field observations, or collateral information, because regions cannot be identified by image information alone. As the interpretation is refined, the analyst may combine some photomorphic regions, or subdivide others.

Delineation of photomorphic regions is a powerful interpretation tool, but one that must be applied with caution. Photomorphic regions do not always correspond neatly to the categories of interest to the interpreter. The appearance of one region may be dominated by factors related to geology and topography, whereas that of another region on the same image may be controlled by the vegetation pattern. And the image appearance of a third region may be the result of the interaction of several other factors.

## 5.11. Image Interpretation Equipment

Image interpretation can usually be conducted with relatively simple, inexpensive equipment, although some tasks may require expensive items. Typically, an image interpretation laboratory is equipped for storage and handling of images both as paper prints and as film transparencies. Paper prints are most frequently 9 in. × 9 in. contact prints, often stored in sequence in a standard file cabinet. Larger prints and indices must be stored flat in a map cabinet. Transparencies are available as individual 9 in. × 9 in. frames, but often they are stored as long rolls of film wound on spools and sealed in canisters as protection from dust and moisture.

### *Light Tables*

A light table is simply a translucent surface illuminated from behind to permit convenient viewing of film transparencies. In its simplest form, the light table is simply a box-like frame with a frosted glass surface. The viewing area can be desk-size in its largest form or as small as a briefcase. If roll film is to be used, light tables must be equipped with special brackets to hold the film spools and rollers at the edges to permit the film to move freely without damage. More elaborate models have dimmer switches to control intensity of the lighting, high-quality lamps (to control spectral properties of the illumination), and sometimes power drives to wind and unwind long spools of film.

### *Measurement of Length*

Ordinary household rulers are not satisfactory for photointerpretation. Analysts should use an engineer's scale or a ruler with accurate graduations. Both SI units (to at least 1 mm) and English units (to at least 1/20 in.) are desirable. It is convenient for both measurement and calculation if English units are subdivided into decimal divisions.

### *Measurement of Area*

Areas on maps or remote sensing imagery can be measured using any of several techniques. At one time, the *dot grid* was the standard technique for measuring areas. An array of equally spaced dots printed on a tranparent overlay was superimposed over the area to be measured. The analyst could then count the number of dots within a delineated area and apply a formula that related the density of dots on the grid, the number of dots counted, and the scale of the image to estimate the area in question. If applied properly, the dot grid was simple, inexpensive, and reasonably accurate under most circumatances. However, it is time-consuming and impractical for complex maps. The *polar planimeter* is a compact instrument with a movable arm that can be used to trace the outline of an area; a dial at the base of the arm records the area outlined by the perimeter, usually as the map area (e.g., square centimeters or square inches), which can then be converted to ground area. Planimeters are simple and reliable. However, accurate results require averaging from several repetitions, so this method is cumbersome when complex maps must be analyzed.

Today most analysts measure areas using some form of *electronic digitizer* (Chapter 15), which records an electronic version of the outline traced by the analyst. From this electronic record, a microprocessor can compute areas, and apply corrections for image scale. A special, less expensive, version of the elctronic digitizer is the *electronic planimeter* (Figure 5.8). The analyst can trace outlines with a movable arm. Coordinates, areas,

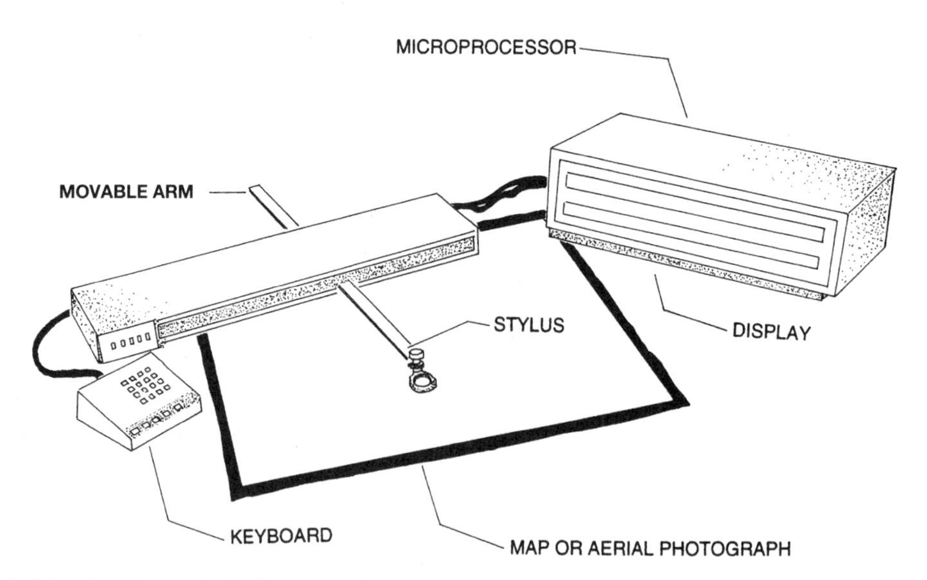


**FIGURE 5.8.** Electronic planimeter. As the operator traces an outline using the movable arm, the microprocesser records and displays coordinates, distances, and areas, as required. The keyboard permits the operator to identify specific lines and regions within identifying codes.

and distances can be read from the display or transferred to a computer if desired. The keyboard permits the analyst to enter comands to the microprocessor and to enter identifying codes as areas are digitized.

### *Stereoscopes*

Stereoscopes are devices that facilitate stereoscopic viewing of aerial photographs. The simplest and most common is the *pocket stereoscope.* Its compact size and low cost make it one of the most widely used remote sensing instruments. Typically, the pocket stereoscope (Figure 5.9) consists of a body holding two low-power lenses attached to a set of collapsible legs that can be folded so the entire instrument is only a bit larger than a deck of playing cards. The body is usually formed from two separate pieces, each holding one of the two lenses that can be adjusted to control the spacing between the two lenses to accommodate the individual user. Use of this instrument is described later.

Other kinds of stereoscopes include the *mirror stereoscope* (Figure 5.10), which permits stereoscopic viewing of large areas, usually at low magnification, and the *binocular stereoscope* (Figure 5.11), designed primarily for viewing transparencies on light tables. Often the binocular stereoscope has adjustable magnification that enables enlargement of portions of the image up to 20 or 40 times.

### *Magnification*

Image analysts almost always wish to examine images using magnification, although the exact form depends on individual preference and the nature of the task at hand. A simple, hand-held reading glass is satisfactory in many circumstances. *Tube magnifiers* (Figure 5.12) are low-power lenses (2× to about 8×) mounted in a transparent tube-like stand. The base may include a reticule calibrated in units as small as 0.001 ft. or 0.1 mm, to permit accurate measurement of small scale images of objects Sometimes it is necessary to use

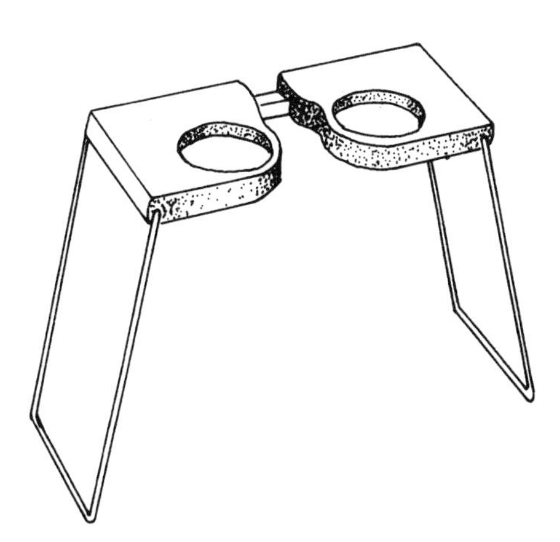

**FIGURE 5.9.** Pocket stereoscope.

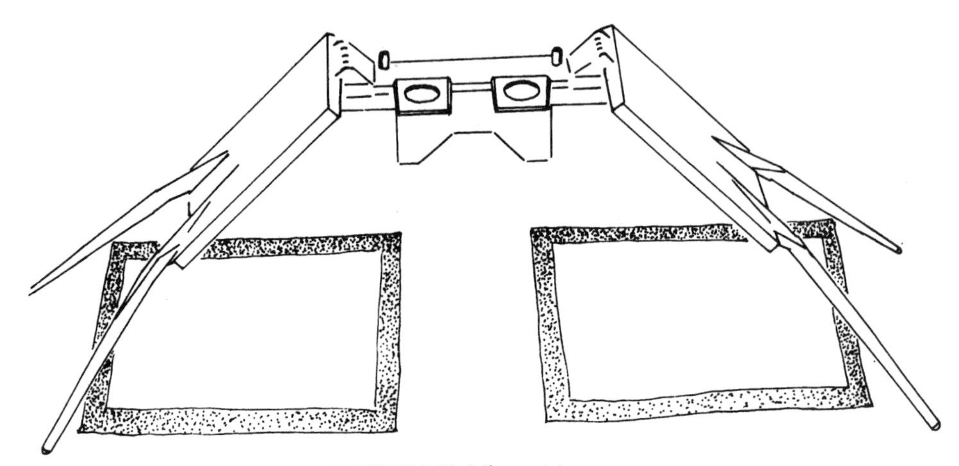

**FIGURE 5.10.** Mirror stereoscope.

the much more expensive binocular microscopes (Figure 5.13) for examination of film transparencies; such instruments may have adjustable magnification to as much as 40×, which will approach or exceed the limits of resolution for most images.

### *Densitometry*

*Densitometry* is the science of making accurate measurements of film density. In the context of remote sensing the objective is often to reconstruct estimates of brightness in the original scene, or sometimes merely to estimate relative brightnesses on the film (Chapter

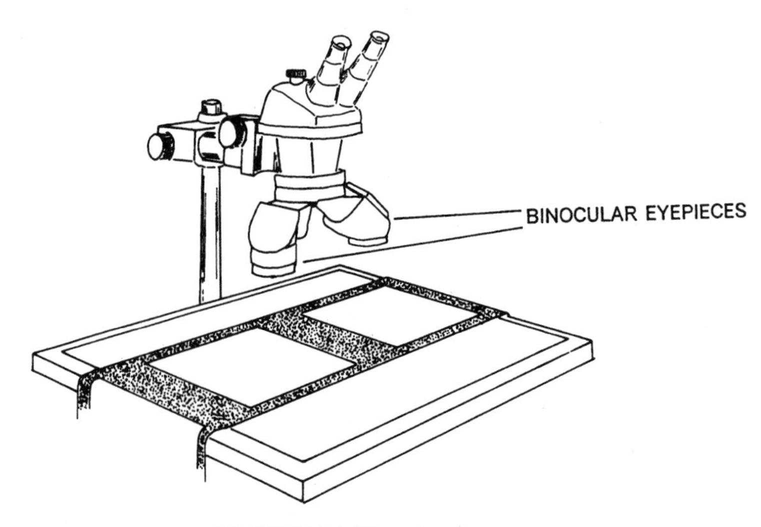


**FIGURE 5.11.** Binocular stereoscope.

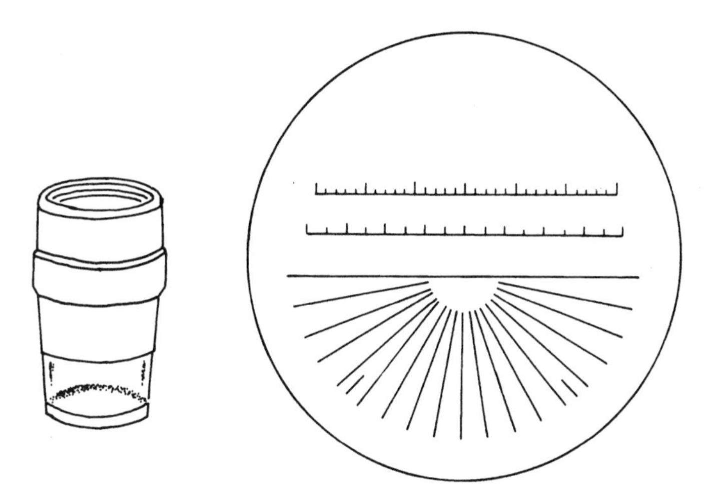

**FIGURE 5.12.** Tube magnifier (left) and reticule (right).

3). A densitometer (Figure 5.14) is an instrument that measures image density by directing a light of known brightness through a small portion of the image, then measuring its brightness as altered by the film. Typically the light beam might pass through an opening perhaps 1 mm in diameter; use of smaller openings (measured sometimes in micrometers) is known as *microdensitometry.* Such instruments find densities for selected regions within an image; an interpreter might use a densitometer to make quantitative measurements of image tone. For color or CIR images, filters are used to make three measurements, one for each of the three additive primaries.

In principle, densitometric measurements can be used to estimate brightnesses in the original scene. However, several factors intervene to make such estimates difficult. The

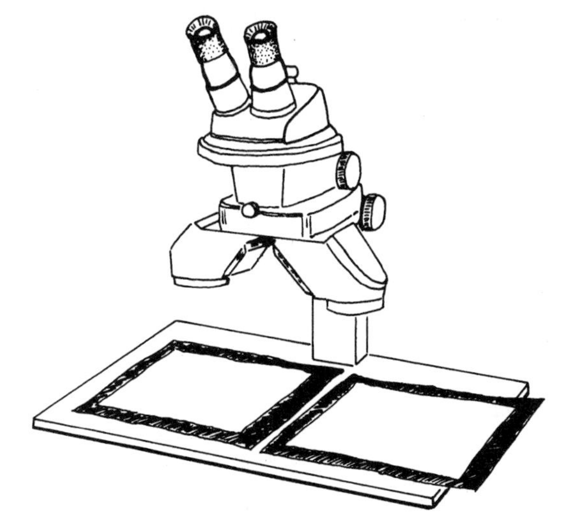

**FIGURE 5.13.** Binocular microscope.

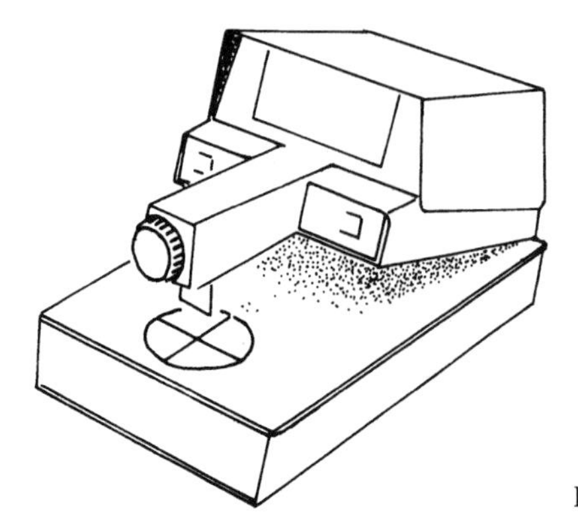

**FIGURE 5.14.** Densitometer.

densitometer must be carefully calibrated; areas of known brightness must be represented on the film and subjected to the same processing as the image to be examined. Measurements of densities that fall in the nonlinear portion of the characteristic curve cannot of course be related to the brightness of the original scene. For these reasons and others, it is difficult to make reliable estimates of scene brightness by densitometry.

### *Image Scanning*

Paper or film images can be scanned for digital representation as outlined in Chapter 4. Among the most accurate instruments for image scanning are *scanning densitometers* (Figure 5.15), designed to accurately and precisely measure image density by systematically scanning across an image, creating an array of digital values to represent the image pattern. Such instruments have been designed to precisely measure position and image density, possibly using resolutions of a few micrometers or so. Although instruments designed for such precise tolerances are expensive, many other less precise but still serviceable scanners can sometimes be used for tasks that do not require high accuracy. Desktop scanners, although designed for office use, can be used to scan maps and images for visual analysis (Carstensen & Campbell, 1991).

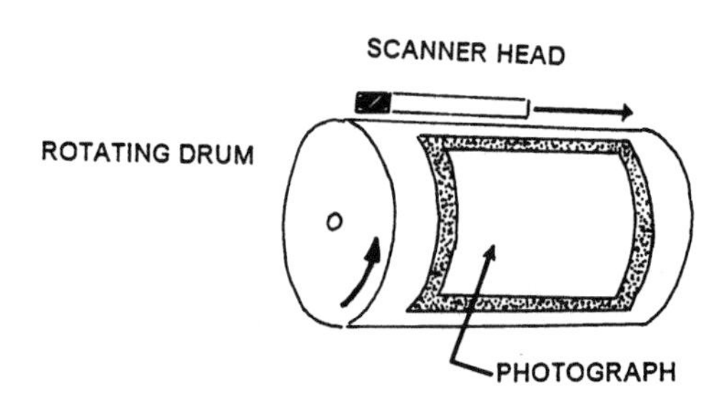


**FIGURE 5.15.** Scanning densitometer.

### *Height Finder*

The height finder (Figure 5.16) is an instrument designed for use with a stereoscope; it permits estimation of topographic elevation or of the heights of features from stereo aerial photographs. The height finder is a bar that attaches at the base of the stereoscope; the bar holds two plastic tabs, one under each lens of the stereoscope. Both tabs are marked with a small black dot, but one tab is fixed in position whereas the other can be moved from side to side along a scale that measures its movement left to right parallel to the bar.

The operator aligns the photographs for stereoscopic viewing after marking their principal points and conjugate principal points and the flight line. The stereoscope is positioned for stereoscopic viewing as normal (see below). Then the interpreter views the scene stereoscopically, positioning the movable dot over the feature of interest. Adjustment of the movable dot causes it to appear to float above the terrain surface; when it is positioned so it appears to rest on the terrain surface, a reading of the scale gives the parallax measurement for that point. A parallax factor is found by using a set of tables. Readings for several points, then simple calculations, permit determination of elevation differences between the individual points (in feet or meters for example). If the photographs show a point of known elevation, these points can be assigned elevations with reference to a datum (such as mean sea level); otherwise they provide only relative heights.

### *Data Transfer*

Analysts often need to transfer information from one map or image to another to ensure accurate placement of features, to update superseded information, or to bring several kinds of information into a common format. A variety of devices are available for such tasks. The simplest are designed to change scale and project an image onto a working surface where it can be traced onto an overlay that registers to another image. More sophisticated instruments (such as the Bausch and Lomb Zoom Transfer Scope) are able not only to change scale but to selectively enlarge or reduce portions of the image to correct for tilt and other geometric errors. Finally, other instruments permit imagery to be viewed in stereo, for detail to be digitized and matched to other data, with computational adjustments for positional errors (Figure 5.17).

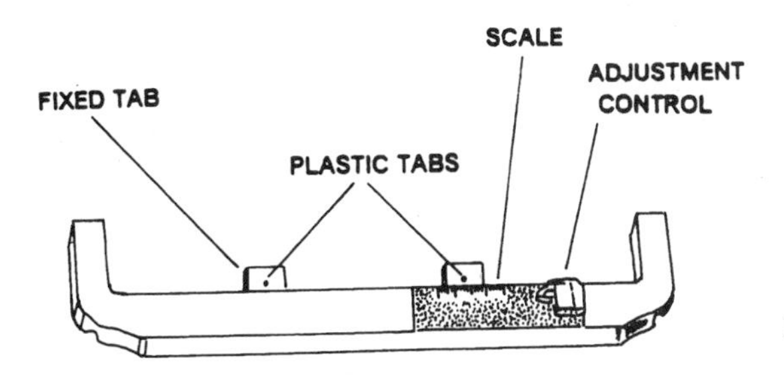


**FIGURE 5.16.** Height finder.

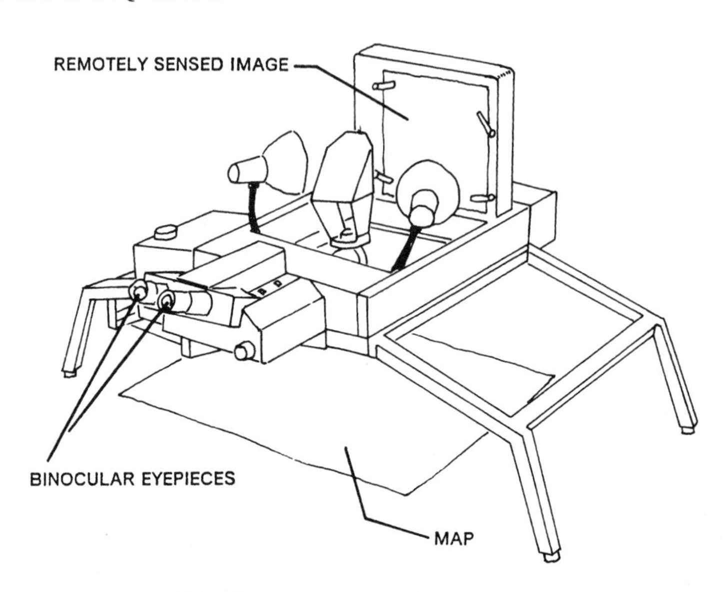


**FIGURE 5.17.** Image transfer equipment.

## 5.12. Preparation for Interpretation

In general, the interpreter should be able to work at a large, well-lighted desk or work table with convenient access to electrical power. Often it is useful to be able to control lighting with blackout shades or dimmer switches. Basic equipment and materials, in addition to those described above, include a supply of translucent drafting film and an engineer's scale, together with protractors, triangles, dividers, and so on. Maps, reference books, and other supporting material should be available as required.

If the interpretation is made from paper prints, special attention must be devoted to prevent folding, tearing, or rough use that will cause the prints to become worn. Usually it is best to mark an overlay registered to the print rather than the print itself. The drafting tape used to attach drafting film must be selected specifically for its weak adhesive quality, which will not tear the emulsion; the stronger adhesive used on many of the popular brands of paper tape will damage paper prints.

If transparencies are used, special care must be given to handling and storage. The surface of the transparency must be protected by a transparent plastic sleeve or handled only with clean cotton gloves Moisture and oils naturally present on unprotected skin may damage the emulsion, and dust and dirt will scratch the surface. Even if such damage is invisible to the unaided eye, it will later emerge as a major problem when high-power magnification is used.

For most interpretations, images should be oriented so that the shadows fall toward the analyst. Otherwise most individuals will see an apparent reversal in the relative elevations of terrain features—ridges will seem to be valleys and valleys will appear as ridge lines.

## 5.13. Use of the Pocket Stereoscope

The pocket stereoscope is one of the most frequently used interpretation aids. As a result, every student should be proficient in its use. Although many students will require the assistance of the instructor as they learn to use the stereoscope, the following paragraphs may provide some assistance for beginners.

First, stereo photographs must be aligned so that the flight line passes left to right (as shown in Figure 5.18). Check the photo numbers to be sure that the photographs have been selected from adjacent positions on the flight line. Usually (but not always) the numbers and annotations on photos are placed on the leading edge of the image—the edge of the image nearest the front of the aircraft at the time the image was taken. Therefore, these numbers should usually be oriented in sequence from left to right, as shown in Figure 5.18. If the overlap between adjacent photos does not correspond to the natural positions of objects on the ground, the photographs are incorrectly oriented.

Next, the interpreter should identify a distinctive feature on the image within the zone of stereoscopic overlap. The photos should then be positioned so that the duplicate images of this feature (one on each image) are approximately 64 mm (2.5 in.) apart. This distance represents the distance between the two pupils of a person of average size (it is referred to as the "interpupillary distance"), but for many it may be a bit too large or too small, so the spacing of photographs may require adjustment as the interpreter follows the procedure outlined here.

The pocket stereoscope should be opened so that the legs are locked in place to position the lens at their correct height above the photographs. The two segments of the body of the stereoscope should be adjusted so that the centers of the eyepieces are about 64 mm

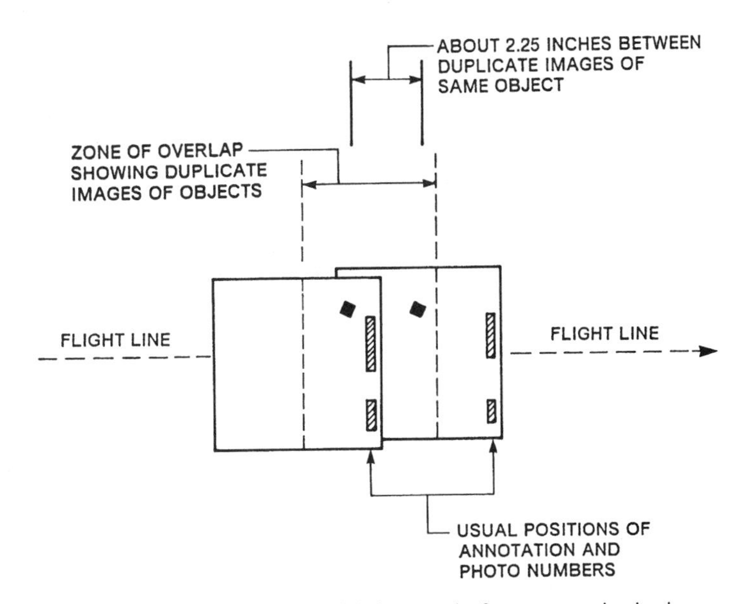


**FIGURE 5.18.** Positioning aerial photographs for stereoscopic viewing

apart (or a slightly larger or smaller distance, as mentioned above). Then the stereoscope should be positioned so that the centers of the lenses are positioned above the duplicate images of the distinctive feature selected previously. Looking through the two lenses the analyst sees two images of this feature; if the images are properly positioned, the two images will appear to "float" or "drift." The analyst can with some effort control the apparent positions of the two images so that they fuse into a single image; as this occurs, the two images should merge into a single image which is then visible in three dimensions. Usually aerial photos show exaggerated heights, due to the large separation (relative to distance to the ground) between successive photographs as they were taken along the flight line. Although exaggerated heights can prevent convenient stereo viewing in regions of high relief, it can be useful in interpretations of subtle terrain features that might not otherwise be noticeable.

The student who has successfully used the stereoscope to examine a section of the photo should then practice moving the stereoscope over the image to view the entire region within the zone of overlap. As long as the axis of the stereoscope is oriented parallel to the flight line, it is possible to retain stereo vision while moving the stereoscope. If the stereoscope is twisted with respect to the flight line, the interpreter loses stereo vision. By lifting the edge of one of the photographs, it is possible to view the image regions near the edges of the photos.

## 5.14. Image Scale Calculations

Scale is a property of all images. As a result, all who work with aerial images and maps must be skilled in computation of image scale. Knowledge of image scale is essential for making measurements from images and for understanding the geometric errors present in all remotely sensed images. Work with image scale is not difficult, although it is easy to make simple mistakes that can result in serious errors.

Scale is simply an expression of the relationship between the *image distance* between two points and the *actual distance* between the two corresponding points on the ground. This relationship can be expressed in several ways. The *word statement* sets a unit distance on the map or photograph equal to the correct corresponding distance on the ground. For example, "One inch equals one mile," or, just as correctly, "One centimeter equals 5 kilometers." The first unit in the statement in the expression specifies the map distance, the second, the corresponding ground distance. A second method of specifying scale is the bar scale, which simply labels a line with subdivisions that show ground distances.

The third method, the *representative fraction* (RF), is more widely used and often forms the preferred method of reporting image scale. The RF is the ratio between image distance and ground distance. It usually takes the form "1:50,000" or "1/50,000," with the numerator set equal to 1 and the denominator equal to the corresponding ground distance. The RF has meaning in any unit of length, as long as both the numerator and the denominator are expressed in the same units. Thus, "1:50,000" can mean "one inch on the image equals 50,000 inches on the ground," or "one centimeter on the image equals 50,000 centimeters on the ground."

A frequent source of confusion is converting the denominator into the larger units that we find more convenient to use for measuring large ground distance. With metric units, the conversion is usually simple; in the example given above, it is easy to see that 50,000 cm is equal to 0.50 km, and that 1 cm on the map represents 0.5 km on the ground. With English units, the same process is not quite so easy. It is necessary to convert inches to miles to derive "1 in. equals 0.79 mi." from 1:50,000. For this reason, it is useful to know that 1 mi. equals 63,360 in. Thus, 50,000 in. is equal to 50,000/63,360 = 0.79 mi.

A typical scale problem requires estimation of the scale of an individual photograph. One method is to use the focal length and altitude method (Figure 5.19):

$$\text{RF} = \frac{\text{Focal length}}{\text{Altitude}} \qquad \text{(Eq. 5.1)}$$

Both values must be expressed in the same units. Thus, if a camera with a 6-in. focal length is flown at 10,000 ft., the scale is 0.5/10,000 = 1:20,000. (Altitude always specifies the flying height above the terrain—*not* above sea level. Because a given flying altitude is seldom the *exact* altitude at the time the photography was taken, and because of the several sources that contribute to scale variations within a given photograph (Chapter 3), we must always regard the results of such calculations as an approximation of the scale of any specific portion of the image. Often such values are referred to as the *nominal* scale of an image, meaning that it is recognized that the stated scale is an approximation and that image scale will vary within any given photograph.

A second method is the use of a *known ground distance*. We identify two points on the aerial photograph that are also represented on a map. For example, in Figure 5.20, the image distance between points *A* and *B* is measured to be approximately 2.2 in. (5.6 cm).

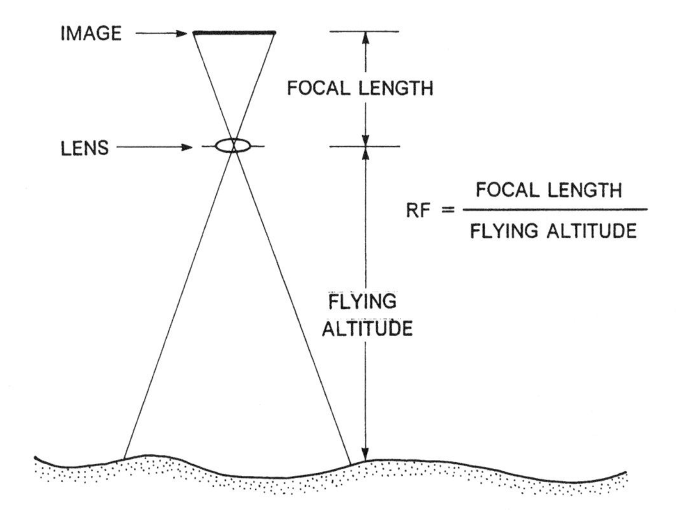


**FIGURE 5.19.** Estimating image scale by focal length and altitude.

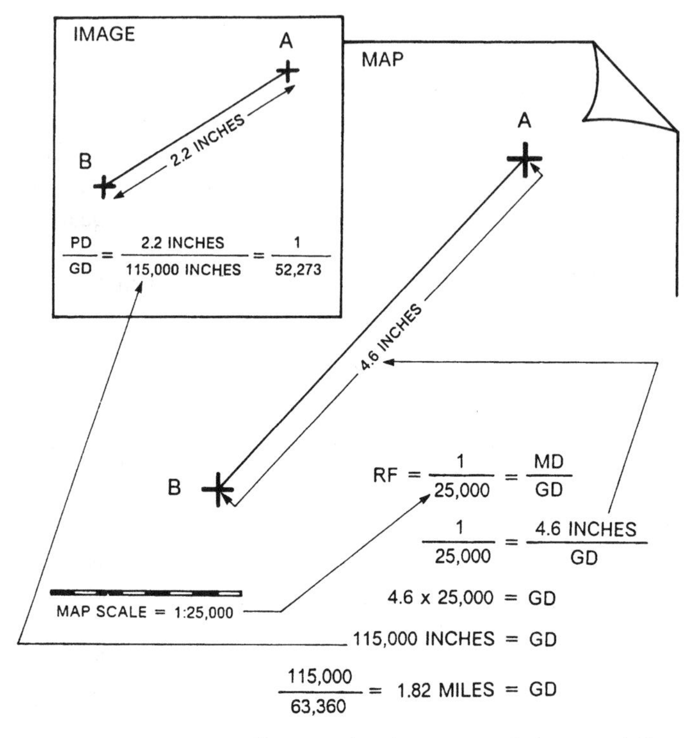


**FIGURE 5.20.** Measurement of image scale using a map to derive ground distance.

From the map, the same distance is determined to correspond to a ground distance (GD) of 115,000 in. (about 1.82 mi.). Thus the scale is found to be

$$\text{RF} = \frac{\text{Image distance}}{\text{Ground distance}} = \frac{2.2 \text{ in.}}{1.82 \text{ mi.}} = \frac{2.2 \text{ in.}}{115{,}000 \text{ in.}} \qquad \text{(Eq. 5.2)}$$

$$= \frac{1}{52{,}273}$$

In instances in which accurate maps of the area represented on the photograph may not be available, the interpreter may not know focal length and altitude. Then an approximation of image scale can be made if it is possible to identify an object or feature of known dimensions. Such features might include a baseball diamond or football field; measurement of a distance from these features as they are shown on the image provides the image distance value needed to use the relationship given above. The ground distance is derived from our knowledge of the length of a football field, or the distance between bases on a baseball diamond. Some photointerpretation manuals provide tables of standard dimensions of features commonly observed on aerial images, to include sizes of athletic fields (soccer, field hockey, etc.), lengths of railroad boxcars, distances between telephone poles, and so on, as a means of using the known ground distance method.

A second kind of scale problem is the use of a known scale to measure a distance on the photograph. Such a distance might separate two objects on the photographs but not be represented on the map, or the size of a feature has changed since the map was compiled. For example, we know that image scale is 1:15,000. A pond not shown on the map is measured on the image as 0.12 in. in width. Therefore, we can estimate the actual width of the lake to be

$$\frac{1}{15{,}000} = \frac{\text{Image distance}}{\text{Ground distance}} \qquad \text{(Eq. 5.3)}$$

$$\frac{0.12 \text{ in.}}{\text{Unknown GD}} = \frac{\text{Image distance}}{\text{Ground distance}}$$

$$\text{GD} = 0.12 \times 15{,}000 \text{ in.}$$

$$\text{GD} = 1{,}800 \text{ in., or } 150 \text{ ft.}$$

This example can illustrate two other points. First, because image scale varies throughout the image, we cannot be absolutely confident that our distance for the width of the pond is accurate; it is simply an estimate, unless we can have high confidence in our measurements and in the image scale at this portion of the photo. Second, measurement of short image distances is likely to have errors simply because we are unable to make accurate measurements of short distances (such as the 0.12-in. distance measured above). As distances become shorter, our errors constitute a greater proportion of the estimated length. Thus, an error of 0.005 in. is 0.08% of a distance of 6 in., but 4% of our distance of 0.12 in. mentioned above. Therefore, the interpreter should exercise a healthy skepticism regarding measurements made from images, unless care has been taken to ensure maximum accuracy and consistency.

## 5.15. Reporting Results of an Interpretation

An image interpreter possesses a specialized knowledge and experience that are not usually understood by those who use the information derived from remotely sensed images. These individuals may not appreciate the value and the reliability of the information, nor may they realize that the same information may have certain errors and limitations. Thus the interpreter must be skilled not only in analysis of images but also in the reporting of the results of the interpretation in a form that can be clearly understood by those who are not specialists in remote sensing.

Although the character of a report will vary greatly depending on the topic, purpose, and audience, it is possible to suggest a few general features that might be important in most circumstances. The report itself might include six separate sections that outline (1) objectives, (2) equipment and materials, (3) regional setting, (4) procedure, (5) results, and (6) conclusions.

### *Objectives*

The *objectives* of the interpretation should be specified clearly so that there is no opportunity for misunderstanding of the scope or detail of the project. Examples are given in

**TABLE 5.1. Objectives for Image Interpretation**

A. For a land use interpretation:
To classify the land use area of Montgomery County using five categories: Forest, Urban Land, Agricultural Land, Open Water, and Barren Land. Aerial photographs at 1:20,000 dated April 1984 are to be used, with field surveys as necessary. Results are to be reported as a map at 1:50,000 and as a table reporting areas in each category.

B. For an inventory of housing units:
To inventory housing units in Washington County, as of January 1985, then to report results by census subdivisions. To identify each structure on each image either as a dwelling unit or as a nondwelling unit. All structures must be assigned to categories that can be defined solely on the basis of their appearance on the image. Classification of dwelling units must be defined, then tested in early phases of the study, using field observations as necessary.

Table 5.1. Note that example *A* in Table 5.1 requires that the results be plotted on a accurate map for measurement of areas, whereas example *B* presents no such requirement, although it must be possible to mark census divisions boundaries on the photographs. Officials who are to use the results may not realize the implications of these requirements, but the image analyst must anticipate problems that arise from these objectives (i.e., the requirements to register data to the map and to the census units, respectively) and devise suitable procedures to ensure that the objectives can be met.

### *Equipment and Materials, and Regional Setting*

*Equipment and materials* can be reported in a simple list as shown in Table 5.2. The *regional setting* should briefly outline the geographic setting of the area considered in the interpretation, both for the benefit of the person who will use the information and as a means of preparing the interpreter for the conditions that will be encountered within the image. This section could include brief descriptions of local climate, topography, agriculture, soils, industry, population and other factors that might pertain to interpretation of images of the region (Table 5.3).

**TABLE 5.2. Equipment and Materials**

Light tables

9 in. × 9 in. black-and-white aerial photographs (positive transparencies at 1:24,000 (dated Winter 1985)

Black-and-white paper enlargements at 1:8,000 (dated Winter 1985)

Descriptions of census division boundaries

Mylar drafting film

Magnifiers

Pens, pencils, drafting type, paper, etc.

Stereoscopes

IBM Personal Computer with dBASE III

**TABLE 5.3. Example of Regional Setting Section Report**

*Landscape and climate*

Topography is characterized by a series of parallel valleys and ridges. Valleys, oriented northeast–southwest, are underlain by limestones and shales, and are separated by steep ridges underlain by sandstones and shales. Topography in valleys is rolling, with local relief of 3 to 30 m (10 to 100 ft.). . . .

In valleys, soils are generally deep and well drained, with moderate to gentle slopes. . . .

Climate here is characterized by an average annual rainfall of about 1,016 mm (40 in.), an average annual temperature of about 13°C (55°F), and about 200 frost-free days each year. Maximum precipitation is received during the summer months although late-summer moisture deficits are common. . . .

*Economic and social characteristics*

Population in this region is largely rural; about 75% of the land is farms, with only 2% to 3% urban land. . . .

In addition to agriculture, important industries include light manufacturing, recreation and tourism, manufacture of forest products. . . .

### *Procedures*

The *procedures* give a step-by-step description of the interpretation process in detail (Table 5.4). This section serves three purposes. First, it ensures that the analyst has thought through the interpretation process in a way that is likely to anticipate problems before they arise. Second, a written description of procedures is especially important if several people are working together on the same project—to ensure that results are consistent, all must follow the same procedure. Although written procedures are not in themselves sufficient to maintain high standards of uniformity, they are necessary to clearly describe. Finally, the written procedures form a record of how the interpretation was conducted, so it is possible to answer questions that arise after the interpretation is complete even though the individuals who actually conducted the interpretation may have moved to different jobs assignments jobs in the interval since the project was completed.

### *Results*

*Results* are reported in a brief narrative section that is supported by tables, maps, and diagrams as appropriate. Frequently the interpretation will result in an inventory or enumeration of specific items or categories that can be reported in a detailed tabulation (Table 5.5). These tabulations should of course match the requirements of those who will use the information, so, for example, in part B of Table 5.5 the categories should correspond to those of interest to the users, and the values should be reported in appropriate units and with suitable detail to satisfy the intended use. The narrative can discuss points that may not be obvious from the table alone, to address topics such as relative accuracies or precision of values for the individual categories, for example.

**TABLE 5.4. Example of Procedure for Image Interpretation**

(This example corresponds to Objective B, Table 5.1.)

1. Plot census boundaries on photographs or on an overlay that registers to the photographs.
2. Each structure is interpreted separately and assigned a unique identifying number within its census division.
3. Interpretations usually will be made from the positive transparency, using the light table and the tube magnifier. Stereoscopes will be used to interpret paper prints in regions where detailed interpretation of multistory buildings is required. In urbanized areas, large-scale images are to be used to supplement the smaller scale photos.
4. As interpretations are made, the classifications are recorded on an overlay that registers to the enlarged photograph. On this overlay the interpreter outlines each structure and marks it with the identifying number and a symbol that matches to one of the previously defined categories.
5. When interpretation of a photograph is complete, data are recorded on the computer. Each record will specify structure number, category, census identifier, photo identification, and the interpreter. The computer file permits recording of changes in interpretation, but all earlier interpretations are to be preserved.
6. Overlays to the large-scale photographs form the authoritative graphic record of the interpretation and are marked to record the latest change, if changes in classification are necessary.
7. For field use, the large-scale overlays are to be duplicated, then carried to the field and marked as necessary to record changes. Changes are to be recorded later on the computer (preserving all earlier interpretations). Field sheets are to be marked with a date and worker's initials and are to be retained for the duration of the project.

Interpreters to work in teams. One examines the image using magnification, the other records the results on the overlay. Periodically tasks are rotated. Each team participates in recording information on the computer and in field verification work.

The data base system permits retrieval of information by census unit, by dwelling unit type, by photo, or by interpreter. It also enables tabulation of data by census units at different levels and retrieval of data for each photo or each interpreter, if necessary to correct errors late in the process.

## 5.16. Interpretation of Digital Imagery

Image interpretation was once practiced entirely within the realm of photographic prints and transparencies, using equipment and techniques outlined in preceding sections. Digital analyses have increased in significance, as have interpretations on imagery as represented on computer displays. Although such interpretations are based on the same principles outlined above for traditional imagery, digital data have their own characteristics which require special treatment in the context of visual interpretation. Haack and Jampoler (1995), for example, have investigated issues relating to the visual interpretation of multispectral data as displayed on color computer screens.

### *Image Enhancement*

Image enhancement is the process of improving the visual appearance of digital images. Image enhancement has increasing significance in remote sensing because of the growing

**TABLE 5.5. Example of Results Section of Report**

A. For inventory of land use and land cover:

| Category | Area (ha) |
|---|---|
| Forested land | 261 |
| Coniferous | 113 |
| Deciduous | 85 |
| Mixed | 63 |
| Urban land | 446 |
| Residential | 217 |
| Commercial | 93 |
| Industrial | 105 |
| Transportation | 31 |
| Agricultural | 1139 |
| Pasture | 318 |
| Cropland, irrigated | 97 |
| Cropland, nonirrigated | 453 |
| Orchards | 85 |
| Market gardens | 101 |

B. For count of housing units:

| Category | Number |
|---|---|
| Single-family dwellings | 251 |
| Multiunit dwellings | 23 |
| Hotels and motels | 12 |
| Mobile home units | 86 |

significance of digital analyses. Although some aspects of digital analysis may seem to reduce or replace traditional image interpretation, many of these procedures require analysis to examine images on computer displays, tasks that require many of the skills outlined in earlier sections of this chapter.

Most image enhancement techniques are designed to improve the visual appearance of an image, often as evaluated by narrowly defined criteria. Therefore, it is important to remember that enhancement is often an arbitrary exercise—what is successful for one purpose may be unsuitable for another image or another purpose. In addition, image enhancement is conducted without regard for the integrity of the original data—the original brightness values will be altered in the process of improving their visual qualities and will lose their relationships to the original brightnesses on the ground. Therefore, enhanced images should not be used as input for additional analytical techniques; rather, any further analysis should use the original values as input.

### *Contrast Enhancement*

*Contrast* refers to the range of brightness values present on an image. Contrast enhancement is required because, as outlined in Chapter 4, digital data usually have brightness ranges that do not match the capabilities of the human visual system or those of photo-

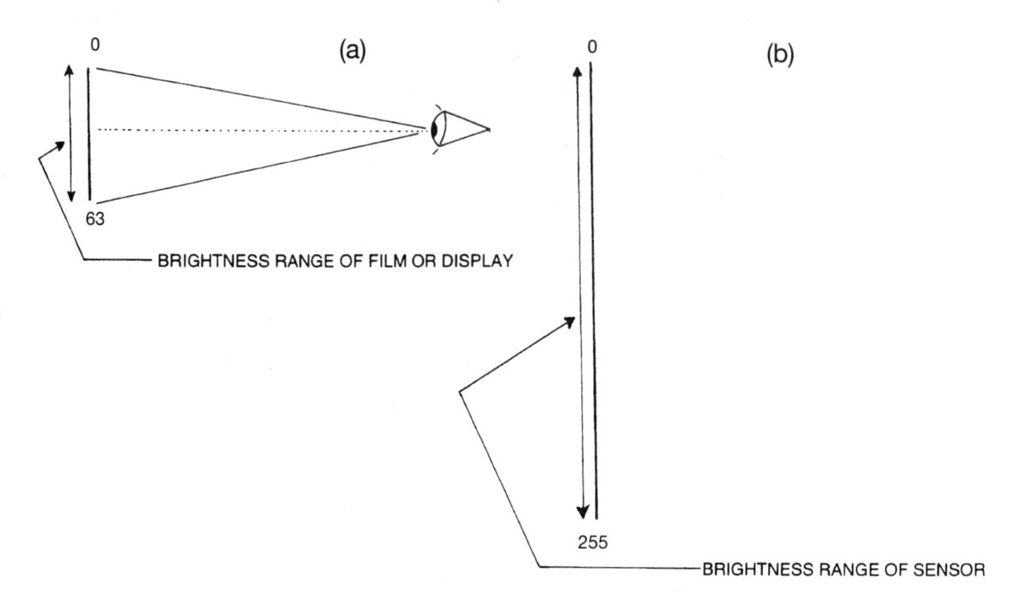


**FIGURE 5.21.** Loss of visual information in display of digital imagery.

graphic films (Chapter 3). Therefore, for analysts to view the full range of information conveyed by digital images, it is usually necessary to rescale image brightnesses to ranges that can be accommodated by human vision, photographic films, and computer displays.

For example, if the maximum possible range of values is 0 to 255 (i.e., 8 bits), but the display can show only the range from 0 to 63 (6 bits), the image will have poor contrast and important detail may be lost in the values that cannot be shown on the display (Figure 5.21a). Contrast enhancement alters each pixel value in the old image to produce a new set of values that exploits the full range of 256 brightness values (Figure 5.21b). Several approaches can be applied to enhance contrast.

*Linear stretch* converts the original digital values into a new distribution, using new minimum and maximum values specified. The algorithm then matches the old minimum to the new minimum and the old maximum to the new maximum. All the old intermediate values are scaled proportionately between the new minimum and maximum values (Figure 5.22). Piecewise linear stretch means that the original brightness range was divided

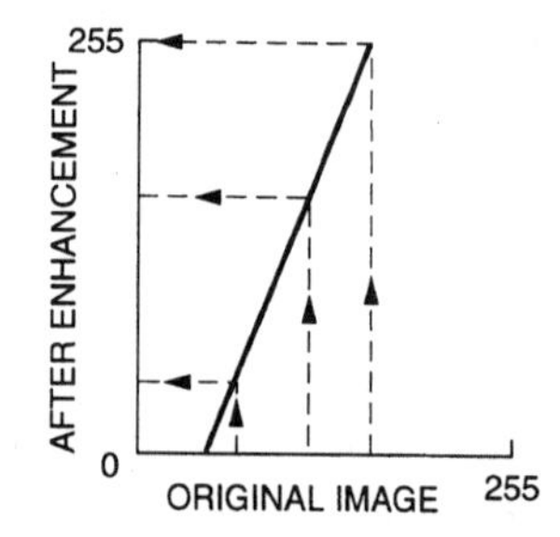


**FIGURE 5.22.** Linear stretch.

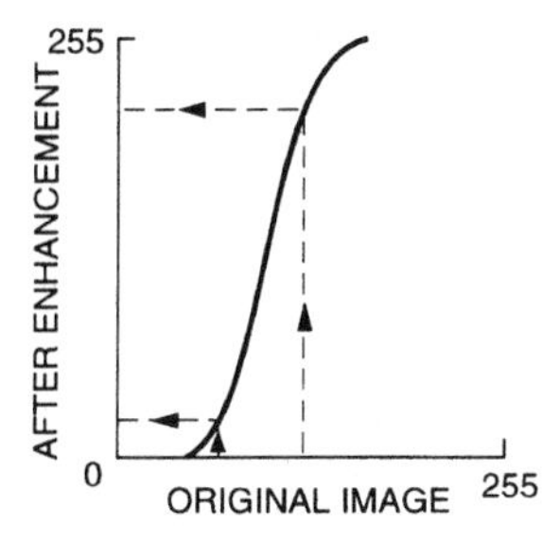


**FIGURE 5.23.** Histogram equalization.

into segments before each segment was stretched individually. This variation permits the analyst to emphasize certain segments of the brightness range that might have more significance for a specific application.

*Histogram equalization* reassigns digital values in the original image such that brightnesses in the output image are equally distributed among the range of output values (Figure 5.23). This procedure is often effective in improving the appearance of a single band of imagery.

*Density slicing* is accomplished by arbitrarily dividing the range of brightnesses in a single band into intervals, then assigning each interval to a color (Figure 5.24 and Plate 9). Density slicing may have the effect of emphasizing certain features that may be represented in vivid colors but, of course, does not carry any more information than the single image used as the source.

### *Edge Enhancement*

*Edge enhancement* is an effort to reinforce the visual boundaries between regions of contrasting brightness. Typically the human interpreter prefers sharp edges between adjacent parcels, whereas the presence of noise, coarse resolution, and other factors often tends to blur, or weaken, the distinctiveness of these edges. Edge enhancement is in effect the strengthening of local contrast—enhancement of contrast within a local region. A typical edge enhancement algorithm consists of a window that is systematically moved through the image, centered successively on each pixel. At each position, it is then possible to calculate a local average of values within the window; the central value can be compared to the averages of the adjacent pixels. If the value exceeds a specified difference from this

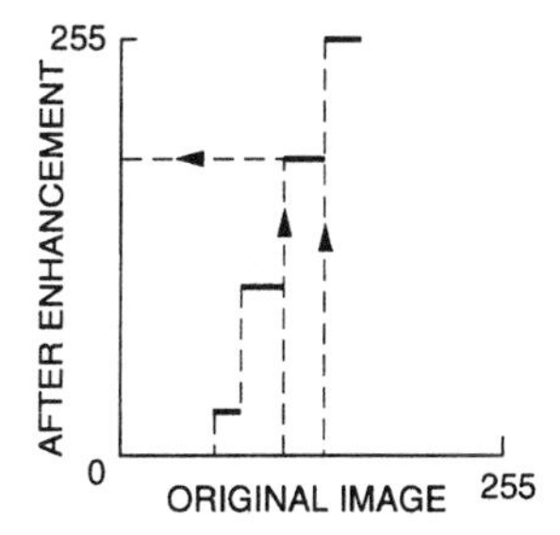


**FIGURE 5.24.** Density slicing.

average, the value can be altered to accentuate the difference in brightness between the two regions.

Rohde et al. (1978) describe an edge enhancement procedure used at the EROS Data Center. A new *(output)* digital value is calculated using the original *(input)* value and the local average of five adjacent pixels. A constant can be applied to alter the effect of the enhancement as necessary in specific situations. The output value is the difference between twice the input value and the local average. Use of this procedure reveals that its effect is to increase the brightness of those pixels that are already brighter than the local average and to decrease the brightnesses of pixels that are already darker than the local average. Thus the effect is to accentuate differences in brightnesses, especially at places ("edges") where a given value differs greatly from the local average. Note that the use of an average value from adjacent pixels reduces the effect of a single value upon the output values.

## 5.17. Summary

Despite the increasing significance of digital analysis in all aspects of remote sensing, image interpretation still forms a key component in the way that humans understand images. Analysts must evaluate imagery, either as paper prints or as displays on a computer monitor, using the skills outlined in this chapter. The fundamentals of manual image interpretation were developed for application to aerial photographs at an early date in the history of aerial survey, although it was not until the 1940s and 1950s that they were formalized into their present form. Since then, these techniques have been applied, without substantial modification, to other kinds of remote sensing imagery. As a result, we have a long record of experience in their application and knowledge of their advantages and limitations.

Interesting questions remain. In what ways might image interpretation skills be modified in the context of interpretation using computer monitors? What new skills might be necessary? How have analysts already adjusted to new conditions? How might equipment and software be improved to facilitate interpretation in this new context?

## Review Questions

1. A vertical aerial photograph was acquired using a camera with a 9-in. focal length at an altitude of 15,000 ft. Calculate the nominal scale of the photograph.

2. A vertical aerial photograph shows two objects to be separated by 6¾ in. The corresponding ground distance is 9½ mi. Calculate the nominal scale of the photograph.

3. A vertical aerial photograph shows two features to be separated by 4.5 in. A map at 1:24,000 shows the same two features to be separated by 9.3 in. Calculate the scale of the photograph.

4. Calculate the area represented by a 9 in. × 9 in. vertical aerial photograph taken at an altitude of 10,000 ft. using a camera with a 6-in. focal length.

5. You plan to acquire coverage of a county using a camera with 6-in. focal length, and a 9

in. × 9 in. format. You require an image scale of 4 in. equal to 1 mi., 60% forward overlap, and sidelap of 10%. Your country is square in shape, measuring 15.5 mi. on a side. How many photographs are required? At what altitude must the aircraft fly to acquire these photos?

6. You have a flight line of 9 in. × 9 in. vertical aerial photographs taken by camera with a 9-in. focal length at an altitude of 12,000 ft. above the terrain. Forward overlap is 60%. Calculate the distance (in miles) between ground nadirs of successive photographs.

7. You require complete stereographic coverage of your study area, which is a rectangle measuring 1.5 mi. × 8 mi. How many 9 in. × 9 in. vertical aerial photographs at 1:10,000 are required?

8. You need to calculate the scale of a vertical aerial photograph, so you compile the following data:

| Ground distance (as derived by map measurement) | Measured photo distance |
|---|---|
| 3.03 mi. | 8.0 in. |
| 0.81 mi. | 2.3 in. |
| 0.21 mi. | 0.43 in. |
| 2.31 mi. | 0.51 ft. |
| 1.29 mi. | 0.29 ft. |

You are not convinced that all values are equally reliable but must use these data, as they are derived from the easily identifiable points visible on the photograph. After examining the data, you proceed to estimate the scale of the photo. Give your estimate, the procedure you used, and explain the rationale for your procedure.

9. You have very little information available to estimate the scale of a vertical aerial photograph, but are able to recognize a baseball diamond among features in an athletic complex. You use a tube magnifier to measure the distance between first and second base to be 0.006 ft. What is your estimate of the scale of the photo?

10. Assume you can easily make an error of 0.001 in your measurement for question 9. Recalculate the image scale to estimate the range of results produced by this level of error. Now return to question 3 and assume that the same measurement error applies (do not forget to consider the different measurement units in the two questions). Calculate the effect on your estimates of image scale. The results should illustrate why it is always better whenever possible to use long distances to estimate image scale.

## References

Avery, T. E., and G. L. Berlin. 1992. *Fundamentals of Remote Sensing and Airphoto Interpretation.* New York: Macmillan, 472 pp.

Campbell, J. B. 1978. A Geographical Analysis of Image Interpretation Methods. *The Professional Geographer,* Vol. 30, pp. 264–269.

Carstensen, L. W., and J. B. Campbell. 1991. Desktop Scanning for Cartographic Digitization and Spatial Analysis. *Photogrammetric Engineering and Remote Sensing,* Vol. 57, pp. 1437–1446.

Chavez, P. S., G. L. Berlin, and W. B. Mitchell. 1977. Computer Enhancement Techniques of Landsat MSS Digital Images for Land Use/Land Cover Assessment. In *Proceedings of the Sixth Annual Remote Sensing of Earth Resources Conference,* pp. 259–276.

Cihlar, J., and R. Protz. 1972. Perception of Tone Differences from Film Transparencies. *Photogrammetria,* Vol. 8, pp. 131–140.

Colwell, R. N. (ed.). 1960. *Manual of Photographic Interpretation.* Falls Church, VA: American Society of Photogrammetry, 868 pp.

Coiner, J. C. 1972. *SLAR Image Interpretation Keys for Geographic Analysis.* Technical Report 177-19. Lawrence, KS: Center for Research, 110 pp.

Erdman, C., K. Riehl, L. Mayer, J. Leachtenauer, E. Mohr, J. Odenweller, R. Simmons, and D. Hothem. 1994. Quantifying Multispectral Imagery Interpretability. *International Symposium on Spectral Sensing Research,* Vol. 1, pp. 468–476.

Estes, J. E., E. J. Hajic, and L. R. Tinney, et al. 1983. Fundamentals of Image Analysis: Analysis of Visible and Thermal Infrared Data. Chapter 24 in *Manual of Remote Sensing* (R. N. Colwell, ed.). Falls Church, VA: American Society of Photogrammetry, pp. 987–1124.

Haack, B., and S. Jampoler. 1995. Colour Composite Comparisons for Agricultural Assessments. *International Journal of Remote Sensing,* Vol. 16, pp. 1589–1598.

*Image Interpretation Handbook.* 1967. Departments of the Army, Navy, and Air Force. TM 30-245, NAVAIR 10-35-685, AFM 200-50.

Lueder, D. R. 1959. *Aerial Photographic Interpretation: Principles and Applications.* New York: McGraw-Hill. 462 pp.

Nunnally, N. R. 1969. Integrated Landscape Analysis With Radar Imagery. *Remote Sensing of Environment,* Vol. 1, pp. 1–6.

Philipson, Warren (ed.). 1996 (in press). *Photointerpretation.* Bethesda, MD: American Society for Photogrammetry and Remote Sensing.

*Photo Interpretation Handbook.* 1954. Deptartments of the Army, Navy, and Air Force. TM 30-245, NAVAIR 10-35-610, AFM 200-50. Washington, DC: U.S. Government Printing Office, 303 pp.

Rohde, W. G., J. K. Lo, and R. A. Pohl. 1978. EROS Data Center Landsat Digital Enhancement Techniques and Imagery Availability, 1977. *Canadian Journal of Remote Sensing,* Vol. 4, pp. 63–76.

Stone, Kirk S. 1964. A Guide to the Interpretation and Analysis of Aerial Photos. *Annals of the Association of American Geographers,* Vol. 54, pp. 318–328.

Taranick, James V. 1978. *Principles of Computer Processing of Landsat Data for Geological Applications.* USGS Open File Report 78-117, 50 pp.

CHAPTER SIX

# Satellites

## 6.1. Satellite Remote Sensing

Today, several nations operate or plan to operate satellite remote sensing systems specifically designed for observation of earth resources, including crops, forests, water bodies, land use, and minerals. Satellite sensors offer several advantages over aerial photography; they provide a synoptic view (observation of large areas in a single image), as well as fine detail and systematic, repetitive coverage. Such capabilities are well suited to creating and maintaining a worldwide cartographic infrastructure and to monitoring many of the broad-scale environmental problems that the world faces today, to list two of many pressing concerns.

Because of the number of satellite observation systems in use, or proposed for use in the near future, and the rapid changes in their design, this chapter cannot attempt to provide complete listings or descriptions of all systems planned or in use. It can, however, provide the basic framework needed to understand key aspects of earth observation satellites in general as a means of preparing readers to acquire knowledge of specific satellite systems as they become available. Therefore, this chapter outlines essential characteristics of the most important systems—past, present, and future—as a guide to learning about the systems that are not discussed here.

Today's land observation satellites have evolved from earlier systems that, although tailored for other purposes, provided design and operational experience necessary for their successful operation. The first earth observation satellite, Television and Infrared Observation Satellite (TIROS), was launched in April 1960 as the first of a series of experimental weather satellites designed to monitor cloud patterns. TIROS formed a prototype for the operational programs that now provide meteorological data for daily weather forecasts throughout much of the world. Successors to the original TIROS vehicle have seen long service in several programs designed to acquire meteorological data. Although data from meteorological satellites have been used to study land resources (as discussed in Chapters 16, 19, and 20), this chapter focuses specifically on satellite systems specifically tailored for observation of land resources.

## 6.2. Landsat Origins

Early meteorological sensors had limited capabilities for land resources observation. Although these sensors were valuable for observing cloud patterns, most had rather coarse resolution, so they could provide only the coarsest level of detail about land resources. Even the most rudimentary patterns of land use, vegetation, and drainage were not consistently visible. Landsat ("land satellite") was designed in the 1960s and launched in 1972 as the first satellite tailored specifically for broad-scale observation of the earth's land areas—to accomplish for land resource studies what meteorological satellites had accomplished for meteorology and climatology. Today Landsat is important both in its own right—as a remote sensing system that has contributed greatly to earth resources studies—and as an introduction to similar land observation satellites operated by other organizations.

Landsat was proposed by scientists and administrators in the U.S. government who envisioned application of the principles of remote sensing to broad-scale, repetitive surveys of the earth's land areas. The first Landsat sensors recorded energy in the visible and near infrared spectrum. Although these portions of the spectrum had long been used for aircraft photography, it was by no means certain that they would also prove practical for observation of earth resources from satellite altitudes—scientists could not be completely confident that the sensors would work as planned, that they would prove to be reliable, that detail would be satisfactory, or that a sufficient proportion of scenes would be free of cloud cover. Although many of these problems were experienced, the feasibility of the basic concept was demonstrated, and Landsat formed the model for similar systems operated by other organizations.

The Landsat system consists of spacecraft-borne sensors that observe the earth and transmit information by microwave signals to ground stations that receive, and then process, data for dissemination to a community of data users. Early landsat vehicles carried two sensor systems; the return beam vidicon (RBV) and the multispectral scanner subsystem (MSS) (Table 6.1). The RBV was a camera-like instrument designed to provide, relative to the MSS, high geometric accuracy but lower spectral and radiometric de-

**TABLE 6.1. Landsat Missions**

| Satellite | Launched | Retired[a] | Principal sensors[b] |
|---|---|---|---|
| Landsat 1 | 23 June 1972 | 6 January 1978 | MSS, RBV |
| Landsat 2 | 22 January 1975 | 27 July 1983 | MSS, RBV |
| Landsat 3 | 5 March 1978 | 7 September 1983 | MSS, RBV |
| Landsat 4 | 16 July 1982 | — | TM, MSS |
| Landsat 5 | 1 March 1984 | — | TM, MSS |
| Landsat 6 | 5 October 1993 | Destroyed at launch | ETM |
| Landsat 7 | Planned for 1998 | — | ETM+ |
| Landsat 7 Follow-on | Under design | — | LATI |

[a]Satellite systems typically operate on an intermittent or stand-by basis for considerable periods prior to formal retirement from service.

[b]Sensors are discussed in the text. MSS, multispectral scanner subsystem; RBV, return beam vidicon; TM, thematic mapper; ETM, enhanced TM; ETM+, enhanced TM plus; LATI, Landsat Advanced Technology Instrument.

tail. That is, positions of features would be accurately represented but without fine detail concerning their colors and brightnesses. In contrast, the MSS was designed to provide finer detail concerning spectral characteristics of the earth but less positional accuracy. Because technical difficulties restricted RBV operation, the MSS soon became the primary Landsat sensor. The second generation of Landsat vehicles (Landsats 4 and 5) carried the MSS, as well as the thematic mapper—a more sophisticated version of the MSS.

## 6.3. Satellite Orbits

Satellites are placed into orbits tailored to match the capabilities of the sensors they carry and to the objectives of each satellite mission. For simplicity, this section describes *normal* orbits based on the assumption that the earth's gravity field is spherical, although in fact satellites actually follow *perturbed* orbits due to the earth's irregular shape and to distortion of the earth's gravity field by lunar and solar gravity, tides, solar wind, and other influences. A normal orbit forms an ellipse with the center of the earth at one foci, characterized by an *apogee* (point farthest from the earth), *perigee* (point closest to the earth), *ascending node* (AN) (point where the satellite crosses the equator moving south to north), *descending node* (DN) (point where the satellite crosses the equator north to south). The *inclination* (*i*) defines the angle that the satellite track forms with respect to the equator at the descending node (Figure 6.1).

A satellite's *period* (time to complete one orbit) increases with altitude; at an altitude of 36,000 km a satellite has the same period as the earth, so (if positioned in the equatorial plane) it remains stationary with respect to the earth's surface—a *geostationary* orbit. Geostationary orbits are ideal for meteorological or communications satellites designed to maintain a constant position with respect to a specific portion of the earth's surface. Earth observation satellites, however, are designed to satisfy other objectives that outweigh advantages offered by geostationary orbits.

Ideally, all satellite images would be acquired under conditions of uniform illumina-

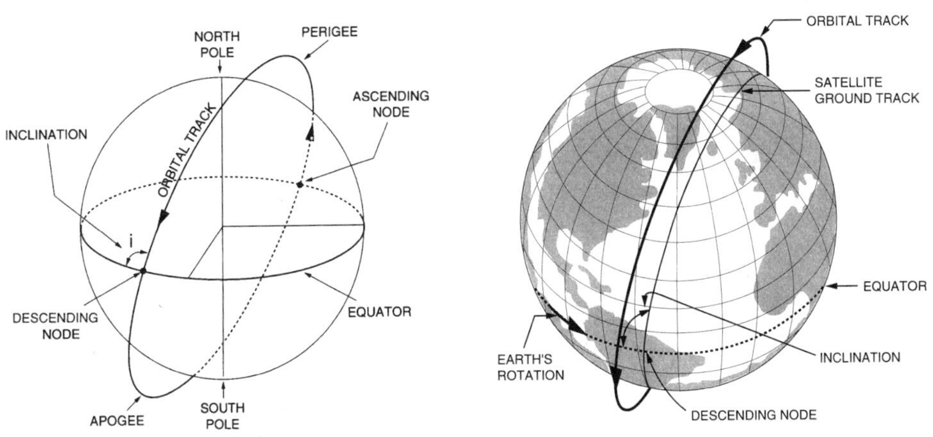


**FIGURE 6.1.** Satellite orbits. Left: Definitions. Right: Schematic representation of a sun-synchronous orbit.

tion, so that appearances of features observed within each scene would indicate changes in conditions on the ground rather than changes in conditions of observation. In reality, satellite images differ greatly because differences in latitude, time of day, and season lead to variations in the nature and intensity of light that illuminates each scene.

*Sun-synchronous* orbits are designed to remove one source of variation in illumination—that caused by differences in time of day, which arises from the fact that the earth rotates inside the solar beam. The hour angle ($h$) describes the difference in longitude between a point of interest and that of the direct solar beam (Figure 6.2). The value of $h$ (measured in degrees of longitude) can be found using

$$h = [(\mathrm{GMT} - 12.0) \times 15] - \text{longitude} \qquad \text{(Eq. 6.1)}$$

where GMT represents Greenwich Mean Time, and longitude represents the longitude of the point in question. Because $h$ varies with longitude, to maintain uniform local sun angle, it is necessary to design satellite orbits that acquire each scene at the same local sun time. The optimum local sun time varies with the objectives of each project; most earth observation satellites are placed in orbits designed to acquire imagery between 9:30 and 10:30 A.M., local sun time—a time that minimizes cloud cover in tropical regions and provides optimum illumination for many interpolations.

A satellite in a sun-synchronous orbit has an inclination that carries the satellite track westward at a rate that compensates for the change in local sun time as the satellite moves from north to south. Therefore, the satellite observes each scene at the same local sun time (constant $h$), removing time of day as a source of variation in illumination.

Illumination also varies due to seasonal effects caused by the tilt of the earth's orbit with respect to the plane of the ecliptic. This tilt tips the north pole 23.5° toward the sun on the northern hemisphere's summer solstice (about 21 June) and tips the south pole toward the sun on 21 December. This effect causes the intensity of the solar beam (as observed at the earth's surface) to vary with day of the year and latitude. Between 23.5°N

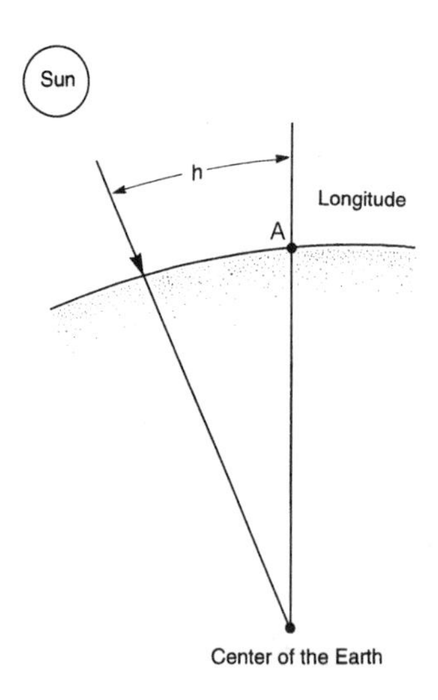


**FIGURE 6.2.** Hour angle. The hour angle ($h$) measures the difference in longitude between a point of interest ($A$) of known longitude and the longitude of the direct solar beam.

and 23.5°S the solar beam will be directly overhead at noon at least once each year; the latitude of the direct solar beam is known as the sun's *declination* ($\delta$) (the latitude of the subsolar point). Solar declination for specific dates can be calculated or found in almanacs and astronomical tables.

Satellites in sun-synchronous orbits pass from north to south on the sunlit side (the *descending node*) and from south to north on the shadowed side (the *ascending node*). Sensors that depend on reflected solar radiation acquire data only during the descending pass, although radar and thermal sensors (Chapters 7 and 8) can acquire data independently of solar illumination, thereby observing the earth's surface during both passes.

## 6.4. The Landsat System

Although the first generation of Landsat sensors is no longer in service, they have acquired a large library of images that are available as a baseline reference of environmental conditions for land areas throughout the world. Therefore, knowledge of these early Landsat images is important both as an introduction to later satellite systems and as a basis for work with the historical archives of images from Landsats 1, 2, and 3.

From 1972 to 1983, various combinations of Landsats 1, 2, and 3 orbited the earth in sun-synchronous orbits every 103 minutes—14 times each day. After 251 orbits—completed every 18 days—Landsat passed over the same place on the earth to produce repetitive coverage (Figure 6.3). When two satellites were both in service their orbits were tailored to provide repetitive coverage every 9 days. Sensors were activated to acquire images only at scheduled times, so this capability was not always used. In addition, equipment malfunctions and cloud cover sometimes prevented acquisition of planned coverage.

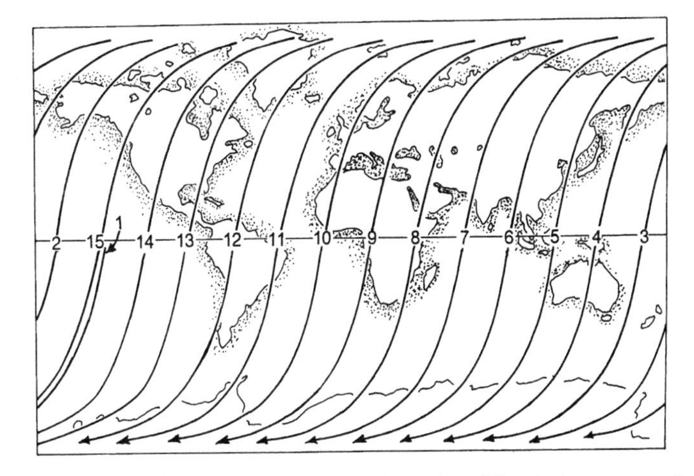


**FIGURE 6.3.** Coverage cycle, Landsats 1, 2, and 3. Each numbered line designates a northeast-to-southwest pass of the satellite. In a single 24-hour interval the satellite completes 14 orbits; the first pass on the next day (orbit 15) is immediately adjacent to pass 1 on the preceding day.

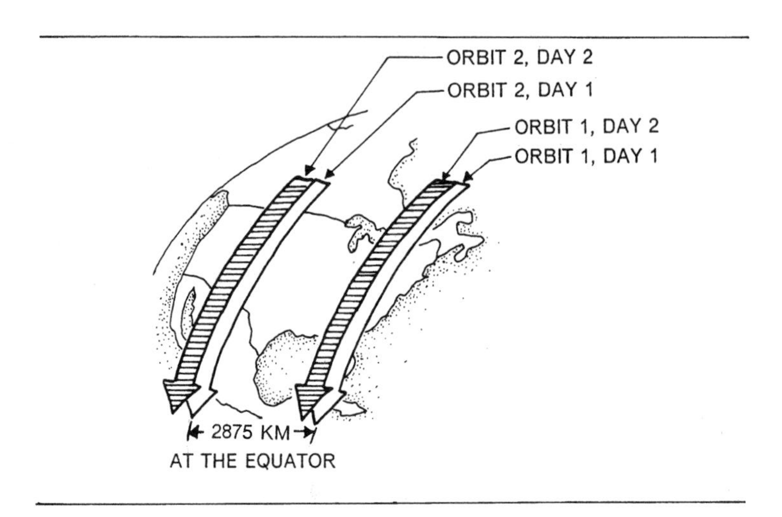


**FIGURE 6.4.** Incremental increases in Landsat coverage. On successive days, orbital tracks begin to fill in the gaps left by the displacement of orbits during the preceding day. After 18 days, progressive accumulation of coverage fills in all gaps left by coverage acquired on day one.

As a function of the earth's rotation on its axis from west to east, each successive north-to-south pass of the Landsat platform was offset to the west by 2,875 km (1,786 mi.) at the equator (Figure 6.4). Because the westward longitudinal shift of adjacent orbital tracks at the equator was approximately 159 km (99 mi.), gaps between tracks were incrementally filled during the 18-day cycle. Thus, on day 2, orbit number one was displaced 159 km to the west of the path of orbit number one on day 1. On the eighteenth day, orbit number one was identical to that of orbit number one, day 1. The first orbit on the nineteenth day coincided with that of orbit number one on day 2, and so on. Therefore, the entire surface of the earth between 81°N and 81°S latitude was subject to coverage by Landsat sensors once every 18 days (every 9 days, if two satellites were in service).

### *Support Subsystems*

Although our interest here is primarily the sensors that these satellites carried, it is important to briefly mention the support subsystems—units that are necessary to maintain proper operation of the sensors. Although this section refers specifically to the Landsat system, remember that all earth observation satellites require similar support systems, so it describes general features as well as specifics. The *attitude control subsystem* (ACS) maintained orientation of the satellite with respect to the earth's surface, and with respect to the orbital path. The *orbit adjust subsystem* (OAS) maintained the orbital path within specified parameters after the initial orbit was attained. The OAS also made adjustments throughout the life of the satellite to maintain the planned repeatable coverage of imagery. The *power subsystem* supplied electrical power required to operate all satellite systems by means of two solar array panels and eight batteries. The batteries were charged by energy provided by the solar panels while the satellite was on the sunlit side of the earth, then provided power when the satellite was in the earth's shadow. The *thermal control subsystem* controlled the temperatures of satellite components by means of heaters, passive radiators (to dissipate excess heat), and insulation. The *communications and data handling subsystem* provided microwave communications with ground stations for transmitting data from the sensors, commands to satellite subsystems, and information regarding satellite status and location.

Data from sensors were transmitted, in digital form, by microwave signal to ground stations equipped to receive and process data. Direct transmission from the satellite to the ground station as the sensor acquired data was possible only when the satellite had direct line-of-sight view of the ground antenna (a radius of about 1,800 km from the ground stations). In North America, stations at Greenbelt, Maryland; Fairbanks, Alaska; Goldstone, California; and Prince Albert, Saskatchewan, provided this capability for most of the United States and Canada. Elsewhere a network of ground stations has been established over a period of years through agreements with other nations (Figure 6.5). Areas outside the receiving range of a ground station could be imaged only by use of the two tape recorders on board each of the early Landsats. Each tape recorder could record about 30 minutes of data; then, as the satellite moved within range of a ground station, the tape recorders could transmit the stored data to a receiving station. Thus these satellites had, within the limits of their orbits, capability for worldwide coverage. Unfortunately, the tape

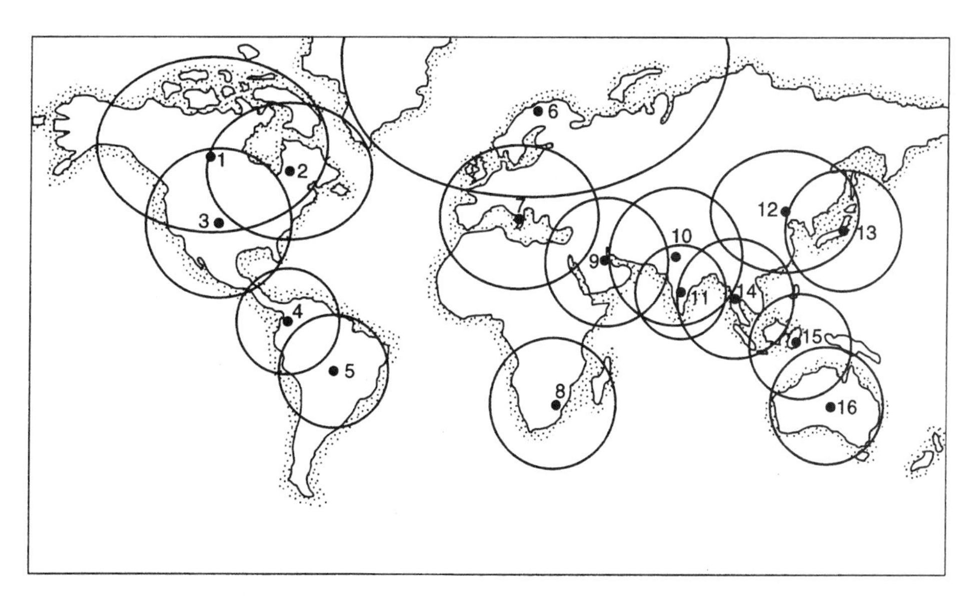


**FIGURE 6.5.** Landsat ground stations. Each circle shows the approximate range for direct communication with the satellite. Because of map projection distorts sizes of areas at high latitudes, ranges for northern stations appear here as large, unevenly shaped regions. On a globe, all would have the same size and shape. Key to Landsat ground stations, current as of 1995: (1) Prince Albert, Canada; (2) Gatineau, Canada; (3) Norman, Oklahoma, United States; (4) Cotopaxi, Ecuador; (5) Cuiaba, Brazil; (6) Kiruna, Sweden; (7) Fucino, Italy; (8) Johannesburg, South Africa; (9) Riyadh, Saudi Arabia; (10) Islamabad, Pakistan; (11) Shadnagar, India; (12) Beijing, China; (13) Hatoyama, Japan; (14) Bangkok, Thailand; (15) Parepare, Indonesia; (16) Alice Springs, Australia.

recorders proved to be one of the most unreliable elements of the Landsat system, so when they failed the system was unable to image areas beyond the range of the ground stations. These unobservable areas became smaller as more ground stations were established, but there still remained areas that Landsat could not observe. Later sections in this chapter explain how Landsats 4 and 5 were able to avoid this problem by use of relay satellites.

### *Return Beam Vidicon*

The return beam vidicon camera system generated high-resolution television-like images of the earth's surface. On Landsats 1 and 2, the RBV system consisted of three independent cameras which operated simultaneously, each sensing a different segment of the spectrum (Table 6.2). All three instruments were aimed at the same region beneath the satellite, so the images they acquired registered to one another to form a three-band multi-

**TABLE 6.2. Landsat Sensors**

| Sensor | Band | Spectral sensitivity |
|---|---|---|
| | | Landsats 1 and 2 |
| RBV | 1 | 0.475–0.575 μm (green) |
| RBV | 2 | 0.58–0.68 μm (red) |
| RBV | 3 | 0.69–0.83 μm (near infrared) |
| MSS | 4 | 0.5–0.6 μm (green) |
| MSS | 5 | 0.6–0.7 μm (red) |
| MSS | 6 | 0.7–0.8 μm (near infrared) |
| MSS | 7 | 0.8–1.1 μm (near infrared) |
| | | Landsat 3 |
| RBV | | 0.5–0.75 μm (panchromatic response) |
| MSS | 4 | 0.5–0.6 μm (green) |
| MSS | 5 | 0.6–0.7 μm (red) |
| MSS | 6 | 0.7–0.8 μm (near infrared) |
| MSS | 7 | 0.8–1.1 μm (near infrared) |
| MSS | 8 | 10.4–12.6 μm (far infrared) |
| | | Landsats 4 and 5[a] |
| TM | 1 | 0.45–0.52 μm (blue-green) |
| TM | 2 | 0.52–0.60 μm (green) |
| TM | 3 | 0.63–0.69 μm (red) |
| TM | 4 | 0.76–0.90 μm (near infrared) |
| TM | 5 | 1.55–1.75 μm (mid infrared) |
| TM | 6 | 10.4–12.5 μm (far infrared) |
| TM | 7 | 2.08–2.35 μm (mid infrared) |
| MSS | 1 | 0.5–0.6 μm (green) |
| MSS | 2 | 0.6–0.7 μm (red) |
| MSS | 3 | 0.7–0.8 μm (near infrared) |
| MSS | 4 | 0.8–1.1 μm (near infrared) |

[a]On Landsats 4 and 5 MSS bands were renumbered although the spectral definitions remain the same.

spectral representation of a 185 km × 170 km ground area, known as a *Landsat scene* (Figure 6.6). This area matched the area represented by the corresponding MSS scene.

The RBV shutter was designed to open briefly, in the manner of a camera shutter, to view the entire scene simultaneously (Figure 6.7a). Solar radiation reflected from the ground passed through a lens and was focused on the photosensitive surface of the vidicon camera. After the shutter closed, the photosensitive surface was then scanned by an electron beam to convert the scene's reflectance values into a video signal for transmission to a ground station. The RBV design is therefore analogous to that of a camera but with a photosensitive plate substituted in the focal plane in the position that would normally hold the film. The RBV system could acquire one image every 25 seconds to produce a series of images with a small forward overlap. Because the RBV was designed with an optical geometry analogous to that of a camera, scientists hoped to be able to use RBV imagery as the source of accurate cartographic measurements, much in the same way that aerial photographs had long been used to compile accurate maps. Thus the function of the RBV was primarily to provide the basis for accurate measurements of position and distance.

Because of electrical failures on Landsat 1, the RBV was turned off less than 1 month after launch. A similar malfunction caused the Landsat 2 RBV to be shut down shortly after launch, so the first two satellites acquired relatively little RBV imagery. Some scientists found the increased detail of the Landsat 3 RBV imagery to be of interest,

**FIGURE 6.6.** Schematic diagram of Landsat scene.

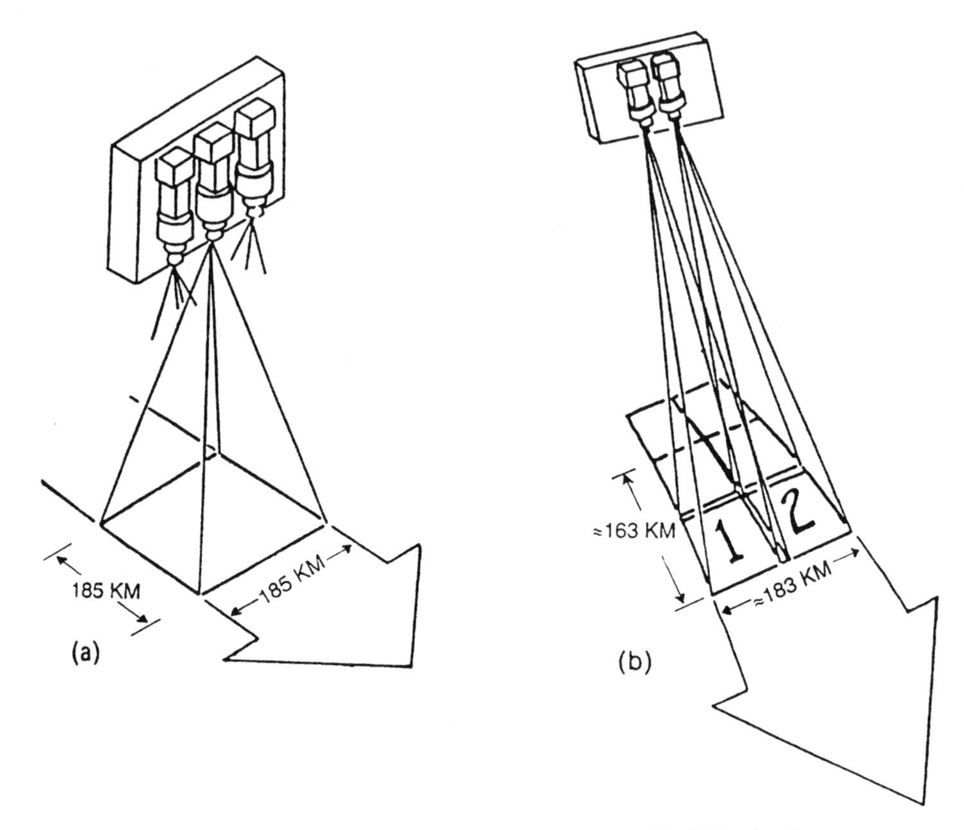


**FIGURE 6.7.** RBV configuration. (*a*) Landsats 1 and 2. (*b*) Landsat 3.

but the sensor experienced technical problems that seemed to preclude routine acquisition of high quality imagery. As a result, Landsat RBV imagery was not used in the role that was envisioned at the start of the Landsat program.

### *Multispectral Scanner Subsystem*

As a result of malfunctions in the RBV sensors early in the missions of Landsats 1 and 2, the multispectral scanner subsystem became the primary Landsat sensor. Whereas the RBV was designed to capture images with known geometric properties, the MSS was tailored to provide multispectral data without as much concern for positional accuracy. In general, MSS imagery and data were found to be of good quality—much better than many expected—and clearly demonstrated the merits of satellite observation for acquiring earth resources data. The economical, routine availability of MSS digital data has formed the foundation for a sizable increase in the number and sophistication of digital image processing capabilities available to the remote sensing community. A version of the MSS was placed on Landsats 4 and 5, and later systems have been designed with an eye toward maintaining continuity of MSS data.

The MSS (Figure 6.8) is a scanning instrument utilizing a flat, oscillating mirror to scan from west to east to produce a ground swath of 185 km (100 nautical miles) perpendicular to the orbital track. The satellite motion along the orbital path provides the along-track dimension to the image. Solar radiation reflected from the earth's surface is directed by the mirror to a telescope-like instrument that focuses the energy onto fiber-optic bundles located in the focal plane of the telescope. The fiber-optic bundles then transmit energy to detectors sensitive to four spectral regions (Table 6.2).

Each west-to-east scan of the mirror represents a strip of ground approximately 185 km long in the east–west dimension and 474 m wide in the north–south dimension. The 474 m distance corresponds to the forward motion of the satellite during the interval required for the west-to-east movement of the mirror and its inactive east-to-west retrace to return to its starting position. The mirror returned to start another active scan just as the satellite reached the position to record another line of data at a ground position immediately adjacent to the preceding scan line. Each motion of the mirror acquired six lines of data on the image because the fiber optics split the energy from the mirror into six contiguous segments.

The *instantaneous field of view* (IFOV) of a scanning instrument can be informally defined as the ground area viewed by the sensor at a given instant in time. The nominal IFOV for the MSS is 79 m × 79 m. Slater (1980) provides a more detailed examination of MSS geometry that shows the IFOV to be approximately 76 m × 76 m (about 0.58 hectares, or 1.4 acres), although differences exist between Landsats 1, 2, and 3 and within orbital paths as satellite altitude varies. The brightness from each IFOV is displayed on the image as a *pixel* (picture element) formatted to correspond to a ground area said to be approximately 79 m × 57 m in size (about 0.45 hectares, or 1.1 acres). In everyday terms, the ground area cor-

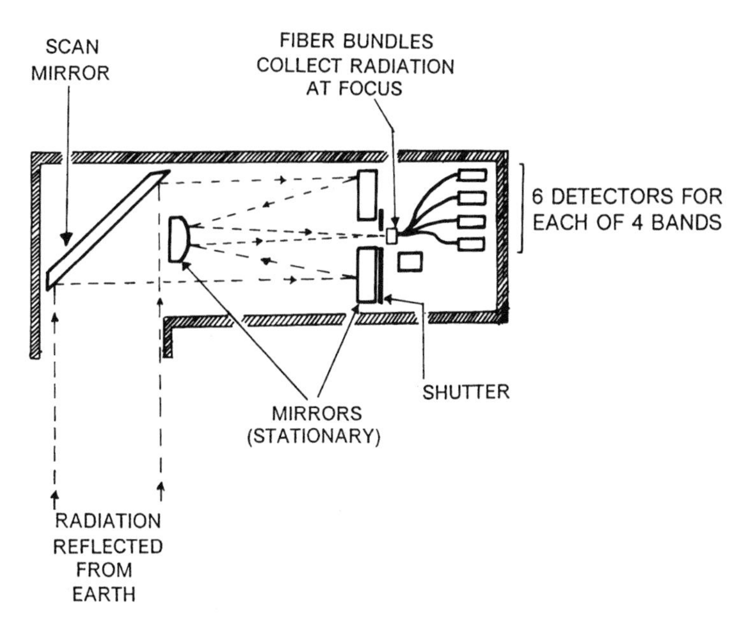


**FIGURE 6.8.** Schematic diagram of Landsat multispectral scanner.

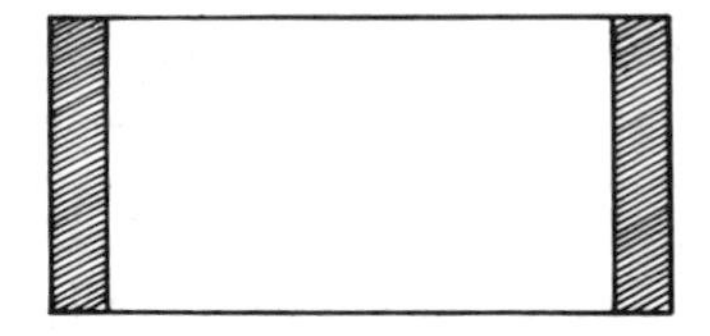

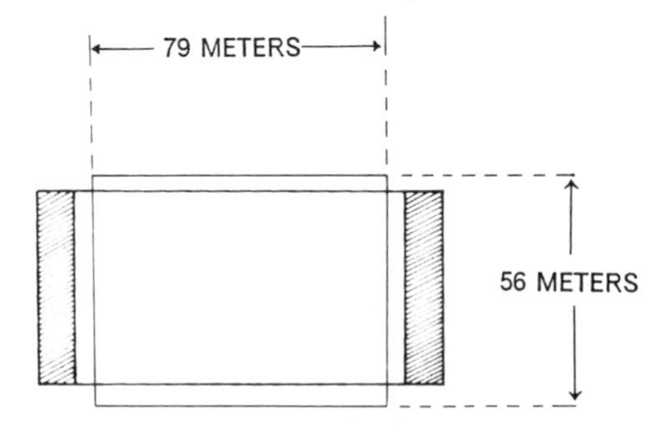


**FIGURE 6.9.** Nominal MSS pixel shown in relation to a U.S. football field. (Here the *orientation* of the pixel is rotated 90° to match to the usual rendition of a football field; on an image, the narrow ends of MSS pixels are oriented aproximately north and south.)

responding to a MSS pixel can be said to be somewhat less than that of a U.S. football field (Figure 6.9). Those who work routinely with MSS data know that, as a practical matter, images matching the stated levels of detail are observed only under ideal circumstances.

For the MSS instruments on board Landsat 1 and Landsat 2, the four spectral channels were located in the green, red, and infrared portions of the spectrum:

- *Band 1:* 0.5–0.6 µm (green)
- *Band 2:* 0.6–0.7 µm (red)
- *Band 3:* 0.7–0.8 µm (near infrared)
- *Band 4:* 0.8–1.1 µm (near infrared)

The Landsat 3 MSS included an additional band in the far infrared from 10.4 to 12.6 µm. Because this band included only two detectors, energy from each mirror movement was subdivided into two segments, each 234 m wide. The IFOV for the thermal band was therefore 234 m × 234 m, producing much coarser resolution than MSS images for the other bands.

## 6.5. MSS Images

### *The MSS Scene*

The MSS scene is defined as an image representing a ground area approximately 185 km in the east–west (across-track) direction, and 178 km in the north–south (along-track) direction (Figure 6.10). The across-track dimension is defined by the side-to-side motion of the MSS; the along-track dimension is provided by the forward motion of the satellite along its orbital path. If the MSS were operated continuously for an entire descending pass, it would

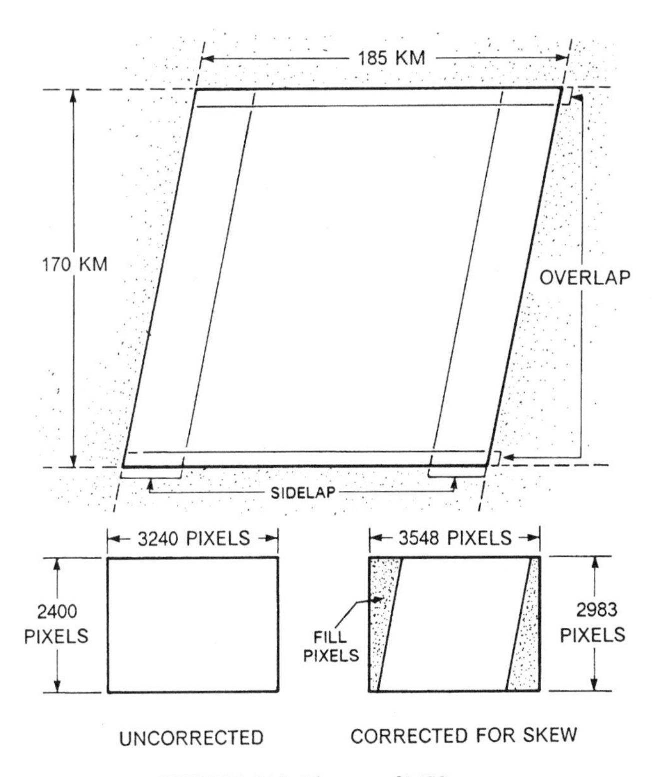


**FIGURE 6.10.** Diagram of MSS scene.

provide a continuous strip of imagery representing an area 185 km wide. The 178 km north–south dimension simply divides this strip into segments of convenient size.

The MSS scene, then, is an array of pixel values (in each of four bands) consisting of about 2,400 scan lines, each composed of 3,240 pixels (Figure 6.10). Although center points of scenes acquired at the same location at different times are intended to register with each other, there is often, in fact, a noticeable shift (the problem of *temporal registration*) in ground locations of center points due to uncorrected drift in the orbit.

There is a small overlap (about 5%, or 9 km) between scenes to the north and south of a given scene. This overlap is generated by repeating the last few lines from the preceding image, not by stereoscopic viewing of the earth. Overlap with scenes to the east and west depends on latitude; sidelap will be a minimum of 14% (26 km) at the equator and increases with latitude to 57% at 60°, then to 85% at 80° north and south latitude. Because this overlap is created by viewing the same area of the earth from different perspectives, the area within the overlap can be viewed in stereo. At high latitudes, this area can constitute an appreciable portion of a scene.

## *Image Format*

MSS data are available in several image formats that have been subjected to different forms of processing to adjust for geometric and radiometric errors. The following section describes some of the basic forms for MSS data, which provide a general model for other kinds of satellite imagery, although specifics vary with each kind of data.

In its initial form a digital satellite image consists of a rectangular array of pixels in each of four bands (Figure 6.10). In this format, however, no compensation has been made for the combined effects of spacecraft movement and rotation of the earth as the sensor acquires the image (this kind of error is known as *skew*). When these effects are removed, the image assumes the shape of a parallelogram (Figure 6.10). For convenience in recording data, *fill* pixels are added to preserve the correct shape of the image. Fill pixels, of course, convey no information, as they are simply assigned values of zero as necessary to attain the desired shape.

Each of the spectral channels of a multispectral image forms a separate image, each emphasizing landscape features that reflect specific portions of the spectrum (Figures 6.11 and 6.12). These separate images, then, record in black-and-white form the spectral reflectance in the green, red, and infrared portions of the spectrum, for example. The

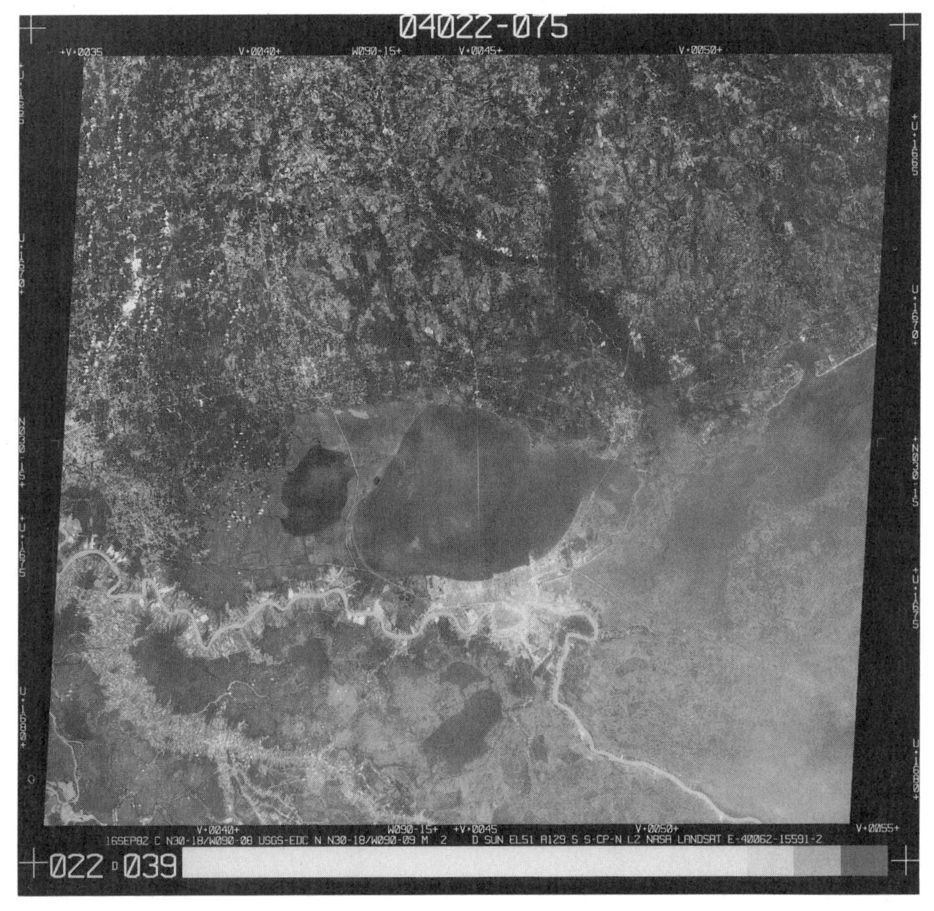


**FIGURE 6.11.** Landsat MSS image, band 2. New Orleans, Louisiana, 16 September 1982. Scene ID 40062-15591-2. Image reproduced by permission of EOSAT. (MSS band 2 was formerly designated as band 5. See Table 6.2.)

green, red, and infrared bands can be combined into a single color image (Plate 5), known as a *false-color composite.* The near infrared band is projected onto color film through a red filter, the red band 2 through a green filter, and the green band through a blue filter. The result is a false-color rendition that uses the same assignment of colors used in conventional color infrared aerial photography (Chapter 3). Strong reflectance in the green portion of the spectrum is represented as blue on the color composite, red as green, and infrared as red. Thus, living vegetation appears bright red, turbid water as a blue color, and urban areas as gray or sometimes pinkish-gray.

On photographic prints of MSS images, the annotation block at the lower edge gives essential information concerning the identification, date, location, and characteristics of the image. During the interval since the launch of Landsat 1, the content and form of the annotation block have been changed several times, but some of the basic information can be shown in a simplified form to illustrate key items (Figure 6.13). The *date* has obvious meaning. The *format center* and *ground nadir* give, in degrees and minutes of latitude and

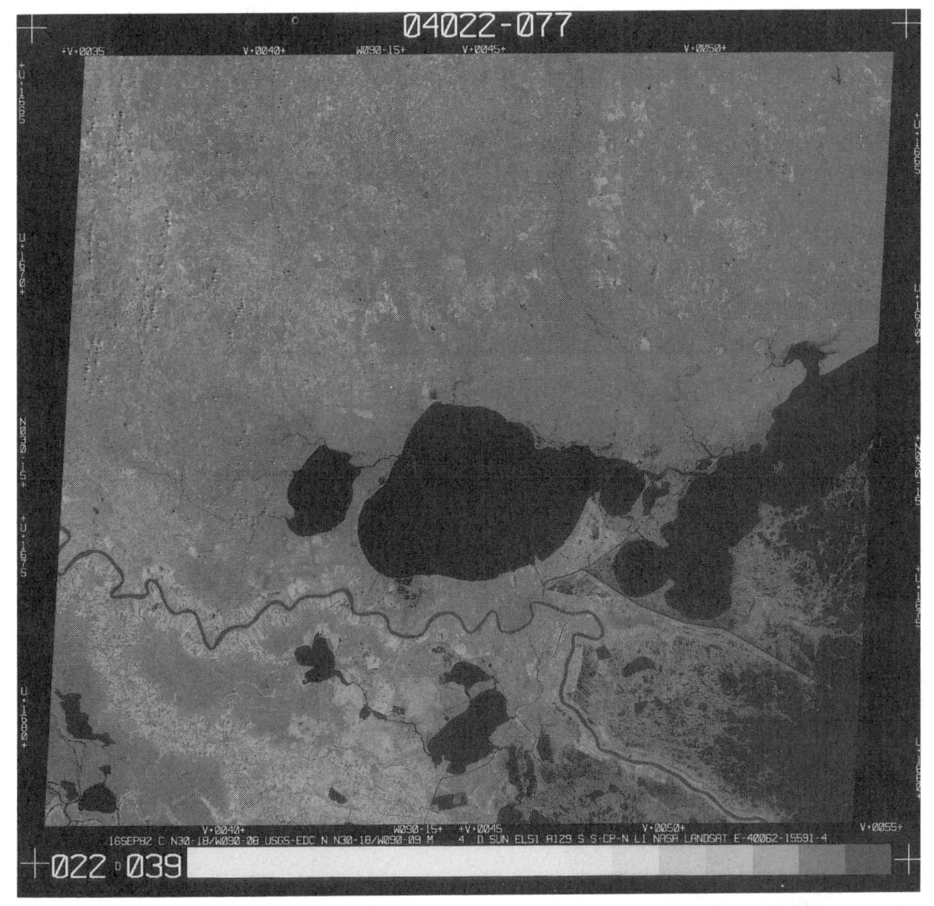


**FIGURE 6.12.** Landsat MSS image, band 4. New Orleans, Louisiana, 16 September 1982. Scene ID 40062-15591-7. Image reproduced by permission of EOSAT. (MSS band 4 was formerly designated as band 7. See Table 6.2.)

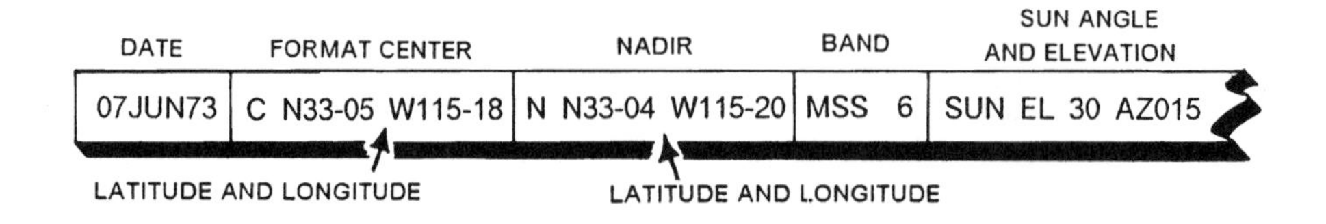


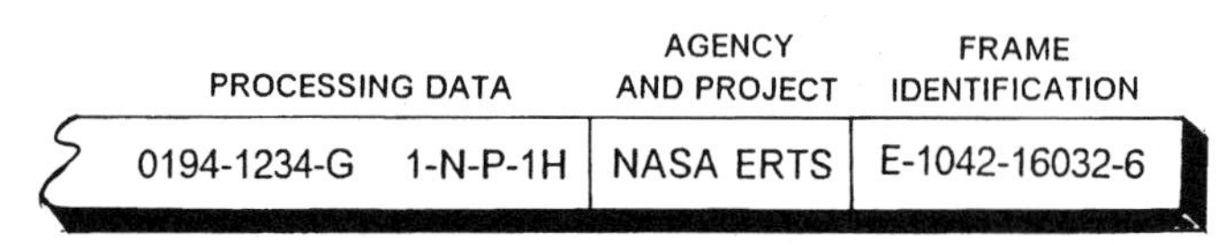


**FIGURE 6.13.** Landsat annotation block. The annotation block has been changed several times, but all show essentially the same information.

longitude, the ground location of the center point of the image. The spectral band is given in the form "MSS 1" (meaning multispectral scanner, band 1). *Sun angle* and *sun elevation* designate, in degrees, the solar elevation (above the horizon) and the azimuth of the solar beam from true north at the center of the image. Of the remaining items on the annotation block, the most important for most users is the *scene ID,* a unique number that specifies the scene and band. The scene ID uniquely specifies any MSS scene, so it is especially useful as a means of cataloging and indexing MSS images as explained below. MSS imagery can be purchased from Earth Observation Satellite Company (EOSAT) (4300 Forbes Blvd., Lanham, MD 20706; 301-552-3762 or 800-344-9933).

### *Worldwide Reference System*

The worldwide reference system (WRS) is a concise designation of nominal center points of Landsat scenes used to index Landsat scenes by location. The reference system is based on a coordinate system in which there are 233 north–south paths corresponding to orbital tracks of the satellite, and 119 rows representing latitudinal center lines of Landsat scenes. The combination of a path number and a row number uniquely identifies a nominal scene center (Figure 6.14). Because of the drift of satellite orbits over time, actual scene centers may not match exactly to the path–row locations, but the method does provide a convenient and effective means of indexing locations of Landsat scenes. Landsat MSS data are available from EROS Data Center (Mundt Federal Building, Sioux Falls, SD 57198; 605-594-6511).

## 6.6. Landsat Thematic Mapper

Even before Landsat 1 was launched, it was recognized that existing technology could improve the design of the MSS, and efforts were made to incorporate improvements into a new instrument modeled in the basic design of the MSS. Landsats 4 and 5 carried a re-

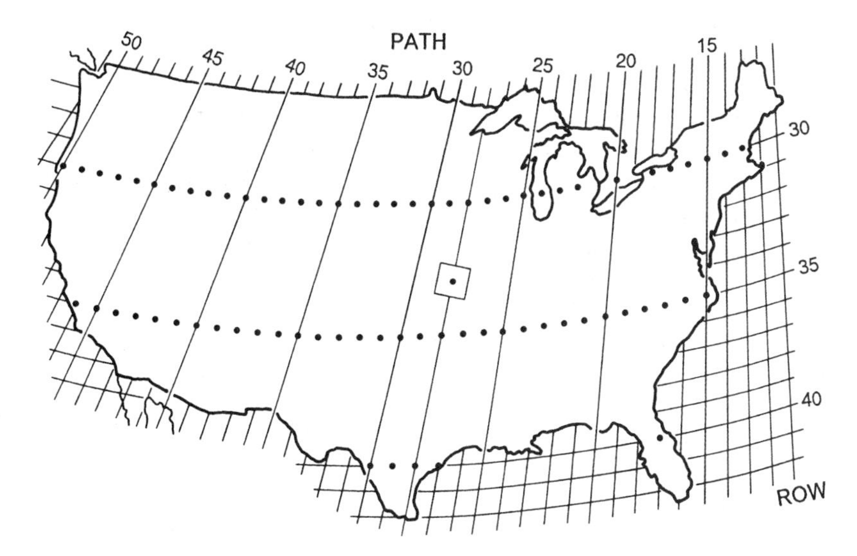


**FIGURE 6.14.** WRS path–row coordinates for the United States.

placement for MSS known as the *thematic mapper* (TM), which can be considered as an upgraded MSS. In these satellites, both the TM and an MSS were carried on an improved platform (Figure 6.15) that maintained a high degree of stability in orientation as a means of improving geometric qualities of the imagery. Landsats 4 and 5 completed a coverage cycle in 233 revolutions (16 days).

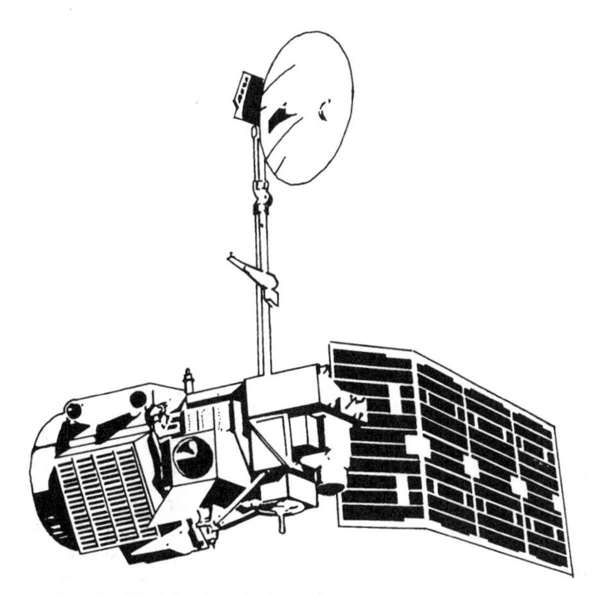

**FIGURE 6.15.** Platform for Landsats 4 and 5.

The design of the TM was based on the same principles as the MSS but with a more complex design. It provides finer spatial resolution, improved geometric fidelity, greater radiometric detail, and more detailed spectral information in more precisely defined spectral regions. Improved satellite stability and an orbit adjust subsystem are designed to improve positional and geometric accuracy. The objectives of the second generation of Landsat instruments were to assess the performance of the TM, to provide continued availability of MSS data, and to continue foreign data reception.

Despite the historical relationship between the TM and the MSS, the two sensors are distinct. Whereas the MSS has four broadly defined spectral regions, the TM records seven spectral bands (Table 6.3). TM band designations do not follow the sequence of the spectral definitions (the band with the longest wavelength is band 6, rather than band 7) because design changes occurred so late in the engineering process that it was not feasible to relabel the bands to follow a logical sequence.

TM spectral bands were tailored to record radiation of interest to specific scientific investigations, rather than the arbitrary definitions used for the MSS. Spatial resolution is about 30 m (about 0.09 hectare; 0.22 acre), compared to the 76 m IFOV of the MSS. (TM band 6 had coarser spatial resolution of about 120 m.) The finer spatial resolution provided a noticeable increase (relative to the MSS) in spatial detail recorded by each TM image (Figure 6.16). Digital values are quantized at 8 bits (256 brightness levels), which provide

**TABLE 6.3. Summary of TM Sensor Characteristics**

| Band | Resolution | Spectral definition | Some applications[a] |
|---|---|---|---|
| 1 | 30 m | Blue-green 0.45–0.52 μm | Penetration of clear water; bathymetry; mapping of coastal waters; chlorophyll absorption; distinction between coniferous and deciduous vegetation |
| 2 | 30 m | Green 0.52–0.60 μm | Records green radiation reflected from healthy vegetation; assesses plant vigor; reflectance from turbid water |
| 3 | 30 m | Red 0.63–0.69 μm | Chlorophyll absorption important for plant-type discrimination |
| 4 | 30 m | Near infrared 0.76–0.90 μm | Indicator of plant cell structure; biomass; plant vigor; complete absorption by water facilitates delineation of shorelines |
| 5 | 30 m | Mid infrared 1.55–1.75 μm | Indicative of vegetation moisture content; soil moisture soil mapping; differentiating snow from clouds; penetration of thin clouds |
| 6 | 120 m | Far infrared 10.4–12.5 μm | Vegetation stress analysis; soil moisture discrimination; thermal mapping; relative brightness temperature; soil moisture; plant heat stress |
| 7 | 30 m | Mid infrared 2.08–2.35 μm | Discrimination of rock types; alteration zones for hydrothermal mapping; hydroxyl ion absorption |

[a]Sample applications listed here; these are not the only applications.

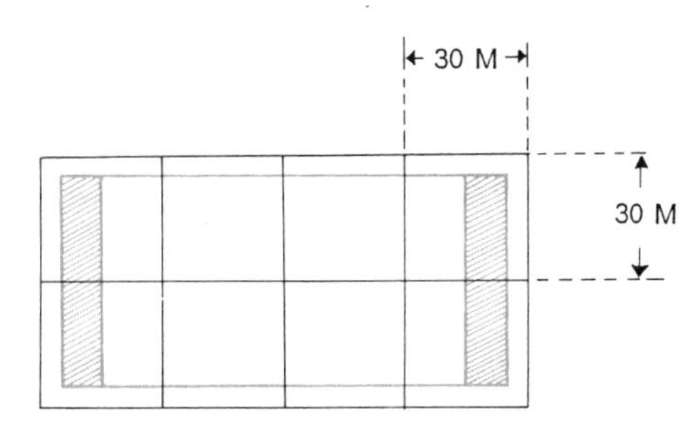


**FIGURE 6.16.** Nominal size of TM pixel.

(relative to the MSS) a much larger range of brightness values. These kinds of changes produced images with much finer detail than those of the MSS (Figures 6.17 and 6.18).

Each scan of the TM mirror acquired 16 lines of data. Unlike the MSS, the TM scan acquired data as it moved in both the east–west and west–east directions. This feature permitted engineers to design a slower speed of mirror movement, thereby improving the length of time the detectors could respond to brightness in the scene. However, this design required additional processing to reconfigure image positions of pixels to form a geometrically accurate image. TM detectors are positioned in an array in the focal plant (it does not use the fiber optics employed in the MSS to ensure perfect spatial registration of the MSS bands); as a result, there may be a slight misregistration of TM bands.

TM imagery is analogous to MSS imagery with respect to aerial coverage and organization of data into several sets of multispectral digital values that overlay to form an image. In comparison with MSS images, TM imagery has a much finer spatial and radiometric resolution, so TM images show relatively fine detail of patterns on the earth's surface.

The use of seven rather than four spectral bands, as well as a smaller pixel size within the same image area, means that TM images consist of many more data values than do MSS images. As a result, each analyst must determine those TM bands that are likely to provide the required information. Because the "best" combinations of TM bands vary according to the purpose of each study, season, geographic region, and other factors, a single selection of bands is unlikely to be equally effective in all circumstances.

A few combinations of TM bands appear to be effective for general-purpose use. The use of TM bands 2, 3, and 4 creates an image that is analogous to the usual false-color aerial photograph. TM bands 1 (blue-green), 2 (green), and 3 (red) form a natural color composite, approximately equivalent to a color aerial photograph in rendition of colors. Experiments with other combinations have shown that bands 2, 4, and 5; 2, 4, and 7; and 2, 4, and 5 are also effective for visual interpretation. Of course, there are many other combinations of the seven TM bands that may be useful in specific circumstances. TM imagery can be purchased from EOSAT (4300 Forbes Blvd., Lanham, MD 20706; 301-552-3762 or 800-344-9933).

### *Orbit and Ground Coverage: Landsats 4 and 5*

Landsats 4 and 5 were placed into orbits resembling those of earlier Landsats. Sun-synchronous orbits bring the satellites over the equator at about 9:45 A.M., thereby maintain-

**FIGURE 6.17.** TM image (band 3). This image shows New Orleans, Louisiana, 16 September 1982. Scene ID: 40062-15591-4. Image reproduced by permission of EOSAT.

ing approximate continuity of solar illumination with imagery from Landsats 1, 2, and 3. Data were collected as the satellite passed northeast to southwest on the sunlit side of the earth. The image swath remained at 185 km. In these respects, coverage was compatible with that of the first generation of Landsat systems.

However, there are important differences. The finer spatial resolution of the TM is achieved in part by a lower orbital altitude, which requires several changes in the coverage cycle. Earlier Landsats produced adjacent image swaths on successive days. However, Landsats 4 and 5 acquire coverage of adjacent swaths at intervals of 7 days. Landsat 4 completed a coverage cycle in 16 days. Successive passes of the satellite are separated at the equator by 2,752 km; gaps between successive passes are filled in over an interval of

**FIGURE 6.18.** TM image (band 4). This image shows New Orleans, Louisiana, 16 September 1982. Scene ID: 40062-15591-5. Image reproduced by permission of EOSAT.

16 days. Adjacent passes are spaced at 172 km. At the equator, adjacent passes overlap by about 7.6%; overlap increases as latitude increases.

A complete coverage cycle is achieved in 16 days—233 orbits. Because this pattern differs from earlier Landsats, Landsats 4 and 5 required a new WRS indexing system for labeling paths and rows (Figure 6.19). Row designations remained the same as before, but a new system of numbering paths was required. In all there are 233 paths, and 248 rows, with row 60 positioned at the equator. Communications for Landsats 4 and 5 were linked with a network of tracking and data relay satellites (TDRSS). TDRSS are in geosynchronous orbits that permitted direct transmission of sensor data directly to a central ground receiving station near White Sands, New Mexico.

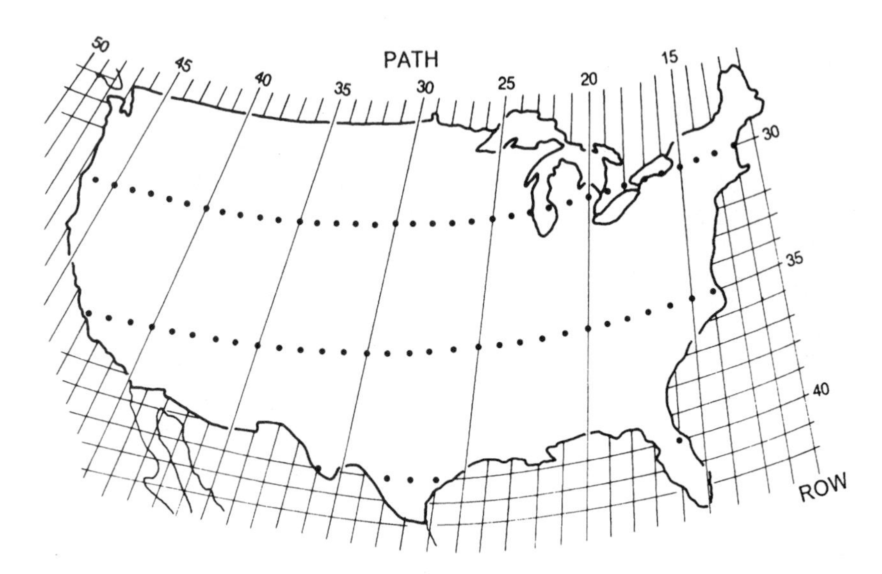


**FIGURE 6.19.** WRS for Landsats 4 and 5.

## 6.7. Administration of the Landsat Program

Landsat was originally operated by the NASA as a part of its mission to develop and demonstrate applications of new technology related to aerospace engineering. Although NASA initially had primary responsibility for Landsat, other federal agencies, including the U.S. Geological Survey and the National Oceanographic and Atmospheric Administration (NOAA) contributed to several aspects of the program. Although these federal agencies operated the Landsat system for many years, Landsat was officially considered "experimental" because of NASA's mission to develop new technology (rather than assume responsibility for routine operations) and because other federal agencies were unwilling to assume responsibility for a program of such cost and complexity. In 1983, NOAA assumed administrative responsibility for the Landsat system, although NASA continued to provide some of the facilities and personnel and technical services required for daily operation.

Over a period of many years, both the Carter and Reagan administrations and the U.S. Congress participated in debates concerning the future of many federal services, including Landsat. In essence, the debate centered on the merits of public operation (by agencies of the federal government) compared to operation by a private corporation. Those favoring private operation emphasized the prospects for more efficient operation of the system and more aggressive pursuit of new applications and new technologies. Those who favored continued government operation stressed the need for supervision to maintain continuity of data flow and data format, the importance of providing for public access to data, and the significance of a public archive of data as an historical record.

In June 1984, the U.S. Congress passed the Land Remote Sensing Act of 1984, which established the process by which the Landsat system passed from public to private operation. Title I of the act summarizes the rationale for the transfer to private operation and established certain constraints including assurances of public access to data. It also defines the fundamental distinction between "unenhanced" data—raw data that have been subjected to only minimal processing and "enhanced" data that have been processed and analyzed to derive usable information. Title II established the Department of Commerce (the parent organization for NOAA) as the monitor for supervising the performance of a private contractor and provides for selection of a private contractor to market unenhanced data.

Title III ensured continuity of data as private contractors assume responsibility for operation of the existing system and development of new components to the system. Title IV establishes government regulation and licensing of private remote sensing systems operated in space. Title V provides for continued research and development by agencies of the federal government, and Title VI outlined general provisions of the act, including maintenance of an image archive by the federal government. Finally, Title VII prohibited commercialization of weather satellites—an issue that assumed political significance during the congressional debate.

In 1985, the first phases of this policy were implemented by the transfer of Landsat from NOAA to Earth Observation Satellite Company, a private firm formed as a partnership between Hughes Aircraft Company and the Radio Corporation of America (RCA). EOSAT operated Landsat components, received and disseminated data and images, conducted research to define new applications for Landsat data, and designed new spacecraft and instruments to continue development of Landsat.

In October 1992, Congress passed the Land Remote Sensing Policy Act. This act and related agreements among federal agencies are described by Sheffner (1994) and Salomonson (1995). It has many provisions, including a step to maintain continuity in the collection of data, establishment of the basis for a pricing policy for image data, and an outline of the nature of additional instruments to be designed for future satellites. The following section describes the *Enhanced Thematic Mapper* (ETM), the sensor designed for Landsat 6, which was lost at launch, and the two replacement instruments now planned for future Landsat missions, *Enhanced Thematic Mapper Plus* (ETM+) and the *Landsat Advanced Technology Instrument* (LATI).

### *Enhanced Thematic Mapper*

The ETM, planned as a replacement for Landsat 6, was designed to acquire imagery over a 185-km swath, with a 16-day coverage cycle. It was designed to acquire seven channels of data in the visible, near infrared, shortwave infrared, and thermal regions. The thermal band would have had a resolution of 120 meters, the others 30 m, plus an additional panchromatic band at 15 m. The ETM was lost with the failed launch of Landsat 6 in September 1993.

At the time of this writing, planning for Landsat 7 is under way, with launch planned for 1998. NASA will design the satellite and ground support system and NOAA will operate the satellite after launch and process the data. The principal sensor for Landsat 7 will be the ETM+, intended to extend the design of the Landsats 4 and 5 TM, with some modest improvements. In the visible, near infrared, and mid infrared its spectral channels are to duplicate those of earlier TMs; the thermal channel will have 60 m (improved from the

120 m of earlier TMs), and it will have a 15-m panchromatic channel. Swath width will remain at 185 km, and the system will experience some improvement with respect to accuracy of calibration, data transmission, and other characteristics.

Planning is also under way for an additional sensor to extend the Landsat program after Landsat 7: the Landsat Advanced Technology Instrument. Current plans call for two identical pushbroom sensors to be mounted side by side in the spacecraft. Together the two instruments would image the 185 swath of current Landsat sensors (93 km each, with overlap of 1 km). Spectral channels would include the current TM channels in the visible, near infrared, and mid infrared at 30 m resolution. The plan does not now include a thermal channel, but the instrument would include an additional band in the mid infrared (at 30-m resolution), a panchromatic band (at 10-m resolution), and additional bands designed to provide data for atmospheric corrections.

Both ETM+ and LATI are intended to extend the continuity of earlier Landsat data (from both MSS and TM) by maintaining consistent spectral definitions, resolutions, and scene characteristics while taking advantage of improved technology, improving calibration, and providing for more efficient transmission of data.

## 6.8. SPOT

### *SPOTs 1, 2, and 3*

The success of Landsat in the early 1970s stimulated interests in other nations. The most significant of these earth observation systems is the SPOT program designed by a French organization in collaboration with other European nations. Laidet (1983) reported that French users were "strongly impressed" with the Landsat system, and that "the French space program has been strongly under the influence" of the U.S. program as they designed their own system.

SPOT—*Le Système Pour l'Observation de la Terre* ("Earth Observation System")—was initiated in 1977 and began operations in 1986, with the launch of SPOT 1. SPOT was conceived and designed by the *Centre National d'Etudes Spatiales* (CNES) in Paris, with the cooperation of other European organizations. SPOT 1, launched on 22 February 1986, was followed by SPOT 2 (21 January 1990) and SPOT 3 (23 September 1993), which carry sensors identical to those of SPOT 1. Two additional satellites, SPOT 4 (planned for launch in 1997) and SPOT 5 (planned for the year 2000), will carry sensors modified from those used on the earlier systems, to be described later.

The SPOT system is designed to provide data for land-use studies, assessment of renewable resources, exploration of geologic resources, and cartographic work at scales of 1:50,000 to 1:100,000. It is the first commercial remote sensing satellite designed to provide high-quality service and data for an operational user community worldwide. SPOT data, like those from Landsat, promise to meet the needs of a wide variety of customers with diverse technical, scientific, and commercial needs. Design requirements included provision for complete world coverage, rapid dissemination of data, a stereo capability, high spatial resolution, and sensitivity in spectral regions responsive to reflectance from vegetation.

The SPOT *bus* is the basic satellite vehicle, designed to be compatible with a variety of sensors (Figure 6.20). The bus provides basic functions related to orbit control and sta-

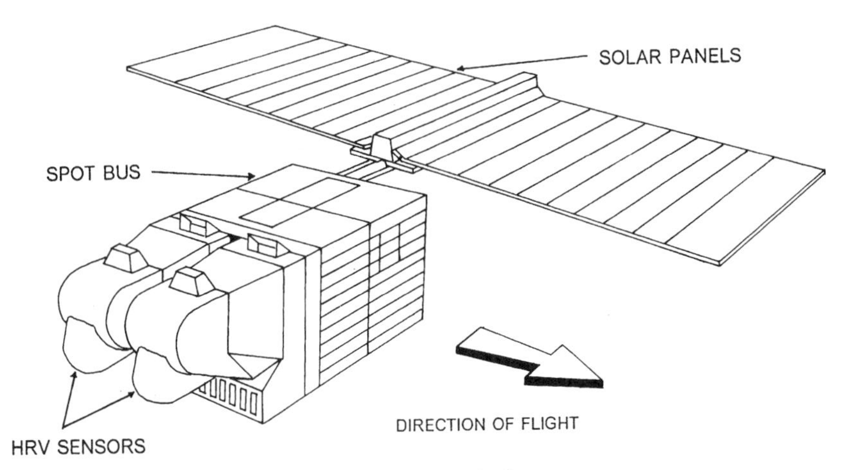


**FIGURE 6.20.** SPOT bus.

bilization, reception of commands, telemetry, monitoring of sensor status, and so on. The bus, with its sensors, is placed in a sun-synchronous orbit at about 832 km, with a 10:30 A.M. equatorial crossing time. For vertical observation, successive passes occur at 26-day intervals, but because of the ability of SPOT sensors to view areas at the oblique, successive imagery can be acquired, on the average, at 2½-day intervals. (The exact interval for repeat coverage varies with latitude.)

The SPOT payload consists of two identical sensing instruments, a telemetry transmitter, and magnetic tape recorders. The two sensors are known as HRV ("high-resolution visible") instruments. HRV sensors use pushbroom scanning, based on charge-coupled devices (CCDs) as discussed in Chapter 4, which simultaneously images an entire line of data in the cross-track axis (Figure 4.6). SPOT linear arrays consist of 6,000 detectors for each scan line in the focal plane; the array is scanned electronically to record brightness values (8 bits; 256 brightness values) in each line. Radiation from the ground is reflected to the two arrays by means of a movable plane mirror. An innovative feature of the SPOT satellite is the ability to control the orientation of the mirror by commands from the ground—a capability that enables the satellite to acquire oblique images, as described below.

The HRV can be operated in either of two modes. In the *panchromatic* (PN) mode the sensor is sensitive across a broad spectral band from 0.51 to 0.73 μm. It images a 60-km swath with 6,000 pixels per line for a spatial resolution of 10 m (Figure 6.21a). In this mode the HRV instrument provides fine spatial detail but records a rather broad spectral region. In the PN mode the HRV instrument provides coarse spectral resolution but fine spatial resolution.

In the other mode, the *multispectral* (×S) configuration (Plate 6), the HRV instrument senses three spectral regions:

- *Band 1:* 0.50–0.59 μm (green)
- *Band 2:* 0.61–0.68 μm (red; chlorophyll absorption)
- *Band 3:* 0.79–0.89 μm (near infrared; atmospheric penetration)

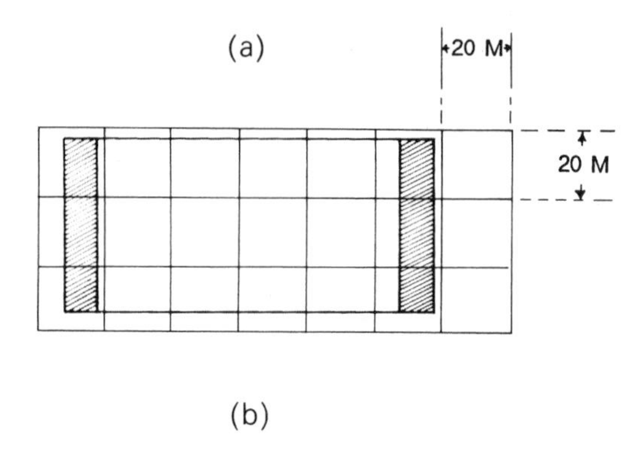


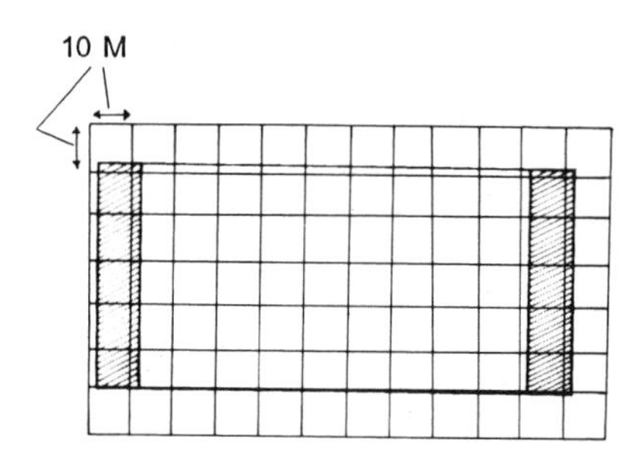


**FIGURE 6.21.** Nominal sizes of SPOT HRV pixels. (*a*) The pixel size of the HRV employed in the multispectral mode. (*b*) The panchromatic mode achieves finer resolution. An outline of a U.S. football field is shown for approximate scale.

In this mode the sensor images a strip 60 km in width using 3,000 samples for each line at a spatial resolution of about 20 m (Figure 6.21b). Thus, in the ×S mode, the sensor records fine spectral resolution but coarse spatial resolution. The three images from the ×S mode can be used to form false-color composites, in the manner of CIR, MSS, and TM images. In some instances, it is possible to "sharpen" the lower spatial detail of multispectral images by superimposing them on the fine spatial detail of high-resolution PN imagery of the same area.

With respect to sensor geometry, each of the HRV instruments can be positioned in either of two configurations (Figure 6.22). For nadir viewing (Figure 6.22a) both sensors are oriented in a manner that provides coverage of adjacent ground segments. Because the two 60-km swaths overlap by 3 km, the total image swath is 117 km. At the equator, centers of adjacent satellite tracks are separated by a maximum of only 108 km, so in this mode the satellite can acquire complete coverage of the earth's surface.

An off-nadir viewing capability is possible by pointing the HRV field of view as much as 27° relative to the vertical in 45 steps of 0.6° each (Figure 6.22b) in a plane perpendicular to the orbital path. Off-nadir viewing is possible because the sensor observes the earth through a pointable mirror that can be controlled by command from the ground. (Note that although mirror orientation can be changed on command, it is not a scanning mirror, as used by the MSS and the TM.) With this capability, the sensors can observe any

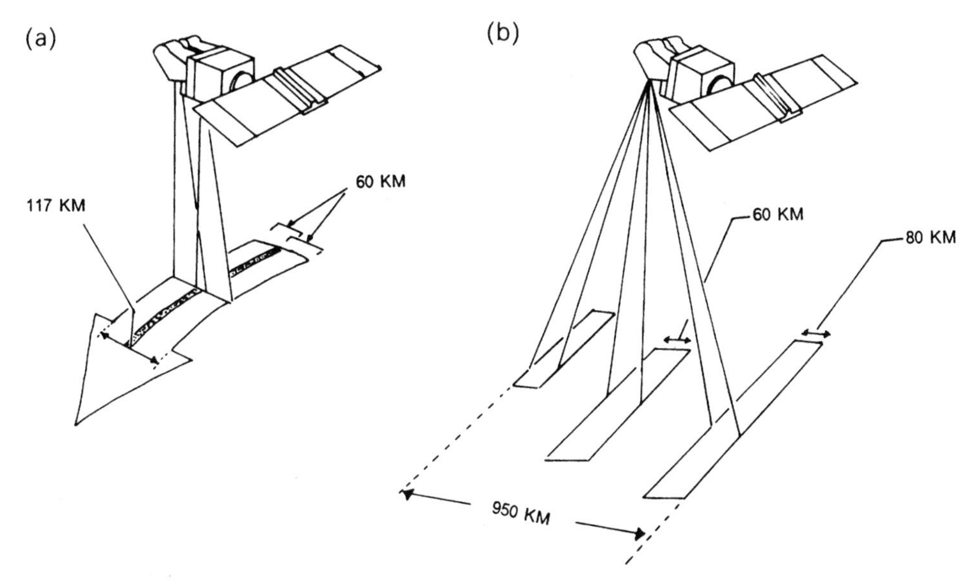


**FIGURE 6.22.** Geometry of SPOT imagery. (*a*) Nadir viewing. (*b*) Off-nadir viewing.

area within a 950-km swath centered on the satellite track. The pointable mirror can position any given off-nadir scene center at 10-km increments within this swath. At the maximum off-nadir viewing angle, HRV pixels increase in size to about 13.5 m in PN mode and 27 m in ×S mode, although at level 1b processing (explained below) apparent pixel sizes of 10 m (PN) and 20 m (×S) are present.

When SPOT uses off-nadir viewing, the swath width of individual images varies from 60 to 80 km, depending on viewing angle. Alternatively, the same region can be viewed from separate positions (different satellite passes) to acquire stereo coverage. (Such stereo coverage depends, of course, on cloud-free weather during both passes.) The twin sensors are not required to operate in the identical configuration; that is, one HRV can operate in the vertical mode while the other images obliquely. Using the off-nadir viewing capability, SPOT can acquire repeat coverage at intervals of 1 to 5 days, depending on latitude.

Ground control is provided by a station in Toulouse, France, with a number of ground stations around the world providing relay and backup capability as required. Direct reception of image data is possible within a 2,600-km radius of participating ground receiving stations. In addition, the tape recorders provide the capability for recording data acquired beyond this range. An on-board computer provides the capability for programming a sequence of viewing modes and angles for later execution. Reception of SPOT data is possible at a network of ground stations positioned throughout the world, in a pattern similar to that shown in Figure 6.5. In fact, many of the stations listed in Figure 6.5 are equipped to receive SPOT data.

CNES processes SPOT data in collaboration with the *Institut Geographique National* (IGN); image archives are maintained by the *Centre de Rectification des Images Spa-*

*tiales* (CRIS), also operated by CNES and IGN. Four levels of processing are listed below in ascending order of precision:

- *Level 1:* Basic geometric and radiometric adjustments
  - *Level 1a:* Sensor normalization
  - *Level 1b:* Level 1a processing, with the addition of simple geometric corrections
- *Level 2:* Use of ground control points to correct image geometry; no correction for relief displacement
- *Level 3:* Further corrections using digital elevation models

In the United States, SPOT imagery can be purchased from SPOT Image Corporation (1897 Preston White Drive, Reston, VA 22091-4368; 703-620-2200).

### *SPOTs 4 and 5*

SPOT 4, planned for launch in 1997, and SPOT-5, scheduled for the year 2000, continue the SPOT program. A principal feature of the SPOT-4 mission will be the high-resolution visible and infrared (HRVIR) instrument, a modification of the HRV used for SPOTs 1–3.

HRVIR resembles the HRV, with the addition of a mid infrared band (1.58 to 1.75 μm), designed to provide capabilities for geologic reconnaissance, vegetation surveys, and survey of snow cover. Whereas the HRV's 10-m resolution band covered a panchromatic range from 0.51 to 0.73 μm, the HRVIR's 10-m band is positioned to provide spectral coverage identical to band 2 (0.61–0.68 μm). In addition, the 10-m band will be registered to match data from band 2, facilitating uses of the two levels of resolution in the same analysis.

SPOT 4 will carry two identical HRVIR instruments, each with the ability to point 27° to either side of the ground track, providing a capability to acquire data within a 460-km swath for repeat coverage or stereo. In its monospectral (M) mode, HRVIR will provide data in band 2's spectral range at 10-m resolution. In multispectral (×S) mode, the HRVIR will acquire four bands of data (1, 2, 3, and mid infrared) at 20-m resolution. Both ×S and M data are compressed on board the satellite, then decompressed on the ground to provide 8 bits.

SPOT 4 will carry auxiliary instruments, of which the VEGETATION (VGT) instrument, a joint project of several European nations, is the most significant for our discussion. VGT is a wide-angle radiometer designed for high radiometric sensitivity, broad areal coverage, and repetitive coverage to detect changes in spectral responses of vegetated surfaces. Swath width is 2,200 km, with repeat coverage on successive days at latitudes above 35°, and 3 out of every 4 days at the Equator.

The VGT instrument has a CCD linear array sensitive in four spectral bands designed to be compatible with the HRVIR. In direct (regional) mode, it provides a resolution of 1 km at nadir. In recording (worldwide observation) mode, each pixel corresponds to four of the 1-km pixels, formed by on-board processing. Worldwide observations mode can acquire data within the region between 60°N and 40°S latitude. Data from HRVIR and VGT can be added or mixed, using on-board processing, as required to meet requirements for specific projects.

SPOT 5, planned for launch in the year 2000, will carry an upgraded version of the HRVIR, is planned to acquire data at 5-m resolution, and can provide a capability for along-track stereo imagery. Stereo imagery at 5-m resolution is intended to provide the basis for compilation of large-scale topographic maps and data. The new instrument, the high-resolution geometrical (HRG), will also have the flexibility to acquire data using the same bands and resolutions as the SPOT 4 HRVIR, thereby providing continuity with earlier systems.

## 6.9. India Remote Sensing

After operating two coarse-resolution remote sensing satellites in the 1970s and 1980s, India began to develop multispectral remote sensing programs in the style of the Landsat system. During the early 1990s, two India remote sensing (IRS) satellites were in service, with additional vehicles planned for 1995 and 1996. IRS-1A (launched in 1988) and IRS-1B (launched in 1991) carry the LISS-I and LISS-II pushbroom sensors (Table 6.4). These instruments collect data in four bands: blue (0.45–0.52 μm), green (0.52–0.59 μm), red (0.62–0.68 μm), and near infrared (0.77–0.86 μm), creating images of 2,400 lines in each band. LISS-I provides resolution of 72.5 m in a 148-km swath, and LISS-II has 36.25-m resolution. Two LISS-II cameras acquire data from 74-km wide swaths positioned within the field of view of LISS-I (Figure 6.23), so that four LISS-II images cover the area imaged by LISS-I, with an overlap of 1.5 km in the cross-track direction, and about 12.76 km in the along-track direction. Repeat coverage is 22 days at the equator, with more frequent revisit capabilities at higher latitudes. Image data can be acquired within range of receiving stations in Shadnagar, India, and Norman, Oklahoma. In the United States, IRS imagery can be purchased from EOSAT (4300 Forbes Blvd., Lanham, MD 20706; 301-552-3762 or 800-344-9933).

LISS III is planned for use on the IRS-1C and IRS-1D missions, planned for initial launch in the interval 1995–1999. These systems will carry tape recorders, permitting acquisition of data outside the ranges of the receiving stations mentioned above. It will ac-

**TABLE 6.4. Spectral Limits for LISS-I and LISS-II (IRS-1A and IRS-1B)**

| | | Resolution | |
|---|---|---|---|
| Band | Spectral limits | LISS-I | LISS-II |
| 1 | Blue-green 0.45–0.52 μm | 72.5 m | 36.25 m |
| 2 | Green 0.52–0.59 μm | 72.5 m | 36.25 m |
| 3 | Red 0.62–0.68 μm | 72.5 m | 36.25 m |
| 4 | Near infrared 0.77–0.86 μm | 72.5 m | 36.25 m |

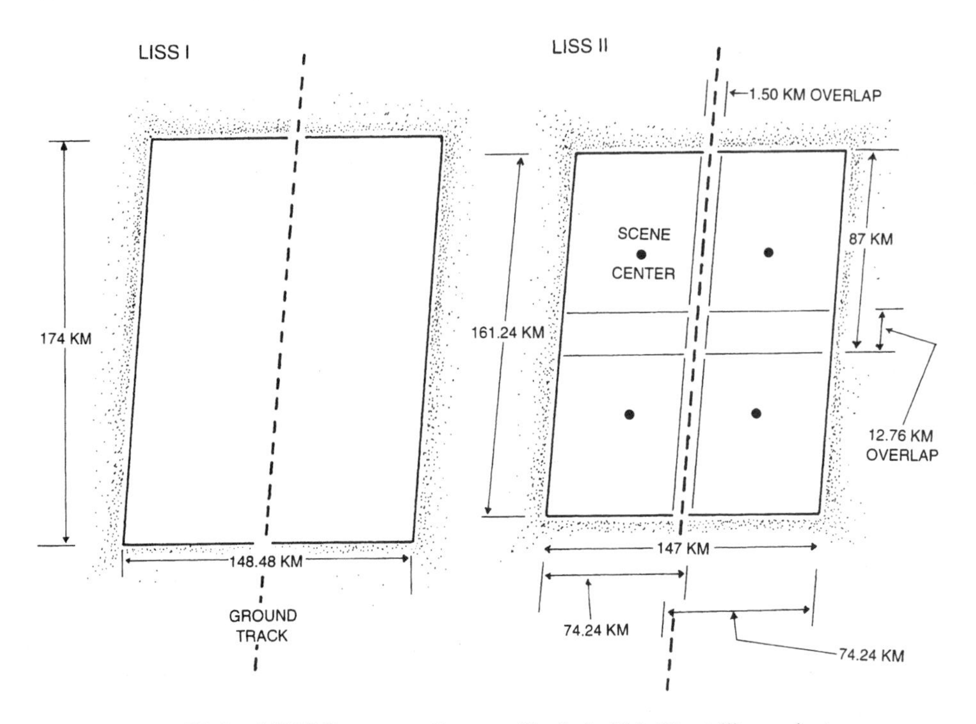


**FIGURE 6.23.** LISS-I and LISS-II coverage diagrams. The India IRS-1B satellite carries two sensors, a camera (LISS-I), and a pushbroom scanner (LISS-II) that each collect data in four spectral bands. Left: Coverage diagram for LISS-I. Right: Coverage diagram for LISS-II. Here four scenes represent an area slightly smaller than an LISS-I scene; the diagram depicts the four LISS-II scene centers (represented as dots) and the overlap between the four images. Diagram based on information provided by EOSAT.

quire data in four bands: green (0.52–0.59 µm), red (0.62–0.68 µm), near infrared (0.77–0.86 µm), and the shortwave infrared (1.55–1.70 µm). LISS-III will provide 23-m resolution for all bands, except for the shortwave infrared, which will have 70-m resolution (Table 6.5). A panchromatic band (0.5–0.75 µm) is planned to collect 10-m data within a 70-km swath. Swath width will be 142 km for bands 2, 3, and 4, and 148 km for band 5. The satellite will provide the capability for 24-day repeat coverage at the equator.

**TABLE 6.5. Characteristics of LISS-III (IRS-1C and IRS-1D)**

| Band | Spectral limits | Resolution |
|---|---|---|
| 1[a] | Blue — | |
| 2 | Green 0.52–0.59 µm | 23 m |
| 3 | Red 0.62–0.68 µm | 23 m |
| 4 | Near infrared 0.77–0.86 µm | 23 m |
| 5 | Mid infrared 1.55–1.70 µm | 70 m |

[a]Band 1 is not included in this instrument, although the numbering system from earlier satellites is maintained to provide continuity.

## 6.10. Computer Searches

Because of the unprecedented amount of data generated by earth observation satellite systems, it is necessary to use computerized data bases to index data by area, data, quality, cloud cover, and other qualities. Usually customers can access to this data base through requests submitted by mail, facsimile, phone, or electronic mail. In general, a user can obtain information from most vendors, free of charge, regarding availability of imagery although payment is required to acquire actual images.

The search of the image data base uses customer-supplied information concerning the area of interest, desirable dates of coverage, and the minimum quality of coverage (Figure 6.24). The result is a computer listing that provides a tabulation of all coverage meeting the constraints specified by the user. Some organizations can provide *quick-look* images—simplified versions of specific images that permit customers to examine an image to verify, prior to ordering data, the coverage, quality, and cloud cover of specific scenes. Some organizations provide access to the image archive and quick-look scenes by Internet (Chapter 4) and opportunities to establish users accounts that will permit electronic assess (for a fee) to image data.

A list of results of a search of the image data base will vary in length according to the nature of the user's request. A conservative, restrictive request closely tailored to the user's needs will result in a much shorter listing than will a loosely defined, broad request. Unless the user suspects that there is very little coverage of an area, it is usually best to define the request to closely match his or her needs, as it can be very tedious to search through a long list to identify the best images.

An example (Figure 6.25) illustrates information on the computer listing. In general, scenes are usually grouped geographically, with codes signifying the amount of cloud cover, image quality, and other characteristics. The analyst must become proficient in interpreting such a list to identify those scenes that will best serve a specific purpose. Because it is usually necessary to make compromises between competing qualities (date, coverage, quality, etc.), it is important that an analyst be intimately familiar not only with the technical qualities of the satellite system but also with the purposes and objectives of the specific project for which they will be used. Note that to prepare the request for a geographic computer search, and to interpret the resulting computer listing, it is necessary to have access to medium- or large-scale maps with accurate representation of latitude and longitude.

Because tabulations such as that shown in Figure 6.25 are difficult to interpret, analysts must plot information on a map, so that image coverage can be examined in relation to local landmarks (Figure 6.26). Computer programs have been written to automatically plot coverages of images in formats such as that shown in Figure 6.26; such programs greatly facilitate the examination of long lists of images.

Descriptive records such as those describing coverages of satellite images illustrate a class of data known as *metadata*. Metadata consist of descriptive summaries of other data sets (in this instance, the satellites images themselves). As increasingly large volumes of remotely sensed data are accumulated, the ability to search, compare, and examine metadata has become increasingly significant (Mather & Newman, 1995). In the context of remote sensing, metadata usually consist of text describing images—dates, spectral regions, quality ratings, cloud cover, geographic coverages, and so forth. An increasing number of

# SPOT Catalog Inquiry

| Licensee ID | Licensee Name |
|---|---|
| Your Reference Number (Optional) | Date |

**Instructions**

1. This form will be used by SPOT Image Corporation to conduct computer searches of the SPOT Catalog for available SPOT scenes meeting the Licensee's scene image parameters.
2. A list of available scenes will be mailed to the Licensee.
3. A geographic search can be made by identifying the latitudes and longitudes of a central point or area rectangle or by specifying the Grid Reference System (GRS) K,J coordinates of a central point or area rectang
4. Mail the completed "SPOT Catalog Inquiry" form to: SPOT Image Corporation
   1897 Preston White Drive
   Reston, VA 22091-4326

**I. SPOT SCENE IMAGE PARAMETERS***

| Time Period of Acquisition | Spectral Mode | Maximum Cloud Cover |
|---|---|---|
| ☐ Earliest Date: ____ / ____ / ____ | ☐ Multispectral | ☐ 10% |
| ☐ Latest Date: ____ / ____ / ____ | ☐ Panchromatic | ☐ 25% |
| ☐ All Coverage | | ☐ > 25% |

* These parameters will be used for all searches defined below.

**II. LATITUDE/LONGITUDE-SPECIFIED SEARCHES**

Single Point Coverage

| Point 1 | Point 2 | Point 3 |
|---|---|---|
| Latitude ____ ° ____ ' N / S | Latitude ____ ° ____ ' N / S | Latitude ____ ° ____ ' N / S |
| Longitude ____ ° ____ ' E / W | Longitude ____ ° ____ ' E / W | Longitude ____ ° ____ ' E / W |

Rectangle Coverage

(Rectangle with corners 1, 2, 3, 4)

| | |
|---|---|
| **1** Latitude ____ ° ____ ' N / S<br>Longitude ____ ° ____ ' E / W | **3** Latitude ____ ° ____ ' N / S<br>Longitude ____ ° ____ ' E / W |
| **2** Latitude ____ ° ____ ' N / S<br>Longitude ____ ° ____ ' E / W | **4** Latitude ____ ° ____ ' N / S<br>Longitude ____ ° ____ ' E / W |

**III. GRS COORDINATE-SPECIFIED SEARCHES**

Single Scene Searches

| 1 | 2 | 3 | 4 |
|---|---|---|---|
| K ____ , J ____ | K ____ , J ____ | K ____ , J ____ | K ____ , J ____ |

Contiguous Scene Searches

| 1 | 2 | 3 | 4 |
|---|---|---|---|
| K ____ to K ____ | K ____ to K ____ | K ____ to K ____ | K ____ to K ____ |
| J ____ to J ____ | J ____ to J ____ | J ____ to J ____ | J ____ to J ____ |

**FIGURE 6.24.** Request for computer search.

```
                DATA FORMAT FOR SICORP CATALOG QUERY

     S#  GRS(K,J)     A-DATE       A-TIME      H#  SM     CC      TQ     GAIN
     ORIEN         INCID        AZIM         ELEV                        CTR

                   UL CNR                                                UR CNR
                   LL CNR                                                LR CNR

       NUMBER OF CATALOG SCENES THAT MET INQUIRY CRITERIA  =  6


1 617     275    86  08        07  16       12     50       1     X     2121 E NNN
10.3        -08.4        -136.6       -063.9        N0372052             W0801250
N0373936          W0802835                          N0373342             W0794755
N0370759          W0803749                          N0370208             W0795726

1 617     275    87  04        08  16       19     17       2     X     1100 E NNN
11.1        +03.6         150.1        056.1        N0372052             W0802243
N0373945          W0803733                          N0373324             W0795738
N0370815          W0804733                          N0370156             W0800755

1 617     276    86  11        29  16       20     28       2     X     0000 E NNN
11.1        +03.6         166.5        030.7        N0365128             W0803331
N0371021          W0804817                          N0370402             W0800837
N0363849          W0805810                          N0363233             W0801846

1 617     276    87  04        08  16       19     26       2     X     0011 E NNN
11.0        +03.6         149.6        056.9        N0365128             W0803208
N0371021          W0804653                          N0370402             W0800714
N0363849          W0805648                          N0363233             W0801724

1 618     275    86  06        26  16       20     25       1     X     0012 E NNN
11.7        +08.6         132.8        071.0        N0372052             W0793649
N0373952          W0795144                          N0373311             W0791110
N0370826          W0800204                          N0370147             W0792146

1 619     275    86  07        12  16       12     43       1     X     2221 E NNN
10.6        +02.3         130.2        068.6        N0372052             W0791722
N0373939          W0793226                          N0373335             W0785234
N0370805          W0794204                          N0370204             W0790228
```

**FIGURE 6.25.** Results of computer search for satellite imagery.

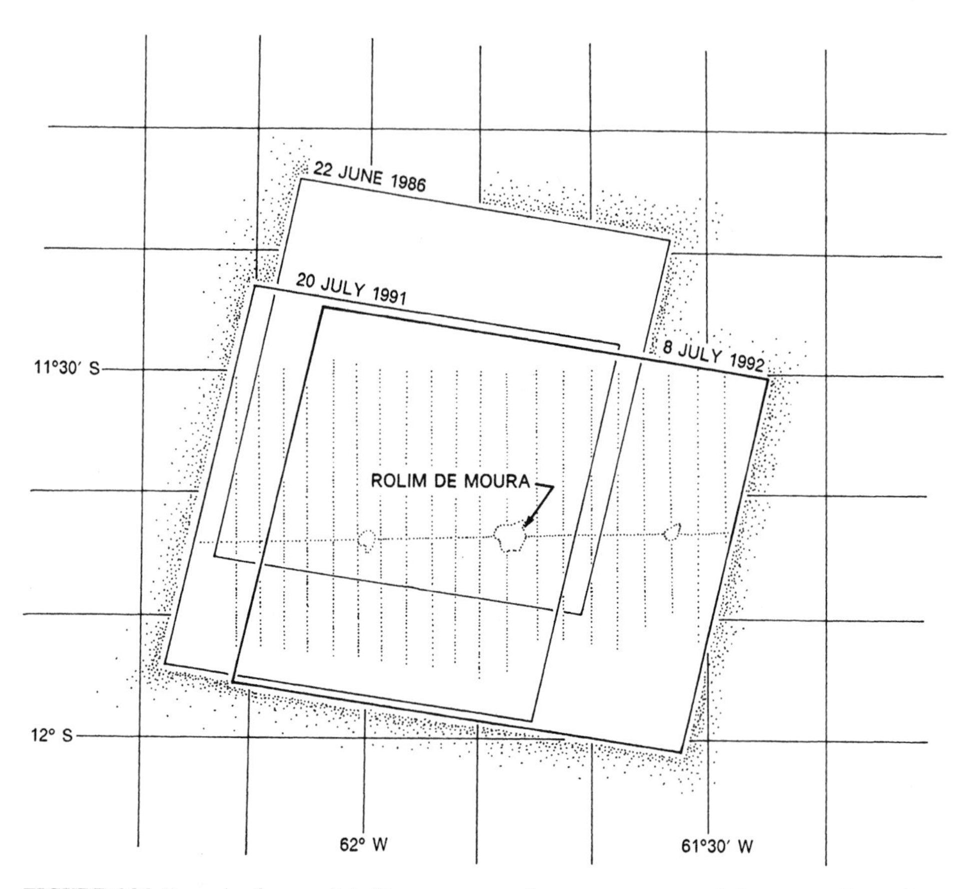


**FIGURE 6.26.** Example of a map plot of image coverage from a computer search. Image coordinates from a listing such as that shown in Figure 6.25 have been plotted in map form to show coverages of several images in relation to local landmarks (in this instance, western Brazil). This kind of display of data, now available using computer programs, assists efficient evaluation of alternative images for a given region.

efforts are under way to develop computer programs and communication systems that will link data bases together to permit searches of multiple archives.

## 6.11. Microsatellites

Landsat, SPOT, and other systems established the technical and commercial value of the land observation satellite concept in the 1970s and 1980s. But they also revealed the high costs and technical challenges of designing and operating a system intended to provide general-purpose data for a broad community of users who may have diverse requirements. During the late 1980s and early 1990s, interest increased among those who believed it would be feasible to operate smaller, special-purpose earth observation satellites focused on the requirements of specific groups of users, thereby decreasing the costs of operation.

For example, the broad areal and multispectral coverage of Landsat and SPOT data might be sacrificed for much finer resolution (perhaps as fine as 1 m) in a single panchromatic band focused on a small area selected by a specific customer. In addition, it might be possible to link imaging technology with global positioning systems (Chapter 12) and data relay systems to integrate image data with observations acquired directly in the field. This kind of approach to satellite remote sensing generated a multitude of proposals for small satellite systems tailored for specific markets and purposes. At the time of this writing it is not yet clear which of these concepts will prove to be feasible, but if only a few of the ideas that have been discussed are implemented, the nature of satellite remote sensing will be changed greatly.

## 6.12. CORONA

CORONA is the project designation for the satellite reconnaissance system operated by the United States during the interval 1960–1972. CORONA provided photographic imagery that was interpreted to provide strategic intelligence on the activities of Soviet industry and strategic forces. For many years this imagery and details of the system were closely held as national security secrets. By 1995, when the Soviet threat was no longer present, and CORONA had been replaced by more advanced systems, President Clinton announced that CORONA imagery would be declassified and released to the public. MacDonald (1995) and Ruffner (1995) provide detailed descriptions of the system. Curran (1985) provides an overview based on open sources available prior to the declassification.

### *Historical Context*

When Dwight Eisenhower became President of the United States in 1953, he was distressed to learn of the rudimentary character of the nation's intelligence estimates concerning the military stature of the Soviet Union. Within the Soviet sphere, closed societies and rigid security systems denied Western nations the information required to prepare reliable estimates of their military capabilities. As a result, Eisenhower feared the nation faced the danger of developing policies based on speculation or political dogma, rather than on estimates of the actual situation.

His concern led to a priority program to build the U-2 system, a high-altitude reconnaissance aircraft designed to carry a camera system of unprecedented capabilities. The first flight occurred in the summer of 1956. Although U-2 flights imaged only a small portion of the Soviet Union, they provided information that greatly improved estimates of Soviet capabilities. The U-2 program was intended only as a stopgap measure, as it was anticipated that the Soviet Union would develop countermeasures that would prevent long-term use of the system. In fact, the flights continued for about 4 years, ending in 1960 when a U-2 was shot down near Sverdlovsk, in the Soviet Union. This event ended use of the U-2 over the Soviet Union, although this aircraft continued to form a valuable asset for observing other regions of the world and later for collecting imagery for environmental analyses in the civilian sphere.

At the time of the U-2 incident, work was already under way to design a satellite system that would provide photographic imagery from orbital altitudes, thereby avoiding risks to pilots and the controversy arising from overflights of another nation's territory. Although CORONA had been conceived beforehand, the Soviet Union's launch of the first artificial earth-orbiting satellite in October 1957 increased the urgency of the effort.

The satellite was designed and constructed as a joint effort of the U.S. Air Force and the Central Intelligence Agency (CIA), in collaboration with private contractors who designed and manufactured launch systems, satellites, cameras, and films. Virtually every element of the system extended technical capabilities beyond their known limits. Each major component encountered difficulties and failures that were, in time, overcome to eventually produce an effective, reliable system. During the interval June 1959–December 1960, the system experienced a succession of failures, first with the launch system, then with the satellite and recovery system, and finally with the cameras and films. Each problem was identified and solved within a remarkably short time, so that CORONA was, in effect, operational by August 1959, only 3 months after the end of the U-2 flights over the Soviet Union. Today, CORONA's capabilities may seem commonplace, as we are familiar with artificial satellites, satellite imagery, and high-resolution films. However, at the time, CORONA extended reconnaissance capabilities far beyond even what technical experts thought might be feasible.

As a result, the existence of the program was a closely held secret for many years. Although the Soviet Union knew the effort was under way, it chose not to publicize the fact that it had knowledge of the system. Details of the system, the launch schedule, nature of the imagery, and the ability of interpreters to derive information from the imagery were closely guarded secrets, in part to deny Soviets the capabilities of the system, which might have permitted development of effective measures for deception. In time, the existence of the program became more widely known in the United States, although details of the system's capabilities were still secret.

CORONA is the name given to the satellite system, whereas the camera systems carried KEYHOLE (KH) designations that were familiar to those who used the imagery. The designations KH-1, KH-4, refer to different models of the cameras used for the CORONA program. Most of the imagery was acquired with KH-4 and KH-5 systems (including KH-4A, and KH-4B).

### *Satellite and Orbit*

The reconnaissance satellites were placed into near-polar orbits, with, for example, an inclination of 77°, apogee of 502 mi., and perigee of 116 mi. Initially missions would last only 1 day; by the end of the program, missions extended for 16 days. Unlike the other satellite systems described here, images were returned to earth by a capsule (satellite recovery vehicle) ejected from the principal satellite at the conclusion of each mission. The recovery vehicle was designed to withstand the heat of reentry into the earth's atmosphere and to deploy a parachute at an altitude of about 60,000 ft. The capsule was then recovered in the air by specially designed aircraft. Recoveries were planned for the Pacific Ocean near Hawaii. Capsules were designed to sink if not recovered within a few days, to prevent recovery by other nations in the event of malfunction. Several capsules were in fact lost when this system failed. Later models of the satellite were designed with two capsules, which extended the length of CORONA missions to as long as 16 days.

### *Cameras*

Camera designs varied as problems were solved and new capabilities were added. MacDonald (1995) and Ruffner (1995) provide details of the evolution of the camera systems; here the description focuses on the main features of the later models that acquired much of the imagery in the CORONA archive. The basic camera design, the KH-3, manufactured by the Itek Corporation, was a vertically oriented panoramic camera with a 24-in. focal length. The camera's 70° panoramic view was acquired by a mechanical motion of the system at right angles to the line of flight. Image motion compensation (Chapter 3) was employed to correct effects of satellite motion relative to the earth's surface during exposure.

The KH-4 camera acquired most of the CORONA imagery (Figure 6.27). The KH-4B (sometimes known as the *MURAL* system) consisted of two KH-3 cameras oriented to observe the same area from different perspectives, thereby providing a stereo capability (Figure 6.28). One pointed 15° forward along to flight path, the other was aimed 15° aft. A small-scale index image provided the context for proper orientation of the panoramic imagery, stellar cameras viewed the pattern of stars, and horizon cameras viewed the earth's horizon to ensure correct orientation of the spacecraft and the panoramic cameras.

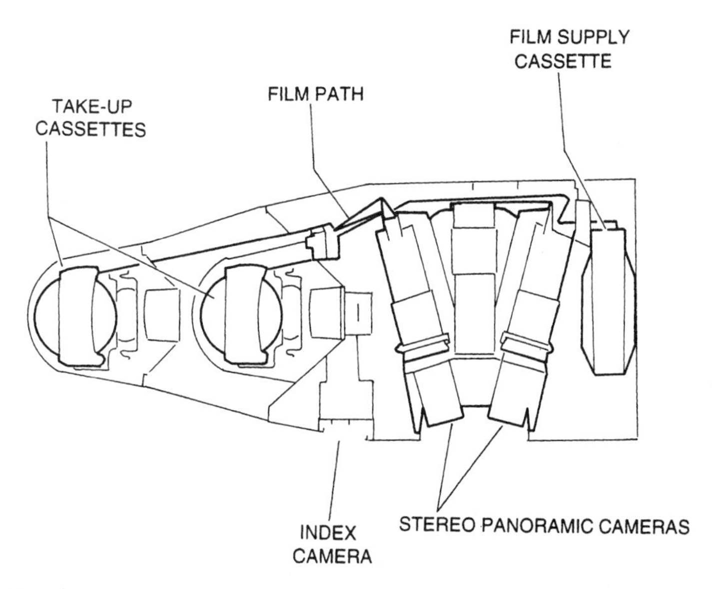


**FIGURE 6.27.** Line drawing of major components of the KH-4B camera (based on MacDonald, 1995). The two panoramic cameras are pointed forward and aft along the ground track to permit acquisition of stereo coverage as shown in Figure 6.28. Each of the two take-up cassettes could be ejected sequenctially to return exposed film to earth. The index camera, as shown in Figure 6.28, provided a broad-scale overview to assist in establishing the context for coverage from the panoramic cameras.

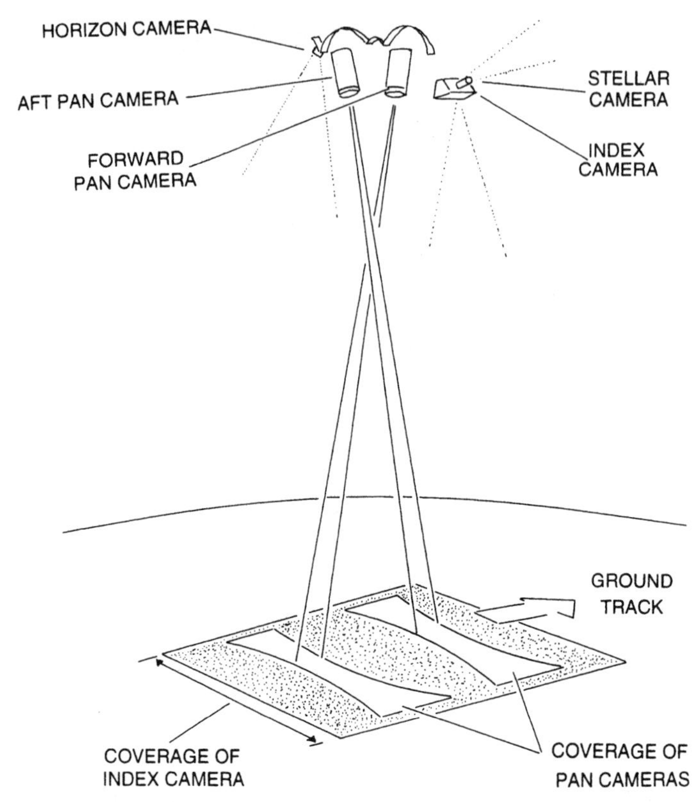


**FIGURE 6.28.** Coverage of the KH-4B camera system. The two stereo panoramic cameras point forward and aft along the ground track, so a given area can be photographed first by the aft camera (pointing forward), then by the forward camera (pointing aft) (see Figure 6.27). Each image from the panoramic camera represents an area about 134.8 mi. (216.8 km) long and about 9.9 mi. (15.9 km) wide at the image's greatest width. The index camera provides a broad-scale overview of the coverage of the region; the stellar and horizon cameras provide information to maintain satellite stability and orientation. Based on MacDonald (1995).

A related system, ARGON, acquired photographic imagery to compile accurate maps of the Soviet Union, using cameras with 3-in. and 1.5-in. focal lengths.

### *Imagery*

The CORONA archive consists of about 866,000 images, totaling about 400 mi. of film, acquired during the interval August 1960–May 1972. Although film format varied, most imagery is recorded on 70-mm film in strips about 25 or 30 in. long. Actual image width is typically either 4.4 in. wide or 2.5 in. wide. Almost all CORONA imagery is recorded on black-and-white panchromatic film, although a small portion used a color infrared emulsion and some is in a natural color emulsion. The earliest imagery is said to have a resolution of about 40 ft.; by the end of the program in 1972, the resolution was as fine as

6 ft., although details varied greatly depending on atmospheric effects, illumination, and the nature of the target (Figure 6.29).

The imagery is indexed to show coverage and dates. About 50% is said to be obscured by clouds, although the index does not record which scenes are cloud covered. Most of the early coverage was directed at the Soviet Union (Figure 6.30), although later coverage often focused on other areas of current or potential international concern.

### *National Photographic Interpretation Center*

Interpretation of CORONA imagery was largely centralized at the National Photographic Interpretation Center (NPIC), maintained by the CIA in a building in the Navy Yard in southeastern Washington, DC. Here teams of photointerpreters from the CIA, armed forces, and allied governments were organized to examine imagery to derive intelligence pertaining to the strategic capabilities of the Soviet Union and its satellite nations. Film from each mission was immediately interpreted to provide immediate reports on topics of current significance.

Then imagery was reexamined to provide more detailed analyses of less urgent developments. In time, the system accumulated an archive that provided retrospective record of the development of installations, testing of weapons systems, and operations of units in the field. By this means image analysts could trace the development of individual military and industrial installations, recognize unusual activities, and develop an understanding of

**FIGURE 6.29.** Severodvinsk shipyard (on the White Sea coastline, near Archangel, Russia, formerly USSR) as imaged by KH-4B camera 10 February 1969. This image is much enlarged from the original image. Image courtesy of the CIA.

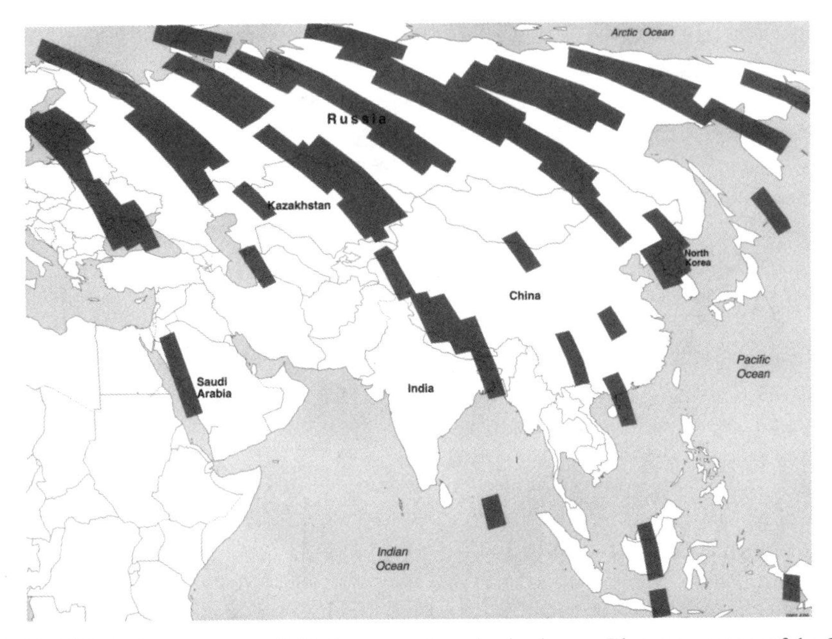


**FIGURE 6.30.** Typical coverage of KH-4B camera, Eurasian land mass. Diagram courtesy of the CIA.

the strategic infrastructure of the Soviet Union. Although the priority was on understanding the strategic capabilities of Soviet military forces, CORONA imagery was also used to examine agricultural and industrial production, environmental problems, mineral resources, population patterns, and other facets of the Soviet Union's economic infrastructure.

Reports from NPIC photointerpreters formed only one element of the information considered by analysts as they prepared intelligence estimates, based not only on photography but also on such other sources as electronic signals, reports from observers, and press reports. However, CORONA photographs came to assume an important role because of ready availability, reliability, and timeliness.

### *Availability*

Until 1995, the CIA maintained the CORONA archive, which was released to the National Archives and Records Administration and to the US. Geological Survey after declassification. The National Archives retain the original negatives, and provide copies for purchase by the public, through the services of contractors. The USGS will have duplicate negatives to provide imagery to the public and will maintain a digital index as part of the Global Land Information System (Chapter 19), accessible by the Internet (Chapter 4):

http://edcwww.cr.usgs.gov/declass/dclass.html

Other organizations may decide to purchase imagery for resale to the public, possibly in digitized form, or to sell selected scenes of special interest.

### *Uses*

CORONA provided the first satellite imagery of the earth's surface, and thus extended the historical record of satellite imagery into the late 1950s, about 10 years before Landsat. Therefore, it may be able to assist in assessment of environmental change, trends in human use of the landscape, and similar phenomena. Plate 7 shows the Aral Sea as observed over the interval 1962–1994. The left-hand image shows the Aral Sea (formerly in the Soviet Union, now at the border between Kazakhstan and Uzbekistan) as recorded by CORONA imagery in 1962. The right-hand image shows the same area, at the same scale, as recorded by AVHRR imagery (Chapter 16) in 1994, clearly revealing the dramatic difference in extext. Located in an arid region, the Aral Sea receives little moisture from local sources; it is fed by melting snow and ice in mountains far to the south. Since 1962, increased withdrawal of irrigation water from the rivers that flow into the Aral Sea has produced the dramatic changes seen here.

Further, CORONA imagery records many of the pivotal events of the Cold War and therefore may form an important source for historians who wish to examine and reevaluate these events more closely than was possible before imagery was released. Ruffner (1995) provides examples that show archaeological and geologic applications of CORONA imagery, and other applications may not yet be identified.

However, today's interpreters of this imagery may face some difficulties. Many of the original interpretations depended not only on the imagery itself but also on the skill and experience of the interpreters (who often were specialized in identification of specific kinds of equipment), access to collateral information, and availability of specialized equipment. Further, many critical interpretations depended on the expertise of interpreters with long experience in analysis of images of specific weapons systems—experience that is not readily available to today's analysis. Thus, it may be difficult to reconstruct or reevaluate interpretations of earlier analysts.

## 6.13. Summary

Satellite observation of the earth has greatly altered the field of remote sensing. Since the launch of Landsat 1, a larger and more diverse collection of scientists than ever before have conducted remote sensing research an applications. Public knowledge of, and interest in, remote sensing has increased. Digital data for satellite images has contributed greatly to the growth of image processing, pattern recognition, and image analysis (Chapters 10–13). Satellite observation systems have increased international cooperation through joint construction and operation of ground receiving stations and through collaboration in training scientists.

This chapter forms an important part of the development of topics presented in subsequent chapters. Much of the information presented here in relation to Landsat is equally important as a basis for understanding other satellite systems that operate in the microwave (Chapter 7) and far infrared (Chapter 8) regions of the spectrum.

Finally, it can be noted that the discussion thus far has emphasized acquisition of satellite data. Little has been said about analysis of these data and their applications to specific fields of study. Both topics are covered in subsequent chapters (Chapters 9–12 and 13–17, respectively).

## Review Questions

1. Outline the procedure for identifying and ordering SPOT, IRS, or Landsat images for a study area near your home. Identify for each step the *information* and the *materials* (maps, etc.) necessary to complete that step and proceed to the next. Can you anticipate some of the difficulties you might encounter?

2. In some instances it may be necessary to form a mosaic of several satellite scenes, by matching several images together at the edges. List some of the problems you expect to encounter as you prepare such a mosaic.

3. What are some of the advantages (relative to use of aerial photography) of using satellite imagery? Can you identify disadvantages?

4. Manufacture, launch, and operation of earth observation satellites is a very expensive undertaking—so large that it requires the resources of a national government to support the many activities necessary to continue operation. Many people question whether it is necessary to spend government funds for earth resource observation satellites and have other ideas for use of these funds. What arguments can you give to justify the costs of such programs?

5. Why are orbits of land observation satellites so low relative to those of communications satellites?

6. Would it be feasible to design an earth observation satellite with a sun-synchronous orbit to provide coverage of the poles? Explain.

7. Discuss problems that would arise as engineers attempt to design multispectral satellite sensors with smaller and smaller pixels. How might some of these problems be avoided?

8. Can you suggest some of the factors that might be considered as scientists select the observation time (local sun time) for a sun-synchronous earth observation satellite?

9. Earth observation satellites do not continuously acquire imagery, but only those individual scenes as instructed by mission control. List factors that might be considered in planning scenes to be acquired during a given week. Design a strategy for acquiring satellite images worldwide, specifying rules for deciding which scenes are to be given priority.

10. Using information given in the text, calculate the number of pixels for a single band of an MSS scene, for a TM scene, and for a SPOT HRV image. (For SPOT assume the image is acquired at nadir.) Recompute the numbers to include all bands available for each sensor.

11. Estimate the number of aerial photographs at a scale of 1:15,470 that would be required to

show the land area represented on a single SPOT scene. Assume end lap of 20% and side lap of 10%.

12. Explain why a satellite image and an aerial mosaic of the same ground area are not equally useful, even though image scale might be the same.
13. How many pixels are required to represent a complete SPOT scene (one band only)? A single band of a TM scene?
14. Prepare a template showing (at the correct scale) the dimensions of MSS, TM, and SPOT pixels for an aerial photograph of a nearby area, or other images provided by your instructor. Position the template at various sites throughout the aerial photograph, and assess the effectiveness of the sensors in recording various components of the landscape, including forested land, agricultural land, urban land, etc. (If pixels are composed of only a single category or feature, they tend to be recorded more effectively than if pixels are composed of two or more classes.)
15. On a small-scale map (such as a road map, or similar map provided by your instructor) plot at the correct scale the outlines of an MSS scene centered on a nearby city. How many different counties are covered by this area?

## References

American Society for Photogrammetry and Remote Sensing. 1995. *Conference Proceedings, Land Satellite Information in the Next Decade.* Bethesda, MD: Author, 149 pp.

Arnaud, M. 1995. The SPOT Programme. Chapter 2 in *TERRA 2: Understanding the Terrestrial Environment* (P. M. Mather, ed.). New York: Wiley, pp. 29–39.

Begni, Gerard. 1982. Selection of the Optimum Spectral Bands for the SPOT Satellite. *Photogrammetric Engineering and Remote Sensing,* Vol. 48, pp. 1613–1620.

Brugioni, Dino. A. 1991. *Eyeball to Eyball: The Inside Story of the Cuban Missle Crisis.* New York: Random House, 622 pp.

Brugioni, Dino. A. 1996. The Art and Science of Photoreconnaissance. *Scientific American,* Vol. 274, pp. 78–85.

Chevrel, M., M. Courtois, and G. Weill. 1981. The SPOT Satellite Remote Sensing Mission. *Photogrammetric Engineering and Remote Sensing,* Vol. 47, pp. 1163–1171.

Curran, Paul J. 1985. *Principles of Remote Sensing.* New York: Longman, 282 pp.

*EOSAT Notes.* Newsletter published by EOSAT, 4300 Forbes Blvd., Lanham, MD 20706-9954.

General Electric Company. no date. *Data Users Handbook.* Philadelphia: Space Division, Author.

General Electric Company. no date. *Landsat 3 Reference Manual.* Philadelphia: Space Division, Author.

MacDonald, Robert A. 1995. CORONA: Success for Space Reconnassance, a Look into the Cold War, and a Revolution for Intelligence. *Photogrammetric Engineering and Remote Sensing,* Vol. 61, pp. 689–720.

Mack, Pamela. 1990. *Viewing the Earth: The Social Constitution of the Landsat Satellite System.* Cambridge, MA: MIT Press, 270 pp.

Mather, P. M., and I. A. Newman. 1995. U.K. Global Change Federal Metadata Network. Chapter 9 in *TERRA 2: Understanding the Terrestrial Environment* (P. M. Mather, ed.). New York: Wiley, pp. 103–111.

McClain, E. Paul. 1980. Environmental Satellites. Entry in *McGraw-Hill Encyclopedia of Environmental Science.* New York: McGraw-Hill.

Morain, Stanley A., and A. M. Budge. 1995. *Earth Observing Platforms and Sensors CD-ROM.* Bethesda, MD: American Society for Photogrammetry and Remote Sensing.

Ruffner, Kevin C. (ed.). 1995. *CORONA: America's First Satellite Program.*Washington, DC: Center for the Study of Intelligence, 360 pp.

Salomonson, V. V., J. R. Irons, and D. L. Williams. 1995. The Future of Landsat: Implications for Commercial Development. *American Institute for Physics,* Proceedings 325 (M. El-Genk and R. P. Whitten, eds.), pp. 353–359.

Satellite Technology Serving Earth. *Aviation Week and Space Technology,* 11 October 1977, pp. 15–30.

Slater, Philip N. 1979. A Re-examination of the Landsat MSS. *Photogrammetric Engineering and Remote Sensing,* Vol. 45, pp. 1479–1485.

Sheffield, Charles. 1981. *Earth Watch: A Survey of the World from Space.* New York: Macmillan, 160 pp.

Sheffield, Charles. 1983. *Man on Earth: How Civilization and Technology Changed the Face of the World—A Survey from Space.* New York: Macmillan, 166 pp.

Sheffner, Edwin, J. 1994. The Landsat Program: Recent History and Prospects. *Photogrammetric Engineering and Remote Sensing,* Vol. 60, pp. 735–744.

Short, Nicholas M. 1976. *Mission to Earth: Landsat Views the World.* Washington, DC: NASA, 459 pp.

Slater, Philip N. 1980. *Remote Sensing: Optics and Optical Systems.* Reading, MA: Addison-Wesley, 575 pp.

*SPOTLIGHT.* Newsletter published by SPOT Image Corporation, 1897 Preston White Drive, Reston, VA 22091-4368.

Taranick, James V. 1978. *Characteristics of the Landsat Multispectral Data System.* USGS Open File Report 78-187, 76 pp.

CHAPTER SEVEN

# Active Microwave

## 7.1. Introduction

The microwave region of the electromagnetic spectrum extends from wavelengths of about 1 mm to about 1 m. This region is, of course, far removed from those in and near the visible spectrum where our direct sensory experience can assist in interpretation of images and data. Thus, formal understanding of the concepts of remote sensing is vital to understanding imagery acquired in the microwave region. As a result, the study of microwave imagery is often a difficult subject for beginning students and requires more attention than is usually necessary for the study of other regions of the spectrum. The family of sensors discussed here (Figure 7.1) includes *active* microwave sensors (imaging radars carried by either aircraft or satellites). These are active sensors that illuminate the ground with their own energy, then record a portion of the energy reflected back to the instrument. *Passive* microwave sensors (instruments sensitive to microwave energy emitted from the earth's surface) are discussed in Chapter 8.

### *Active Microwave Sensors*

Active microwave sensors are *radars*—instruments that transmit a microwave signal, then receive its reflection as the basis for forming images of the earth's surface. The rudimentary components of an imaging radar system include a transmitter, receiver, antenna array, and recorder (Figure 7.2). A *transmitter is* designed to transmit repetitive pulses of microwave energy at a given frequency. A *receiver* accepts the reflected signal as received by the antenna, then filters and amplifies it as required. An *antenna array* is designed to transmit a narrow beam of microwave energy. Such an array is composed of *waveguides*—devices that control the propagation of an electromagnetic wave such that the waves follow a path defined by the physical structure of the guide. (A simple waveguide might be formed from a hollow metal tube.) Usually the same antenna is used both to transmit the radar signal and to receive its echo from the terrain. Finally, a *recorder* records and/or displays the signal as an image. Numerous refinements and variations of these basic components are possible; a few are described here in greater detail.

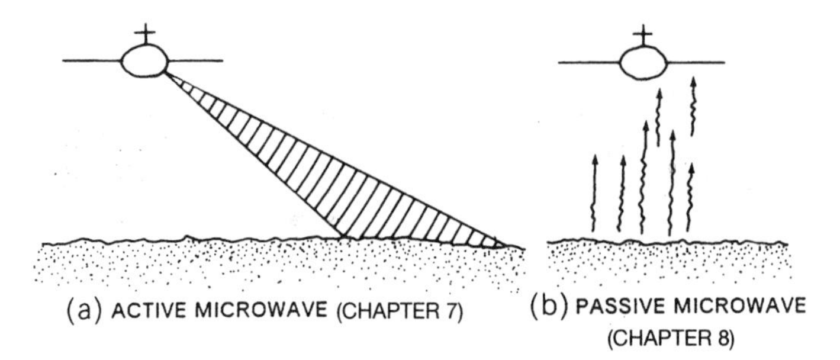


**FIGURE 7.1.** Active and passive microwave remote sensing. (*a*) Active microwave sensing, using energy generated by the sensor, as described in this chapter. (*b*) Passive microwave sensing, which detects energy emitted by the earth's surface (Chapter 8).

### *Side-Looking Airborne Radar*

*Radar* is an acronym for "radio detection and ranging." The "ranging" capability is achieved by measuring the time delay from the time a signal is transmitted to the terrain until its echo is received. Through its ranging capability, possible only with active sensors, radar can accurately measure the distance from the antenna to features on the ground. A second unique capability, also a result of radar's status as an active sensor, is its ability to detect frequency and polarization shifts. Because the sensor transmits a signal of known wavelength, it is possible to compare the received signal with the transmitted signal. From such comparisons imaging radars detect changes in frequency that form the basis of capabilities not possible with other sensors.

*Side-looking airborne radar* (SLAR) imagery is acquired by an antenna array aimed to the side of the aircraft, so that it forms an image of a strip of land parallel to, and at some distance from, the ground track of the aircraft (Figure 7.3). The resulting image geometry differs greatly from those of other remotely sensed images. These qualities es-

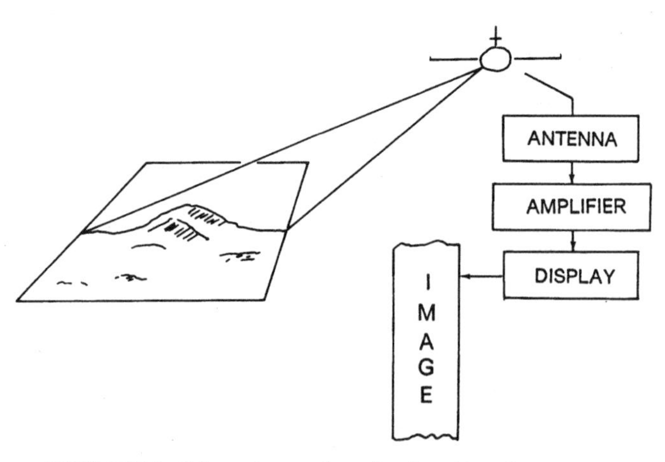


**FIGURE 7.2.** Schematic overview of an imaging radar system.

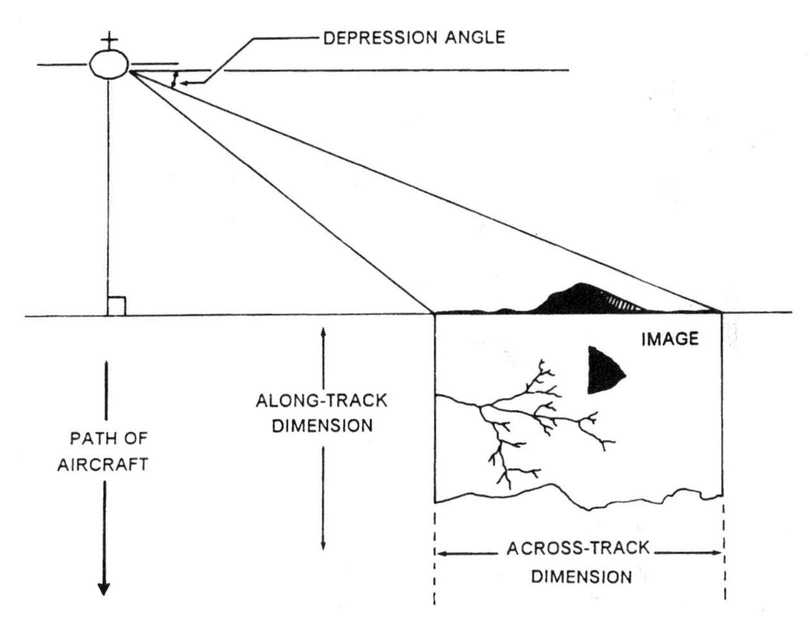


**FIGURE 7.3.** Schematic diagram of a radar image.

tablish radar imagery as a distinctive form of remote sensing imagery. One of SLAR's most unique and useful characteristics is its ability to function during inclement weather. SLAR is often said to possess an "all-weather" capability, meaning that it can acquire imagery in all but the most severe weather conditions. The microwave energy used for SLAR imagery is characterized by wavelengths long enough to escape interference from clouds and light rain (although not necessarily from heavy rainstorms). Because SLAR systems are independent of solar illumination, missions can be scheduled at night or during early morning or evening hours when solar illumination might be unsatisfactory for acquiring aerial photography. This advantage is especially important for imaging radars carried by the earth-orbiting satellites described later in this chapter.

Radar images typically provide crisp, clear representations of topography and drainage (Figure 7.4). Despite the geometric errors described later, radar images typically provide good positional accuracy, and in some areas they have formed the basis for small-scale base maps depicting major drainage and terrain features. Analysts have registered TM and SPOT data to radar images to form composites of the two quite different kinds of information. Some of the most successful operational applications of SLAR imagery have occurred in tropical climates, where persistent cloud cover has prevented acquisition of aerial photography, and where shortages of accurate maps create a context in which radar images may provide cartographic information superior to that on existing conventional maps. Another important characteristic of SLAR imagery is its synoptic view of the landscape. SLAR's ability to clearly represent the major topographic and drainage features within relatively large regions at moderate image scales forms a valuable addition to our selection of remote sensing imagery. Furthermore, because it is acquired in the microwave spectrum, it may show detail and information that differ greatly from that of sensors operating in the visible and near infrared spectra.

**FIGURE 7.4.** Radar image of region near Chattanooga, Tennessee, September 1985 (X-band, HH polarization). This image has been processed to produce pixels of about 11.5 m in size. Image courtesy of U.S. Geological Survey.

### *Origins and History*

The foundations for imaging radars were laid by scientists who first investigated the nature and properties of microwave and radio energy. James Clerk Maxwell (1831–1879) first defined essential characteristics of electromagnetic radiation; his mathematical descriptions of the properties of magnetic and electrical fields prepared the way for further theoretical and practical work. In Germany, Heinrich R. Hertz (1857–1894) confirmed much of Maxwell's work, and further studied properties and propagation of electromagnetic energy in microwave and radio portions of the spectrum. The Hertz, the unit for designation of frequencies (Chapter 2), is named in his honor. Hertz was among the first to demonstrate the reflection of radio waves from metallic surfaces and thereby begin research that led to development of modern radios and radars. In Italy, Guglielmo M. Marconi (1874–1937) continued the work of Hertz and other scientists, in part by devising a practical antenna suitable for transmitting and receiving radio signals. In 1895 he demonstrated the practicability of the wireless telegraph. After numerous experiments over shorter distances, he demonstrated in 1901 the feasibility of long-range communications by sending signals across the Atlantic and, in 1909, he shared the Nobel prize in physics. Later he proposed that ships could be detected by the reflection of radio waves, but there is no evidence that his suggestion influenced the work of other scientists.

The formal beginnings of radar date from 1922, when A. H. Taylor and L. C. Young, civilian scientists working for the U.S. Navy, were conducting experiments with high-frequency radio transmissions. Their equipment was positioned near the Anacostia River near

Washington, D.C., with a transmitter on one bank of the river and a receiver on the other. They observed that the passage of a river steamer between the transmitter and the receiver interrupted the signal in a manner that clearly revealed the potential of radio signals as a means of detecting the presence of large objects (Figure 7.5). Taylor and Young recognized the significance of their discovery for marine navigation in darkness and inclement weather and, in a military context, the potential for detection of intruding vessels. Initial efforts to implement this idea depended on the placement of transmitters and receivers at separate locations, so that a continuous microwave signal was reflected from an object, then recorded by a receiver placed some distance away. Designs evolved so that a single instrument contained both the transmitter and the receiver at a single location, integrated in a manner that permitted use of a pulsed signal to be reflected from the target back to the same antenna that transmitted the signal. Such instruments were first devised during the interval 1933–1935 more or less simultaneously in the United States (by Young and Taylor), Great Britain, and Germany. Many credit Sir Robert Watson-Watt, a British inventor, as the inventor of the first radar system, although others have given credit to Young and Taylor.

Subsequent improvements were based mainly on refinements in the electronics required to produce high-power transmissions over narrow wavelength intervals, to carefully time short pulses of energy, and to amplify the reflected signal. These and other developments led to rapid evolution of radar systems during the years preceding World War II, and during the war, due to the profound military significance of radar technology. During the postwar years, research with imaging radars began to assume greater significance. Experience with conventional radars during World War II revealed that radar reflection from oceans and ground surfaces (*ground clutter*) varied greatly according to terrain, season, settlement patterns, and so on, and to winds and waves. Ground clutter was undesirable for radars designed to detect aircraft and ships, but later systems were designed specifically to record the differing patterns of reflection from the ground surfaces. The side-looking characteristic of SLAR was desirable because of the experiences of reconnaissance pilots during the war, who often were required to fly low-level missions along predictable flight paths in lightly armed aircraft to acquire the aerial photography required for battlefield intelligence. The side-looking capability provided a capability for acquiring information at a distance using an aircraft flying over friendly territory. The situation posed by the emergence of the Cold War in Europe, with clearly defined frontiers and strategic requirement for information within otherwise closed borders, also provided an incentive for development of sensors with SLAR's capabilities. The experience of the war years also must have provided an intense interest in development of SLAR's all-weather capabilities, because Allied intelligence efforts were several times severely restricted by the absence of aerial reconnaissance during inclement weather. Thus, the development of both imaging radars

**FIGURE 7.5.** Beginnings of radar. Schematic diagram of the situation that led to the experiments by Young and Taylor.

is linked to military and strategic reconnaissance, even though many current applications focus upon civilian requirements.

## 7.2. Geometry of the Radar Image

The basics of the geometry of a SLAR image are illustrated in Figure 7.6. Here the aircraft is viewed head on, with the radar beam represented in vertical cross section as the fan-shaped figure at the side of the aircraft. The upper edge of the beam forms an angle with a horizontal line extended from the aircraft; this angle is designated as the *depression angle* of the far edge of the image. Upper and lower edges of the beam, as they intersect with the ground surface, define the edges of the radar image; the forward motion of the aircraft (toward the reader, out of the plane of the illustration) forms what is usually the "long" dimension of the strip of radar imagery. The smallest depression angle forms the *far-range* side of the image. The *near-range* region is the edge nearest to the aircraft. Intermediate regions between the two edges are sometimes referred to as *mid-range* portions of the image. Steep terrain may hide areas of the imaged region from illumination by the radar beam, causing radar shadow. Note that radar shadow depends on topographic relief and the direction of the flight path in relation to topography. Within an image, radar shadow depends also on depression angle, so that (given equivalent topographic relief) radar shadow will be more severe in the far-range portion of the image (where depression angles are smallest), or for those radar systems that use shallow depression angles. (A specific radar system is usually characterized by a fixed range of depression angles.)

Radar systems measure distance to a target by timing delay for a transmitted signal to be returned to the antenna. Because the speed of electromagnetic energy is a known constant, the measure of time translates directly to a measure of distance from the antenna. Microwave energy travels in a straight path from the aircraft to the ground—a path that defines the *slant-range* distance, as if one were to stretch a string from the aircraft to a specific point on the ground as a measure of distance (Figure 7.7). Image interpreters prefer images to be presented in *ground-range* format, with distances portrayed in their correct relative positions on the earth's surface. Because radars collect all information in the

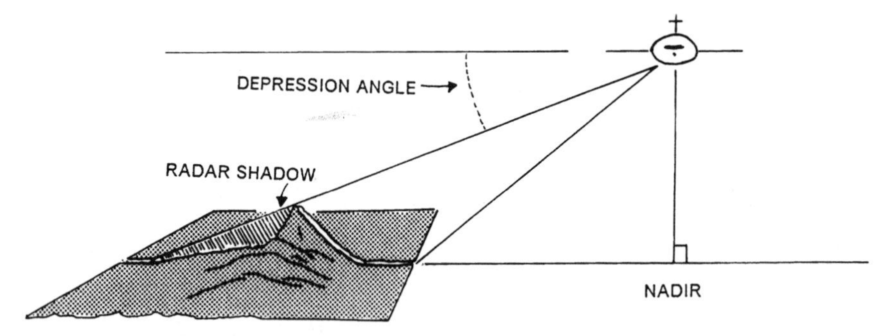


**FIGURE 7.6.** Geometry of an imaging radar system. The radar beam illuminates a strip of ground parallel to the flight path of the aircraft; the reflection and scattering of the microwave signal from the ground forms the basis for the image.

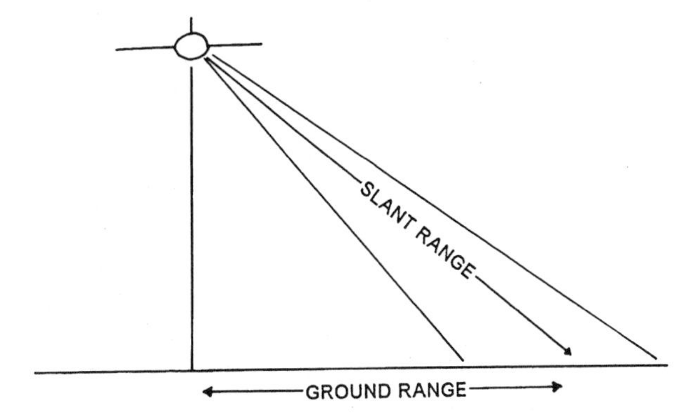


**FIGURE 7.7.** Slant range and ground range. Slant range is the direct distance from the antenna to an object on the ground, as measured by the time delay from transmission of the signal to reception of its echo. Ground range represents the correct scaling of distances as we would measure them on a map.

slant-range domain, radar images inherently contain geometric artifacts, even though the image display ostensibly matches a ground-range presentation.

One such error is *radar layover* (Figure 7.8). At near range, the top of a tall object is closer to the antenna than is its base. As a result, the echo from the top of the object reaches the antenna before it receives the echo from the base. Because radar measures all distances with respect to time elapsed between transmission of a signal and the reception of its echo, the top of the object appears (in the slant-range domain) to be closer to the antenna

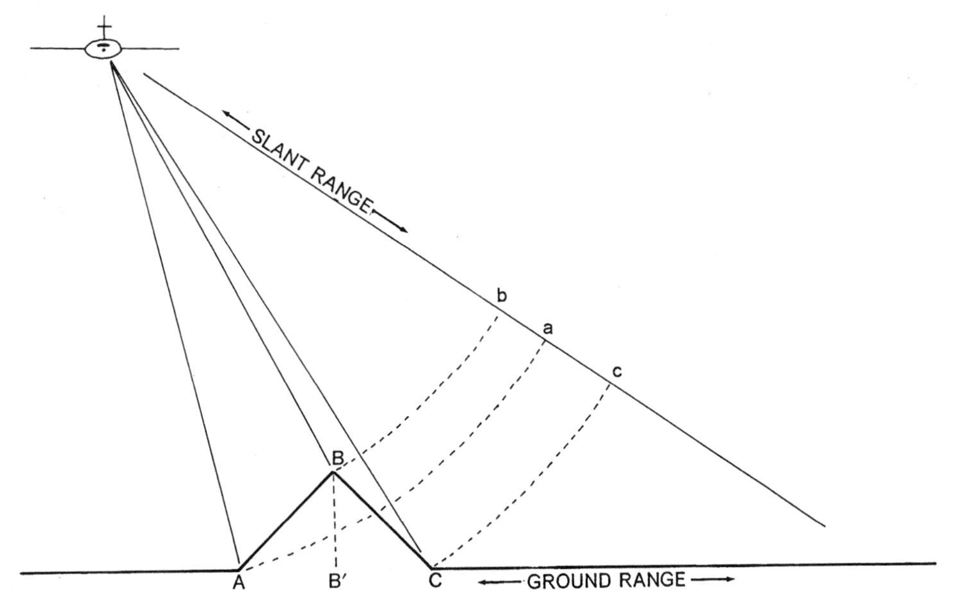


**FIGURE 7.8.** Radar layover. In the ground-range domain *AB* and *BC* are equal. Because the radar can measure only slant range distances, *AB* and *BC* are projected onto the slant-range domain, represented by the line *bac*. The three points are not shown in their correct relationship because the slant-range distance from the antenna to the points does not match to their ground-range distances. Point *B* is closer to the antenna than is *A*, so it is depicted on the image as closer to the edge of the image.

than does its base. And, in fact, it is closer, if only the slant-range domain is considered. However, in the ground-range domain (the context for correct positional representation and for accurate measurement) both the top and the base of the object occupy the same geographic position. In the slant-range domain of the radar image, they occupy different image positions—a geometric error analogous perhaps to relief displacement on aerial photography.

Radar foreshortening is depicted in Figure 7.9. Here the topographic feature *ABC* is shown with *AB* = *BC* in the ground-range representation. However, because the radar can position *A*, *B*, and *C* only by the time delay with relation to the antenna, it must perceive the relationships between *A*, *B*, and *C* as shown in the slant range (image plane). Here *A* and *B* are reversed from their ground-range relationships, so that *ABC* is now *bac*, due to the fact that the echo from *B* must be received before the echo from *A*.

A second form of geometric error, *radar foreshortening*, occurs in terrain of modest to high relief depicted in the mid- to far-range portion of an image (Figure 7.9). Here the slant-range representation depicts *ABC* in their correct relationship *abc,* but the distances between them are not accurately shown. Whereas *AB* = *BC* in the ground-range domain, when they are projected into the slant range, *ab* < *bc*. Radar foreshortening tends to cause images of a given terrain feature to appear to have steeper slopes than they do in nature on the near-range side of the image, and to have shallower slopes than they do in nature on

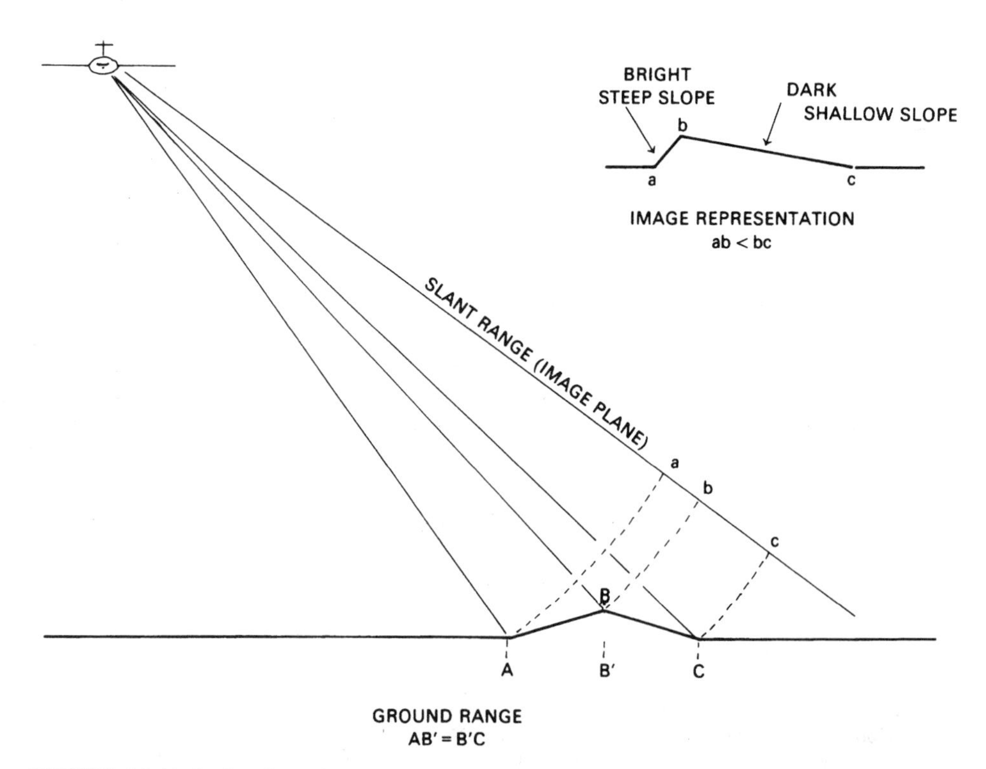


**FIGURE 7.9.** Radar foreshortening. Projection of *A*, *B*, and *C* into the slant-range domain distorts the representations of *AB* and *BC*, so that *ab* appears shorter, steeper, and brighter than it should be in a faithful rendition, and *bc* is longer, shallower in slope, and darker than it should be.

the far-range side of the feature (Figure 7.10). Thus a terrain feature with equal fore and back slopes may be imaged to have shorter, steeper, and brighter slopes than it would in a correct representation, and the image of the back slope would appear to be longer, shallower, and darker than it should in a correct representation. Because depression angle varies with position on the image, the amount of radar foreshortening in the image of a terrain feature depends not only on the steepness of its slopes but also on its position on the radar image. As a result, apparent terrain slope and shape on radar images are not necessarily accurate representations of their correct character in nature, and care should be taken in interpretation of these features from radar images.

## 7.3. Wavelength

Imaging radars normally operate within a small range of wavelengths with the rather broad interval defined at the beginning of this chapter. Table 7.1 lists primary subdivisions of the

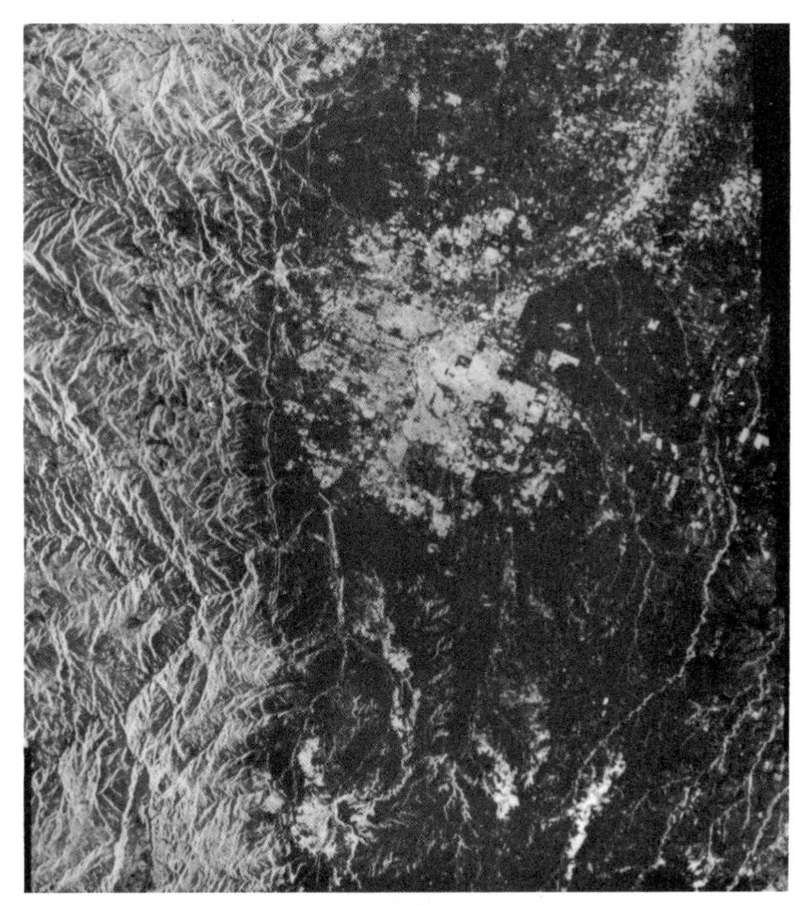

**FIGURE 7.10.** Image illustrating radar foreshortening. Note the unnatural appearance of the topography in the near-range (left-hand) portion of the image. Denver, Colorado, region. Image courtesy of NASA and Jet Propulsion Laboratory, Pasadena, California.

**TABLE 7.1. Radar Frequency Designations**

| Band | Wavelengths |
|---|---|
| P-band | 107–77 cm |
| UHF | 100–30 cm |
| L-band | 30–15 cm |
| S-band | 15–7.5 cm |
| C-band | 7.5–3.75 cm |
| X-band | 3.75–2.40 cm |
| Ku-band | 2.40–1.67 cm |
| K-band | 1.67–1.18 cm |
| Ka-band | 1.18–0.75 cm |

active microwave region, as commonly defined in the United States. These divisions and their designations have an arbitrary, illogical flavor that is the consequence of their origin during the development of military radars when it was important to conceal the use of specific frequencies for given purposes. In the context of military security, the designations were designed as much to confuse unauthorized parties as to provide convenience for authorized personnel. In the interval since then, these designations have become established in everyday usage, even though there is no longer a requirement for secrecy.

Although experimental radars can often change frequency, or sometimes even use several frequencies (for a kind of "multispectral radar"), operational systems are generally designed to use a single wavelength band. Airborne imaging radars have frequently used L-, C-, K-, and X-bands. As described elsewhere in this chapter, imaging radars used for satellite observation frequently use C- and L-band frequencies.

Choice of a specific microwave band has several implications for the nature of the radar image. For real aperture imaging radars, spatial resolution improves as wavelength becomes shorter with respect to antenna length (i.e., for a given antenna length, resolution is finer with use of shorter wavelength). Penetration of the signal into the soil is in part a function of wavelength; for given moisture conditions, penetration is greatest at longer wavelengths. The longer wavelengths of microwave radiation (e.g., relative to visible radiation) mean that imaging radars are insensitive to the usual problems of atmospheric attenuation—usually only heavy rain will interfere with transmission of microwave energy.

## 7.4. Penetration of the Radar Signal

In principle, radar signals are capable of penetrating what would normally be considered solid features, including vegetative cover and the soil surface. In practice, it is very difficult to assess the existence or amount of radar penetration in the interpretation of specific images. Penetration is assessed by specifying the skin depth—the depth to which the strength of a signal is reduced to $1/e$ of its surface magnitude, or about 37%. Separate features are subject to differing degrees of penetration; specification of skin depth, measured in standard units of length, provides a means of designating variations in the ability of radar signals to penetrate various substances.

Skin depth increases with increasing wavelength and in the absence of moisture.

Thus, optimum conditions for observing high penetration would be in arid regions, using long-wavelength radar systems. Penetration is also related to surface roughness and to incidence angle; penetration is greater at steeper angles and decreases as incidence angle increases. We should therefore expect maximum penetration at the near-range edge of the image and minimum penetration at the far-range portion of the image.

As a practical matter, the difficulties encountered in the interpretation of an image that might record penetration of the radar signal would probably prevent practical use of any information that might be conveyed to the interpreter. There is no clearly defined means by which an interpreter might be able to recognize the existence of penetration and to separate its effects from the many other variables that contribute to radar backscatter. For radar systems operating near X-band and K-band, empirical evidence suggests that the radar signal is generally scattered from the first surface it strikes, probably foliage in most instances. L-band signals, at much longer wavelengths, are believed to be capable of a higher degree of penetration, to reach branches, trunks, and terrain surfaces below the canopy.

## 7.5. Polarization

The polarization of a radar signal denotes the orientation of the field of electromagnetic energy emitted and received by the antenna. Radar systems can be configured to transmit either horizontally or vertically polarized energy and to receive either horizontally or vertically polarized energy as it is scattered from the ground. Unless otherwise specified, an imaging radar usually transmits horizontally polarized energy and receives the horizontally polarized echo from the terrain. However, some radars are designed to transmit horizontally polarized signals but to separately receive the horizontally and vertically polarized reflections from the landscape. Such systems produce two images of the same landscape (Figure 7.11). One is the image formed by the transmission of a horizontally polarized signal and the reception of the horizontally polarized return signal. This is often referred to as the *HH image* or the *like-polarized* mode. A second image is formed by the transmission of a horizontally polarized signal, and the reception of only the vertically polarized return; this is the *HV image* or the *cross-polarized* mode.

By comparing the two images, the interpreter can identify features and areas that represent regions on the landscape that tend to depolarize the signal. Such areas will reflect the incident horizontally polarized signal back to the antenna as vertically polarized energy—that is, they change the polarization of the incident microwave energy. Such areas can be identified on the images as bright regions on the HV image and as dark or dark gray regions on the corresponding HH image. Their appearance on the HV image is much brighter due to the effect of depolarization—the polarization of the energy that would have contributed to the brightness of the HH image has been changed, so it creates instead a bright area on the HV image. Comparison of the two images, therefore, permits detection of those areas that are good depolarizers.

Causes of depolarization are related to physical and electrical properties of the ground surface. A rough surface (with respect to the wavelength of the signal) may depolarize the signal. Another cause of depolarization is volume scattering from an inhomogeneous medium; such scatter might occur if the radar signal is capable of penetrating be-

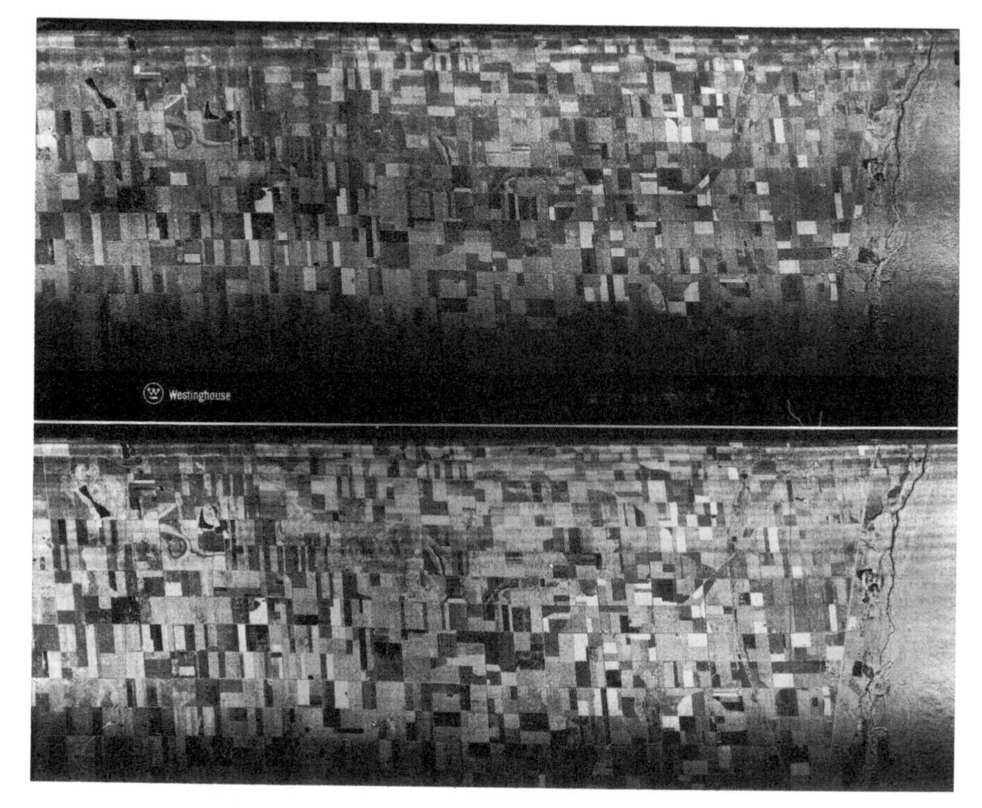


**FIGURE 7.11.** Radar polarization. Upper image: HH image. Lower image: HV image. Agricultural landscape in western Kansas. Image courtesy of Westinghouse.

neath the soil surface (as might be conceivabe in some desert areas where vegetation is sparse and the soil dry enough for significant penetration to occur) where it might encounter such subsurface inhomogeneities as buried rocks.

## 7.6. Look Direction and Look Angle

### *Look Direction*

The direction at which the radar signal strikes the landscape is important in both natural and man-made landscapes. In natural landscapes, look direction is especially important when terrain features display a preferential alignment. Look directions perpendicular to topographic alignment will tend to maximize radar shadow, whereas look directions parallel to topographic orientation will tend to minimize radar shadow. In regions of small or modest topographic relief, radar shadow may be desirable as a means of enhancing microtopography or revealing the fundamental structure of the regional terrain. The extent of

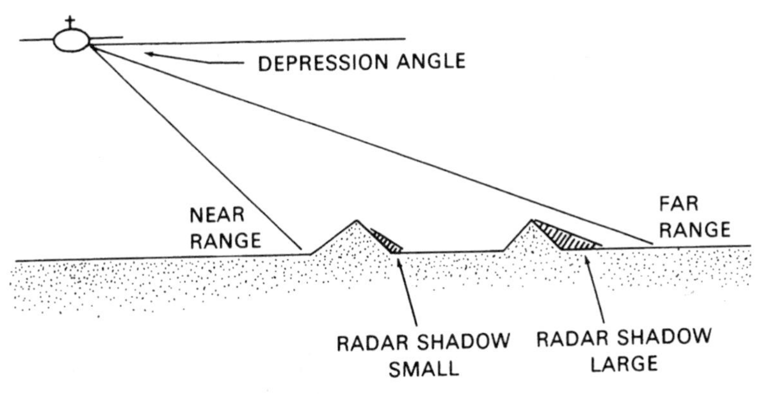


**FIGURE 7.12.** Radar shadow.

radar shadow depends not only on local relief but also on the positions of features relative to the flight path; those features positioned in the near-range portion (other factors being equal) will have the smallest shadows, whereas those at the far-range edge of the image will cast larger shadows (Figure 7.12). In areas of high relief, radar shadow is usually undesirable as it masks large areas from observation.

In landscapes that are heavily altered by mankind's activities, the orientation of structure and land-use patterns is often a significant influence on the character of the radar return and, therefore, on the manner in which given landscapes appear on radar imagery. For instance, if an urban area is viewed at a look angle that maximizes the scattering of the radar signal from, for example, urban structures aligned along a specific axis, it will have an appearance quite different from that of an image acquired at a look direction that tends to minimize reflection from such features.

### *Look Angle*

The depression angle of the radar varies across an image, from relatively steep at the near-range side of the image to relatively shallow at the far-range side (Figure 7.13). The exact values of the look angle vary with the design of specific radar systems, but some broad generalizations are possible concerning the effects of varied look angles. First, the basic geometry of a radar image ensures that the resolution of the image must vary with look angle; at steeper depression angles, a radar signal illuminates a smaller area than does the same signal at shallow depression angles. Therefore, the spatial resolution, at least in the across-track direction, varies with respect to depression angle.

It has been shown that the sensitivity of the signal to ground moisture is increased as depression angle becomes steeper. Furthermore, the slant-range geometry of a radar image means that all landscapes are viewed (by the radar) at oblique angles. As a result, the image tends to record reflections from the sides of features. The obliqueness, and therefore the degree to which we view sides rather than tops of features, varies with look angle. In some landscapes, the oblique view may be different from the overhead view to which we are accustomed in the use of other remotely sensed imagery (Figure 7.14). Such varia-

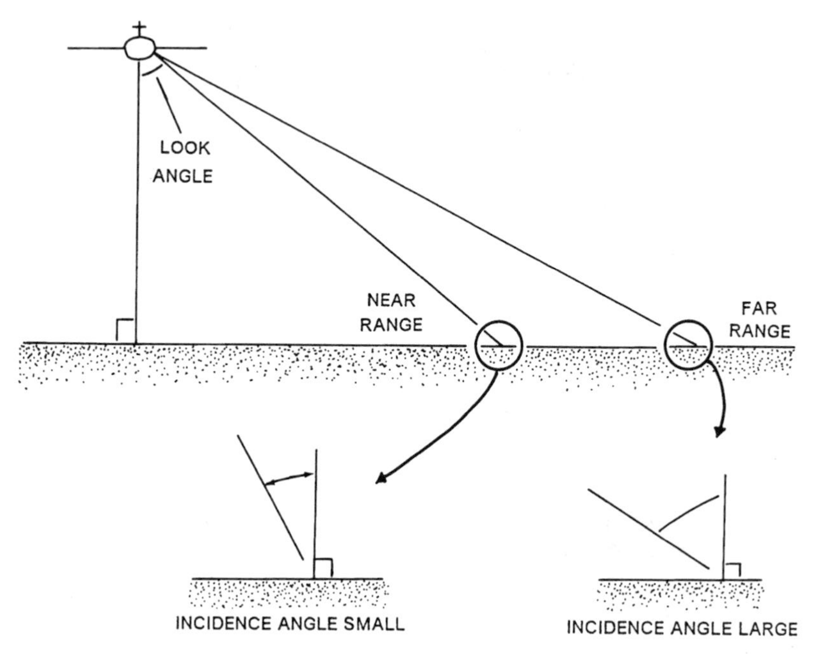


**FIGURE 7.13.** Look angle.

tions in viewing angle may contribute to variations in the appearance on radar imagery of otherwise similar landscapes.

## 7.7. Real Aperture Systems

*Real aperture* SLAR systems (sometimes referred to as *brute force* systems), one of the two strategies for acquiring radar imagery, are the oldest, simplest, and least expensive of imaging radar systems. They follow the general model described earlier for the basic configuration of a SLAR system (Figure 7.15). The transmitter generates a signal at a specified wavelength and of a specified duration. The antenna directs this signal toward the ground, then receives its reflection. The reflected signal is amplified, filtered, and displayed on a cathode ray tube, where a moving film records the radar image line by line as it is formed by the forward motion of the aircraft.

The resolution of such systems is controlled by several variables. One objective is to cause the transmitted signal to illuminate as small an area as possible on the ground, as it is the size of this area that determines spatial detail recorded on the image. If the area illuminated is large, reflections from diverse features may be averaged together to form a single graytone value on the image, and their distinctiveness is lost. If the area is small, individual features are recorded as separate features on the image, and their identities are preserved.

The size of the area illuminated is controlled by several variables. One is antenna length in relation to wavelength. A long antenna length permits the system to focus en-

**FIGURE 7.14.** Perspective as related to depression angle. The side-looking nature of imaging radars means that radars tend to view the landscape from an oblique angle, which means that the image tends view the sides of objects, rather than the overhead view that we are accustomed to in aerial photography.

ergy on a small ground area. Thus, real aperture systems require long antennas in order to achieve fine detail; usually the ability of an aircraft to carry long antennas forms a practical limit to the resolution of radar images. This restriction on antenna length forms a barrier for use of real aperture systems on spacecraft, as small antennae would provide very coarse resolution from spacecraft altitudes, but practical limitations prevent use of large antennas.

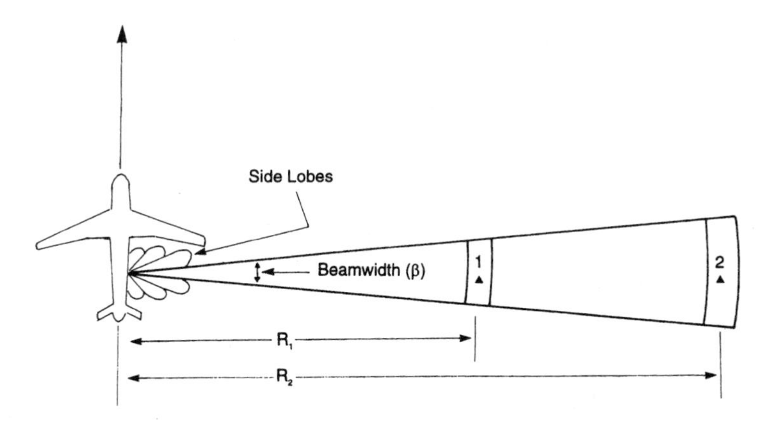


**FIGURE 7.15.** Azimuth resolution. For real aperture radar, the ability of the system to acquire fine detail in the a long-track axis derives from the ability of the system to focus the radar beam to illuminate a small area. A long antenna, relative to wavelength, permits the system to focus energy on a small strip of ground, improving detail recorded in the along-track dimension of the image. The *beamwidth (β)* measures this quality of an imaging radar. Beamwidth, in relation to range ($R$), determines detail—region 1, at range $R_1$, will be imaged in greater detail than region 2, at greater range $R_2$. Also illustrated here are *side lobes,* smaller beams of microwave energy created because the antenna cannot be perfectly effective in transmitting a single beam of energy.

The area illuminated by a real aperture SLAR system can be considered analogous to the spot illuminated by a flashlight aimed at the floor. As the flashlight is aimed straight down, the spot is small and nearly circular in shape. As it is aimed at a spot on the floor at increasingly large distances, the spot becomes larger and dimmer and assumes a more irregular shape. By analogy, the near-range portions of a radar image will have finer resolution than the far-range portions. Thus, antenna length in relation to wavelength determines the angular resolution of a real aperture system—the ability of the system to separate two objects in the along-track dimension of the image (Figure 7.15).

The relationship between resolution and antenna length and wavelength is given by the expression

$$\beta = \lambda / A \qquad \text{(Eq. 7.1)}$$

where $\beta$ represents the beam width, $\lambda$ is the wavelength, and $A$ is the antenna length. Real aperture systems can be designed to attain finer along-track resolution by increasing the length of the antenna and decreasing wavelength. Therefore, these qualities form the design limits on the resolution of real aperture systems—antenna length is constrained by practical limits of aeronautical design and operation, and shorter wavelengths increase susceptibility to atmospheric effects.

Radar systems have another, unique means of defining spatial resolution. The length of the radar pulse determines the ability of the system to resolve the distinction between two objects in the cross-track axis of the image (Figure 7.16). Long pulses strike two nearby features at the same time, thereby recording the two objects as a single reflection and as a single feature on the image. Shorter pulses are each reflected separately from nearby features and can record the distinctive identities and locations of the two objects.

## 7.8. Synthetic Aperture Systems

The alternative design for an imaging radar is the *synthetic aperture radar* (SAR), which is based on principles and technology differing greatly from those of real aperture radars. SAR systems are much more complex and much more expensive to manufacture and operate than are real aperture systems. But, they can overcome some of the limitations inherent to real aperture systems and therefore can be applied in a wider variety of applications (including observation from earth-orbiting satellites) and may provide imagery of higher quality in some instances.

Consider an aircraft with a SAR that images the landscape depicted in Figure 7.17. At *1* the aircraft is positioned so that the objects shown on the landscape are just barely inside the region illuminated by the SAR. At *2* they are more fully within the area of illumination, and at *3* they are again just at the edge of the illuminated area. Finally (*4*), the aircraft moves so that the objects are not illuminated by the radar beam. The principle of the SAR depends on the fact that objects within a scene experience illumination by the radar over an interval of time, as the aircraft moves along its flight path. A SAR system receives the scattered signal from the landscape during this interval and saves the complete history of reflections from each object. Knowledge of this history permits later reconstruction of

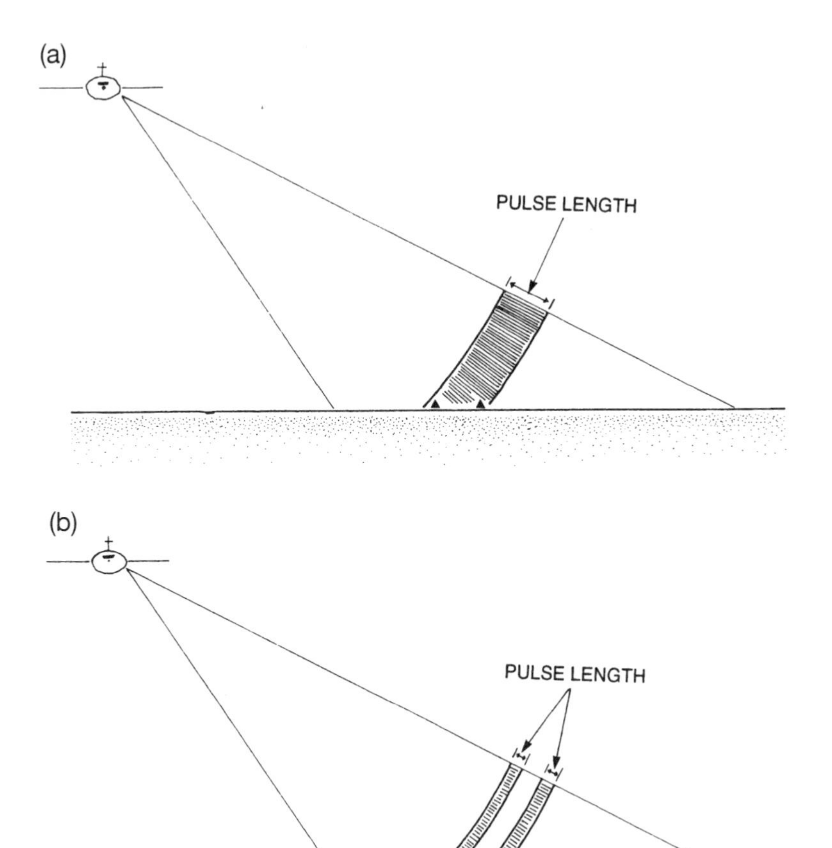


**FIGURE 7.16.** Effect of pulse length. (*a*) Longer pulse length means that the two objects shown here are illuminated by a single burst of energy, creating a single echo that cannot reveal the presence of two separate objects. (*b*) Shorter pulse length illuminates the two objects with separate pulses, creating separate echos for each object. Pulse length determines resolution in the cross-track dimension of the image.

the reflected signals as though they were received by a single antenna occupying physical space *abc*, even though they were in fact received by a much shorter antenna that was moved in a path along distance *1234*. (Thus, the term *synthetic aperture* denotes the artificial length of the antenna, in contrast to the "real" aperture based on the actual physical length of the antenna used with real aperture systems.)

In order to implement this strategy, it is necessary to define a practical means of assigning separate components of the reflected signal to their correct positions as the spatial representation of the landscape is re-created on the image. This process is, of course, extraordinarily complex if each such assignment must be considered an individual problem in unraveling the complex signal history of the radar reflection at each of a multitude of antenna positions.

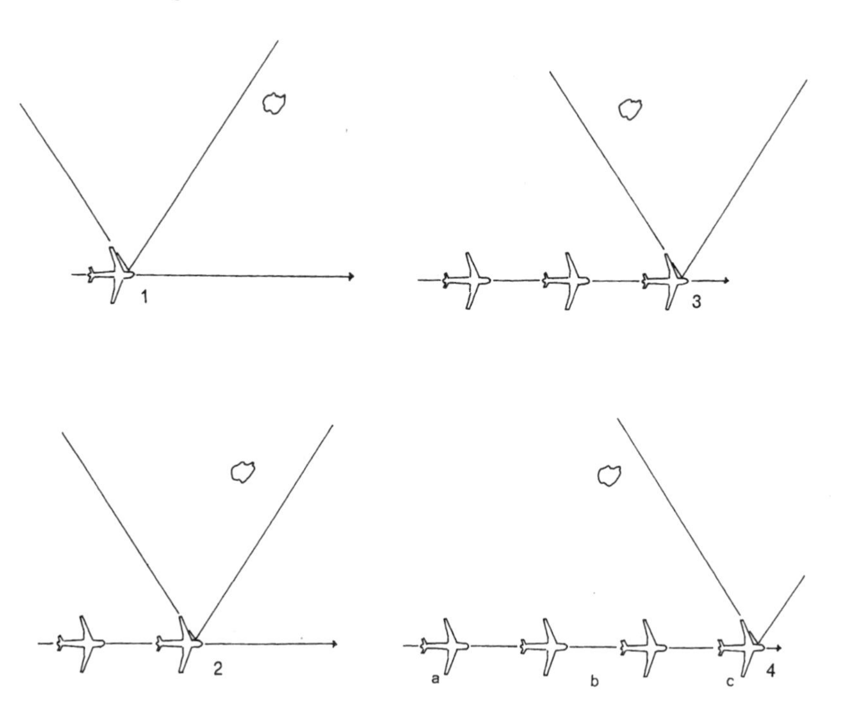


**FIGURE 7.17.** Synthetic aperture imaging radar. Synthetic aperture systems accumulate a history of backscattered signals from the landscape as the antenna moves along path *abc*.

This problem can be solved in a practical manner because of the systematic changes in frequency experienced by the radar signal as it is scattered from the landscape. Objects within the landscape experience different frequency shifts in relation to their distances from the aircraft track. At a given instant, objects at the leading edge of the beam reflect a pulse with an increase in frequency (relative to the transmitted frequency) due to their position ahead of the aircraft, and those at the trailing edge of the antenna experience a decrease in frequency (Figure 7.18). This is the *Doppler effect*, often explained by analogy to the change in pitch of a train whistle heard by a stationary observer as a train passes at high speed. As the train approaches, the pitch appears higher than that of a stationary whistle due to the increase in frequency of sound waves. As the train passes the observer, then recedes into the distance, the pitch appears lower due to the decrease in frequency. Radar, as an active remote sensing system, is operated with full knowledge of the frequency of the transmitted signal; as a result, it is possible to compare the frequencies of transmitted and reflected signals to determine the nature and amount of frequency shift. Knowledge of frequency shift permits the system to assign reflections to their correct relative positions on the image and to synthesize the effect of a long antenna.

Nonetheless, practical difficulties of recording such complex signals and assigning them their correct meaning in the image are immense. One method of solving this prob-

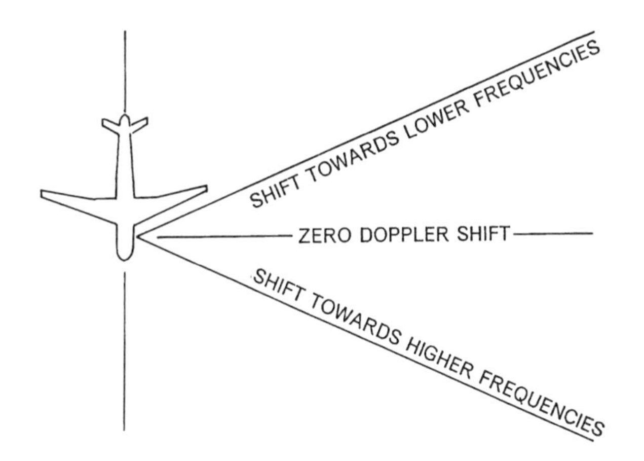


**FIGURE 7.18.** Frequency shifts experienced by features within the field of view of the radar system.

lem is to record information concerning the reflected signal as an interference pattern, recorded on the *signal film* (sometimes called the *data film*) (Figure 7.19). The signal film records on photographic film the interference pattern between the reference (transmitted) signal and its reflection from the landscape. Such a record does not present an intelligible record of landscape features, as all information is recorded in the frequency domain. A given object on the landscape appears on the signal film as a broken line; its distance from the edge of the film is proportional to the distance of the actual object from the flight path of the aircraft.

The signal film is generated as the aircraft flies a mission and the sensor records the microwave energy scattered from the ground. After the aircraft lands, the signal film is removed from the aircraft and taken to a processing facility for processing. The signal film forms the basis for a reconstruction of the data into a conventional image of the landscape. This reconstruction transforms the frequency information in the signal film into its spatial counterpart on the final image by using the signal film as the basis for what might

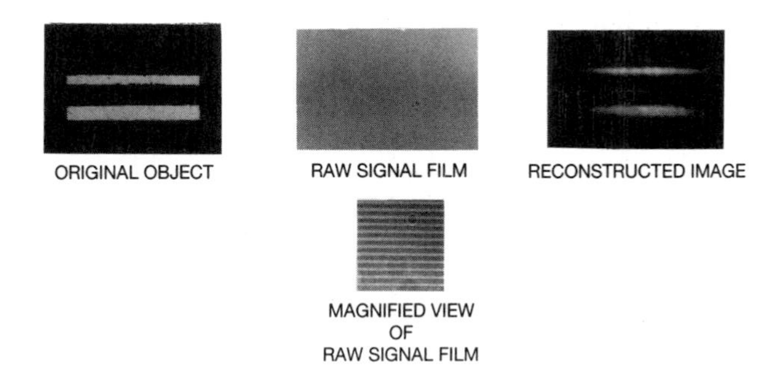


**FIGURE 7.19.** Signal film. This image is formed by the interference pattern between the reference (transmitted) signal from the radar and the signal as reflected from the terrain. All information is presented in the frequency domain, so it is not susceptible to visual interpretation.

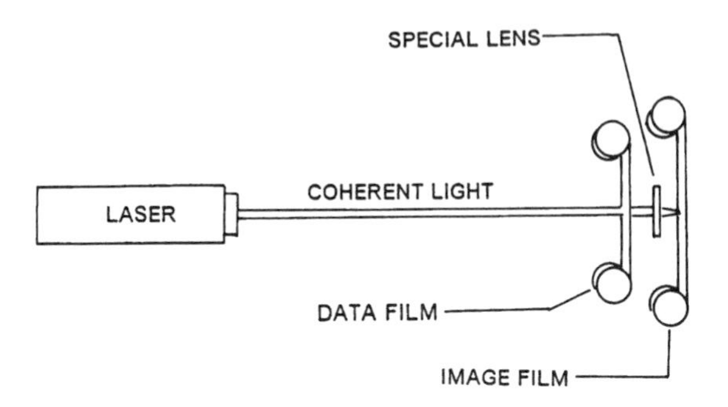


**FIGURE 7.20.** Optical processing of signal film. The signal film (Figure 7.19) is illuminated with coherent light (i.e., a laser) to form a hologram that reconstructs the image in the spatial domain, forming the image that we are accustomed to interpreting.

be called a microwave hologram (Figure 7.20). Remember that the signal film records the interference pattern between a microwave signal of known frequency and its reflection from the landscape; the reflected signal differs from the reference signal with respect to both frequency and phase. Because the reference signal is in effect coherent microwave energy (i.e., energy that is "pure" with respect to wavelength and that is in phase), the signal film can be regarded as a record of a microwave hologram. By directing a beam of coherent light through the signal film using special lenses to correct image geometry (Jensen et al., 1977), the image is reconstructed in the visual spectrum to form an image in the spatial domain. If the image data have been recorded in digital, rather than optical, form, a magnetic tape forms a counterpart to the signal film, and the image reconstruction can be performed digitally.

This final image is comparable to images generated by real aperture systems except that it can be free of some of the limitations that constrain the spatial resolution of real aperture systems. Often real aperture radar images of homogeneous regions sometimes display a characteristic speckled appearance caused by in-phase reflectance from the terrain. Because the techniques available to remove visual effects of the speckle degrade the image resolution, the speckle is usually retained.

## 7.9. Interpreting Brightness Values

Each radar image is composed of many image elements of varying brightness (Figure 7.21). Variations in image brightness correspond, at least in part, to place-to-place changes within the landscape; through knowledge of this correspondence the image interpreter has a basis for predictions, or inferences, concerning landscape properties.

Unlike passive remote sensing systems, active systems illuminate the land with radiation of known, and carefully controlled, properties. Therefore, in principle, the interpreter should find a firm foundation for deciphering the meaning of the image because the only "unknowns" of the many variables that influence image appearance are the ground conditions—the object of study.

However, in practice the interpreter of a radar image is faced with many difficult obstacles for a rigorous interpretation of a radar image. First, returned signals from a terrain span a broad range of magnitudes from very low to very high; the ranges in values are often so large that they exceed the ability of photographic emulsions to accurately portray

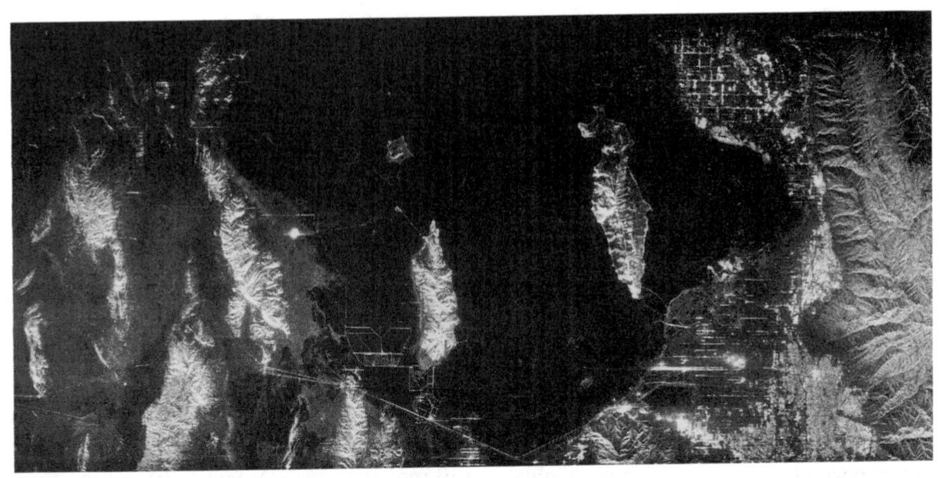

**FIGURE 7.21.** Varied brightnesses and tones within a radar image. Landscape near Salt Lake City, Utah. Image courtesy of National Space Science Data Center.

the actual range of values (Chapter 3). Very dark and very bright landscape elements may be portrayed in the nonlinear portions of the characteristic curve and therefore lack consistent relationships with their representations on the image. Furthermore, the features that compose even the simplest landscapes have complex shapes and arrangements and are formed from diverse materials of contrasting electrical properties.

As a result, there are often few detailed models of the kinds of backscattering that should in principle be expected from separate classes of surface materials. Direct experience and intuition are not always reliable guides to interpretation of images acquired outside the visible spectrum. In addition, many SLAR images observe the landscape at shallow depression angles. As most interpreters gain experience from observations at ground level, or from overhead aerial views, the oblique radar view from only a few degrees above the horizon may be difficult to interpret.

### *The Radar Equation*

The fundamental variables influencing the signal returned from the terrain are formally given by the *radar equation*:

$$P_r = \frac{\sigma G^2 P_t \lambda^2}{(4\pi)^3 R^4} \qquad \text{(Eq. 7.2)}$$

Here $P_r$ designates the power returned to the antenna from the ground surface; $R$ specifies the range to the target from the antenna; $P_t$ the transmitted power; $\lambda$ the wavelength of the energy; and $G$ the antenna gain (a measure of the system's ability to focus the transmitted energy). All these variables are determined by the design of the radar system and are therefore known or controlled quantities. The one variable in the equation not thus far identified is $\sigma$, the backscattering coefficient; $\sigma$ is of course not controlled by the radar system but by the specific characteristics of the terrain surface represented by a specific region on the image. Whereas $\sigma$ is often an incidental factor for the radar engineer, it is, of course, the primary focus of study for the image interpreter as it is this quantity that carries information about the landscape.

The value of $\sigma$ conveys information concerning the amount of energy scattered from a specific region on the landscape as measured by $\sigma^\circ$, the radar cross section. It specifies the corresponding area of an isotropic scatterer that would provide the same power returned as does the observed signal. The backscattering coefficient ($\sigma^\circ$) expresses the observed scattering from a large surface area as a dimensionless ratio between two areal surfaces; it measures the average radar cross section per unit area. $\sigma^\circ$ varies over such wide values that it must be expressed as a ratio rather than an absolute value.

Ideally, radar images should be interpreted with the objective of relating observed $\sigma^\circ$ (varied brightnesses) to properties within the landscape. It is known that backscattering is related to specific *system variables*, including wavelength, polarization, and azimuth, in relation to landscape orientation and depression angle. In addition, *landscape parameters* are important, including surface roughness, soil moisture, vegetative cover, and microtopography. Because so many of these characteristics are interrelated, detailed interpretations of individual variables are usually very difficult, in part because of the extreme complexity of landscapes, which normally are intricate compositions of diverse natural and man-made features. Often many of the most useful landscape interpretations of radar images have attempted to recognize integrated units defined by assemblages of several variables rather than separate individual components. The notion of "spectral signatures" is difficult to apply in the context of radar imagery because of the high degree of variation in image tone as incidence angle and look direction change.

### *Moisture*

Moisture in the landscape influences the backscattering coefficient through changes in the dielectric constant of landscape materials. (The dielectric constant is a measure of the ability of a substance to conduct electrical energy, an important variable determining the response of a substance that is illuminated with microwave energy.) Although natural soils and minerals vary in their ability to conduct electrical energy, these properties are difficult to exploit as the basis for remote sensing because the differences between dielectric properties of separate rocks and minerals in the landscape are overshadowed by the effects of even small amounts of moisture, which greatly change the dielectric constant. As a result, the radar signal is sensitive to the presence of moisture both in the soil and in vegetative tissue; this sensitivity appears to be greatest at steep depression angles.

The presence of moisture also influences effective skin depth; as the moisture content of surface soil increases, the signal tends to scatter from the surface. As moisture content decreases, skin depth increases and the signal may be scattered from a greater thickness of soil.

### *Roughness*

A radar signal that strikes a surface will be reflected in a manner that depends on both characteristics of the surface and properties of the radar wave, as determined by the radar

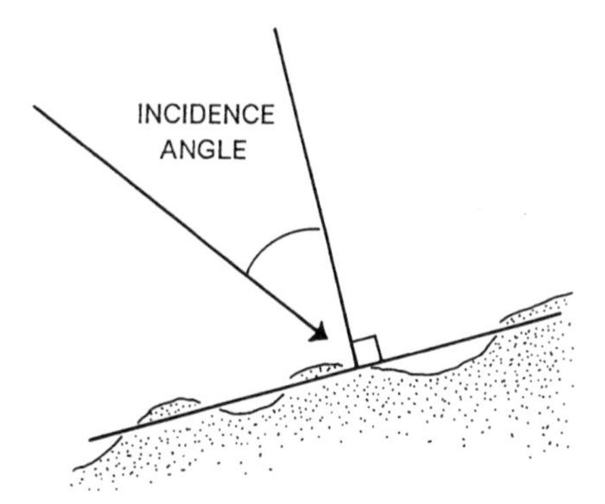


**FIGURE 7.22.** Incidence angle.

system, and the conditions under which it is operated. The *incidence angle* (θ) is defined as the angle between the axis of the incident radar signal and a perpendicular to the surface that the signal strikes (Figure 7.22). If the surface is homogeneous with respect to its electrical properties and is "smooth" with respect to the wavelength of the signal, the reflected signal will be reflected at an angle equal to the incidence angle, with most of the energy directed in a single direction (i.e., specular reflection).

For "rough" surfaces, reflection will not depend as much on incidence angle, and the signal will be scattered more or less equally in all directions (diffuse, or isotropic, scattering). For radar systems, the notion of a rough surface is defined in a manner considerably more complex than that of everyday experience, as roughness depends not only on the physical configuration of the surface but also on the wavelength of the signal and its incidence angle. Consider the physical configuration of the surface to be expressed by the standard deviation of the heights of individual facets (Figure 7.23). Although definitions of surface roughness vary, one common definition defines a rough surface as one in which the standard deviation of surface height ($S_h$) exceeds one-eighth of the wavelength (λ) divided by the cosine of the incidence angle (cos θ):

$$S_h > \lambda/(8 \cos \theta) \quad \text{(Eq. 7.3)}$$

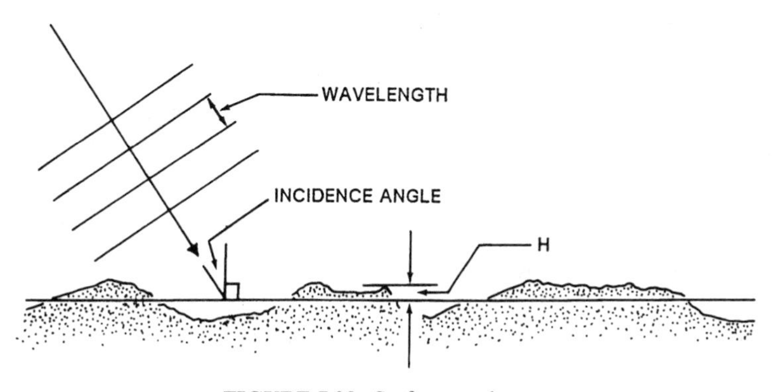


**FIGURE 7.23.** Surface roughness.

**TABLE 7.2. Surface Roughness Defined for Several Wavelengths**

| Roughness category | K-band ($\lambda$ = 0.86 cm) | X-band ($\lambda$ = 3 cm) | L-band ($\lambda$ = 25 cm) |
|---|---|---|---|
| Smooth | $h < 0.05$ cm | $h < 0.17$ cm | $h < 1.41$ cm |
| Intermediate | $h$ = 0.05–0.28 cm | $h$ = 0.17–0.96 cm | $h$ = 1.41–8.04 cm |
| Rough | $h > 0.28$ cm | $h > 0.96$ cm | $h > 8.04$ cm |

*Note*. Data from Jet Propulsion Laboratory (1982).

where $h$ is the average height of the irregularities. In practice, this definition means that a given surface appears rougher as wavelengths become shorter (see Table 7.2). Also, for a given wavelength, surfaces will act as smooth scatterers as incidence angle becomes greater (i.e., equal terrain slopes will appear as smooth surfaces as depression angle becomes smaller, as occurs in the far-range portions of radar images).

### *Corner Reflectors*

The return of the radar signal to the antenna can be influenced not only by moisture and roughness but also by the broader geometric configuration of targets. Objects that have complex geometric shapes, such as those encountered in an urban landscape, can create radar returns that are much brighter than would be expected based on size alone. This effect is caused by the complex reflection of the radar signal directly back to the antenna in a manner analogous to a ball that bounces from the corner of a pool table directly back to the player. This behavior is caused by objects classified as *corner reflectors*, which often are corner-shaped features (such as the corners of buildings and alleyways between them in a dense urban landscape) but are also formed by other objects of complex shape.

Corner reflectors are common in urban areas due to the abundance of concrete, masonry, and metal surfaces constructed in complex angular shapes (Figure 7.24). Corner reflectors can also be found in rural areas, perhaps formed sometimes by natural surfaces but more commonly by metallic roofs of farm buildings, agricultural equipment, and items such as powerline pylons and guardrails along divided highways (Figure 7.25).

Corner reflectors are important in interpretation of the radar image. They form a characteristic feature of the radar signatures of urban regions and identify other features such as powerlines, highways, and railroads. It is important to remember that the image of a corner reflector is not shown in proportion to its actual size—the returned energy forms a star-like burst of brightness that is proportionately much larger than the size of the object that caused it. Thus they can be valuable in interpreting the radar image but do not appear on the image in their correct relative sizes.

## 7.10. Satellite Imaging Radars

Scientists working with radar remote sensing have for years held an interest in the possibility of observing the earth by means of imaging radars carried by earth satellites. Whereas real aperture systems cannot be operated at satellite altitudes without unaccept-

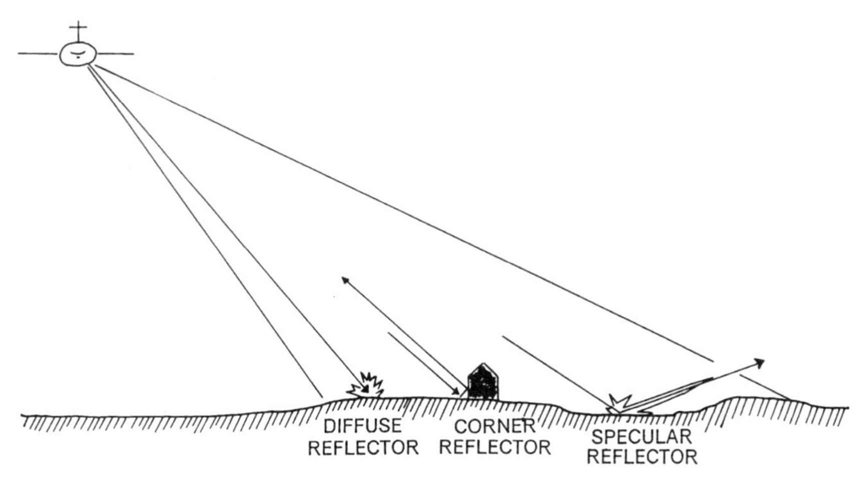


**FIGURE 7.24.** Three classes of features important for interpretation of radar imagery. See Figure 7.25 for examples.

ably coarse spatial resolution (or use of impractically large antennas), the synthetic aperture principle permits relatively compact radar systems to acquire imagery of fine spatial detail at high altitudes. This capability, combined with the capability of imaging radars to acquire imagery in darkness, through cloud cover, and during inclement weather, provides the opportunity for development of a powerful remote sensing capability, with potential for observing large areas of the earth's ocean and land areas that might otherwise be unobservable because of remoteness and atmospheric conditions.

Some difficulties have been encountered in designing orbital radar systems suitable for operational use. Large electrical power requirements have caused some problems; possibly these kinds of difficulties can be solved by improved design and engineering of future systems. Other problems may be more difficult to solve within the near future as they limit our ability to derive information from the image. Some of these problems can be resolved only by a more sophisticated and detailed knowledge of the interaction between the radar signal and fundamental properties of the landscape. As shown in some of the following sections, current research with satellite imaging radars is designed to acquire more knowledge of these basic qualities and more experience in deriving information from a range of diverse terrains. In addition, it may be possible to combine information derived from imaging radars with that from satellite sensors operating in the visible and near infrared spectrum to provide images and data that portray the best features of both kinds of sensors.

Several satellite missions have used imaging radars; all have been experimental programs to further develop the concepts and experience necessary for longer-term efforts. These satellite imaging radars have been designed primarily to acquire a library of imagery representative of diverse terrains and to develop more experience and knowledge of the interpretation and analysis of radar images. In addition, it may be possible to further develop our ability to combine information derived from radar imagery from that collected by satellites observing the visible and near infrared wavelengths to enhance the capabilities of both kinds of sensors.

**FIGURE 7.25.** Airborne SAR image of region near Tucson, Arizona. X-band SAR image, HH polarization, acquired 15 July 1976, shown here at a scale of about 1:100,000. The upper right-hand portion of the image is dominated by an urban and suburban landscape composed chiefly of corner reflectors—the tiny bright points formed as the radar signal is reflected by angular structures of the built environment (Figure 7.24). An interstate highway and railroad pass left to right through the center of the image. Below these features is a river floodplain, and in the lower left of the image, a landscape composed of range and brush, with some residential development. The rangeland generally forms a diffuse surface at this wavelength and appears in medium gray tones. On this image, black surfaces are bare fields or open land with sparse or low, grassy vegetation, which forms a smooth surface at this wavelength, creating specular reflection. Image courtesy of Goodyear Aerospace Corporation and U.S. Air Force Systems Command.

### *Seasat SAR*

Seasat (Figure 7.26) was a satellite specifically tailored to observe the earth's oceans by means of several sensors designed to monitor winds, waves, temperature, and topography. Many of the sensors detected active and passive microwave radiation, although one radiometer operated in the visible and near infrared spectrum. (Seasat carried three microwave radiometers, one imaging radar, and a radiometer that operated both in the visible and in the infrared. Radiometers are described in Chapter 8.) Some sensors were capable of observing 95% of the earth's oceans every 36 hours. Specifications for the Seasat SAR are as follows:

- *Launch:* 28 June 1978
- *Electrical system failure:* 10 October 1978
- *Orbit:* nearly circular at 108° inclination; 14 orbits each day
- *Frequency:* L-band (1.275 GHz)
- *Wavelength:* 23 cm
- *Look direction:* looks to starboard side of track
- *Swath width:* 100 km, centered 20° off nadir
- *Ground resolution:* 25 m × 25 m
- *Polarization:* HH

Our primary interest here is the synthetic aperture radar, designed to observe ocean waves, sea ice, and coastlines. The satellite orbit was designed to provide optimum cover-

**PLATE 1.** Color and color infrared aerial photographs, Torch Lake, Michigan. Top: natural color aerial photograph. Bottom: color infrared aerial photograph of approximately the same region (see text, pp. 69–70). Photograph courtesy of U.S. Environmental Protection Agency.

**PLATE 2.** High oblique aerial photograph, Anchorage, Alaska, 17 September 1983 (see text, p. 71). Photograph courtesy of North Pacific Aerial Surveys.

**PLATE 3.** Low oblique aerial photograph (see text, p. 71). Photograph courtesy of North Pacific Aerial Surveys.

**PLATE 4.** High-altitude aerial photograph, Corpus Christi, Texas, 31 January 1995. Photograph courtesy of U.S. Geological Survey.

**PLATE 5.** TM color composite (bands 1, 3, and 4). This image forms a color composite providing spectral information approximately equivalent to a color infrared aerial photograph (cf. Plates 1 and 4). It shows the same region depicted by Figures 6.17 and 6.18: New Orleans, Louisiana, 16 September 1982. Scene ID: 40062-15591. Image reproduced by permission of EOSAT.

**PLATE 6.** SPOT color composite. SPOT 3 HRV image of the Strait of Bonifacio, separating Corsica (on the north) from Sardinia (on the south), 27 September 1993. Copyright CNES/SPOT Image. Reproduced by permission.

**PLATE 7.** Aral Sea shrinkage, 1962–1994 (see text, p. 197). The left image is a mosaic of CORONA images from 1962, revealing the extent of the Aral Sea; the right image, acquired by AVHRR in 1994, has been prepared to the same scale and orientation as the CORONA image. Images courtesy of U.S. Central Intelligence Agency and U.S. Geological Survey.

**PLATE 8.** Shuttle Imaging Radar images (SIR-C/X-SAR) of the region near Mount Pinatubo, Philippines (see text, pp. 231–232). The left image was acquired 14 April 1994; the right image 5 October 1994. Images courtesy of NASA and Jet Propulsion Laboratory, Pasadena, California.

**PLATE 9.** Thermal images of a residential structure showing thermal properties of separate elements of a wooden structure (March 1979, 11 P.M.; air temperature = 2°C). Top: the original thermal image (white = hot; dark = cool). Bottom: a color-coded version of the same data, with each change in color representing a change in temperature of about 1.5°C (see text, p. 257). Images courtesy of Daedalus Enterprises, Inc., Ann Arbor, Michigan.

**PLATE 10.** Landscape near Erfurt, Germany, as observed by a thermal scanner. See Chapter 8 (pp. 260–261) for extended discussion. Images courtesy of EUROSENSE-BELFOTOP N.W. and Daedalus Enterprises, Inc., Ann Arbor, Michigan.

**PLATE 11.** Applications Explorer Mission 1 (HCMM) image of northeastern United States and southeastern Canada, 11 May 1978 (see text, p. 262). Image courtesy of NASA.

**PLATE 12.** Example of a classified image. A portion of a Landsat TM scene of Topeka, Kansas, after image classification. Each pixel has been assigned to a land cover class and coded with a color, as discussed in Chapter 11 (p. 313).

**PLATE 13.** Two examples of AVIRIS images displayed in image cube format. The top portion shows the image as recorded at the shortest wavelength, in conventional format with a perspective view. Sides of the cubes represent edges of each image as recorded in 220 channels, with the longest wavelength at the bottom (see Figure 14.4, p. 403). Left: Moffett Field, California. Right: Rogers Dry Lake, California. Images courtesy of NASA and Jet Propulsion Laboratory, Pasadena, California.

**PLATE 14.** Agricultural scene, Columbia River, eastern Oregon, as viewed by the SPOT HRV, August 1993 (see text, pp. 453–454). Copyright CNES/SPOT Image. Reproduced by permission.

**PLATE 15.** Example of a ratio image. The left image shows a portion of a SPOT HRV image of a landscape in Rondonia, Brazil. Red indicates forested regions; blue, dry grasses of open pasture; lighter reds and oranges, brushy regrowth in open land. The right image shows the same image represented using the normalized difference vegetation index, as discussed in the text (pp. 461–463). White indicates pixels dominated by healthy vegetation; black, sparsely vegetated pixels; gray, pixels with varying amounts of brushy regrowth. Copyright CNES/SPOT Image. Reproduced by permission.

**PLATE 16.** Band ratios used to study lithologic differences, Cuprite, Nevada, mining district. Variations in color represent variations in ratios between spectral bands, as discussed in the text (pp. 503–504). Image courtesy of GeoSpectra Corporation. From Vincent et al. (1984, p. 226).

**PLATE 17.** Chesapeake Bay, as photographed 8 June 1991 from the Space Shuttle. Land areas appear in silhouette due to underexposure to record hydrographic features. Visible here are the coastal marshes discussed in Chapter 18 (Section 18.10, Figures 18.18–18.21). Photo courtesy of NASA (STS-40-614-047).

**PLATE 18.** Belgian port of Zeebrugge as observed by the Daedalus digital multispectral scanner using 12 channels in the visible, near infrared, and thermal spectra. Reds and yellows indicate high sediment content; blues and greens, clearer water. The left image shows conditions at low tide; the right image, high tide (see text, p. 546). Images courtesy of EUROSENSE-BELFOTOP N.W. and Daedalus Enterprises, Inc., Ann Arbor, Michigan.

**PLATE 19.** Landsat TM quarter scene depicting Santa Rosa del Palmar, Bolivia (northwest of Santa Bruise), July 1992 (TM bands 3, 4, and 5). The landscape in the southeast is dominated by broad-scale mechanized agriculture. The upper right is a mountainous area occupied by Indians who practice slash-and-burn agriculture, visible as dispersed patches of light green. The Bolivian government has encouraged broad-scale clearing of forest for agriculture; in the upper left, these clearings appear as light green spots aligned northwest to southeast; at the center of each patch is a central facility providing colonists with staples. In the southwest, the complex field pattern reflects an established agricultural landscape (see text, p. 550). GEOPIC image courtesy of Earth Satellite Corporation, Rockville, Maryland, Reproduced by permission.

**PLATE 20.** Images depicting global remote sensing data. Top: AVHRR data representing sea-surface temperatures, July 1984. The warmest surfaces are shown as red; in decreasing sequence, oranges, yellows, green, and finally blues represent the coolest temperatures. Bottom: composite data from two separate NASA satellites combined to represent the earth's biosphere. In the oceans, colors represent phytoplankton concentrations (highest values in browns and yellows, lower values in purples and dark blues). On land, greens represent dense vegetation; yellows, sparsely vegetated regions (see text, p. 584). Images courtesy of NASA Goddard Space Flight Center.

**PLATE 1**

**PLATE 2**

PLATE 3

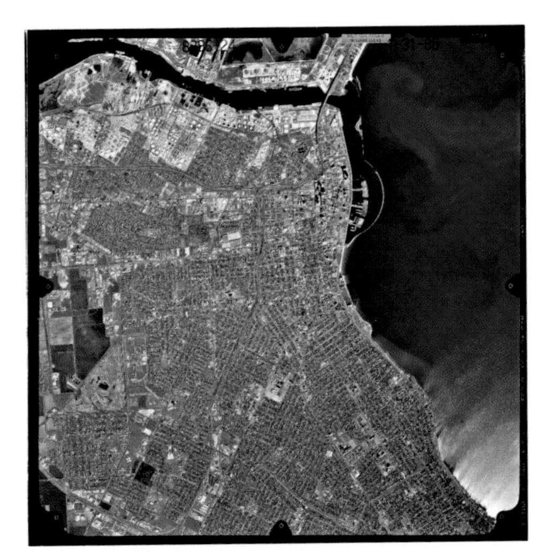

PLATE 4

PLATE 5

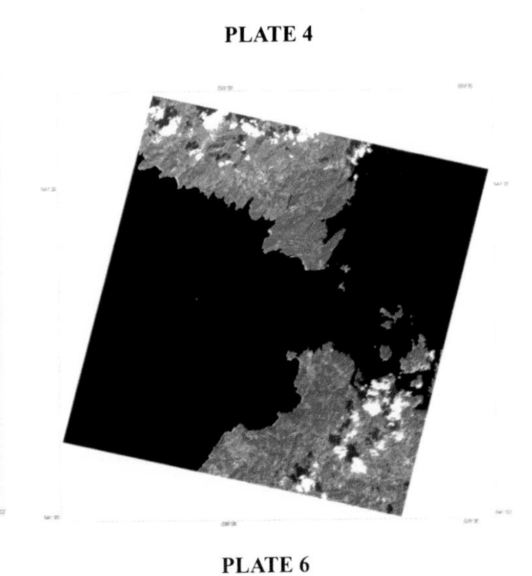

PLATE 6

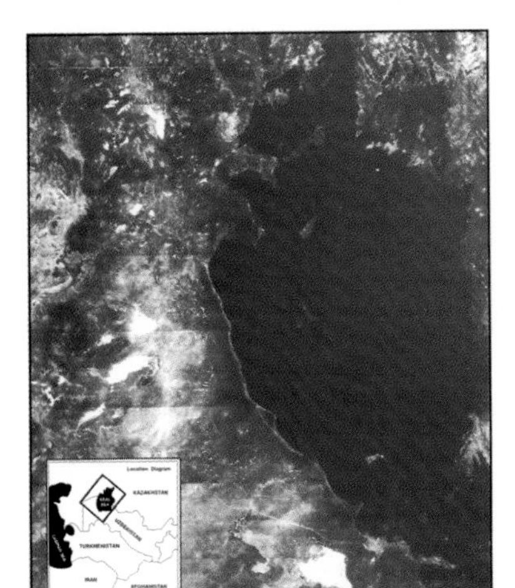


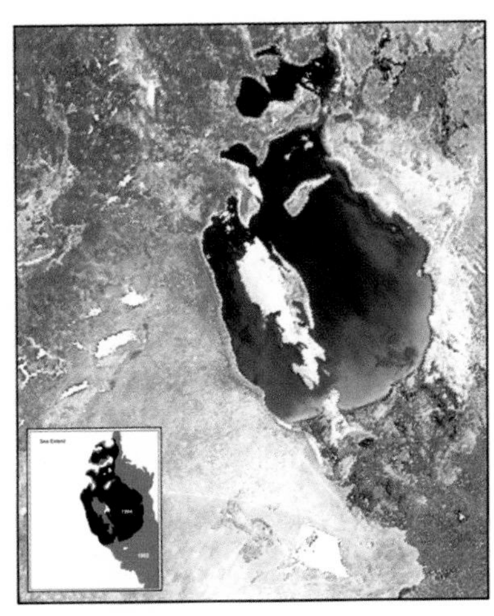


PLATE 7

**PLATE 11**

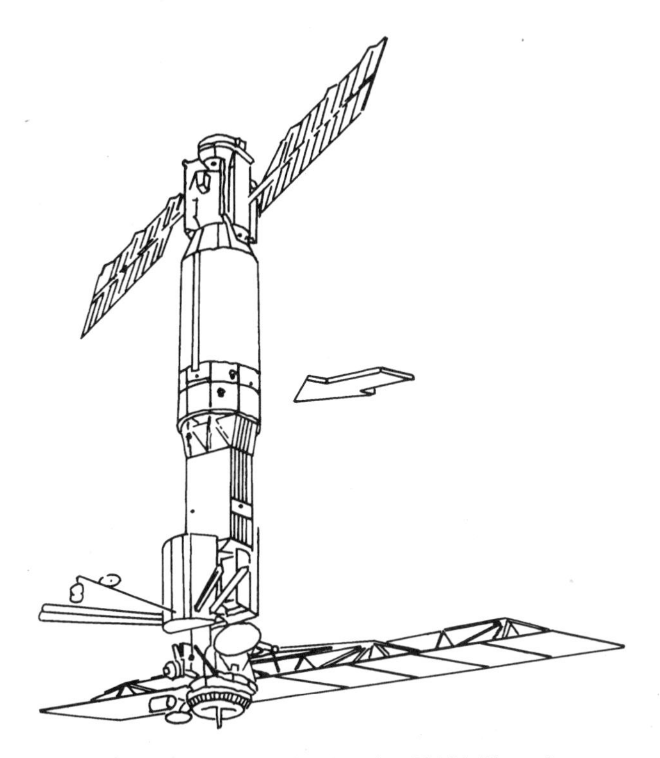

**FIGURE 7.26.** Seasat SAR. Based on NASA illustration.

age of oceans, but its track did cross land areas, thereby offering the opportunity to acquire radar imagery of the earth's surface from satellite altitudes (Figure 7.27). The SAR, of course, had the capability for operation during both daylight and darkness, and during inclement weather, so (unlike most other satellite sensors) the Seasat SAR could acquire data on both ascending and descending passes.

The high transmission rate required to convey data of such fine resolution means that SAR data cannot be recorded on board for later transmission to ground stations but must be immediately transmitted to a ground station. Therefore, data could be acquired only when the satellite was within line of sight of one of the five ground stations equipped to receive Seasat data, located in California, Alaska, Florida, Newfoundland, and England. The SAR was first turned on in early July 1978 (day 10); in all, it was in operation for 98 days, and acquired some 500 passes of data. The average duration of each pass was about 5 minutes, but the longest SAR track covered about 4,000 km in ground distance. Data are available in digital form, as either digitally or optically processed imagery. All land areas covered are in the Northern Hemisphere, including portions of North America and Western Europe.

Seasat SAR data have formed the basis for important oceanographic studies. In addition, Seasat data have provided the basis for the study of radar applications within the earth sciences and for the examination of settlement and land-use patterns. Because data are acquired on both ascending and descending passes, the same ground features can be observed from differing look angles under conditions that hold most other factors constant, or almost constant.

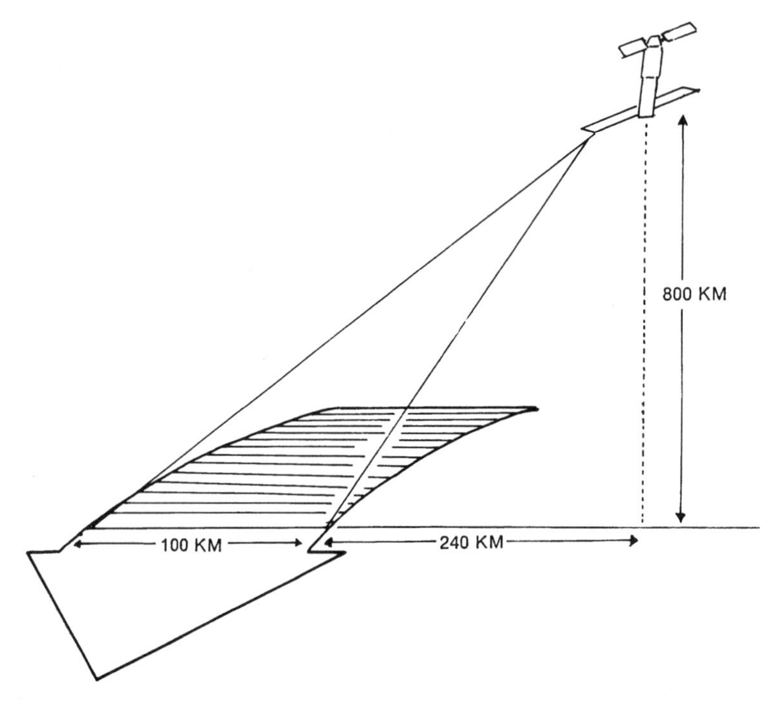


**FIGURE 7.27.** Seasat SAR geometry.

### *Shuttle Imaging Radar-A*

The Shuttle Imaging Radar (SIR) is a synthetic aperture imaging radar carried by the Shuttle Transportation System. (NASA's Space Shuttle Orbiter has the ability to carry a variety of scientific experiments; the SIR is one of several.) The first scientific payload carried on board the shuttle (during the second flight of Columbia) was SIR-A, the first of the imaging radars carried by the shuttle. SIR-A was operated for about 54 hours during the Columbia flight in November 1981. Although this time was shorter than planned, due to the reduction in length of the mission from original plans, SIR-A was able to acquire images of almost 4,000,000 mi.$^2$ of the earth's surface during this interval. Additional details of SIR-A are given below:

- *Launch:* 12 November 1981
- *Land:* 14 November 1981
- *Altitude:* 259 km
- *Frequency:* L-band (1.278 GHz)
- *Wavelength:* 23.5 cm
- *Depression angle:* 40°
- *Swath width:* 50 km
- *Ground resolution:* about 40 m × 40 m
- *Polarization:* HH

**FIGURE 7.28.** SIR-A image, southwestern Virginia. Image courtesy of National Space Science Data Center.

Geometric qualities were fixed; there was no provision for changing depression angle All data were recorded on magnetic tape and signal film on board the shuttle, then after the Shuttle landed, physically carried to ground facilities for processing into image products. First-generation film images were typically 12 cm (5 in.) in width at a scale of about 1:5,250,000 (Figure 7.28). About 1,400 ft. of film were processed; most has been judged to be of high quality.

SIR-A completed five passes over the United States (during which SIR-A was in operation for about 50 minutes). Portions of all continents (except Antarctica) were imaged, to provide images of a wide variety of environments differing with respect to climate, vegetation, geology, land use, and other qualities. Analyses of these data have led to important scientific studies of several subjects, including the geology of remote areas (Sabins, 1983) and further knowledge of the capabilities of radar imagery.

### *Shuttle Imaging Radar-B*

SIR-B was the second imaging radar experiment for the Space Shuttle. It was similar in design to SIR-A, except that it provided greater flexibility in acquiring imagery at varied depression angles (Figure 7.29), as shown below:

- *Orbital altitude:* 225 km
- *Orbital inclination:* 57°
- *Frequency:* 1.28 GHz
- *Wavelength:* 23 cm
- *Swath width:* 40 to 50 km
- *Resolution:* about 25 m × 17 m (at 60° depression), or
  about 25 m × 58 m (at 15° depression)

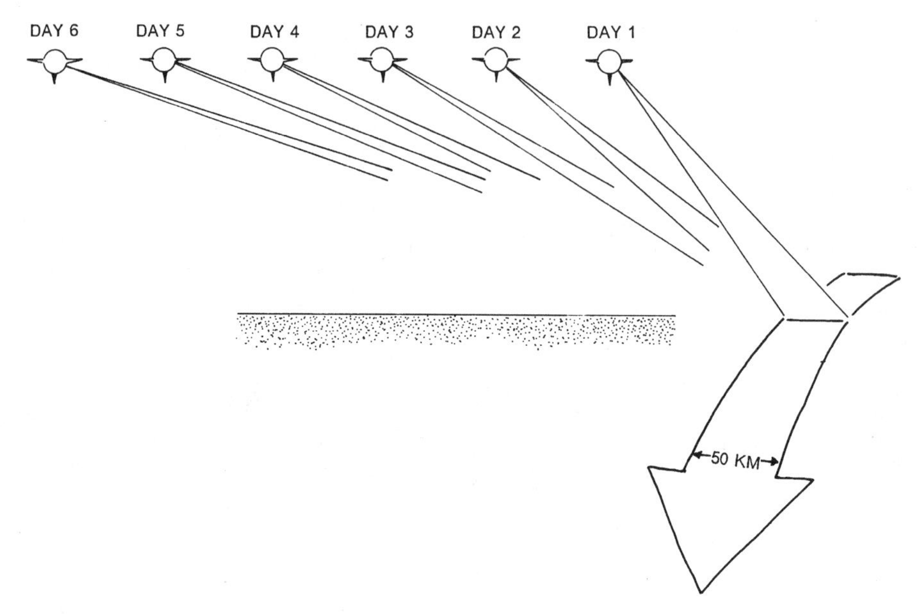


**FIGURE 7.29.** SIR-B geometry. Depression angle can be changed to view the same area on different dates.

SIR-B used horizontal polarization at L-band (23 cm). Azimuth resolution was about 25 m; range resolution varied from about 17 m at an angle of 60° to 58 m at 15°. The shuttle orbited at 225 km; the orbital track drifted eastward 86 km each day (at 45° latitude). In contrast to the fixed geometric configuration of the Seasat SAR and SIR-A, the radar pallet onboard the shuttle permited control of the angle of observation, which thereby enabled operation of the SIR in any of several modes. A given image swath of 40 to 50 km could be imaged repeatedly, at differing depression angles, by changing the orientation of the antenna as spacecraft position changes with each orbit. This capability provided the ability to acquire stereo imagery. Or, the antenna could be oriented to image successive swaths and thereby build up coverage of a larger region than can be covered in any single pass. A mosaic composed of such images would be acquired at a consistent range of depression angles, although angles would, of course, vary within individual images.

The shuttle orbit also provided the opportunity to examine a single study area at several look directions. Varied illumination directions and angles permit imaging of a single geographic area at varied depression angles and azimuths and thereby enable (in concept) derivation of information concerning the influence of surface roughness and moisture content on the radar image and information concerning the interactions of these variables with look direction and look angle. Thus, an important function of SIR-B was its ability to advance understanding of the radar image itself and its role in remote sensing of the earth's landscapes.

The significance of SIR imagery arises in part from the repetitive nature of the orbital coverage, with an accompanying opportunity to examine temporal changes in such phenomena as soil moisture, crop growth, land use, and so on. Because of the all-weather capability of SAR, the opportunities to examine temporal variation with radar imagery may be even more significant than it was with Landsat imagery. Second, costs of the SIR are said to be low relative to those of a satellite dedicated specifically to acquiring radar imagery. Third, the flexibility of the shuttle platform and the SIR experiment permits examination of a wide range of configurations for satellite radars, over a broad range of geographic regions, by many subject areas in the earth sciences. Although a number of practical applications will no doubt be discovered as SIR imagery is examined, it seems clear that the primary role of SIR-B will be to investigate optimum configurations and applications for use in the design of future satellites tailored specifically to acquire radar imagery of the earth's surface.

### *SIR-C/X-SAR Radar System*

In April and August 1994, the Space Shuttle conducted a third experiment with imaging radars. SIR-C is an SAR operating at both L-band (23 cm) and C-band (6 cm), with the capability for transmitting and receiving both horizontally and vertically polarized radiation. SIR-C, designed and manufactured by the Jet Propulsion Laboratory (Pasadena, CA) and Ball Communications Systems Division, is one of the largest and most complex items built for flight on the shuttle. In addition to the use of two microwave frequencies and its dual polarization, the antenna has the ability to electronically aim the radar beam, to supplement the capability of the Shuttle to aim the antenna by manuvering the spacecraft. Data from SIR-C can be recorded on board using tape storage or can be trensmitted by microwave to the Tracking and Data Relay Satellite System (TDRSS) link to ground stations (Chapter 6).

X-SAR was designed and built in Europe as a joint German–Italian project for flight on the shuttle, to be used independently or in coordination with SIR-C. X-SAR was an X-band SAR (VV polarization) with the ability to create highly focused radar beams. It was mounted in a manner that permits it to be aligned with the L- and C-band beams of the SIR-C. Together, the two systems provided the ability to gather data at three frequencies and two polarizations. The SIR-C/X-SAR experiment was coordinated with field experiments that collected ground data at specific sites devoted to studies of ecology, geology, hydrology, agriculture, oceanography, and other topics. These instruments continue the effort to develop a clearer understanding of the capabilities of SAR data to monitor key processes at the earth's surface.

- *Frequencies:* X-band (3 cm)<br>C-band (6 cm)<br>L-band (23 cm)
- *Ground swath:* 15 to 90 km, depending on orientation of the antenna
- *Resolution:* 10 to 200 m

Plate 8 provides an example of the interpretation of data from SIR-C. The region near Mount Pinatubo, an active volcano in the Philippines, is shown by a pair of images, each depicting the same 40 km × 65 km (25 mi. × 40 mi.) region. The left-hand image was acquired 14 April 1994, the right-hand image on 5 October 1994. Both images were acquired using the same viewing geometry and displayed so that the L-band HH shows as red, the L-band HV data appear as green, and the C-band HV data appear as blue.

Red areas reveal surfaces that depolarized the signal (the horizontally transmitted signal is scattered as a vertically polarized signal); here these surface are formed chiefly from recent ash deposits from the 1991 eruption of the volcano. The elongated dark drainage features radiating from the main peak of the volcano are mudflows caused by redistribution of the ash deposits during the heavy rains of the summer months. During the interval between the two images, the summer rains greatly increased the area influenced by the mudflows.

Mudflows, and many other changes to the rivers, lakes, and groundwater in this region, are closely watched by geologists and hydrologists because of the large danger they pose to nearby communities, which have suffered major loss of life and property damage since the 1991 eruption, and the many aftereffects caused by erosion and deposition of the volcanic deposits.

### *ERS SAR*

The European Space Agency (ESA), a joint organization of several European nations, designed a remote sensing satellite with several sensors configured to conduct both basic and applied research. Here our primary interest is the SAR for European Remote Sensing Satellites 1 and 2 (ERS-1, launched in 1991, and ERS-2, launched in 1995). One of the satellite's primary missions is to use several of its sensors to derive wind and wave information within 5 km × 5 km SAR scenes positioned within the SAR's 80-km swath width. Nonetheless, the satellite has acquired a library of images of varied land and maritime scenes (Figure 7.30).

Because of the SAR's extremely high data rate, SAR data cannot be stored on board and thus must be acquired within range of a ground station. Although the primary ground control and receiving stations are located in Europe, receiving stations throughout the world are equipped to receive ERS-1 data. ERS-1 has a sun-synchronous, nearly polar orbit that crosses the equator at about 10:30 A.M. local sun time. Other characteristics include:

- *Altitude:* 785 km
- *Frequency:* 5.3 GHz
- *Wavelength:* C-band
- *Incidence angle:* 23° at midrange
- *Swath width:* 100 km
- *Spatial resolution:* 30 m
- *Polarization:* VV

ERS data are available from EOSAT (800-343-9933) and from RADARSAT International (604-244-0400).

### *RADARSAT SAR*

RADARSAT is a joint project of Canadian federal and provincial governments, the United States government, and private corporations. The United States contribution, through NASA, provides launch facilities and the services of a receiving station in Alaska. Canada

**FIGURE 7.30.** ERS-1 SAR image, Sault St. Marie, Michigan–Ontario, showing forested areas as bright tones, rough texture. Within water bodies, dark areas are calm surfaces; lighter tones are rough surfaces caused by winds and currents. Copyright 1991 by ESA. Received and processed by Canada Centre for Remote Sensing; distributed by RADARSAT International.

has special interests in use of radar sensors because of its large territory, the poor illumination prevailing during much of the year at high latitudes, unfavorable weather conditions, interest in monitoring sea ice in shipping lanes, and many other issues arising from assessing natural resources, especially forests and mineral deposits. Radar sensors, particularly those carried by satellites, provide capabilities tailored to address many of these concerns.

RADARSAT's C-band SAR was launched on 4 November 1995 into a sun-synchronous orbit at 98.6° inclination, with a 16-day repeat cycle and equatorial crossings at 0600 and 1800 hours. An important feature of the RADARSAT SAR is the flexibility to select among a wide range of trade-offs between area covered and spatial resolution, and to use a wide variety of incidence angles (Figure 7.31). This flexibility increases opportunities for practical applications. Critical specifications include:

- *Altitude:* 793–821 km
- *Frequency:* 5.3 GHz
- *Wavelength:* C-band (5.6 cm)
- *Incidence angle:* varies; 10°–60°
- *Swath width:* varies; 45–510 km
- *Spatial resolution:* varies; 100 m × 100 m to 9 m × 9 m
- *Polarization:* HH

Data are available from RADARSAT International (3851 Shell Road, Suite 200, Richmond, British Columbia V6X 2W2, Canada; 604-244-0400 voice; 604-244-0404 fax).

Figure 7.32 shows a RADARSAT SAR image of Cape Breton Island, Canada, on 28 November 1995; this image represents a region extending about 132 km east to west and 156 km north to south. At the time the RADARSAT SAR acquired this image, this region was in darkness and was experiencing high winds and inclement weather, so it illustrates well radar's ability to operate under conditions of unfavorable weather and low solar illumination. Sea conditions near Cape Breton Island's Cape North and Aspy Bay are recorded as variations in image tone. Ocean surfaces appearing black or dark are calmer waters, often sheltered from winds and currents by the configuration of the coastline or by the islands themselves. Brighter surfaces are rougher seas, usually on windward coastlines, formed as winds and currents interact with subsurface topography. The urban area of Sidney is visible as the bright area in the lower center of the image.

### *JERS-1*

JERS-1 is a satellite launched in February 1992 by the National Space Development Agency of Japan carrying a C-band SAR. In addition, it carries an optical CCD sensor sensitive in seven spectral regions from 0.52 to 2.40 μm, with resolution of about 18 m.

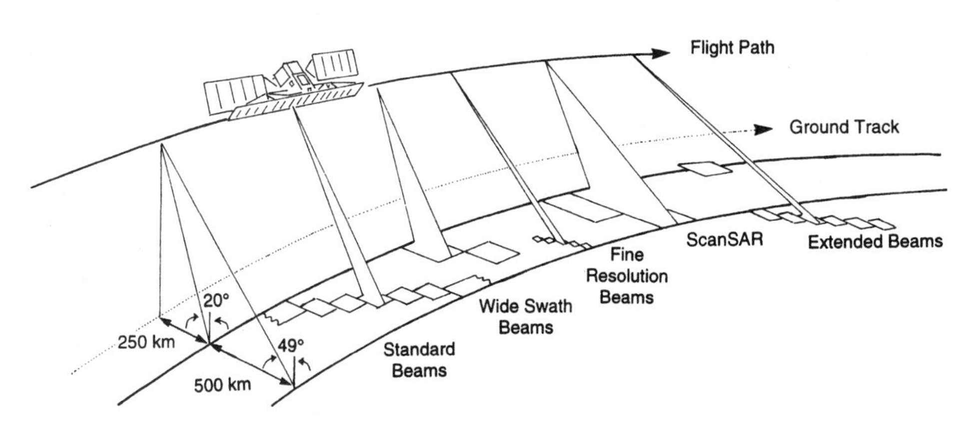


**FIGURE 7.31.** RADARSAT SAR geometry.

**FIGURE 7.32.** RADARSAT SAR Image, Cape Breton Island, Canada. RADARSAT's first SAR image, acquired 28 November 1995. Image copyright 1995 by Canadian Space Agency. Received by the Canada Centre for Remote Sensing; processed and distributed by RADARSAT International.

One band in the near infrared is collected with a forward-looking sensor; when combined with data from the nadir-aimed instruments, it can provide a stereo capability.

The JERS-1 SAR has the following characteristics:

- *Altitude:* 568 km
- *Orbit:* sun-synchronous
- *Wavelength:* L-band (23 cm)
- *Incidence angle:* 35°
- *Swath width:* 75 km
- *Spatial resolution:* 30 m
- *Polarization:* HH

## 7.11. Summary

Radar imagery is especially useful because it complements many of the characteristics of images acquired in other portions of the spectrum. For example, aerial photography depends on proper solar illumination and tends to record information concerning the biologic components of plant tissues. In contrast, radar imagery is independent of solar radiation and tends to portray information about the physical structure of plant communities. Likewise, the two forms of imagery have different geometric characteristics. Therefore, for many purposes, radar images supplement, rather than replace, the more common forms of imagery from the visible and infrared regions.

There is now a large and growing archive of high-quality radar imagery that has contributed to an increased understanding of the information conveyed by radar data. To exploit the potential of radar imagery, it is necessary to understand its performance in diverse environments that can permit unraveling of the complex interactions between frequency, look angle, look direction, soil moisture, vegetation structure, and the many other variables that influence radar images. Fortunately, the availability of high-quality radar data for regions throughout the world is providing the basis for developing this understanding. The growing number of satellite radar systems forms evidence of the value of radar data in many spheres of the practice of remote sensing.

## Review Questions

1. List advantages for the use of radar images, relative to aerial photography and Landsat MSS. Can you identify disadvantages?

2. Imaging radars may not be equally useful in all regions of the earth. Can you suggest certain geographic regions where they might be most effective? Are there other geographic zones where imaging radars might be less effective?

3. Radar imagery has been combined with data from other imaging systems, such as the Landsat MSS, to produce composite images. Because these composites are formed from data from two widely separated portions of the spectrum, together they convey much more information than either image can alone. Perhaps you can suggest (from information already given in Chapters 3 and 6) some of the problems encountered in forming and interpreting such composites.

4. Why might radar images be more useful in many less developed nations than in industrialized nations? Can you think of situations in which radar images might be especially useful in the industrialized regions of the world?

5. A given object or feature will not necessarily have the same appearance on all radar images. List some of the factors that will determine the texture and tone of an object as it is represented on a radar image.

6. Seasat was, of course, designed for observation of the earth's oceans. Why are the steep depression angles of the Seasat SAR inappropriate for many land areas? Can you think of advantages for use of steep depression angles in some regions?

7. What problems would you expect to encounter if you attempted to prepare a mosaic from several radar images?
8. Why are synthetic aperture radars required for radar observation of the earth by satellite?
9. Why is the shuttle imaging radar so important in developing a more complete understanding of interpretation of radar imagery?
10. Compare and contrast satellite sensors that use active microwave energy and those that depend upon energy in the visible and near infrared imagery. Consider scale, resolution, orbit, timing, wavelengths used, satellite altitude, and area represented on a given image.

## References

Born, G. H., J. A. Dunne, and D. B. Lane. 1979. Seasat Mission Overview. *Science,* Vol. 204, pp. 1405–1406.
Brown, W. M., and L. J. Porcello. 1969. An Introduction to Synthetic Aperture Radar. *IEEE Spectrum,* Vol. 6, No. 9, pp. 52–62.
Bryan, M. Leonard, and F. M. Henderson. 1983. *Radar Remote Sensing from Space* (40 35-mm slides with text and descriptive material). Lincoln, NE: GPN (Box 80669; 402-472-2007).
Elachi, C. et al. 1982. Shuttle Imaging Radar Experiment. *Science,* Vol. 218, pp. 996–1003.
Elachi, C. et al. 1982. Subsurface Valleys and Geoarcheology of the Eastern Sahara. *Science,* Vol. 218, pp. 1004–1007.
Ford, J. P., R. G. Blom, M. L. Bryan, M. I. Daly, T. H. Dixon, C. Elachi, and E. C. Xenos. 1980. Seasat Views North America, the Caribbean, and Western Europe with Imaging Radar. JPL Publication 80-67. Pasadena, CA: Jet Propulsion Laboratory, 141 pp.
Ford, J. P., J. B. Cimino, and C. Elachi. 1983. *Space Shuttle Columbia Views the World with Imaging Radar: The SIR-A Experiment.* JPL Publication 82-95. Pasadena, CA: Jet Propulsion Laboratory, 179 pp.
Jensen, H., L. C. Graham, L. J. Porcello, and E. M. Leith. 1977. Side-Looking Airborne Radar. *Scientific American,* Vol. 237, pp. 84–95.
Jet Propulsion Laboratory (JPL). 1982. *The SIR-B Science Plan.* JPL Publication 82-78. Pasadena, CA: Author, 90 pp.
Page, Robert M. 1962. *The Origin of Radar.* New York: Doubleday, 169 pp.
Sabins, F. F. 1983. Geologic Interpretation of Space Shuttle Radar Images of Indonesia. *AAPG Bulletin,* Vol. 67, pp. 2076–2099.
Simpson, Robert. B. 1966. Radar, Geographic Tool. *Annals, Association of American Geographers.* Vol. 56, pp. 80–96.
Settle, Mark, and J. V. Taranick. 1982. Use of the Space Shuttle for Remote Sensing Research: Recent Results and Future Prospects. *Science,* Vol. 218, pp. 993–995.
Southworth, C. Scott. 1984. The Side-Looking Airborne Radar Program of the U.S. Geological Survey. *Photogrammetric Engineering and Remote Sensing,* Vol. 50, pp. 1467–1470.

**PROJECT RADAM**

Project RADAM (for "Radar-Amazon") began in 1971 as a joint project of the Brazilian government and the U.S. firms Goodyear Aerospace Corporation and Aero Service Corporation.

Prevailing weather in tropical regions has long presented problems for acquisition of aerial photography. In tropical regions, cloud cover is by no means continuous or perpetual, but its extent and occurrence are sufficient to prevent routine acquisition during those times of day when sun angles are most favorable for aerial photography. In addition, the Amazon basin is so remote, and so extensive (larger than the 48 contiguous United States) that even under favorable weather conditions, acquisition of photographic coverage would be a formidable task. Much of the region is covered by dense forest, is without good landmarks, and before Project RADAM was not covered by reliable maps.

Thus, the problem of surveying the Amazon basin seemed to be tailor-made for the capabilities of imaging radars. The Goodyear–Aero Service team equipped a Carvelle jet with the latest navigational equipment, and with the synthetic aperture radar developed by Goodyear. Flights were planned along north–south flight lines with the radar looking to the west; terrain shadowing therefore would approximate that of Landsat imagery (which in 1971 had not yet been launched) as a means of facilitating comparisons of the two kinds of imagery. Each radar image covered a 20-mi. swath, with about 20% sidelap to permit mosaicking. Some images were flown with 60% overlap to permit stereo viewing.

During the period 1971–1975 virtually all the Amazon basin was covered by imagery acquired by Project RADAM. Over 100,000 square miles in Venezuela, 1,000,000 mi.$^2$ in central Brazil, then later another 386,000 mi.$^2$ in southeastern Brazil were covered by the radar imagery. The radar images were acquired at rather small scales by the standards of developed nations; original images were at 1:400,000. From these images, semicontrolled mosaics at 1:250,000 were constructed for use in image interpretation. Coverages of individual mosaic sheets were tailored to match boundaries of quadrangles of the International Map of the World (published at 1:1,000,000); several hundred sheets were required to cover the Amazon region.

Initial interpretations were made as overlays to the 1:250,000 mosaics. Scientists from many disciplines (including geologists, pedologists, botanists, and foresters) worked to produce a variety of maps. The completed maps were published at 1:500,000 and 1:1,000,000, with detail reduced to an level appropriate for the smaller scales of the published sheets.

The maps produced by Project RADAM provided the basis for important discoveries. Previous information was so sketchy that the accurate maps revealed the courses of river systems previously unknown in detail. In addition, geologists working with radar imagery discovered important mineral deposits. Foresters, botanists, and pedologists were able to compile reconnaissance maps to provide the basis for planning of resource use. The total cost of the project was over $37 million.

CHAPTER EIGHT

# Thermal Radiation

## 8.1. Introduction

The literal meaning of infrared is "below the red," indicating that its frequencies are lower than those in the red portion of the visible spectrum. With respect to wavelength, however, the infrared spectrum is "beyond the red," having wavelengths longer than those of red radiation. The infrared is often defined from about 0.76 μm to about 1,000 μm (1 mm), there are great differences between properties of radiation within this range, as noted below.

The infrared spectrum was discovered in 1800 by Sir William Herschel (1738–1822), a British astronomer who was searching for the relationship between heat sources and visible radiation. Later, in 1847, two Frenchmen, A. H. L. Fizeau (1819–1896) and J. B. L. Foucault (1819–1868), demonstrated that infrared radiation has optical properties similar to those of visible light with respect to reflection, refraction, and interference patterns.

The infrared portion of the spectrum extends beyond the visible region to wavelengths about 1 mm (Figure 8.1). The shorter wavelengths of the infrared spectrum, near the visible, behave in a manner analogous to visible radiation—this region forms the *reflective infrared* spectrum, extending from about 0.7 μm to 3.0 μm. Many of the same kinds of films, filters, lenses, and cameras that we use in the visible portion of the spectrum can also be used, with minor variations, for imaging in the near infrared. The very longest infrared wavelengths are in some respects similar to the shorter wavelengths of the microwave region.

This chapter discusses use of radiation from about 7 to 18 μm for remote sensing of landscapes. This spectral region is often referred to as the *emissive* or *thermal* infrared. Of course, emission of thermal energy occurs over many wavelengths, so some prefer to use the term *far infrared* for this spectral region. In addition, passive remote sensing of shorter microwave radiation, which resembles thermal remote sensing in some respects, is also considered.

Remote sensing in the mid and far infrared is based on a family of imaging devices that differ greatly from the cameras and films used in the visible and near infrared. Interaction with the atmosphere is also quite different from that of shorter wavelengths. The far infrared regions are free from the scattering that is so important in the ultraviolet and visible regions, but absorption by atmospheric gases restricts uses of the mid and far infrared

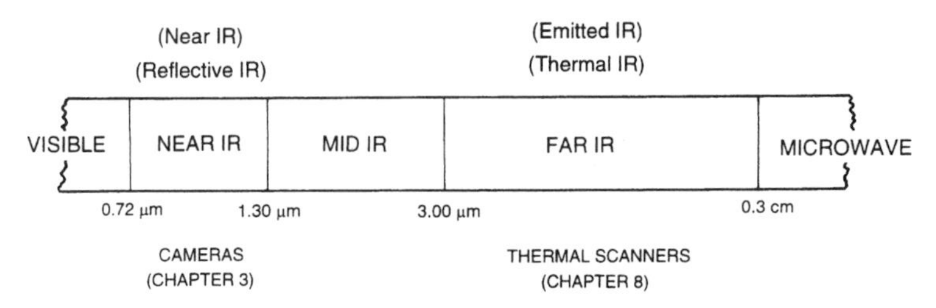


**FIGURE 8.1.** Infrared spectrum.

spectrum to specific atmospheric windows. Also, the kinds of information acquired by sensing the far infrared differ from those acquired in the visible and near infrared. Variations in emitted energy in the far infrared provide information concerning surface temperature and thermal properties of soils, rocks, vegetation, and man-made structures. Inferences based on thermal properties lead to inferences of the identities of surface materials.

## 8.2. Thermal Detectors

Before the 1940s, the absence of suitable instruments limited use of thermal infrared radiation for aerial reconnaissance. Aerial mapping of thermal energy depends on use of a sensor that is sufficiently sensitive to thermal radiation that variations in apparent temperature can be detected from by an aircraft moving at considerable speed high above the ground. Detectors in use prior to the 1940s, known as *thermal detectors,* responded to differences in thermal radiation through changes in electrical resistance or in other physical properties of the detector. Such instruments are widely used in the laboratory setting but are not sufficiently sensitive for use in the context of remote sensing. They respond slowly to changes in temperature and cannot be used at distances required for remote sensing applications.

During the 1940s, the development of photo detectors provided a practical means for use of the thermal portion of the spectrum in remote sensing. Such detectors are capable of responding directly to incident photons by changes in electrical resistance, providing a sensitivity and speed of response suitable for use in reconnaissance instruments. Since the 1940s a family of photo detectors has been developed to provide the basis for electro-optical instruments used in several portions of the spectrum.

Detectors are devices formed from substances known to respond to energy over a defined wavelength interval, generating a weak electrical signal with a strength related to the radiances of the features in the field of view of the sensor. The electrical current is amplified, then used to generate a digital signal that can be used to form a pictorial image, roughly similar in overall form to an aerial photograph (Figure 8.2).

Detectors have been designed with sensitivities for many of the spectral intervals of interest in remote sensing, including regions of the visible, near infrared, and ultraviolet spectra. Detectors sensitive in the thermal portion of the spectrum are formed from rather exotic materials, such as InSb (indium antimonide) and Ge:Hg (mercury-doped germanium). InSb has a peak sensitivity near 5 μm, in the mid infrared, and Ge:Hg has a peak

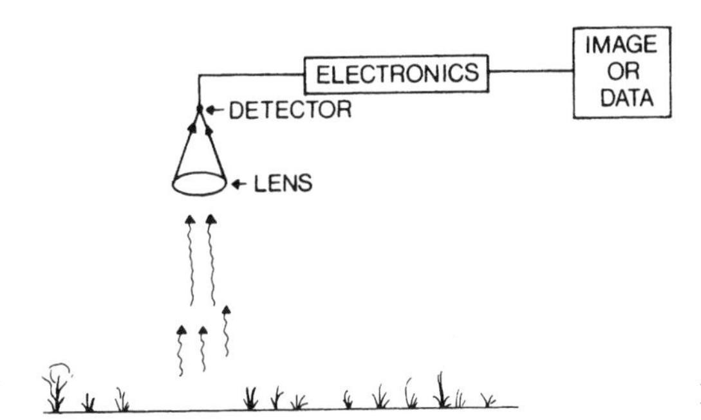


**FIGURE 8.2.** Use of thermal detectors.

sensitivity near 10 μm, in the far infrared spectrum (Figure 8.3). Mercury cadmium telluride (MCT) is sensitive over the range 8–14 μm. To maintain maximum sensitivity, such detectors must be cooled to very low temperatures (–196° C or –243° C) using liquid nitrogen or liquid helium.

The sensitivity of the detector can be seen to be a significant variable in the design and operation of the system. A low sensitivity means that only large differences in brightness are recorded ("coarse radiometric resolution") and most of the finer detail in the scene is lost. High sensitivity means that finer differences in scene brightness are recorded ("fine radiometric resolution"). The signal-to-noise ratio expresses this concept (Chapter 4). The "signal" in this context refer to differences in image brightness caused by actual variations in scene brightness. "Noise" designates variations unrelated to scene brightness. Such variations may be the result of unpredictable variations in the performance of the system. (There may also be random elements contributed by the landscape and the atmosphere, but here "noise" refers specifically to those contributed by the sensor.) If noise is large relative to the signal, the image does not provide a reliable representation of the feature of interest. Clearly, high noise levels will prevent imaging of subtle features, and even if noise levels are low, there must be minimum contrast between a feature and its background (i.e., a minimum magnitude for the signal) for the feature to be imaged. Also, note that increasing fineness of spatial resolution decreases the energy inci-

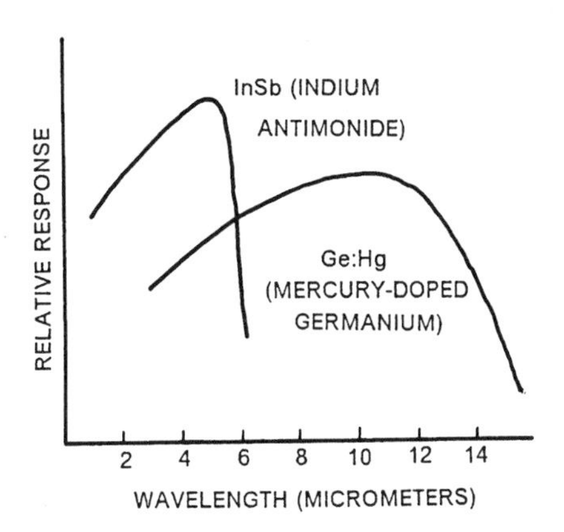


**FIGURE 8.3.** Sensitivity of some common thermal detectors.

dent upon the detector, with the effect of decreasing the strength of the signal. For many detectors, noise levels may remain constant even though the level of incident radiation decreases; if so, the increase in spatial resolution may be accompanied by decreases in radiometric resolution, as suggested in Chapter 3.

## 8.3. Thermal Radiometry

A *radiometer* is a sensor that measures the intensity of radiation received within a specified wavelength interval and within a specific field of view. Figure 8.4 gives a schematic view of a radiometer. A lens or mirror gathers radiation from the ground, then focuses it upon a detector positioned in the focal plane. A field stop may restrict the field of view, and filters may be used to restrict the wavelength interval that reaches the detector. A characteristic feature of radiometers is that radiation received from the ground is compared to a reference source of known radiometric qualities.

A device known as a *chopper* is capable of interrupting the flow radiation that reaches the detector. The chopper consists of a slotted disk, or similar device, rotated by an electrical motor so that as the disk rotates it causes the detector to alternately view the target and the reference source of radiation. Because the chopper rotates very fast, the signal from the detector consists of a stream of data that alternately measures the radiance of the reference source and radiation from the ground. The amplitude of this signal can be used to determine the radiance difference between the reference and the target. Because the reference source has known radiance, radiance of the target can then be estimated. The detector is alternately exposed to radiance from the target and the reference, producing an alternating current (AC). This allows the use of AC amplifiers, which are inherently more stable than direct current amplifiers.

Although there are many variations on this design, this description identifies the

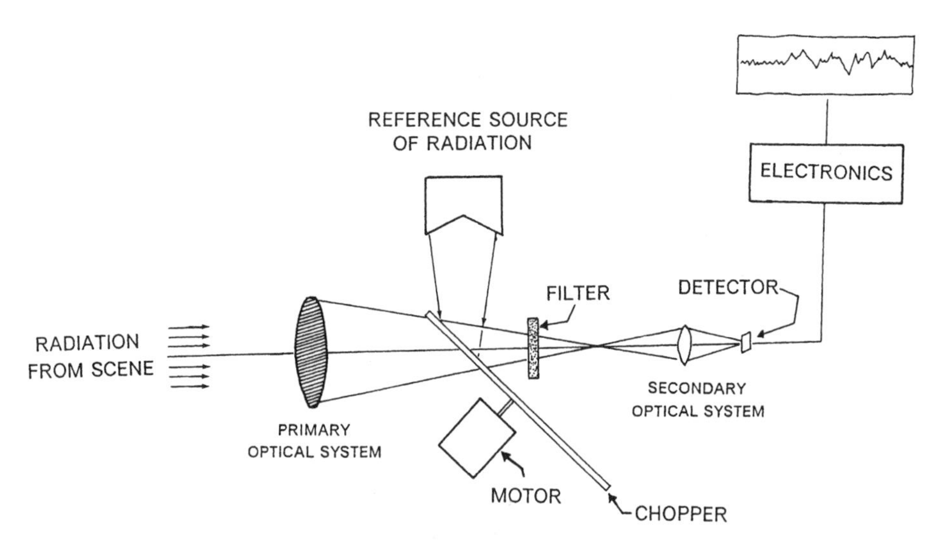


**FIGURE 8.4.** Schematic diagram of a radiometer.

most important components of radiometers. Related instruments include *photometers,* which operate at shorter wavelengths and often lack the internal reference source, and *spectrometers,* which very precisely examine radiance over a range of wavelengths, often extending from the infrared to the ultraviolet. By carefully tailoring the sensitivity of radiometers, scientists have been able to design instruments that are very useful in studying atmospheric gases and cloud temperatures. Conventional radiometers used for earth resource applications are often configured to view only a single trace along the flight path; the output signal then consists of a single stream of data that varies in response to differences in radiances of features along the flight line (Figure 8.5). A scanning radiometer can gather data from a corridor beneath the aircraft; output from such a system resembles those from some of the scanning sensors discussed in earlier chapters.

Spatial resolution of a radiometer is determined by an instantaneous field of view (IFOV), which is in turn controlled by the sensor's optical system, the detector, and flying altitude. Radiometers often have relatively coarse spatial resolution, for example (satellite-borne radiometers may have spatial resolutions of 100 km or more), in part because of the desirability of maintaining high radiometric resolution. To ensure that the sensor receives enough energy to make reliable measurements of radiance, the IFOV is defined to be a rather large; a smaller IFOV would mean that less energy would reach the detector, that the signal would be much too small with respect to system noise, and that the measurement of radiance would be much less reliable.

The IFOV can be defined informally as the area viewed by the sensor if the motion of the instrument were to be suspended so that it records radiation from only a single patch of ground. The IFOV can be more formally expressed as the angular field of view ($\beta$) of the optical system (Figure 8.6). The projection of this field of view onto the ground

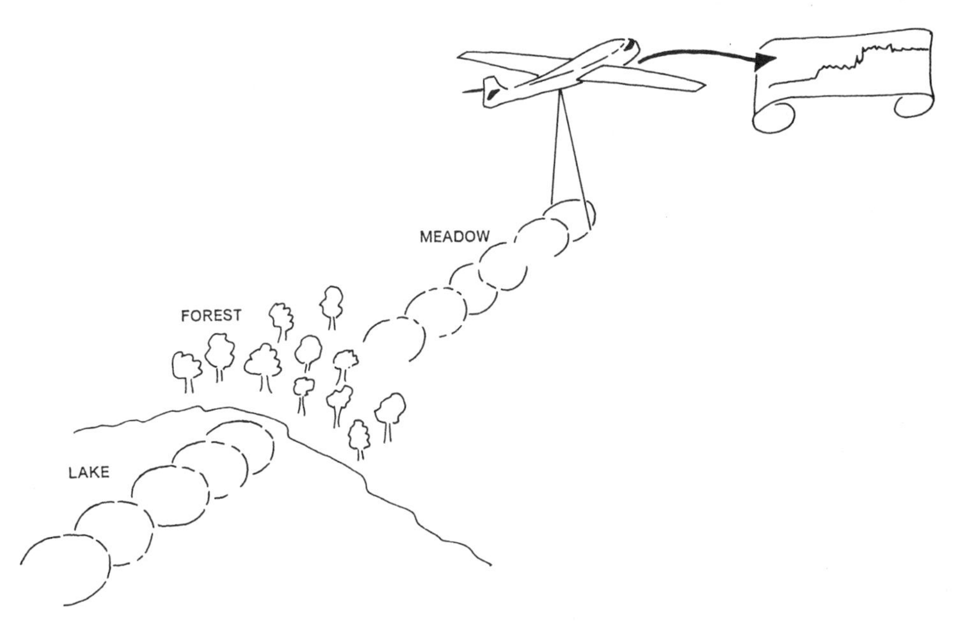


**FIGURE 8.5.** Single stream of data from a radiometer.

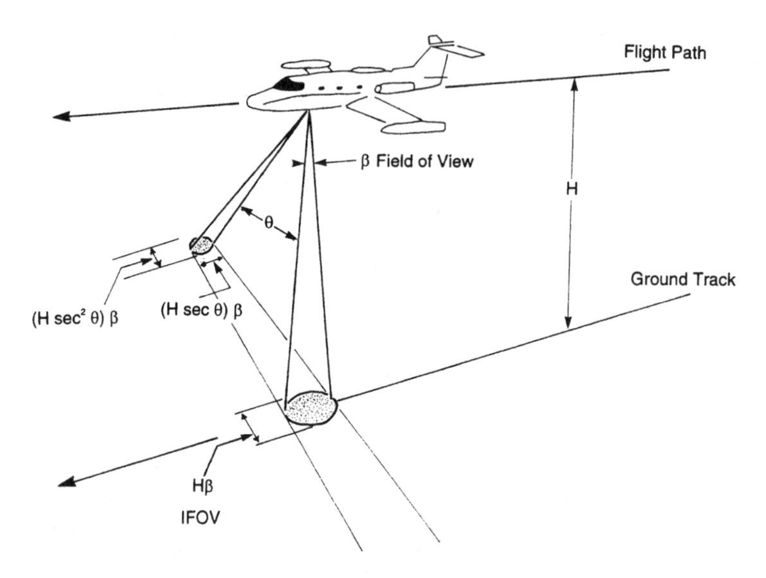


**FIGURE 8.6.** Instantaneous field of view. At nadir, the diameter of the IFOV is given by flying altitude (*H*) and the instrument's field of view (β). As the instrument scans to the side, it observes at angle θ; in the off-nadir position, the IFOV becomes larger and unevenly shaped.

surface defines the circular area that contributes radiance to the sensor. Usually β for a particular sensor is expressed in radians; to determine the IFOV for a particular image it is necessary to know the flying altitude (*H*) and to calculate the size of the circular area viewed by the detector.

From elementary trigonometry, it can be seen that the diameter of this area (*D*) is given as

$$D = H\beta \qquad \text{(Eq. 8.1)}$$

as illustrated by Figure 8.6. Thus, for example, if the angular field of view is 1.0 milliradians (mr) (1 mr = 0.001 radian) and the flying altitude (*H*) is 400 m above the terrain, then:

$$D = H\beta$$

$$D = 400 \times 1.0 \times 0.001$$

$$D = 0.40 \text{ m}$$

Because a thermal scanner views a landscape over a range of angles as it scans from side to side, the IFOV varies in size depending on the angle of obsevation ($\theta$). Near the nadir (ground track of the aircraft), the IFOV is relatively small; near the edge of the image, the IFOV is large. This effect is beneficial in one sense, becuse it compensates for effects of the increased distance from the sensor to the landscape, thereby providing consistent radiometric sensitivity across the image. Other effects are more troublesome.

Equation 8.1 defines the the IFOV at nadir as $H\beta$. At angle $\theta$, the IFOV measures $H$ sec $\theta\beta$ in the direction of flight and $H$ sec$^2$ $\theta\beta$ along the scan axis. Thus, near the nadir the IFOV is small and symetrical; near the edge of the image, it is larger and elongated along the axis of the scan. The variation in the shape of the IFOV creates geometric errors in the representations of features, discussed in subsequent sections. The variation in size means that radiance from the scene is averaged over a larger area and can be influenced by the presence of small features of contrasting temperature. Although small features can influence data for IFOVs of any size, such effects are more severe when the IFOV is large.

## 8.4. Microwave Radiometers

Microwave emissions from the earth convey some of the same information carried by thermal (far infrared) radiation. Even though their wavelengths are much longer than those of thermal radiation, microwave emissions are related to temperature and emissivity in much the same manner as is thermal radiation. Microwave radiometers are sensitive instruments tailored to receive and record radiation in the range from about 0.1 mm to 3 cm. Whereas the imaging radars discussed in Chapter 7 illuminate the terrain with their own energy and therefore are active sensors, microwave radiometers are passive sensors that receive microwave radiation naturally emitted by the environment. The strength and wavelength of such radiation are largely a function of the temperature and emissivity of the target. Thus, although microwave radiometers, like radars, use the microwave region of the spectrum, they are functionally most closely related to the thermal sensors discussed in this chapter.

In the present context, we are concerned with microwave emissions from the earth, which indirectly provide information pertaining to vegetation cover, soil moisture status, and surface materials. Other kinds of studies, peripheral to the field of remote sensing, derive information from microwave emissions from the earth's atmosphere or from extraterrestrial objects. In fact, the field of microwave radiometry originated with radio astronomy, and some of its most dramatic achievements have been in the reconnaissance of extraterrestrial objects.

A microwave radiometer consists of a sensitive receiving instrument typically in the form of a horn or dish-shaped antenna that observes a path directly beneath the aircraft; the signal gathered by the antenna is electronically filtered and amplified and displayed as a stream of digital data or, in the instance of scanning radiometers, as an image (Figure 8.7). As with thermal radiometers, microwave radiometers have a reference signal from an object of known temperature. The received signal is compared with the reference as a means of deriving the radiance of the target.

Examination of data from a microwave radiometer can be complex due to many factors that contribute to a given observation. The component of primary interest is usually energy radiated by the features within the IFOV; of course, variations within the IFOV are lost, as the sensor can detect only the average radiance within this area. The atmosphere also radiates energy, so it contributes radiance, depending on moisture content and temperatures. In addition, solar radiation in the microwave radiation can be reflected from the surface to the antenna.

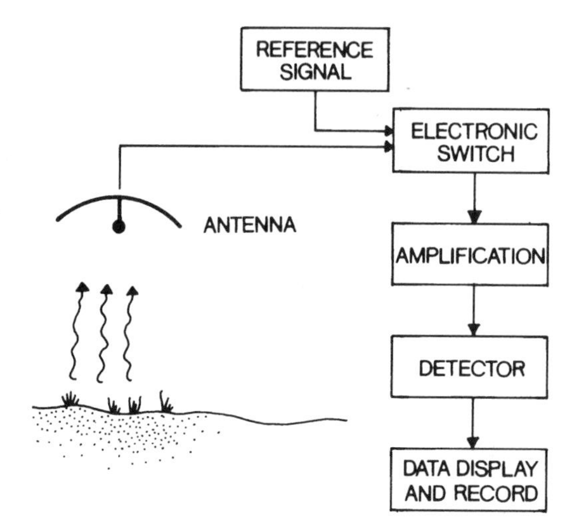


**FIGURE 8.7.** Microwave radiometer.

## 8.5. Thermal Scanners

The most widely used imaging sensors for thermal remote sensing are known as *thermal scanners*. Thermal scanners sense radiances of features beneath the aircraft flight path and produce digital and/or pictorial images of the terrain beneath the aircraft. There are several designs for thermal scanners. *Object-plane scanners* view the landscape by means of a moving mirror that oscillates at right angles to the flight path of the aircraft, generating a series of parallel (or perhaps overlapping) scan lines that together image a corridor directly beneath the aircraft. *Image-plane scanners* use a wider field of view to collect a more comprehensive image of the landscape; this image is then moved, by means of a moving mirror, relative to the detector. In either instance, the instrument is designed as series of lenses and mirrors configured to acquire energy from the ground and to focus it on the detector.

An infrared scanning system consists of a scanning unit, with gyroscopic roll connection unit, infrared detectors (connected to a liquid nitrogen cooling unit), and an amplification and control unit (Figure 8.8). A magnetic tape unit records data for later display as a video image; some systems may provide film recording of imagery as it is acquired. Together these units might weigh about 91 kg (200 lb.), and are usually mounted in an aircraft specially modified to permit the scanning unit to view the ground through an opening in the fuselage.

Infrared energy is collected by a scanning mirror that scans side to side across the flight path, in a manner similar to that described earlier for the MSS. The typical field of view might be as wide as 77°, so an aircraft flying at an altitude of 300 m (1,000 ft.) could record a strip as wide as 477 m (1,564 ft.). The forward motion of the aircraft provides the long dimension of the imagery. The mirror might make as any as 80 scans per second, with each scan representing a strip of ground about 46 cm (18 in.) in width (assuming the 300-m altitude mentioned above).

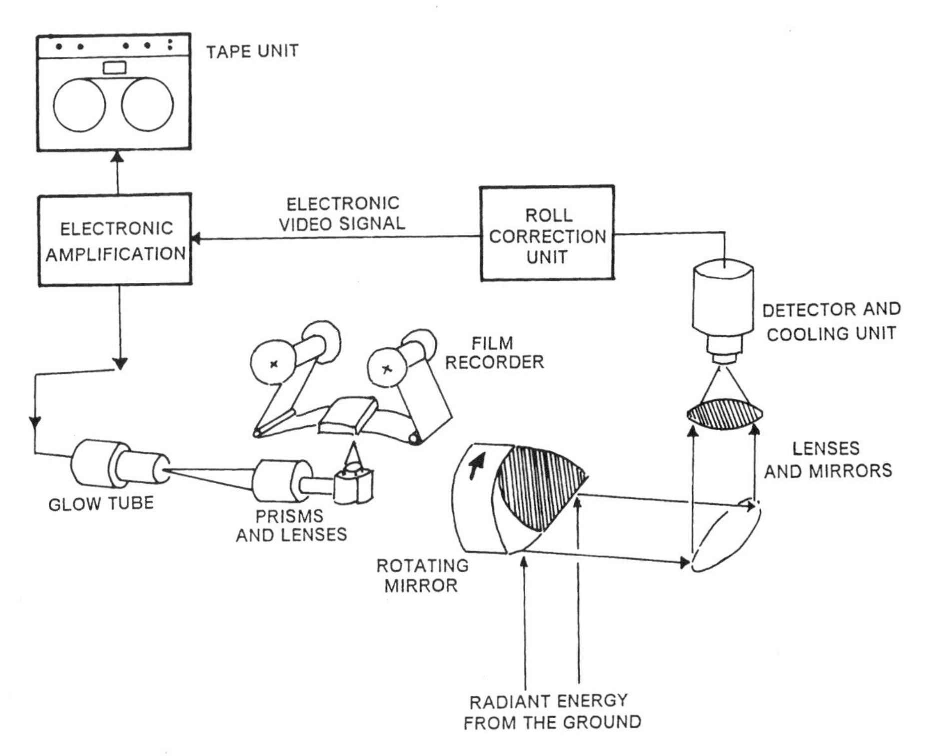


**FIGURE 8.8.** Thermal scanner.

Energy collected by the scanning mirror is focused on a parabolic mirror and then on a flat, stationary, mirror that focuses energy on the infrared detector unit. The infrared detector unit consists of one of the detectors mentioned above, confined in a vacuum container cooled by liquid nitrogen (to reduce electronic noise and the enhance sensitivity of the detector). On some units the detector can be easily changed by removing a small unit and replacing it with another. The detector generates an electrical signal that varies in strength in proportion to the radiation received by the mirror and focused on the detector. The signal from the detector is, however, very weak, so it must be amplified before it is recorded by the magnetic tape unit that is connected to the scanner. A *roll correction unit*, consisting in part of a gyroscope, senses side-to-side motion of the aircraft and sends a signal that permits the electronic control unit to correct the signal to reduce geometric errors caused by aircraft instability.

After the aircraft has landed, the magnetic tape from the sensor is removed from the aircraft and taken to a specially equipped laboratory. There, data from the tape can be displayed on a cathode ray tube; brightnesses on the screen are proportional to infrared energy received by the scanning mirror. Photographic representations of such images are prepared to form the scenes used for interpretation, or often the digital data may be analyzed directly as described in Chapters 9 through 13.

## 8.6. Thermal Properties of Objects

All objects at temperatures above absolute zero emit thermal radiation, although the intensity and peak wavelength of such radiation varies with the temperature of the object, as specified by the radiation laws outlined in Chapter 2. For remote sensing in the visible and near infrared, we examine contrasts in the abilities of objects to reflect direct solar radiation to the sensor. For remote sensing in the far infrared spectrum, we sense differences in the abilities of objects and landscape features to absorb shortwave visible and near infrared radiation, then to emit this energy as longer wavelengths in the mid and far infrared regions.

Thus, except for geothermal energy, man-made thermal sources, and range and forest fires, the immediate source of emitted thermal infrared radiation is shortwave solar energy. Direct solar radiation (with a peak at about 0.5 $\mu$m in the visible spectrum) is received and absorbed by the landscape (Chapter 2). The amount and spectral distribution of energy emitted by landscape features depends on the thermal properties of these features, as discussed below. The contrasts in thermal brightness, observed as varied gray tones on the image, are used as the basis for identification of features.

A *blackbody* is an object that acts as a perfect absorber and emitter of radiation; it absorbs and reemits all energy that it receives. Although the blackbody is an idealized concept, it is useful in describing and modeling the thermal behavior of actual objects, and it is possible to approximate the behavior of blackbodies in laboratory experiments.

As explained in Chapter 2, as the temperature of a blackbody increases, the wavelength of peak emission decreases in accordance with Wien's displacement law. The Stefan–Boltzmann law describes mathematically the increase in total radiation emitted (over a range of wavelengths) as the temperature of a blackbody increases.

Emissivity ($\epsilon_\lambda$) is a ratio between emittance from an object in relation to emittance from a blackbody at the same temperature:

$$\epsilon_\lambda = \frac{\text{Radiant emittance of an object}}{\text{Radiant emittance of a blackbody at the same temperature}} \quad \text{(Eq. 8.2)}$$

The subscript ($\lambda$) sometimes used with $\epsilon$ signifies that $\epsilon$ has been measured for specific wavelengths. Emissivity therefore varies from 0 to 1, with 1 signifying a substance with a thermal behavior identical to that of a blackbody. Table 8.1 lists emissivities for some common materials. Note that many of the substances commonly present in the landscape (e.g., soil and water) have emissivities rather close to 1. Note, however, that emissivity can vary with temperature, wavelength, and angle of observation.

### *Graybodies*

An object that has an emissivity less than 1.0 but has constant emissivity over all wavelengths is known as a *graybody* (Figure 8.9). A *selective radiator* is an object with an emissivity that varies with respect to wavelength. If two objects in the same setting are at the same temperature but have different emissivities, the one having the higher emissivity will radiate more strongly. Because the sensor detects radiant energy (apparent tempera-

**TABLE 8.1. Emissivities of Some Common Materials**

| Material | Temperature (°C) | Emissivity[a] |
|---|---|---|
| Polished copper | 50–100 | 0.02 |
| Polished brass | 200 | 0.03 |
| Polished silver | 100 | 0.03 |
| Steel alloy | 500 | 0.35 |
| Graphite | 0–3,600 | 0.7–0.8 |
| Lubricating oil (thick film on nickel base) | 20 | 0.82 |
| Snow | –10 | 0.85 |
| Sand | 20 | 0.90 |
| Wood (planed oak) | 20 | 0.90 |
| Concrete | 20 | 0.92 |
| Dry soil | 20 | 0.92 |
| Brick (red common) | 20 | 0.93 |
| Glass (polished plate) | 20 | 0.94 |
| Wet soil (saturated) | 20 | 0.95 |
| Distilled water | 20 | 0.96 |
| Ice | –10 | 0.96 |
| Carbon lamp black | 20–400 | 0.96 |
| Lacquer (matte black) | 100 | 0.97 |

*Note.* Data from Hudson (1969) and Weast (1986).
[a]Measured at normal incidence over a range of wavelengths.

ture) rather than the kinetic ("true") temperature, precise interpretation of an image requires knowledge of emissivities of features shown on the image.

### *Heat*

*Heat* is the internal energy of a substance arising from the motion of its component atoms and molecules. *Temperature* measures the relative warmth or coolness of a substance. It is the kinetic temperature or average thermal energy of molecules within a substance. *Kinetic temperature,* sometimes known as the *true temperature*, is measured using the usual

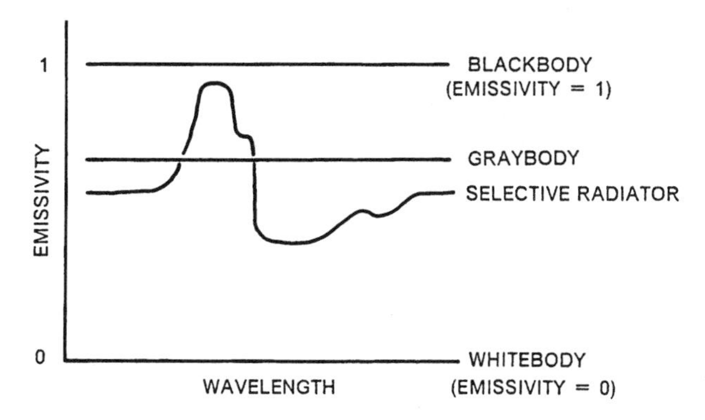


**FIGURE 8.9.** Blackbody, graybody, whitebody.

temperature scales, most notably the Fahrenheit, Celsius (Centigrade), and Kelvin (absolute) scales. *Radiant* (or *apparent*) *temperature* measures the emitted energy of an object. Photons from the radiant energy are detected by the thermal scanner.

*Heat capacity* is the ratio of the change in heat energy per unit mass to the corresponding change in temperature (at constant pressure). For example, we can measure the heat capacity of pure water to be 1 calorie per gram, meaning that 1 calorie is required for each gram to raise its temperature by 1°C. The specific heat of a substance is the ratio of its heat capacity to that of a reference substance. Because the reference substance typically is pure water, specific heat is often numerically equal to its heat capacity. Because a calorie is defined as the amount of heat required to raise by 1°C the temperature of 1 gram of pure water, use of water as the reference means that heat capacity and specific heat will be numerically equivalent. In this context, specific heat can be defined as the amount of heat (measured in calories) required to raise the temperature of one gram of a substance 1°C.

Thermal conductivity is a measure of the rate that a substance transfers heat. Conductivity is measured as calories per centimeter per second per degree Celsius, so it measures calories required to transfer a change in temperature over specified intervals of length and time.

Some of these variables can be integrated into a single measure, called *thermal inertia (P),* defined as

$$P = \sqrt{KC\rho} \qquad \text{(Eq. 8.3)}$$

where $K$ is the thermal conductivity (cal · $cm^{-1}$ · $sec^{-1}$ · $°C^{-1}$); $C$ is the heat capacity (cal · $g^{-1}$ · $°C^{-1}$); and $\rho$ is the density (g · $cm^{-3}$). *Thermal inertia, P,* is then measured in cal · $cm^{-2}$ · $°C^{-1}$ · $sec^{-1/2}$. Thermal intertia measures the tendency of a substance to resist changes in temperature, or more precisely, the rate of heat transfer at the contact between two substances. Table 8.2 gives thermal inertia values for a number of common materials. A substance with low thermal inertia is one with low density, low conductivity, and low specific heat and therefore resists changes in temperature deep within the material, although surface temperatures may change dramatically. Such substances (e.g., wood, glass, or cork) are insensitive to changes in temperature. In contrast, substances with high values of $P$ will heat and cool quickly. Such substances as silver, copper, and lead have high densities, high conductivity, and high specific heat.

These measures, and the terminology, are established by long-standing convention, even though they cause confusion. Many people find it confusing that a low value for thermal inertia indicates high resistance to temperature change because they interpret low inertia to mean the opposite. Therefore, the *thermal parameter,* $P^{-1}$, is perhaps easier to interpret, as high values indicate high resistance to change and low values indicate low resistance to thermal change.

## 8.7. Geometry of Thermal Images

Thermal scanners, like all remote sensing systems, generate geometric errors as they gather data. These errors mean that representations of positions and shapes of features de-

TABLE 8.2. Thermal Properties of Some Common Substances

| | $K$ | $\rho$ | $C$ | $P$ | $P^{-1}$ |
|---|---|---|---|---|---|
| Basalt | 0.0050 | 2.8 | 0.20 | 0.053 | 19 |
| Clay soil (moist) | 0.0030 | 1.7 | 0.35 | 0.042 | 24 |
| Dolomite | 0.012 | 2.6 | 0.18 | 0.075 | 13 |
| Granite | 0.0065 | 2.6 | 0.16 | 0.052 | 19 |
| Limestone | 0.0048 | 2.5 | 0.17 | 0.045 | 22 |
| Sandy soil | 0.0014 | 1.8 | 0.24 | 0.024 | 41 |
| Shale | 0.0030 | 2.3 | 0.17 | 0.034 | 29 |
| Slate | 0.0050 | 2.8 | 0.17 | 0.049 | 21 |
| Aluminum | 0.538 | 2.69 | 0.215 | 0.544 | 1.81 |
| Copper | 0.941 | 8.93 | 0.092 | 0.879 | 1.14 |
| Pure iron | 0.18 | 7.86 | 0.107 | 0.389 | 2.57 |
| Lead | 0.083 | 11.34 | 0.031 | 0.171 | 5.86 |
| Silver | 1.00 | 10.42 | 0.056 | 0.764 | 1.31 |
| Carbon steel | 0.150 | 7.86 | 0.110 | 0.360 | 2.78 |
| Glass | 0.0021 | 2.6 | 0.16 | 0.029 | 34 |
| Wood | 0.0005 | 0.5 | 0.327 | 0.009 | — |

*Note.* Selected from values given by Janza (1975). Thermal conductivity, $K$, measured in cal · cm$^{-1}$ · sec$^{-1}$ · °C$^{-1}$; density, $\rho$, measured in g · cm$^{-3}$; thermal inertia, $P$, measured in cal · cm$^{-2}$ · °C$^{-1}$ · sec$^{-1/2}$; thermal parameter symbolized by $P^{-1}$.

picted on thermal imagery do not match to their correct planimetric forms. Therefore, images cannot be directly used as the basis for accurate measurements.

Some errors are caused by aircraft instability (Figure 8.10). As the aircraft rolls and pitches, the scan lines lose their correct positional relationships, and, of course, the features they portray are not accurately represented in the image.

Thermal imagery also exhibits relief displacement analogous to that encountered in aerial photography (Figure 8.11). Thermal imagery, however, does not have the single central perspective of an aerial photograph but rather a separate nadir for each scan line. Thus, the focal point for relief displacement is the nadir for each scan line, or in effect the trace of the flight path on the ground. Thus relief displacement is projected from a line that follows the center of the long axis of the image. At the center of the image the sensor views objects from directly overhead and planimetric positions are correct. However, as distance from the center line increases, the sensor tends to view the sides rather than only the tops of features and relief displacement increases. These effects are visible in Figures 8.12 and 8.13; the tanker and the tanks appear to lean outward from a line that passes longitudinally through the center of the image. The effect increases toward the edges of the image.

Figure 8.12 also illustrates other geometric qualities of thermal line scan imagery. Although the scanning mirror rotates at a constant speed, the projection of the IFOV onto the ground surface does not move (relative to the ground) at equal speed because of the varied distance from the aircraft to the ground. At nadir, the sensor is closer to the ground than it is at the edge of the image; in a given interval of time, the sensor scans a shorter distance at nadir than it does at the edge of the image. Therefore, the scanner produces a geometric error that tends to compress features along an axis oriented perpendicular to the flight line, parallel to the scan lines. In Figure 8.12, this effect, known as *tangential*

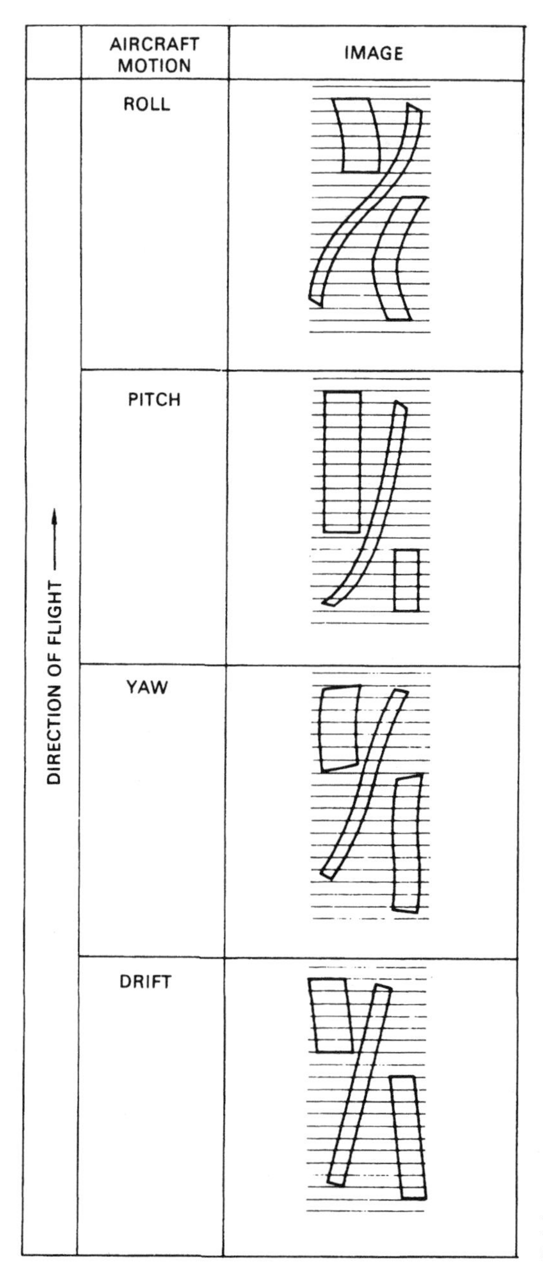


**FIGURE 8.10.** Geometric errors caused by aircraft instability.

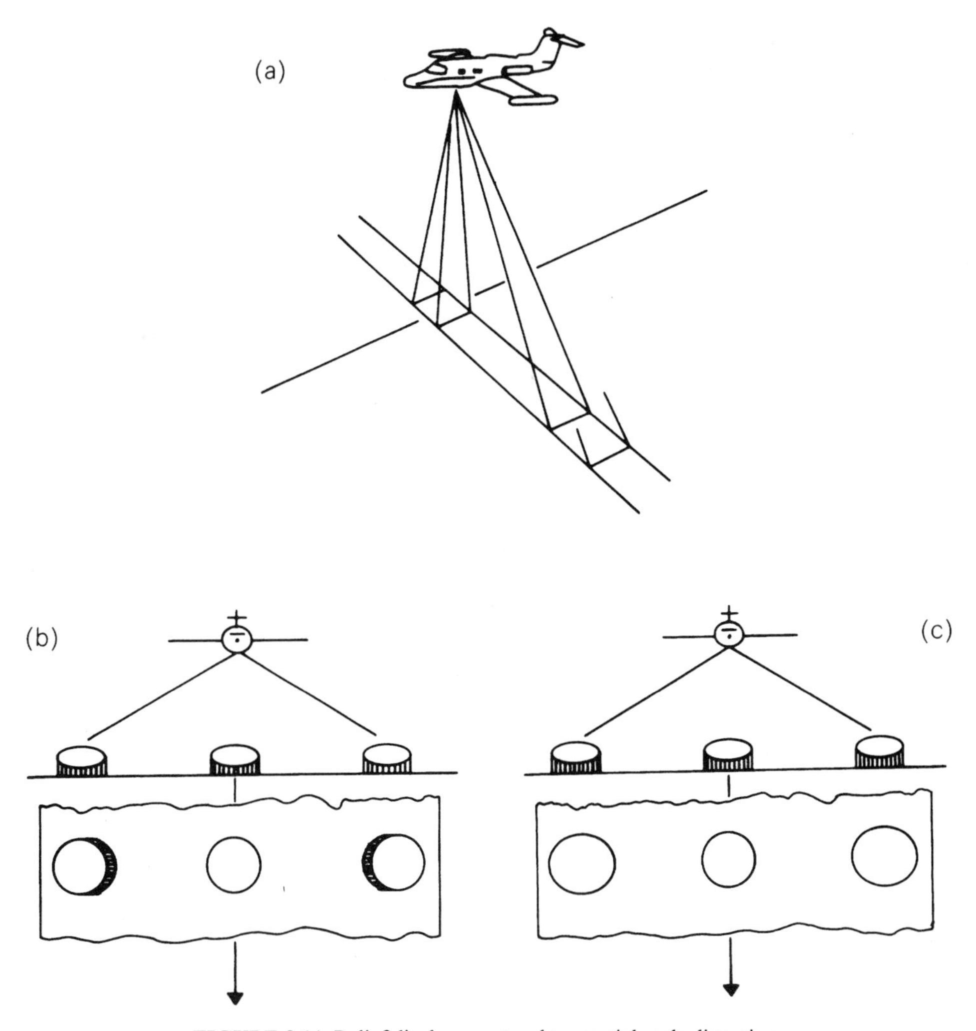


**FIGURE 8.11.** Relief displacement and tangential scale distortion.

*scale distortion*, is visible in the shapes of the cylindrical storage tanks. Images of those nearest the flight line are more circular, whereas shapes of those nearest the edge of the image are compressed along an axis perpendicular to the flight line. Sometimes the worst effects can be removed by corrections applied as the film image is generated, although it is often necessary to avoid use of the extreme edges of the image.

## 8.8. The Thermal Image and Its Interpretation

The image generated by a thermal scanner is a strip of black-and-white film depicting thermal contrasts in the landscape as variations in gray tones (e.g., Figure 8.12). Usually brighter tones (whites and light grays) represent warmer features; darker tones (dark grays

**FIGURE 8.12.** Thermal image of an oil tanker and petroleum storage facilities near the Delaware River, 19 December 1979. This image, acquired at 11:43 P.M., shows discharge of warm water into the Delaware River, and thermal patterns related to operation of a large petrochemical facility. Thermal image provided by Daedalus Enterprises, Inc., Ann Arbor, Michigan.

and blacks) represent cooler features. In some applications the black-and-white image may be subjected to level slicing or other enhancements that assign distinctive hues to specific gray tones as an aid for manual interpretation. Often it is easier for the eye to separate subtle shades of color than the variations in gray on the original image. Such enhancements are simply manipulations of the basic infrared image and do not represent differences in means of acquisition or in the quality of the basic information available for interpretation.

For any thermal infrared image the interpreter must always determine (1) whether the image at hand is a positive or negative image and (2) the time of day the image was ac-

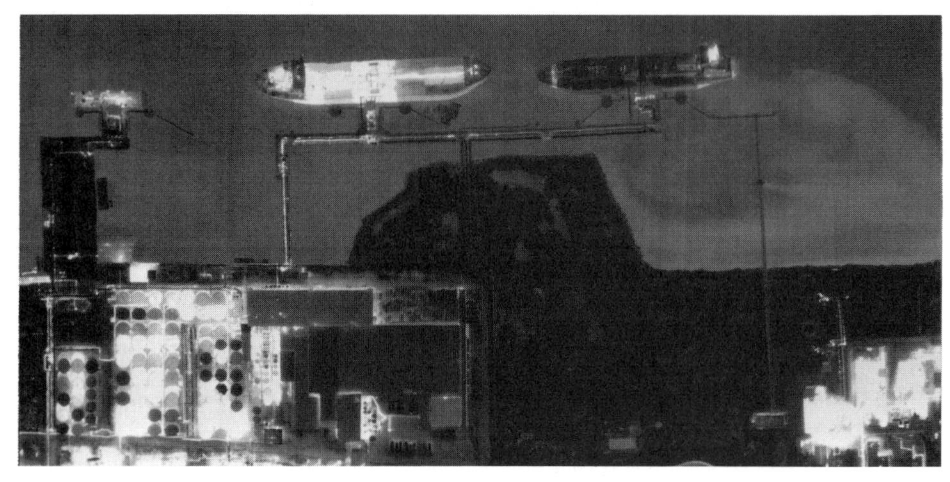

**FIGURE 8.13.** This thermal image shows another section of the same facility imaged in Figure 8.12. Thermal image provided by Daedalus Enterprises, Inc., Ann Arbor, Michigan.

quired. Sometimes it may not be possible to determine the correct time of day from information within the image itself; misinterpretation can alter the meaning of gray tones on the image and render the resulting interpretation useless.

As the sensor views objects near the edge of the image, the distance from the sensor to the ground increases. This relationship means that the IFOV is larger nearer the edges of the image than it is near the flight line.

Thermal scanners are generally uncalibrated, so they show relative radiances rather than absolute measurements of radiances. However, some thermal scanners do include reference sources that are viewed by the scanning system at the beginning and end of each scan. The reference sources can be set at specific temperatures that are related to those expected to be encountered in the scene. Thus, each scan line includes values of known temperature that permit the analyst to estimate temperatures of objects within the image.

In addition, errors caused by the atmosphere and by the system itself prevent precise interpretation of thermal imagery. Typical system errors might include recording noise, variations in reference temperatures, and detector errors. Full correction for atmospheric conditions requires information not usually available in detail, so often it is necessary to use approximations, or value-based samples acquired at a few selected times and places and extrapolated to estimate values elsewhere. Also, the *atmospheric path traveled* by radiation reaching the sensor varies with angle of observation, which changes as the instrument scans the ground surface. These variations in angle lead to errors in observed values in the image.

Even when accurate measures of radiances are available, it is difficult to derive data for kinetic temperatures from the apparent temperature information within the image. Derivation of kinetic temperatures requires knowledge of emissivities of the materials. In some instances, such knowledge may be available, as the survey may be focused on a known area that must be repeatedly imaged to monitor changes over time (e.g., as moisture conditions change). But many other surveys examine areas not previously studied in detail, and information regarding surface materials and their emissivities may not be known.

*Emissivity* is a measure of the effectiveness of an object in translating temperature into emitted radiation (and in converting absorbed radiation into a change in observed temperature). Because objects differ with respect to emissivity, observed differences in emitted infrared energy do not translate directly into corresponding differences in temperature. As a result, it is necessary to apply a knowledge of surface temperature, or emissivity variations, to accurately study surface temperature patterns from thermal imagery. Because knowledge of these characteristics assumes a detailed prior knowledge of the landscape, such interpretations should be considered appropriate for examination of a distribution known already in some detail, rather than for reconnaissance of an unknown pattern. (For example, one might know already the patterns of soils and crops at an agricultural experiment station but wish to use the imagery to monitor temperature patterns.) Often estimated values for emissivity are used or assumed values are applied to areas of unknown emissivity.

Also, it should be remembered that the sensor records radiances of the surfaces of objects. Because radiances may be determined at the surface of an object by a layer perhaps as thin as 50 μm, a sensor may record conditions that are not characteristic of the subsurface mass, which is probably the object of the study. For example, evaporation from a water body or a moist soil surface may cool the thin layer of moisture at the contact with the atmosphere. Because the sensor detects radiation emitted at this surface layer, the observed temperature may differ considerably from that of the remaining mass of the soil or water body.

Leckie (1982) estimates that calibration error and other instrument errors are generally rather small, although they may be important in some instances. Errors in estimating emissivity and in attempts to correct for atmospheric effects are likely to be the most important sources of error in quantitative studies of thermal imagery.

In most instances, a thermal image must be interpreted to yield qualitative rather than quantitative information. Although, of course, some applications do require interpretations of quantitative information, there are many others for which qualitative interpretation is completely satisfactory. An interpreter who is well-informed about the landscape represented on the image, the imaging system, the thermal behavior of various materials, and the timing of the flight is prepared to derive considerable information from an image even though it may not be possible to derive precise temperatures from the image.

The thermal landscape is a composite of the familiar elements of surface material, topography, vegetation cover, and moisture. Various rocks, soils, and other surface materials respond differently to solar heating. Thus in some instances the differences in thermal properties tabulated in Table 8.1 can be observed in thermal imagery. However, the thermal behavior of surface materials is also influenced by other factors. For example, slopes that face the sun will tend to receive more solar radiation than slopes that are shadowed by topography. Such differences are, of course, combined with those arising from different surface materials. Also, the presence and nature of vegetation alters the thermal behavior of the landscape. Vegetation tends to heat rather rapidly but can also shade areas, creating patterns of warm and cool.

Water tends to retain heat, to cool slowly at night, and to warm slowly during daytime. In contrast, many soils and rocks (if dry) tend to release heat rapidly at night and to absorb heat quickly during the daytime. Even small or modest amounts of moisture can therefore greatly alter the thermal properties of soil and rock. Therefore, thermal sensors can be effective in monitoring the presence and movement of moisture in the environment. In any given image, the combined influences of surface materials, topography, vegetation, and moisture can combine to cause complex image patterns. However, often it is possible to isolate the effect of some of these variables and therefore to derive useful information concerning, for example, movement of moisture or the patterns of differing surface materials.

Timing of acquisition of thermal imagery is very important. The optimum times vary according to the purpose and subject of the study, so it is is not possible to specify universally applicable rules. Often the greatest thermal contrast occurs during the daylight hours, so sometimes thermal images are acquired in the early afternoon to capture the differences in thermal properties of landscape features. However, in the 3–6 μm range the sensor may record reflected as well as emitted thermal radiation, so daytime missions in this region may not be optimum for thermal information. Also during daytime, the sensor may record thermal patterns caused by shadowing: Although shadows may sometimes be useful in interpretation, they are more likely to complicate analysis of a thermal image, so it is usually best to avoid acquiring images. In a daytime image, water bodies typically appear as cool relative to land, and bare soil, meadow, and wooded areas appear as warm features.

Some of the problems arising from daytime images are avoided by missions planned just before dawn. Shadows are absent, and sunlight, of course, cannot cause reflection (at shorter wavelengths) or shadows. However, thermal contrast is lower, so it may be more difficult to distinguish between broad classes of surfaces based on differences in thermal behavior. On such an image, water bodies would appear as warm relative to land. Forested areas may also appear to be warm. Open meadows and dry, bare soil are likely to appear as cool features.

The thermal images of petroleum storage facilities (Figures 8.12 and 8.13) show thermal contrasts that are especially interesting. A prominent feature on Figure 8.12 is the bright thermal plume discharged by the tributary to the Delaware River. The image clearly shows the sharp contrast in temperature as the warm water flows into the main channel, then disperses and cools as it is carried downstream. Note the contrast between the full and partially full tanks and the warm temperatures of the pipelines that connect the tanker with the storage tanks. Many of the same features are also visible in Figure 8.13, which also shows a partially loaded tanker with clear delineation of the separate storage tanks in the ship.

Thermal imagery has obvious significance for studies of heat loss, thermal efficiency, and effectiveness of insulation. Plate 9 shows ground views of two residential structures. Windows and walls are major avenues for the escape of heat, and the chimney shows as an especially bright feature. Note that walkways, paved roads, and parked automobiles are relatively cool.

In Figure 8.14 two thermal images depict a portion of the Cornell University campus (Ithaca, NY) acquired in January (left) and again the following November (right). Campus buildings are clearly visible, as are losses of heat through vents in the roofs of buildings and at manholes where steampipes for the campus heating system join or change direction. The left-hand image shows a substantial leak in a steam pipe as it passes over the bridge in the right center of the image. On the right, a later image of the same region shows clearly the effects of repair of the defective section.

Figure 8.15 shows Painted Rock Dam, Arizona, as depicted both by an aerial photograph and a thermal infrared image. The aerial photograph (top) was taken at about 10:30 A.M.; the thermal image was acquired at about 7 A.M. the same day. The prominent linear feature is a large earthen dam with the spillway visible at the lower left. On the thermal image, the open water upstream from the dam appears as a uniformly white (warm) region, whereas land areas are dark (cool)—a typical situation for early morning hours, before solar radiation has warmed the earth. On the downstream side of the dam, the white (warm) areas reveal areas of open water or saturated soil. The open water in the spillway is, of course, expected, but the other white areas indicate places where there may be seepage and potentially weak points in the dam structure.

Figure 8.16 shows thermal images of a power plant acquired at four different stages of the tidal cycle. The discharge of warm water by the plant is visible as the bright plume in the upper left of each image. In the top image, acquired at low tide (5:59 A.M.), the warm water is carried downstream, toward the ocean. The second image, acquired at flood tide (8:00 A.M.), shows the deflection of the plume by rising waters. In the third image, acquired at high tide (10:59 A.M.), the plume extends upstream for a considerable distance. Finally, at ebb tide (2:20 P.M.; bottom image), the shape of the plume reflects the reversal of the flow of tide once again.

If imagery or data for two separate times are available, it may be possible to employ a knowledge of thermal inertia as a means of studying the pattern of different materials at the earth's surface. Figure 8.17 illustrates the principle. Two images are acquired at times that permit observation of extremes of temperature, perhaps near noontime and again just before dawn. These two sets of data permit estimation of the ranges of temperature variation for each region on the image. Because these variations are determined

**FIGURE 8.14.** Two thermal images of a portion of the Cornell University campus (Ithaca, NY). Thermal images provided by Daedalus Enterprises, Inc., Ann Arbor, Michigan.

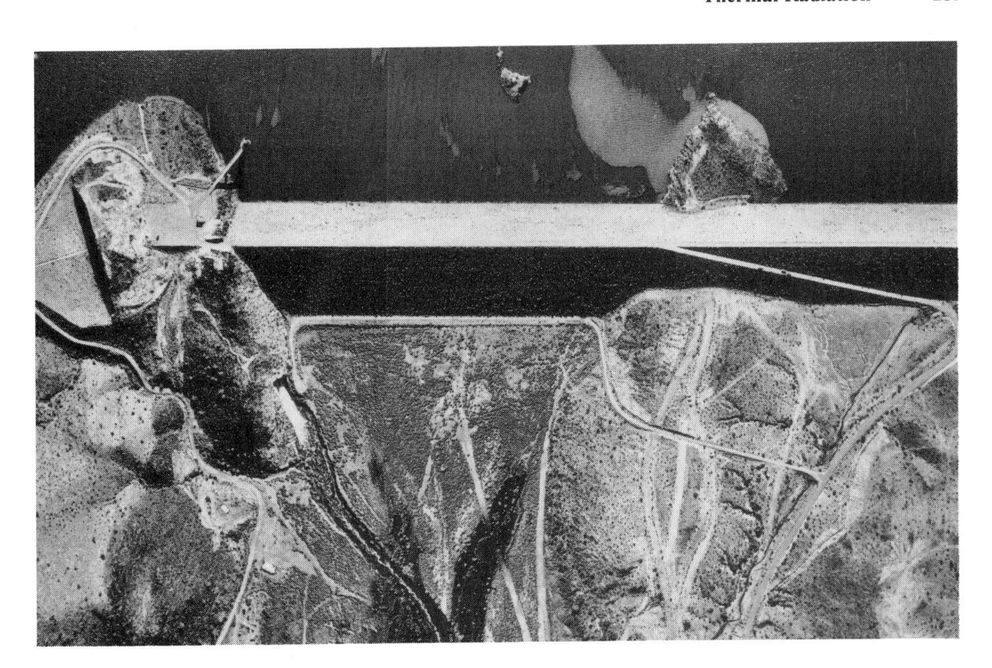

**FIGURE 8.15.** Painted Rock Dam, Arizona, 28 January 1979 aerial photograph (top) and thermal image (below). Thermal image provided by Daedalus Enterprises, Inc., Ann Arbor, Michigan.

by the thermal inertias of the substances, they permit interpretation of features represented by the images. Leckie (1982) notes that misregistration can be a source of error in comparisons of day and night images, although such errors are thought to be small relative to other errors.

In Figure 8.17, images are acquired at noon, when temperatures are highest, and again just before dawn, when temperatures are lowest. By observing the differences in temperature, it is possible to estimate thermal inertia. Thus the lake, composed of a mate-

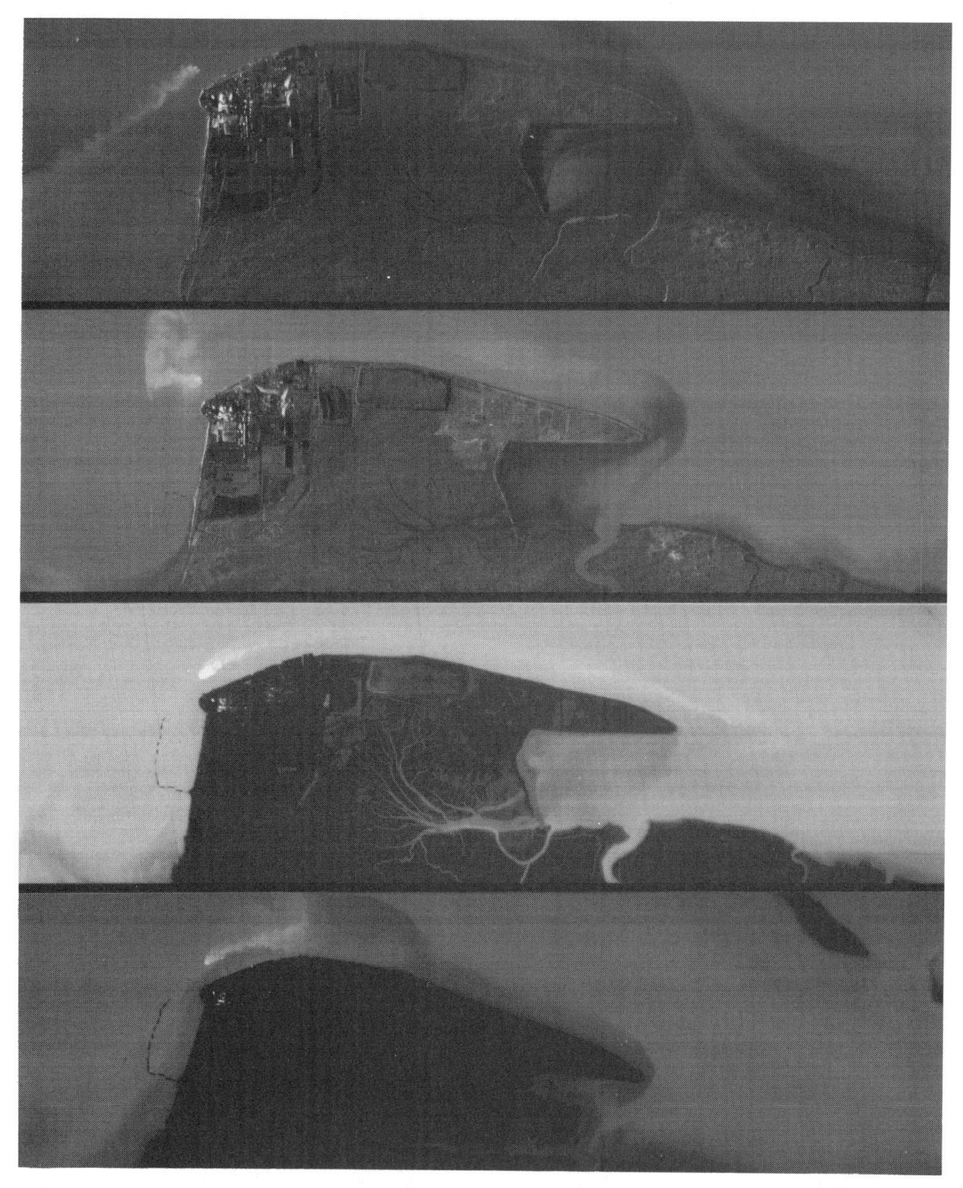

**FIGURE 8.16.** Thermal images of a power plant acquired at different stages of the tidal cycle. Thermal images provided by Daedalus Enterprises, Inc., Ann Arbor, Michigan.

rial that resists changes in temperature, shows rather small changes in temperature, whereas the surface of open sand displays little resistance to thermal changes and exhibits a wider range of temperature.

Plate 10 shows another example of diurnal temperature changes. These two images depict the landscape near Erfurt, Germany, as observed by a thermal scanner in the region between 8.5 and 12 μm. The thermal data have been geometrically and radiometrically

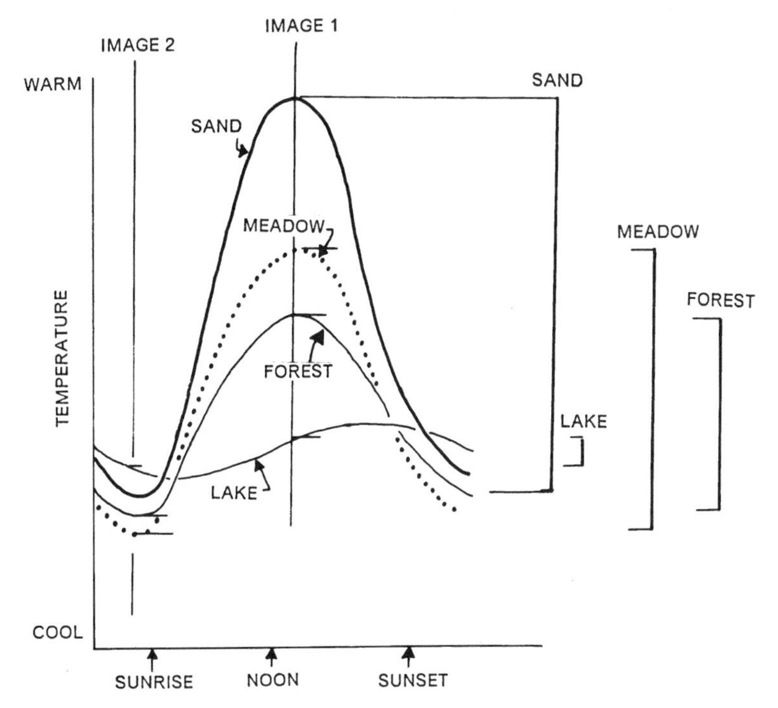


**FIGURE 8.17.** Schematic illustration of diurnal temperature variation of several broad classes of land cover. Modified from Fagerlund et al. (1970). The two solid vertical lines depict timing of two thermal images acquired at an interval timed to maximize thermal contrast over a 24-hour period. The vertical bars at the right are projections from the diagram on the left, so that the lengths of the bars represent thermal variation of each class of land cover.

processed, then superimposed over digital elevation data, with reds and yellows assigned to represent warmer temperatures, and blues and greens cooler temperatures. The city of Erfurt is positioned at the edge of a limestone plateau, which is visible as the irregular topography in the foreground of the image. The valley of the river Gera is visible in the lower left, extending across the image to the upper center region of the image.

The top image represents this landscape as observed just after sunset. The urbanized area and much of the forested topography south of the city show as warm reds and yellows. The bottom image shows the same area observed just before sunrise, when temperatures are at their coolest. The open land of the rural landscape and the areas at the periphery of the urbanized areas have cooled considerably; the forested region on the plateau south of the city is now the warmest surface, due to the thermal effects of the forest canopy.

## 8.9. Heat Capacity Mapping Mission

The Heat Capacity Mapping Mission (HCMM) (in service from April 1978 to September 1980) was a satellite system specifically designed to evaluate the concept that orbital observations of temperature differences at the earth's surface at different points in the daily heating/cooling cycle might provide a basis for estimation of thermal inertia and other thermal properties of surface materials. The satellite was in a sun-synchronous orbit, at an altitude of 620 km, bringing it over the equator at 2 P.M. local sun time. At 40° N latitude the satellite passes overhead at 1:30 P.M., then again at 2:30 A.M. These times provide observations at two points on the diurnal heating/cooling cycle (Figure 8.18). (It would be desirable to use a time later in the morning just before sunrise, in order to provide higher thermal contrast, but orbital constraints dictate the time differences between passes.) The repeat cycle varied with latitude—at mid-latitude locations the cycle was 5 days. The 12-hour repeat coverage is available for some locations at 16-day intervals, but other locations receive only 36-hour coverage. The HCMM radiometer used two channels; one, in the reflective portion of the spectrum (0.5 to 1.1 μm), had a spatial resolution of about 500 m × 500 m. A second channel is available in the thermal infrared region, from 10.5 to 12.5 μm, had a spatial resolution of 600 m × 600 m. The image swath was about 716 km wide. The HCMM was launched on 26 April 1978 and acquired data until 30 September 1980.

Plate 11 shows an HCMM image of a portion of the eastern coast of North America. The image represents a strip of land about 688 km (430 mi.) in width extending from Lake Ontario in the north (top) south toward Cape Hatteras, North Carolina (bottom left). The image uses colors to portray variations in observed temperature so that the sequence purple, blue, green, brown, yellow, orange, red, gray, and white represents increasing temperature.

Black areas represent cloud cover; the coldest areas are the larger water bodies and clouds. Major metropolitan centers including New York, Philadelphia, and Washington are visible as bright white areas near the coastline. Variations in radiances within land areas are mainly caused by differences in land cover, which in some areas (especially southeastern Pennsylvania) closely follows topography and geologic structure.

Estimation of thermal inertia requires that two HCMM images be registered to observe apparent temperatures at different points on the diurnal heating and cooling cycle. Because the day and night images must be acquired on different passes with different inclinations (Figure 8.19) the registration of the two images is often much more difficult than might be the case with similar images. The rather course resolution means that nor-

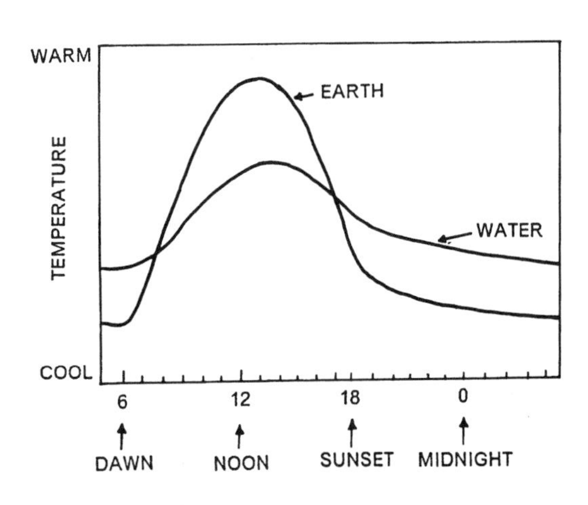


**FIGURE 8.18.** Diurnal temperature cycle.

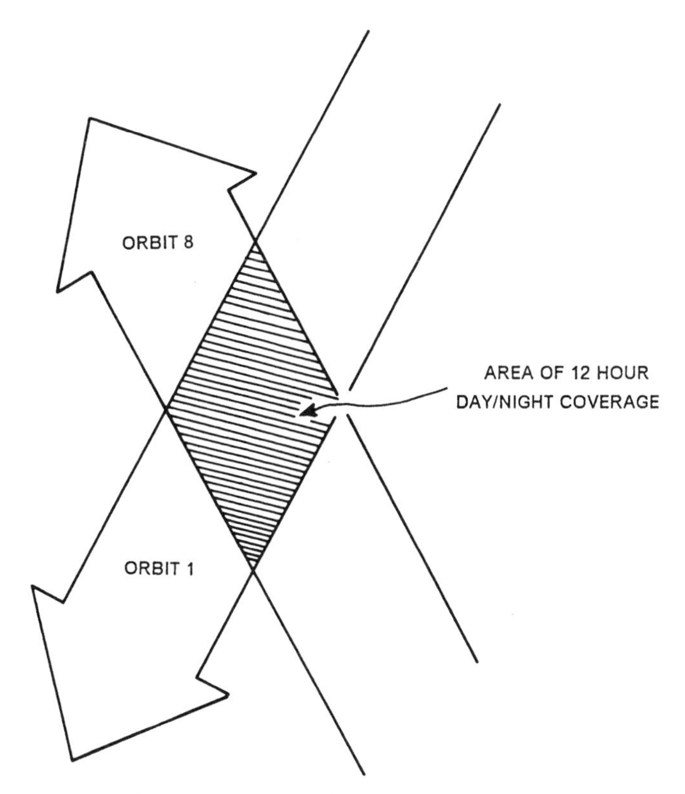


**FIGURE 8.19.** Overlap of two HCMM passes at 40° latitude.

mally distinct land marks are not visible, or more difficult to recognize. Watson et al. (1982) have discussed some of the problems encountered in registering HCMM images.

Goddard Space Flight Center received and processed HCMM data, calibrating the thermal data, and performing geometric registration, when possible, for the two contrasting thermal images. Ideally, these two images permit reconstruction of a thermal inertia map of the imaged area. In fact, atmospheric conditions vary during the intervals between passes, and the measured temperatures can be influenced by variations in cloud cover, wind, evaporation at the ground surface, and atmospheric water vapor. As a result, the measured differences are *apparent thermal inertia* (ATI) rather than a true measure of thermal inertia. Because of a number of approximations employed and scaling of the values, the values for ATI are best considered to be measures of relative thermal inertia.

As an experimental prototype, HCMM was in service for a relatively short interval and acquired data for rather restricted regions of the earth. Some of the concepts developed and tested by HCMM have been employed in the design of other programs, most notably, the the Earth Observing System (Chapter 20).

Data are available in both film and digital formats; HCMM scences show areas about 700 km × 700 km (1,127 mi. × 1,127 mi.) in size. Nonetheless, the HCMM archive, maintained by the National Space Science Data Center (World Data Center-A, Code 601, NASA Goddard Space Flight Center, Greenbelt, MD 20771) provides coverage for many mid-latitude regions (including North America, Australia, Europe, and northern Africa).

## 8.10. MSS and TM Thermal Data

Chapter 6 mentioned, but did not discuss in any detail, the thermal data collected by the TM and by the Landsat 3 MSS. The Landsat 3 MSS, unlike earlier Landsats, included a channel (MSS band 5, formerly designated as band 8) sensitive to thermal data in the region 10.4–12.6 µm. Due to decreased sensitivity of the thermal detectors, spatial resolution of this band is significantly coarser than that of the visible and near infrared channels (237 m × 237 m compared to 79 m × 79 m). Each of the MSS band 5 pixels therefore corresponds to about nine pixels from one of the other four bands. Landsat 3 band 5 was designed to detect apparent temperature differences of about 15° C. Although initial images were reported to meet this standard, progressive degradation of image quality was observed as the thermal sensors deteriorated over time. In addition, the thermal band was subject to other technical problems. Because of these problems, the MSS thermal imaging system was turned off about one year after Landsat 3 was launched.

As a result, relatively few images were acquired using the MSS thermal band and relatively little analysis has been attempted. Lougeay (1982) examined a Landsat 3 thermal image of an alpine region in Alaska with the prospect that the extensive glaciation in this region could contribute to sufficient thermal contrast to permit interpretation of terrain features. The coarse spatial resolution, solar heating of south-facing slopes, and extensive shadowing due to rugged topography and low sun angle contributed to difficulties in interpretation. Although he was able to recognize major landscape features, his analysis forms more of a suggestion of potential applications of the MSS thermal data, rather than an illustration of their usefulness.

Price (1981) also studied the Landsat MSS thermal data. He concluded that the thermal band provided new information not conveyed by the other MSS channels. Rather broad land-cover classes (including several classes of open water, urban and suburban land, vegetated regions, and barren areas) could readily be identified on the scene he examined. The time of the Landsat overpass is too early to record maximum thermal contrast (which occurs in early afternoon), and analyses of temperatures are greatly complicated by topographic and energy balance effects at the ground surface and the influence of the intervening atmosphere. Despite the possibility of adding new information to the analysis, Price advised against routine use of the thermal channel in the usual analyses of Landsat MSS data. In general, the experience of other analysts has confirmed this conclusion and these data have not found widespread use.

The Landsat TM includes a thermal band, usually designated as TM band 6, sensitive in the region 10.4–12.5 µm (Figure 8.20). It has lower radiometric sensitivity, and coarser spatial resolution (about 120 m) relative to other TM bands. Preliminary studies (e.g., Toll, 1985) suggest that the TM thermal band does not significantly add to the accuracy of the usual land cover analysis, probably due to some of the same influences noted in the Price study. However, it seems likely that continued research of characteristics of these data will reveal more information about their characteristics and lead to further research.

**FIGURE 8.20.** Thematic mapper (band 6), showing thermal image of New Orleans, Louisiana, 16 September 1982. This image shows the same area, and was acquired at the same time, as Figures 6.11 and 6.12. Resolution of TM band 6 is coarser than the other bands, so detail is not as sharp. Here dark image tone represent relatively cool areas, and bright areas represent relatively warm areas. The horizontal banding near the center of the image is caused by a defect in the operation of the TM discussed in Chapter 10. Image reproduced by permission of EOSAT.

## 8.11. Summary

Thermal imagery is a valuable asset for remote sensing because it conveys information not easily derived from other forms of imagery. The thermal behavior of different soils, rocks, and construction materials can permit derivation of information not present in other images. The thermal properties of water contrast with those of many other landscape materials, so that thermal images can be sensitive to the presence of moisture in the environment. And the presence of moisture is itself often a clue to the differences between different classes of soil and rock. Modern thermal imagery is of high quality and usually provides fine detail.

Of course, use of data from the far infrared region can present its own problems. Like all images, thermal imagery has its own geometric errors. The analyst cannot derive detailed quantitative interpretations of temperatures unless detailed knowledge of emissivity is at hand. Timing of image acquisition can be critical. Atmospheric effects can be a serious problems, especially from satellite altitudes. Because the thermal landscape differs so greatly from the visible landscape, it may often be necessary to use aerial photography to locate familiar landmarks while interpreting thermal images. Existing archives of thermal imagery are not comparable in scope to those for aerial photography or satellite data (such as those of Landsat or SPOT), so it may be difficult to acquire thermal data, unless it is feasible to purchase custom-flown imagery.

## Review Questions

1. Explain why choice of time of day is so important in planning acquisition of thermal imagery.
2. Would you expect season of the year to be important in acquiring thermal imagery of, for example, a region in Pennsylvania? Explain.
3. In your new job as an analyst for an institution that studies environmental problems in coastal areas, it is necessary for you to prepare a plan to acquire thermal imagery of a tidal marsh. List the important factors you must consider as you plan the mission.
4. Many beginning students are surprised to find that geothermal heat and man-made heat are of such modest significance in the earth's energy balance and in determining information presented on thermal imagery. Explain, then, what it is that thermal imagery does depict, and why thermal imagery is so useful in so many disciplines.
5. Can you understand why thermal infrared imagery is considered so useful to so many scientists even though it does not usually provide measurements of actual temperatures? Can you identify situations in which it might be important to be able to determine actual temperatures from imagery?
6. Fagerlund et al. (1982) report that it is possible to judge from thermal imagery if storage tanks for gasoline and oil are empty, full, or partially full. Examine Figure 8.12 and find examples of each. How can you confirm from evidence on the image itself that full tanks are warm (bright) and empty tanks are cool (dark)?
7. Examine Figure 8.12. Is the tanker empty, partially full, or full? Examine Figure 8.13; are these tankers empty, partially full, or full? From your inspection of the imagery can you determine something about the construction of tankers and the procedures used to empty or fill tankers?
8. What information would be necessary to plan an aircraft mission to acquire thermal imagery to study heat loss from residential area in the northeastern United States?
9. In what ways would thermal imagery be important in agricultural research?

10. Outline ways in which thermal imagery would be especially useful in studies of the urban landscape.

## References


Colcord, J. E. 1981. Thermal Imagery Energy Surveys. *Photogrammetric Engineering and Remote Sensing,* Vol. 47, pp. 237–240.

Fagerlund, E., B. Kleman, L. Sellin, and H. Svenson. 1970. Physical Studies of Nature by Thermal Mapping, *Earth Science Reviews,* Vol. 6, pp. 169–180.

Gillespie, A. R. and A. B. Kahle. 1978. Construction and Interpretation of a Digital Thermal Inertia Image. *Photogrammetric Engineering and Remote Sensing,* Vol. 43, pp. 983–1000.

Goddard Space Flight Center. 1978. *Data Users Handbook, Heat Capacity Mapping Mission (HCMM), for Applications Explorer Mission-A (AEM).* Greenbelt, MD: NASA.

Goward, S. N. 1981. Longwave Infrared Observation of Urban Landscapes. *Technical Papers, American Society of Photogrammetry 47th Annual Meeting.* Washington, DC.

Hatfield, J. L., J. P. Millard, and R. C. Goettelman. 1982. Variability of Surface Temperature in Agricultural Fields of Central California. *Photogrammetric Engineering and Remote Sensing,*Vol. 48, pp. 1319–1325.

Hudson, R. D. 1969. *Infrared System Engineering.* New York: Wiley, 642 pp.

Janza, Frank J. 1975. Interaction Mechanisms. Chapter 4 in *Manual of Remote Sensing* (R. G. Reeves, ed). Falls Church, VA: American Society of Photogrammetry, pp. 75–179.

Leckie, Donald G. 1982. An Error Analysis of Thermal Infrared Line-Scan Data for Quantitative Studies. *Photogrammetric Engineering and Remote Sensing,* Vol 48, pp. 945–954.

Lougeay, R. 1982. Landsat Thermal Imaging of Alpine Regions. *Photogrammetric Engineering and Remote Sensing,* Vol. 48, pp. 269–273.

Lowe, Donald S. (ed.). 1975. Imaging and Nonimaging Sensors. Chapter 8 in *Manual of Remote Sensing* (R. G. Reeves, ed.). Falls Church, VA: American Society of Photogrammetry, pp. 367–398.

Moore, Richard K. (ed.). 1975. Microwave Remote Sensors. Chapter 9 in *Manual of Remote Sensing* (R. G. Reeves, ed.). Falls Church, VA: American Society of Photogrammetry, pp. 399–538.

Pratt, D. A., and C. D. Ellyett. 1979. The Thermal Inertia Approach to Mapping of Soil Moisture and Geology. *Remote Sensing of Environment,* Vol. 8, pp. 151–168.

Price, John C. 1981. The Contribution of Thermal Data in Landsat Multispectral Classification. *Photogrammetric Engineering and Remote Sensing,* Vol. 47, pp. 229–236.

Price, John C. 1978. Thermal Inertia Mapping—A New View of the Earth. *Journal of Geophysical Research,* Vol. 82, pp. 2582–2590.

Sabins, F. 1969. Thermal Infrared Imaging and its Application to Structural Mapping, Southern California. *Geological Society of America Bulletin,* Vol. 80, pp. 397–404.

Schott, John R., and W. J. Volchok. 1985. Thematic Mapper Infrared Calibration. *Photogrammetric Engineering and Remote Sensing,* Vol. 51, pp. 1351–1358.

Short, Nicholas M., and L.M. Stuart. 1982. *The Heat Capacity Mapping Mission (HCMM) Anthology.* NASA Special Publication 465. Washington, DC: U.S. Government Printing Office, p. 264.

Toll, David L. 1985. Landsat-4 Thematic Mapper Scene Characteristics of a Suburban and Rural Area. *Photogrammetric Engineering and Remote Sensing,* Vol. 51, pp. 1471–1482.

Watson, K., S. Hummer-Miller, and D. L. Sawatzky. 1982. Registration of Heat Capacity Mapping Mission Day and Night Images. *Photogrammetric Engineering and Remote Sensing,* Vol. 48, pp. 263–268.

Weast, R. C. (ed.). 1986. *CRC Handbook of Chemistry and Physics.* Boca Raton, FL: CRC Press.

CHAPTER NINE

# Image Resolution

## 9.1. Introduction and Definitions

In very broad terms, resolution refers to the ability of a remote sensing system to record and display fine detail. A working knowledge of resolution is essential for understanding both practical and conceptual aspects of remote sensing. Our understanding, or lack of understanding, of resolution may be the limiting factor in our efforts to use remotely sensed data, especially at coarse spatial resolution.

For scientists with interests in instrument design and performance, measurement of resolution is of great significance in determining the optimum design and configuration of individual elements (e.g., specific lenses or photographic emulsions) of a remote sensing system. Here our interest focuses on understanding image resolution of entire remote sensing systems as they are used to image landscapes. Regardless of our specific interests in elements of the landscape, whether they may focus on soil patterns, geology, water quality, land use, or vegetation distributions, a knowledge of image resolution is a prerequisite for understanding the information recorded on the images we examine.

The purpose of this chapter is to discuss image resolution as a separate concept in recognition of its significance throughout the field of remote sensing. Thus, it is an effort to outline some of the generally applicable concepts without ignoring the special and unique factors that apply in any given instance.

Estes and Simonett (1975) define resolution as "the ability of an imaging system . . . to record fine detail in a distinguishable manner" (p. 879). This definition includes several key concepts. The emphasis on the imaging *system* is significant because in most practical situations it makes little sense to focus attention on the resolving power of a single element of the system (e.g., the film) if another element (perhaps the camera lens) forms the limit on the resolution of the final image. "Fine" detail is of course a relative concept, as is the specification that detail be recorded in a "distinguishable" manner. Both of these aspects of the definition emphasize that resolution can be clearly defined only by operational definitions applicable under specified conditions, as discussed later.

For the present, it is sufficient to note that there is a practical limit to the level of detail that can be acquired from a given aerial or satellite image. This limit we define informally as the "resolution" of the remote sensing system, although it must be recognized that image detail also depends on the character of the scene that has been imaged, atmospheric conditions, illumination, and the experience and ability of the image interpreter.

Usually most individuals think of resolution as *spatial resolution*—the fineness of the spatial detail visible in an image. Fine detail in this sense means that small objects can be identified on an image. Other forms of resolution are equally important. *Radiometric resolution* can be defined as the ability of an imaging system to record many levels of brightness. Coarse radiometric resolution would record a scene using only a few brightness levels (i.e., at very high contrast), whereas fine radiometric resolution would record the same scene using many brightnesses. *Spectral resolution* denotes the ability of a sensor to define fine wavelength intervals. The thematic mapper, for example has fine spectral resolution relative to the Landsat MSS, as the spectral bands are more narrowly defined and are focused more on spectral regions of interest to specific scientific disciplines.

In many situations, there are clear trade-offs between different forms of resolution. For example, in traditional photographic emulsions, increases in spatial resolving power are based on decreased size of film grain, which produces accompanying decreases in radiometric resolution (i.e., the decreased sizes of grains in the emulsion portray a lower range of brightnesses). In other systems there are similar trade-offs. In scanning systems, increasing spatial detail requires a smaller instantaneous field of view; energy reaching the sensor has been reflected from a smaller ground area. If all other variables have been held constant, this must translate to decreased energy reaching the sensor; lower levels of energy mean that the sensor may record less "signal" and more "noise," thereby reducing the usefulness of the data. This effect can, of course, be compensated for by broadening the spectral window to pass more energy (i.e, decreasing spectral resolution) or by dividing the energy into fewer brightness levels (decreasing radiometric resolution). Of course, overall improvements can be achieved by improved instrumentation or by altering operating conditions (flying at a lower altitude). The general situation, however, seems demand costs in one form of resolution for benefits achieved in another.

## 9.2. Target Variables

Observed spatial resolution in a specific image depends greatly on the character of the scene that has been imaged. In complex natural landscapes identification of the essential variables influencing detail observed in the image may be difficult to separate, although many of the key factors can be enumerated. *Contrast* is clearly one of the most important influences on spatial and radiometric resolution. Contrast can be defined as the difference in brightness between an object and its background. (Note the importance in this discussion of distinguishing between contrast in the scene and contrast in the image of that scene; the two may be related, but not necessarily in a direct manner, as discussed in Section 3.4 of this volume). Other factors held constant, high contrast favors recording of fine spatial detail; low contrast produces coarser detail. A black automobile imaged against a black asphalt background will be more difficult to observe than white vehicle observed under the same conditions.

The significance of contrast as an influence on spatial resolution illustrates the interrelationships between the various forms of resolution and emphasizes that no single element of system resolution can be considered in isolation from the others. It is equally important to distinguish between contrast in the original scene and that recorded on the image of that scene; the two may be related but not necessarily in a direct fashion. Also, it should be noted that contrast in the original scene is a dynamic quality that, for a given landscape, varies greatly from season to season (with changes in vegetation, snow cover, etc.) and within a single day (as angle and intensity of illumination change).

The *shape* of an object or feature is significant. *Aspect ratio* refers to the length of a feature in relation to its width. Usually long, thin features, such as highways, railways, and rivers, tend to be visible on aerial imagery, even when their widths are much less than the nominal spatial resolution of the imagery. *Regularity of shape* favors recording of fine detail. Features with regular shapes, such as cropped agricultural fields, tend to be recorded in fine detail, whereas complex shapes will be imaged in rather coarse detail.

The *number* of objects in a pattern also controls the level of detail recorded by a sensor. For example, the pattern formed by the number and regular arrangement of tree crowns in an orchard favors the imaging of the entire pattern in fine detail. Under similar circumstances, the crown of a single isolated tree might not be visible on the imagery.

*Extent and uniformity of background* contribute to resolution of fine detail in many distributions. For example, a single automobile in a large, uniform parking area or a single tree positioned in a large cropped field will be imaged in detail not achieved under other conditions.

## 9.3. System Variables

From earlier chapters, remember that resolution of individual sensors depends in part on the design of that sensor, and upon its operation at a given time. In any specific situation these considerations must be acknowledged, and studies to determine their role in defining resolution should be performed. For example, resolution of an aerial photograph (Chapter 3) is determined by the quality of the camera lens, the choice of film, and the design of the aerial camera. For scanning systems such as the Landsat MSS/TM (Chapter 6) or thermal scanners (Chapter 8), the IFOV determines many of the qualities of image resolution. The IFOV depends, of course, on the optical system (the angular field of view) and operating altitude. Speed of the scanning motion and movement of the vehicle that carries the sensor will also have their effect on image quality. For active microwave sensors (Chapter 7), image resolution is determined by beamwidth (antenna gain), angle of observation, wavelength, and other factors as discussed previously.

## 9.4. Operating Conditions

For all remote sensing systems, the operating conditions, including flying altitude and ground speed, are important elements influencing the level of detail in the imagery. Atmospheric conditions can be included as important variables, especially for satellite and high-altitude imagery.

## 9.5. Measurement of Resolution

### *Ground Resolved Distance*

Perhaps the simplest measure of spatial resolution is *ground resolved distance* (GRD), defined simply as the dimensions of the smallest objects recorded on an image. One might speak of the resolution of an aerial photograph as being "2 m," meaning that objects of that size and larger could be detected and interpreted from the image in question. Smaller objects presumably would not be resolved and therefore would not be interpretable.

Such measures of resolution may have utility as a rather rough suggestion of usable detail but must be recognized as having only a subjective meaning. The objects and features that compose the landscape vary greatly in size, shape, contrast with background, and pattern; usually we have no means of relating a given estimate of GRD to a specific problem of interest. For example, the spatial resolution of the U.S. Department of Agriculture (USDA) 1:20,000 black-and-white aerial photography is often said to be "about one meter," yet typically one can easily detect on these photographs the painted white lines in parking lots and highways; these lines may be as narrow as 6 to 9 in. Does this mean that the resolution of this photography should be assessed as 6 in. rather than 1 m? Only if we are interested in the interpretation of long, thin features in high contrast with their background could we accept such an estimate as useful. Similarly, the estimate of 1 m may be inappropriate for many applications.

### *Line Pairs per Millimeter*

*Line pairs per millimeter* (LPM) is a means of standardizing the characteristics of targets used to assess image resolution. Essentially, it is a means of quantifying, under controlled conditions, the estimate of GRD by using a standard target, positioned on the ground, which is imaged by the remote sensing system under specified operating conditions.

Although many targets have been used, the resolution target designed by the U.S. Air Force (USAF) has been a standard for a variety of studies (Figure 9.1). This target consists

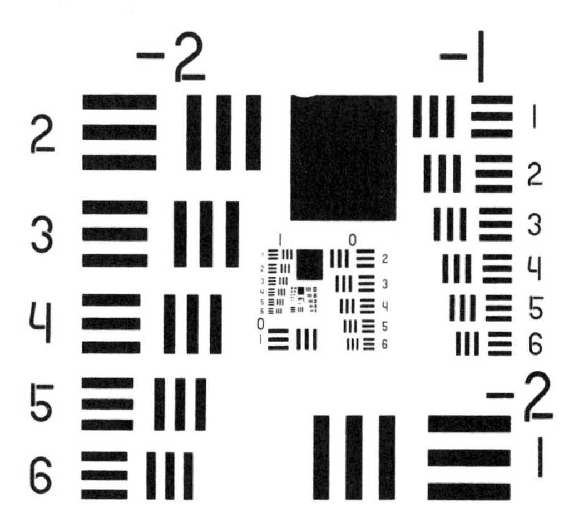


**FIGURE 9.1.** Bar target used in resolution studies.

of parallel black lines positioned against a white background. The width of spaces between lines is equal to that of the lines themselves; their length is five times their width. As a result, a block of three lines and the two white spaces that separate them form a square. This square pattern is reproduced at varied sizes to form an array consisting of bars of differing widths and spacings. Sizes are controlled to produce changes in spacing of the bars (spatial frequency) of 12%. Repetition of the pattern at differing scales assures that the image of the pattern will include at least one pattern so small that individual lines and their spaces will not be fully resolved.

If images of two objects are visually separated, they are said to be "spatially resolved." Images of the resolution target are examined by an interpreter to find that smallest set of lines in which the individual lines are all completely separated along their entire length. The analyst measures the width of the image representation of one "line pair" (i.e., the width of the image of one line and its adjacent white space is measured) (Figure 9.2). This measurement provides the basis for the calculation of the number of line pairs per millimeter (or any other length we may choose; LPM is standard for many applications). For example, in Figure 9.2 the width of a line and its adjacent gap is measured to be 0.04 mm. From 1 line pair / 0.04 mm we find a resolution of 25 LPM.

For aerial photography this measure of resolution can be translated into GRD by the relationship

$$\text{GRD} = \frac{H}{(f)\,(R)} \quad \text{(Eq. 9.1)}$$

where GRD represents ground resolved distance in meters; $H$, the flying altitude above the terrain, in meters; $f$, the focal length, in millimeters; and $R$, the system resolution, in LPM.

Such measures have little predictable relationship to the actual sizes of landscape features that might be interpreted in practical situations because seldom will the features of interest have the same regularity of size, shape, and arrangement, and the high contrast of the resolution target used to derive the measures. They are, of course, valuable as comparative measures for assessing the performance of separate systems under the same operating conditions or of a single system under different conditions.

Although the Air Force target has been widely used, other resolution targets have been developed; a colored target has been used to assess the spectral fidelity of color films (Brooke, 1974), and bar targets have been constructed with contrast ratios somewhat closer to conditions observed during actual applications. The USGS target array is a large array painted on the roof of the USGS National Center, Reston, Virgina, as a means

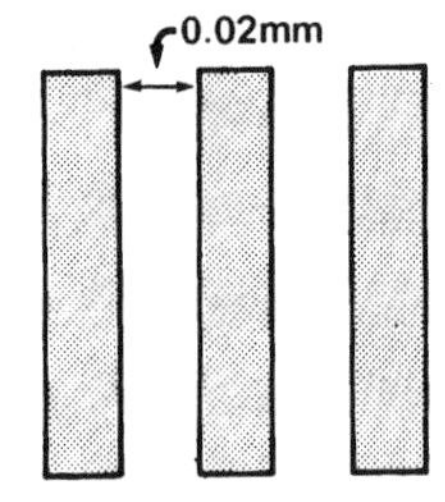


**FIGURE 9.2.** Use of bar target to find LPM.

of assessing aerial imagery under operational conditions from high altitudes. The array is formed from large bar targets, about 100 ft. in length, of known contrast, and a star target (about 140 ft. in diameter) designed for assessment of the resolution of nonphotographic sensors.

### *Modulation Transfer Functions*

The *modulation transfer function* (MTF) records system response to a target array with elements of varying spatial frequency (i.e., unlike the bar targets described above, targets used to find MTFs are spaced at varied intervals). Often the target array is formed from bars of equal length spaced against a white background at intervals that produce a sinusoidal variation in image density along the axis of the target.

*Modulation* refers to changes in the widths and spacings of the target. *Transfer* denotes the ability of the imaging system to record these changes on the image—that is, to "transfer" these changes from the target to the image. Because the target is explicitly designed with spatial frequencies too fine to be recorded on the image, some frequencies (the high frequencies at the closest spacings) cannot be imaged. The "function" then shows the degree to which the image records specified frequencies (Figure 9.3).

Although the MTF is probably the "best" measure of the ability of an imaging system, or of a single component of a system to record spatial detail, the complexity of the method prevents routine use in many situations. The MTF can be estimated using simpler, and more readily available targets, including the USAF target described above (Welch, 1971).

## 9.6. Mixed Pixels

As spatial resolution interacts with the fabric of the landscape, a special problem is created by those pixels that are not completely occupied by a single, homogeneous category. The subdivision of a scene into discrete pixels acts to average brightnesses over the entire pixel area. If a uniform, or relatively uniform, land area occupies the pixel, then similar brightnesses are averaged and the resulting digital value forms a reasonable representation of the brightnesses within the pixel. That is, the average value does not differ greatly from

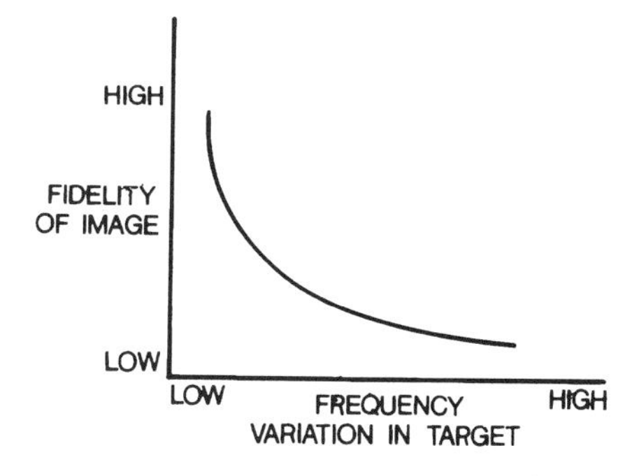


**FIGURE 9.3.** Modulation transfer function.

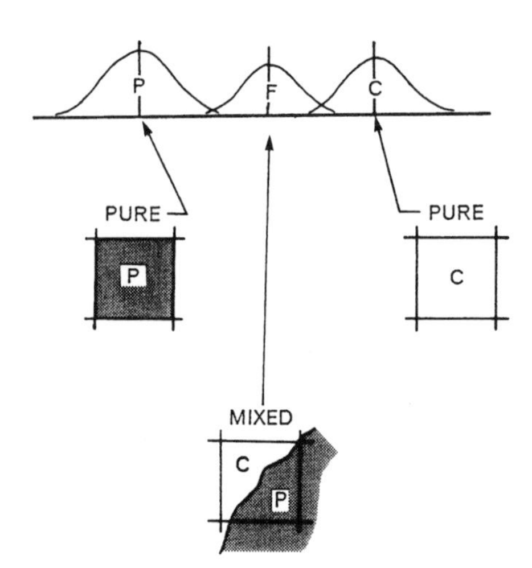


**FIGURE 9.4.** False resemblance of mixed pixels to a third category.

the values that contribute to the average. However, when a pixel area is composed of two or more areas that differ greatly with respect to brightness, the average is composed of several very different values, and the single digital value that represents the pixel may not accurately represent any of the categories present (Figure 9.4).

An important consequence of the occurrence of mixed pixels is that pure spectral responses of specific features are mixed together with the pure responses of other features. The mixed response sometimes known as a *composite signature* does not match to the pure signatures that we wish to use to map the landscape. Sometimes composite signatures can be useful because they permit us to map features that are too complex to resolve individually.

However, mixed pixels are also a source of error and confusion. In some instances, the digital values from mixed pixels may not resemble any of the several categories in the scene; in others, the value formed by a mixed pixel may resemble those from other categories in the scene but not present within the pixel—an especially misleading kind of error.

Mixed pixels occur often at the edges of large parcels, or along long linear features such as rivers or highways, where contrasting brightnesses are immediately adjacent to one another (Figure 9.5). The edge, or border, pixels then form opportunities for errors in

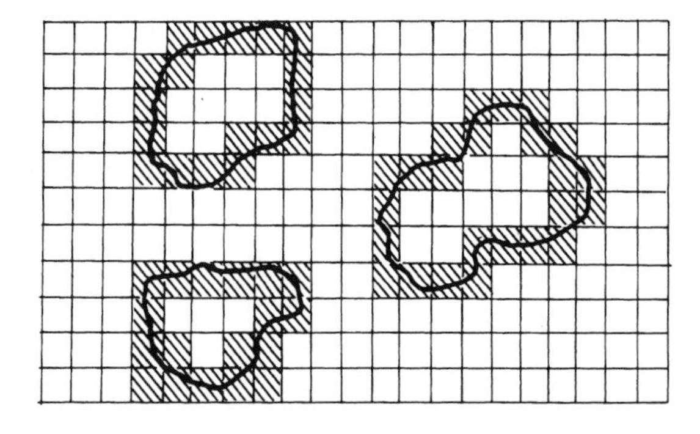

**FIGURE 9.5.** Edge pixels.

digital classification. Scattered occurrences of small parcels (such as farm ponds observed at the resolution of the Landsat MSS) may produce special problems because they may be represented *only* by mixed pixels, and the image analyst may not be aware of the presence of the small areas of high contrast because they occur at subpixel sizes. An especially difficult situation can be created by landscapes composed of many parcels that are small relative to the spatial resolution of the sensor. A mosaic of such parcels will create an array of digital values, *all* formed by mixed pixels (Figure 9.6).

It is interesting to examine the relationships between the numbers of mixed pixels in a given scene and the spatial resolution of the sensor. Studies have documented the increase in numbers of mixed pixels that occurs as spatial resolution decreases. Because the numbers, sizes, and shapes, of landscape parcels vary greatly with season, and geographic setting, there can be no generally applicable conclusions regarding this problem. Yet examination of a few simple examples may help us understand the general character of the problem.

Consider the same contrived scene that is examined at several different spatial resolutions (Figure 9.7). This scene consists of two contrasting categories with two parcels of one superimposed against the more extensive background of the other. This image is then examined at four levels of spatial resolution; for each level of detail, pixels are categorized as "background," "interior," or "border." (Background and interior pixels consist only of a single category; border pixels are those composed of two categories.) A tabulation of proportions of the total in each category reveals a consistent pattern (Table 9.1). As resolution becomes coarser, the number of mixed pixels increases (naturally) at the expense of the pure background and interior pixels. In this example, interior pixels experience the larger loss, but this result is the consequence of the specific circumstances of this example, and is unlikely to reveal any generally applicable conclusions.

If other factors could be held constant, it would seem that fine spatial resolution would offer many practical advantages, including capture of fine detail. Note however the substantial increases in the total numbers of pixels required to achieve this advantage; increases in the numbers of pixels produce compensating disadvantages, including increased costs. Also, this example does not consider other effects often encountered as spatial resolution is increased. Finer detail may resolve features not recorded at coarser detail,

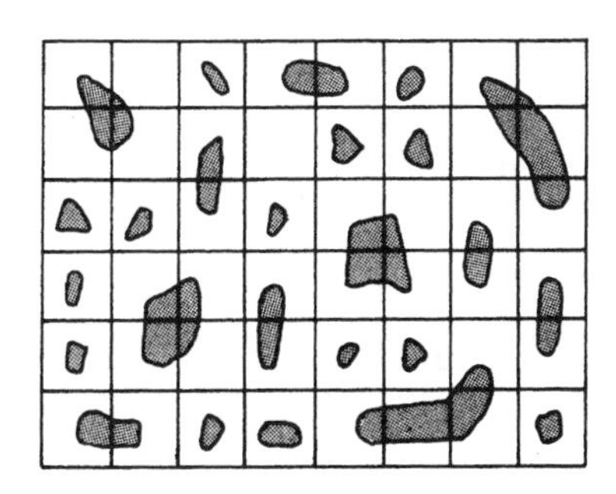


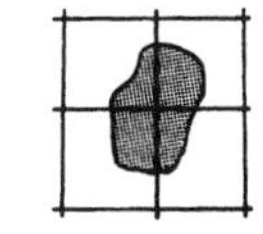


**FIGURE 9.6.** Mixed pixels generated by image of landscape composed of small parcels.

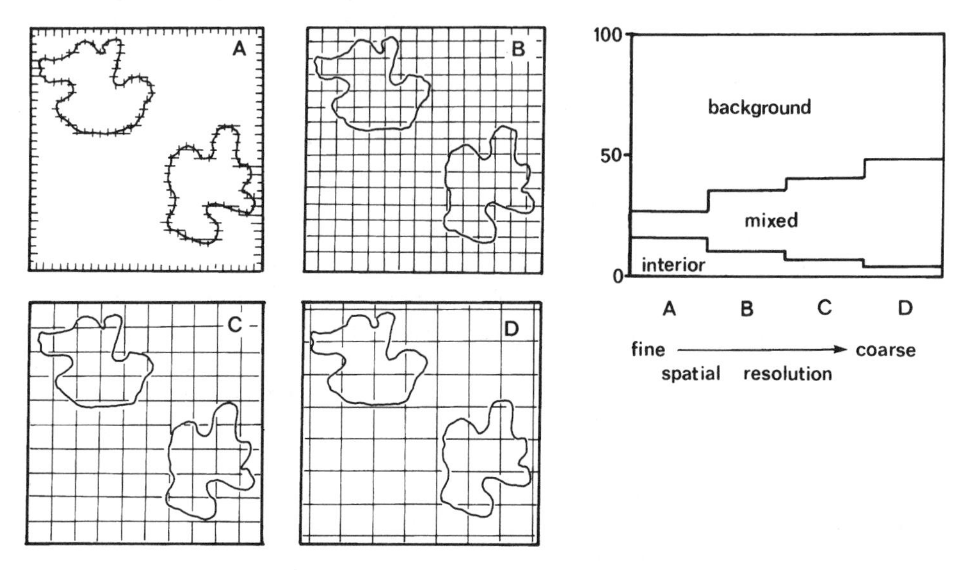


**FIGURE 9.7.** Influence of spatial resolution upon proportions of mixed pixels.

thereby altering the criteria by which the mixed pixels are evaluated. Although mixed pixels decrease with respect to initial classes (those defined at coarse resolution), increased detail reveals new classes, thereby changing the basis for the evaluation. This effect may explain some of the results observed by Sadowski and Sarno (1976), who found that classification accuracy decreased as spatial resolution became finer.

Marsh et al. (1980) have reviewed strategies for resolving the percentages of components that compose the ground areas with mixed pixels. The measured digital value for each pixel is determined by the brightnesses of distinct categories within that pixel area projected on the ground, as integrated by the sensor over the area of the pixel. For example, the projection of a pixel on the earth's surface may encompass areas of open water ($W$) and forest ($F$). Assume that we know that: (1) the digital value for such a pixel is "mixed," not "pure," (2) the mean digital value for water in all bands is $i$ ($W_i$), (3) the mean digital value for forest in all bands is ($F_i$), and (4) the observed value of the mixed pixel in all spectral bands is ($M_i$). We wish then to find the areal percentages $P_W$ and $P_F$ that contribute to the observed value $M_i$.

Marsh et al. outline several strategies for estimating $P_W$ and $P_F$ under these conditions; the simplest, if not the most accurate, is the weighted average method:

**TABLE 9.1. Summary of Data Derived from Figure 9.7**

| Spatial resolution | Total | Mixed | Interior | Background |
|---|---|---|---|---|
| Fine A | 900 | 109 | 143 | 648 |
| B | 225 | 59 | 25 | 141 |
| C | 100 | 34 | 6 | 60 |
| Coarse D | 49 | 23 | 1 | 25 |

$$P_W = (M_i - F_i)/(W_i - F_i) \quad \text{(Eq. 9.2)}$$

An example can be shown using the following data:

| | Band | | | |
|---|---|---|---|---|
| | 1 | 2 | 3 | 4 |
| Means for the mixed pixel ($M_i$): | 16 | 12 | 16 | 18 |
| Means for forest ($F_i$): | 23 | 16 | 32 | 35 |
| Means for water ($W_i$): | 9 | 8 | 0 | 1 |

Using equation 9.2, the areal proportion of the mixed pixel composed of the water category can be estimated as follows:

Band 1: $P_W = (16 - 23)/(9 - 23) = -7/-14 = 0.50$
Band 2: $P_W = (12 - 16)/(8 - 16) = -4/-8 = 0.50$
Band 3: $P_W = (16 - 32)/(0 - 32) = -16/-32 = 0.50$
Band 4: $P_W = (18 - 35)/(1 - 35) = -17/-34 = 0.50$

Thus, the mixed pixel is apparently composed of about 50% water and 50% forest. Note that in practice we may not know which pixels are mixed and the categories that might contribute to the mixture. Note also that this procedure may yield different estimates for each band. Other procedures, too lengthy for concise description here, may give more suitable results in some instances (Marsh et al., 1980).

## 9.7. Spatial and Radiometric Resolution: A Simple Example

Some of these effects can be illustrated by an artificial example. The contrived scene in Figure 9.8 is composed of two water bodies (W), several areas of forest (F), a large area of pasture (P), and a cultivated region (A) composed of a pattern of agricultural fields, some with mature crops, others composed of plowed bare ground. A digital representation of this scene might resemble Figure 9.9, the product of an imaginary sensor with fine spatial and radiometric resolution operating in the near infrared portion of the spectrum.

This hypothetical sensor records the scene at 10 brightness levels, from "0" (very dark) to "9" (very bright). In digital representation the water bodies are very dark, mainly "0's" and "1's;" forest is brighter, "3's" and "5's." Pasture has an intermediate brightness of "2." And agricultural land is represented as either "0" (the dark areas of bare soil), or "9" (the brighter areas of living vegetation).

Figure 9.10a represents the same scene as portrayed at high spatial resolution but very low radiometric resolution; the sizes of the pixels are the same as in Figure 9.9, but only two levels of brightness are used. (Pixels darker than 5 in Figure 9.9 are coded 0 in Figure 9.10a; those brighter that 5 are coded as 1.) At such coarse radiometric resolution, few features in the original scene can be recognized even though the spatial resolution remains constant from one to the other.

Figure 9.10b shows the original scene recorded at finer radiometric resolution. Three

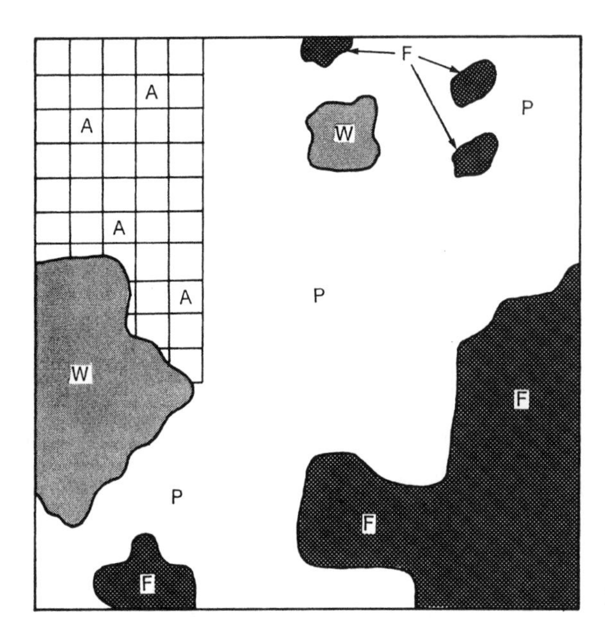


**FIGURE 9.8.** Hypothetical landscape.

brightness levels are formed by representing the original 0's, 1's, and 2's as 0's; 3's, 4's, 5's, and 6's as 1's; and 7's, 8's, and 9's as 2's. In this image the distinction between pasture and forest is evident, but the contrast between water and pasture is insufficient for these two categories to be represented as separate features.

Figures 9.10c and 9.10d show effects of changing spatial resolution while retaining essentially the same level of radiometric resolution. In Figure 9.10c, spatial resolution has been decreased to show one fourth of the detail visible in Figure 9.9; each pixel is formed by the average of four of the smaller original pixels. Here many of the major features in

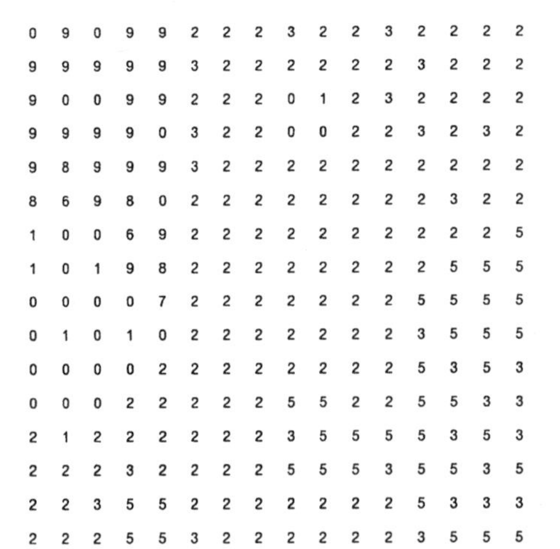

| 0 | 9 | 0 | 9 | 9 | 2 | 2 | 2 | 3 | 2 | 2 | 3 | 2 | 2 | 2 | 2 |
|---|---|---|---|---|---|---|---|---|---|---|---|---|---|---|---|
| 9 | 9 | 9 | 9 | 9 | 3 | 2 | 2 | 2 | 2 | 2 | 2 | 3 | 2 | 2 | 2 |
| 9 | 0 | 0 | 9 | 9 | 2 | 2 | 2 | 0 | 1 | 2 | 3 | 2 | 2 | 2 | 2 |
| 9 | 9 | 9 | 9 | 0 | 3 | 2 | 2 | 0 | 0 | 2 | 2 | 3 | 2 | 3 | 2 |
| 9 | 8 | 9 | 9 | 9 | 3 | 2 | 2 | 2 | 2 | 2 | 2 | 2 | 2 | 2 | 2 |
| 8 | 6 | 9 | 8 | 0 | 2 | 2 | 2 | 2 | 2 | 2 | 2 | 2 | 3 | 2 | 2 |
| 1 | 0 | 0 | 6 | 9 | 2 | 2 | 2 | 2 | 2 | 2 | 2 | 2 | 2 | 2 | 5 |
| 1 | 0 | 1 | 9 | 8 | 2 | 2 | 2 | 2 | 2 | 2 | 2 | 2 | 5 | 5 | 5 |
| 0 | 0 | 0 | 0 | 7 | 2 | 2 | 2 | 2 | 2 | 2 | 2 | 5 | 5 | 5 | 5 |
| 0 | 1 | 0 | 1 | 0 | 2 | 2 | 2 | 2 | 2 | 2 | 2 | 3 | 5 | 5 | 5 |
| 0 | 0 | 0 | 0 | 2 | 2 | 2 | 2 | 2 | 2 | 2 | 2 | 5 | 3 | 5 | 3 |
| 0 | 0 | 0 | 2 | 2 | 2 | 2 | 2 | 5 | 5 | 2 | 2 | 5 | 5 | 3 | 3 |
| 2 | 1 | 2 | 2 | 2 | 2 | 2 | 2 | 3 | 5 | 5 | 5 | 5 | 3 | 5 | 3 |
| 2 | 2 | 2 | 3 | 2 | 2 | 2 | 2 | 5 | 5 | 5 | 3 | 5 | 5 | 3 | 5 |
| 2 | 2 | 3 | 5 | 5 | 2 | 2 | 2 | 2 | 2 | 2 | 2 | 5 | 3 | 3 | 3 |
| 2 | 2 | 2 | 5 | 5 | 3 | 2 | 2 | 2 | 2 | 2 | 2 | 3 | 5 | 5 | 5 |

**FIGURE 9.9.** Digital representation of Figure 9.8.

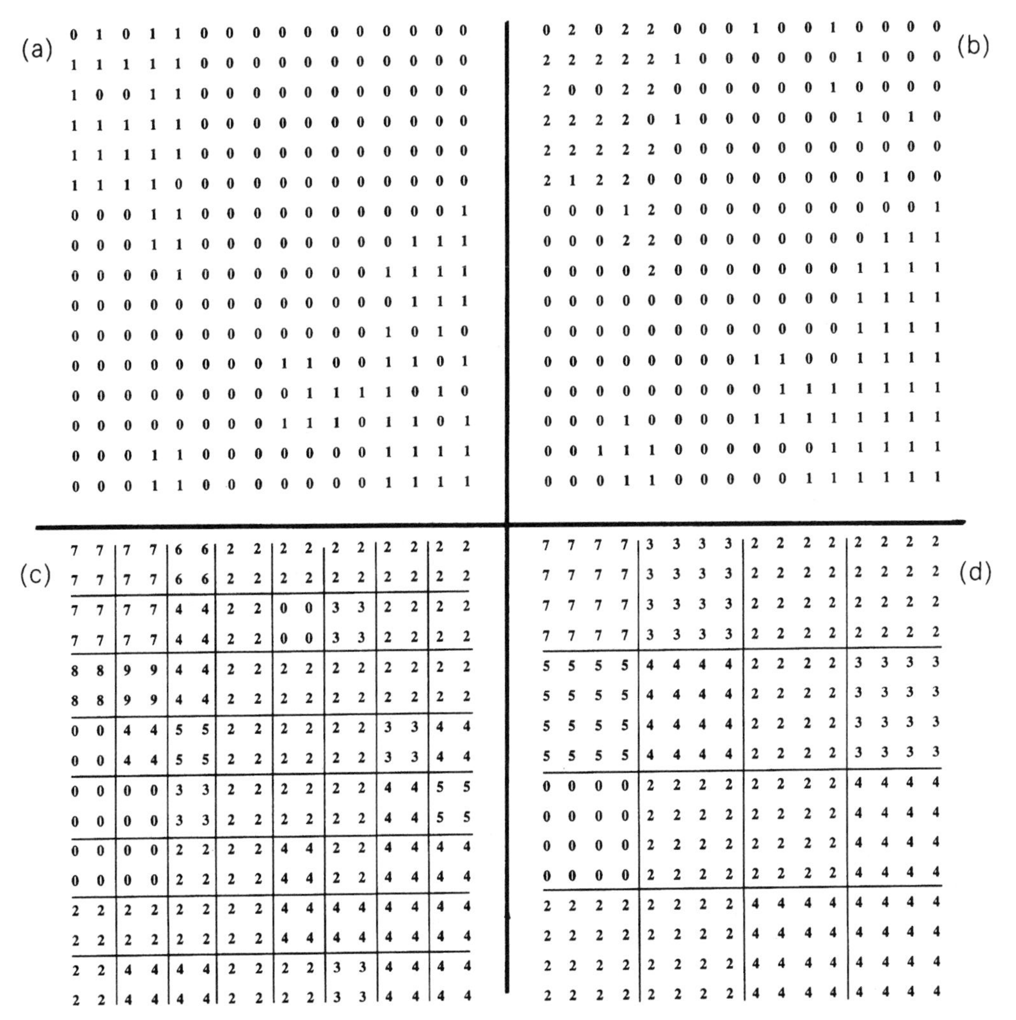


**FIGURE 9.10.** Figure 9.9 represented at (*a*) coarse radiometric resolution, (*b*) modest rdiometric resolution, (*c*) modest level of spatial resolution, and (*d*) coarse spatial resolution.

the original scene are recognizable, although sizes, shapes, and positions are indistinct. Figure 9.10d shows the scene at extremely coarse spatial resolution; these values are formed by averaging brightness values over areas 16 times as large as those of the original pixels. Many of the major differences in brightness are distinguishable, although the pattern is greatly altered from the original distribution in Figure 9.9.

## 9.8. Interactions with the Landscapes

Although most discussions of image resolution tend to focus on sensor characteristics, understanding the significance of image resolution in the application sciences requires assessment of the effect of specific resolutions on images of specific landscapes, or classes of landscapes. For example, relatively low resolution may be sufficient for recording the essential features of landscapes with rather coarse fabrics (such as the broad-scale patterns of the agricultural fields of the North American Great Plains) but inadequate for imaging complex landscapes composed of many small parcels with low contrast.

Podwysocki's studies (1976, 1977) of field sizes in the major grain producing regions of the world is an excellent example of the systematic investigation of this topic. His research can be placed in the context of the widespread interest in accurate forecasts of the world wheat production in the years that followed the large international wheat purchases by the Soviet Union in 1972. Computer models of biophysical processes of crop growth and maturation could provide accurate estimates of yields (given suitable climatological data), but estimates of total production require accurate estimates of planted acreage. Satellite imagery would seem to provide the capability to derive the required estimates of area plowed and planted. Podwysocki attempted to define the extent to which the spatial resolution of the Landsat MSS would be capable of providing the detail necessary to provide the required estimates.

He examined Landsat MSS scenes of the United States, China, Soviet Union, Argentina, and other wheat-producing regions, sampling fields for measurements of length, width, and area. His data are summarized by frequency distributions of field sizes for samples for each of the world's major wheat-producing regions. (He used his samples to evaluate the Guassian distributions for each of them, so he was able to extrapolate the frequency distributions to estimate frequencies at sizes smaller than the resolution of the MSS data.) Cumulative frequency distributions for his normalized data reveal the percentages of each sample that equal or exceed specific areas (Figure 9.11). For example, the

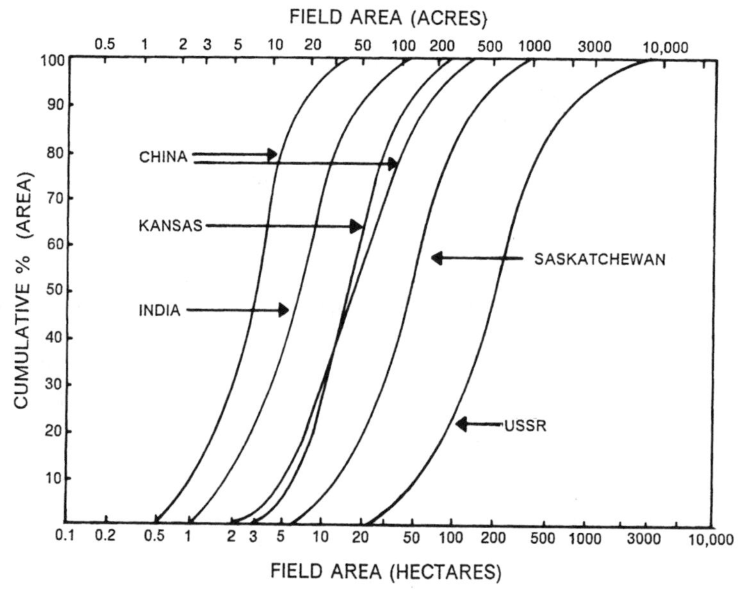


**FIGURE 9.11.** Field size distributions for selected wheat-producing regions. From Podwysocki (1976).

curve for India reveals that 99% or more of this sample were at least 1 hectare in size, that all were smaller than about 100 ha (247 acres), and that we can expect the Indian fields to be smaller than those in Kansas. These data, and others presented in his study, provide the basis for evaluating the effectiveness of a given resolution in monitoring features of specified sizes. This example is especially instructive because it emphasizes not only the differences in average field size in the different regions (shown in Figure 9.11 by the point where each curve crosses the 50% line) but also the differences in variation of field size between the varied regions (shown in Figure 9.11 by the slopes of the curves).

In a different analysis of relationships between sensor resolution and landscape detail, Simonett and Coiner (1971) examined 106 sites in the United States, each selected to represent a major land use region. Their study was conducted prior to the launch of Landsat 1, with the objective of assessing the effectiveness of MSS spatial resolution in recording differences between major land use regions in the United States. Considered as a whole, their sites represent a broad range of physical and cultural patterns in the 48 coterminous states.

For each site they simulated the effects of imaging with low-resolution imagery by superimposing grids over aerial photographs, with grid dimensions corresponding to ground distances of 800, 400, 200, and 100 ft. Samples were randomly selected within each site. Each sample consisted of the *number* of land use categories within cells of each size and thereby formed a measure of landscape diversity, as considered at several spatial resolutions. For example, those landscapes that show only a single land use category at the 800-ft. resolution have a very coarse fabric and would be effectively imaged at the low resolution of satellite sensors. Those landscapes that have many categories within the 100-ft. grid are so complex that very fine spatial resolution would be required to record the pattern of landscape variation. Their analysis grouped sites according to their behavior at various resolutions. One of their conclusions reported that natural landscapes appeared to be more susceptible than man-made landscapes to analysis at the coarse resolutions of satellite sensors.

Welch and Pannell (1982) examined Landsat MSS (bands 2 and 4) and Landsat 3 RBV images (in both pictorial and digital formats) to evaluate their suitability as sources of landscape information at levels of detail consistent with a map scale of 1:250,000. Images of three study areas in China provided a variety of urban and agricultural landscapes representing a range of spatial detail and a number of geographical settings. Their analysis of modulation transfer functions reveals that the RBV imagery represents an improvement in spatial resolution of about 1.7 over the MSS imagery, and that Landsat 4 TM provides an improvement of about 1.4 over the RBV (for target:background contrasts of about 1.6:1).

Features appearing on each image were evaluated with corresponding representations on 1:250,000 maps in respect to size, shape, and contrast. A numerical rating system provided scores for each image based on the numbers of features represented and the quality of the representations on each form of imagery. MSS images portrayed about 40–50% of the features shown on the usual 1:250,000 topographic maps. MSS band 2 was the most effective for identification of airfields; band 4 performed very well for identification and delineation of water bodies. Overall, the MSS images achieved scores of about 40–50%. RBV images attained higher overall scores (50–80%), providing considerable improvement in representation of high-contrast targets but little improvement in imaging of detail in fine-textured urban landscapes. The authors concluded that spatial resolutions of MSS and RBV images are inadequate for compilation of usual map detail at 1:250,000.

## 9.9. Summary

This chapter highlights the significance of image resolution as a concept that extends across many aspects of remote sensing. Although the special and unique elements of any image must always be recognized and understood, many of the general aspects of image resolution can assist us in understanding how to interpret remotely sensed images.

Although there has long been an intense interest in measuring image resolution, especially in photographic systems, it is clear that much of our more profound understanding has been developed through work with such satellite scanning systems as the Landsat MSS. Such data were of much coarser spatial resolution than any studied previously, and as more and more attention was focused on their analysis and interpretation (Chapters 11 and 12) it was necessary to develop a better understanding of image resolution and its significance for specific tasks. Now much finer resolution data are available, but we can continue to develop and apply our knowledge of image resolution to maximize our ability to understand and interpret these images.

## Review Questions

1. Most individuals are quick to appreciate advantages of fine resolution. However, there may well be *disadvantages* to fine resolution data, relative to data of coarser spatial, spectral, and radiometric detail. Suggest what some of these effects might be.

2. Imagine that the spatial resolution of the digital remote sensing system is increased from about 80 m to 40 m. List some of the consequences, assuming that image coverage remains the same. What would be some of the consequences of *decreasing* detail from 80 m to 160 m?

3. You examine an image of the U.S. Air Force resolution target and determine that the image distance between the bars is the smallest pair of lines is 0.01 mm. Find the LPM for this image. Find the LPM for an image in which you measure the distance to be 0.04 mm. Which image has finer resolution?

4. For each object or feature listed below, discuss the characteristics that will be significant in our ability to resolve the object on a remotely sensed image. Categorize each as "easy" or "difficult" to resolve clearly. Explain.
   a. A white car parked alone in an asphalt parking lot.
   b. A single tree in a pasture.
   c. An orchard.
   d. A black cat in a snow-covered field.
   e. Painted white lines on a crosswalk across an asphalt highway.
   f. Painted white lines on a crosswalk across a concrete highway.
   g. A pond.
   h. A stream.

5. Write a short essay describing how spatial resolution, spectral resolution, and radiometric resolution are interrelated. Is it possible to increase one kind of resolution without influencing the others?

6. Review Chapters 1–8 to identify the major features that influence spatial resolution of images collected by the several kinds of sensors described. Prepare a table to list these factors in summary form.

7. Explain why some objects might be resolved clearly in one part of the spectrum yet resolved poorly in another portion of the spectrum.

8. Although the U.S. Air Force resolution target is very useful for evaluating some aspects of remotely sensed images, it is not necessarily a good indication of the ability of a remote sensing system to record patterns that are significant for environmental studies. List some of the reasons this might be true.

9. Describe ideal conditions for achieving maximum spatial resolution.

## References


Badhwar, G. B. 1984. Automatic Corn–Soybean Classification Using Landsat MSS Data. II. Early Season Crop Proportion Estimation. *Remote Sensing of Environment,* Vol. 14, pp. 31–37.

Badhawr, G. D. 1984. Use of Landsat-Derived Profile Features for Spring Small-Grains Classification. *International Journal of Remote Sensing,* Vol. 5, pp. 783–797.

Badhwar, G. D., J. G. Carnes, and W. W. Austen. 1982. Use of Landsat-Derived Temporal Profiles for Corn–Soybean Feature Extraction and Classification. *Remote Sensing of Environment,* Vol. 12, pp. 57–79.

Badhwar, G. D., C. E. Garganti, and F. V. Redondo. 1987. Landsat Classification of Argentina Summer Crops. *Remote Sensing of Environment,* Vol. 21, pp. 111–117.

Brooke, Robert K. 1974. *Spectral/Spatial Resolution Targets for Aerial Imagery.* Technical Report ETL-TR-74-3. Ft. Belvior, VA: U.S. Army Engineer Topographic Laboratories, 20 pp.

Chhikara, Raj. S. 1984. Effect of Mixed (Boundary) Pixels on Crop Proportion Estimation. *Remote Sensing of Environment,* Vol. 14, pp. 207–218.

Crapper, P. F. 1980. Errors Incurred in Estimating an Area of Uniform Land Cover Using Landsat. *Photogrammetric Engineering and Remote Sensing,* Vol. 46, pp. 1295–1301.

Ferguson, M. C., G. D. Badhwar, R. S. Chhikara, and D. E. Pitts. 1986. Field Size Distributions for Selected Agricultural Crops in the United States and Canada. *Remote Sensing of Environment,* Vol. 19, pp. 25–45.

Hall, F. G. and G. D. Badhwar. 1987. Signature Extendable Technology: Global Space-Based Crop Recognition. *IEEE Transactions on Geoscience and Remote Sensing,* Vol. GE-25, pp. 93–103.

Hallum, Ceal R., and C. R. Perry. 1984. Estimating Optimal Sampling Unit Sizes for Satellite Surveys. *Remote Sensing of Environment,* Vol. 14, pp. 183–196.

Hyde, Richard F., and N. J. Vesper. 1983. Some Effects of Resolution Cell Size on Image Quality. *Landsat Data Users Notes,* Issue 29, pp. 9–12.

Irons, J. R., and D. L. Williams. 1982. Summary of Research Addressing the Potential Utility of Thematic Mapper Data for Renewable Resource Applications. *Harvard Computer Graphics Week.* Cambridge, MA: Harvard University Graduate School of Design.

Latty, R. S., and R. M. Hoffer. 1981. Computer-Based Classification Accuracy Due to the Spatial Resolution Using Per-Point versus Per-Field Classification Techniques. In *Proceedings, 7th International Symposium on Machine Processing of Remotely Sensed Data.* West Lafayette, IN: LARS, pp. 384–393.

MacDonald, D. E. 1958. Resolution as a Measure of Interpretability. *Photogrammetric Engineering,* Vol. 24, No. 1, pp. 58–62.

Markham, B. L., and J. R. G. Townshend. 1981. Land Cover Classification Accuracy as a Function of Sensor Spatial Resolution. In *Proceedings, 15th International Symposium on Remote Sensing* of *Environment.* Ann Arbor: University of Michigan, pp. 1075–1090.

Marsh, S. E., P. Switzer, and R. J. P. Lyon. 1980. Resolving the Percentage of Component Terrains within Single Resolution Elements. *Photogrammetric Engineering and Remote Sensing,* Vol. 46, 1079–1086.

Pitts, David E., and G. Badhwar. 1980. Field Size, Length, and Width Distributions Based on LACIE Ground Truth Data. *Remote Sensing of Environment,* Vol. 10, pp. 201–213.

Podwysocki, M. H. 1976. *An Estimate of Field Size Distribution for Selected Sites in the Major Grain Producing Countries.* X-923-76-93. Greenbelt, MD: Goddard Space Flight Center, 34 pp.

Podwysocki, M.H. 1977. *Analysis of Field Size Distributions: LACIE Test Sites 5029, 5033, 5039, Anwhei Province, People's Republic of China.* X-923-76-145. Greenbelt, MD: Goddard Space Flight Center, 8 pp.

Potdar, M. B. 1993. Sorghum Yield Modelling Based on Crop Growth Parameters Determined from Visible and Near-IR Channel NOAA AVHRR Data. *International Journal of Remote Sensing,* Vol. 14, pp. 895–905.

Sadowski, F., and J. Sarno. 1976. *Forest Classification Accuracy as Influenced by Multispectral Scanner Spatial Resolution.* Report for Contract NAS9-14123:NASA. Houston, TX: LBJ Space Center.

Salmonowicz, P. H. 1982. USGS Aerial Resolution Targets. *Photogrammetric Engineering and Remote Sensing,* Vol. 48, pp. 1469–1473.

Simonett, D. S., and J. C. Coiner. 1971. Susceptibility of Environments to Low Resolution Imaging for Land Use Mapping. In *Proceedings, 7th International Symposium on Remote Sensing of Environment.* Ann Arbor: University of Michigan, pp. 373–394.

Tucker, C. J. 1980. Radiometric Resolution for Monitoring Vegetation: How Many Bits Are Needed? *International Journal of Remote Sensing,* Vol. 1, pp. 241–254.

Welch, R. 1971. Modulation Transfer Functions. *Photogrammetric Engineering,* Vol. 47, pp. 247–259.

Welch, R., and C. W. Pannell. 1982. Comparative Resolution of Landsat 3 MSS and RBV Images of China. *Photogrammetric Record,* Vol. 10, pp. 575–586.

Wehde, M. E. 1979. Spatial Quantification of Maps or Images; Cell Size or Pixel Size Implication. In *Joint Proceedings, American Society of Photogrammetry and American Congress of Surveying and Mapping,* pp. 45–65.

PART TWO

# ANALYSIS

CHAPTER TEN

# Preprocessing

## 10.1. Introduction

In the context of digital analysis of remotely sensed data, *preprocessing* refers to those operations that are preliminary to the main analysis. Typical preprocessing operations could include (1) radiometric preprocessing to adjust digital values for the effect of a hazy atmosphere and/or (2) geometric preprocessing to bring an image into registration with a map or another image. Once corrections have been made, the data can then be subjected to the primary analyses described in subsequent chapters. Thus, preprocessing forms a preparatory phase that, in principle, improves image quality as the basis for later analyses that will extract information from the image.

It should be emphasized that, although certain preprocessing procedures are frequently used, there can be no definitive list of "standard" preprocessing steps because each project requires individual attention and some preprocessing decisions may be a matter of personal preference. Furthermore, the quality of image data vary greatly, so some data may not require the preprocessing that would be necessary in other instances. Preprocessing changes data. We assume that such changes are beneficial, but the analyst should remember that preprocessing may create artifacts that are not immediately obvious. As a result, the analyst should tailor preprocessing to the data at hand, the needs of specific projects, using only those preprocessing operations essential to obtain a specific result.

Preprocessing includes a wide range of operations, from the very simple to extremes of abstractness and complexity. Most can be categorized into one of three groups: (1) feature extraction, (2) radiometric corrections, and (3) geometric corrections. Although there are far too many preprocessing methods to discuss in detail here, it will be possible to illustrate some of the principles important for each group.

## 10.2. Feature Extraction

In the context of image processing, the term *feature extraction* (or *feature selection*) has specialized meaning. "Features" are not geographical features, visible on an image but, rather, "statistical" characteristics of image data—individual bands, or combinations of band values, that carry information concerning systematic variation within the scene. Thus, feature extraction could also be known as "information extraction"—isolation of components within multispectral data that are most useful in portraying the essential elements of an image. In principle, discarded data contain noise and errors present in original data. Thus feature extraction may increase accuracy. In addition, feature extraction reduces the number of spectral channels, or bands, that must be analyzed. After feature selection is complete, the analyst works with fewer channels, but each of the individual channels is more potent. The reduced data set may convey almost as much information as does the complete data set. Thus, feature selection may increase speed and reduce costs of analysis.

Multispectral data, by their nature, consist of several channels of data. Although some images may have as few as 3, 4, or 7 channels (Chapter 6), other image data may have many more, possibly 200 or more channels (Chapter 15). With so much data, processing of even modest-sized images requires considerable time and inconvenience. In this context, feature selection assumed considerable practical significance, as image analysts wish to reduce the amount of data while retaining effectiveness and/or accuracy.

Our examples here are based on TM data, which provide enough channels to illustrate the concept but are compact enough to be reasonably concise (Table 10.1). A variance–covariance matrix shows interrelationships between pairs of bands; some pairs show rather strong correlations—for example, bands 1 and 3 and 2 and 3 both show correlations above 0.9. High correlation between pairs of bands means that the values in the two channels are closely related. Thus, as values in channel 2 rise or fall, so do those in channel 3; one channel tends to duplicate information in the other. Feature selection attempts to identify and remove such duplication so that the data set can include maximum information using the minimum number of channels.

For example, for data represented by Table 10.1, bands 3, 5, and 6 might include almost as much information as the entire set of seven channels because band 3 is closely related to bands 1 and 2, band 5 is closely related to bands 4 and 7, and band 6 carries infor-

**TABLE 10.1. Similarity Matrices for Seven Bands of a TM Scene**

| | Covariance matrix | | | | | | |
|---|---|---|---|---|---|---|---|
| | 1 | 2 | 3 | 4 | 5 | 6 | 7 |
| 1. | 48.8 | 29.2 | 43.2 | 50.0 | 76.5 | 0.9 | 44.9 |
| 2. | 29.2 | 20.3 | 29.0 | 48.6 | 65.4 | 1.5 | 32.8 |
| 3. | 43.2 | 29.0 | 46.4 | 59.9 | 101.2 | 0.6 | 53.5 |
| 4. | 49.9 | 48.6 | 59.9 | 327.8 | 325.6 | 12.4 | 104.32 |
| 5. | 76.5 | 65.4 | 101.2 | 325.6 | 480.5 | 10.2 | 188.5 |
| 6. | 0.9 | 1.5 | 0.6 | 12.5 | 10.2 | 14.0 | 1.1 |
| 7. | 45.0 | 32.8 | 53.5 | 104.3 | 188.5 | 1.1 | 90.8 |

| | Correlation matrix | | | | | | |
|---|---|---|---|---|---|---|---|
| | 1 | 2 | 3 | 4 | 5 | 6 | 7 |
| 1. | 1.00 | | | | | | |
| 2. | 0.92 | 1.00 | | | | | |
| 3. | 0.90 | 0.94 | 1.00 | | | | |
| 4. | 0.39 | 0.59 | 0.48 | 1.00 | | | |
| 5. | 0.49 | 0.66 | 0.67 | 0.82 | 1.00 | | |
| 6. | 0.03 | 0.08 | 0.02 | 0.18 | 0.12 | 1.00 | |
| 7. | 0.67 | 0.76 | 0.82 | 0.60 | 0.90 | 0.02 | 1.00 |

mation largely unrelated to any others. Therefore, each of the discarded channels (1, 2, 4, and 7) resembles one of the channels that have been retained. So, a simple approach to feature selection discards unneeded bands, thereby reducing the number of channels. Although such a selection can be used as a kind of rudimentary feature extraction, feature selection is a more typically complex process based on statistical interrelationships between channels.

A common approach to feature selection applies a method of data analysis called *principal components analysis* (PCA) (Davis, 1986). This text will offer only a superficial description, as a more complete explanation requires the level of detail provided by Davis (1986) and others. In essence, PCA identifies the optimum linear combination of the original channels that can account for variation of pixels values in an image. Linear combinations are of the form

$$A = C_1 X_1 + C_2 X_2 + C_3 X_3 + C_4 X_4 \quad \text{(Eq. 10.1)}$$

where $X_1$, $X_2$, $X_3$, and $X_4$ are pixel values in four spectral channels, and $C_1$, $C_2$, $C_3$, and $C_4$ are coefficients applied individually to the values in the respective channels. $A$ represents a transformed value for the pixel. Assume, as an example, that $C_1 = 0.35$, $C_2 = -0.08$, $C_3 = 0.36$, and $C_4 = 0.86$. For a pixel with $X_1 = 28$, $X_2 = 29$, $X_3 = 21$, $X_4 = 54$, the transformation assumes a value of 61.48. Optimum values for coefficients are calculated by a procedure that ensures that the values they produce account for maximum variation within the entire data set. Thus, this set of coefficients provides maximum information that can be conveyed by any single channel formed by a linear combination of the original channels. If we make an image from all the values formed by applying this procedure to an entire image, we generate a single band of data that provides an optimum depiction of the information present within the four channels of the original scene.

Effectiveness of this procedure depends, of course, on calculation of the optimum coefficients. Here our description must be, by intention, abbreviated because calculation of the coefficients is accomplished by methods requiring full explanations, such as those given by upper-level statistics texts or discussions such as those of Davis (1986) and Gould (1967). For the present, the important point is that PCA permits identification of a set of coefficients that concentrates maximum information in a single band.

The same procedure also yields a second set of coefficients that will yield a second set of values (we could represent this as the *B* set, or image) that will be a less effective conveyor of information but will still represent variation of pixels within the image. In all, the procedure will yield seven sets of coefficients (one set for each band in the original image), and therefore produces seven sets of values, or bands (here denoted as *A, B, C, D, E, F,* and *G*), each in sequence conveying less information than the preceding band. Thus, in Table 10.2 transformed channels I and II (each formed from linear combinations of the seven original channels) together account for about 93% of the total variation in the data, whereas channels III–VII together account for only about 7% of the total variance. The analyst may be willing to discard the variables that convey 7% the variance as a means of reducing the number of channels. The analyst still retains 93% of the original information in a much more concise form. Thus, feature selection reduces the size of the data set by eliminating replication of information.

The effect is easily seen in Figure 10.1, which shows transformed data for a subset of a

**TABLE 10.2. Results of Principal Components Analysis of Data in Table 10.1**

| | Component | | | | | | |
|---|---|---|---|---|---|---|---|
| | I | II | III | IV | V | VI | VII |
| | | | | Eigenvectors | | | |
| % var.: | 82.5% | 10.2% | 5.3% | 1.3% | 0.4% | 0.3% | 0.1% |
| EV: | 848.44 | 104.72 | 54.72 | 13.55 | 4.05 | 2.78 | 0.77 |
| | 0.14 | 0.35 | 0.60 | 0.07 | –0.14 | –0.66 | –0.20 |
| | 0.11 | 0.16 | 0.32 | 0.03 | –0.07 | –0.15 | –0.90 |
| | 0.37 | 0.35 | 0.39 | –0.04 | –0.22 | 0.71 | –0.36 |
| | 0.56 | –0.71 | 0.37 | –0.09 | –0.18 | 0.03 | –0.64 |
| | 0.74 | 0.21 | –0.50 | 0.06 | –0.39 | –0.10 | 0.03 |
| | 0.01 | –0.05 | 0.02 | 0.99 | 0.12 | 0.08 | –0.04 |
| | 0.29 | 0.42 | –0.08 | –0.09 | 0.85 | 0.02 | –0.02 |
| | | | | Loadings | | | |
| Band 1 | 0.562 | 0.519 | 0.629 | 0.037 | –0.040 | –0.160 | –0.245 |
| Band 2 | 0.729 | 0.369 | 0.529 | 0.027 | –0.307 | –0.576 | –0.177 |
| Band 3 | 0.707 | 0.528 | 0.419 | –0.022 | –0.659 | –0.179 | –0.046 |
| Band 4 | 0.903 | –0.401 | 0.150 | –0.017 | 0.020 | 0.003 | –0.003 |
| Band 5 | 0.980 | 0.098 | –0.166 | 0.011 | –0.035 | –0.008 | –0.001 |
| Band 6 | 0.144 | –0.150 | 0.039 | 0.969 | 0.063 | 0.038 | –0.010 |
| Band 7 | 0.873 | 0.448 | –0.062 | –0.033 | 0.180 | 0.004 | –0.002 |

Landsat TM scene. Images PCI and PCII are the most potent; PCII, PCIV, PCVI, and PCVII show the decline in information content such that the final images record (one assumes) artifacts of system noise, atmospheric scatter, and other undesirable contributions to image brightness. If these two channels are excluded from subsequent analysis, it is likely that accuracy can be retained (relative to the entire set of four channels) while also reducing costs.

This method is not the only means for feature selection but does illustrate the objectives of this step—to reduce the number of channels to be examined while simultaneously retaining as much information as possible and reducing contributions of noise and error.

## 10.3. Subsets

Because of the very large sizes of many remotely sensed images, analysts typically work with those segments of full images that specifically pertain to the task at hand. Therefore, to minimize computer storage and the analyst's time and effort, one of the first tasks in each project is to prepare *subsets*—portions of larger images, selected to show only the region of interest.

Although selecting subsets would not appear to be one of remote sensing's most challenging tasks, it is more difficult than one might first suppose. Often subsets must match to other data, or to other projects, so it is necessary to find distinctive landmarks in both sets of data to assure that coverages match. Second, time and computational effort devoted

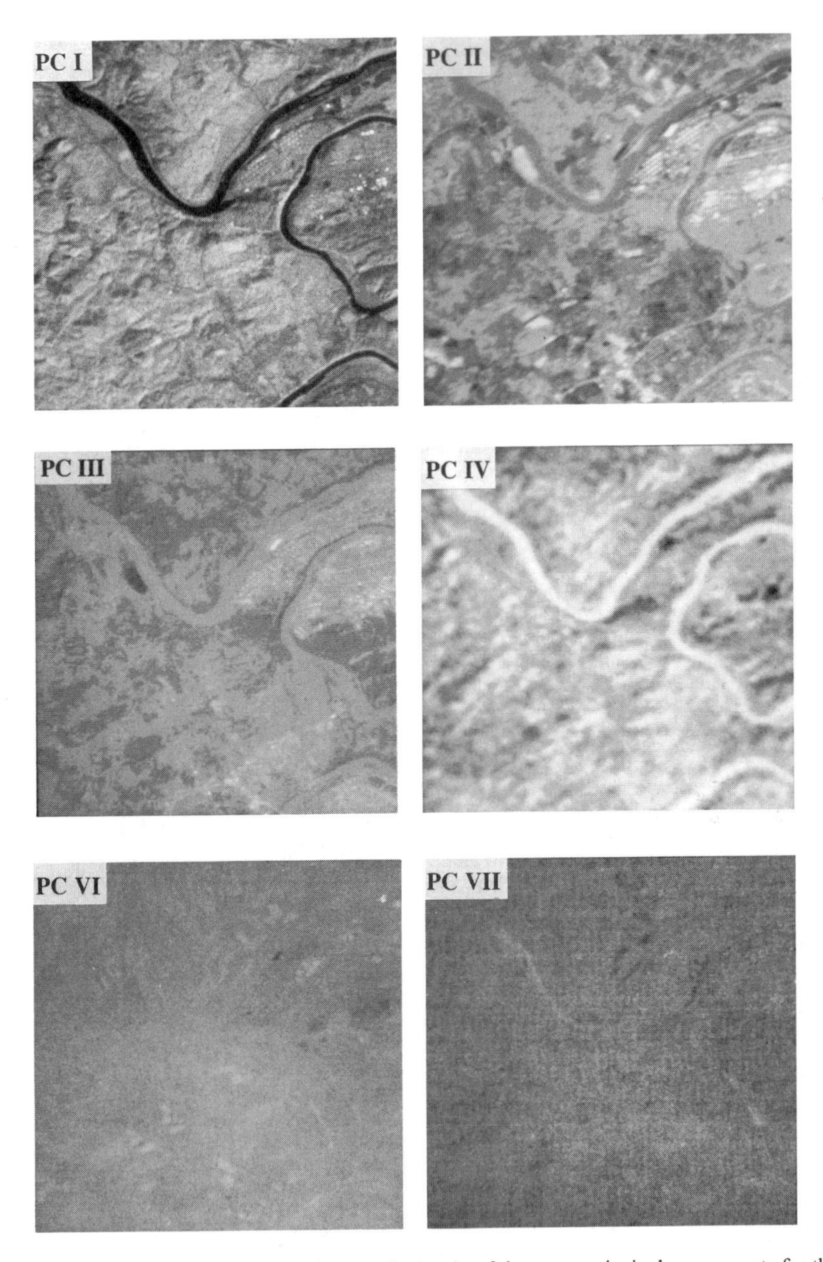


**FIGURE 10.1.** Feature selection. These images depict six of the seven principal components for the image described by Tables 10.1 and 10.2. The first principal component image (PC I), formed from a linear combination of data from all seven original bands, accounts for 82.5% of the total variation of the image data. PC II and PC III present 10.2% and 5.3% of the total variation, respectively. The higher components (e.g., PC VI and PC VII) account for very low proportions of the total variation and convey mainly noise and error, as is clear by the image patterns they show.

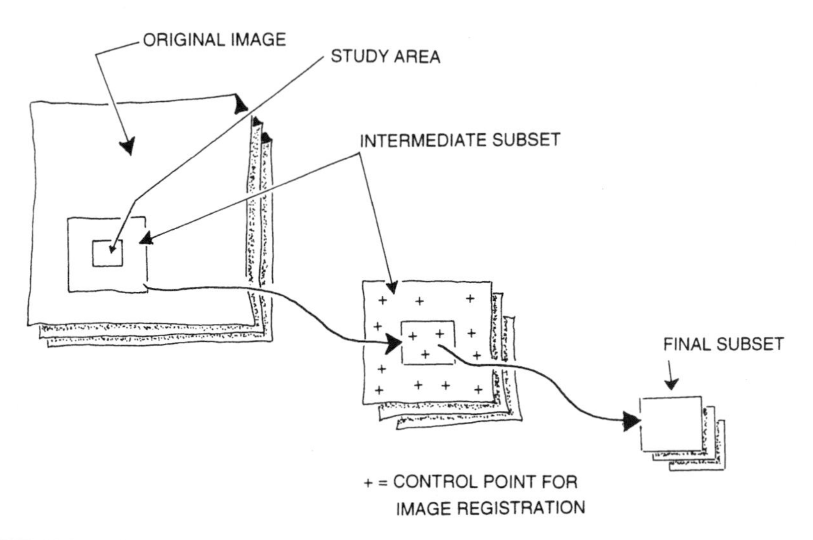


**FIGURE 10.2.** Subsets. Sometimes a subset of a particular area is too small to encompass sufficient points to allow the subset to be matched accurately to an accurate map (to be discussed in Section 10.6). Selection of an intermediate, temporary subset permits accurate registration using an adequate number of control points. After the temporary subset has been matched to the map, the study area can be more precisely subset without concern for the distribution of control points.

to matching images to maps or other images (as described below) increase with large images, so it is often convenient to prepare subsets before registration. Yet if the subset is too small, it may be difficult to identify sufficient landmarks for efficient registration. Therefore, it may be useful to prepare a preliminary subset, large enough to conduct the image registration effectively, before selecting the final, smaller subset for analytical use (Figure 10.2).

The same kinds of considerations apply in other steps of an analysis. Subsets should be large enough to provide the context required for the specific analysis. For example, it may be important to prepare subsets large enough to provide sufficient numbers of training fields for image classification (Chapter 11), or a sufficient set of sites for accuracy assessment (Chapter 13).

## 10.4. Radiometric Preprocessing

Many preprocessing operations fall into the category of image restoration (Estes et al., 1983)—the effort to remove the undesirable influence of atmospheric interference, system noise, and sensor motion. By applying a knowledge of the nature of these effects, it is possible to estimate their magnitude and remove, or minimize, their influence on the data used in later steps of the analysis. By removing these effects, the data are said to be "restored" to their (hypothetical) correct condition, although we can, of course, never know what the correct values might be and must always remember that attempts to correct data may themselves introduce errors. So, the analyst must decide whether the errors removed are likely to be greater than those that might be introduced. Typically, image restoration includes efforts to correct for both radiometric and geometric errors.

Radiometric preprocessing influences the brightness values of an image to correct for sensor malfunctions or to adjust the values to compensate for atmospheric degradation. Any sensor that observes the earth's surface using visible or near-visible radiation will record a mixture of two kinds of brightnesses. One brightness is due to the reflectance from the earth's surface—the brightnesses that are of interest for remote sensing. But the sensor also observes the brightnesses of the atmosphere itself—the effects of scattering (Chapter 2). Thus an observed digital brightness value ("56") might be in part the result of surface reflectance (perhaps "45") and partially the result of atmospheric scattering ("11"). Of course, we cannot immediately distinguish the two brightnesses, so one objective of atmospheric correction is to identify and separate these two components so the main analysis can focus on examination of correct surface brightness (the "45" in this example). Ideally, atmospheric correction should find a separate correction for each pixel in the scene; in practice, we may apply the same correction to an entire band, or a single factor to a local region within the image.

Preprocessing operations to correct for atmospheric degradation fall into three rather broad categories. First are those procedures based on efforts to model the physical behavior of the radiation as it passes through the atmosphere. Application of such models permits observed reflectances to be adjusted to approximate true values that might be observed under a clear atmosphere, thereby improving image quality and accuracies of analyses. Physical models (i.e., models that attempt to model the physical process of scattering at the level of individual particles and molecules) have important advantages with respect to rigor, accuracy, and applicability to a wide variety of circumstances. But they have significant disadvantages. Often they are very complex, usually requiring use of intricate computer programs. A more important limitation is the requirement for detailed meteorological information pertaining to atmospheric humidity and the concentrations of atmospheric particles. Such data may be difficult to obtain in the necessary detail and may apply only to a few points within, for example, a Landsat scene. Also, atmospheric conditions vary with altitude, and the radiosonde data that depict changes with altitude are routinely collected only at a few locations. Although the availability of data collected by meteorological satellites offers the possibility for increased use of such methods, routine application of such models is not now practicable.

Among the most widely used and widely available of the physical models are the LOWTRAN 7 (Kneizys et al., 1988) and MODTRAN (Berk et al., 1989) models developed by the U.S. Air Force and available in versions tailored for personal computers. LOWTRAN calculates atmospheric transmittance and atmospheric background radiance for a variety of atmospheric conditions (Table 10.3 and Figure 10.3). LOWTRAN applies relatively low spectral resolution (2.0 cm, in increments as small as 0.50 cm, beginning at 0.2 μm), whereas MODTRAN uses somewhat finer spectral resolution (0.2 cm). These programs estimate atmospheric absorption and emission by atmospheric gasses ($H_2O$, $NO_2$, O, $O_3$, $CH_4$, $CO_2$, $N_2O$, CO, $O_2$, $NH_3$, and $SO_2$). These models accommodate differing atmospheric conditions, including seasonal and geographic variations, cloud conditions, rain, and haze. The model considers all possible atmospheric paths.

A second group of atmosphere correction procedures is based on examination of re-

**TABLE 10.3. Sample Output from LOWTRAN**

| | RADIANCE (WATTS/CM2-STER-XXX) | | | | |
|---|---|---|---|---|---|
| FREQ (CM-1) | WAVLEN (MICRN) | ATMOS (CM-1) | RADIANCE (MICRN) | INTEGRAL (CM-1) | TOTAL TRANS |
| 740. | 13.514 | 1.29E-05 | 7.07E-04 | 3.23E-05 | .0001 |
| 745. | 13.423 | 1.28E-05 | 7.13E-04 | 9.65E-05 | .0001 |
| 750. | 13.333 | 1.28E-05 | 7.18E-04 | 1.60E-04 | .0004 |
| 755. | 13.245 | 1.27E-05 | 7.22E-04 | 2.24E-04 | .0010 |
| 760. | 13.158 | 1.23E-05 | 7.24E-04 | 2.88E-04 | .0023 |
| 765. | 13.072 | 1.24E-05 | 7.26E-04 | 3.48E-04 | .0043 |
| 770. | 12.987 | 1.23E-05 | 7.28E-04 | 4.10E-04 | .0063 |
| 775. | 12.903 | 1.22E-05 | 7.31E-04 | 4.70E-04 | .0083 |
| 780. | 12.821 | 1.21E-05 | 7.35E-04 | 5.31E-04 | .0095 |
| 785. | 12.739 | 1.20E-05 | 7.40E-04 | 5.91E-04 | .0103 |
| 790. | 12.658 | 1.19E-05 | 7.44E-04 | 6.50E-04 | .0112 |
| 795. | 12.579 | 1.18E-05 | 7.48E-04 | 7.10E-04 | .0117 |
| 800. | 12.500 | 1.17E-05 | 7.50E-04 | 7.68E-04 | .0138 |
| 805. | 12.422 | 1.16E-05 | 7.49E-04 | 8.26E-04 | .0178 |
| 810. | 12.346 | 1.14E-05 | 7.46E-04 | 8.83E-04 | .0232 |
| 815. | 12.270 | 1.11E-05 | 2.40E-04 | 9.39E-04 | .0297 |
| 820. | 12.195 | 1.09E-05 | 7.34E-04 | 9.93E-04 | .0363 |
| 825. | 12.121 | 1.08E-05 | 7.33E-04 | 1.05E-03 | .0408 |
| 830. | 12.048 | 1.06E-05 | 7.30E-04 | 1.10E-03 | .0457 |
| 835. | 11.976 | 1.05E-05 | 7.30E-04 | 1.15E-03 | .0488 |
| 840. | 11.905 | 1.04E-05 | 7.34E-04 | 1.20E-03 | .0497 |
| 845. | 11.834 | 1.03E-05 | 7.36E-04 | 1.26E-03 | .0518 |
| 850. | 11.765 | 1.02E-05 | 7.39E-04 | 1.31E-03 | .0525 |
| 855. | 11.696 | 1.01E-05 | 7.38E-04 | 1.36E-03 | .0555 |
| 860. | 11.62B | 9.90E-06 | 7.32E-04 | 1.41E-03 | .0620 |
| 865. | 11.561 | 9.73E-06 | 7.28E-04 | 1.46E-03 | .0667 |
| 870. | 11.494 | 9.56E-06 | 7.24E-04 | 1.50E-03 | .0717 |
| 875. | 11.429 | 9.44E-06 | 7.23E-04 | 1.55E-03 | .0749 |
| 880. | 11.364 | 9.35E-06 | 7.24E-04 | 1.60E-03 | .0759 |
| 885. | 11.299 | 9.22E-06 | 7.22E-04 | 1.64E-03 | .0794 |
| 890. | 11.236 | 9.05E-06 | 7.17E-04 | 1.69E-03 | .0847 |
| 895. | 11.173 | 8.88E-06 | 7.11E-04 | 1.73E-03 | .0901 |
| 900. | 11.111 | 8.71E-06 | 7.05E-04 | 1.78E-03 | .0958 |
| 905. | 11.050 | 8.59E-06 | 7.04E-04 | 1.82E-03 | .0991 |
| 910. | 10.989 | 8.51E-06 | 7.05E-04 | 1.86E-03 | .1001 |
| 915. | 10.929 | 8.41E-06 | 7.04E-04 | 1.90E-03 | .1020 |
| 920. | 10.870 | 8.31E-06 | 7.03E-04 | 1.95E-03 | .1043 |
| 925. | 10.811 | 8.18E-06 | 7.00E-04 | 1.99E-03 | .1085 |
| 930. | 10.753 | 8.03E-06 | 6.94E-04 | 2.03E-03 | .1134 |
| 935. | 10.695 | 7.92E-06 | 6.92E-04 | 2.07E-03 | .1158 |
| 940. | 10.638 | 7.83E-06 | 6.92E-04 | 2.11E-03 | .1170 |
| 945. | 10.582 | 7.76E-06 | 6.93E-04 | 2.14E-03 | .1174 |
| 950. | 10.528 | 7.68E-06 | 6.93E-04 | 2.18E-03 | .1183 |
| 955. | 10.471 | 7.60E-06 | 6.93E-04 | 2.22E-03 | .1196 |
| 960. | 10.417 | 7.51E-06 | 6.92E-04 | 2.26E-03 | .1206 |
| 965. | 10.363 | 7.44E-06 | 6.93E-04 | 2.30E-03 | .1200 |
| 970. | 10.309 | 7.38E-06 | 6.93E-04 | 2.33E-03 | .1183 |
| 975. | 10.256 | 7.29E-06 | 6.93E-04 | 2.37E-03 | .1183 |
| 980. | 10.204 | 7.13E-06 | 6.85E-04 | 2.40E-03 | .1216 |
| 985. | 10.152 | 6.97E-06 | 6.76E-04 | 2.44E-03 | .1238 |

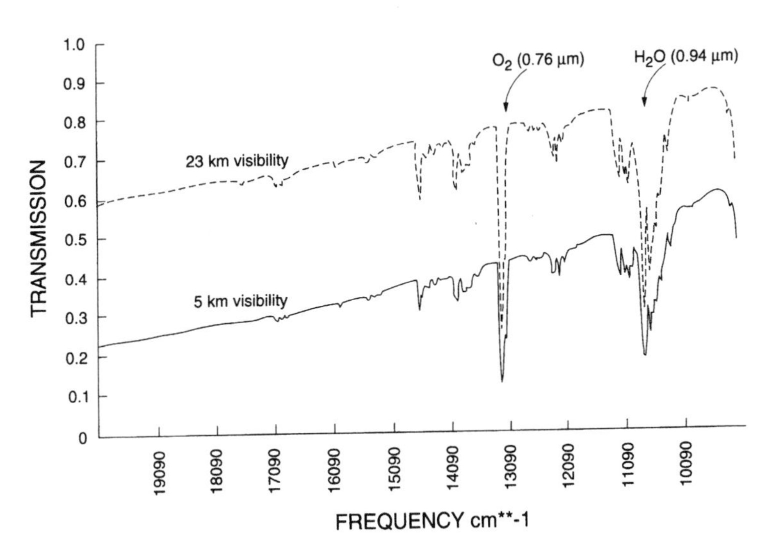


**FIGURE 10.3.** LOWTRAN output.

flectances from objects of known, or assumed, brightness recorded by multispectral imagery. From basic principles of atmospheric scattering, it is known that scattering is related to wavelength, sizes of atmospheric particles, and their abundance. If a known target is observed using a set of multispectral measurements, the relationships between values in the separate bands can help assess atmospheric effects.

Ideally, the target consists of a natural or man-made feature that can be observed with airborne or ground-based instruments at the time of image acquisition, so the analyst could know from measurements independent of the image, the true brightness of the object when the image was acquired. However, in practice we seldom have such measurements and therefore must look for features of known brightness that commonly, or fortuitously, appear within an image. In its simplest form, this strategy can be implemented by identifying a very dark object or feature within the scene. Such an object might be a large water body or possibly shadows cast by clouds or by large topographic features. In the infrared portion of the spectrum, both water bodies and shadows should have brightness at or very near zero because clear water absorbs strongly in the near infrared spectrum and because very little infrared energy is scattered to the sensor from shadowed pixels. Analysts who examine such areas, or the histograms of the digital values for a scene, can observe that the lowest values (for dark areas, such as clear water bodies) are not zero but some larger value. Typically this value will differ from one band to the next, so, for example, for Landsat band 1 the value might be 12, band 2 = 7, band 3 = 2, band 4 = 2 (Figure 10.4). These values are then assumed to represent the value contributed by atmospheric scattering for each band and subtracted from all digital values for that scene and that band. Thus, the lowest value in each band is set to zero, the dark black color assumed to be the correct tone for a dark object in the absence of atmospheric scattering. This procedure forms one of the simplest, most direct methods for adjusting digital values for atmospher-

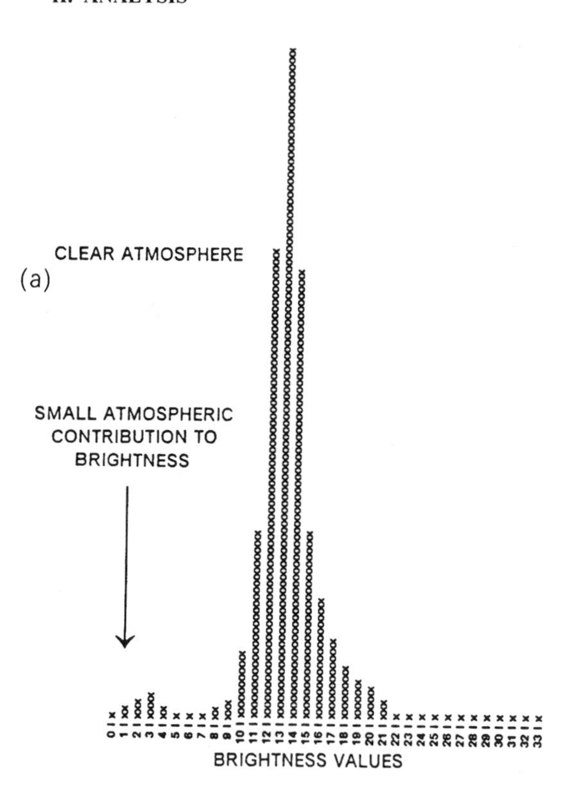


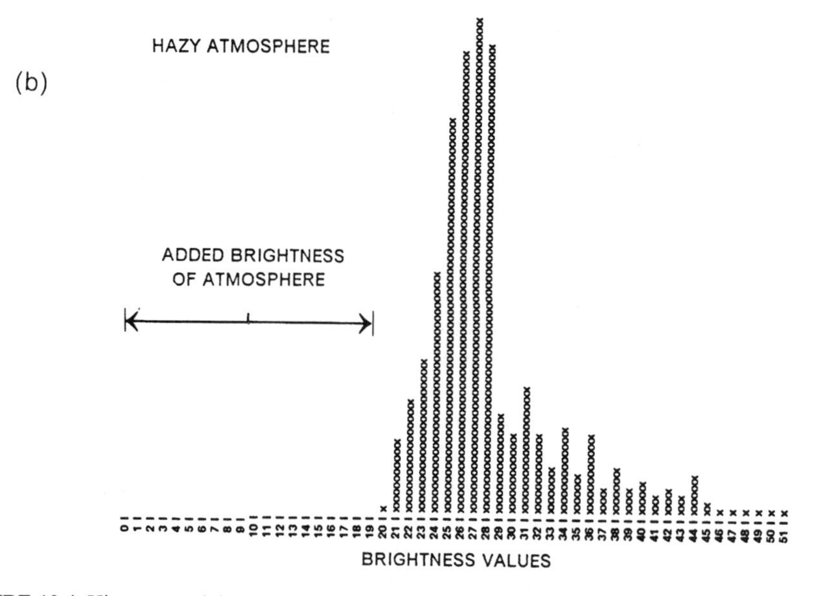


**FIGURE 10.4**. Histogram minimum method for correction of atmospheric effects. The lowest brightness value in the scene is assumed to reveal the added brightness of the atmosphere and is then subtracted from all pixels in a specific band.

ic degradation (Chavez, 1975), known sometimes as the histogram minimum method (HMM).

This procedure has the advantages of simplicity, directness, and almost universal applicability, as it exploits information present within the image itself. Yet it must be considered as an approximation; atmospheric effects change not only the position of the histogram on the axis but also its shape (i.e., not all brightnesses are affected equally). (Chapter 2 explained that the atmosphere can cause dark pixels to become brighter and bright pixels to become darker, so application of single correction to all pixels will provide only a rough adjustment for atmospheric effects.) In addition, in arid regions observed at high sun angles, shadows, clouds, and open water may be so rare, or of such small areal extent, that the method cannot be applied.

A more sophisticated approach retains the idea of examining brightness of objects within each scene but attempts to exploit knowledge of interrelationships between separate spectral bands. Chavez (1975) devised a procedure that paired values from each band with values from band 7. The *Y* intercept of the regression line is then taken as the correction value for the specific band in question. Whereas the HMM is applied to entire scenes, or to very large areas, the regression technique can be applied to local areas (of possibly only 100 to 500 pixels each), ensuring that the adjustment is tailored to conditions important within specific regions. An extension of the regression technique is to examine the variance–covariance matrix (i.e., the set of variances and covariances between all band pairs on the data). (This is the covariance matrix method [CMM] described by Switzer et al., 1981.) Both procedures assume that within a specified image region, variations in image brightness are due to topographic irregularities and the reflectivity is constant (i.e., land cover reflectivity in several bands is uniform for the region in question). Therefore, variations in brightness are caused by small-scale topographic shadowing, and the dark regions reveal the scattering term mentioned previously. Although these assumptions may not always be strictly met, the procedure, if applied with care and with knowledge of the local geographic setting, seems to be a robust and often satisfactory procedure. Campbell et al. (1984) modified this strategy by using an automatic grouping procedure (as described in Chapter 11) to ensure that regions are defined in such a way that their reflectances are in fact uniform.

How can the analyst decide whether atmospheric corrections are necessary? This may be a difficult decision, as effects of atmospheric degradation are not always immediately obvious from casual inspection. The analyst should always examine summary statistics for each scene, inspecting means, variances, and frequency histograms for suggestions of poor image quality, and the absence of dark values (especially if the image is known to show larger water bodies) (Figure 10.5).

Of course, inspection of the image may reveal evidence suggesting a requirement for correction. Loss of resolution and low contrast may indicate poor atmospheric conditions. Sometimes the image date may itself suggest the nature of atmospheric quality. In the central United States, summer dates often imply high humidity, haze, and poor visibility, whereas winter, autumn, and spring dates are often characterized by clearer atmospheric conditions. Thus, the image date may provoke further investigation by the analyst to determine whether corrections are necessary. Finally, the analyst should examine summary statistics for the scene and especially the frequency histograms for each band.

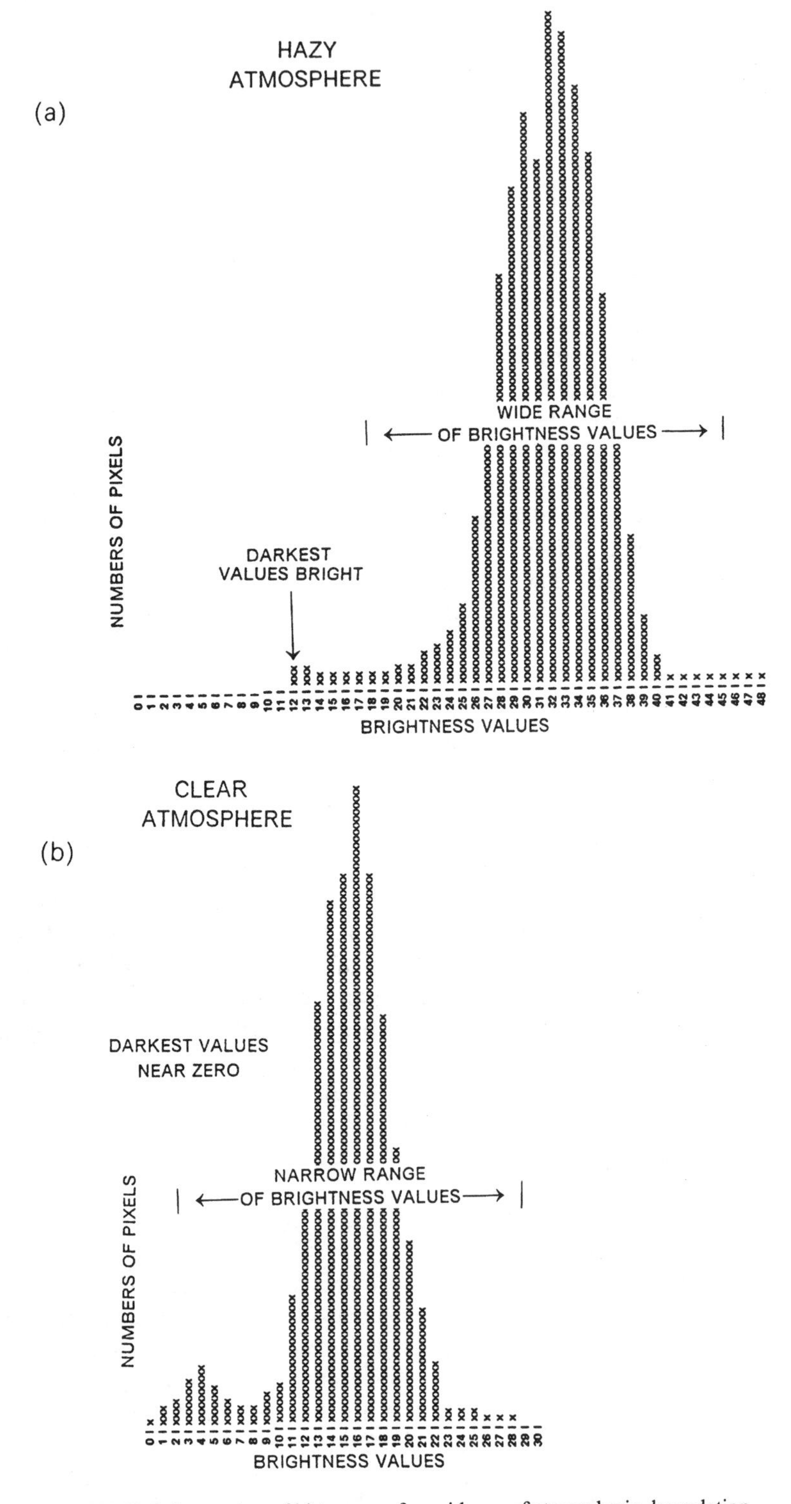


**FIGURE 10.5.** Inspection of histograms for evidence of atmospheric degradation.

### *Destriping*

Landsat MSS data sometimes exhibit a kind of radiometric error known as *sixth-line striping,* caused by small differences in the sensitivities of detectors within the sensor. Within a given band, such differences appear on images as a horizontal banding, or "striping," because individual scan lines exhibit unusually brighter or darker brightness values that contrast noticeably with the background brightnesses of the "normal" detectors. Because MSS detectors are positioned in arrays of six (Chapter 6), anomalous detector response appears as linear banding at intervals of six lines. Striping may appear on only one or two bands of a multispectral image or may be severe for only a portion of a band. Other forms of digital imagery sometimes exhibit similar effects, all caused by difficulties in maintaining consistent calibration of detectors within a single sensor. Campbell and Liu (1995) found that striping and paneling in digital data seemed to have minimal impact on the character of the data—much less than might be suggested by the visual appearance of the imagery.

Although sixth-line striping is often clearly visible as an obvious banding (Figure 10.6), it may also be present in a more subtle form that may escape casual visual inspection. *Destriping* refers to the application of algorithms to adjust incorrect brightness values to values thought to be near the correct values. Some image processing software provides special algorithms to detect striping. Such procedures search through an image line by line to look for systematic differences in average brightnesses of lines spaced at intervals of six; examination of the results permits the analyst to have objective evidence of the

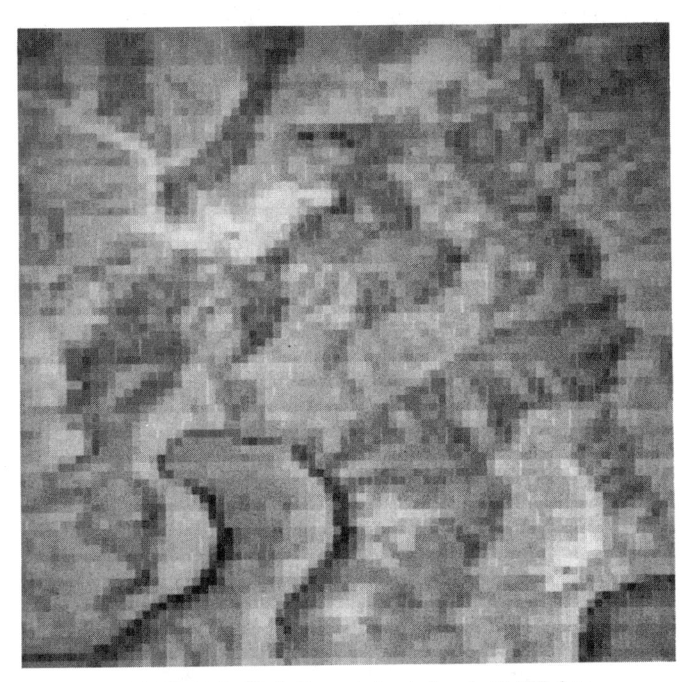

**FIGURE 10.6.** Sixth-line striping in Landsat MSS data.

existence or absence of sixth-line striping. If striping is present, the analyst must make a decision. If no correction is applied, the analysis must proceed with brightness values that are known to be incorrect. On the other hand, efforts to correct for such serious errors may yield rather rudimentary approximations of the (unknown) true values. Often striping may be so severe that it is obvious the bad lines must be adjusted.

A variety of destriping algorithms have been devised. All identify the values generated by the defective detectors by searching for lines that are noticeably brighter or darker than the remainder of the scene. These lines are presumably the bad lines caused by the defective detectors (especially if they occur at intervals of six lines). Then the destriping estimates corrected values for the bad lines. There are many different estimation procedures; most belong to one of two groups. One approach is to replace bad pixels with values based on the average of adjacent pixels not influenced by striping; this approach is based on the notion that the missing value is probably quite similar to the pixels that are nearby (Figure 10.7a). A second strategy is to replace bad pixels with new values based on the mean and standard deviation of the band in question, or upon statistics developed for each detector (Figure 10.7b). This second approach is based on the assumption that the overall statistics for the missing data must, because there are so many pixels in the scene, resemble those from the good detectors.

The algorithm described by Rohde et al. (1978) combines elements of both strategies. Their procedure attempts to bring all values in a band to a normalized mean and variance based on overall statistics for the entire band. Because brightnesses of individual regions within the scene may vary considerably from these overall values, a second algorithm can be applied to perform a local averaging to remove remaining influences of striping. Some destriping algorithms depend entirely on local averaging; because they tend to degrade image resolution, and introduce statistical dependencies between adjacent brightness values, it is probably best to be cautious in their application if alternatives are available.

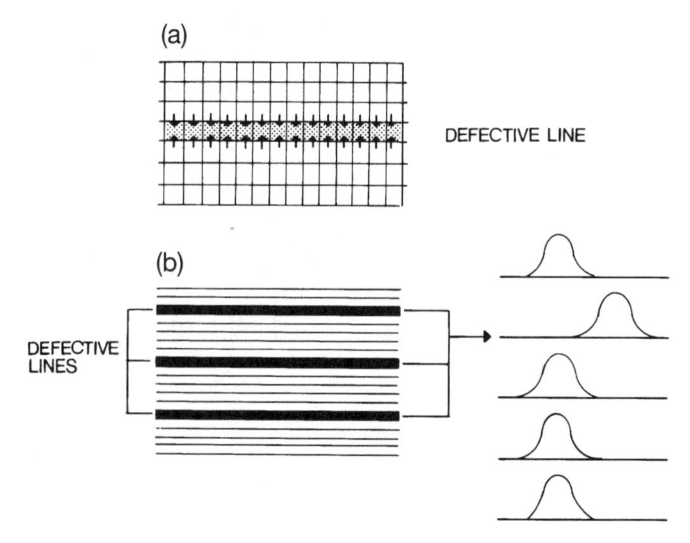


**FIGURE 10.7.** Two strategies for destriping. (*a*) Local averaging. (*b*) Normalization.

## 10.5. Image Matching

Image matching is the process of superimposing two images of the same area and moving them to find the position at which they best match. Visually, two transparencies have a registration point where the image detail matches.

Digitally there are many separate procedures; some are very complex. One of the simplest and most widely used strategies is to digitally overlay the two images and calculate a correlation for the area where the two images overlap (or for some smaller zone). Matched positions are systematically shifted, pixel by pixel, until all possible matches have been attempted. At each position a new correlation value is calculated and saved. The optimum position, presumably the correct registration, is the position that yields the highest correlations. This kind of image-matching procedure is important in many automated procedures, including matching digital stereo images.

A much more common problem is the registration of two images, meaning that two images are brought to registration, but one is usually altered using one of the procedures described below. (In image matching, the two images are not changed—we simply find the position at which the two match.) Typically we wish to register a Landsat image to a planimetrically correct map, or to register two images from different dates, or to register Landsat and Seasat images of the same area. In each instance, we have the same problem—we must rearrange pixels to match to the new map, or we must estimate pixel values at points that match the new distribution.

The most difficult but most rigorous approach to image registration is to apply knowledge of sensor geometry and motion to derive accurate coordinates for each pixel. To apply this approach, which we will designate the *analytical* approach, the analyst must know the satellite altitude, its trajectory, the shape of the earth's surface, its motion relative to the satellite, and the motion of the sensor scanner. Although for satellites such as Landsat factors are known with some precision, analytical image correction can correct only some of the geometric errors in images.

## 10.6. Geometric Correction by Resampling

A second approach to image registration treats the problem in a completely different manner. No effort is made to apply our knowledge of system geometry; instead, the images are treated simply as an array of values that must be manipulated to create another array with the desired geometry.

This can be seen essentially as an interpolation problem similar to those routinely considered in cartography and other disciplines. In Figure 10.8, the input image is represented as an array of open dots, each representing the center of a pixel in the uncorrected image. Superimposed over this image is a second array, symbolized by the solid dots, which shows the centers of pixels in the image as transformed to have the desired geometric properties (the "output" image).

The locations of the output pixels are derived from locational information provided by *ground control points* (GCPs)—sites on the input image that can be located with precision on the ground and on planimetrically correct maps. (Or, if two images are to be registered, GCPs must be easily recognized on both images.) The locations of these points establish

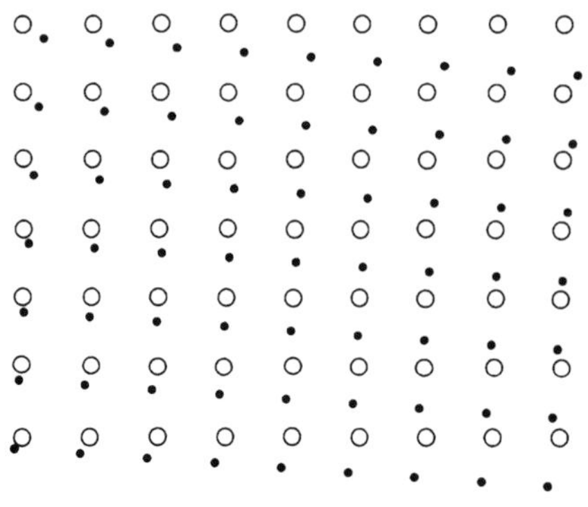

**FIGURE 10.8.** Resampling. Open circles represent the reference grid of known values in the input image. Black dots represent the regular grid of points to be estimated to form the output image. Each resampling method employs a different strategy to estimate values at the output grid, given known values for the input grid.

the geometry of the output image and its relationship to the input image. Thus, this first step establishes the framework of pixel positions for the output image using the GCPs.

The problem then is to decide how to best estimate the values at pixels in the corrected image based on information in the uncorrected image. The simplest strategy from a computational perspective is simply to assign each "corrected" pixel the value from the nearest "uncorrected" pixel. This is the *nearest-neighbor* approach to resampling (Figure 10.9). It has the advantages of simplicity and the ability to preserve original values in the unaltered scene. On the other hand, it may create noticeable positional errors, which may be severe in linear features where the realignment of pixels is obvious. In Kovalick's (1983) study, the nearest-neighbor method was computationally the most efficient of the three methods studies.

A second, more complex approach to resampling is *bilinear interpolation* (Figure 10.10). Bilinear interpolation calculates a value for each output pixel based on a weighted average of the four nearest input pixels. In this context, "weighted" means that nearer pixel values are given greater influence in calculating output values than are more distant pixels. Because each output value is based on several input values, the output image will not have the unnaturally blocky appearance of some nearest-neighbor images. The image therefore has a more "natural" look.

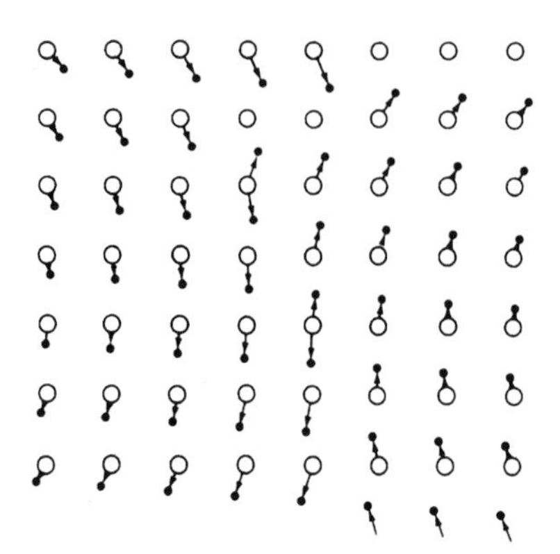

**FIGURE 10.9.** Nearest-neighbor resampling. Each estimated value (●) receives its value from the nearest point on the reference grid (○).

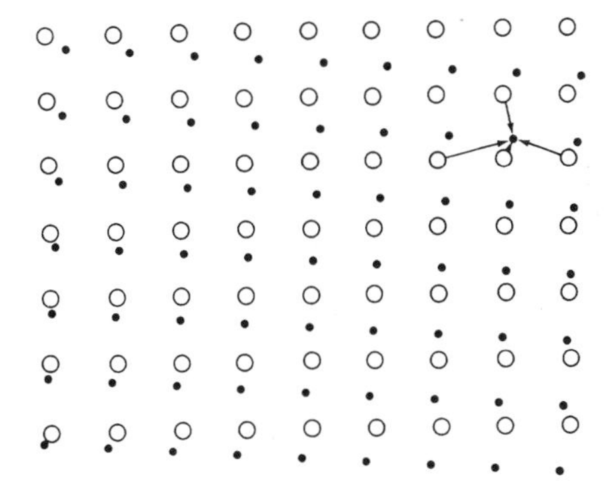

**FIGURE 10.10.** Bilinear interpolation. Each estimated value in the output image is formed by calculating a weighted average of the four nearest neighbors in the input image. Each estimated value is weighted according to its distance from the known values in the input image.

Yet there are important changes. Because bilinear interpolation creates new pixel values, the brightness values in the input image are lost, and the analyst may find that the range of brightness values in the output image differs from those in the input image. Such changes to digital brightness values may be significant in later processing steps. Second, because the resampling is conducted by averaging over areas (i.e., blocks of pixels), it decreases spatial resolution by a kind of "smearing" that will be caused by averaging the small features with their background.

Finally, the most sophisticated, most complex, and (possibly) most widely used resampling method is *cubic convolution* (Figure 10.11). Cubic convolution uses a weighted average of values within a neighborhood that extends about two pixels in each direction, usually encompassing 16 pixels. Typically the images produced by cubic convolution resampling are much more attractive than are those of other procedures, but the data are altered much more drastically than those of nearest neighbor or bilinear interpolation.

### *Identification of Ground Control Points*

A practical problem in applying image registration procedures is the selection of control points (Figure 10.12). GCPs are features that can be located with precision and accuracy

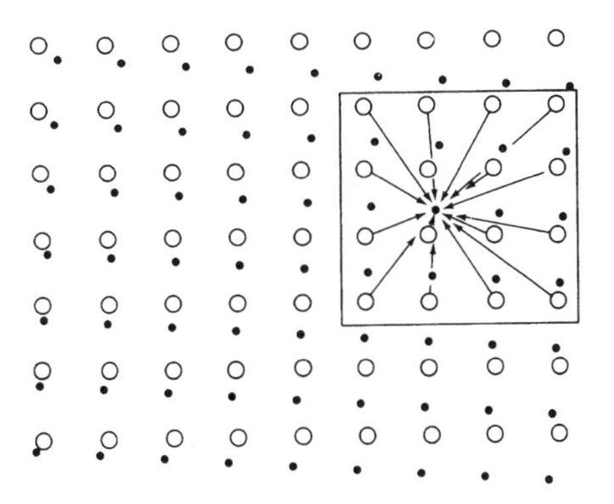

**FIGURE 10.11.** Cubic convolution. Each estimated value in the output matrix is found by assessing values within a neighborhood of 16 pixels in the input image.

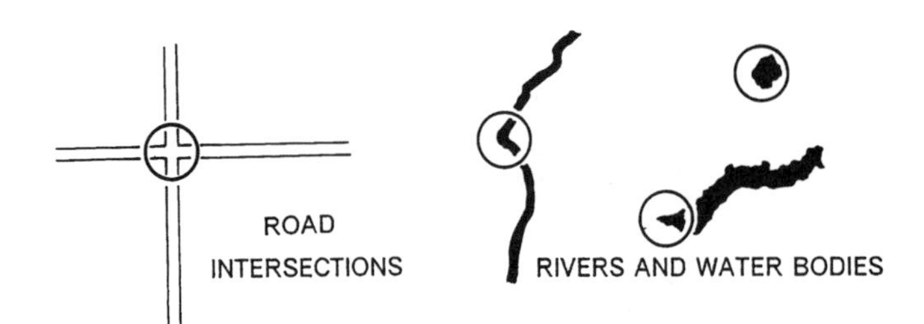


**FIGURE 10.12**. Selection of ground control points.

on accurate maps yet are also easily located on digital images. Ideally, GCPs could be as small as a single pixel, if one could be easily identified against its background. In practice, most GCPs are likely to be spectrally distinct areas as small as a few pixels. Examples might include intersections of major highways, distinctive water bodies, edges of land cover parcels, stream junctions, and similar features. Although identification of such points may seem to be an easy task, in fact, difficulties that might emerge during this step can form a serious roadblock to the entire analytical process, as another procedure may depend on completion of an accurate registration.

Typically it is relatively easy to find a rather small or modest-sized set of control points. However, in some scenes, the analyst finds it increasingly difficult to expand this set, as one has less and less confidence in each new point added to the set of GCPs. Thus there may be a rather small set of "good" GCPs—points the analyst can locate with confidence and precision both on the image and on an accurate map of the region.

The locations may also be a problem. In principle, GCPs should be dispersed throughout the image, with good coverage near edges. Obviously there is little to be gained from a large number of GCPs if they are all concentrated in a few regions of the image. Analysts who attempt to expand areal coverage to ensure good dispersion are forced to consider points in which it is difficult to locate them with confidence. Therefore the desire to select "good" GCPs and the desire to gain good dispersion may work against each other so the analyst finds it difficult to select a judicious balance. Analysts should anticipate difficulties in selecting GCPs as they prepare subsets early in the analytical process. If subsets are too small, or do not encompass important landmarks, the analysts may later find that the subset region of the image does not permit selection of a sufficient number of high-quality GCPs.

Bernstein et al. (1983) present information that shows how registration error decreases as the number of GCPs is increased. Obviously it is better to have more, rather than fewer, GCPs. But the quality of GCP accuracy may decrease as their number increases because the analyst probably picks the best points first. Bernstein et al. recommend that 16 GCPs may be a reasonable number if each can be located with an accuracy of one third of a pixel. This number may not be sufficient if the GCPs are poorly distributed, or if the nature of the landscape prevents accurate placement.

Some image-processing systems permit analysts to anticipate the accuracy of the registration by reporting errors observed at each GCP if a specific registration has been applied. The standard measure of the location error is the root mean square (rms) error, which reports the standard deviation of differences between actual values of GCPs and their calculated values (i.e., after registration). These differences are known as the *residuals*. Usually rms is reported in units of image pixels for both north–south and east–west directions; small rms values indicate more precise estimates of correct values (see Figure 13.1).

Table 10.4 illustrates an evaluation of an image registration. Each GCP is listed by its sequential identification and its image coordinates (*X* pixel and *Y* pixel), together with corresponding residuals, and the rms for the registration. From these data, it is possible to calculate a total error for each GCP (a measure of its proportionate share of the total error for the set of GCPs), and then to use this information to eliminate those points (e.g., numbers 18 and 4) that contribute most to total error. By repeated evaluation of errors (after each point is deleted), it is then possible to select those GCPs that provide accurate resampling. Note that this procedure estimates errors of GCPs, which may not reflect error in the broader population of image pixels.

## 10.7. Map Projections for Representing Satellite Images and Ground Tracks

When mapping any distribution on the earth's surface that exceeds a few square miles, a map projection is required because of the fundamental problem of projecting locations on the earth's spherical surface onto the flat surface of the map. A map projection is a system of transformations that enables locations on the spherical earth to be represented systematically on a flat map. Because of the inherent difference between the two surfaces, there is always some sacrifice in accurate representation of area, shape, scale, or direction on maps relative to gloves. However, such errors may be confined to a few of these characteristics or may be very small in certain portions of a map.

Continuous mapping of a region viewed by Landsat or other earth observation satellites presents serious problems for those using conventional map projections. Both the satellite and the earth's surface are moving as imagery is acquired, so, at best, usual maps are inconvenient. Therefore, users of early Landsat data encountered problems in representing coverage of Landsat scenes on conventional maps. The ground tracks of sun-synchronous satellites trace curved lines on the usual map projections (Chapter 6), greatly complicating the representation of satellite paths and coverage. Snyder (1981) addressed this problem by devising projections specially designed to represent ground tracks of sun-synchronous satellites as straight lines. For global tracking his map is based on a cylindrical projection, and for larger-scale maps of continents or areas of similar size, he uses a map based on a conic projection. To show the ground track as a straight line, it is of course necessary to sacrifice other map qualities. Thus, these maps cannot show accurate shapes and area everywhere. Furthermore, representation of Landsat data as a rectangular array of pixels does not place them in their correct positions relative to the earth's surface. Convenient representation of correct positions of pixels requires projects specifically designed to capture the complex geometry of the Landsat image.

Alden P. Colvocoresses is credited with defining the Space Oblique Mercator (SOM) projection tailored for use with Landsat data, although others contributed to its develop-

**TABLE 10.4. Sample Tabulation of Data for GCPs**

| Point No. | Image $X$ pixel | $X$ pixel residual | Image $Y$ pixel | $Y$ pixel residual |
|---|---|---|---|---|
| 1 | 1269.75 | –0.2471E+00 | 1247.59 | 0.1359E+02 |
| 2 | 867.91 | –0.6093E+01 | 1303.90 | 0.8904E+01 |
| 3 | 467.79 | –0.1121E+02 | 1360.51 | 0.5514E+01 |
| 4 | 150.52 | 0.6752E+02 | 1413.42 | –0.8580E+01 |
| 5 | 82.20 | –0.3796E+01 | 163.19 | 0.6189E+01 |
| 6 | 260.89 | 0.2890E+01 | 134.23 | 0.5234E+01 |
| 7 | 680.59 | 0.3595E+01 | 70.16 | 0.9162E+01 |
| 8 | 919.18 | 0.1518E+02 | 33.74 | 0.1074E+02 |
| 9 | 1191.71 | 0.6705E+01 | 689.27 | 0.1127E+02 |
| 10 | 1031.18 | 0.4180E+01 | 553.89 | 0.1189E+02 |
| 11 | 622.44 | –0.6564E+01 | 1029.43 | 0.8427E+01 |
| 12 | 376.04 | –0.5964E+01 | 737.76 | 0.6761E+01 |
| 13 | 162.56 | –0.7443E+01 | 725.63 | 0.8627E+01 |
| 14 | 284.05 | –0.1495E+02 | 1503.73 | 0.1573E+02 |
| 15 | 119.67 | –0.8329E+01 | 461.59 | 0.4594E+01 |
| 16 | 529.78 | –0.2243E+00 | 419.11 | 0.5112E+01 |
| 17 | 210.42 | –0.1558E+02 | 1040.89 | –0.1107E+01 |
| 18 | 781.85 | –0.2915E+02 | 714.94 | –0.1521E+03 |
| 19 | 1051.54 | –0.4590E+00 | 1148.97 | 0.1697E+02 |
| 20 | 1105.95 | 0.9946E+01 | 117.04 | 0.1304E+02 |

$X$ rms error = 18.26133
$Y$ rms error = 35.33221

Total rms error = 39.77237

| Point No. | Error | Error contribution by point |
|---|---|---|
| 1 | 13.5913 | 0.3417 |
| 2 | 10.7890 | 0.2713 |
| 3 | 12.4971 | 0.3142 |
| 4 | 68.0670 | 1.7114 |
| 5 | 7.2608 | 0.1826 |
| 6 | 5.9790 | 0.1503 |
| 7 | 9.8416 | 0.2474 |
| 8 | 18.5911 | 0.4674 |
| 9 | 13.1155 | 0.3298 |
| 10 | 12.6024 | 0.3169 |
| 11 | 10.6815 | 0.2686 |
| 12 | 9.0161 | 0.2267 |
| 13 | 11.3944 | 0.2865 |
| 14 | 21.6990 | 0.5456 |
| 15 | 9.5121 | 0.2392 |
| 16 | 5.1174 | 0.1287 |
| 17 | 15.6177 | 0.3927 |
| 18 | 154.8258 | 3.8928 |
| 19 | 16.9715 | 0.4267 |
| 20 | 16.3982 | 0.4123 |

ment and mathematical definition (Colvocoresses, 1974). Research to develop the SOM was conducted by the USGS specifically for the purpose of defining a map projection to provide constant scale for the ground track of an earth observation satellite for the entire coverage cycle. Although the SOM was designed for use with landsat, it is equally applicable to other land observation satellites if tailored to their specific orbital characteristics. For Landsat, the projection must depict coverage from 81° N to 81° S, through a coverage cycle of 251 orbits. The projection shows only a narrow strip parallel to the ground track of the satellite—essentially that area subject to view by the MSS.

The SOM is based on the map projection devised by Gerhard Kramer (1512–1594), a Dutch scholar and cartographer. The Latinized version of his name, Gerhardus Mercator, has been given to the map projection he devised for an atlas he published in 1569. The Mercator projection was the first projection to attain widespread use, primarily because of its utility for marine navigation.

The Mercator projection can be envisioned as a transformation of the network of lines of latitude and longitude (known as the *graticule*) onto a flat surface such that the meridians of longitude form equally spaced vertical lines and the parallels of latitude form horizontal lines intersecting the meridians at right angles. The creation of this projection is envisioned as wrapping a transparent globe in a cylinder tangent at the equator with a light inside the globe projecting the graticule onto the cylinder, which is then opened up to form a flat, map-like, surface.

On a globe, lines of latitude are parallel to one another, but lines of longitude converge near the poles. The Mercator projection differs significantly from a globe because it shows the converging meridians as parallel lines. But, simply described, the essence of Mercator's projection is that this error is, in part, compensated for by his method of spacing lines of latitude. Although the globe shows lines of latitude as equally spaced. Mercator's map increases this spacing so that as distance from the equator increases and as divergence of the meridians from their true spacing increases, so does spacing between lines of latitude. Thus the interval between lines of latitude increases dramatically toward the poles to compensate for converging meridians near the poles. This compensation preserves correct shape in cartographic representation of the earth's features. The poles cannot be shown on the classic (i.e., centered on the equator) Mercator as parallels of latitude would have to be spaced at an infinite distance from one another at the poles.

If a Mercator projection is centered on the equator, the portion nearest the equator (where the hypothetical cylinder is tangent to the globe) is correct in representation of distance, shape, area, and direction; all are shown accurately or with only minor errors (Figure 10.13a). As distance from the equator increases, errors become increasingly severe. Area in particular is shown with large errors at high latitudes. However, shape is shown accurately throughout the map. Furthermore, this projection has an important feature that has been especially valuable throughout its history. A line of constant compass bearing (known as a loxodrome) is shown on a Mercator projection as a straight line—a feature that has made the projection especially valuable for navigational use, both now and in the past.

For our present concerns, the Mercator projection is significant in its modified forms,

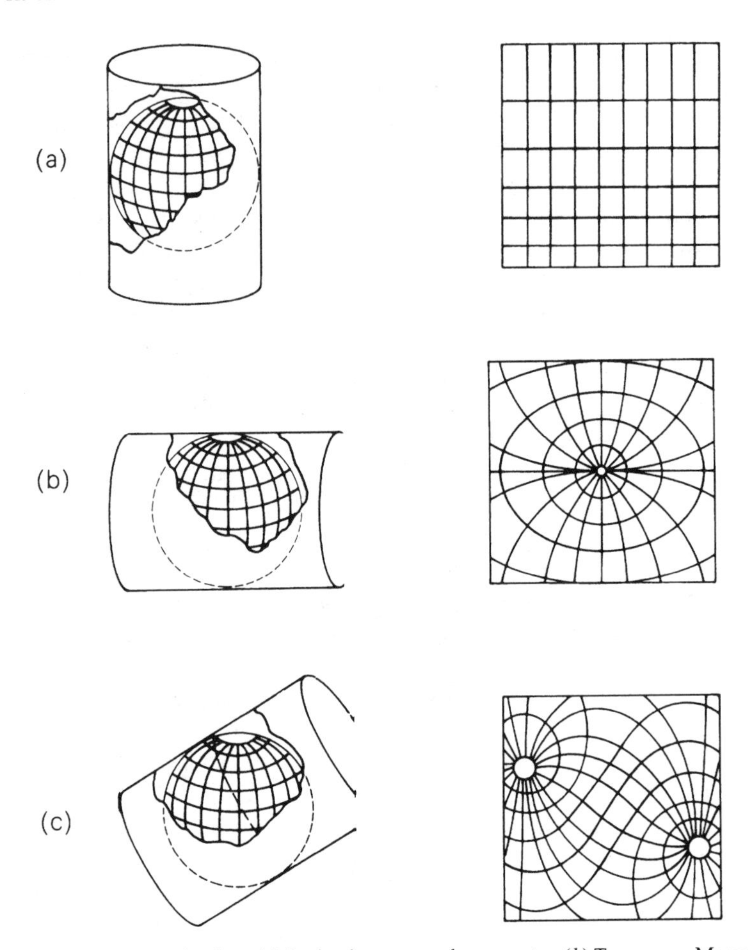


**FIGURE 10.13**. Mercator projection. (*a*) Projection centered on equator. (*b*) Transverse Mercator projection. (*c*) Oblique Mercator. From Snyder (1982).

the *transverse Mercator* and the *oblique Mercator* projections. The transverse Mercator can be visualized as a cylinder tangent, not at the equator but at a meridian of longitude (Figure 10.13b). When the cylinder is unfolded, only the most accurate strip centered at the meridian of tangency is selected for use—the outer sections of the map, where the map is inaccurate, are discarded. If the process is repeated successively for different meridians, each time retaining only the accurate center strip, accurate maps of large areas of the earth can be constructed. This process forms the basis for the *Universal Transverse Mercator* (UTM) geographic reference system (described in Chapter 15), which is based on such a projection.

For the oblique Mercator projection, the line of tangency is shifted from a meridian to a great circle on the earth oriented at some angle (other than 90°) to the meridians (Figure 10.13c). On oblique Mercator projections the graticule appears as a set of curved lines.

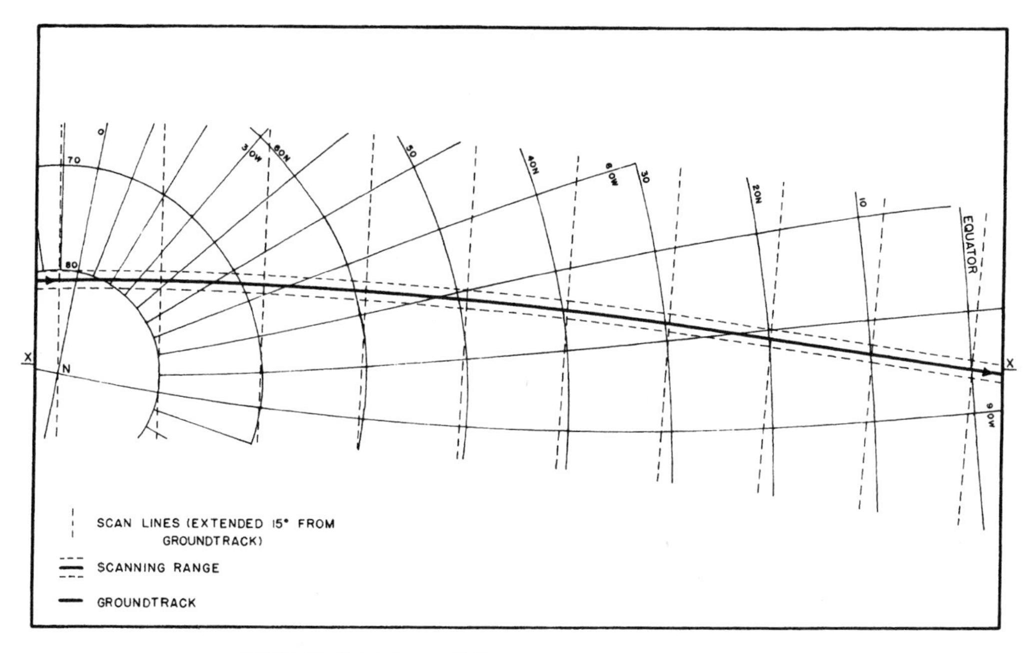


**FIGURE 10.14.** Space Oblique Mercator. From Snyder (1982).

For the SOM (Figure 10.14), the line of tangency is defined not as a circle on the earth's surface but as a curve that follows the ground track of the satellite. For Landsat and other satellites in sun-synchronous orbits, this line is a nearly sinusoidal curve.

The scan lines, representing the area depicted by MSS imagery, intersect the ground track at 86° at the equator, but at 90° where the track is closest to the poles. This difference is caused by variations in *skew* (0 to 4°) caused by interaction of mirror scan motion with movement of the earth's surface as it rotates beneath the moving satellite. Like other Mercator projections the SOM has errors in representation of distance and area, but errors are very small near the ground track.

## 10.8. Effects of Preprocessing

There is indirect evidence that preprocessing operations sometimes have unwanted effects on digital values, and therefore on the accuracies of classifications based on such data. It seems clear from a basic understanding of the preprocessing operations that the data are in fact altered from their original values, and that such operations alter both the mean values for each band as well as the variances and covariances. Presumably other qualities, such as correlations between bands, are also influenced. Yet, there have been few studies that have systematically examined actual effects of preprocessing operations upon final results of analyses.

Kovalick (1983) examined a Landsat MSS scene of a forested wetland in North Carolina to study the effects of resampling and destriping on accuracies of image classification. He found that resampling tended to reduce class means (of the training data he examined) and to increase variances relative to the original data.

## 10.9. Summary

It is important to recognize that many of the preprocessing operations used today have been introduced into the field of remote sensing from the related fields of pattern recognition and image processing. In such disciplines, the emphasis is usually on detection or recognition of objects as portrayed on digital images. In this context, the digital values have much different significance than they do in remote sensing. Often, the analysis requires recognition simply of contrasts between different objects or the study of objects against their backgrounds, detection of edges, and reconstruction of shapes from the configuration of edges and lines. Digital values can be manipulated freely to change image geometry or to enhance images without concern that their fundamental information content will be altered.

However, in remote sensing we usually are concerned with much more subtle variations in digital values and are concerned when preprocessing operations alter the digital values. Such changes may alter spectral signatures, contrasts between categories, or variances and covariances of spectral bands.

## Review Questions

1. How can an analyst know if preprocessing is advisable? Suggest how you might make this determination.
2. How can analyst determine if specific preprocessing procedures have been effective?
3. Can you identify situations in which application of preprocessing might be inappropriate? Explain.
4. Discuss the merits of preprocessing techniques that improve the visual appearance of an image but do not alter its basic statistical properties. Are visual qualities important in the context of image analysis?
5. Examine images and maps of your region to identify prospective GCPs. Evaluate the pattern of GCPs; is the *pattern* even, or is it necessary to select questionable points to attain an even distribution?
6. Are optimum decisions regarding preprocessing likely to vary accordingly to the subject of the investigation? For example, would optimum preprocessing decisions for a land-cover analysis differ from those for a hydrologic or geologic analysis?
7. Assume for the moment that sixth-line striping in MSS data has a purely visual impact, with no effect on the underling statistical qualities of the image. In your judgment, should preprocessing procedures by applied? Why or why not?

8. Can you identify analogies for preprocessing in other contexts?

9. Suppose an enterprise offers to sell images with preprocessing already completed. Would such a product be attractive to you? Why or why not?

## References


Berk, A., L. S. Bernstein, and D. C. Robertson. 1989. *MODTRAN: A Moderate Resolution Model for LOWTRAN 7.* Hanscom Air Force Base, MA: Air Force Geophysics Laboratory, 38 pp.

Bernstein, Ralph et al. 1983. Image Geometry and Rectification. Chapter 21 in *Manual of Remote Sensing* (R. N. Colwell, ed.). Falls Church, VA: American Society of Photogrammetry, pp. 873–922.

Brach, E. J., A. R. Mack, and V. R. Rao. 1979. Normalization of Radiance Data for Studying Crop Spectra over Time with a Mobile Field Spectro-Radiometer. *Canadian Journal of Remote Sensing,* Vol. 5, pp. 33–42.

Campbell, J. B. 1993. Evaluation of the Dark-Object Subtraction Method of Adjusting Digital Remote Sensing Data for Atmospheric Effects. In *Digital Image Processing and Visual Communications Technologies in the Earth and Atmospheric Sciences II* (M. J. Carlotto, ed.). *SPIE Proceedings,* Vol. 1819, pp. 176–188.

Campbell, J. B., R. M. Haralick, and S. Wang. 1984. Interpretation of Topographic Relief from Digital Multispectral Imagery. In *Remote Sensing* (P. N. Slater, ed.). *SPIE Proceedings,* Vol. 475, pp. 98–116.

Campbell, J. B., and L. Ran. 1993. CHROM: A C Program to Evaluate the Application of the Dark Object Subtraction Technique to Digital Remote Sensing Data. *Computers and Geosciences,* Vol. 19, pp. 1475–1499.

Campbell, J. B., and X. Liu. 1994. Application of Dark Object Subtraction to Multispectral Data. *Proceedings, International Symposium on Spectral Sensing Research (ISSSR '94),* pp. 375–386.

Campbell, J. B., and X. Liu. 1995. Chromaticity Analysis in Support of Multispectral Remote Sensing. In *Proceedings, ACSM/ASPRS Annual Convention and Exposition.* Bethesda, MD: American Society for Photogrammetry and Remote Sensing, pp. 724–932.

Chavez, P. S. 1975. Atmospheric, Solar, and M.T.F. Corrections for ERTS Digital Imagery. *Proceedings, American Society of Photogrammetry,* pp. 69–69a.

Chavez, P. S., G. L. Berlin, and W. B. Mitchell. 1977. Computer Enhancement Techniques of Landsat MSS Digital Images for Land Use/Land Cover Assessment. *Proceedings of the Sixth Annual Remote Sensing of Earth Resources Conference,* pp. 259–276.

Colvocoresses, Alden P. 1974. Space Oblique Mercator, a New Map Projection of the Earth. *Photogrammetric Engineering and Remote Sensing,* Vol. 40, pp. 921–926.

Davis, J. C. 1986. *Statistics and Data Analysis in Geology.* New York: Wiley, 646 pp.

Deetz, Charles H., and Oscar S. Adams. 1945. *Elements of Map Projection.* U.S. Department of Commerce, Special Publication 68, 226 pp.

Edwards, K., and P. A. Davis. 1994. The Use of Intensity–Hue–Saturation Transformation for Producing Color Shaded Relief Images. *Photogrammetric Engineering and Remote Sensing,* Vol. 56, pp. 1369–1374.

Estes, J. E., E. J. Hajic, L. R. Tinney, et al. 1983. Fundamentals of Image Analysis: Analysis of Visible and Thermal Infrared Data. Chapter 24 in *Manual of Remote Sensing* (R. N. Colwell, ed.). Falls Church, VA: American Society of Photogrammetry, pp. 987–1124.

Franklin, S. E., and P. T. Giles. 1995. Radiometric Processing of Aerial Imagery and Satellite Remote-Sensing Imagery. *Computers and Geosciences,* Vol. 21, pp. 413–423.

Gould, P. 1967. On the Geographical Interpretation of Eigenvalues. *Transactions, Institute of British Geographers,* Vol. 42, pp. 53–86.

Heller, D. L., B. K. Quirk, and J. L. Hood. 1992. A Technique for the Reduction of Banding in Thematic Mapper Images. *Photogrammetric Engineering and Remote Sensing,* Vol. 58, pp. 1425–1431.

Holben, B., E. Vermote, Y. J. Kaufman, D. Taré, and V. Kalb. 1992. Aerosol Retrieval over Land from AVHRR Data. *IEEE Transactions on Geoscience and Remote Sensing,* Vol. 30, pp. 212–222.

Jensen, John R. 1996. *Introductory Digital Image Processing.* Upper Saddle River, NJ: Prentice Hall, 316 pp.

Kneizys, F. X., E. P. Shettle, L. W. Abreu, J. H. Chetwynd, G. P. Anderson, W. O. Gallery, J. E. A. Selby, and S. A. Clough. 1988. *Users Guide to LOWTRAN 7.* Hanscom Air Force Base, MA: Air Force Geophysics Laboratory, 137 pp.

Kovalick, William M. 1983. *The Effect of Selected Preprocessing Procedures upon the Accuracy of a Landsat-Derived Classification of a Forested Wetland.* Master's thesis. Blacksburg: Virginia Polytechnic Institute, 109 pp.

Lam, Nina Siu-Ngan. 1983. *Spatial Interpolation Methods: A Review. The American Cartographer*, Vol. 10, pp. 129–149.

Lambeck, P. F., and J. F. Potter. 1979. Compensation for Atmospheric Effects in LANDSAT Data. In *The LACIE Symposium: Proceedings of Technical Sessions.* Vol. II. Houston, TX: NASA-JSC, pp. 723–738.

Lavreau, J. 1991. De-Hazing Landsat Thematic Mapper Images. *Photogrammetric Engineering and Remote Sensing,* Vol. 57, pp. 1297–1302.

Mausel, P. W., W. J. Kramber, and J. Lee. 1990. Optimum Band Selection for Supervised Classification of Multispectral Data. *Photogrammetric Engineering and Remote Sensing,* Vol. 56, pp. 55–60.

Pitts, David E., Williams E. McAllum, and Alyce E. Dillinger. 1974. The Effect of Atmospheric Water Vapor on Automatic Classification of ERTS Data. In *Proceedings of the Ninth International Symposium on Remote Sensing of Environment.* Ann Arbor: University of Michigan, Institute of Science and Technology, pp. 483–497.

Potter, J. F. 1984. The Channel Correlation Method for Estimating Aerosol Levels from Multispectral Scanner Data. *Photogrammetric Engineering and Remote Sensing,* Vol. 50, pp. 43–52.

Rohde, W. G., J. K. Lo, and R. A. Pohl. 1978. EROS Data Center Landsat Digital Enhancement Techniques and Imagery Availability, 1977. *Canadian Journal of Remote Sensing,* Vol. 4, pp. 63–76.

Snyder, John P. 1978. The Space Oblique Mercator Projection. *Photogrammetric Engineering and Remote Sensing,* Vol. 44, pp. 585–596.

Snyder, John P. 1981. Map Projections for Satellite Tracking. *Photogrammetric Engineering and Remote Sensing,* Vol. 47, pp. 205–213.

Snyder, John P. 1982. *Map Projections Used by the U.S. Geological Survey.* U.S. Geological Survey Bulletin 1532. Washington, DC: USGS, 313 pp.

Switzer, Paul, W. S. Kowalick, and R. J. P. Lyon. 1981. Estimation of Atmospheric Path-Radiance by the Covariance Matrix Method. *Photogrammetric Engineering and Remote Sensing,* Vol. 47, pp. 1469–1476.

Taranick, James V. 1978. *Principles of Computer Processing of Landsat Data for Geological Applications.* USGS Open File Report 78-117, 50 pp.

Vishnubhatla, S. S. 1977. *Radiometric Correction of Landsat 1 and Landsat II MSS Data.* Technical Note 77-1. Ottawa: Canada Centre for Remote Sensing, 9 pp.

Westin, Torbjorn. 1990. Precision Rectification of SPOT Imagery. *Photogrammetric Engineering and Remote Sensing,* Vol. 56, pp. 247–253.

CHAPTER ELEVEN

# Image Classification

## 11.1. Introduction

Digital image classification is the process of assigning pixels to classes. Usually each pixel is treated as an individual unit composed of values in several spectral bands. By comparing pixels to one another and to those of known identity, it is possible to assemble groups of similar pixels into classes that match the informational categories of interest to users of remotely sensed data. These classes form regions on a map or an image; after classification, the digital image is presented as a mosaic of uniform parcels, each identified by a color or symbol (Figure 11.1 and Plate 12). In principle, these classes are homogeneous—pixels within classes are more similar to one another than they are to pixels in other classes. In practice, of course, each class will display some diversity, as each scene will exhibit some variability within classes.

Image classification has formed an important part of the fields of remote sensing, image analysis, and pattern recognition. In some instances, the classification itself may form the object of the analysis. For example, classification of land use from remotely sensed data (Chapter 19) produces a map-like image that forms the final product of the analysis. In other instances, the classification may form only an intermediate step in a more elaborate analysis. For example, in a study of water quality using remote sensing (Chapter 18), an initial step may use image classification to identify all water pixels within the scene; later steps may then focus on more detailed study of these pixels to map water quality. Image classification therefore forms an important tool for examination of digital images—sometimes to produce a final product, other times as one of several analytical procedures applied to derive information from an image.

The term *classifier* refers loosely to a computer program that implements a specific procedure for image classification. Over the years scientists have devised many classification strategies. From these alternatives the analyst must select the classifier that will best accomplish a specific task. At present it is not possible to state that a given classifier is "best" for all situations because characteristics of each image and the circumstances for each study vary so greatly. Therefore, it is essential that analysts understand the alternative strategies for image classification so that they may be prepared to select the most appropriate classifier for the task at hand.

The simplest form of digital image classification is to consider each pixel individual-

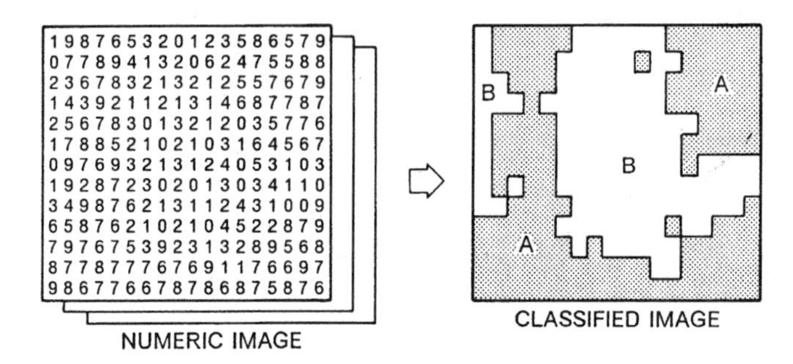


**FIGURE 11.1.** Numeric image and classified image. The classified image (right) is defined by examining the numeric image, then grouping together those pixels that have similar spectral values. Here class A is formed from bright pixels (values of 6, 7, 8, and 9), and B is formed from dark pixels (values of 0, 1, 2, and 3). Usually there are many more classes and at least three or four spectral bands.

ly, assigning it to a class based on its several values measured in separate spectral bands (Figure 11.2). Sometimes such classifiers are referred to as *spectral* or *point* classifiers because they consider each pixel a "point" observation (i.e., as values isolated from their neighbors). Although point classifiers offer the benefits of simplicity and economy, they are not capable of exploiting information contained in relationships between each pixel and those that neighbor it. Human interpreters, for example, could derive little information using the point-by-point approach, as they derive less information from the brightnesses of individual pixels than they do from the patterns of brightnesses formed by groups of adjacent pixels. This is the same quality we designate *image texture* in the context of manual image interpretation (Chapter 5).

As an alternative, more complex classification processes consider groups of pixels within their spatial setting within the image as a means of using the textural information

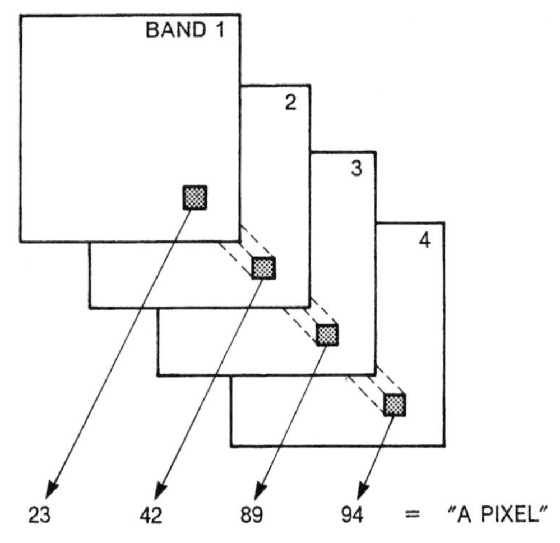


**FIGURE 11.2.** Point classifiers operate on each pixel as a single set of spectral values considered in isolation from its neighbors.

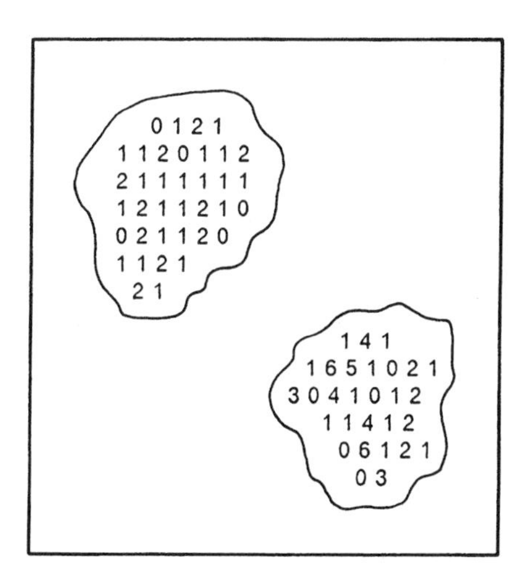


**FIGURE 11.3.** Image texture—the basis for neighborhood classifiers. A neighborhood classifier considers values within a region of the image in defining class membership. Here two regions within the image differ with respect to average brightness but also with respect to "texture"—the uniformity of pixels within a small neighborhood.

so important for the human interpreter. These are spatial, or neighborhood, classifiers, which examine small areas within the image using both spectral and textural information to classify the image (Figure 11.3). Spatial classifiers are often much more difficult to program, and much more expensive to use, than are point classifiers. In some situations spatial classifiers have demonstrated improved accuracy, but few have found their way into routine use for remote sensing image classification.

Another kind of distinction between classifiers separates supervised classification from unsupervised classification. Supervised classification procedures require considerable interaction with the analyst, who must guide the classification by identifying areas on the image that are known to belong to each category. Unsupervised classification, on the other hand, proceeds with only minimal interaction with the analyst, in a search for natural groups of pixels present within the image. The distinction between supervised and unsupervised classification is useful, especially for students who are first learning about image classification. But the two strategies are not as clear as these definitions suggest, for some methods do not clearly fit into either category. These are known as *hybrid classifiers,* which share characteristics of both supervised and unsupervised methods.

## 11.2. Informational Classes and Spectral Classes

Informational classes are the categories of interest to the users of the data. Informational classes are, for example, the different kinds of geological units, forest, or land use that convey information to planners, managers, administrators, and scientists who use information derived from remotely sensed data. These classes form the information that we wish to derive from the data—they are the object of our analysis. Unfortunately, these classes are not directly recorded on remotely sensed images—we can derive them only indirectly, using the evidence contained in brightnesses recorded by each image. For example, the image cannot directly show geological units but, rather, only the differences in topography, vegetation, soil color, shadow, and other factors that lead the analyst to conclude that certain geological conditions exist in specific areas.

Spectral classes are groups of pixels that are uniform with respect to the brightnesses in their several spectral channels. The analyst can observe spectral classes within remotely sensed data; if it is possible to define links between the spectral classes on the image and the informational classes that are of primary interest, then the image forms a valuable source of information. Thus, remote sensing classification proceeds by matching spectral categories to informational categories. If the match can be made with confidence, the information is likely to be reliable. If spectral and informational categories do not correspond, the image is unlikely to be a useful source for that particular form of information. Seldom can we expect to find exact, one-to-one matches between informational and spectral classes. Any informational class includes spectral variations arising from natural variations within the class. For example, a region of the informational class *forest* is still "forest," even though it may display variations in age, species composition, density, and vigor, which all lead to differences in the spectral appearance of a single informational class. Furthermore, other factors such as variations in illumination and shadowing may produce additional variations even within otherwise spectrally uniform classes.

Thus, informational classes are typically composed of numerous spectral subclasses—spectrally distinct groups of pixels that together may be assembled to form an informational class (Figure 11.4). In digital classification, we must often treat spectral subclasses as distinct units during classification but then display several spectral classes under a single symbol for the final image or map to be used by planners or administrators (who are, after all, interested only in the informational categories, not in the intermediate steps required to generate them).

In subsequent chapters we will be interested in several properties of spectral classes. For each band, each class is characterized by a mean, or average, value that, of course, represents the typical brightness of each class. In nature, all classes exhibit some variability around their mean values—some pixels are darker than the average; others a bit

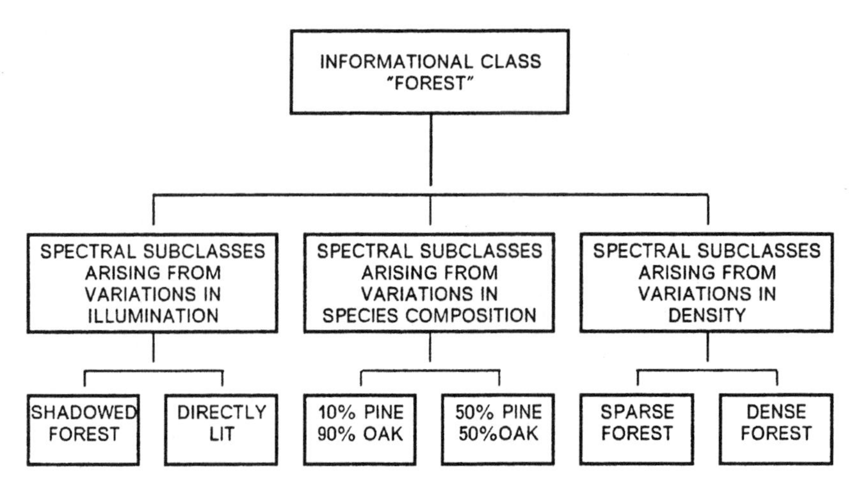


**FIGURE 11.4.** Spectral subclasses.

brighter. These departures from the mean are measured as the variance, or sometimes by the standard deviation (the square root of the variance).

Often we wish to assess the distinctiveness of separate spectral classes, perhaps to determine whether they really are separate classes or whether they should be combined to form a single, larger class. A crude measure of the distinctiveness of two classes is simply the difference in the mean values; presumably classes that are very different should have big differences in average brightness, whereas classes that are similar to one another should have small differences in average brightness. This measure is a bit too simple because it does not take into account the differences in variability between classes.

Another simple measure of distinctiveness is the normalized difference (ND), found by dividing the difference in class means by the sum of their standard deviations. For two classes A and B, with means $\bar{x}_a$ and $\bar{x}_b$, and standard deviations $s_a$ and $s_b$, the ND is defined as

$$\text{Normalized difference} = \frac{|\bar{x}_a - \bar{x}_b|}{s_a + s_b} \qquad \text{(Eq. 11.1)}$$

As an example, Table 11.1 summarizes properties of several classes and uses these properties to show the ND between these classes in several spectral channels. Note that some pairs of categories are distinct relative to others. Also note that some spectral channels are much more effective than others in separating categories from one another.

## 11.3. Unsupervised Classification

Unsupervised classification can be defined as the identification of natural groups, or structures, within multispectral data. The notion of the existence of natural, inherent groupings of spectral values within a scene may not be intuitively obvious, but it can be

**TABLE 11.1. Normalized Difference**

| | MSS band | | | | MSS band | | | |
|---|---|---|---|---|---|---|---|---|
| | 4 | 5 | 6 | 7 | 4 | 5 | 6 | 7 |
| | Water | | | | Forest | | | |
| $x$ | 37.5 | 31.9 | 22.8 | 6.3 | 26.9 | 16.6 | 55.7 | 32.5 |
| $s$ | 0.67 | 2.77 | 2.44 | 0.82 | 1.21 | 1.49 | 3.97 | 3.12 |
| | Crop | | | | Pasture | | | |
| $x$ | 37.7 | 38.0 | 52.3 | 27.3 | 28.6 | 22.0 | 53.4 | 32.9 |
| $s$ | 3.56 | 5.08 | 4.13 | 4.42 | 1.51 | 5.09 | 13.16 | 3.80 |

$$\text{Water – forest (band 6): } \frac{55.7 - 22.8}{2.44 + 3.97} = \frac{32.9}{6.41} = 5.13$$

$$\text{Crop – pasture (band 6): } \frac{52.3 - 53.4}{4.13 + 13.16} = \frac{1.1}{17.29} = 0.06$$

demonstrated that remotely sensed images are usually composed of spectral classes that internally are reasonably uniform in respect to brightnesses in several spectral channels. Unsupervised classification is the definition, identification, labeling, and mapping of these natural classes.

### *Advantages*

Advantages of unsupervised classification (relative to supervised classification) can be enumerated as follows:

- *No extensive prior knowledge of the region is required.* Or, more accurately, the nature of knowledge required for unsupervised classification differs from that required for supervised classification. To conduct supervised classification, detailed knowledge of the area to be examined is required to select representative examples of each class to be mapped. To conduct unsupervised classification, no detailed prior knowledge is required, but knowledge of the region is required to interpret the meaning of the results produced by the classification process.
- *The opportunity for human error is minimized.* To conduct unsupervised classification, the operator may perhaps specify only the number of categories desired (or possibly, minimum and maximum limits on the number of categories), and sometimes constraints governing the distinctness and uniformity of groups. Many of the detailed decisions required for supervised classification are not required for unsupervised classification, so the analyst is presented with less opportunity for error. If analysts have inaccurate preconceptions regarding the region, they will have little opportunity to influence the classification. The classes defined by unsupervised classification are often much more uniform, with respect to spectral composition, than those generated by supervised classification.
- *Unique classes are recognized as distinct units.* Such classes, perhaps of very small areal extent, may remain unrecognized in the process of supervised classification and could inadvertently be incorporated into other classes, generating error and imprecision throughout the entire classification.

### *Disadvantages and Limitations*

Disadvantages and limitations arise primarily from a reliance on "natural" groupings and difficulties in matching these groups to the informational categories that are of interest to the analyst.

- Unsupervised classification identifies spectrally homogeneous classes within the data; these classes do not necessarily correspond to the informational categories that are of interest to the analyst. As a result, the analyst is faced with the problem of matching spectral classes generated by the classification to the informational classes that are required by the ultimate user of the information. Seldom is there a simple one-to-one correspondence between the two sets of classes.

• The analyst has limited control over the menu of classes and their specific identities. If it is necessary to generate a specific menu of informational classes (e.g., to match to other classifications for other dates or adjacent regions), the use of unsupervised classification may be unsatisfactory.

• Spectral properties of specific informational classes will change over time (on a seasonal basis, as well as over the years). As a result, relationships between informational classes and spectral classes are not constant and relationships defined for one image can seldom be extended to others. Although detailed knowledge of the spectral characteristics of specific classes can permit extension of training data over space and time, such efforts require carefully considered preconditions.

### *Distance Measures*

Some of the basic elements of unsupervised classification can be illustrated using data presented in Tables 11.2 and 11.3. These values can be plotted on simple diagrams constructed using brightnesses of two spectral bands as orthogonal axes (Figures 11.5 and 11.6). Plots of band pairs 4 and 5 and band pairs 5 and 6 are sufficient to illustrate the main points of significance here. These two-dimensional plots illustrate principles that can be extended to include additional variables (to analyze all four MSS bands simultane-

**TABLE 11.2. Landsat MSS Digital Values for February**

| | MSS band | | | | | MSS band | | | |
|---|---|---|---|---|---|---|---|---|---|
| | 4 | 5 | 6 | 7 | | 4 | 5 | 6 | 7 |
| 1. | 19 | 15 | 22 | 11 | 21. | 24 | 24 | 25 | 11 |
| 2. | 21 | 15 | 22 | 12 | 22. | 25 | 25 | 38 | 20 |
| 3. | 19 | 13 | 25 | 14 | 23. | 20 | 29 | 19 | 3 |
| 4. | 28 | 27 | 41 | 21 | 24. | 28 | 29 | 18 | 2 |
| 5 | 27 | 25 | 32 | 19 | 25. | 25 | 26 | 42 | 21 |
| 6. | 21 | 15 | 25 | 13 | 26. | 24 | 23 | 41 | 22 |
| 7. | 21 | 17 | 23 | 12 | 27. | 21 | 18 | 12 | 12 |
| 8. | 19 | 16 | 24 | 12 | 28. | 25 | 21 | 31 | 15 |
| 9. | 19 | 12 | 25 | 14 | 29. | 22 | 22 | 31 | 15 |
| 10. | 28 | 29 | 17 | 3 | 30. | 26 | 24 | 43 | 21 |
| 11. | 28 | 26 | 41 | 21 | 31. | 19 | 16 | 24 | 12 |
| 12. | 19 | 16 | 24 | 12 | 32. | 30 | 31 | 18 | 3 |
| 13. | 29 | 32 | 17 | 3 | 33. | 28 | 27 | 44 | 24 |
| 14. | 19 | 16 | 22 | 12 | 34. | 22 | 22 | 28 | 15 |
| 15. | 19 | 16 | 24 | 12 | 35. | 30 | 31 | 18 | 2 |
| 16. | 19 | 16 | 25 | 13 | 36. | 19 | 16 | 22 | 12 |
| 17. | 24 | 21 | 35 | 19 | 37. | 30 | 31 | 18 | 2 |
| 18. | 22 | 18 | 31 | 14 | 38. | 27 | 23 | 34 | 20 |
| 19. | 21 | 18 | 25 | 13 | 39. | 21 | 16 | 22 | 12 |
| 20. | 21 | 16 | 27 | 13 | 40. | 23 | 22 | 26 | 16 |

*Note.* These are raw digital values for a forested area in central Virginia as acquired by the Landsat 1 MSS in February 1974. These values represent the same area as those in Table 11.3, although individual pixels do not correspond.

**TABLE 11.3. Landsat MSS Digital Values for May**

| | MSS band | | | | | MSS band | | | |
|---|---|---|---|---|---|---|---|---|---|
| | 4 | 5 | 6 | 7 | | 4 | 5 | 6 | 7 |
| 1. | 34 | 28 | 22 | 6 | 21. | 26 | 16 | 52 | 29 |
| 2. | 26 | 16 | 52 | 29 | 22. | 30 | 18 | 57 | 35 |
| 3. | 36 | 35 | 24 | 6 | 23. | 30 | 18 | 62 | 28 |
| 4. | 39 | 41 | 48 | 23 | 24. | 35 | 30 | 18 | 6 |
| 5. | 26 | 15 | 52 | 31 | 25. | 36 | 33 | 24 | 7 |
| 6. | 36 | 28 | 22 | 6 | 26. | 27 | 16 | 57 | 32 |
| 7. | 28 | 18 | 59 | 35 | 27. | 26 | 15 | 57 | 34 |
| 8. | 28 | 21 | 57 | 34 | 28. | 26 | 15 | 50 | 29 |
| 9. | 26 | 16 | 55 | 30 | 29. | 26 | 33 | 24 | 27 |
| 10. | 32 | 30 | 52 | 25 | 30. | 36 | 36 | 27 | 8 |
| 11. | 40 | 45 | 59 | 26 | 31. | 40 | 43 | 51 | 27 |
| 12. | 33 | 30 | 48 | 24 | 32. | 30 | 18 | 62 | 38 |
| 13. | 28 | 21 | 57 | 34 | 33. | 28 | 18 | 62 | 38 |
| 14. | 28 | 21 | 59 | 35 | 34. | 36 | 33 | 22 | 6 |
| 15. | 36 | 38 | 48 | 22 | 35. | 35 | 36 | 56 | 33 |
| 16. | 36 | 31 | 23 | 5 | 36. | 42 | 42 | 53 | 26 |
| 17. | 26 | 19 | 57 | 33 | 37. | 26 | 16 | 50 | 30 |
| 18. | 36 | 34 | 25 | 7 | 38. | 42 | 38 | 58 | 33 |
| 19. | 36 | 31 | 21 | 6 | 39. | 30 | 22 | 59 | 37 |
| 20. | 27 | 19 | 55 | 30 | 40. | 27 | 16 | 56 | 34 |

*Note.* These are raw digital values for a forested area in central Virginia as acquired by the Landsat 1 MSS in May 1974. These values represent the same area as those in Table 11.2, although individual pixels do not correspond.

ously, for example). The additional variables create three- or four-dimensional plots of points in multidimensional data space, which are difficult to depict in a diagram but can be envisioned as groups of pixels represented by swarms of points with depth in several dimensions (Figure 11.7).

The general form for such diagrams is illustrated by Figure 11.8. A diagonal line of points extends upward from a point near the origin. This pattern occurs because a pixel that is dark in one band will often tend to be dark in another band, and as brightness increases in one spectral region, so also does it tend to increase in others. (This relationship often holds for spectral measurements in the visible and near infrared; it is not necessarily

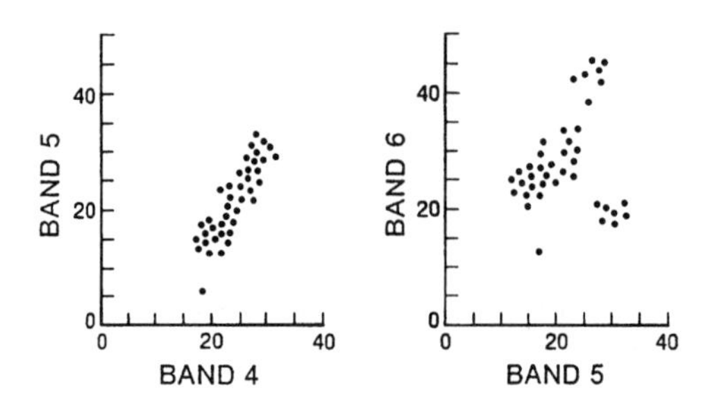


**FIGURE 11.5.** Two-dimensional scatter diagrams, February (Table 11.2). Compare with Figure 11.6.

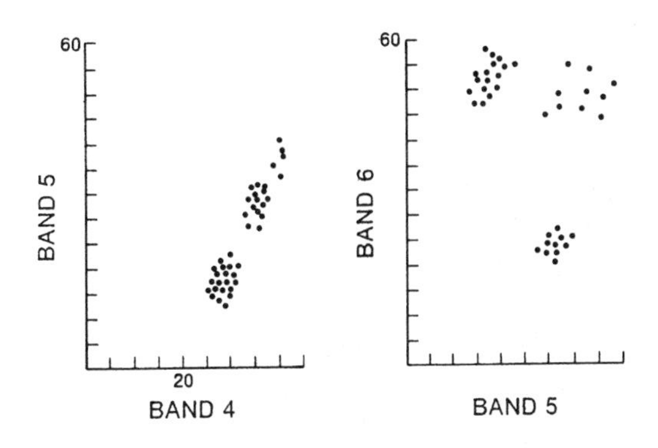


**FIGURE 11.6.** Two-dimensional scatter diagrams, May (Table 11.3) These pixels represent the same categories shown in Figure 11.5 but have different brightnesses because of difference in date.

observed in data from widely separated spectral regions.) There are 40 values plotted in each of the diagrams in Figure 11.6. Specific groupings, or clusters, are evident; these clusters may correspond to informational categories of interest to the analyst. Unsupervised classification is the process of defining such clusters in multidimensional data space and (if possible) matching them to informational categories.

When we consider large numbers of pixels, the clusters are not usually as distinct because pixels of intermediate values tend to fill in the gaps between groups (Figure 11.8). We must therefore apply a variety of methods to assist us in the identification of those groups that may be present within the data but not necessarily obvious to visual inspection. Over the years, image scientists and statisticians have developed a wide variety of procedures for identification of such clusters—procedures that vary greatly in complexity and effectiveness. The general model for unsupervised classification can, however, be illustrated using one of the simplest classification strategies—one that can form the basis for understanding more complex approaches.

Figure 11.9 shows two pixels, each with measurements in several spectral channels, plotted in multidimensional data space in the same manner as those illustrated in Figure 11.6. For ease of illustration, only two bands are shown here, although the principles illustrated extend to as many bands as may be available.

Unsupervised classification of an entire image must consider many thousands of pixels. But the classification process is always based on the answer to the same question: "Do the two pixels belong to the same group?" For this example, the question is, "Should pix-

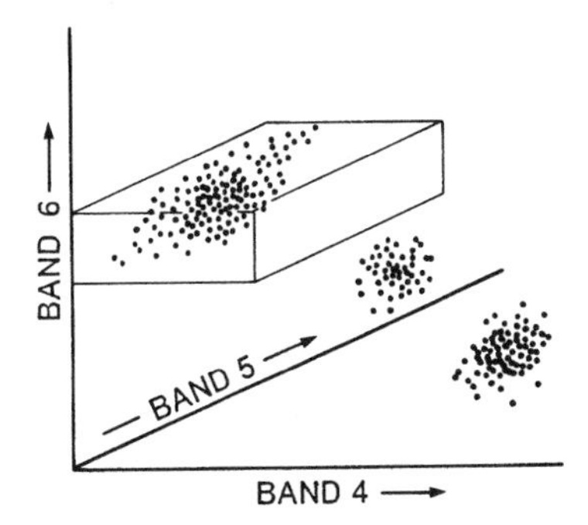


**FIGURE 11.7.** Sketch illustrating multidimensional scatter diagrams. Here three bands of data are shown.

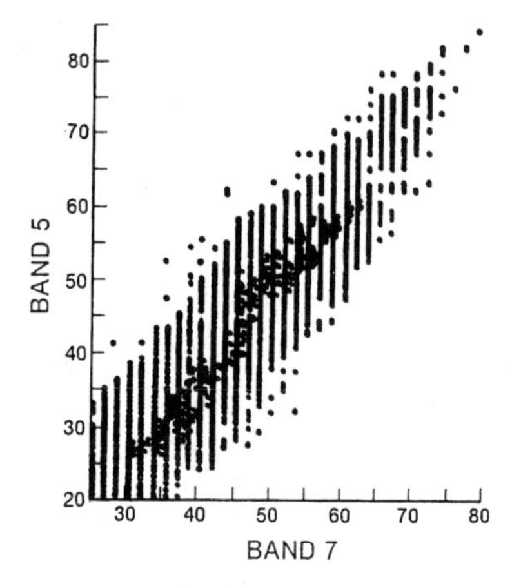


**FIGURE 11.8.** Scatter diagram. Data from two Landsat MSS bands illustrate the general form of the relationship between spectral measurements in contiguous regions of the spectrum. This diagram shows several hundred points; when so many values are shown, the distinct clusters visible in Figure 11.6 are not visible. The groups may still be present, although they can be detected only by application of classification algorithms that can simultaneously consider values in many spectral bands. From Todd et al. (1980, p. 511). Copyright 1980 by the American Society for Photogrammetry and Remote Sensing. Reproduced by permission.

el C be grouped with A, or with B?" (See the lower part of Figure 11.9.) This question can be answered by finding the distance between pairs of pixels. If the distance between A and C is greater than that from B to C, B and C are said to belong to the same group and A may be defined as a member of a separate class.

There are thousands of pixels in a remotely sensed image; if they are considered individually as prospective members of groups, the distances to other pixels can always be used to define group membership. How can such distances be calculated? A number of methods for finding distances in multidimensional data space are available. One of the simplest is *Euclidean distance*:

$$D_{ab} = \left[ \sum_{i=1}^{n} (a_i - b_i)^2 \right]^{1/2} \quad \text{(Eq. 11.2)}$$

where $i$ represents one of $n$ spectral bands, $a$ and $b$ are pixels, and $D_{ab}$ is the distance between the two pixels. The distance calculation is based on the Pythagorean theorem (Figure 11.10):

$$c = \sqrt{a^2 + b^2} \quad \text{(Eq. 11.3)}$$

In this instance we are interested in distance $c$; $a$, $b$, and $c$ are measured in units of the two spectral channels.

$$c = D_{ab} \quad \text{(Eq. 11.4)}$$

To find $D_{ab}$ we need to find distances $a$ and $b$. Distance $a$ is found by subtracting values of A and B in MSS channel 7 ($a = 38 - 15 = 23$). Distance $b$ is found by finding the difference between A and B with respect to channel 6 ($b = 30 - 10 = 20$).

$$D_{ab} = c = \sqrt{20^2 + 23^2}$$

$$D_{ab} = \sqrt{400 + 529} = \sqrt{929}$$

$$D_{ab} = 30.47$$

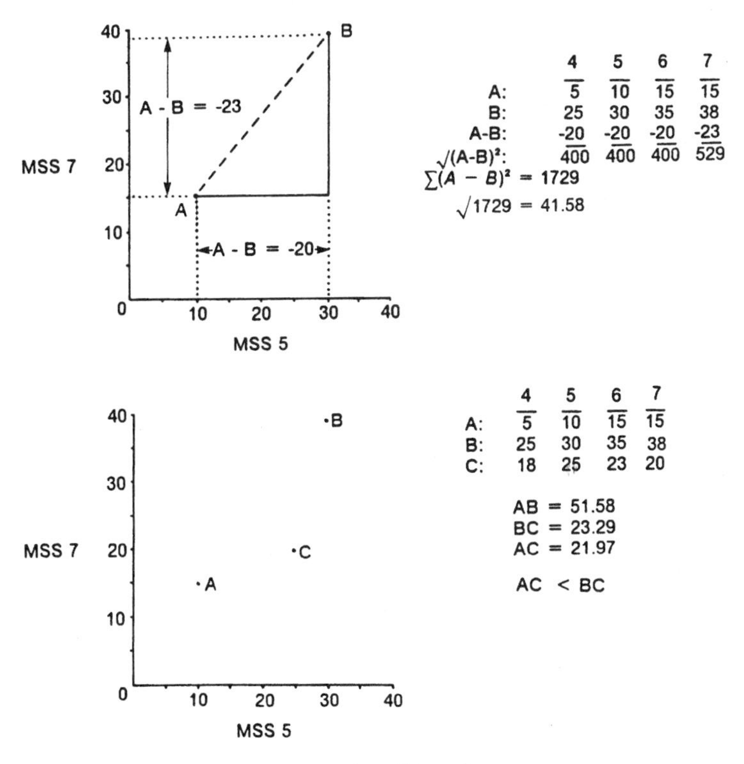


**FIGURE 11.9.** Illustration of Euclidean distance measure.

This measure can be applied to as many dimensions (spectral channels) as might be available, by addition of distances. For example:

| | Landsat MSS band | | | |
|---|---|---|---|---|
| | 1 | 2 | 3 | 4 |
| Pixel A | 34 | 28 | 22 | 6 |
| Pixel B | 26 | 16 | 52 | 29 |
| Difference | 8 | 12 | –30 | –23 |
| (Difference)$^2$ | 64 | 144 | 900 | 529 |
| Total of (differences)$^2$ = 1,637 | | | | |
| $\sqrt{\text{total}}$ = 40.5 | | | | |

The lower part of Figure 11.9 shows another worked example.

Thus, the Euclidean distance between A and B is equal to 40.45 distance units. This value in itself has little significance, but in relation to other distances it forms a means of defining similarities between pixels. For example, if we find that distance *ab* = 40.45 and distance *ac* = 86.34, we know that pixel A is closer (i.e., more nearly similar) to B than it is to C, and that we should form a group from A and B rather than A and C.

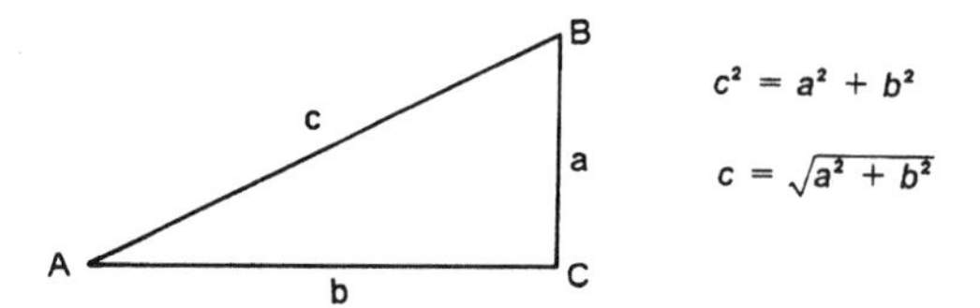


**FIGURE 11.10.** Definition of symbols used for Euclidean distance.

Unsupervised classification proceeds by making thousands of distance calculations as a means of determining similarities for the many pixels and groups within an image. Usually the analyst does not actually know any of these many distances that must be calculated for unsupervised classification, as the computer presents only the final classified image, without the intermediate steps necessary to derive the classification. Nonetheless, distance measures are the heart of unsupervised classification.

But not all distance measures are based on Euclidean distance. Another simple measure of distance is the $L_1$ *distance*, the sum of the absolute differences between values in individual bands (Swain & Davis, 1978). For the example given above, the $L_1$ distance is 73 (73 = 8 + 12 + 30 + 23). Other distance measures have been defined for unsupervised classification; many are rather complex methods of scaling distances to promote effective groupings of pixels.

Unsupervised classification often proceeds in an interactive fashion to search for an optimal allocation of pixels to categories, given constraints specified by the analyst. A computer program for unsupervised classification includes an algorithm for calculation of distances as described above (sometimes the analyst might be able to select between several alternative distance measures) and a procedure for finding, testing, and revising classes according to limits defined by the analyst. The analyst may be required to specify limits on the number of clusters to be generated, to constrain the diversity of values within classes, or to require that classes exhibit a specified minimum degree of distinctness with respect to neighboring groups. Specific classification procedures may define distances differently; for example, it is possible to calculate distances to the centroid of each group, or to the closest member of each group, or perhaps to the most densely occupied region of a cluster. Such alternatives represent refinements of the basic strategy outlined here and may offer advantages in certain situations. Also, there are many variations in the details of how each classification program may operate; because so many distance measures must be calculated for classification of a remotely sensed image, most programs use variations of these basic procedures that accomplish the same objectives with improved computational efficiency.

### *Sequence for Unsupervised Classification*

A typical sequence might begin with the analyst specifying minimum and maximum numbers of categories to be generated by the classification algorithm. These values might be based on the analyst's knowledge of the scene, or on the user's requirements that the final classification display a certain number of classes. The classification starts with a set of arbitrarily selected pixels as cluster centers; often these are selected at random to ensure that the analyst cannot influence the classification and that the selected pixels are representative of values found throughout the scene. The classification algorithm then finds distances (as described above) between pixels and forms initial estimates of cluster centers as permitted by constraints specified by the analyst. The class can be represented by a single point, known as the *class centroid,* which can be thought of as the center of the cluster of pixels for a given class, even though many classification procedures do not always define it as the exact center of the group. At this point, classes consist only of the arbitrarily selected pixels chosen as initial estimates of class centroids. In the next step, all the remaining pixels in the scene are assigned to the nearest class centroid. The entire scene has now been classified, but this classification forms only an estimate of the final result, as the classes formed by this initial attempt are unlikely to be the optimal set of classes and may not meet the constraints specified by the analyst.

To begin the next step, the algorithm finds new centroids for each class, as the addition of new pixels to the classification means that the initial centroids are no longer accurate. Then the entire scene is classified again, with each pixel assigned to the nearest centroid. And again new centroids are calculated; if the new centroids differ from those found in the preceding step, the process repeats until there is no significant change detected in locations of class centroids and the classes meet all constraints required by the operator.

Throughout the process, the analyst generally has no interaction with the classification, so it operates as an "objective" classification within the constraints provided by the analyst. Also, the unsupervised approach identifies the "natural" structure of the image in the sense that it finds uniform groupings of pixels that form distinct classes without the influence of preconceptions regarding their identities or distributions. The entire process, however, cannot be considered "objective," as the analyst has made decisions regarding the data to be examined, the algorithm to be used, the number of classes to be found, and, possibly, the uniformity and distinctness of classes. Each of these decisions influences the character and the accuracy of the final product, so it cannot be regarded as a result isolated from the context in which it was made.

Many different procedures for unsupervised classification are available; despite their diversity, most are based on the general strategy just described. Although some refinements are possible to improve computational speed and efficiency, this approach is in essence a kind of wearing down of the classification problem by repetitive application assignment and reassignment of pixels to groups. Key components to any unsupervised classification algorithm are effective methods of measuring distances in data space, identifying class centroids, and testing the distinctness of classes. There are many different strategies for accomplishing each of these tasks; an enumeration of even the most widely used methods is outside the scope of this text, but some are described in the articles listed in the references.

### *AMOEBA*

A useful variation on the basic strategy for unsupervised classification has been described by Bryant (1979). The AMOEBA classification operates in the manner of usual unsupervised classification, with the addition of a contiguity constraint that considers the locations of values as spectral classes are formed. The analyst specifies a tolerance limit that governs the diversity permitted as classes are formed. As a class is formed by a group of neighboring pixels, adjacent pixels belonging to other classes are considered prospective members of the class if they occur as a small region within a larger, more homogeneous

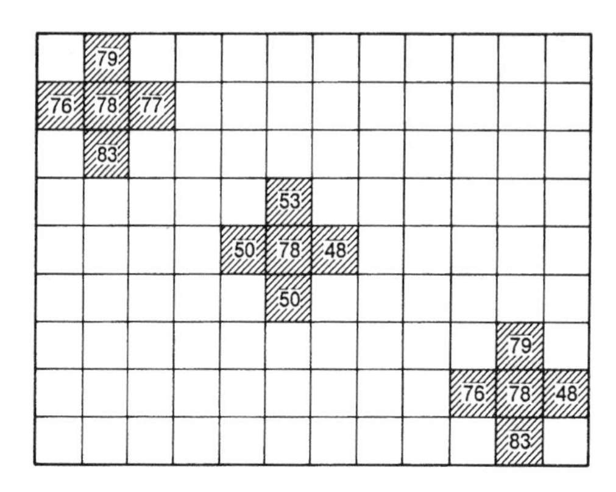


**FIGURE 11.11.** AMOEBA spatial operator. This contrived example shows AMOEBA as it considers classification of three pixels, all with the same digital value of 78. Upper left: The central pixel has a value similar to those of all neighbors; it is classified together with the surrounding pixels. Center: The central pixel differs from its neighbors, but due to its position within a background of contrasting but homogeneous pixels, it is classified with the darker pixels. Lower left: The central pixel is classified with the darker pixels at the top, left, and bottom. In all three situations, classification is based on values at adjacent locations.

background (Figure 11.11). If the candidate pixel has values that fall within the tolerance limits specified by the analyst, the pixel is accepted as a member of the class despite the fact that it differs from other members. Thus, locations, as well as spectral properties, of pixels form classification criteria for AMOEBA and similar algorithms. Although this approach increases the spectral diversity of classes—normally an undesirable quality—the inclusion of small areas of foreign pixels as members of a more extensive region may satisfy our notion of the proper character of a geographic region. Often, in the manual preparation of maps, the analyst will perform a similar form of generalization as a means of eliminating small inclusions within larger regions. This kind of generalization presents cartographic data in a form suitable for presentation at small scale. The AMOEBA classifier was designed for application to scenes composed most of large, homogeneous regions (e.g., the agricultural landscapes of the North American prairies). For such scenes, it seems to work well, although it may not be as effective in more complex landscapes composed of smaller parcels (Story et al., 1984).

### *Assignment of Spectral Categories to Informational Categories*

The classification results described thus far provide uniform groupings of pixels—classes that are uniform with respect to the spectral values that compose each pixel. These spectral classes are of interest only to the extent that they can be matched to one or more informational classes that are of interest to the user of the final product. These spectral classes may sometimes correspond directly to informational categories. For example, in Figure 11.12 the pixels that compose the group closest to the origin correspond to "open water." This identification is more or less obvious from the spectral properties of the class—few other classes will exhibit such dark values in both spectral channels. However, seldom can we depend on a clear identification of spectral categories from spectral values alone. Often it is possible to match spectral and informational categories by examining patterns on the image; many informational categories are recognizable by the positions, sizes, and shapes of individual parcels and their spatial correspondence with areas of known identity.

Often, however, spectral classes do not match informational classes directly. In some instances, informational classes may occur in complex mixtures and arrangements. Perhaps forested patches are scattered in small areas against an more extensive background

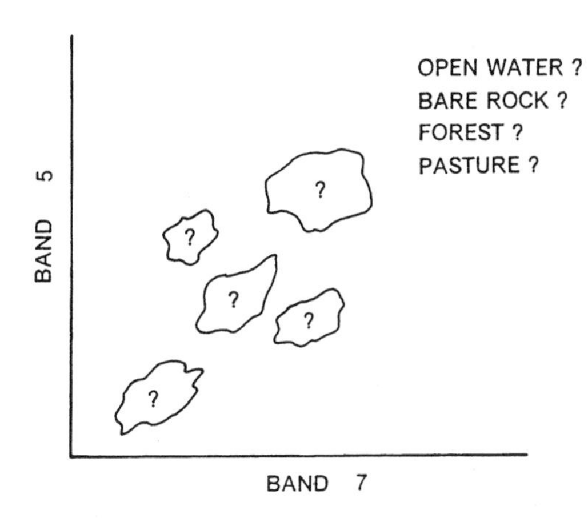


**FIGURE 11.12.** Assignment of spectral categories to image categories. Unsupervised classification defines the clusters defined schematically on the scatter diagram. The analyst must decide which, if any, match the list of informational categories that form the object of the analysis.

of grassland. If these forested areas are small relative to the spatial resolution of the sensor, the overall spectral response for such an area will differ from either "forest" or "grassland," due to the effect of mixed pixels on the spectral response. This region may be assigned then to a class separate from either the forest or the grassland classes (Chapter 9). It is the analyst's task to use knowledge of the region and the sensor to identify the character of the problem and apply an appropriate remedy.

Or, because of spectral diversity, even nominally uniform informational classes may manifest themselves as a set of spectral classes. For example, a forested area may be recorded as several spectral clusters, due perhaps to variations in density, age, aspect, shadowing, and other factors that alter the spectral properties of a forested region, but it does not alter the fact that the region belongs to the informational class *forest* (Figure 11.4). The analyst must therefore examine the output to match spectral categories from the classification with the informational classes of significance to those who will use the results.

Thus, a serious practical problem with unsupervised classification is that clear matches between spectral and informational classes are not always possible; some informational categories may not have direct spectral counterparts, and vice versa. Furthermore, the analyst does not have control over the nature of the categories generated by unsupervised classification. If a study is conducted to compare results with those from an adjacent region, or, for example, from a different date, it may be necessary to have the same set of informational categories on both maps or the comparison cannot be made. If unsupervised classification is to be used, it may be difficult to generate the same sets of informational categories on both images.

## 11.4. Supervised Classification

Supervised classification can be defined informally as the process of using samples of known identity (i.e., pixels already assigned to informational classes) to classify pixels of unknown identity (i.e., to assign unclassified pixels to one of several informational classes). Samples of known identity are those pixels located within *training areas* (or *training fields*). The analyst defines training areas by identifying regions on the image that can be clearly matched to areas of known identity on the image. Such areas should typify spectral properties of the categories they represent and, of course, must be homogeneous in respect to the informational category to be classified. That is, training areas should not include unusual regions, nor should they straddle boundaries between categories. Size, shape, and position must favor convenient identification both on the image and on the ground. Pixels located within these areas form the training samples used to guide the classification algorithm to assign specific spectral values to appropriate informational classes. Clearly, the selection of these training data is a key step in supervised classification.

### *Advantages*

Advantages of supervised classification, relative to unsupervised classification, can be enumerated as follows: The analyst has control of a set, selected menu of informational categories tailored to a specific purpose and geographic region. This quality may be vitally important if it is necessary to generate a classification for the specific purpose of comparison with another classification of the same area at a different date, or if the classification must be compatible with those of neighboring regions. Under such circumstances, the unpredictable qualities of categories generated by unsupervised classification (i.e., unpredictability with respect to number, identity, size, and pattern) may be inconvenient or unsuitable. Second, supervised classification is tied to specific areas of known identity, as provided through the process of selecting training areas. Furthermore, the analyst using supervised classification is not faced with the problem of matching spectral categories on the final map with the informational categories of interest (this task has, in effect, been addressed during the process of selecting training data). Finally, the operator may be able to detect serious errors by examining training data to determine whether they have been correctly classified—inaccurate classification of training data indicates serious problems in the classification (or in the selection of training data), although correct classification of training data does not always indicate correct classification of other data.

### *Disadvantages and Limitations*

Disadvantages of supervised classification are numerous. The analyst, in effect, imposes a classification structure upon the data (recall that unsupervised classification searches for "natural" classes). These operator-defined classes may not match the natural classes that may exist within the data and therefore may not be distinct or well-defined in multidimensional data space. Second, training data are often defined primarily on informational categories and only secondarily on spectral properties. A training area that is 100% forest may be accurate with respect to the *forest* designation but very diverse with respect to density, age, shadowing, and so on, and therefore forms a poor training area. Third, training data selected by the analyst may not be representative of conditions encountered throughout the image. This may be true despite the best efforts of the analyst, especially if the area to be classified is large, complex, or inaccessible.

Conscientious selection of training data may be a time-consuming, expensive, and tedious undertaking, even if ample resources are at hand. The analyst may experience problems in matching prospective training areas as defined on maps and aerial photographs to the image to be classified. Finally, supervised classification may not be able to recognize and represent special or unique categories not represented in the training data, possibly because they are not known to the analyst or because they occupy very small areas on the image.

### *Training Data*

Training fields are areas of known identity delineated on the digital image, usually by specifying the corner points of a rectangular or polygonal area using line and column numbers within the coordinate system of the digital image. The analyst must, of course, know the correct class for each area. Usually the analyst begins by assembling maps and aerial photographs of the area to be classified. (Here we assume that the analyst has some basic familiarity with the specific area to be studied and the particular problem the study is to address and has conducted the necessary field observations prior to initiating the actual selection of training data.) Specific training areas are identified for each informational category following the guidelines outlined below. The objective is to identify a set of pixels that accurately represents spectral variation present within each informational region (Figure 11.13).

### *Key Characteristics of Training Areas*

#### NUMBER OF PIXELS

An important concern is the overall number of pixels selected for each category; as a general guideline, the operator should ensure that several individual training areas for each category provide a total of at least 100 pixels for each category.

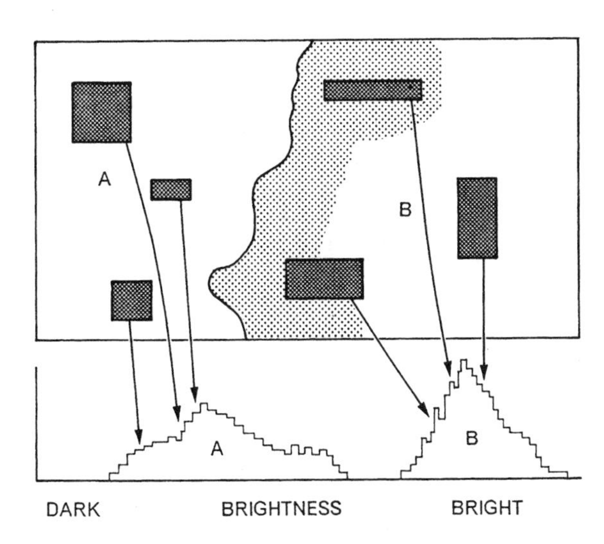


**FIGURE 11.13.** Training fields and training data. Training fields, each composed of many pixels, sample the spectral characteristics of informational categories. Here the shaded figures represent training fields, each positioned carefully to estimate spectral properties of each class, as represented by the histograms. This information provides the basis for classification of the remaining pixels not within the training fields.

### SIZE

Sizes of training areas are important. They must each be large enough to provide accurate estimates of the properties of each informational class. Therefore, they must as a group include enough pixels to form reliable estimates of the spectral characteristics of each class (hence, the minimum figure of 100 suggested above). Individual training fields should not, on the other hand, be too big, as large areas include undesirable variation. (Because the total number of pixels in the training data for each class can be formed from many separate training fields, each individual area can be much smaller than the total number of pixels required for each class.) Joyce (1978) recommends that individual training areas be at least 4 ha (10 acres) in size at the absolute minimum and preferably include about 16 ha (40 acres). Small training fields are difficult to locate accurately on the image, and to accumulate an adequate total number of training pixels, the analyst must devote more time to definition and analysis of the additional training fields. On the other hand, use of large training fields increases the opportunity for including of spectral inhomogeneities. Joyce (1978) suggests 65 ha (160 acres) as the maximum size for training fields.

Joyce specifically refers to Landsat MSS data, so in terms of numbers of pixels, he is recommending from 10 to about 40 pixels for each training field. For TM or SPOT data, of course, his recommendations would specify different numbers of pixels, as the resolutions of these sensors differ from those of the MSS. Also, one would expect optimum sizes of training fields to vary according to the heterogeneity of each landscape and each class, so each analyst should develop his or her own guidelines as experience is acquired in specific circumstances.

### SHAPE

Shapes of training areas are not important provided that shape does not prohibit accurate delineating and positioning of outlines of regions on digital images. Usually it is easiest to define rectangular or polygonal areas, as such shapes minimize the number of vertices that must be specified, which is usually the most bothersome task for the analyst.

### LOCATION

Location is important, as each informational category should be represented by several training areas positioned throughout the image. Training areas must be positioned in locations that favor accurate and convenient transfer of their outlines from maps and aerial photographs to the digital image (Figure 11.14). As the training data are to represent variation within the image, they must not be clustered in favored regions of the image, which may not typify conditions encountered throughout the image as a whole. As it is desirable for the analyst to use direct field observations in the selection of training data, the requirement for an even distribution of training fields often conflicts with practical constraints, as it may not be practical to visit remote or inaccessible sites that may seem to form good areas for training data. Often aerial observation, or use of good maps and aerial photographs, can provide the basis for accurate delineation of training fields that cannot be inspected in the field. Although such practices are often sound, it is important to avoid a cavalier approach to the selection of training data that depends completely on indirect evidence when direct observation is feasible.

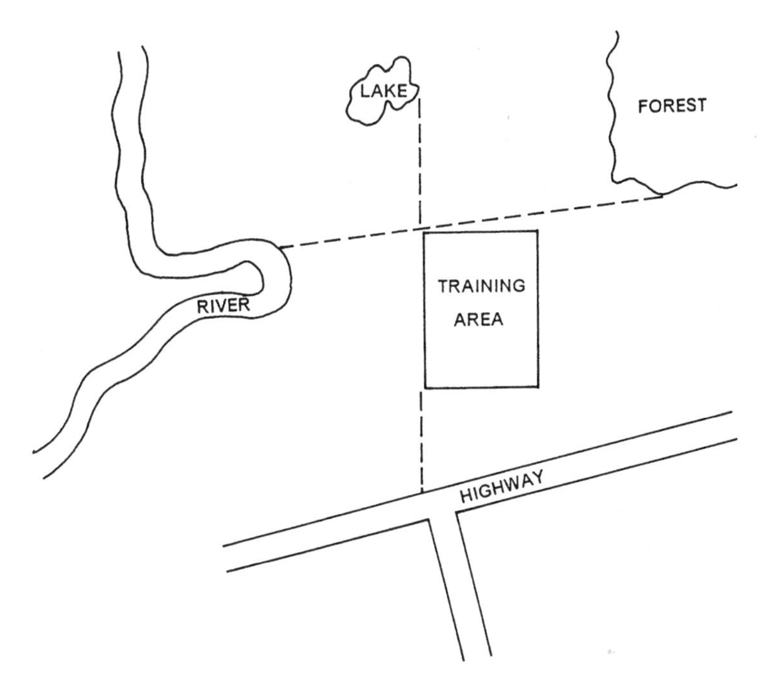


**FIGURE 11.14.** Location of training fields with reference to landmarks. Training fields must be positioned with reference to features easily recognizable on the ground, on maps and aerial photographs, and on the digital imagery.

NUMBER

The optimum number of training areas depends on the number of categories to be mapped, their diversity, and the resources that can be devoted to delineating training areas. Ideally, each informational category, or each spectral subclass, should be represented by a number (perhaps 5 to 10 at a minimum) of training areas to ensure that spectral properties of each category are represented. Because informational classes are often spectrally diverse, it may be necessary to use several sets of training data for each informational category, due to the presence of spectral subclasses. Selection of multiple training areas is also desirable because later in the classification process it may be necessary to discard some training areas if they are discovered to be unsuitable. Experience indicates that it is usually better to define many small training fields than to use only a few large fields.

PLACEMENT

Placement of training areas may be important. Training areas should be placed within the image in a manner that permits convenient and accurate location with respect to distinctive features, such as water bodies, or boundaries between distinctive features on the image. They should be distributed throughout the image so that they provide a basis for representation of diversity present within the scene. Boundaries of training fields should be placed well away from the edges of contrasting parcels so that they do not encompass edge pixels (Figure 11.15).

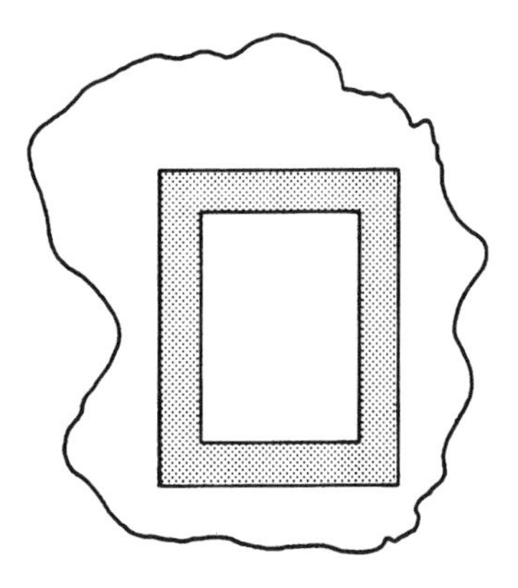

FIGURE 11.15. Positioning training fields. Here the irregular line represents the edge of a parcel visible on a digital image. The inner rectangle represents the training field surrounded by a "buffer" of pixels within the parcel but not included in the training data. These unused pixels help ensure that mixed pixels at the edge of the parcel are not included within the training field. (Remember that the analyst may not have absolute confidence in his or her ability to match map or photo information to the digital image.)

### UNIFORMITY

Perhaps the most important property of a good training area is its uniformity, or homogeneity. Data within each training area should exhibit a unimodal frequency distribution for each spectral band to be used (Figure 11.16). Prospective training areas that exhibit bimodal histograms should be discarded if their boundaries cannot be adjusted to yield more uniformity. Training data provide values that estimate the means, variances, and covariances of spectral data measured in several spectral channels. For each class to be mapped, these estimates approximate the mean values for each band, variability of each band, and interrelationships between bands. In the ideal, these values represent the conditions present within each class within the scene and thereby form the basis for classification of the vast majority of pixels within each scene that do not belong to training areas. In practice, of course, scenes vary greatly in complexity, and individual analysts differ in their knowledge of a region and in their ability to define training areas that accurately represent spectral properties of informational classes. Some informational classes are not spectrally uniform and cannot be conveniently represented by a single set of training data.

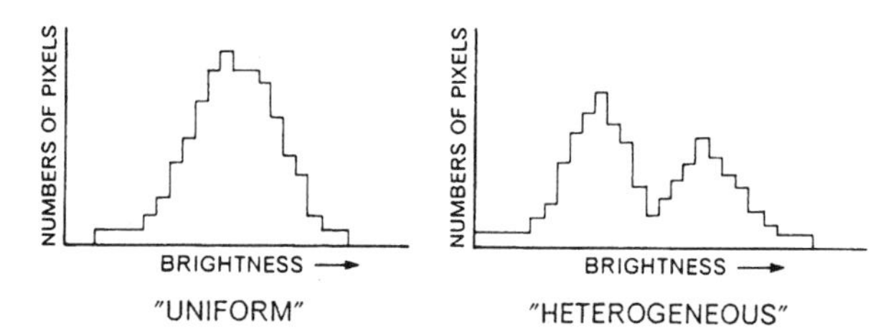


FIGURE 11.16. Uniform and heterogeneous training data. On the left, the histogram of the training data has a single peak, indicating a degree of spectral homogeneity. Data from such training fields form a suitable basis for image classification. On the right, a second set of training data displays a bimodal histogram that reveals that this area encompasses two, rather than one, spectral classes. This training area is not satisfactory for image classification and must be discarded or redefined.

### *Significance of Training Data*

Scholz et al. (1979) and Hixson et al. (1980) discovered that selection of training data may be as important as, or even more important than, choice of classification algorithm in determining classification accuracies of agricultural areas in the central United States. They concluded that differences in the selection of training data were more important influences on accuracy than were differences among some five different classification procedures.

The results of this study show little difference in the classification accuracies achieved by the five classification algorithms that were considered if the same training statistics were used. However, in one part of the study, a classification algorithm given two alternative training methods for the same data produced significantly different results. This finding suggests that the choice of training method, in at least some instances, is as important as the choice of classifier. Scholz et al. (1979, p. 4) concluded that the most important aspect of training is that all cover types in the scene must be adequately represented by a sufficient number of samples in each spectral subclass.

A study by Campbell (1981) examined the character of the training data as it influences accuracy of the classification. His examples showed that adjacent pixels within training fields tended to have similar values; as a result, the samples that compose each training field may not be independent samples of the properties within a given category. Training samples collected in contiguous blocks may tend to underestimate the variability within each class and overestimate the distinctness of categories. His examples also show that the degree of similarity varies between land cover categories from band to band and from date to date. If training samples are selected randomly within classes, rather than as blocks of contiguous pixels, effects of high similarity are minimized and classification accuracies improve. Also, his results suggest that it is probably better to use a rather large number of small training fields rather than a few large areas.

### *Idealized Sequence for Selecting Training Data*

Specific circumstances for conducting supervised classification vary greatly, so it is not possible to discuss in detail the procedures to follow in selecting training data, which will be determined in part by equipment and software available at a given facility. However, it is possible to outline an idealized sequence as a suggestion of the key steps in the selection and evaluation of training data.

1. Assemble information, including maps and aerial photographs of the region to be mapped. Conduct field studies as necessary to acquire firsthand information of the region to be studied.
2. Conduct field studies, to acquire firsthand information regarding the area to be studied. The amount of effort devoted to field studies varies depending on the analyst's familiarity with the region to be studied. If the analyst is intimately familiar with the region and has access to up-to-date maps and photographs, additional field observations may not be necessary. However, in the vast majority of situations fieldwork will be required; analysts typically overestimate their knowledge of a specific region.

3. Carefully plan collection of field observations, with a route designed to observe all regions of the study area. Maps and images should be taken into the field in a form that permits the analyst to annotate them as observations are made in the field (e.g., images may be prepared with overlays; photocopies of maps can be marked with colored pens). Although time and access may be limited, it is important to observe all classes of terrain encountered within the study area, as well as all regions. Observations should not be limited to a few easily observed segments of the area. The analyst should keep good notes, carefully keyed to annotations on the image, and may find it useful to take photographs as a permanent record of conditions observed. In very remote areas, aerial observation may be the only practical means of observing the study region. Ideally, field observations should be timed to coincide with image acquisition or, or when this is not possible, should at least be made during the season images were acquired.

4. Conduct a preliminary examination of the digital scene. Determine landmarks that may be useful in positioning training fields. Assess image quality. Examine frequency histograms of the data and determine whether preprocessing is necessary.

5. Identify prospective training areas, using guidelines proposed by Joyce (1978) and outlined here. Sizes of prospective areas must be assessed in the light of scale differences between maps or photographs and the digital image. Locations of training areas must be defined with respect to features easily recognizable on the image and on the maps and photographs used as collateral information.

6. Display the digital image and then locate and delineate training areas on the digital image. Be sure to place training area boundaries well inside parcel boundaries to avoid including mixed pixels within training areas. At the completion of this step, all training areas should be identified with respect to row and column coordinates within the image.

7. For each training field, display and inspect frequency histograms of all spectral bands to be used in the classification. If possible, examine means, variance, divergence measures, covariances, and so on, to assess the usefulness of training data.

8. Modify boundaries of the training field to eliminate bimodal frequency distributions; if necessary, discard those areas that are not suitable. If necessary, return to step 1 to define new areas to replace those that have been eliminated.

9. Incorporate training data information into a form suitable for use in the classification procedure and proceed with the classification process as described in subsequent sections of this chapter.

## *Specific Methods for Supervised Classification*

A variety of different methods have been devised to implement the basic strategy of supervised classification. All use information derived from the training data as a means of classifying those pixels not assigned to training fields. The following sections outline only a few of the many methods of supervised classification.

### PARALLELEPIPED CLASSIFICATION

*Parallelepiped classification,* sometimes also known as *box decision rule,* or *level slice procedures,* is based on the ranges of values within the training data to define regions

TABLE 11.4. Data for Example Shown in Figure 11.17

| | Group A | | | | Group B | | | |
|---|---|---|---|---|---|---|---|---|
| | Band | | | | Band | | | |
| | 4 | 5 | 6 | 7 | 4 | 5 | 6 | 7 |
| | 34 | 28 | 22 | 3 | 28 | 18 | 59 | 35 |
| | 36 | 35 | 24 | 6 | 28 | 21 | 57 | 34 |
| | 36 | 28 | 22 | 6 | 28 | 21 | 57 | 30 |
| | 36 | 31 | 23 | 5 | 28 | 14 | 59 | 35 |
| | 36 | 34 | 25 | 7 | 30 | 18 | 62 | 28 |
| | 36 | 31 | 21 | 6 | 30 | 18 | 62 | 38 |
| | 35 | 30 | 18 | 6 | 28 | 16 | 62 | 36 |
| | 36 | 33 | 24 | 2 | 30 | 22 | 59 | 37 |
| | 36 | 36 | 27 | 10 | 27 | 16 | 56 | 34 |
| High | 34 | 28 | 18 | 10 | 27 | 14 | 56 | 28 |
| Low | 36 | 36 | 27 | 3 | 30 | 22 | 62 | 38 |

*Note.* These data have been selected from a larger data set to illustrate parallelepiped classification.

within a multidimensional data space. The spectral values of unclassified pixels are projected into data space; those that fall within the regions defined by the training data are assigned to the appropriate categories.

An example can be formed from data presented in Table 11.4. Here Landsat MSS bands 5 and 7 are selected from a larger data set to provide a concise, easily illustrated example. In practice, four or more bands can be used. The ranges of values with respect to band 5 can be plotted on the horizontal axis in Figure 11.17. The extremes of values in band 7 training data are plotted on the vertical axis, then projected to intersect with the ranges from band 5. The polygons thus defined (Figure 11.17) represent regions in data

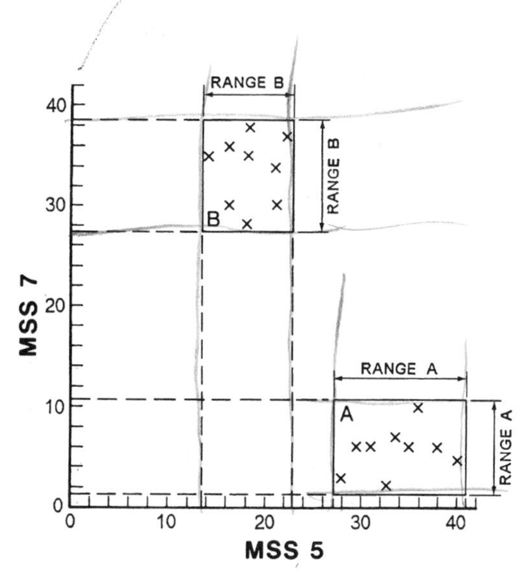


**FIGURE 11.17.** Parallelepiped classification. Ranges of values within training data (Table 11.4) define decision boundaries. Here only two bands are shown, but the principle extends to several spectral bands. Other pixels, not from the training fields, are classified as a given category if their positions fall within the polygons defined by the training data.

space that are assigned to categories in the classification. As pixels of unknown identity are considered for classification, those that fall within these regions are assigned to the category associated with each polygon, as derived from the training data. The procedure can be extended to as many bands, or as many categories, as necessary. In addition, the decision boundaries can be defined by the standard deviations of the values within the training areas rather than their ranges. This kind of strategy is useful because fewer pixels will be placed in an "unclassified" category (a special problem for parallelepiped classification), but it also increases the opportunity for classes to overlap in spectral data space.

Although this procedure for classification has the advantages of accuracy, directness, and simplicity, some of its disadvantages are obvious. Spectral regions for informational categories may intersect. Training data may underestimate actual ranges of classification and leave large areas in data space and on the image, unassigned to informational categories. Also, the regions as defined in data space are not uniformly occupied by pixels in each category; those pixels near the edges of class boundaries may belong to other classes. Also, if training data do not encompass the complete range of values encountered in the image (as is frequently the case), large areas of the image remain unclassified, or the basic procedure described here must be modified to assign these pixels to logical classes.

This strategy was among the first used in the classification of Landsat data and is still used, although it may not always be the most effective choice for image classification.

### MINIMUM DISTANCE CLASSIFICATION

Another approach to classification uses the central values of the spectral data that form the training data as a means of assigning pixels to informational categories. The spectral data from training fields can be plotted in multidimensional data space in the same manner illustrated previously for unsupervised classification. Values in several bands determine the positions of each pixel within the clusters that are formed by training data for each category (Figure 11.18). These clusters may appear to be the same as those defined earlier for unsupervised classification. However, in unsupervised classification, these clusters of pixels were defined according to the "natural" structure of the data. Now, for

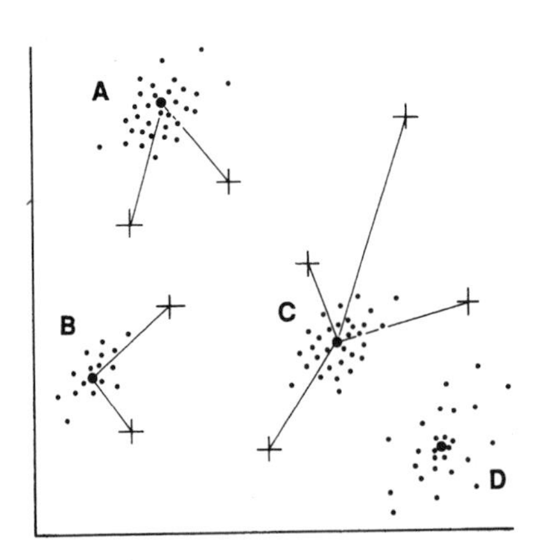


**FIGURE 11.18.** Minimum distance classifier. Here small dots represent pixels from training fields, and the large unclassified pixels from elsewhere in the image are represented by the crosses; each of these pixels is assigned to the class with the closest centroid, as measured using the distance measures discussed in the text.

supervised classification, these groups are formed by values of pixels within the training fields defined by the analyst.

Each cluster can be represented by its centroid, often defined as its mean value. As unassigned pixels are considered for assignment to one of the several classes, the multidimensional distance to each cluster centroid is calculated and the pixel is then assigned to the closest cluster. Thus, the classification proceeds by always using the "minimum distance" from a given pixel to a cluster centroid defined by the training data as the spectral manifestation of an informational class.

Minimum distance classifiers are direct in concept and in implementation but are not widely used in remote sensing work. In its simplest form, minimum distance classification is not always accurate; there is no provision for accommodating differences in variability of classes, and some classes may overlap at their edges. It is possible to devise more sophisticated versions of the basic approach just outlined by using different distance measures and different methods of defining cluster centroids.

### ISODATA

The ISODATA classifier (Duda & Hart, 1973) can be considered a variation on the minimum distance method; however, it produces results that are often considered superior to those derived from the basic minimum distance approach. ISODATA is often seen as a form of supervised classification, although it differs appreciably from the classical stereotype of supervised classification presented at the beginning of this chapter. It is a good example of a technique that shares characteristics of both supervised and unsupervised methods (i.e., "hybrid" classification) and forms evidence that the distinction between the two approaches is not as clear as idealized descriptions imply.

ISODATA starts with the training data selected as previously described; these data can be envisioned as clusters in multidimensional data space. All unassigned pixels are then assigned to the nearest centroid. Thus far, the approach is the same as described previously for the minimum distance classifier. Then, new centroids are found for each group, and the process of allocating pixels to the closest centroid is repeated if the centroid changes in position. The process is repeated until there is no change, or only small change, in class centroids from one iteration to the next. A step-by-step description follows:

1. Choose initial estimates of class means. These can be derived from training data; in this respect, ISODATA resembles supervised classification.
2. Assign all other pixels in the scene to the class with the closest mean, as considered in multidimensional data space.
3. Recompute class means to include effects of those pixels that may have been reassigned in step 2.
4. If any class mean changes in value from step 2 to step 3, return to step 2 and repeat the assignment of pixels to the closest centroid. Otherwise, the result at the end of step 3 is final.

Steps 2, 3, and 4 use the methodology of unsupervised classification, although the requirement for training data identifies the method essentially as supervised classification. It could probably best be considered a hybrid rather than a clear example of either approach.

### MAXIMUM LIKELIHOOD CLASSIFICATION

In nature, the classes that we classify exhibit natural variation in their spectral patterns, and further variability is added by the effects of haze, topographic shadowing, system noise, and mixed pixels. As a result, remote sensing images seldom record spectrally pure classes; typically, they display a range of brightnesses in each band. The classification strategies considered thus far do not consider variation that may be present within spectral categories and do not address problems that arise when frequency distributions of spectral values from separate categories overlap. For example, for application of a parallelepiped classifier, the overlap of classes is a serious problem because spectral data space cannot then be neatly divided into discrete units for classification. This kind of situation arises frequently because often our attention is on classifying those pixels that tend to be spectrally similar rather than those that are distinct enough to be easily and accurately classified by other classifiers.

As a result, the situation depicted in Figure 11.19 is common. Assume that we examine a digital image representing a region composed of three fourths forested land and one fourth cropland. The two classes, *forest* and *cropland,* are distinct with respect to average brightness, but extreme values (very bright forest pixels, or very dark crop pixels) are similar in the region in which the two frequency distributions overlap. (For clarity, Figure 11.19 shows data for only a single spectral band, although the principle extends to values observed in several bands and to more than the two classes shown here.) Brightness value "45" falls into the region of overlap, where we cannot make a clear assignment to either "forest," or "crop." Using the kinds of decision rules mentioned above, we cannot decide which group should receive these pixels unless we place the decision boundary arbitrarily.

In this situation, an effective classification would consider the relative likelihoods of "45 as a member of forest" and "45 as a member of crop." We could then choose the class that would maximize the probability of a correct classification given the information in the training data. This kind of strategy is known as *maximum likelihood* classification—it uses the training data as a means of estimating means and variances of the classes, which

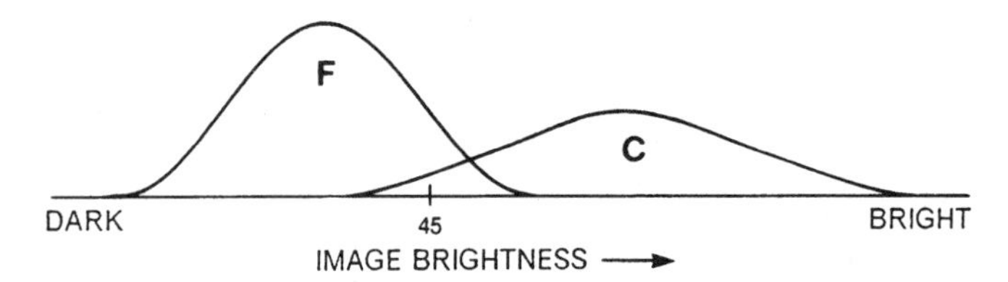


**FIGURE 11.19.** Maximum likelihood classification. These frequency distributions represent pixels from two training fields; the zone of overlap depicts pixel values common to both categories. The relation of the values within the overlap region to the overall frequency distribution for each class forms the basis for assigning pixels to classes.

are then used to estimate the probabilities. Maximum likelihood classification considers not only the mean, or average, values in assigning classification but also the variability of brightness values in each class.

The maximum likelihood decision rule, implemented quantitatively to consider several classes and several spectral channels simultaneously, forms a powerful classification technique. It requires intensive calculations so it has the disadvantage of requiring more computer resources than do most of the simpler techniques mentioned above. Also, it is sensitive to variations in the quality of training data, even more so than most other supervised techniques. Computation of the estimated probabilities is based on the assumption that both training data and the classes themselves display multivariate normal (Gaussian) frequency distributions. (This is one reason that training data should exhibit unimodal distributions, as discussed above.) Data from remotely sensed images often do not strictly adhere to this rule, although the departures are often small enough that the usefulness of the procedure is preserved. Nonetheless, training data that are not carefully selected may introduce error.

### *Bayes's Classification*

This classification problem can be expressed more formally by stating that we wish to estimate the "probability of forest, given that we have an observed digital value 45," and the "probability of crop, given that we have an observed digital value 45." These questions are a form of conditional probabilities, written as "$P(F|45)$," and "$P(C|45)$," and read as "the probability of encountering category forest, given that digital value 45 has been observed at a pixel," and "the probability of encountering category crop, given that digital value 45 has been observed at a pixel." That is, they state the probability of one occurrence (finding a given category at a pixel) given that another event has already occurred (the observation of digital value 45 at that same pixel). Whereas estimation of the probabilities of encountering the two categories at random (without a conditional constraint) is straightforward (here $P(F) = 0.50$, and $P(C) = 0.50$, as mentioned above), conditional probabilities are based on two separate events. From our knowledge of the two categories as estimated from our training data, we can estimate $P(45|F)$ ("the probability of encountering digital value 45, given that we have category forest"), and $P(45|C)$ ("the probability of encountering digital value 45, given that we have category crop"). For this example, $P(45|F) = 0.75$, and $P(45|C) = 0.25$.

However, what we want to know are values for probabilities of "forest, given that we observed digital value 45" [$P(F|45)$] and "crop, given that we have observed digital value 45" [$P(C|45)$], so that we can compare them to choose the most likely class for the pixel. These probabilities cannot be found directly from the training data, and from a purely intuitive examination of the problem, there would seem to be no way to estimate these probabilities.

But, in fact, there is a way to estimate $P(F|45)$ and $P(C|45)$ from the information at hand. Thomas Bayes (1702–1761) defined the relationship between the unknowns $P(F|45)$ and $P(C|45)$, and the known $P(F)$, $P(C)$, $P(45|F)$, and $P(45|C)$. His relationship, now known as *Bayes's law*, is expressed as follows for our example:

$$P(\mathrm{F}|45) = \frac{P(\mathrm{F})\,P(45|\mathrm{F})}{P(\mathrm{F})P(45|\mathrm{F}) + P(\mathrm{C})P(45|\mathrm{C})} \qquad \text{(Eq. 11.5)}$$

$$P(\mathrm{C}|45) = \frac{P(\mathrm{C})P(45|\mathrm{C})}{P(\mathrm{C})P(45|\mathrm{C}) + P(\mathrm{F})P(45|\mathrm{F})} \qquad \text{(Eq. 11.6)}$$

In a more general form, Bayes's rule can be written:

$$P(b_1|a_1) = \frac{P(b_1)\,P(a|b_1)}{P(b_1)P(a_1|b_1) + P(b_2)P(a_1|b_2) + \ldots} \qquad \text{(Eq. 11.7)}$$

where $a_1$ and $a_2$ represent alternative results of the first stage of the experiment, and where $b_1$ and $b_2$ represent alternative results for the second stage.

For our example, Bayes's theorem can be applied as follows:

$$P(\mathrm{F}|45) = \frac{P(\mathrm{F})P(45|\mathrm{F})}{P(\mathrm{F})P(45|\mathrm{F}) + P(\mathrm{C})P(45|\mathrm{C})} \qquad \text{(Eq. 11.8)}$$

$$= \frac{\frac{1}{2} \times \frac{3}{4}}{(\frac{1}{2} \times \frac{3}{4}) + (\frac{1}{2} \times \frac{1}{4})} = \frac{\frac{3}{8}}{\frac{4}{8}} = \frac{3}{4}$$

$$P(\mathrm{C}|45) = \frac{P(\mathrm{C})P(45|\mathrm{C})}{P(\mathrm{C})P(45|\mathrm{C}) + P(\mathrm{F})P(45|\mathrm{F})}$$

$$= \frac{\frac{1}{2} \times \frac{1}{4}}{(\frac{1}{2} \times \frac{1}{4}) + (\frac{1}{2} \times \frac{3}{4})} = \frac{\frac{1}{8}}{\frac{4}{8}} = \frac{1}{4}$$

So we conclude that this pixel is more likely to be forest than crop. Usually data for several spectral channels are considered, and usually we wish to choose from more than two categories, so this example is greatly simplified. We can extend this procedure to as many bands, or as many categories, as may be necessary, although the expressions become more complex than can be discussed here.

For remote sensing classification, application of Bayes's rule is especially effective when classes are indistinct or overlap in spectral data space. It can also form a convenient vehicle for incorporating ancillary data (Chapter 17) into the classification, as the added information can be expressed as a conditional probability. In addition, it can provide a means of introducing costs of misclassification into the analysis. (Perhaps an error in misassignment of a pixel to forest is more serious than a misassignment to crop.) Furthermore, we can combine Bayes's rule with other classification procedures, so, for example, most of the pixels can be assigned using a parallelepiped classifier, and a Baysean classifier can be used for those pixels that are not within the decision boundaries or within a region of overlap. Some studies have shown that such classifiers are very accurate (Story et al., 1984).

Thus, Bayes's theorem is an extremely powerful means of using information at hand to estimate probabilities of outcomes related to the occurrence of preceding events. The weak point of the Bayesean approach to classification is the selection of the training data. If the probabilities are accurate, Bayes's strategy must give an effective assignment of observations to classes. Of course, from a purely computational point of view, the procedure will give an answer with any values. But to make accurate classification, the training data must have a sound relationship to the categories they represent. For the multidimensional case, with several spectral bands, it is necessary to estimate for each category a mean brightness for each band and a variance–covariance matrix to summarize the variability of each band and its relationships with other bands. From these data we extrapolate to estimate means, variances, and covariances of entire classes. Usually this extrapolation is made on the basis that the data are characterized by multivariate normal frequency distributions. If such assumptions are not justified, the classification results may not be accurate.

If classes and subclasses have been judiciously selected, and if the training data are accurate, Bayes's approach to classification should be as effective as any that can be applied. If the classes are poorly defined, and the training data are not representative of the classes to be mapped, the results can be no better than those for other classifiers applied under similar circumstances. Use of Bayes's approach to classification forms a powerful strategy because it includes information concerning the relative diversities of classes as well as the means and ranges used in the previous classification strategies. The simple example used here is based on only a single channel of data and on a choice between only two classes. The same approach can be extended to consider several bands of data and several sets of categories.

This approach to classification is extremely useful and flexible and, under certain conditions, provides what is probably the most effective means of classification given the constraints of supervised classification. Note, however, that in most applications this strategy is limited by the quality of the estimates of the probabilities required for the classification; if these are accurate, the results can provide optimal classification; if they are makeshift values conjured up simply to provide numbers for computation, the results may have serious defects.

### ECHO

Kettig and Landgrebe (1975) describe a classification approach known as ECHO (*extraction and classification of homogeneous objects*). The ECHO approach is not as widely used as the other methods discussed here, but it illustrates an important classification strategy. ECHO classifies a digital image into fields of spectrally similar pixels before the pixels are assigned to categories. Classification is then conducted using the fields, rather than individual pixels, as the features to be classified.

ECHO searches for neighboring pixels that are spectrally similar and enlarges these groups to include adjacent pixels that have spectral values that resemble those of the core group. For example, the algorithm can first search for neighborhoods of four contiguous pixels (Figure 11.20a). For each group, it then tests members for homogeneity, perhaps using distance measures analogous to those described above. The degree of similarity required for group membership can be set by the analyst, as can other variables that guide the formation of groups. Pixels that are not similar to their neighbors are rejected from the group under the assumption that they are border pixels or belong to very small parcels of distinct categories. These pixels are classified separately using the traditional point classifiers.

Each of the homogeneous patches is then compared to each of its neighbors, and if

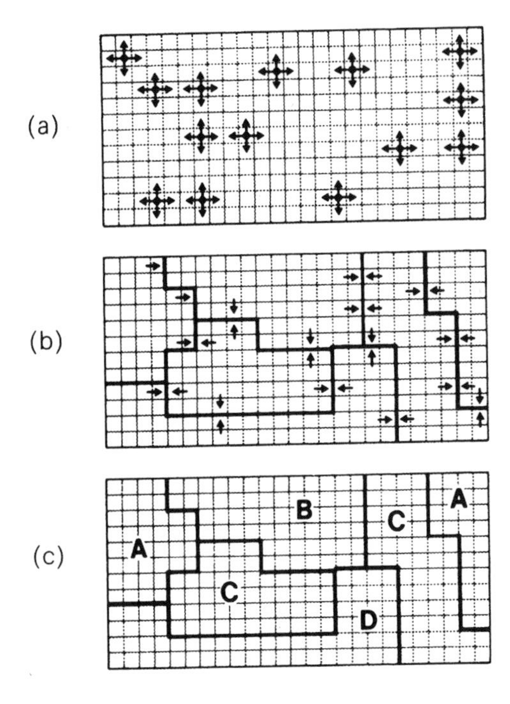


**FIGURE 11.20.** Schematic representation of the process used by ECHO. (*a*) ECHO grows uniform regions by searching outwards from groups of pixels throughout the image. (For clarity only a few groups are depicted.) (*b*) Growing stops when region edges encounter edges of other regions; the image is now segmented but not classified. (*c*) Regions defined in (*b*) are now classified using usual classification strategies. This diagram does not show the regions that might occur as mixed pixels are isolated from the main groups.

similar patches border each other, they are merged to form a larger patch. Patches are allowed to grow until they meet the edges of contrasting patches; when all patches reach their maximum extents within the constraints defined by the operator, the growing process stops (Figure 11.20b). At this point, the image has been segmented into distinct regions but none have been assigned to categories. Thus, the next step is to find the average values within each region and use these values as the basis for classification as in the usual image classification process. The difference is that the units of classification are not individual pixels but the regions defined in the earlier steps (Figure 11.20c). For the classification step, Kettig and Landgrebe apply a maximum likelihood classifier, but other approaches could also be applied, if desired, as both supervised and unsupervised versions have been written.

ECHO is a good example of a classifier that operates on fields of pixels rather than on each pixel in isolation. However, it performs the classification on the average brightness of each patch, so it does not attempt to use image texture, as mentioned at the beginning of this chapter and discussed in the following section. Textural classifiers attempt to use the variation of values within neighborhoods, rather than just the average value, as the basis for classification. ECHO is reported to be an efficient and accurate classifier when used with training site data.

### *Idealized Sequence for Conducting Supervised Classification*

The practice of supervised classification is considerably more complex than unsupervised classification. The analyst must evaluate the situation at each of several steps and return to an earlier point if it appears that refinements or corrections are necessary to ensure an accurate result.

1. *Preparation of the mean of categories to be mapped.* Ideally these categories correspond to those of interest to the final users of the maps and data. But, these requirements may not be clearly defined, or the user may require the assistance of the analyst to prepare a suitable plan for image classification.

2. *Selection and definition of training data as outlined above.* This step may be the most expensive and time-consuming phase of the project and may be conducted in several stages as the analyst attempts to define a satisfactory set of training data. The analyst must outline the limits of each training site within the image, usually by using a mouse or trackball to define a polygon on a computer screen that displays a portion of the image to be classified.

3. *Modification of categories and training fields as necessary to define homogeneous training data.* As the analyst examines the training data, it may be necessary to delete some categories from the menu, combine others, or subdivide still others into spectral subclasses. If training data for the modified categories meet the requirements of size and homogeneity, they can be used to enter the next step of the classification. Otherwise, the procedure (at least for certain categories) must start again at step 1.

4. *Conduct classification.* Each image analysis system will require a different series of commands to conduct a classification, but in essence the analyst must provide the program with access to the training data (often written to a specific computer file) and must identify the image that is to be examined. The results can be displayed on a video display terminal.

5. *Evaluation of classification performance.* Finally, the analyst must conduct an evaluation using the procedures discussed in Chapter 13.

This sequence outlines the basic steps required for supervised classification; details may vary from individual to individual and with different image processing systems, but the basic principles and concerns remain the same—accurate definition of homogeneous training fields.

## 11.5. Textural Classifiers

Image classification presents special problems for digital analysis because of the diverse spectral characteristics of the landscape. Idealized landscape regions consist of spectrally homogeneous patches on the earth's surface—accurate mapping of such regions could be accomplished by the relatively straightforward process of matching spectral categories to the spectral "signatures" of informational categories, as described above.

Actual landscape regions are, of course, usually formed by assemblages of spectrally diverse features. Low-density residential land, for example, as viewed from above in detail, is composed largely of tree crowns, rooftops, lawns, and paved streets, driveways, and parking lots. We are interested in classification of the composite of these many features rather than an inventory of the many components that in themselves may be of little interest. Thus, in the ideal, the classification should focus on the overall pattern of variation that characterizes each category rather than on the average brightness, which may not reveal much about the essential differences between categories. Although human interpreters can intuitively recognize such complex patterns, many digital classification algorithms encounter serious problems in accurate classification of such scenes because they are designed to classify each separate spectral region as a separate informational category.

Textural classifiers attempt to measure image texture—distinctive spatial and spectral relationships between neighboring pixels. For example, the standard deviation of brightness values within a neighborhood of specified size, systematically moved over the entire image, may provide a rough measure of the spectral variability over short distances as a measure of image texture. Such a measure may permit the analyst to classify such composite categories as the one mentioned above.

Usually more sophisticated measures of texture are required to produce satisfactory results. For example, other textural measures examine relationships between brightness values at varying distances and directions from a central pixel, which is systematically moved over the image (Haralick et al., 1973; Haralick, 1979).

Jensen (1979) found the use of textural measures improved his classification of Level II and III suburban and transitional land using band 5 of the Landsat MSS. Improvements were confined to certain categories and were perhaps of marginal significance when considered in the context of increased costs.

Textural measures seem to work best when relatively large neighborhoods are defined (perhaps as large as 64 × 64 pixels). Such large neighborhoods may cause problems when they straddle boundaries between categories. In addition, such large neighborhoods may decrease the effective spatial resolution of the final map, as they must form the smallest spatial elements on the final map.

## 11.6. Ancillary Data

Ancillary data refer to the use of non-image information as an aid for classification of spectral data (Tom et al., 1978). In the process of manual interpretation of images, there is a long tradition of both implicit and explicit use of such information, including data from maps, photographs, field observations, reports, and personal experience. For digital analysis, ancillary data often consist of data available in formats consistent with the digital spectral data, or in forms that can be conveniently transformed into usable formats. Examples include digital elevation data or digitized soil maps (Anuta, 1976).

Ancillary data can be used in either of two ways. They can be "added to" or "superimposed over" the spectral data to form a single multiband image; the ancillary data are treated simply as additional channels of data. Or, the analysis can proceed in two steps using a layered classification strategy (as described below); the spectral data are classified in the first step; then the ancillary data form the basis for reclassification and refinement of the initial results.

A serious limitation to practical application of digital ancillary data is the problem of incompatibility with the remotely sensed data. Physical compatibility (matching digital formats, etc.) is, of course, one such problem of obvious practical significance, but logical compatibility may form a more subtle but equally important problem. Seldom are ancillary data collected for use with a specific remote sensing problem; usually they are derived from archived data collected for another purpose as a means of reducing costs in time and money. For example, digital terrain data gathered by the U.S. Geological Survey and the Defense Mapping Agency for topographic quadrangles are frequently used as ancillary data for remote sensing studies; seldom if ever could remote sensing projects absorb the costs of digitizing, then editing and correcting, these data for specific use in a single study. One consequence of this practice is that the ancillary data are seldom compatible with the remotely sensed data with respect to scale, resolution, date, and accuracy. Some differences can be minimized by preprocessing the ancillary data to reduce effects of different measurement scales, resolutions, and so on. In other situations, unresolved incompatibilities, possibly quite subtle, may detract from the potential effectiveness of the ancillary data.

Choice of ancillary variables may be critical. In the mountainous regions of the western United States, elevation data have been effective as ancillary data for mapping vegetation patterns with digital MSS data, due in part to the large ranges in local elevation and the close associations of vegetation distributions with elevation, slope, and aspect. In other settings in which elevation differences may have more subtle influences on vegetation distributions, elevation data may not form effective ancillary variables. Although some scientists advocate use of all available ancillary data, in the hope of deriving whatever advantage might be possible, common sense would seem to favor careful selection of those variables with conceptual and practical significance to the mapped distributions.

## 11.7. Layered Classification

Layered classification refers to use of a hierarchical process in which two or more steps form the basis for classification. The usual single-step classification algorithm uses all available information in a single process to produce the entire menu of informational categories. Layered classification uses subsets of data in a series of separate steps, presumably by applying each form of information in its most effective context. The hierarchical structure permits the most difficult classification decisions to be made in a context that isolates the problem categories from others and focuses the most effective variables on that classification decision.

For example, Jensen (1979) devised a layered classification strategy that separated vegetated and nonvegetated regions early in the classification by using vegetation indices (analogous to those discussed by Tucker, 1979), as a measure of extent of living biomass with each pixel. This basic distinction separated built-up and heavily vegetated regions at an early point in the classification, preventing confusion between categories later in the classification process.

Layered classification can be useful only if the classification logic can be structured in a way that minimizes errors at the upper levels of the decision tree. If errors are made at the first stage, they are carried to lower levels and will appear in the final product regardless of the soundness of subsequent decisions.

## 11.8. Fuzzy Clustering

Fuzzy clustering addresses a problem implicit in much of the preceding material—pixels must be assigned to a single discrete class. Although such classification attempts to maxi-

**TABLE 11.5. Partial Membership in Fuzzy Classes**

| Class | Pixel | | | | | | |
|---|---|---|---|---|---|---|---|
| | A | B | C | D | E | F | G |
| Water | 0.00 | 0.00 | 0.00 | 0.00 | 0.00 | 0.00 | 0.00 |
| Urban | 0.00 | 0.01 | 0.00 | 0.00 | 0.00 | 0.00 | 0.85 |
| Transportation | 0.00 | 0.35 | 0.00 | 0.00 | 0.99 | 0.79 | 0.14 |
| Forest | 0.07 | 0.00 | 0.78 | 0.98 | 0.00 | 0.00 | 0.00 |
| Pasture | 0.00 | 0.33 | 0.21 | 0.02 | 0.00 | 0.05 | 0.00 |
| Cropland | 0.92 | 0.30 | 0.00 | 0.00 | 0.00 | 0.15 | 0.00 |

mize correct classifications, the logical framework allows only for direct one-to-one matches between pixels and classes. We know, however, that many processes contribute to prevent clear matches between classes and pixels, as noted in Chapter 9, and by the work of Robinove (1981) and Richards and Kelly (1984). Therefore, even though errors may be minimized, the effort to define a discrete match between the model and reality ensures that many pixels will be in error, at least to a degree.

Fuzzy logic (Kosko & Isaka, 1993) has applications in many fields but special significance for remote sensing. Fuzzy logic permits partial membership, a property that is especially significant in the field of remote sensing as it translates closely to the problem of mixed pixels (Chapter 9). Whereas traditional classifiers must label pixels as either forest or water, for example, a fuzzy classifier is permitted to assign a pixel a membership grade of 0.3 for water and 0.7 for forest in recognition that the pixel may not be properly assigned to a single class. Membership grades typical vary from 0 (non membership) to 1.0 (full membership), with intermediate values signifying partial membership in one or more other classes (Table 11.5).

A fuzzy classifier assigns membership to pixels based on a membership function (Figure 11.21). Membership functions for classes are determined either by general rela-

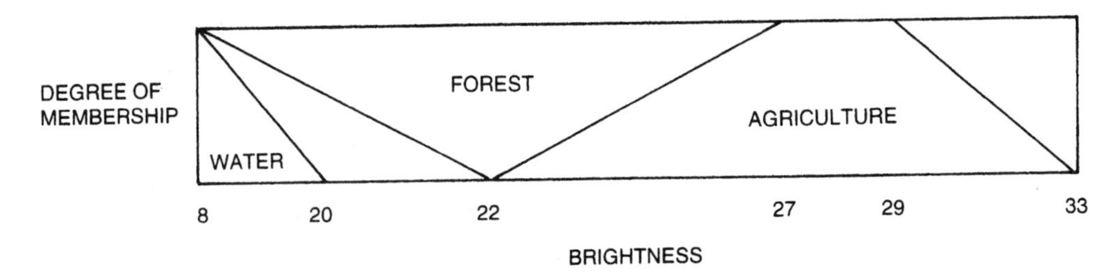


**FIGURE 11.21.** Membership functions for fuzzy clustering. This example illustrates membership functions for the simple instance of three classes considered for a single band, although the method can be applied to multiple bands. The horizontal axis represents pixel brightness; the vertical axis represents degree of membership, from low at the bottom to high near the top. The class "water" consists of pixels darker than brightness 20, although only pixels darker than 8 are likely to be completely covered by water. The class "agriculture" can include pixels as dark as 22 and as bright as 33, although pure agriculture pixels are found only in the range 27 to 29. A pixel of brightness 28, for example, can be only "agriculture," although a pixel of brightness 24 could be partially forested, partially in agriculture. Unlabeled areas on this diagram are not occupied by any of the classes in this classification.

TABLE 11.6. "Hardened" Classes for Example Shown in Table 11.5

| | Pixel | | | | | | |
|---|---|---|---|---|---|---|---|
| Class | A | B | C | D | E | F | G |
| Water | 0.00 | 0.00 | 0.00 | 0.00 | 0.00 | 0.00 | 0.00 |
| Urban | 0.00 | 0.00 | 0.00 | 0.00 | 0.00 | 0.00 | 1.00 |
| Transportation | 0.00 | 1.00 | 0.00 | 0.00 | 1.00 | 1.00 | 0.00 |
| Forest | 0.00 | 0.00 | 1.00 | 1.00 | 0.00 | 0.00 | 0.00 |
| Pasture | 0.00 | 0.00 | 0.00 | 0.00 | 0.00 | 0.00 | 0.00 |
| Cropland | 1.00 | 0.00 | 0.00 | 0.00 | 0.00 | 0.00 | 0.00 |

tionships or by definitional rules, describing the relationships between data and classes. Or, as is more likely in the instance of remote sensing classification, membership functions are derived from experimental (i.e., training) data for each specific scene to be examined. In the instance of remote sensing data, a membership function describes the relationship between class membership and brightness in several spectral bands (Figure 11.21).

Figure 11.21 provides a simple example showing membership grades. (Actual output from a fuzzy classification is likely to form an image that shows varied levels of membership for specific classes.) Membership grades can be hardened (Table 11.6) by setting the highest class to 1.0, and all others to 0.0. Hardened classes are equivalent to traditional classifications—each pixel is labeled with a single label and the output is a single image labeled with the identify of the hardened class. Programs designed for remote sensing applications (Bezdek et al., 1984) provide the ability to adjust the degree of fuzziness, and thereby the structures of classes and the degree of continuity, in the classification pattern.

Fuzzy clustering has been judged to improve results, at least marginally, with respect to traditional classifiers, although the evaluation is difficult because the usual methods require the discrete logic of traditional classifiers. Thus, the improvements noted for hardened classifications are probably conservative as they do not reveal the full power of fuzzy logic.

## 11.9. Artificial Neural Networks

*Artificial neural networks* (ANNs) are computer programs designed to simulate human learning processes through establishment and reinforcement of linkages between input data and output data. It is these linkages, or pathways, that form the analogy with the human learning process in that repeated associations between input and output in the training process reinforce linkages that can then be employed to link input and output, in the absence of training data.

ANNs are often represented as composed of three elements. An *input layer* consists of the source data, which in the context of remote sensing is the multispectral observations, perhaps in several bands and dates. ANNs are designed to work with large volumes of data, including perhaps many bands and dates of multispectral observations, together with related ancillary data. The *output layer* consists of the classes required by the analyst. There are few restrictions on the nature of the output layer, although the process will be most reliable when the number of output labels is small, or modest, with respect to the number of input channels. Included are training data in which the association between output labels and input data is clearly established. During the training phase, an ANN establishes association between input and output data by establishment of weights within one or more *hidden layers* (Figure 11.22). In the context of remote sensing, repeated associations between classes and digital values, as expressed in the training data, strengthen weights within hidden layers that permit the ANN to assign correct labels when given spectral values in the absence of training data.

Further, ANNs can also be trained by *back propagation* (BP). If establishment of the usual training data for conventional image classification can be thought of as "forward propagation," BP can be thought of as a retrospective examination of the links between input and output data in which differences between expected and actual results can be used to adjust weights. This process establishes *transfer functions*, quantitative relationships between input and output layers that assign weights to emphasize effective links between input and output layers. (For example, such weights might acknowledge that some band combinations may be very effective in defining certain classes and others effective for other classes.) In BP, hidden layers note errors in matching data to classes and adjusts the weights to minimize errors.

ANNs are designed using less severe statistical assumptions than many of the usual classifiers (e.g., maximum likelihood), especially those based on the assumption of normally distributed data (such as the maximum likelihood classifier). In practice, successful application requires careful application. ANNs have been found to be accurate in the classification of remotely sensed data, although improvements in accuracies have generally been small or modest. ANNs require considerable effort to train and use computer time intensively.

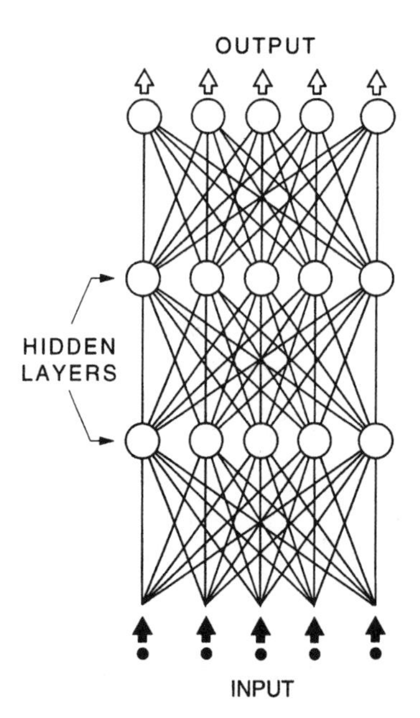


**FIGURE 11.22.** Artificial neural net. Input data are represented as the dots at the bottom of the diagram. Within the classification program several hidden layers define relationships between input data and output. Some pathways connecting input data and outcomes are preferentially selected by repeated use as the system classifies the training data. As data from outside training areas are examined, the system uses those pathways most effective in classifying training data to classify the remaining data in the scene.

## 11.10. Contextual Classification

Contextual information is derived from spatial relationships among pixels within a given image. Whereas texture usually refers to spatial interrelationships among unclassified pixels within a window of specified size, context is determined by positional relationships between pixels, either classified or unclassified, anywhere within the scene (Swain, 1984; Gurney & Townshend, 1983).

Although contextural classifiers can operate on either classified or unclassified data, it is convenient to assume that some initial processing has assigned a set of preliminary classes on a pixel-by-pixel basis without using spatial information. The function of the contextual classifier is then to operate on the preliminary classification to reassign pixels as appropriate in the light of contextual information.

Context can be defined in several ways, as illustrated in Figure 11.23a. In each instance the problem is to consider the classification of a pixel, or a set of pixels (represented by the shaded pattern), using information concerning the classes of other, related pixels. Several kinds of links define the relationships between the two groups. The simplest link is distance. Perhaps the unclassified pixels are agricultural land, which is likely to be "irrigated cropland" if positioned within a certain distance of a body of open water; if the distance to water exceeds a certain threshold, the area might be more likely to be assigned to rangeland or unirrigated cropland. The second example in Figure 11.23 illustrates the use of both distance and direction. Contiguity (Figure 11.23c) may be an important classification aid. For example, in urban regions specific land use regions may be found primarily in locations adjacent to a specific category. Finally (Figure 11.23d), specific categories may be characterized by their positions within other categories.

Contextual classifiers are efforts to simulate some of the higher-order interpretation

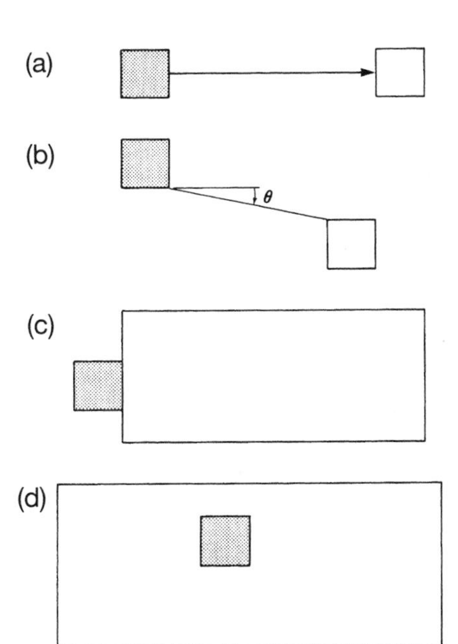


**FIGURE 11.23.** Contextual classification from Gurney and Townshend (1983). The shaded regions depict pixels to be classified; the open regions represent other pixels used in the classification decision. (*a*) Distance is considered. (*b*) Direction is considered. (*c*) Contiguity forms a part of the classification process. (*d*) Inclusion is considered. Copyright 1983 by the American Society for Photogrammetry and Remote Sensing. Reproduced by permission.

processes used by human interpreters in which the identity of an image region is derived in part from its location in relation to other regions of specified identity. For example, a human interpreter considers sizes and shapes of parcels in identifying land use, as well as the identities of neighboring parcels. The characteristic spatial arrangement of the central business district, industrial, residential, and agricultural land in an urban region permits the interpreter to identify parcels that might be indistinct, if considered with conventional classifiers.

Contextual classifiers can also operate on data to reclassify erroneously classified pixels, or to reclassify isolated pixels (perhaps correctly classified) that form regions so small and so isolated that they are of little interest to the user. Some uses may be considered essentially cosmetic operations, but they may be helpful in editing the results for final presentation.

## 11.11. Relative Accuracy of Classification Procedures

Although this chapter discusses only a few of the many classification strategies available, it should be clear that the analyst has many choices as the classification process is considered. For the individual analyst, the range of choices is limited by the specific algorithms available at a given laboratory. Today, however, almost all image processing facilities provide a selection of several classification procedures. Some of the larger systems may have as many as 8 to 12 classification procedures. Sometimes the choice may be determined by convenience, cost, or speed, but usually the analyst chooses the most accurate classification approach.

Although individual analysts have accumulated considerable experience in applications of alternative classification techniques, there are only a few systematic investigations of relative classification accuracies. Typically, studies of accuracy examine the same scene using different classification strategies while holding other factors (e.g., training data) constant. Presumably, the differences in classification accuracy (measured using the methods described in Chapter 13) can be attributed to differences in the classification procedure.

Scholz et al. (1979) and Hixson et al. (1980) concluded that selection of training data had a stronger influence on accuracy than did choice of classification algorithm. It should be noted that they examined only a few classification procedures applied to data from a rather restricted set of landscapes—mainly agricultural regions of the central United States—that may not represent typical conditions for many applications of image classification. Story et al. (1984) studied accuracies of a different selection of classification strategies as applied to a much more complex land cover pattern near Roanoke, Virginia. Their results, although preliminary in nature, revealed that choice of classification procedure did influence accuracy of their classifications. A parallelepiped classifier provided high accuracy but did not classify all pixels because of the basic limitations of the method, as described above. The most accurate classification of the entire scene was achieved by an approach that combined a parallelepiped classifier with a Bayes's classification for those pixels that did not fall within the decision boundaries defined by the parallelepiped.

The study of the effectiveness of different classification strategies has just st
there is little that can be stated with confidence. Yet it does seem clear that it is
that a single classification strategy will be "best" for all purposes. Rather, certain approaches may be more effective for certain classes of landscapes whereas others may be more effective in different settings. It will require extensive research to define what these specific combinations of landscape and classifier might be.

## 11.12. Summary

This chapter has described a few specific classifiers as a means of introducing the student to the variety of classification strategies available today. Possibly the student may have the opportunity to use some of these procedures, so these descriptions may form the first step in a more detailed learning process. It is more likely, however, that the student who uses this text will never use many of the classifiers described here. Nonetheless, those procedures that are available for student use are likely to be based on the same principles outlined here using specific classifiers as examples. Therefore, this chapter should not be regarded as a complete catalogue of image classification methods but rather as an effort to illustrate some of the primary methods of image classification. Specific details and methods will vary greatly, but if the student has mastered the basic strategies and methods of image classification, it will be easy to see the unfamiliar methods as variations on the fundamental approaches described here.

## Review Questions

1. This chapter mentions only a few of the many strategies available for image classification. Why have so many different methods been developed? Why not use just one?

2. Why might the decision to use or not to use preprocessing be especially significant for image classification?

3. Image classification is not necessarily equally useful for all fields. For a subject of interest to you (geology, forestry, etc.) evaluate the significance of image classification by citing examples of how classification might be used. Also list some subjects for which image classification might be more or less useful.

4. Review Chapters 6 and 11. Speculate on the course of further developments in image classification. Can you suggest relationships between sensor technology and the design of image classification strategies?

5. The table on page 352 lists land cover classes for the pixels given in Tables 11.2 and 11.3. (Note that individual pixels do not correspond for the two dates.) Select a few pixels from Table 11.3; conduct a rudimentary feature selection "by inspection" (i.e., by selecting from the set of four bands). Can you convey most of the ability to distinguish between classes by choosing a subset of the four MSS bands?

| | Feb. | May | | Feb. | May | | Feb. | May | | Feb. | May |
|---|---|---|---|---|---|---|---|---|---|---|---|
| (1) | F | W | (11) | P | C | (21) | F | C | (31) | F | C |
| (2) | F | F | (12) | F | C | (22) | P | P | (32) | W | F |
| (3) | C | W | (13) | W | P | (23) | W | P | (33) | P | F |
| (4) | P | C | (14) | F | P | (24) | W | W | (34) | F | W |
| (5) | P | F | (15) | F | C | (25) | P | W | (35) | W | C |
| (6) | C | W | (16) | F | W | (26) | P | F | (36) | F | C |
| (7) | F | F | (17) | P | F | (27) | F | F | (37) | W | F |
| (8) | F | P | (18) | C | W | (28) | C | F | (38) | P | C |
| (9) | C | F | (19) | F | W | (29) | C | P | (39) | F | P |
| (10) | W | C | (20) | F | F | (30) | P | W | (40) | C | F |

F, forest; C, crop; P, pasture; W, water

6. Using the information given above and in Table 11.3, calculate normalized differences between water and forest, bands 5 and 7. (For this question, and those that follow, the student should have access to a calculator.) Identify those classes easiest to separate, and the bands likely to be most useful.

7. For both February and May, calculate normalized differences between forest and pasture, band 7. Give reasons to account for differences between the results for the two dates.

8. The results for questions 5 through 7 illustrate that normalized difference values vary greatly, depending upon data, bands, and classes considered. Discuss some of the reasons why this is true, both in general and for specific classes in this examples.

9. Refer to Table 11.1. Calculate Euclidean distance between means of the four classes given in the table. Again, explain why there are differences from date to date and from band to band.

## References

### General


Anuta, P. E. 1976. Digital Registration of Topographic Data and Satellite MSS Data for Augmented Spectral Analysis. In *Proceedings, 42nd Annual Meeting, American Society of Photogrammetry*. Falls Church, VA: American Society of Photogrammetry, pp. 180–187.

Bryant, J. 1979. On the Clustering of Multidimensional Pictorial Data. *Pattern Recognition*, Vol. 11, pp. 115–125.

Duda, R. O., and P. E. Hart. 1973. *Pattern Classification and Scene Analysis*. New York: Wiley, 482 pp.

Gurney, C. M., and J. R. Townshend. 1983. The Use of Contextual Information in the Classification of Remotely Sensed Data. *Photogrammetric Engineering and Remote Sensing*, Vol. 49, pp. 55–64.

Hord, Richard. 1982. *Digital Image Processing of Remotely Sensed Data*. New York: Academic Press, 221 pp.

Jensen, J. R., and D. L. Toll. 1982. Detecting Residential Land Use Development at the Urban Fringe. *Photogrammetric Engineering and Remote Sensing*, Vol. 48, pp. 629–643.

Kettig, R. L., and D. A. Landgrebe. 1975. Classification of Multispectral Image Data by Extraction and Classification of Homogeneous Objects. In *Proceedings, Symposium on Machine Classification of Remotely Sensed Data*. West Lafayette, IN: Laboratory for Applications in Remote Sensing, pp. 2A-1–2A-11.

Moik, Johannes G. 1980. *Digital Processing of Remotely Sensed Images*. National Aeronautics and Space Administration Special Publication 431. Washington, DC: U.S. Government Printing Office, 330 pp.

Richards, J. A., and D. J. Kelly. 1984. On the Concept of the Spectral Class. *International Journal of Remote Sensing,* Vol. 5, pp. 987–991.

Robinove, Charles J. 1981. The Logic of Multispectral Classification and Mapping of Land. *Remote Sensing of Environment,* Vol. 11, pp. 231–244.

Schowengerdt, Robert A. 1983. *Techniques for Image Processing and Classification in Remote Sensing.* New York: Academic Press, 272 pp.

Strahler, Alan H. 1980. The Use of Prior Probabilities in Maximum Likelihood Classification of Remotely Sensed Data. *Remote Sensing of Environment,* Vol. 10, pp. 135–163.

Swain, Philip H. 1984. Advanced Computer Interpretation Techniques For Earth Data Information Systems. In *Proceedings of the Ninth Annual William H. Pecora Remote Sensing Symposium.* Silver Spring, MD: IEEE, pp. 184–189.

Swain, P. H., and S. M. Davis (eds.). 1978. *Remote Sensing: The Quantitative Approach.* New York: McGraw-Hill, 396 pp.

Todd, W. J., D. G. Gehring, and J. F. Haman. 1980. Landsat Wildland Mapping Accuracy. *Photogrammetric Engineering and Remote Sensing,* Vol. 46, pp. 509–520.

Tom, C., L. D. Miller, and J. W. Christenson. 1978. *Spatial Land-Use Inventory, Modeling, and Projection: Denver Metropolitan Area, with Inputs from Existing Maps, Airphotos, and Landsat Imagery.* NASA Technical Memorandum 79710. Greenbelt MD: Goddard Space Flight Center, 225 pp.

Tucker, C. J. 1979. Red and Photographic Infrared Linear Combinations for Monitoring Vegetation. *Remote Sensing of Environment*, Vol. 8, pp. 127–150.

### Training Data

Campbell, J. B. 1981. Spatial Correlation Effects upon Accuracy of Supervised Classification of Land Cover. *Photogrammetric Engineering and Remote Sensing,* Vol. 47, pp. 355–363.

Joyce, A. T. 1978. *Procedures for Gathering Ground Truth Information for a Supervised Approach to Computer-Implemented Land Cover Classification of Landsat-Acquired Multispectral Scanner Data.* NASA Reference Publication 1015. Houston, TX: National Aeronautics and Space Administration, 43 pp.

McAffrey, T. M., and Steven F. Frankin. 1993. Automated Training Site Selection for Large-Area Remote-Sensing Image Analysis. *Computers and Geosciences,* Vol. 10, pp. 1413–1428.

### Image Texture

Haralick, R. M. et al. 1973. Textural Features for Image Classification. *IEEE Transactions on Systems, Man, and Cybernetics,* SMC-3, pp. 610–622.

Haralick, R. M. 1979. Statistical and Structural Approaches to Texture. *Proceedings of the IEEE,* Vol. 67, pp. 786–804.

Jensen, J. R. 1979. Spectral and Textural Features to Classify Elusive Land Cover at the Urban Fringe. *The Professional Geographer,* Vol. 31, pp. 400–409.

### Comparisons of Classification Techniques

Chen, K. S., Y. C. Tzeng, C. F. Chen, and W. L. Kao. 1995. Land-Cover Classification of Multispectral Imagery Using a Dynamic Learning Neural Network. *Photogrammetric Engineering and Remote Sensing,* Vol. 61, pp. 403–408.

Gong, Peng, and P. J. Howarth. 1992. Frequency-Based Contextual Classification and Gray-Level Vector Reduction for Land-Use Identification. *Photogrammetric Engineering and Remote Sensing,* Vol. 58, pp. 423–437.

Hixson, M., D. Scholz, and N. Fuhs. 1980. Evaluation of Several Schemes for Classification of Remotely Sensed Data. *Photogrammetric Engineering and Remote Sensing,* Vol. 46, pp. 1547–1553.

Scholz, D., N. Fuhs, and M. Hixson. 1979. An Evaluation of Several Different Classification Schemes,

Their Parameters, and Performance. In *Proceedings, Thirteenth International Symposium on Remote Sensing of the Environment.* Ann Arbor: University of Michigan, pp. 1143–1149.

Story, Michael H., J. B. Campbell, and G. Best. 1984. An Evaluation of the Accuracies of Five Algorithms for Machine Classification of Remotely Sensed Data. In *Proceedings of the Ninth Annual William T. Pecora Remote Sensing Symposium.* Silver Spring, MD: IEEE, pp. 399–405.

**Fuzzy Clustering**

Bezdek, J. C., R. Ehrilich, W. Full. 1984. FCM: The Fuzzy c-Means Clustering Algorithm. *Computers and Geosciences,* Vol. 10, pp. 191–203.

Fisher, P. F., and S. Pathirana. 1990. The Evaluation of Fuzzy Membership of Land Cover Classes in the Suburban Zone. *Remote Sensing of Environment,* Vol. 34, pp. 121–132.

Kent, J. T., and K. V. Marida. 1988. Spatial Classification Using Fuzzy Membership Models. *IEEE Transactions on Pattern Analysis and Machine Intelligence,* Vol. 10, 659–671.

Kosko, Bart, and S. Isaka. 1993. Fuzzy Logic. *Scientific American,* Vol. 271, pp. 76–81.

Wang, Fangju. 1990a. Improving Remote Sensing Image Analysis through Fuzzy Information Representation. *Photogrammetric Engineering and Remote Sensing,* Vol. 56, pp. 1163–1169.

Wang, Fangju. 1990b. Fuzzy Supervised Classification of Remote Sensing Images. *IEEE Transactions on Geoscience and Remote Sensing,* Vol. 28, pp. 194–201.

**Artificial Neural Networks**

Bischof, H. W., W. Schneider, and A. J. Pinz. 1992. Multispectral Classification of Landsat Images Using Neural Networks. *IEEE Transactions on Geoscience and Remote Sensing,* Vol. 28, pp. 482–489.

Chen, K. S., Y. C. Tzeng, C. F. Chen, and W. L. Kao. 1995. Land-Cover Classification of multispectral imagery Using a Dynamic Learning Neural Network. *Photogrammetric Engineering and Remote Sensing,* Vol. 61, pp. 403–408.

Miller, D. M., E. D. Kaminsky, and S. Rana. 1995. Neural Network Classification of Remote-Sensing Data. *Computers and Geosciences,* Vol. 21, pp. 377–386.

CHAPTER TWELVE

# Field Data

## 12.1. Introduction

Every application of remote sensing must apply, if only implicitly, field observations in one form or another. Each analyst must define relationships between image data and conditions at corresponding points on the ground. Although use of field data appears, at first glance, to be self-evident, in fact there are numerous practical and conceptual difficulties to be encountered and solved in the course of even the most routine work (Steven, 1987).

Field data consist of observations collected at or near ground level in support of remote sensing analysis. Although it is a truism that characteristics of field data must be tailored for the specific study at hand, the inconvenience and expense of collecting accurate field data often lead analysts to apply field data collected for one purpose to other, perhaps different, objectives. Therefore, a key concern is not only to acquire field data but to acquire data suitable for a specific task. Accurate field data permit the analyst to match points or areas on imagery to corresponding regions on the ground surface and thereby establish with confidence relationships between the image and conditions on the ground. In their simplest form, field data may simply permit an analyst to identify a specific region as *forest.* In more detailed circumstances, field data may identify a region as a specific form of forest or specify height, volume, leaf area index, photosynthetic activity, or other properties according to specific purposes of a study.

## 12.2. Kinds of Field Data

Field data typically serve one of three purposes. They can be used to verify, evaluate, or assess the results of remote sensing investigations (Chapter 13). A second function is to provide reliable data to guide the analytical process, such as training fields to support supervised classification (Chapter 11). This distinction introduces the basic difference between the realms of qualitative (nominal labels) and quantitative analyses, discussed later in this chapter. Lastly, field data can provide information used to model the spectral behavior of specific landscape features (e.g., plants, soils, or water bodies). For example, analyses of the spectral properties of forest canopies can be based on quantitative models of the behavior of radiation within the canopy—models that require detailed, precise, and accurate field observations (Treitz et al., 1992).

Despite its significance, sound procedures for field data collection have not been discussed in a manner that permits others to discover the principles that underlie them. Discussions of field data collection procedures typically include some combination of the following components:

- Data to record ground-level, or low-level, spectral data for specific features and estimates of atmospheric effects (e.g., Lintz et al., 1976; Curtis & Goetz, 1994; Deering, 1989).
- Broad guidelines for collection of data for use with specific kinds of imagery (e.g., Lintz et al., 1976).
- Discipline-specific guidelines for collection and organization of data (e.g., Roy et al., 1991; Congalton & Biging, 1992).
- Guidelines for collection of data for use with specific digital analyses (Joyce, 1978; Foody, 1990).
- Highlighting of special problems arising from methodologies developed in unfamiliar ecological settings (Wikkramatileke, 1959; Campbell & Browder, 1995).
- Evaluation of relative merits of quantitative biophysical data in relation to qualitative designations (Treitz et al., 1992).

The work by Joyce (1978) probably comes the closest to defining field data collection as a process to be integrated into a study's analytical plan. He outlined procedures for systematic collection of field data for digital classification for multispectral satellite data. Although his work focused on specific procedures of analysis of Landsat MSS data, he implicitly stressed the need for the data collection plan to be fully compatible with the resolution of the sensor, spatial scale of the landscape, and the kinds of digital analyses to be employed. Further, preparation of a specific document devoted to field data collection in itself emphasized the significance of the topic.

Field data must include at least three kinds of information. The first consists of attributes or measurements that describe ground conditions at a specific place. Examples might include identification of a specific crop or land use or measurement of soil moisture content. Second, these observations must be linked to locational information so the attributes can be correctly matched to corresponding points in image data. Third, observations must also be described with respect to the time and date. These three elements—attributes, location, time—form the minimum information for useful field data. Complete field data also include such other information as records of weather and illumination, identities of the persons who collected the data, calibration information for instruments, and other components as required for specific projects.

## 12.3. Nominal Data

Nominal labels consist of qualitative designations applied to regions delineated on imagery that convey basic differences from adjacent regions. Simple examples include *forest, urban land, turbid water,* and so on. Despite their apparent rudimentary character, accurate and precise nominal labels convey significant information. Those that are most precise—*evergreen forest, single-family residential housing, maize,* and *winter wheat,* for example—convey more information than broader designations and therefore are often more valuable.

Nominal labels originate from several alternative sources. One is an established classification system such as that proposed by Anderson et al. (1976) (Chapter 19) and comparable systems established in other disciplines. These offer advantages of acceptance by other workers, comparability with other studies, established definitions, and defined relationships between classes. In other instances, classes and labels have origins in local terminology or in circumstances that are specific to a given study. These may have the advantage of serving well the immediate purposes of the study, but they limit comparison with other studies. Collection of data in the field is usually expensive and inconvenient, so there are incentives to use a set of field data for as many purposes as possible; this effect means that it is often difficult to anticipate the ultimate application of a specific set of data. Therefore, ad hoc classifications that are unique to a specific application often prove unsatisfactory if the scope of the study expands beyond its initial purpose. Nonetheless, analysts must work to address the immediate issues without attempting to anticipate every potential use for their data—the objective should be to maintain flexibility rather than to cover every possibility.

In the field, nominal data are usually easy to collect at points or for small areas; difficulties arise as one attempts to apply the labeling system to larger areas. For these reasons, it is usually convenient to annotate maps or aerial photographs in the field as a means of relating isolated point observations to areal units. As the system is applied to broader regions, requiring the work of several observers, it becomes more important to train observers in the consistent application of the system and to evaluate the results to detect inconsistencies and errors at the earliest opportunity. Because physical characteristics of some classes vary seasonally, it is important to ensure that the timing of field observations match those of the imagery to be examined.

## 12.4. Documentation of Nominal Data

The purpose of field data is to permit reconstruction, in as much detail as possible, of ground conditions at the place and time that imagery was acquired. Therefore, field data require careful and complete documentation as they may be required later to answer questions that had not been anticipated at the time they were collected. Nominal data can be recorded as field sketches or annotated aerial photographs or maps (Figure 12.1). The inherent variability of landscapes within nominal labels means that careful documentation of field data forms an important component for the field data collection plan. Field observations should be collected using standardized procedures designed specifically for each project to ensure that uniform data are collected at each site (Figure 12.2). Reliable documentation includes careful notes and sketches and ground photographs, sometimes including Polaroid prints and videotapes (Figure 12.3). A log must be kept to identify photographs in relation to field sites and to record dates, times, weather, orientations, and related information. Photographs must be cross-indexed to field sketches and notes. In some instances, small-format aerial photographs might be useful if feasible within the constraints of the project (p. 84).

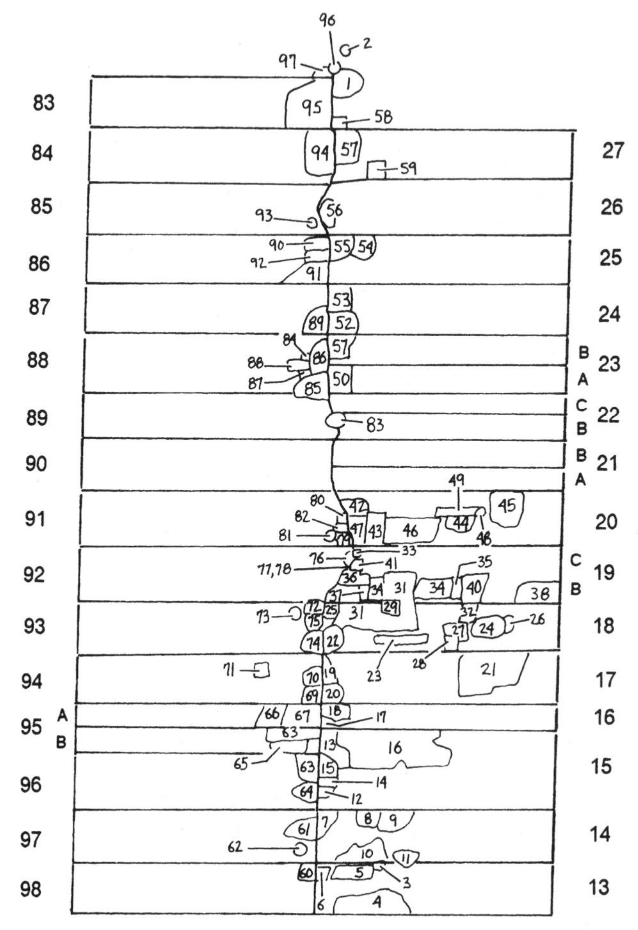


**FIGURE 12.1.** Record of field sites. The rectangular units represent land holdings within the study area in Brazil studied by Campbell and Browder (1995). Each colonist occupied a 0.5 km × 2.0 km strip (numbered at the left and right). Within each colonist's lot, the irregular parcels outlined and numbered here show areas visited by field teams which prepared notes and photographs. See Figure 12.2. Copyright 1995 by the International Society for Remote Sensing. Reproduced by permission.

## 12.5. Biophysical Data

Biophysical data consist of measurements of physical characteristics collected in the field describing, for example, the type, size, form, and spacing of plants that form the vegetative cover. Or, as another example, biophysical data might record the texture and mineralogy of the soil surface. The exact character of biophysical data collected for a specific study depends on the purposes of the study, but typical data might include such characteristics as leaf area index (LAI), biomass, soil texture, soil moisture. Many such measurements vary over time, so they often must be coordinated with image acquisition; in any event, careful records must be made of time, date, location, and weather.

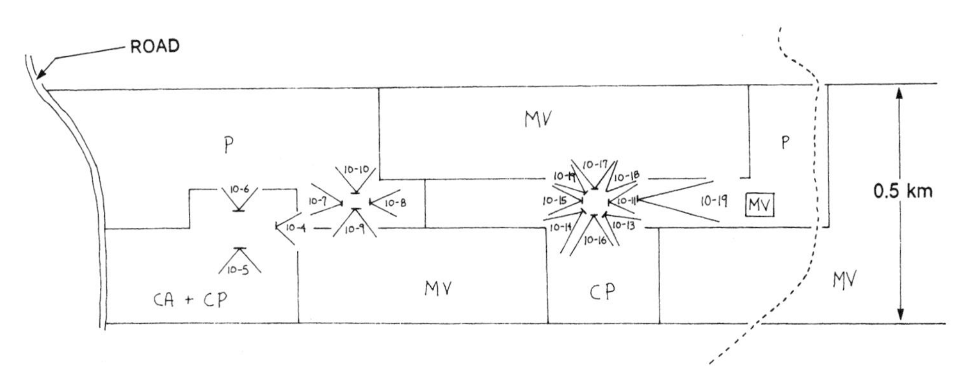


**FIGURE 12.2.** Field sketch illustrating collection of data for nominal labels. The V-shaped symbols represent the fields of view of photographs documenting land cover within cadastral units such as those outlined in Figure 12.1, keyed to notes and maps not shown here. From Campbell and Browder (1995). Copyright 1995 by the International Society for Remote Sensing. Reproduced by permission

GROUND DATA INVENTORY FORM  Data record 7 / 12 / and 7 / 15 / 96

Kansas, Morton County

| Field # | Acreage | Land use Crop Code | | | Irrigated | Fertilized | Planting date Month/Day | |
|---|---|---|---|---|---|---|---|---|
| 1 | 160.4 | 7 | 0 | 0 | YES | NO | | WELLSITE SOUTH SIDE |
| 2 | 163.3 | 7 | 0 | 0 | | NO | | |
| 3 | 158.8 | 4 | 0 | 4 | NO | NO | 9-20-95 | HARVESTED |
| 4 | 77.7 | 4 | 0 | 2 | NO | NO | 9-7-95 | GRAZED - STUBBLED PLOWED |
| 5 | 81.9 | 7 | 0 | 0 | YES | NO | | WELLSITE EAST SIDE |
| 6 | 19.6 | 7 | 0 | 0 | | | | |
| 7 | 72.4 | 4 | 0 | 2 | NO | NO | SEPT 95 | HARVESTED |
| 8 | 65.2 | 7 | 0 | 0 | | | | |
| 9 | 153.7 | 4 | 0 | 2 | NO | NO | 9-16-95 | HARVESTED - YIELD 14 BL? - SPRAYED |
| 10 | 156.6 | 7 | 0 | 0 | | | | |
| 11 | 154.6 | 7 | 0 | 0 | | | | WELLSITE IN SW 40 AC |
| 12 | 152.8 | 4 | 0 | 0 | NO | NO | SEPT 95 | STUBBLE BEING PLOWED |
| 13 | 156.8 | 7 | 0 | 0 | NO | NO | SEPT 95 | STUBBLE BEING PLOWED - WELLSITE E SIDE |
| 14 | 73.0 | 4 | 0 | 2 | | | | |
| 15 | 89.4 | 4 | 0 | 0 | NO | YES | 9-15-95 | STUBBLE PLOWED |
| 16 | 162.8 | 4 | 0 | 0 | NO | NO | SEPT 95 | GRAZED JAN-APRIL 96; RESIDUE PLOWED |
| 17 | 120.0 | 5 | 0 | 0 | | | SEPT 95 | DESTROYED BY GREENBUGS; RESIDUE PLOWED |
| 18 | 35.3 | 7 | 0 | 0 | | | | WELLSITE NW CORNER |
| 19 | 163.2 | 6 | 1 | 6 | | | SEPT 95 | HARVESTED - YIELD 16 |
| 20 | 94.8 | 7 | 0 | 0 | YES | YES | 5-28-96 | SPRAYED IN JULY |

**FIGURE 12.3.** Log recording field obsevations documenting agricultural land use.

Biophysical data typically apply to points, so often they must be linked to areas by averaging of values from several observations within an area. Further, biophysical data must often be associated with nominal labels, so they do not replace nominal data but rather document the meaning of nominal labels. For example, biophysical data often document the biomass or structure of vegetation within a nominal class rather than completely replacing a nominal label. In their assessment of field data recording forest sites, Congalton and Biging (1992) found that purely visual assessment provided accurate records of species present, but that plot and field measurements were required for accurate assessment of dominant-size classes.

## 12.6. Field Radiometry

Radiometric data permit the analyst to relate brightnesses recorded by the aerial sensor to corresponding brightnesses near the ground surface. A field spectrometer consists of a measuring unit with a hand-held probe connected to the measuring unit by a fiber-optic cable (Figure 12.4). The measuring unit consists of an array of photosensitive detectors with filters or diffraction gratings to separate radiation into several spectral regions. Radiation received by the probe can therefore be separated into separate spectral regions and then projected onto detectors similar to those discussed in Chapter 4. Brightnesses can usually be presented as either radiances or reflectances. Some instruments have the ability to match their spectral sensitivities to such specific sensors as SPOT, MSS, or TM. The results can be displayed as spectral plots or as an array of data. Many units interface with notebook computers, which permits the analyst to record spectra in the field and write data to mini disks, which can then be transferred to office or laboratory computers for analysis or incorporation into data bases. Typically, a field spectrometer fits within a portable case designed to withstand inclement weather and other rigors of use in the field.

Analysts can select from several designs for probes with differing fields of view or for immersion in liquids (e.g., to record radiation transmitted by water bodies). Some-

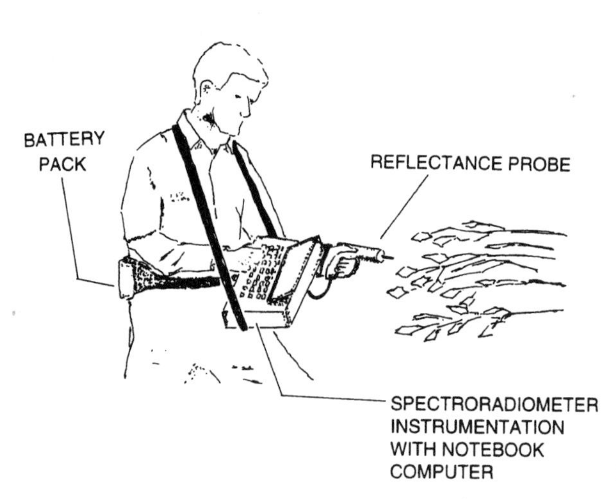


**FIGURE 12.4.** Field radiometry.

FIGURE 12.5 Boom truck equipped for radiometric measurements in the field.

times measuring units are on truck-mounted booms that can be used to acquire overhead views that simulate the viewing perspective of airborne or satellite sensors (Figure 12.5).

Field spectrometers must be used carefully with specific attention devoted to the field of view in relation to the features to be observed, background surfaces, direction of solar illumination, and the character of diffuse light from nearby objects (Figure 12.6). Diffuse light (Chapter 2) is influenced by surrounding features, so measurements should be acquired at sites well clear of features of contrasting radiometric properties. Likewise, the instrument may record indirect radiation reflected from the operator's clothing, so special attention must be devoted to orientation of the probe with respect to surrounding objects.

The most useful measurements are those coordinated with acquisition of aircraft or

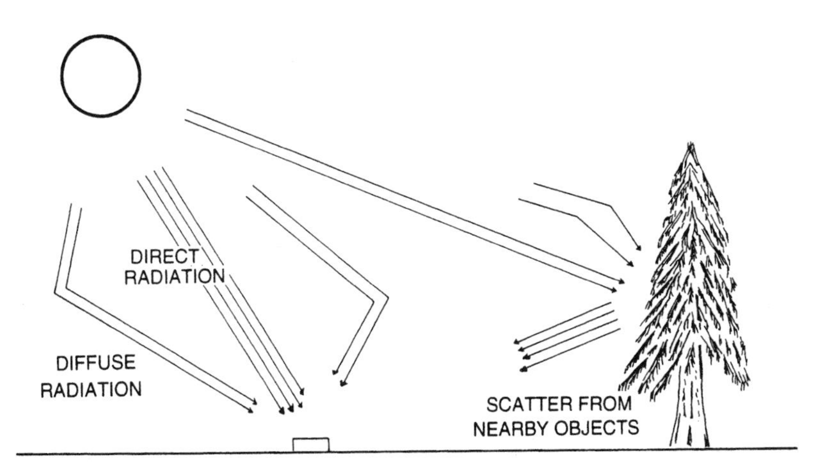


FIGURE 12.6. Careful use of the field radiometer with respect to nearby objects. Redrawn from Curtis and Goetz (1994).

satellite data. Operators must carefully record the circumstances of each set of measurements to include time, date, weather, conditions of illumination, and location. Some instruments permit the analyst to enter such information through the laptop's keyboard as annotations to the radiometric data. As data are collected, the operator should supplement spectral data with photographs or videos to permit clear identification of the region represented by the radiometric observations.

In the field it is often difficult to visualize the relationship between ground features visible at a given location and their representations in multispectral data. Not only is it sometimes difficult to relate specific points in the field to the areas visible in overhead views, but often it is equally difficult to visualize differences between representations in visible and nonvisible regions of the spectrum and at the coarse resolutions of some sensors. Therefore, analysts must devote effort to matching point observations to the spectral classes that will be of interest during image analysis. Robinove (1981) and Richards and Kelley (1984) emphasize difficulties in defining uniform spectral classes.

## 12.7. Locational Information

Locational data permit attributes or measurements gathered in the field to be matched to imagery of the same region. Locations of distinctive landmarks, visible both in the field and within the image data, permit the analyst to clearly match the image with the field data.

Traditionally, many of the principal problems in collection of field data for remote sensing analysis have focused on the difficulty of acquiring locational information that permits field observations to be matched to corresponding points on imagery. Accurate observations have little use unless they can be placed in their correct locations with confidence. Because of the difficulty and inaccuracies in deriving absolute locational information (e.g., latitude and longitude, or comparable coordinates) in the field, most field data have been positioned with reference to distinctive landmarks (e.g., water bodies, road intersections, and topographic features) visible on imagery. Such methods can introduce positional errors that negate effort invested in accurate measurement of other qualities. In well-mapped regions of developed nations, it is often possible to acquire accurate maps to use in positioning field data, but in unmapped or poorly mapped regions, the absence of reliable locational information has formed a major difficulty for the use of field observations.

### *Global Positioning Systems*

In recent years the availability of global positioning system (GPS) technology has permitted convenient, inexpensive, and accurate measurement of absolute location. GPSs have greatly enhanced the usefulness of remote sensing data, especially when it is necessary to integrate image data with field data. These instruments are now inexpensive and easy to use and can be employed in almost any area on the earth's surface. GPS was developed over many years by U.S. military services for military use, but is available for general use worldwide.

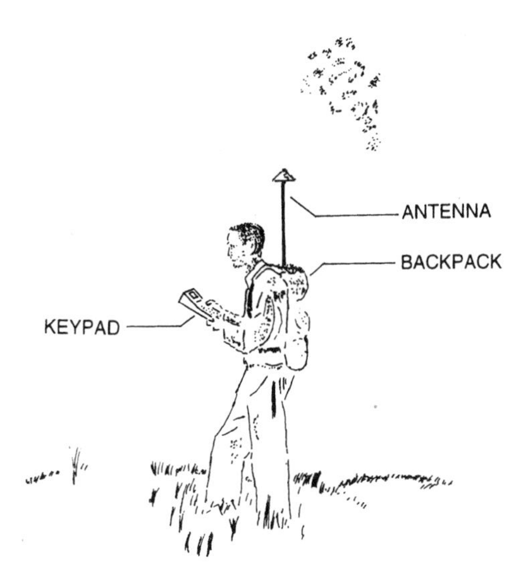


**FIGURE 12.7.** Field portable GPS unit.

A GPS receiver consists of a portable receiving unit (Figure 12.7) sensitive to signals transmitted by a network of earth-orbiting satellites. These satellites are positioned in orbits such that each point on the earth's surface will be in view of at least four, and perhaps as many as nine, satellites at any given time. Today's GPS satellites have evolved from designs developed in the early 1970s. A system of 24 satellites (21 in service, plus 3 spares) is positioned at an altitude of about 11,000 mi. (about 17,600 km), to circle the earth at intervals of 12 hours, spaced in six orbital planes to provide complete coverage of the earth's surface (Figure 12.8).

These satellites continuously broadcast one-way signals at two frequencies within the L-band region of the microwave spectrum (Chapter 7). These signals permit GPS re-

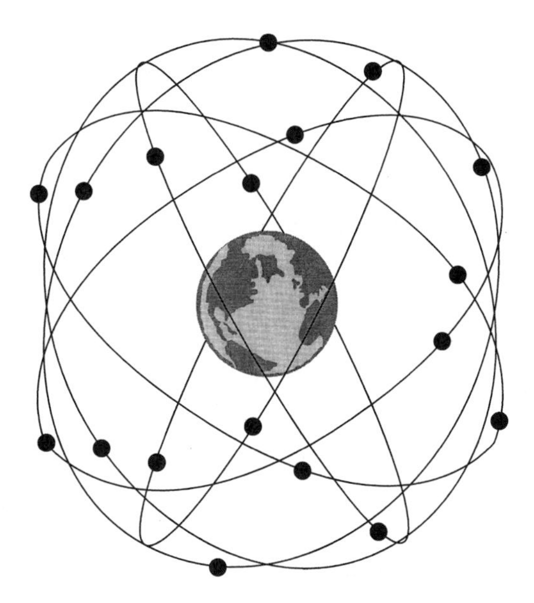

**FIGURE 12.8.** GPS satellites.

ceivers to solve equations to estimate latitude, longitude, and elevation. Although at ground level these signals are weak, they are designed so that they can be detected even under adverse conditions (such as severe weather or interference from other signals). Each of these signals varies in a manner that both identifies the satellite that broadcasts the signal and gives the exact time that the signal was broadcast. A receiver therefore can calculate the time delay for the signal to travel from a specific satellite and accurately estimates the distance from the receiver to a specific satellite. Both L-band signals are modulated to carry a precision P-code; one of the two bands, however, also carries a coarse acquisition (C/A) code that provides less precise information.

One reason that a weak signal is adequate is that the time and identification information each satellite transmits is very simple, and the receiver can listen for long periods to acquire it accurately. Because a receiver is always within range of several satellites, it is possible to combine positional information from two or more satellites to accurately estimate geographic position on the earth's surface. A network of ground stations periodically recomputes and uploads new positional data to the GPS satellites.

### *GPS Receivers*

A GPS receiver consists of an antenna, power supply, electronic clock, and circuitry that can translate the signal into positional information. The receiver typically contains a keypad and a small internal digital computer (Figure 12.7) and may be connected to larger laptop or notebook models. These permit the analyst to enter coding to identify features and to download information to laboratory or office computers for use with image processing systems. Depending on expense and sophistication, receivers can be as small as a portable telephone or as large as several suitcases. Those GPS receivers likely to be used for remote sensing fieldwork might be contained within a small backpack, with an antenna and keypad carried separately (Figure 12.7).

Smaller units achieve portability by conserving power, using less accurate clocks, and receiving signals intermittently. More sophisticated GPS units can continuously receive satellite signals in several channels, use very accurate clocks, and employ more substantial power supplies. These larger, more expensive units can acquire more accurate measurements and can be operated while in motion (installed in boats, aircraft, or field vehicles). Four channels are required, at a minimum, for highly accurate scientific or navigational applications (three channels for positional information and a fourth to ensure that timing information is correct). Most everyday remote sensing applications can be satisfied by the more modest capabilities of two-channel GPS receivers. GPS signals can be used in several ways to estimate location—two of the most important are pseudo-ranging and carrier phase measurement.

### *Pseudo-Ranging*

GPS satellites are equipped with accurate clocks. The system maintains its own time standard, with the capability to update times as signals are received from other satellites. GPS receivers generate codes that match codes generated by GPS satellites at the same time. As GPS units receive signals from satellites, they can estimate the displacement required to synchronize the two times. The amount of displacement forms the basis for an estimate of the distance between the receiver and each of the satellites within range. Pseudo-ranging is less accurate than other methods, but simplicity, convenience, and cost-effectiveness mean that it is widely used for determining positions of fixed points—the kind of information most often required for applications in remote sensing and GISs.

### *Carrier Phase Measurement*

Carrier phase measurement is based on a more detailed examination of signals broadcast by the satellites. GPS receivers detect either or both of the two L-band signals and add them to a signal generated by the receiver. This process is, in essence the application of the Doppler principle (Chapter 7), using observed frequency shifts to derive positional information.

To apply carrier phase measurement, a GPS station must have line-of-sight communication with at least three satellites, and communication with a fourth satellite is required to provide precise timing of signals. If a receiver can acquire data from as many as four satellites, it is possible to estimate both vertical and horizontal positions. In point positioning mode, a single GPS unit can establish horizontal location to within 20 m with 95% reliability and vertical location to within 30 m. With use of multiple carrier phase receivers it may be possible to derive location to within centimeters.

### *Selective Availability*

P-code permits a single, hand-held GPS receiver to estimate location to within 5 m. Because of the implications for national security of universal availability of quick, accurate, locational data, some aspects of the satellite signal can be encrypted, and exploitation of the system's full accuracy (carrier phase measurement, described above) requires access to a cryptographic key provided only to authorized users. Therefore, the system's most accurate data are provided only to military services and to other governmental agencies with requirements for high levels of positional accuracy.

C/A accuracy provides accuracy of 20 to 30 m. But even these capabilities can be degraded by selective availability (SA), achieved by randomly varying the time signals in a manner that reduces the precision of locational estimates. Under SA, using a single receiver, horizontal accuracy can be achieved to about 50 m. SA is designed to provide civilian GPS users with a lower degree of locational precision, while still providing useful information. SA policies are the subject of controversy because some claim SA is not necessary, and others argue that technology and practice (discussed below) allow civilian GPS users to circumvent effects of SA.

In March 1996, the U.S. government announced that its use of SA would be phased out over a period of 4 to 10 years, to be replaced by a system that will permit jamming of accurate GPS signals on a region-by-region basis during periods of military conflict or crisis.

### *Local Differential GPS*

When a GPS receiver can be stationed at a fixed position of known location, it becomes possible to derive estimates of the errors introduced by SA and to apply these estimates to improve the accuracy of GPS locations of points at unknown locations. This process is known as *local differential GPS*, which in effect circumvents the deliberate errors introduced by SA. Differential GPS requires that a GPS receiver be established at a fixed point for which the geographic location is known with confidence—this location forms the *base station*. Information from the base station can be applied to locational information from roving GPS receivers to derive more accurate estimates of location. Data from roving GPS receivers can be transported back to the base station to be processed (*postprocessing differential*) or, alternatively, data from the base station can be transmitted to roving receivers for use in the field (*real-time differential*). Postprocessing differential is likely to be cheaper and more accurate but also more inconvenient and time-consuming. A carefully established differential network, using the highest quality GPS receivers, can permit estimates of location within centimeters of the true position.

Careful use of GPS receivers can provide locational information adequate for use in many if not most remote sensing studies, especially if it is possible to employ differential location. Atmospheric effects (Chapter 2) constitute one of the major sources of error in GPS measurements. Atmospheric interference can delay transmission of signals from the satellites—delays that can be interpreted by the system as a difference in distance. Electrically charged particles in the ionosphere (30 to 300 mi. [48 to 483 km] above the earth's surface) and unusually severe weather in the troposphere (ground level to about 7.5 mi. [12 km]) can combine to cause errors as much as 1 to 5 m. Some of these errors cancel out or can be removed by filtering, whereas others remain to create errors in the estimated positions. Topographic location, presence of structures, and nearby vegetation canopies also can contribute to variations in effective use of GPS receivers.

## 12.8. Using Locational Information

Locational information can serve several purposes. One of the most important is to assist in identification of GCPs (Chapter 10) that allow analysts to resample image data to provide accurate planimetric location and correctly match image detail to maps and other images. Accurate locational information also permits correct positioning of field data with image data so that field data match to the correct regions within the image.

GPSs permit analysts to specify coordinates for distinctive landmarks clearly visible both on the imagery and on the ground. Examples include highway intersections, water bodies, distinctive terrain features, and the like. A network of such points then provides a locational framework that permits specification of locations of other points that might not be so distinctive. In other instances, field teams may acquire data in remote regions, far from distinctive landmarks. Prior to the availability of GPSs, these data could be positioned within imagery only approximately, using relative locations from distant landmarks as the basis for rather rough estimates. GPSs have the capability to specify positions of such observations precisely within the imagery, provided the image itself can be positioned within a coordinate system. That is, even with the accurate information provided by GPSs, it is still important to devote care to establishing the link between the image data and the ground observations.

## 12.9. Geographic Sampling

Although, in the ideal, it would be useful to acquire complete information concerning a study area, practical constraints require that analysts sample the patterns they study. Even modest-sized images can show such large areas that complete inventories are not feasible, and large areas of the study area may be inaccessible. As a result, field data must usually be collected for a sample of pixels from the image. Here the word *observation* signifies the selection of a specific cell or pixel; a *sample* is used here to designate a set of observations that will be used to construct an error matrix.

Any effort to sample a map or any other spatial pattern must consider three separate decisions: (1) determination of the number of observations to be used, (2) choice of the sampling pattern to position observations within an image, and (3) the selection for a given sampling pattern of the spacing of observations.

### *Numbers of Observations*

Clearly, determination of the number of observations to be collected is fundamental. The issue is much more complex than it might seem to be at first consideration, and scientists have employed different strategies to estimate the numbers of observations.

However, it is possible to outline the main considerations:

1. The number of observations determines the confidence interval of an estimate of the accuracy of a classification. Thus, from elementary statistics, we know that a large sample size decreases the width of the confidence interval of our estimate of a statistic—that is, we can have greater confidence that our estimate is close to the actual value when we employ larger numbers of samples.
2. For most purposes, we must consider not only the observations for the entire image but those that are allocated to each class, as it is necessary to have some minimum number of observations assigned to each class.

### *Sampling Pattern*

Sampling patterns specify the arrangement of observations. The most commonly used sampling patterns include the *simple random sampling pattern*, the *stratified random pattern*, *systematic* patterns, and the *systematic stratified unaligned* pattern. A simple random sample (Figure 12.9) positions observations such that the choice of any specific location as the site for an observation is independent of the selection of any other location as an observation. Randomness ensures that all portions of a region are equally subject to selection for the sample, thereby yielding data that accurately represent the area examined and satisfying one of the fundamental requirements of inferential statistics.

Typically the *simple random* sample of a geographic region is defined by first dividing the region to be studied into a network of cells. Each row and column in the network is numbered, then a random number table is used to select values that, taken two at a time,

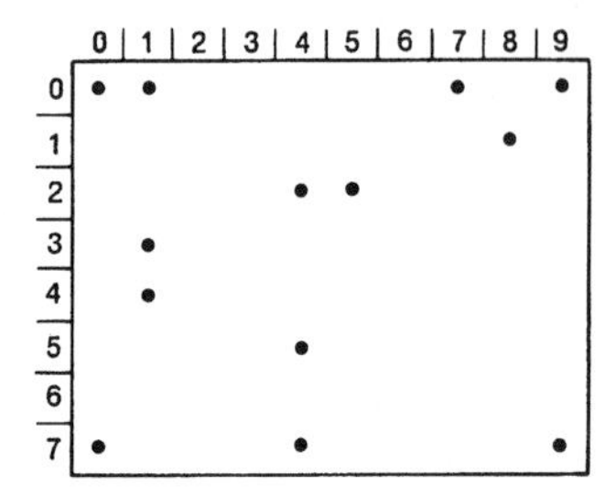


**FIGURE 12.9.** Simple random sampling pattern.

form coordinate pairs for defining the locations of observations. Because the coordinates are selected at random, the locations they define should be positioned at random.

The random sample is probably the most powerful sampling strategy available as it yields data that can be subjected to analysis using inferential statistics. However, in a geographic context, it has some shortcomings. Often a random sample is not uniform but instead positions observations in patterns that tend to cluster together. Also, observations are unlikely to be proportionately distributed among the categories on the image, and classes of small areal extent may not be represented at all in the sample. In addition, observations may fall at or near the boundaries of parcels; in the context of accuracy assessment, such observations should probably be avoided as they may record registration errors rather than classification errors. Thus, the pure random sample, despite its advantages in other situations, may not be the best choice comparing two images. It is, however, important to include an element of randomness in any sampling strategy, as described above.

A *stratified sampling pattern* assigns observations to subregions of the image to ensure that the sampling effort is distributed in a rational manner (Figure 12.10). For example, a stratified sampling effort plan might assign specific numbers of observations to each category on the map to be evaluated. This procedure would ensure that every category would be sampled. If a purely random pattern is used, on the other hand, many of the smaller categories might escape the sampling effort.

Often the positions of observations within strata are selected at random—a process

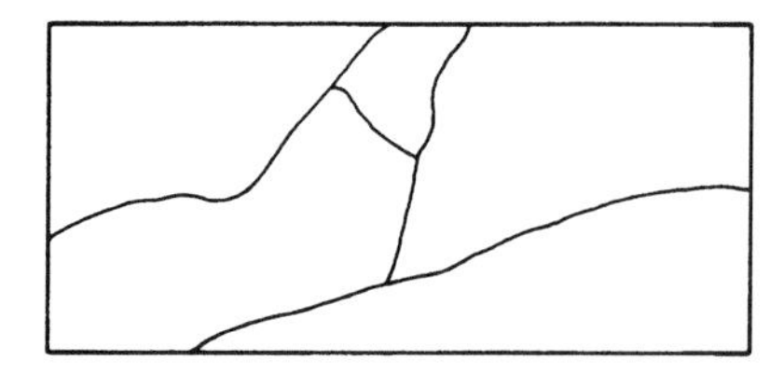

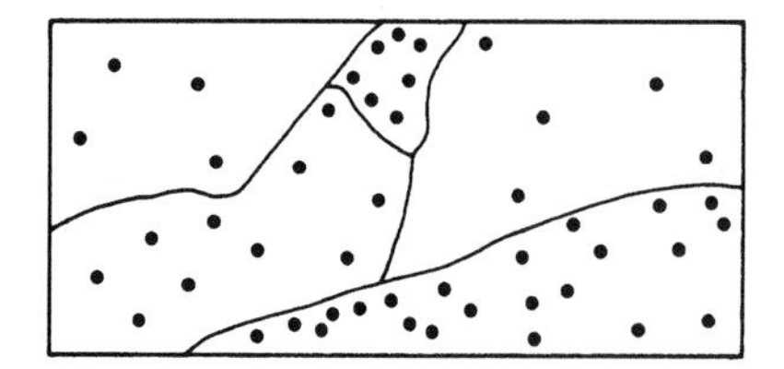

**FIGURE 12.10.** Stratified random sampling pattern. Here a region has been subdivided into subregions *(strata)* shown at the top. Samples are then allocated to each stratum in proportion to its expected signifcance to the study, then positioned randomly within each subarea.

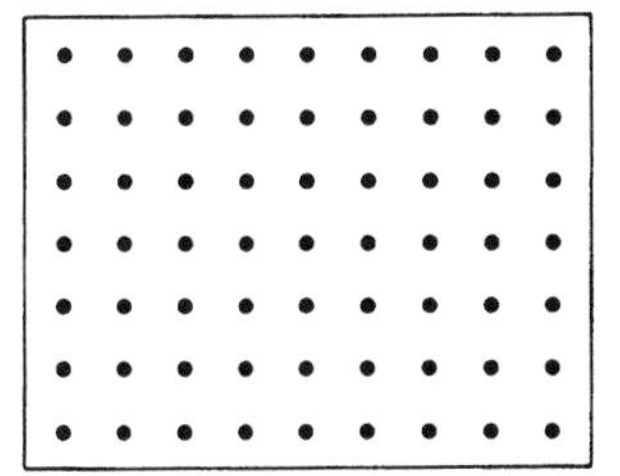

**FIGURE 12.11.** Systematic sampling pattern.

that yields the stratified random sampling pattern. Thus, stratification allocates observations to categories (the strata, in this instance) in proportion to their size significance; the random element ensures that observations are located within categories by chance, thereby ensuring a representative sample within each class.

*Systematic sampling* positions observations at equal intervals according to a specific strategy (Figure 12.11). Because selection of the starting point predetermines positions of all subsequent observations, data derived from systematic samples will not meet requirements of inferential statistics for randomly selected observations.

Another potential problem in the application of systematic samples is the presence of systematic variation in the underlying pattern that is to be sampled. For example, in the central United States, the Township–Range Survey System introduces a systematic component to the sizes, shapes, and positions of agricultural fields. Therefore, a purely systematic sample of errors in the classification of such landscape could easily place all observations at the centers of fields, thereby missing errors that might occur systematically near the edges of fields and giving an inflated estimate of accuracy. Or, systematic samples could oversample the edges and overestimate errors. There may be other, more subtle periodicities in landscapes that could interact with a systematic pattern to introduce other errors. Systematic sampling may nevertheless be useful if it is necessary to ensure that all regions within a study area are represented; a purely random sample, for example, may place observations in clustered patterns and leave some areas only sparsely sampled.

The *stratified systematic nonaligned* sampling pattern combines features of both systematic and stratified samples while simultaneously preserving an element of randomness (Figure 12.12) These features combine to form a sampling strategy that can be especially versatile for sampling geographic distributions.

The entire study area is divided into uniform cells, usually by means of a square grid. The grid cells, of course, introduce a systematic component to the sample and form the basis for stratification; one observation is placed in each cell. An element of randomness

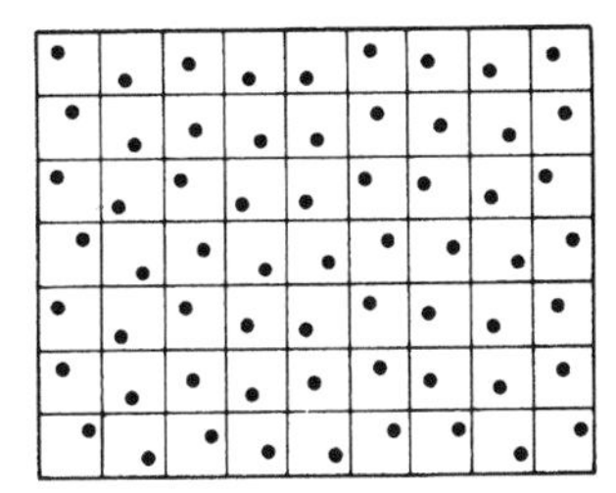

**FIGURE 12.12.** Stratified systematic sampling pattern.

is contributed by the method of placing observations within each cell. A number is randomly selected as the east–west coordinate for placement of the observation within the first cell; this value forms the east–west coordinate not only within the first cell but for all cells in that column (Figure 12.12). A second number is randomly selected to define the north–south coordinate within the first cell; this value also specifies the north–south placement of observations within all cells in the first row of cells. Each row and each column is assigned a new random number that is applied to locating observations within each cell.

The resulting placement of observations within the study area has many favorable characteristics. The use of grid cells as the basis for stratification means that observations are distributed evenly within an image; no segment of the area will be unrepresented. The random element destroys the rigid alignment caused by a purely systematic design and introduces an element of chance that increases the probability that the observations will accurately represent the categories within the image. The primary pitfall is that the analyst must specify an appropriate size for the grid, which determines the number of observations and the spacing between them. These factors determine the effectiveness of a given sample in representing a specific pattern. If the analyst knows enough about the region to make a good choice of grid size, the stratified systematic nonaligned sample is likely to be among the most effective.

In some situations, the sampling effort must be confined to certain sites, perhaps because of difficult access to ground observation of remote areas or because financial restrictions confine the analyst's ability to apply the other sampling patterns mentioned above. Cluster sampling selects points within a study area and uses each point as a center to determine locations of additional "satellite" observations placed nearby, so that the overall distribution of observations forms the clustered pattern illustrated in Figure 12.13. Centers can be located randomly or assigned specifically to certain areas or classes on a reference map. Satellite observations can be located randomly within constraints of distance. Cluster sampling may be efficient with respect to time and resources devoted to the sampling effort. If the pattern to be sampled is known beforehand, it may provide reasonably accurate results. However, if the distribution to be sampled is not known beforehand, it is probably best not to use a clustered pattern if it is at all possible to employ alternatives. Because observations are constrained to be close to one another, there is a distinct danger that the observations will replicate information rather than provide new information. If so, the analyst receives an inflated confidence in the assessment because the number of observations will suggest that more information is available than is actually the case.

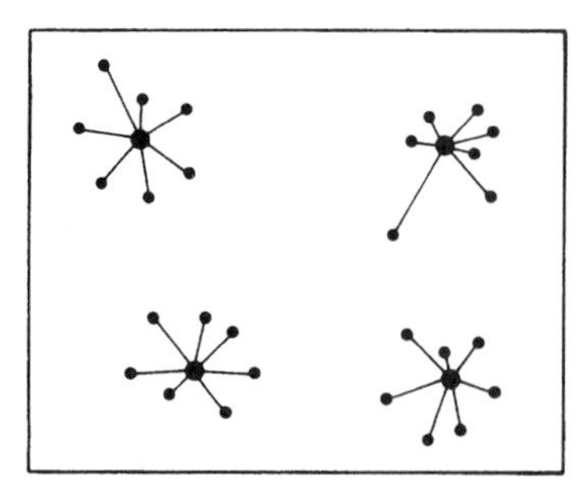

**FIGURE 12.13.** Clustered sampling pattern. Larger dots represent focal points for clusters selected randomly or possibly according to access or availability. Smaller dots represent satellite samples positioned at randomly selected distances and direction from the central point.

### *Spacing of Observations*

All efforts to define geographic sampling patterns must consider the effect of spacing of observations. If observations are spaced too close together, they tend to replicate information already provided by their neighbors and thereby provide the analyst with an unjustified confidence in the reliability of their conclusions. Use of 100 observations, if closely spaced, may in fact convey the information of only 50 evenly spaced observations. Closely spaced samples may underestimate variability and provide a poor indication of the number of categories present.

Examination of only a few images should be sufficient to reveal the significance of sample spacing. Some landscapes have a coarse fabric of landscape parcels—for example, the large agricultural fields of the North American plains can be represented by observations spaced at rather large distances. Other landscapes, typically those of complex topography or densely occupied regions, display a finer, intricate fabric and require more intense sampling patterns.

Spatial autocorrelation describes the tendency of measurements at one location to resemble those at nearby locations. Thus, a measurement of soil moisture at one place is likely to resemble another measurement made 6 in. distant and less likely to resemble one made 6 ft. away. Spatial autocorrelation provides quantitative estimates of the rate that similarities between paired observations change over distance. Spatial autocorrelation assesses the degree to which samples provide independent information.

Without prior knowledge of the pattern to be studied, it is impossible to decide whether a given sampling pattern positions observations too close to one another. However, in some instances we can apply the results of experience acquired in similar situations and avoid using clustered samples unless dictated by other considerations. Even a purely random sample will often place observations in clusters, so many investigators prefer the systematic stratified nonaligned pattern, which retains a random component but forces a degree of uniformity in spacing samples. Even when the most effective sampling patterns are used, it is usually necessary to augment the sample with observations selected at random from the less significant categories (in terms of area) as a means of ensuring that all categories are represented by an adequate number of independent observations.

### *Summary: Sampling*

It is difficult to recommend sampling strategies that might prove best for collecting field data as there has been little study of the comparative advantages of sampling procedures in this context. Congalton (1984) evaluated five sampling patterns to assess their performance in representing agricultural, range, and forested landscapes (as will be shown in Chapter 13). For the agricultural pattern, he concluded that the random sample performed the best. He found that the stratified random sample was best for the rangeland scene and that the systematic random sample provided the most accurate information regarding the forest image. In general, he concluded that the systematic and systematic nonaligned sample performed well on the forest scene but poorly on other scenes.

His results highlight the difficulty in making broad recommendations for application of sampling strategies—the optimum sampling pattern is likely to vary according to the nature of the pattern to be sampled. In some applications, we can examine the pattern to be sampled and select the sampling strategy likely to provide the best results. However, in many instances we are likely to face situations in which it is difficult or impossible to know beforehand the best sampling strategy and will be forced to develop experience on a case-by-case basis.

## 12.10. Summary

Full exploitation of field data can be considered one of the unexplored topics in the field of remote sensing. In recent years the availability of GPSs, field spectroradiometers, and laptop computers (among others) has greatly expanded the nature and accuracy of field data. These developments have provided data that are vastly superior to those of previous eras' data with respect to quantity of accuracy by:

- Effective integration of field data within an image processing system
- Compatibility of varied forms of field data
- Assessment and evaluation
- Coordination of field acquisition with image data
- Evaluation and assessment
- Questions of efficient and accurate sampling
- Access to remote regions

## Review Questions

1. Typically one might expect that a thorough field data collection effort might take much longer than the time required to collect image data. During the interval between the two, elements within the scene are likely to change, and certainly weather and illumination will change. How can field data compensate for these changes?
2. Many kinds of field observations necessarily record characteristics of points. Yet on remotely sensed data we normally examine areas. How can point data be related to areas? How can field observations be planned to accurately record characteristics of areas?
3. Sometimes field data are referred to as "ground truth." Why is this term inappropriate?
4. Prepare a rough estimate of the time and a number of people required to collect field data for a region near your school. Assume they will collect data to support a specific study in your discipline.
5. Prepare an itinerary and schedule for the field crew. Use local road maps or topographic maps to plan a route to observe field sites.
6. Refine your estimates for questions 4 and 5, to include costs of (a) equipment, (b) transportation, (c) wages, and supplies and materials.
7. Further consider your analyses of the questions posed in 4, 5, and 6 to include tabulation and analysis of field data. Develop a plan that can permit you to match field data to image

data. How can you be sure that the two match accurately? How long will it take to prepare the field observations in a form in which they can be used?

8. Devise also a sound sampling plan for field data collection in your discipline and your local area.

9. Some areas are clearly inaccessible, as a practical matter, for field data collection. How then should these areas be treated in a field data collection plan?

10. Some features on the ground are highly variable over time and over distance. How can they be accurately observed for field data collection?

## References

### Field Data: General

Anderson, James R., E. E. Hardy, J. T. Roach, and R. E. Witmer. 1976. *A Land Use and Land Cover Classification for Use with Remote Sensor Data*. U.S. Geological Survey Professional Paper 964. Washington, DC: U.S. Government Printing Office, 28 pp.

Campbell, J. B., and J. O. Browder. 1995. Field Data for Remote Sensing Analysis: SPOT Data, Rondonia, Brazil. *International Journal of Remote Sensing,* Vol. 16, pp. 333–350.

Craven, D., and B. Haack. 1994. Seeking the Truth in Kathmandu; Fieldwork as an Essential Component in Remote Sensing Studies. *International Journal of Remote Sensing,* Vol. 15, pp. 1365–1377.

Congalton, R. G., and G. S. Biging. 1992. A Pilot Study Evaluating Ground Reference Data for Use in Forest Inventory. *Photogrammetric Engineering and Remote Sensing,* Vol. 58, pp. 1701–1703.

Foody, G. M. 1990. Directed Ground Survey for Improved Maximum Likelihood Classification of Remotely Sensed Data. *International Journal of Remote Sensing,* Vol. 11, pp. 1935–1940.

Joyce, A. T. 1978. *Procedures for Gathering Ground Truth Information for a Supervised Approach to Computer-Implemented Land Cover Classification of Landsat-Acquired Multispectral Scanner Data*. NASA Reference Publication 1015. Houston, TX: National Aeronautics and Space Administration, 43 pp.

Lintz, Joseph, P. A. Brennan, and P. E. Chapman. 1976. Ground Truth and Mission Operations. Chapter 12 in *Remote Sensing of Environment* (J. Lintz and D. S. Simonett, eds.). Reading, MA: Addison-Wesley, pp. 412–437.

Peterson, D. L., M. A. Spencer, S. W. Running, and K. B. Teuber. 1987. Relationship of Thematic Mapper Simulator Data to Leaf Area of Temperate Coniferous Forest. *Remote Sensing of Environment,* Vol. 22, pp. 323–341.

Roy, P. S., B. K. Ranganath, P. G. Diwaker, T. P. S. Vophra, S. K. Bhan, I. S. Singh, and V. C. Pandian. 1991. Tropical Forest Type Mapping and Monitoring Using Remote Sensing. *International Journal of Remote Sensing,* Vol. 11, pp. 2205–2225.

Sader, S. A., R. B. Waide, W. T. Lawrence, and A. T. Joyce. 1989. Tropical Forest Biomass and Successional Age Class Relationships to a Vegetation Index Derived from Landsat TM Data. *Remote Sensing of Environment,* Vol. 28, pp. 143–156.

Treitz, P. M., P. J. Howarth, R. C. Suffling, and P. Smith. Application of Detailed Ground Information to Vegetation Mapping with High Resolution Digital Imagery. *Remote Sensing of Environment,* Vol. 42, pp. 65–82.

Steven, M. D. 1987. Ground Truth: An Underview. *International Journal of Remote Sensing,* Vol. 8, pp. 1033–1038.

Wikkramatileke,R. 1959. Problems of Land-Use Mapping in the Tropics: An Example from Ceylon. *Geography,* Vol. 44, pp. 79–95.

### Field Radiometry

*Technical Guide,* Analytical Technical Devices, Inc. Boulder, CO: Author, 47 pp.

Curtis, Brian, and A. Goetz. 1994. Field Spectroscopy: Techniques and Instrumentation. *Proceedings, International Symposium on Spectral Sensing Research (ISSSR '94),* pp. 195–203.

Deering, Donald W. 1989. Field Measurement of Bidirectional Reflectance. Chapter 2 in *Theory and Applications of Optical Remote Sensing* (Ghassem Asar, ed.). New York: Wiley, pp. 14–65.

Milton, E. J., E. M. Rollin, and D. R. Emery. 1995. Chapter 2 in *Advances in Environmental Remote Sensing* (F. M. Danson and S. E. Plummer, eds.). New York: Wiley, pp. 9–32.

## GPS

Hoffman-Wellenhof, B., H. Lichtenegger, and J. Collins. 1994. *Global Positioning Systems: Theory and Practice.* New York: Springer-Verlag, 355 pp.

Hurn, Jeff. 1989. *GPS: A Guide to the Next Utility.* Sunnyvale CA: Trimble Navigation, 76 pp.

Tralli, David M. 1993. A Sense of Where You Are. *The Sciences,* May–June 1993, pp. 14–19.

## Spectral Classes

Richards, J. A., and D. J. Kelley. 1984. On the Concept of the Spectral Class. *International Journal of Remote Sensing,* Vol. 5, p. 987–991.

Robinove, C. J. 1981. The Logic of Multispectral Classification and Mapping of Land. *Remote Sensing of Environment,* Vol. 11, p. 231–244.

## Sampling

Ayeni, O. O. 1982. Optimum Sampling for Digital Terrain Models: A Trend Towards Automation. *Photogrammetric Engineering and Remote Sensing,* Vol. 40, pp. 1687–1694.

Bellhouse, D. R. 1977. Some Optimal Designs for Sampling in Two Dimensions. *Biometrika,* Vol. 64, pp. 605–611.

Campbell, J. B. 1981. Spatial Correlation Effects upon Accuracy of Supervised Classification of Land Cover. *Photogrammetric Engineering and Remote Sensing,* Vol. 47, pp. 355–363.

Cochran, William G. 1961. Comparison of Methods for Determining Stratum Boundaries. *Bulletin of the International Statistical Institute,* Vol. 58, pp. 345–358.

Congalton, R. G. 1984. *A Comparison of Five Sampling Schemes Used in Assessing the Accuracy of Land Cover/Land Use Maps Derived from Remotely Sensed Data.* Ph.D. dissertation. Blacksburg, VA: Virginia Polytechnic Institute and State University, 147 pp.

Congalton, R. G. 1988. A Comparison of Sampling Schemes Used in Generating Error Matrices for Assessing the Accuracy Maps Generated from Remotely Sensed Data. *Photogrammetric Engineering and Remote Sensing,* Vol. 54, pp. 593–600.

Franklin, Steven E., D. P. Peddle, B. A. Wilson, and C. Blodget. 1991. Pixel Sampling of Remotely Sensed Digital Imagery. *Computers and Geosciences,* Vol. 17, pp. 759–775.

Ginevan, M. E. 1979. Testing Land-Use Map Accuracy: Another Look. *Photogrammetric Engineering and Remote Sensing,* Vol. 45, pp. 1371–1377.

Hay, A. 1979. Sampling Designs to Test Land Use Map Accuracy. *Photogrammetric Engineering and Remote Sensing,* Vol. 45, pp. 529–533.

Rosenfield, G. H., K. Fitzpatrick-Lins, & H. S. Ling. 1982. Sampling for Thematic Map Accuracy Testing. *Photogrammetric Engineering and Remote Sensing,* Vol. 48, pp. 131–137.

Rosenfield, George H., and K. Fitzpatrick-Lins. 1986. A Coefficient of Agreement as a Measure of Thematic Classification Accuracy. *Photogrammetric Engineering and Remote Sensing,* Vol. 52, pp. 223–227.

Skidmore, A. K., and B. J. Turner. 1992. Map Accuracy Assessment Using Line Intersect Sampling. *Photogrammetric Engineering and Remote Sensing,* Vol. 58, pp. 1453–1457.

Stehman, Stephen V. 1992. Comparison of Systematic and Random Sampling for Estimating the Accuracy of Maps Generated from Remotely Sensed Data. *Photogrammetric Engineering and Remote Sensing,* Vol. 58, pp. 1343–1350.

Van der Wel, Frans J. M., and L. L. F. Janssen. 1994. A Short Note on Pixel Sampling of Remotely Sensed Digital Imagery. *Computers and Geosciences, Vol. 20, pp. 1263–1264.*

# CHAPTER THIRTEEN

# Accuracy Assessment

## 13.1. Definition and Significance

Prospective users of maps and data derived from remotely sensed images quite naturally ask about the accuracy of the information they will use. Yet questions concerning accuracy are surprisingly difficult to address in a convincing manner. This chapter describes how the accuracy of a map or image can be evaluated and identifies some of the difficulties that remain.

### *Accuracy and Precision*

*Accuracy* defines "correctness"; it measures the agreement between a standard assumed to be correct and a classified image of unknown quality. If the image classification corresponds closely with the standard, it is said to be accurate. There are several methods for measuring the degree of correspondence, described in later sections.

*Precision* defines "detail" (Figure 13.1). The distinction is important because one may be able to increase accuracy by decreasing precision—that is, by being vague in the classification. For example, as the analyst classifies a stand of trees as *forest, coniferous, pine, shortleaf pine,* or *mature shortleaf pine,* detail increases, and so does the opportunity for error. (It is clearly more difficult to be correct in assigning detailed classes than in assigning general classes.) Evaluation of accuracy seldom explicitly considers precision, yet we must always ask whether the precision is appropriate for the purpose at hand. Accuracy of 95% in separating *water* and *forest* is unlikely to be useful if we need to know the distributions of *evergreen* and *deciduous* categories.

In a statistical context, high accuracy means that *bias* is low (that estimated values are consistently close to an accepted reference value) and that the variability of estimates is low (Figure 13.1). The usefulness of a map is related not only to its correctness but also to the precision with which the user can make statements about specific points depicted on the map. A map that offers only general classes (even if correct) enables users to make only vague statements about any given point represented on the map; one that uses detailed classes permits the user to make more precise statements (Webster & Beckett, 1968).

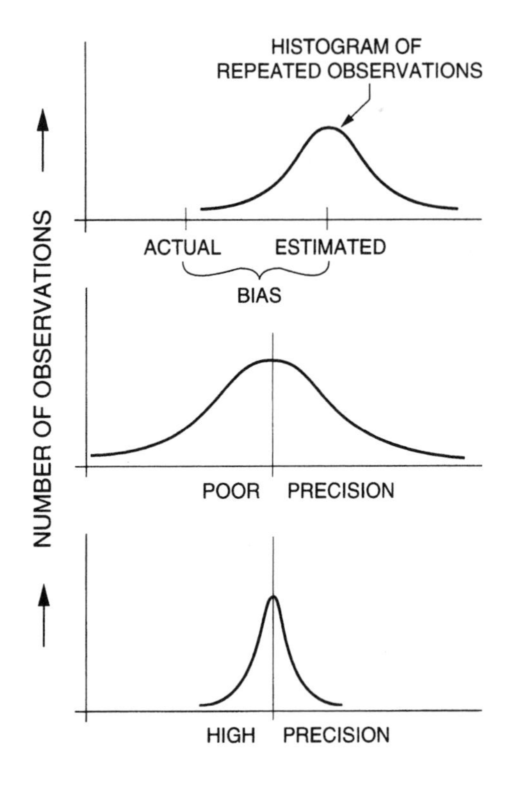


**FIGURE 13.1.** Bias and precision. Accuracy consists of bias and precision. Consistent differences between estimated values and true values create *bias* (top diagram). The lower diagrams illustrate the concept of *precision*: High variability of estimates leads to poor precision; low variability of estimates creates high precision.

### *Significance*

Accuracy has many practical implications. For example, it affects the legal standing of maps and reports derived from remotely sensed data, the operational usefulness of such data for land management, and their validity as a basis for scientific research. Analyses of accuracies of alternative classification strategies have significance for everyday uses of remotely sensed data. There have, however, been few systematic investigations of relative accuracies of manual and machine classifications, of different individuals, of the same interpreter at different times, of alternative preprocessing and classification algorithms, or of accuracies associated with different images of the same area. As a result, accuracy studies are likely to provide the basis for some of the most valuable research in both practical and theoretical aspects of remote sensing.

Often people assess accuracy from the appearance of a map, from past experience, or from personal knowledge of the areas represented. These can all be misleading, as overall accuracy may be unrelated to the map's cosmetic qualities, and often personal experience may be unavoidably confined to a few unrepresentative sites. Instead, accuracy should be evaluated through a well-defined effort to assess the map in a manner that permits quantitative measure of accuracy and comparisons with alternative images of the same area.

Evaluation of accuracies of information derived from remotely sensed images has long been of interest, but concern for accuracies of digital classifications has probably stimulated research on accuracy assessment. As digital classifications were first offered in

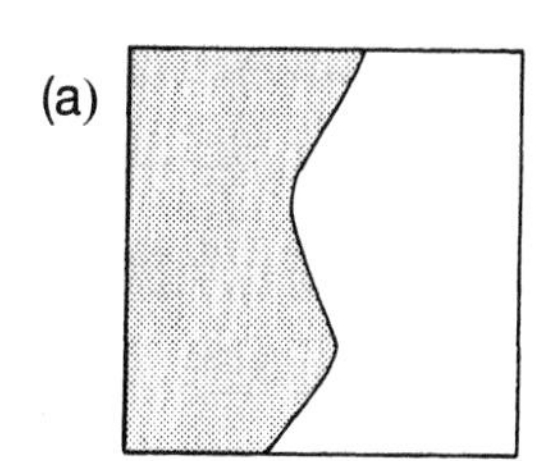


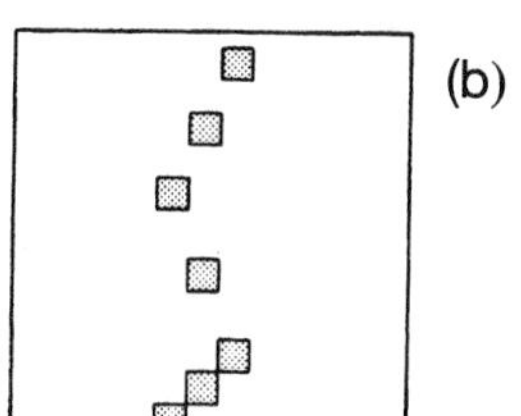


**FIGURE 13.2.** Incorrectly classified border pixels at the edges of parcels. Left: map of edge between two contrasting categories. Right: idealized representation of errors that might occur in a digital classification. The dark cells represent erroneously classified pixels. Usually, of course, we would not be able to know where these errors might occur.

the 1970s as replacements for more traditional products, many found the methods of machine classification to be abstract and removed from the direct control of the analyst; their validity could not be accepted without evidence. This concern formed much of the basis for the research outlined in this chapter.

Users should not be expected to accept the validity of any map, regardless of its origin or appearance, without supporting evidence. We shall see in this chapter how difficult it can be to compile the data necessary to support credible statements of map accuracy.

## 13.2. Sources of Classification Error

Errors are present in any classification. In manual interpretations, errors are caused by misidentification of parcels, excessive generalization, errors in registration, variations in detail of interpretation, and other factors. Perhaps the simplest causes of error are related to the misalignment of informational categories to spectral categories (Chapter 11). Bare granite in mountainous areas, for example, can be easily confused with the spectral response of concrete in urban areas. However, most errors are probably more complex. Mixed pixels occur as resolution elements fall on the boundaries between landscape parcels; these pixels may well have digital values unlike either of the two categories represented and may be misclassified even by the most robust and accurate classification procedures. Such errors may appear in digital classification products as chains of misclassified pixels that parallel the borders of large, homogeneous parcels (Figure 13.2).

In this manner the character of the landscape contributes to the potential for error through the complex patterns of parcels that form the scene. A simple landscape composed of large, uniform, and distinct categories is likely to be easier to classify accurately than one with small, heterogeneous, and indistinct parcels arranged in a complex pattern. Key landscape variables are likely to include:

- Parcel size
- Variation in parcel size
- Parcel identities
- Numbers of categories
- Arrangement of categories
- Number of parcels per category
- Shapes of parcel
- Radiometric and spectral contrast with surrounding parcels

These variables change from one region to another (Podwysocki, 1976; Simonett & Coiner, 1971), and within a given region, from season to season. As a result, errors present in a given image are not necessarily predictable from previous experience in other regions or on other dates.

## 13.3. Error Characteristics

Classification error is the assignment of a pixel belonging to one category (as determined by ground observation) to another category during the classification process. There are few if any systematic studies of geographic characteristics of these errors, but experience and logic suggest that errors are likely to possess at least some of the following characteristics:

- Errors are not distributed over the image at random but display a degree of systematic, ordered occurrence in space. Likewise, errors are not assigned at random to the various categories on the image but are likely to be preferentially associated with certain classes.
- Often, erroneously assigned pixels are not spatially isolated but occur grouped in areas of varied size and shape (Campbell, 1981).
- Errors may have specific spatial relationships to the parcels to which they pertain; for example, they may tend to occur at the edges or the interiors of parcels.

Figure 13.3 shows three error patterns from Landsat classifications given by Congalton (1984). Each image shows an area corresponding to a U.S. Geological Survey 7.5-minute topographic quadrangle; the distributions show errors in land cover classifications derived from Landsat images. These three areas were specifically selected to represent contrasting landscapes of predominantly forested, agricultural, and rangeland land use in the rural United States. Because accurate ground observations were available for each region, it was possible to compare the classification based on the Landsat data with the actual landscapes and to produce a complete inventory of errors for each region on a pixel-by-pixel basis. Dark areas represent incorrectly classified pixels; white areas show correct classification. (Here the scale of the images is so small that we can see only the broad outlines of the pattern without resolving individual pixels.)

We cannot identify sources of these errors. But we can clearly see the contrasting patterns. None of the images show random patterns—misclassified pixels tend to form distinctive patterns. In the forest image, errors form crescent-shaped strips created by shadowed terrain; in the agricultural scene, center-pivot irrigation systems create ring-shaped patches left fallow, or planted to contrasting crops, that have been erroneously classified. On the rangeland image, we clearly see the effect of the systematic Township–Range Survey System, probably through influence of mixed pixels at the edges of rectangular parcels.

In our own image classifications we have no knowledge of the corresponding error patterns because seldom do we have the opportunity to observe the total inventory of classification errors for areas as large as those shown in Congalton's study. Yet these patterns of error do exist, even though we cannot know their abundance or distribution.

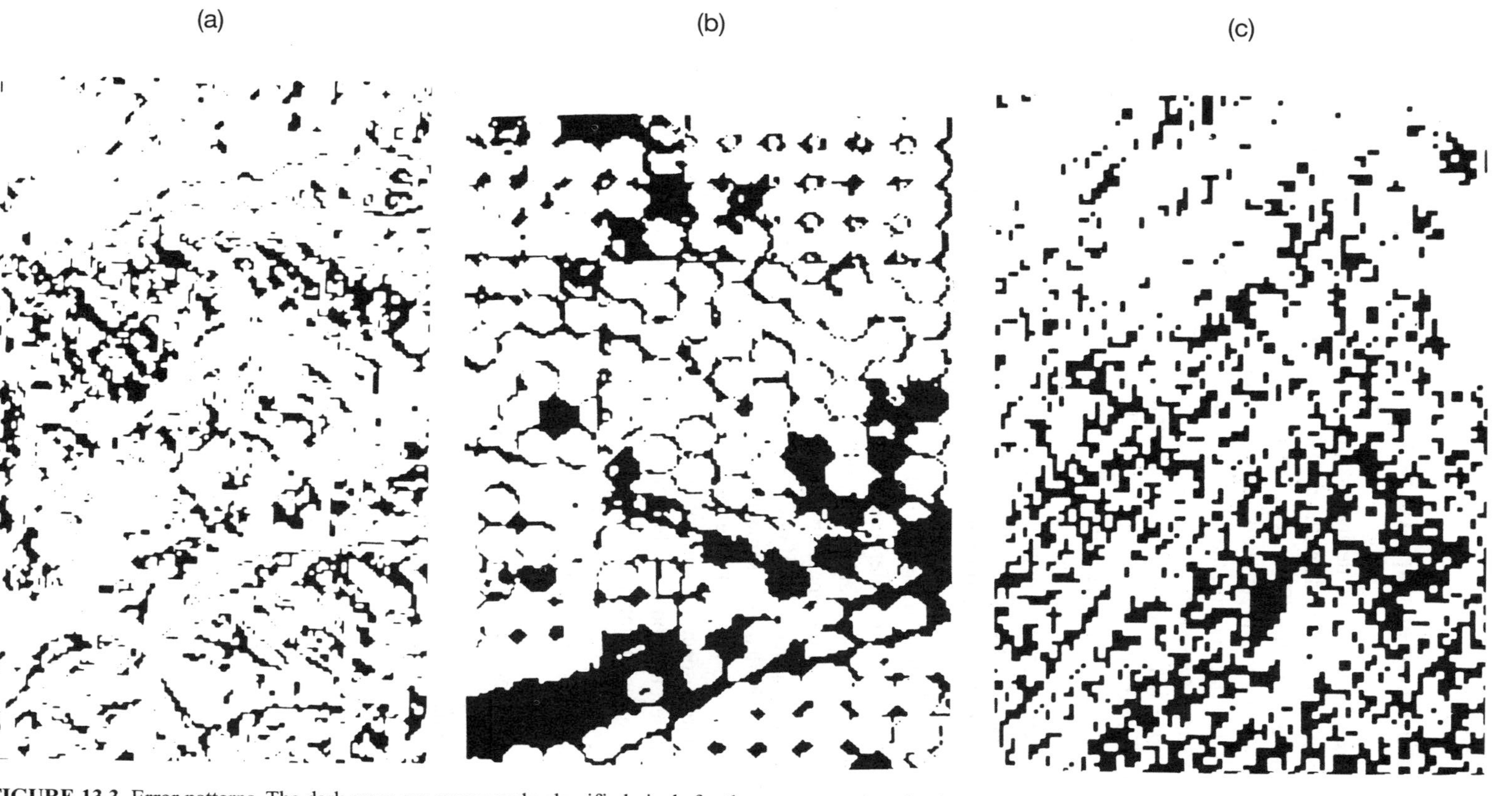


**FIGURE 13.3.** Error patterns. The dark areas are erroneously classified pixels for three areas as given by Congalton (1984): (*a*) Forested region. (*b*) Agricultural scene. (*c*) Rangeland. These images are shown at very small scale; at larger scale, it is possible to recognize the dark areas as individual pixels.

## 13.4. Measurement of Map Accuracy

Accuracy assessment can be defined as the task of comparing two maps, one based on analysis of remotely sensed data (the *map to be evaluated*) and another based on a different source of information. This second map is designated the *reference map,* assumed to be accurate, that forms the standard for comparison. The reference data are of obvious significance; if they are in error, the attempt to measure accuracy will be in error. For information that varies seasonally, or over time, it is important that the reference image shows information that matches with respect to time. Otherwise, differences between images may not be caused solely by inaccuracies in the classification but to a combination of error, changes that have occurred during the interval that elapsed between acquisition of the two images. In other instances we may examine two images simply to decide whether there is a difference, without concluding that one is more accurate than the other. For example, we may compare images of the same area taken by different sensors, or classifications made by different interpretors. In such instances, it is not always necessary to assume that one image is more accurate than the other, as the objective is simply to determine whether the two are different.

Usually, however, the reference map is assumed to be the "correct" map. Both maps must be in the form of "parcel" or "mosaic" maps—that is, maps composed of a network of discrete parcels, each designated by a single label from a set of mutually exclusive categories. To assess the accuracy of one map, it is necessary that the two maps register to one another, that they both use the same classification systems, and that they have been mapped at comparable levels of detail. The strategies described here are not appropriate if the two maps differ with respect to detail, numbers of categories, or meanings of the categories.

The simplest method of evaluation is to compare the two maps with respect to the areas assigned to each category. The result of such a comparison is to report the areal proportions of categories (Figure 13.4). These values report the extent of the agreement between the two maps in respect to total areas in each category but do not take into account compensating errors in misclassification that cause this kind of accuracy measure to be itself inaccurate. For example, underestimation of "forest" in one part of the image can compensate for overestimation of "forest" in another part of the image; serious errors in classification have been made but are not revealed in the simple report of total areas in each category.

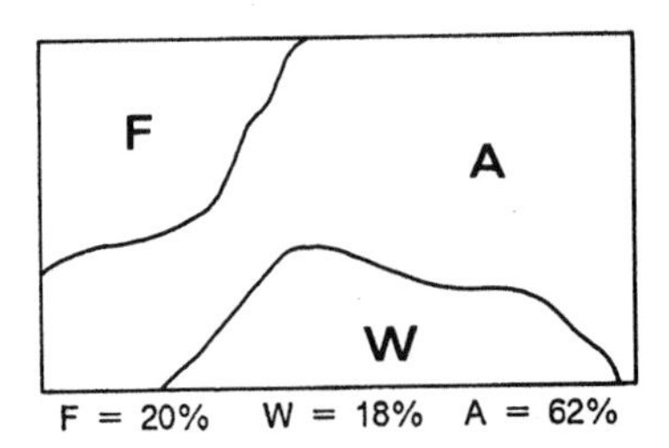


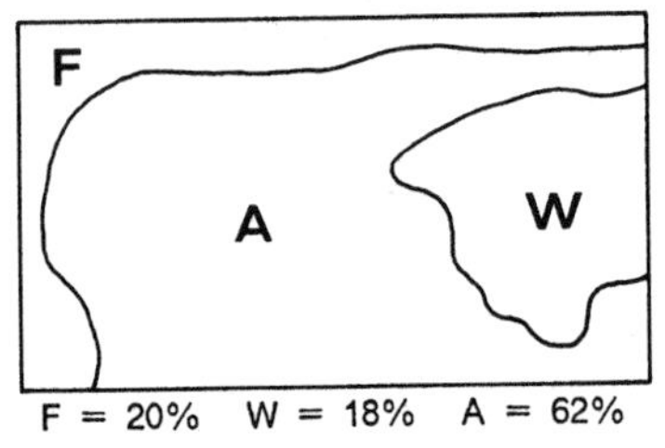


**FIGURE 13.4.** Non-site-specific accuracy. Here two images are compared only on the basis of total areas in each category. Because total areas may be similar even though placement of the boundaries differs greatly, this approach can give misleading results, as shown here.

This form of error assessment is sometimes called *non-site-specific accuracy* because it does not consider agreement between the two maps at specific locations but only the overall figures for the two maps. Figure 13.4 illustrates this point. The two patterns are clearly different, but the difference is not revealed by the overall percentages for each category. Gershmehl and Napton (1982) refer to this kind of error as *inventory error,* as the process considers only the aggregate areas for classes rather than the placement of classes on the map.

### *Site-Specific Accuracy*

The second form of accuracy, *site-specific accuracy,* is based on the detailed assessment of agreement between two maps at specific locations (Figure 13.5). Gershmehl and Napton (1982) refer to this kind of error as *classification error.* In most analyses the units of comparison are simply pixels derived from remotely sensed data, although if necessary a pair of matching maps could be compared using any network of uniform cells. Site-specific accuracy has been measured using several alternative strategies. Sometimes, training data for image classification can be subdivided into two groups, one used to prepare the classification and the other to assess accuracy of the classification.

Hord and Brooner (1976) distinguish between errors in classification and errors in positioning boundaries. Classification errors are misidentifications of the identities of pixels. Boundary errors are caused by misplacement of boundaries between categories. In automated classifications, boundary errors are caused mainly by geometric errors either in the image or in the reference data, so that a correctly classified pixel may be misregistered to the wrong pixel in the reference image and thereby counted as an error.

### *The Error Matrix*

The standard form for reporting site-specific error is the error matrix, sometimes referred to as the *confusion matrix* because it identifies not only overall errors for each category but also misclassifications (due to confusion between categories) by category. Compila-

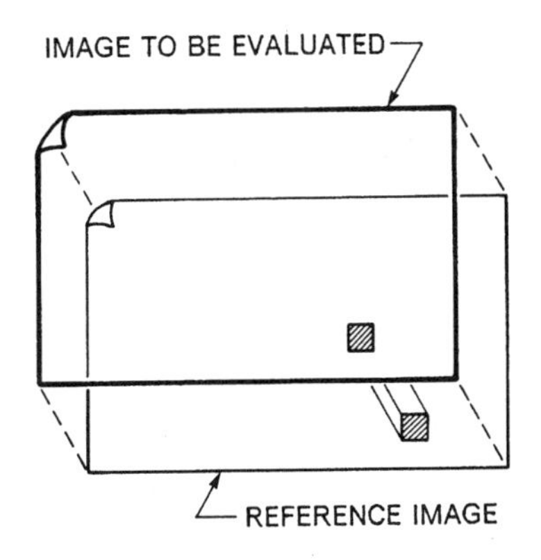


**FIGURE 13.5.** Site-specific accuracy. The two images are compared on a site-by-site ("cell by cell" or "pixel by pixel") basis to accumulate information concerning the correspondence of the two images. Here only a single pair of cells is shown, although the comparison considers the entire images.

**TABLE 13.1. Example of Error Matrix**

| | | Image to be evaluated | | | | | | |
|---|---|---|---|---|---|---|---|---|
| | | Urban | Crop | Range | Water | Forest | Barren | Totals |
| Reference image | Urban | 150 | 21 | 0 | 7 | 17 | 30 | 225 |
| | Crop | 0 | 730 | 93 | 14 | 115 | 21 | 973 |
| | Range | 33 | 121 | 320 | 23 | 54 | 43 | 594 |
| | Water | 3 | 18 | 11 | 83 | 8 | 3 | 126 |
| | Forest | 23 | 81 | 12 | 4 | 350 | 13 | 483 |
| | Barren | 39 | 8 | 15 | 3 | 11 | 115 | 191 |
| | Totals | 248 | 979 | 451 | 134 | 555 | 225 | 1,748 |

*Note.* Percentage correct = sum of diagonal entries/total observations = 1,748/2,592 = 67.4%

tion of an error matrix is required for any serious study of accuracy. It consists of an $n \times n$ array, where $n$ represents the number of categories (Table 13.1, Figure 13.6).

The left-hand side ($y$ axis) is labeled with the categories on the reference (*correct*) classification; the upper edge ($x$ axis) is labeled with the same $n$ categories; these refer to those on the map to be evaluated. (Note that the meanings of the two axes can be reversed in some applications, as the convention is not universal.) Values in the matrix represent numbers of pixels. The matrix reveals the results of a comparison of the evaluated and reference images; sometimes these numbers might constitute the entire image; other times they might only form a sample of the total. (Sometimes the matrix is constructed using percentages rather than absolute values, but here we will show absolute values counted directly from the image.)

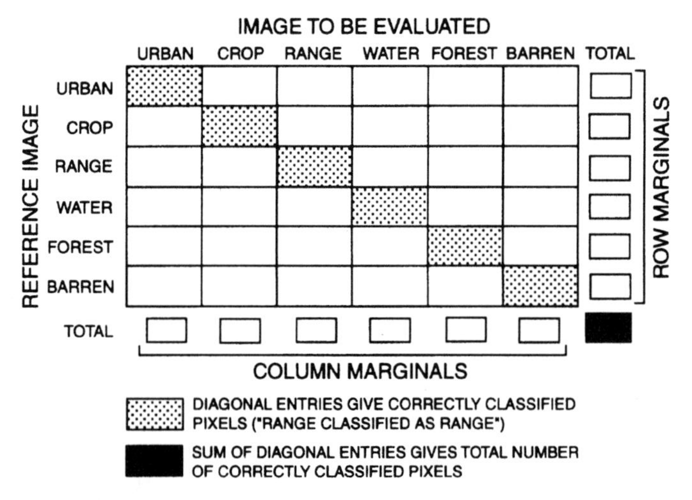


**FIGURE 13.6.** Schematic representation of an error matrix. Here elements of the matrix are represented to correspond to the text and to Table 13.1.

Inspection of the matrix shows how the classification represents actual areas on the landscape. For example, in Table 13.1, there are 225 pixels of urban land (the far right value in the first row). Of the 225 pixels of urban land, 150 were classified as urban (row 1, column 1); these, of course, are the pixels of urban land correctly classified. Reading succeeding values along the first row, next are incorrectly classified urban pixels and the categories to which they were assigned: crop, 21; range, 0; water, 7; forest, 17; and barren land, 30. Reading across each row, we then learn how the classification assigned pixels identified by their actual classes as they occur on the landscape. The diagonal from upper left to lower right gives numbers of correctly classified pixels.

Further inspection of the matrix reveals a summary of other information. Column totals on the far right give the total number of pixels in each class as recorded on the reference image. The bottom row of totals shows numbers of pixels assigned to each class as depicted on the image to be evaluated.

## *Compiling the Error Matrix*

To construct the error matrix, the analyst must compare two images—the reference image and the image to be evaluated—on a point-by-point basis to determine exactly how each site on the reference image is represented in the classification. For this comparison to have meaning, the two images must register to one another. Errors in registration will appear as errors in classification, so registration problems will create errors in the assessment of accuracy. The analyst must establish a network of uniform cells that form the units of comparison (Figure 13.7); these must be small enough to provide enough cells for a statistically valid sample, and small enough to avoid large numbers of mixed cells (caused by cells that fall on the boundaries between parcels), but large enough to avoid the tedium and expense that accompany use of impractically small units.

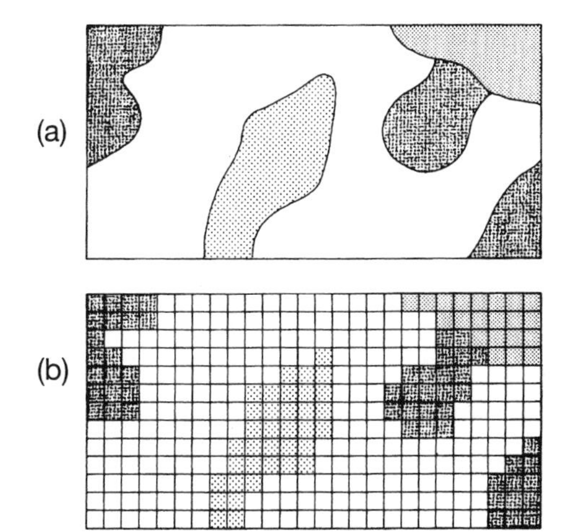


**FIGURE 13.7.** Representation of a category map by uniform cells. The network of uniform cells (bottom) represents the pattern of the original map or image (top). These cells then form the units of comparison for site-specific accuracy assessment.

The two images are then superimposed, either literally if the compilation is to be conducted manually or figuratively if the compilation is conducted digitally, by computer. For manual compilation, the analyst examines the superimposed images systematically on a cell-by-cell basis, tabulating for each cell the predominant category shown on the reference map and the category of the corresponding cell on the image to be evaluated (Table 13.2). The analyst maintains a count of the numbers in each reference category as they are assigned to classes on the map to be evaluated. A summation of this tabulation then forms the basis for construction of the error matrix.

If both images are in digital form, the unit of compilation is likely to be the pixel as defined by the image sensor, or a unit formed from groups of pixels. Matching of the two images, tabulation of the cross-classification, and construction of the error matrix are most difficult when one map (typically the reference map) is in manuscript form and the other, typically the map to be evaluated, is in digital form. It is then necessary to bring them both into the same format, probably by digitizing the manuscript map. Digitization can be time-consuming and expensive, and the analyst then faces questions of optimum registration and resampling (Chapter 10). Thus, although compilation of the error matrix is straightforward, preparation of maps for comparison may be very difficult.

Table 13.3 illustrates the significance of compatibility of the classification used for the two images. Ideally, the two images use the same classification system. In practice, the two images may differ with respect to classification. Sometimes different classifications

**TABLE 13.2. Tabulation of Data for the Error Matrix**

| Worksheet | | |
|---|---|---|
| Cell | Reference map | Classification |
| 1 | F | F |
| 2 | F | U |
| 3 | W | W |
| 4 | W | W |
| 5 | C | C |
| 6 | C | F |
| 7 | C | C |
| . | . | . |
| . | . | . |
| . | . | . |

| Summation | |
|---|---|
| Forest classified as forest | 350 |
| Forest classified as urban | 23 |
| Forest classified as cropland | 81 |
| Forest classified as water | 4 |
| . . . . | . |
| . . . . | . |
| . . . . | . |
| Cropland classified as cropland | 730 |
| Cropland classified as forest | 115 |
| Cropland classified as water | 14 |
| . . . . | . |
| . . . . | . |

**TABLE 13.3. Examples of Compatible and Incompatible Classifications**

| Reference map | Image to be evaluated |
|---|---|
| **Compatible** | |
| Water | Water |
| Urban and built-up land | Urban residential<br>Urban commercial<br>Urban industrial |
| Deciduous forest<br>Coniferous forest<br>Mixed forest | Forest |
| Agricultural land | Cropland<br>Pasture |
| **Incompatible** | |
| Open land | Cropland |
| Dense settlement | Forest |
| Strip development | Cities and towns |
| Water | Lakes and rivers |
| Rough, uneven land | Disturbed land |
| Roads and highways | |

may be compatible—categories can be matched to one another in a manner that permits a valid comparison of the two images. If the differences are simply a matter of detail, often the more detailed categories can be collapsed into the more general classes, as illustrated in Table 13.3. In other instances, the two images may be fundamentally incompatible—the categories may be based on different classification logic, so the two sets of classes cannot be matched (Table 13.3). In some instances two classifications may use differing definitions but the same names, so it is important to examine each image and the supporting information closely before attempting a comparison.

### *Errors of Omission and Errors of Commission*

Examination of the error matrix reveals, for each category, errors of omission and errors of commission (Table 13.4). Errors of omission are, for example, the assignment of errors of forest on the ground to the agricultural category on the map (in other words, an area of "real" forest on the ground has been omitted from the map). Using the same example, an error of commission would be to assign an area of agriculture on the ground to the *forest* category on the map. The analyst's error in this instance has been to actively commit an error by assigning a region of forest to a wrong category. (This error of commission for one category will, of course, also be tabulated as an error of omission for another category.) The distinction is essential because the interpretation could otherwise achieve 100% accuracy in respect to forest by assigning all pixels to forest. Tabulations of errors of commission of reveal such actions to be meaningless because they are balanced by compensating errors.

**TABLE 13.4. Errors of Omission and Commission**

| | | Image to be evaluated | | | | | | | | | |
|---|---|---|---|---|---|---|---|---|---|---|---|
| | | Urban | Crop | Range | Water | Forest | Barren | Totals | PA% | EO% | EC% |
| Reference image | Urban | 150 | 21 | 9 | 7 | 17 | 30 | 234 | 64.1 | 35.9 | 41.9 |
| | Crop | 0 | 730 | 93 | 14 | 115 | 21 | 973 | 75.0 | 25.0 | 25.6 |
| | Range | 33 | 121 | 320 | 23 | 54 | 43 | 594 | 53.9 | 46.1 | 23.6 |
| | Water | 3 | 18 | 11 | 83 | 8 | 3 | 126 | 65.9 | 34.1 | 40.5 |
| | Forest | 23 | 81 | 12 | 4 | 350 | 13 | 483 | 72.5 | 27.5 | 42.4 |
| | Barren | 39 | 8 | 15 | 3 | 11 | 115 | 191 | 60.2 | 39.8 | 57.6 |
| | Totals | 248 | 979 | 460 | 134 | 555 | 225 | 1,748 | | | |
| | CA (%): | 60.5 | 74.6 | 69.6 | 61.9 | 63.1 | 51.1 | | | | |

*Note.* CA, consumer's accuracy; PA, producer's accuracy; EO, errors of omission; EC, errors of commission.

By examining relationships between the two kinds of errors, the map user gains insight into the varied reliabilities of classes on the map, and the analyst learns about the performance of the process that generated the maps. Examined from the user's perspective, the matrix reveals *consumer's accuracy*; examined from the analyst's point of view, the matrix reveals *producer's accuracy.* The difference between the two lies in the base from which the accuracy is assessed. For producer's accuracy, the base is the area in each class on the reference map. Thus, for the example in Table 13.4, producer's accuracy for forest is 350/483, or 73%. For the same class, consumer's accuracy is 350/555, or 63%. Consumer's accuracy forms a guide to the reliability of the map as a predictive device—it tells the user of the map that (in this example), of the area labeled *forest,* 63% actually corresponds to forest on the ground. Producer's accuracy informs the analyst who prepared the classification that, of the actual forested landscape, 73% was correctly classified. In both instances the error matrix permits identification of the classes erroneously labeled *forest,* and forested areas mislabeled as other classes.

## 13.5. Interpretation of the Error Matrix

Table 13.1 and Figure 13.6 show an example of an error matrix. Each of the 2,592 pixels in this scene was assigned to one of six land cover classes. The resulting classification was then compared, pixel by pixel, to a previously existing land use map of the same area, and the differences were tabulated, category by category, to form the data for Table 13.1. The total number of pixels reported by the matrix (in this instance, 2,592) may constitute the entire image or simply a sample selected from the image as explained below. Also, the land use classes here simply form examples; the matrix could be based on other kinds of classes (including forest types, geology, and others) and could be smaller or larger, depending on the number of classes examined.

The column of sums on the right-hand edge of the matrix gives total numbers of pixels in each category on the reference image; the row of sums at the bottom shows total pixels in each category in the classified scene. These are known, respectively, as row and column *marginals* (Figure 13.6). The sequence of values that extends from the upper left corner to the lower right corner is referred to as the *diagonal* (here we are not interested in the opposite diagonal). These diagonal entries show the number of correctly classified pixels—rangeland classified as rangeland, urban as urban, and so on. Nondiagonal values in each row give errors of omission for row-labeled categories. For example, as we read across the third row, we see the numbers of pixel of rangeland that have been placed in other categories: urban land, 33; cropland, 121; water, 23; forest, 54; and barren land, 43. These are errors of omission because the classification has erred by omitting true errors of rangeland from the interpreted image. Specifically, we know that rangeland on the reference map is most likely to be misclassified as cropland on the classified image.

In contrast, nondiagonal values along the columns give errors of commission (for labels at the tops of rows). To continue the example with the rangeland category, errors of commission are caused by active misassignment of other categories to the rangeland class. These errors are found by reading down the third column: urban land, 0; cropland, 93; water, 11; forest, 12; and barren land, 15. These values reveal that the classification of rangeland erred most often by assigning true rangeland to the crop class. Here errors of both omission and commission reveal that rangeland is most often confused with cropland, and that the classified map image is likely to confuse these two categories.

Table 13.4 summarizes errors of omission and commission for another matrix. For example, there were 234 pixels of urban land in the land use map; of these, 150 were correctly classified as urban land, for a percentage of about 64%. The remaining 84 pixels (the sum of the off-diagonal entries from row 1, Table 13.4), forming about 36% of the total urban land, were incorrectly classified, mainly as barren land but also as other categories. These form the error of omission for urban land.

Errors of commission (equal to about 42% of the total of urban land in this area) consist mainly of urban land classified as barren land and rangeland, as shown in Table 13.4. For this example, it is clear that the classifier's confusion of urban land and barren land is a major source of error in the classification of urban land use.

For a contrasting relationship between errors of omission and commission, note that rangeland tends to be classified as urban land whereas seldom is urban land classified as rangeland. Inspection of the error matrix reveals the kinds of errors generated by the classification process, which may in turn permit improved interpretation of the map's reliability and improved accuracy in future classifications.

### *Percentage Correct*

One of the most widely used measures of accuracy is the *percentage correct*—a report of the overall proportion of correctly classified pixels in the image or in the sample used to construct the matrix. The percentage correct, of course, can be easily found; the number correct is the sum of the diagonal entries. Dividing this value by the total number of pixels examined gives the proportion that has been correctly classified (Table 13.1). This value estimates the unknowable "true" value; the closeness of this value to the true values depends, in part, on the sampling strategy, as explained below. The percentage correct can be reported with a confidence interval (Hord & Brooner, 1976).

Often the percentage correct is used alone, without the error matrix, as a simple measure of accuracy. Reported values vary widely. Anderson et al. (1976) state that accuracies of 85% are required for satisfactory land use data for resource management. Fitzpatrick-Lins (1978) reports that accuracies of USGS land cover maps of the central Atlantic coastal region are about 85% (1:24,000), 77% (1:100,000), and 73% (1:250,000). For automated interpretations of land use in the Denver metropolitan area (six level I categories), Tom et al. (1978) achieved 38% accuracy using only MSS data (including band ratios) and about 78% using ancillary data.

By itself, the percentage correct may suggest the relative effectiveness of a classification, but in the absence of an opportunity to examine the full error matrix, it cannot form convincing evidence of the classification's accuracy. A full evaluation must consider the categories used in the classification. For example, it would be easy to achieve high values of percentage correct by classifying a scene composed chiefly of open water—a class that is easy to categorize correctly. Furthermore, variations in the accuracies of specific classes should be noted, as should the precision of the classes. A classification that used only broadly defined classes could achieve high accuracies but would not be useful for someone who required more detail. Finally, later sections of this chapter will show that raw values for the percentage correct can be inflated by chance assignment of pixels to classes—yet another reason for careful inspection of the assessment task.

Hay (1979) stated that it is necessary to consider five questions to thoroughly understand the accuracy of a classification:

- What proportion of the classification decision is correct?
- What proportion of assignments to a given category is correct?
- What proportion of a given category is correctly classified?
- Is a given category overestimated or underestimated?
- Are errors randomly distributed?

Percentage correct can answer only the first of these questions; the others can be answered only by examination of the full error matrix.

### *Quantitative Assessment of the Error Matrix*

After an initial inspection of the error matrix reveals the overall nature of the errors present, there is often a need for a more objective assessment of the classification. For example, we may ask whether the two maps are in agreement—a question that is difficult to answer, as the notion of "agreement" may be difficult to define and implement. The error matrix is an example of a more general class of matrices, known as *contingency tables,* that summarize classifications analogous to those considered here. Some of the procedures that have been developed for analyzing contingency tables can be applied to examination of the error matrix.

Chrisman (1980), Congalton (1983), and Congalton et al. (1983) propose application of techniques described by Bishop et al. (1975) and Cohen (1960) as a means of improving interpretation of the error matrix. A shortcoming of usual interpretations of the error matrix is that even chance assignments of pixels to classes can result in surprisingly good results, as measured by percentage correct. Hood and Brooner (1976) and others have noted that the use of such measures is highly dependent on the samples and, therefore, on the sampling strategy used to derive the observations used in the analysis.

κ (kappa) is a measure of the difference between the observed agreement between two maps (as reported by the diagonal entries in the error matrix) and the agreement that might be attained solely by change matching of the two maps. Not all agreement can be attributed to the success of the classification. κ attempts to provide a measure of agreement that is adjusted for chance agreement. κ is estimated by $\hat{\kappa}$ ("*k* hat"):

$$\hat{\kappa} = \frac{\text{Observed} - \text{expected}}{1 - \text{expected}} \qquad \text{(Eq. 13.1)}$$

This form of the equation, given by Chrisman (1980) and others, is a simplified version of the more complete form given by Bishop et al. (1975). Here *observed* designated the accuracy reported in the error matrix, and *expected* designates the correct classification that can be anticipated by chance agreement between the two images.

"Observed" is the overall value for percentage correct defined previously—the sum of the diagonal entries divided by the total number of samples. "Expected" is an estimate of the contribution of chance agreement to the observed percentage correct. Expected values are calculated using the row and column totals (the "marginals," as explained above). Products of row and column marginals estimate the numbers of pixels assigned to each cell in the confusion matrix, given that pixels are assigned by chance to each category.

The role of marginals in estimating chance agreement is perhaps clearer if we consider the situation in its spatial context (Figure 13.8). Then the marginals can be seen to represent areas occupied by each of the categories on the reference map and the image, and the chance mapping of any two categories is the product of their proportional extents on the two maps.

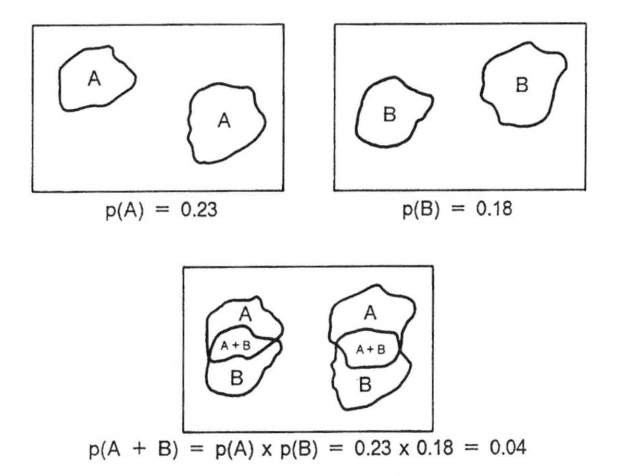


**FIGURE 13.8.** Contrived example illustrating computation of chance agreement of two categories when two images are superimposed. Here $p$(A) and $p$(B) would correspond to row and column marginals; their product computes the agreement expected by chance matching of the two maps.

In tabular form (Table 13.5), the expected correct, based on chance mapping, is found by summing the diagonals, then dividing by the grand total of products of row and column marginals. This estimate for expected correct can be compared to the percentage correct or observed value defined above. Table 13.5 shows a worked example for finding $\hat{\kappa}$.

$\hat{\kappa}$ in effect adjusts the percentage correct measure by subtracting the estimated contribution of chance agreement. Thus, $\hat{\kappa} = 0.83$ can be interpreted to mean that the classification achieved an accuracy that is 83% better than would be expected from chance assignment of pixels to categories. As the percentage correct approaches 100, and as the contribution of chance agreement approaches 0, the value of $\hat{\kappa}$ approaches +1.0, indicating perfect effectiveness of the classification (Table 13.6). On the other hand, as the effect

**TABLE 13.5. Worked Example Illustrating Calculation of $\hat{\kappa}$**

Part 1. Calculation of "observed correct"

| | Image to be evaluated | | | |
|---|---|---|---|---|
| Reference image | 35 | 14 | 11 | 1 |
| | 4 | 11 | 3 | 0 |
| | 12 | 9 | 38 | 4 |
| | 2 | 5 | 12 | 2 |

Grand total = 163
Total correct = 86
"Observed correct" = 86/163 = 0.528
(Percentage correct = 53%)

Part 2. Calculation of "expected correct"

Marginals

| | | | | |
|---|---|---|---|---|
| | | | | 61 |
| | | | | 18 |
| | | | | 63 |
| | | | | 21 |
| 53 | 39 | 64 | 7 | |

Products of row and column marginals

| | | | |
|---|---|---|---|
| 3,233 | 2,379 | 3,904 | 427 |
| 954 | 702 | 1,152 | 126 |
| 3,339 | 2,457 | 4,032 | 441 |
| 1,113 | 819 | 1,344 | 147 |

$$\text{Expected agreement by change} = \frac{\text{Sum of diagonal entries}}{\text{Grand total}} = \frac{8{,}114}{26{,}569} = 0.305$$

Part 3. Calculation of $\hat{\kappa}$ using data from Parts 1 and 2

$$\hat{\kappa} = \frac{\text{Observed} - \text{expected}}{1 - \text{expected}} = \frac{0.528 - 0.305}{1 - 0.305} = \frac{0.223}{0.695} = 0.321$$

**TABLE 13.6. Contrived Matrices Illustrating $\hat{\kappa}$**

| 10 | | | | |
|---|---|---|---|---|
| | 10 | | | |
| | | 10 | | |
| | | | 10 | |
| | | | | 10 |

$\hat{\kappa} = +1.00$

| 10 | | | | |
|---|---|---|---|---|
| | 10 | | | |
| | | | 10 | |
| | | 10 | | |
| | | | | 10 |

$\hat{\kappa} = 0.50$

| | | | | 25 |
|---|---|---|---|---|
| | | | | |
| | | | | |
| | | | | |
| 25 | | | | |

$\hat{\kappa} = -1.00$

| | | | | 10 |
|---|---|---|---|---|
| | | | 10 | |
| | | 10 | | |
| | 10 | | | |
| 10 | | | | |

$\hat{\kappa} = 0.00$

| 2 | 2 | 2 | 2 | 2 |
|---|---|---|---|---|
| 2 | 2 | 2 | 2 | 2 |
| 2 | 2 | 2 | 2 | 2 |
| 2 | 2 | 2 | 2 | 2 |
| 2 | 2 | 2 | 2 | 2 |

$\hat{\kappa} = 0.00$

| | | | | 10 |
|---|---|---|---|---|
| | | | 10 | 10 |
| | | | | |
| 10 | | | | |
| 10 | | | | |

$\hat{\kappa} = -0.25$

of chance agreement increases, and the percentage correct decreases, the value of $\hat{\kappa}$ assumes a negative value.

$\hat{\kappa}$ and related coefficients have been discussed by Rosenfield and Fitzpatrick-Lins (1986), by Foody (1992), and by Ma and Redmond (1995). According to Cohen (1960), any negative value indicates a poor classification, but the possible range of negative values depends on the specific matrix evaluated. Thus, the magnitude of a negative value should not be interpreted as an indication of the relative performance of the classification. Values near zero suggest that the contribution of chance is equal to the effect of the classification, and that the classification process yields no better results than would a chance assignment of pixels to classes. $\hat{\kappa}$ is asymptotically normally distributed.

Congalton (1983) and Congalton et al. (1983) described how the large sample variance can be estimated, then used in a $Z$ test to determine whether individual $\hat{\kappa}$ scores differ significantly from one another. Congalton (1983) gives a FORTRAN computer program that calculates $\hat{\kappa}$ and the necessary values to conduct the test of significance. Students should read the article by Hudson and Ramm (1987) as they consider using these methods. Foody (1992), and Ma and Redmond (1995) have advocated use of the tau statistic ($\tau$) as an improvement over $\hat{\kappa}$. $\tau$ is analogous to $\kappa$, but is based on a priori classification probabilities rather than observed occurrences used in calculation of $\hat{\kappa}$.

### *Comparisons of Error Matrices*

One of the primary purposes of error analyses is to permit comparisons of different interpretations. For a given region, we may have alternative classifications made using images acquired at different dates, classified by different procedures, or produced by different individuals. In such instances, we wish to generate error matrices and ask which classification might be best for a given purpose. Such an analysis can permit identification of the specific imagery, dates, preprocessing procedures, and other procedures that might be best for a given purpose.

For example, Tables 13.7 and 13.8 give error matrices for two classifications of the

**TABLE 13.7. Error Matrix: Euclidean Distance Classification**

| | U | A | R | F | W | WL | B | |
|---|---|---|---|---|---|---|---|---|
| U | 2 | 31 | 1 | | | | | |
| A | 34 | 15 | | | | | | |
| R | | 4 | 44 | | 8 | | | |
| F | 5 | 4 | | 2 | | | | |
| W | | 2 | | | 12 | | | |
| WL | | | 1 | | 2 | 16 | | |
| B | | | | | 1 | | 2 | |
| | | | | | | | | 112 |

$N = 186$

$$\frac{112}{186} = 0.6022 = 60.22\%$$

$\hat{\kappa} = 0.482$

*Note.* U, urban; A, agriculture; R, rangeland; F, forest; W, water; WL, wetlands; B, barren land.

**TABLE 13.8. Error Matrix: Parallelpiped Classification**

| | U | A | R | F | W | WL | B | (UN) |
|---|---|---|---|---|---|---|---|---|
| U | 2 | 3 | | | | | | (5) |
| A | | 12 | 3 | | | | | (60) |
| R | | | 24 | | | 2 | | (34) |
| F | | | | 2 | | | | |
| W | | | | | 3 | | | (12) |
| WL | | | | | | 6 | | (16) |
| B | | | | | | | 2 | |
| | | | | | | | | 51 |

$N = 59$

$$\frac{51}{59} = 0.8644 = 86.44\%$$

$\hat{\kappa} = 0.809$

*Note.* (UN), unclassified. For explanation of other abbreviations, see Table 13.7.

same Landsat subscene near Roanoke, Virginia. One was produced using Euclidean distance classifier; the other a parallelepiped classifier. Because the parallelepiped may classify only a portion of the scene (Chapter 11), the matrix in Table 13.8 is based on fewer samples than in Tables 13.7. Comparison of the matrices is complicated by the differing numbers of observations. Although the percentage correct and $\hat{\kappa}$ both suggest that the parallelepiped classifier is more accurate, it would be useful to be able to compare the matrices item by item. For example, which classifier performed best in classification of agricultural land. From direct inspection of the two original matrices, the comparison is difficult because of the differing values. The sample for the Euclidean distance classifier was much larger; it produced more correctly classified agriculture pixels but also more that were erroneously classified. Especially when there are numerous off-diagonal entries, direct comparison of the two matrices can be difficult.

Congalton (1983) addresses this problem by applying a procedure given by Bishop et al. (1975) for normalization of a square matrix. As applied in this context, normalization is an iterative procedure that brings row and column marginals to a common value of +1.0 by incrementally altering values within the matrix. Thus, the error matrix for a perfect classification would be as shown in Table 13.9; the diagonal and the marginals consist of 1's. All other entries are 0's. Other matrices, from real classifications, will of course have diagonal entries each less than 1. Off-diagonal entries will be greater than 0 in proportion

**TABLE 13.9. Idealized Matrix Illustrating Perfect Classification**

| | | | | | Row marginals |
|---|---|---|---|---|---|
| 1.00 | | | | | 1.00 |
| | 1.00 | | | | 1.00 |
| | | 1.00 | | | 1.00 |
| | | | 1.00 | | 1.00 |
| | | | | 1.00 | 1.00 |
| 1.00 | 1.00 | 1.00 | 1.00 | 1.00 | |
| Column marginals | | | | | |

to the magnitude of the errors in each cell of the original matrix, subject to the constraint that row and column marginals each total +1.00.

The iterative nature of the normalization procedure (called *iterative proportional fitting*) does not lend itself to concise description or calculation by hand. Values in each position in the matrix are successively incremented or decreased to bring row and column marginals closer and closer to +1.0. When they approach within a specified small interval of +1.00, the procedure stops. Congalton (1983) gives a FORTRAN program to calculate a normalized matrix, given an initial error matrix. His program was used to calculate the standardized versions of the two matrices given above (Table 13.10). Examination of the standardized matrices reveals, for example, that the parallelepiped classifier performed better than the Euclidean distance classifier for agricultural land.

These matrices illustrate some other important points. The two matrices given in Tables 13.7 and 13.8 are based on differing classifications of the image. Ideally this region should be large enough to provide adequate numbers of samples in each class. Because of the difficulty of compiling large data sets for accuracy assessment, however, it is expensive to gather sufficient reference data. Therefore, in this example, even though 186 samples sites have been identified, the number of samples for each class are less than necessary to conduct a rigorous assessment of the accuracies of these classifications.

Because the parallelepiped classifier may not classify all pixels (as discussed in Chapter 11), the number of classified pixels for Table 13.8 is less than for Table 13.7, even though both classifications are based on the same data. This example illustrates the difficulty of controlling the numbers of pixels in the accuracy assessment and of comparing different matrices even though ostensibly the comparisons should be straightforward.

These matrices illustrate an undesirable characteristic of the normalization procedure. Error matrices based on remote sensed data typically display numerous entries with zeros, or low values, because some kinds of errors are improbable and occur infrequently. Note that the normalization procedure can force some zero values to assume nonzero values. In some instances the artificial values are so small that they can be regarded as token

**TABLE 13.10. Examples of Standardized Matrices**

| | Euclidean distance classification (Table 13.7) | | | | | | | |
|---|---|---|---|---|---|---|---|---|
| | U | A | R | F | W | WL | B | |
| U | 0.28 | 0.44 | 0.04 | 0.11 | 0.03 | 0.02 | 0.08 | 1.00 |
| A | 0.05 | 0.42 | 0.32 | 0.09 | 0.02 | 0.03 | 0.097 | 1.00 |
| R | 0.02 | 0.04 | 0.58 | 0.06 | 0.02 | 0.23 | 0.05 | 1.00 |
| F | 0.46 | 0.04 | 0.01 | 0.39 | 0.02 | 0.02 | 0.06 | 1.00 |
| W | 0.06 | 0.04 | 0.01 | 0.11 | 0.68 | 0.02 | 0.08 | 1.00 |
| WL | 0.05 | 0.01 | 0.02 | 0.09 | 0.11 | 0.65 | 0.07 | 1.00 |
| B | 0.08 | 0.01 | 0.02 | 0.15 | 0.12 | 0.03 | 0.59 | 1.00 |
| | 1.00 | 1.00 | 1.00 | 1.00 | 1.00 | 1.00 | 1.00 | |
| | Parallelepiped classification (Table 13.8) | | | | | | | |
| | U | A | R | F | W | WL | B | |
| U | 0.45 | 0.23 | 0.03 | 0.08 | 0.08 | 0.05 | 0.08 | 1.00 |
| A | 0.07 | 0.59 | 0.14 | 0.06 | 0.05 | 0.03 | 0.06 | 1.00 |
| R | 0.04 | 0.03 | 0.69 | 0.04 | 0.04 | 0.12 | 0.04 | 1.00 |
| F | 0.12 | 0.04 | 0.04 | 0.54 | 0.09 | 0.06 | 0.11 | 1.00 |
| W | 0.11 | 0.04 | 0.03 | 0.09 | 0.58 | 0.06 | 0.09 | 1.00 |
| WL | 0.09 | 0.03 | 0.03 | 0.08 | 0.07 | 0.62 | 0.08 | 1.00 |
| B | 0.12 | 0.04 | 0.04 | 0.11 | 0.09 | 0.06 | 0.54 | 1.00 |
| | 1.00 | 1.00 | 1.00 | 1.00 | 1.00 | 1.00 | 1.00 | |

*Note.* For explanation of abbreviations, see Table 13.7.

entries that do not alter our interpretation. In other instances, however, the normalized values may differ from the original values sufficiently to alter our conclusions. (Of course, the nonzero entries in the matrix are also altered, but usually the changes in the zero values attract our attention first.) Thus, normalization, like all methods of data analysis, should not be applied without careful consideration of the data and critical interpretation of the results.

## 13.6. Summary

Accuracy assessment is a complex process. This chapter cannot address all the relevant topics in detail, and even the most complete discussion leaves many unresolved issues. Research continues, and despite agreement on many important aspects of accuracy evaluation, other issues are likely to be debated for a long time before they are resolved. For example, there is no clearly superior method of sampling an image for accuracy assessment, and there is disagreement concerning the best way to compare two error matrices. For many problems, there may be no single "correct" way to conduct the analysis, but we may be able to exclude some alternatives and to speak of the relative merits and shortcomings of others.

This chapter provides the background to assess the accuracies of a classification using procedures that, if not perfect, are at least comparable in quality to those in common use today. Furthermore, the student should now be prepared to read some of the current research on this topic, and possibly to contribute to improvements in the study of accuracy assessment. Many of the problems in this field are difficult, but they are not beyond the reach of interested and informed students.

## Review Questions

1. Examine matrix A. Which category has the greatest areal extent on the image? Which category has the greatest areal extent within the ground area shown on the image?

**Matrix A**

| Actual land use | Interpreted land use | | | | | |
|---|---|---|---|---|---|---|
| | Urban | Agriculture | Range | Forest | Water | Total |
| Urban | 510 | 110 | 85 | 23 | 10 | 738 |
| Agriculture | 54 | 1,155 | 235 | 253 | 35 | 1,732 |
| Range | 15 | 217 | 930 | 173 | 8 | 1,343 |
| Forest | 37 | 173 | 238 | 864 | 27 | 1,339 |
| Water | 5 | 17 | 23 | 11 | 265 | 321 |
| Total | 621 | 1,672 | 1,511 | 1,324 | 345 | 3,724 |

**Matrix B**

| Actual land use | Interpreted land use | | | | | |
|---|---|---|---|---|---|---|
| | Urban | Agriculture | Range | Forest | Water | Total |
| Urban | 320 | 98 | 230 | 64 | 26 | 738 |
| Agriculture | 36 | 1,451 | 112 | 85 | 48 | 1,732 |
| Range | 98 | 382 | 514 | 296 | 53 | 1,343 |
| Forest | 115 | 208 | 391 | 539 | 86 | 1,339 |
| Water | 28 | 32 | 68 | 23 | 170 | 321 |
| Total | 597 | 2,171 | 1,315 | 1,007 | 383 | 2,994 |

2. Refer again to matrix A. Which class shows the highest error of commission?
3. Based upon your examination of matrix A, identify the class that was most often confused with agricultural land.
4. For matrix A, which class was most accurately classified? Which class has the lowest accuracy?
5. For matrix A, which class shows the highest number of errors of omission?
6. Compare matrices A and B, which are error matrices for alternative interpretations of the same area, perhaps derived from different images or compiled by different analysts. Each has 5,473 pixels, but they are allocated to classes in different patterns. Which image is the "most accurate"? Can you see problems in responding with a simple answer to this question?
7. Refer again to matrices A and B. If you were interested in accurate delineation of agricultural land, which image would you prefer? If accurate classification of forested land is important, which image should be used?

8. What might be the consequences if we had no effective means of assessing the accuracy of a classification of a remotely sensed image?
9. Data for matrices A and B can be compiled from a complete inventory of each image, or by selecting a sample of pixels from both the reference and classified images. List advantages and disadvantages of each method.
10. Make a list of applications of the accuracy assessment technique. That is, think of different circumstances/purposes in which it is valuable to be able to determine which of two alternative classifications is the more accurate.

## References


Anderson, J. R., E. E. Hardy, J. T. Roach, and R. E. Witmer. 1976. *A Land Use and Land Cover Classification for Use with Remote Sensor Data.* U.S. Geological Survey Professional Paper 964. Washington, DC: U.S. Government Printing Office, 28 pp.

Bishop, Y. M. M., S. E. Fienber and P. W. Holland. 1975. *Discrete Multivariate Analysis: Theory and Practice.* Cambridge, MA: MIT Press, 557 pp.

Campbell, J. B. 1981. Spatial Correlation Effects Upon Accuracy of Supervised Classification of Land Cover. *Photogrammetric Engineering and Remote Sensing,* Vol. 47, pp. 355–363.

Chrisman, N. R. 1980. Assessing Landsat Accuracy: A Geographic Application of Misclassification Analysis. *Second Colloquium on Quantitative and Theoretical Geography,* Trinity Hall, Cambridge, England.

Cohen, J. 1960. A Coefficient of Agreement for Nominal Scales. *Educational and Psychological Measurement,* Vol. 20, No. 1, pp. 37–40.

Congalton, R. G. 1983. *The Use of Discrete Multivariate Techniques for Assessment of Landsat Classification Accuracy.* M.S. thesis. Blacksburg, VA: Virginia Polytechnic Institute and State University, 111 pp.

Congalton, R. G. 1984. *A Comparison of Five Sampling Schemes Used in Assessing the Accuracy of Land Cover/Land Use Maps Derived from Remotely Sensed Data.* Ph.D. dissertation. Blacksburg, VA: Virginia Polytechnic Institute and State University, 147 pp.

Congalton, R. G. 1988. A Comparison of Sampling Schemes Used in Generating Error Matrices for Assessing the Accuracy Maps Generated from Remotely Sensed Data. *Photogrammetric Engineering and Remote Sensing,* Vol. 54, pp. 593–600.

Congalton, R. G. 1988. Using Spatial Autocorrelation Analysis to Explore the Errors in Maps Generated from Remotely Sensed Data. *Photogrammetric Engineering and Remote Sensing,* Vol. 54, pp. 587–592.

Congalton, R. G., and R. A. Mead. 1983. A Quantitative Model to Test for Consistency and Correctness in Photointerpretation. *Photogrammetric Engineering and Remote Sensing,* Vol. 49, pp. 69–74.

Congalton, R. G., R. G. Oderwald, and R. A. Mead. 1983. Assessing Landsat Classification Accuracy Using Discrete Multivariate Analysis Statistical Techniques. *Photogrammetric Engineering and Remote Sensing,* Vol. 49, pp. 1671–1687.

Fisher, Peter F. 1994. Hearing the Reliability in Classified Remotely Sensed Images. *Cartography and Geographic Information Systems,* Vol. 21, pp. 31–36.

Fisher, Peter F. 1994. Visualization of the Reliability in Classified Remotely Sensed Images. *Photogrammetric Engineering and Remote Sensing,* Vol. 54, pp. 905–910.

Fitzpatrick-Lins, K. 1978. Accuracy and Consistency Comparisons of Land Use and Land Cover Maps Made from High-Altitude Photographs and Landsat Multispectral Imagery. *Journal of Research, U.S. Geological Survey,* Vol. 6, pp. 23–40.

Foody, Giles, M. 1992. On the Compensation for Change Agreement in Image Classification Accuracy Assessment. *Photogrammetric Engineering and Remote Sensing,* Vol. 58, pp. 1459–1460.

Franklin, Steven E., D. P. Peddle, B. A. Wilson, and C. Blodget. 1991. Pixel Sampling of Remotely Sensed Digital Imagery. *Computers and Geosciences,* Vol. 17, pp. 759–775.

Gershmehl, Philip J., and D. E. Napton. 1982. Interpretation of Resource Data: Problems of Scale and Transferability. *Practical Applications of Computers in Government, Papers from the Annual Conference of the Urban and Regional Information Systems Association,* pp. 471–482.

Ginevan, M. E. 1979. Testing Land-Use Map Accuracy: Another Look. *Photogrammetric Engineering and Remote Sensing,* Vol. 45, pp. 1371–1377.

Hay, A. 1979. Sampling Designs to Test Land Use Map Accuracy. *Photogrammetric Engineering and Remote Sensing,* Vol. 45, pp. 529–533.

Hord, R. M., and W. Brooner. 1976. Land Use Map Accuracy Criteria. *Photogrammetric Engineering and Remote Sensing,* Vol. 46, pp. 671–677.

Hudson, W. D., and C. W. Ramm. 1987. Correct Formulation of the Kappa Coefficient of Agreement. *Photogrammetric Engineering and Remote Sensing,* Vol. 53, pp. 421–422.

Ma, Z., and R. L. Redmond. 1995. Tau Coefficients for Accuracy Assessment of Classification of Remotely Sensed Data. *Photogrammetric Engineering and Remote Sensing,* Vol. 61, pp. 435–439.

Podwysocki, M. H. 1976. *An Estimate of Field Size Distribution for Selected Sites in Major Grain Producing Countries.* Greenbelt, MD: Goddard Space Flight Center (X-923-76-93), 34 pp.

Quirk, B. K., and F. L. Scarpace. 1980. A Method of Assessing Accuracy of a Digital Classification. *Photogrammetric Engineering and Remote Sensing,* Vol. 46, pp. 1427–1431.

Rosenfield, G. H., and K. Fitzpatrick-Lins. 1986. A Coefficient of Agreement as a Measure of Thematic Map Classification Accuracy. *Photogrammetric Engineering and Remote Sensing,* Vol. 52, pp. 223–227.

Rosenfield, G. H., K. Fitzpatrick-Lins, and H. S. Ling. 1982. Sampling for Thematic Map Accuracy Testing. *Photogrammetric Engineering and Remote Sensing,* Vol. 48, pp. 131–137.

Rosenfield, George H., and K. Fitzpatrick-Lins. 1986. A Coefficient of Agreement as a Measure of Thematic Classification Accuracy. *Photogrammetric Engineering and Remote Sensing,* Vol. 52, pp. 223–227.

Silk, J. 1979. *Statistical Concepts in Geography.* Boston: Allen & Unwin, 276 pp.

Simonett, D. S., and J. C. Coiner. 1971. Susceptibility of Environments to Low Resolution Imaging for Land Use Mapping. In *Proceedings of the Seventh International Symposium on Remote Sensing of Environment.* Ann Arbor: University of Michigan, pp. 373–394.

Skidmore, A. K., and B. J. Turner. 1992. Map Accuracy Assessment Using Line Intersect Sampling. *Photogrammetric Engineering and Remote Sensing,* Vol. 58, pp. 1453–1457.

Stehman, Stephen V. 1992. Comparison of Systematic and Random Sampling for Estimating the Accuracy of Maps Generated from Remotely Sensed Data. *Photogrammetric Engineering and Remote Sensing,* Vol. 58, pp. 1343–1350.

Stehman, Stephen V. 1996. Estimating the Kappa Coefficient and Its Variance under Stratified Random Sampling. *Photogrammetric Engineering and RemoteSensing,* Vol. 62, pp. 401–407.

Story, M., and R. G. Congalton. 1986. Accuracy Assessment—A User's Perspective. *Photogrammetric Engineering and Remote Sensing,* Vol. 52, pp. 397–399.

Todd, W. J., D. G. Gehring, and J. F. Haman. 1980. Landsat Wildland Mapping Accuracy. *Photogrammetric Engineering and Remote Sensing,* Vol. 46, pp. 509–520.

Tom, C., L. D. Miller and J. W. Christenson. 1978. *Spatial Land-Use Inventory, Modeling, and Projection: Denver Metropolitan Area, with Inputs from Existing Maps, Airphotos, and Landsat Imagery.* NASA Technical Memorandum 79710. Greenbelt, MD: Goddard Space Flight Center, 225 pp.

Turk, G. 1979. GT Index: A Measure of the Success of Prediction. *Remote Sensing of Environment,* Vol. 8, pp. 65–75.

Van der Wel, Frans J. M., and L. L. F. Janssen. 1994. A Short Note on Pixel Sampling of Remotely Sensed Digital Imagery. *Computers and Geosciences,* Vol. 20, pp. 1263–1264.

Van Genderen, J., and B. Lock. 1977. Testing Map Use Accuracy. *Photogrammetric Engineering and Remote Sensing,* Vol. 43, pp. 1135–1137.

Verdin, J. 1983. Corrected vs. Uncorrected Landsat 4 MSS Data. *Landsat Data Users Notes,* Issue 27, pp. 4–7.

Webster, R., and P. H. T. Beckett. 1968. Quality and Usefulness of Soil Maps. *Nature,* Vol. 219, pp. 680–682.

CHAPTER FOURTEEN

# Hyperspectral Remote Sensing

## 14.1. Introduction

Remote sensing is practiced by the examination of features as observed in several regions of the electromagnetic spectrum. Conventional remote sensing, as outlined in previous chapters, is based on the use of several rather broadly defined spectral regions—*hyperspectral remote sensing* is based on the examination of many narrowly defined spectral channels. Sensor systems such as SPOT, MSS, and TM provide 3, 4, and 7 spectral channels, respectively. Hyperspectral sensors, described below, can provide as many as 200 or more channels, each only 10 nm wide. Although hyperspectral remote sensing applies the same principles and methods discussed previously, it requires specialized data sets, instruments, field data, and software to the extent that it forms a specialized field of inquiry.

## 14.2. Spectroscopy

Hyperspectral data have detail and accuracy that permit investigation of phenomena and concepts that greatly extend the scope of traditional remote sensing. For example, analysts can begin to match observed spectra to those recorded in data bases and to closely examine radiation as it passes through the atmosphere. Such capabilities present opportunities for much more precise identifications of features than is possible with broad-band sensors, for use of spectral libraries, for detailed investigation of biologic and geologic phenomena, and for self-correction of data in some bands using information from other bands.

These capabilities extend the reach of remote sensing into the field of *spectroscopy,* the science devoted to detailed examination of very accurate spectral data. Classical spectroscopy has origins in experiments conducted by Isaac Newton (1642–1727), who used glass prisms to separate visible light into the spectrum of colors. Later, another English physicist, William Wollaston (1766–1828), noted that spectra displayed dark lines when light is projected through a narrow slit. The meaning of these lines was discovered through the work of Joseph Fraunhofer (1787–1826), a German glassmaker who discovered distinctive lines in spectra of light from the sun and from stars. Dark lines (*absorption spectra*) are observed as radiation passes through gases at low temperature; bright lines (*emission spectra*) form as heated gases (in the sun's atmosphere, for example) emit radiation. These lines have origins in the chemical elements present in the gases—a discovery that has permitted astronomers to investigate differences in chemical compositions of stars and planets. The Danish physicist Neils Bohr (1885–1962) found that the character of Fraunhofer lines is ultimately determined by the atomic structure of gases.

Instruments used in spectroscopy—*spectroscopes, spectrometers, spectrographs*—are designed to collect radiation with a lens and divide it into spectral regions (using prisms or diffraction gratings), which is then recorded on film or measured electronically. These instruments permit application of spectroscopy not only to problems in astronomy but also for laboratory analyses to characterize unidentified materials.

## 14.3. Hyperspectral Remote Sensing

Hyperspectral remote sensing is an application of the practice of spectroscopy to examination of solar radiation reflected from the earth's surface. Although hyperspectral remote sensing can sometimes apply the techniques of classical spectroscopy to the study of atmospheric gasses and pollutants, for example, more commonly it applies the practice of making precise, accurate, and detailed spectral measurements to the study of the earth's surface (*imaging spectrometry*). Such data have accuracy and detail sufficient to begin to match observed spectra to those stored in data bases, known as *spectral libraries.* Instruments for hyperspectral remote sensing differ from those of conventional spectroscopy in that they gather spectra not only for point targets but for areas—not for stars or laboratory samples but for regions of the earth's surface. Instruments for hyperspectral remote sensing differ from other remote sensing instruments in their extraordinarily fine spectral, spatial, and radiometric resolutions and in their careful calibration. Some hyperspectral instruments collect data in 200 or more channels at 10 to 12 bits. Because of their accurate calibration and ability to collect data at fine detail, such instruments greatly extend the reach of remote sensing not only by extending the range of applications but also by defining new concepts and analytical techniques.

Although the techniques of classical spectroscopy can be used in hyperspectral remote sensing to examine, for example, atmospheric gases, hyperspectral remote sensing examines very detailed spectra for images of the earth's surface, applies corrections for atmospheric effects, and matches them to spectra of known features.

## 14.4. AVIRIS

One of the first hyperspectral sensors was designed in the early 1980s by the Jet Propulsion Laboratory (JPL) (Pasadena, CA). The *airborne imaging spectrometer* (AIS) greatly extended the scope of remote sensing by virtue of the number of spectral bands; their fine spatial, spectral, and radiometric detail; and the accuracy of its calibration. AIS collected 128 spectral channels, each about 10 nm wide in the interval 1.2 to 2.4 μm. (Alternatively, a spectral range from 1.9 to 2.1 μm could be used for geological investigations. Spatial resolution was about 8 m.) The term *hyperspectral remote sensing* recognizes the fundamental difference between these data and those of the usual broad-band remote sensing

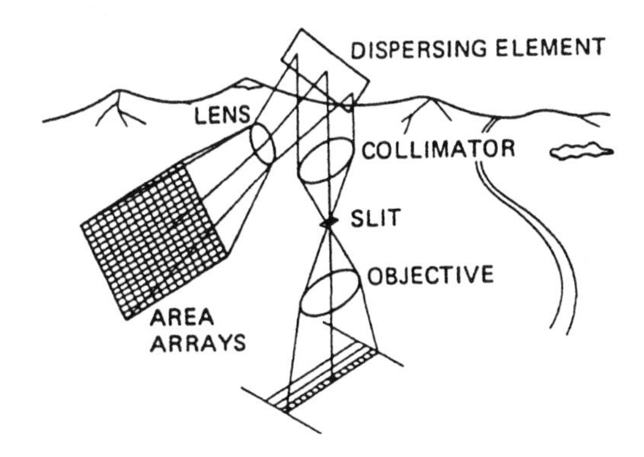


**FIGURE 14.1.** Imaging spectrometer (NASA diagram).

instruments. (Systems with even finer spectral resolution, designed primarily to study atmospheric gases, are known as *ultraspectral* sensors.)

Although several hyperspectral instruments have been constructed or planned, the principal hyperspectral instrument now in use is the *airborne visible/infrared imaging spectrometer* (AVIRIS). AVIRIS was developed by NASA and JPL from the foundations established by AIS. AVIRIS was first tested in 1987 and placed in service in 1989, although the instrument has been modified at intervals to upgrade its reliability and performance. Each year AVIRIS flights collect data to support scientific investigations in oceanography, ecology, geosciences, hydrology, and other fields. It has now acquired hundreds of images of test sites in North America and Europe.

Hyperspectral sensors necessarily employ designs that differ from those of the usual sensor systems. An objective lens collects radiation reflected or emitted from the scene; a collimating lens projects the radiation as a beam of parallel rays through a diffraction grating that separates the radiation into discrete spectral bands (Figure 14.1). Energy in each spectral band is then detected by linear arrays of silicon and indium antimonide (Chapters 4 and 8). Because of the wide spectral range of AVIRIS, detectors are configured in four separate panels (0.4–0.7 μm, 0.7–1.3 μm, 1.3–1.9 μm, and 1.8–2.8 μm) each calibrated independently (Figure 14.2). AVIRIS operates over the spectral range of 400 to 2500 nm (0.4 to 2.45 μm) producing 224 spectral channels, each 10 nm wide (Figure 14.3). After processing, data consist of 210 spectral channels. At its usual operating alti-

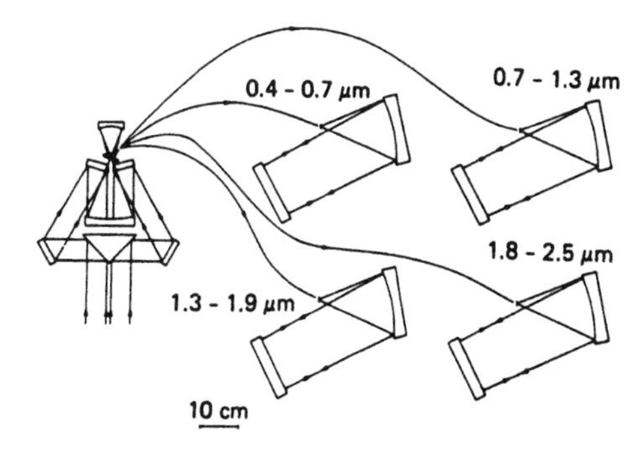


**FIGURE 14.2.** AVIRIS sensor design (NASA diagram).

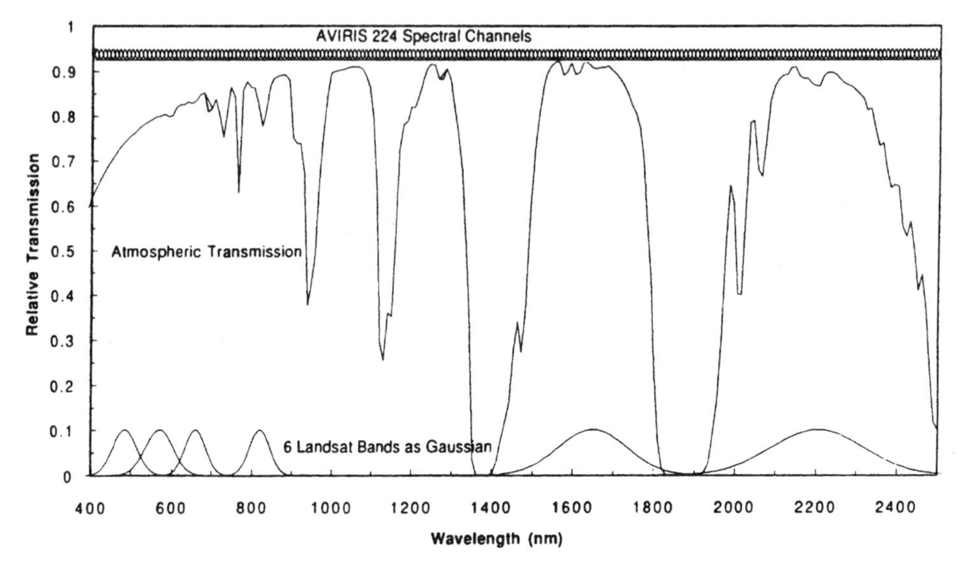


**FIGURE 14.3.** AVIRIS spectral channels compared to Landsat TM spectral channels. The 220 narrow bands at the top represent AVIRIS spectral channels. For comparison, Landsat TM channels are shown at the bottom (Green & Simmonds, 1993).

tude, each image records a strip 11 km wide, processed to forms scenes recording areas about 11 km × 11 km. Each line of data conveys about 614 pixels, each representing a ground area about 20 m on a side.

## 14.5. The Image Cube

The *image cube* refers to the representation of hyperspectral data as a three-dimensional figure, with two dimensions formed by the *x* and *y* axes of the usual map or image display and the third (*z*) formed by the accumulation of spectral data as additional bands are superimposed on each other. In Figure 14.4 and Plate 13, the top of the cube is an image composed of data collected at the shortest wavelength (collected in the ultraviolet), and the bottom formed by the spectral channel collected at the longest wavelength (2.5 μm). Intermediate wavelengths are found as slices through the cube at intermediate positions. Values for a single pixel observed along the edge of the cube form a spectral trace describing the spectra of the surface represented by the pixel.

## 14.6. Data Libraries

The development of hyperspectral remote sensing has been accompanied by the accumulation of detailed spectral data acquired in the laboratory and in the field. These data are organized in *spectral libraries*—data bases maintained primarily by governmental agencies but also by other organizations. These libraries assemble spectra that have been ac-

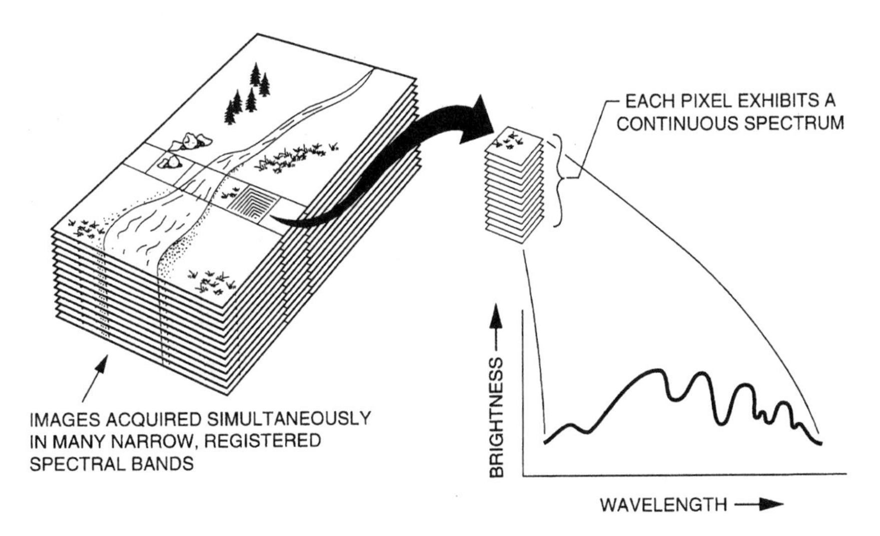


**FIGURE 14.4.** Image cube (based on NASA diagram).

quired at test sites representative of varied terrain and climate zones, observed in the field under natural conditions. Included also are other data describing, for example, construction materials, minerals, vegetation, and fabrics as observed in laboratories under standardized conditions.

Such data are publicly available to the remote sensing community and have been incorporated into software designed for use in hyperspectral remote sensing. Maintenance of a spectral library requires specialized effort to bring data into a common format that can be used by a diverse community of users. Typically spectral data are collected in the field by using varied instruments under a range of illumination conditions (Section 12.6). These differences must be documented and resolved to prepare data in a format that permits use by scientists with differing interests.

Because of the fine spectral, spatial, and radiometric detail of hyperspectral analysis, identification and cataloging form special problems for the design of spectral libraries. Therefore, each spectral record must be linked to detailed information specifying the instruments used, meteorological conditions, nature of the surface, and the circumstances of the measurement. These kinds of ancillary data are required for successful interpretation and analysis of the image data.

## 14.7. Spectral Matching

Figure 14.5, derived from Barr (1994), illustrates a sequence of analysis for hyperspectral data. Analysis begins with acquisition and preprocessing (1) to remove known system errors and ensure accurate calibration. Any of several methods can be used to correct for atmospheric effects (2), as outlined in Chapter 10. The lower portion of Figure 14.5 represents the analytical process. Data are organized to permit convenient display and manipulation of specified bands or combinations of bands.

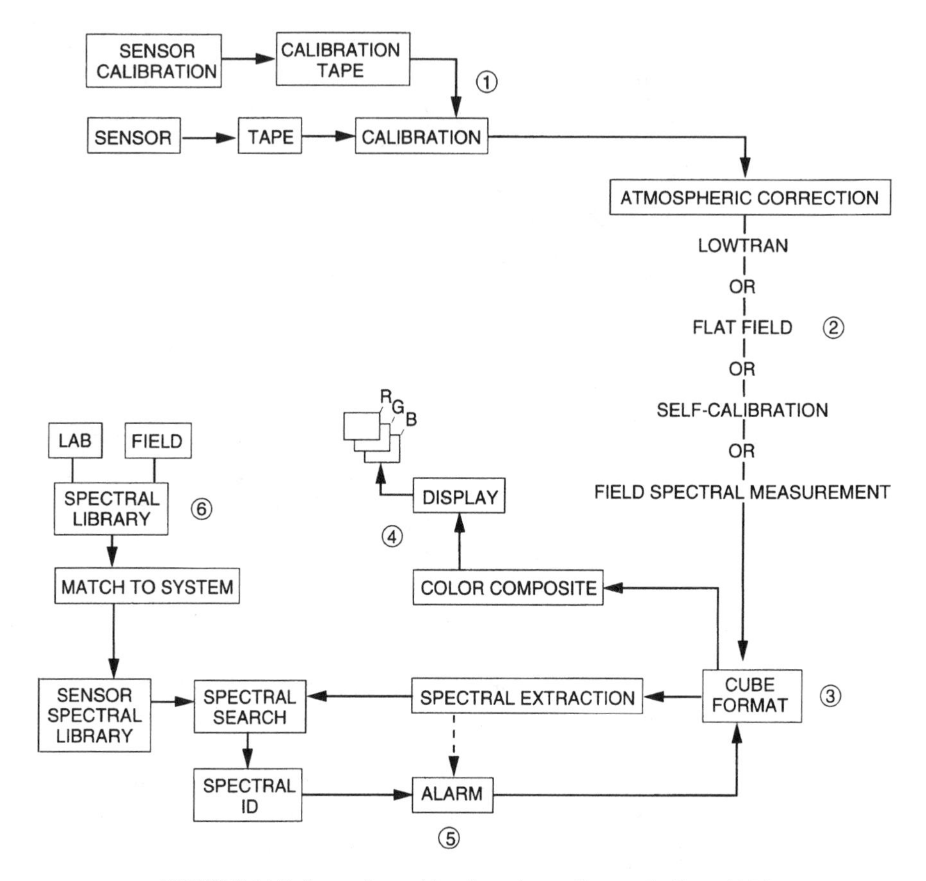


**FIGURE 14.5.** Spectral matching (based on a diagram by Barr, 1994).

Selected bands can be used to prepare color composites for display. *Alarm* (5) means that the analyst can place the cursor on the image to mark a specific pixel or region of pixels and then instruct the computer to highlight (alarm) other regions on the image that are characterized by the same spectral response. Then the analyst can attempt to match the alarmed areas to spectra from a spectral library (6) accessible to the image processing system. If a match can be made, the analyst can examine another region and attempt to match it to entries from the spectral library.

## 14.8. Spectral Mixing Analysis

Fine detail does not overcome the enduring obstacles to the practice of remote sensing. Surface materials recorded by the sensor are not always characteristic of subsurface conditions. Atmospheric effects, shadowing, and topographic variations contribute to observed spectra to confuse interpretations. Even when observed at fine detail, surfaces are often composed of varied materials. Therefore, the sensor observes composite spectra that may not clearly match the pure spectra of spectral libraries. *Linear mixing* refers to additive combinations of several diverse materials that occur in patterns too fine to be resolved by the sensors (Figure 14.6). This is the effect of *mixed pixels* outlined in Chapter 9. As long as the radiation from component patches remains separate until it reaches the sensor, it is possible to estimate proportions of component surfaces from the observed pixel brightness using methods such as those illustrated by Equation 9.2. Linear mixing might occur when components of a composite surface occur in a few compact areas (Figure 14.6b). *Nonlinear mixing* occurs when radiation from several surfaces combines before it reaches the sensor. Nonlinear mixing occurs when component surfaces arise in highly dispersed patterns (Figure 14.6c). Nonlinear mixing cannot be addressed by the techniques described here.

*Spectral mixing analysis* (also known as *spectral unmixing*) is devoted to extracting pure spectra from the complex composites of spectra that by necessity form each image. It assumes that pixels are formed by linear mixing, and that it is possible to identify the components contributing to the mixture. It permits analysts to define key components of a specific scene and forms an essential component in the process of spectral matching discussed next. Analysts desire to match data from hyperspectral images to corresponding laboratory data, to identify surfaces from their spectral data much more precisely than previously was possible. Whereas conventional image classification (Chapter 11) matches pixels to broad classes of features, hyperspectral image matching attempts to make more precise identifications—to specific mineralogies of soils or rocks, for example.

Therefore, spectral matching requires the application of techniques that enable analysts to separate pure pixels from impure pixels. This problem is well matched to the capabilities of *convex geometry,* which examines multidimensional data envisioned in *n* dimensions. Individual points (pixels) within this data space can be examined as linear combinations of an unknown number of pure components. Convex geometry can solve such problems provided the components are linearly weighted, sum to unity, and are positive. We assume also that the data have greater dimensionality (more spectral bands) than the number of pure components.

The illustrations here, for convenience and legibility, show only two dimensions, al-

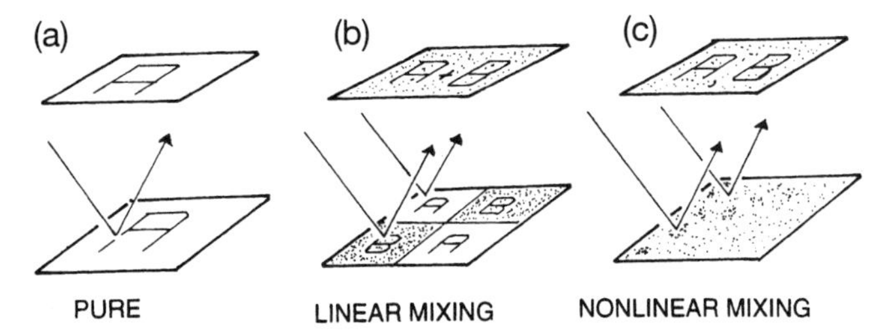


**FIGURE 14.6.** Linear and nonlinear spectral mixing. (*a*) If a pixel represents a uniform ground area at the resolution of the sensor, the pixel represents a pure spectrum. (*b*) If a pixel represents two or more surfaces that occur in patches that are large relative to the sensor's resolution, mixing occurs at the sensor. The pattern of the composite surfaces can never be resolved, but because mixing occurs in a linear manner, proportions of the components can be estimated. (*c*) If the composite occurs at a scale that is fine relative to the resolution of the sensor, mixing occurs before radiation reaches the sensor, and components of the composite cannot be estimated using methods described here.

though the power of the technique is evident only with much higher dimensionality. In Figure 14.7a, the three points *A, B,* and *C* represent three spectral observations at the extreme limits of the swarm of data points, represented by the shaded pattern. That is, the shaded pattern represents values of all the pixels within a specific image or subimage, which is simplified by the triangle shape. (Other shapes can be defined as appropriate to approximate the shape of data swarm, although an objective is to define the simplest shape [*simplex*] that can reasonably approximate the pattern of the data swarm.)

These three points (for the example in Figure 14.7a) form *endmembers,* defined as the pure pixels that contribute to the varied mixtures of pixels in the interior of the data swarm. Once the simple form is defined, the interior pixels can be defined as linear combinations of the pure endmembers. (The discussion of mixed pixels in Chapter 9 anticipates this concept.) In general, interior points can be interpreted as positive unit–sum combinations of the pure variables represented by endmembers at the vertices. In general, a shape defined by $n + 1$ vertices is the simplest shape that encompasses interior points (i.e., for two dimensions ($n = 2$), the simplex is a triangle (i.e., $3 = n + 1$). The faces of the shape are *facets,* and the exterior surface is a *convex hull.* In Figure 14.7, *A, B,* and *C* are the observed approximations of the idealized spectra *A′, B′, and C′* that may not be observed on any specific image.

In the application of convex geometry to hyperspectral data, it is first necessary to

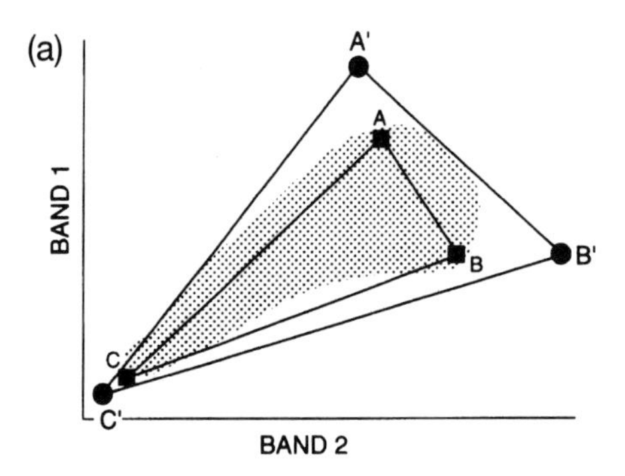


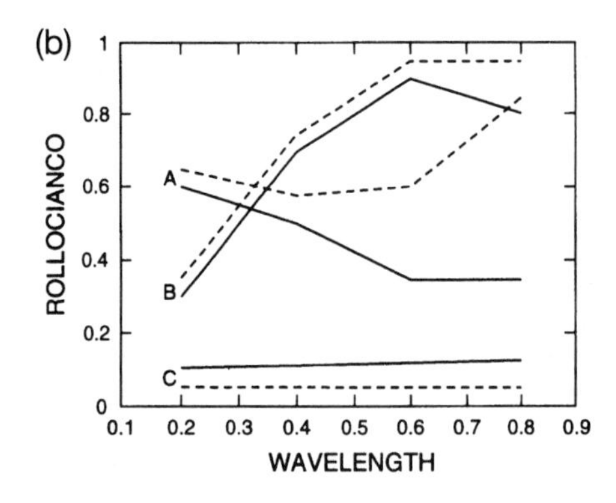


**FIGURE 14.7.** Spectral mixing analysis (Tompkins et al., 1993). (*a*) Simplex. (*b*) Endmembers.

define the dimensionality of the data. An image is typically subset to define a region of relative homogeneity and to remove bands with high variance (assumed to include high levels of error). The original hyperspectral data are converted to ground level reflectances (from radiances) using atmospheric models. The data are condensed by applying a version of principal components analysis (PCA) (Chapter 9). Although hyperspectral data may include many bands (224 for AVIRIS), duplication from one channel to another means that inherent dimensionality is much less (perhaps as few as 3 to 10), depending on the specifics of each scene.

The analyst then examines the transformed data in a data space to define the smallest simplex that fits the data. This process defines the $n + 1$ facets that permit identification of the $n$ endmembers (Figure 14.7a). These vertices, when projected back into the original spectral domain, estimate the spectra of the endmembers. These spectra are represented by Figure 14.7b. The objective is to match these endmembers to spectra from spectral libraries and to prepare maps and images that reveal the varied mixtures of surfaces that contribute to the observed spectra in each image.

Typical endmembers in arid regions have included bare soil, water, partially vegetated surfaces, fully vegetated surfaces, and shadows. Endmembers can be investigated in the field to confirm or revise identifications made by computer. Software for hyperspectral analysis often includes provisions for spectral libraries (and additional spectra as acquired in the field or laboratory) and the ability to search for matches with endmembers identified.

Although it may not always be possible to identify matches in spectral libraries, such analyses can narrow the range of alternatives, and in some instances mathematical models can assist in defining poorly defined endmembers.

## 14.9. Analyses

Other investigations have attempted to understand relationships between spectral data and specific physical or biological processes. For example, Curran (1989) reviews efforts to use hyperspectral data to monitor botanical variables. At high levels of spectral, radiometric, and spatial resolution, observed spectra of laboratory samples of plant tissues are influenced by atomic and molecular structures of water and by concentrations of specific organic compounds (such as cholorophyll, lignin, and cellulose). These relationships have formed the basis for research devoted to examination of hyperspectral data of vegetation canopies to derive estimates of foliar chemistry of plant tissues in situ. Such estimates would support agricultural, forestry, and ecological studies by providing indications of nutrient availability, rates of productivity, and rates of decomposition. Green (1993) and Rivard and Arvidson (1992) report some of the applications of hyperspectral data to lithologic and mineralogic analyses.

## 14.10. Summary

The vast amounts of data collected by hyperspectral systems, and the problems they present for both collection and analysis, prevent routine use of hyperspectral data in the same way that we might collect TM or SPOT data on a regular basis. More likely, hyperspectral data will provide the basis for discovering and refining the knowledge needed to develop improved sensors and analytical techniques that can be applied on a more routine basis. A second important role for hyperspectral data lies in the monitoring of long-term research sites, especially those devoted to study of biophysical processes and other phenomena that change over time. The fine detail of hyperspectral data will provide detailed information for ecological monitoring and for understanding patterns of lower-resolution data recorded on the same sites.

Previous plans to design a hyperspectral sensor for satellite observation have been abandoned (Chapter 20), although strong interest in the use of such an instrument continues.

## Review Questions

1. Discuss *advantages* of hyperspectral remote sensing in relation to more conventional remote sensing instruments.
2. List some *disadvantages* of hyperspectral remote sensing relative to use of systems such as SPOT or Landsat.
3. Prepare a plan to monitor an agricultural landscape in the midwest using *both* hyperspectral data *and* SPOT or Landsat data.
4. How would use of hyperspectral data influence collection of field data, compared to similar studies using SPOT or Landsat imagery?
5. Discuss how *preprocessing* and *image classification* differ with hyperspectral data in comparison to SPOT data.
6. How would *equipment* needs differ for image processing of hyperspectral data, compared to more conventional multispectral data?
7. The question of choosing between broad-scale coverage at coarse detail and focused coverage at fine detail recurs frequently in many fields of study. How does availability of hyperspectral data influence this discussion?
8. Can you think of ways that availability of hyperspectral data will influence the concepts and theories of remote sensing?
9. Hyperspectral data have so much volume that it is unfeasible to accumulate geographic coverage comparable to that of SPOT or Landsat, for example. What value, then, can hyperspectral data have?
10. Discuss the problems that arise in attempting to design and maintain a spectral library. (Consider, for example, the multitude of different materials and surfaces that must be considered, each under conditions of varied illumination.)

## References

Barr, Samuel. 1994. Hyperspectral Image Processing. In *Proceedings of the International Symposium on Spectral Sensing Research.* Ft. Belvior, VA: U.S. Army Corps of Engineers, pp. 447–490.

Curran, P. J. 1989. Remote Sensing of Foliar Chemistry. *Remote Sensing of Environment,* Vol. 29, pp. 271–278.

Curran, P. J. 1994. Imaging Spectrometry. *Progress in Physical Geography,* Vol. 18, pp. 247–266.

Curran, P. J., and J. A. Kupiec. 1995. Imaging Spectrometry: A New Tool for Ecology. Chapter 5 in *Advances in Environmental Remote Sensing* (F. M. Danson, and S. E. Plummer, eds.). New York: Wiley, pp. 71–88.

Goa, B. C., K. B. Heidebrecht, and A. F. H. Goetz. 1993. Derivation of Scaled Surface Reflectances from AVIRIS Data. *Remote Sensing of Environment,* Vol. 44, pp. 165–178.

Green, Robert O. (ed.). 1993. *Summaries of the Fourth Annual JPL Airborne Geoscience Workshop, October 25–29, 1993.* Pasadena, CA: NASA–Jet Propulsion Laboratory, 209 pp.

Green, R. O., and J. J. Simmonds. 1993. A Role for AVIRIS in Landsat and Advanced Land Remote Sensing System Program. In *Summaries of the Fourth Annual JPL Airborne Geoscience Workshop* (R. O. Green, ed.). Pasadena, CA: NASA–Jet Propulsion Laboratory, pp. 85–88.

Lee, C., and D. A. Landgrebe. 1992. Analysing High Dimensional Data. In *Proceedings, International Geoscience and Remote Sensing Symposium (IGARSS '93)* (Vol. 1). New York: Institute of Electrical and Electronics Engineers, pp. 561–563.

Rivard, Benoit, and R. E. Arvidson. 1992. Utility of Imaging Spectrometry for Lithologic Mapping in Greenland. *Photogrammetric Engineering and Remote Sensing,* Vol. 58, pp. 945–949.

Roberts, D. A., M. O. Smith, and J. B. Adams. 1993. Green Vegetation, Nonphotosynthetic Vegetation, and Soils in AVIRIS Data. *Remote Sensing of Environment,* Vol. 44, pp. 255–269.

Tompkins, S., J. F. Mustard, C. M. Pieters, and D. W. Forsyth. 1993. Objective Determination of Image End-Members in Spectral Mixture Analysis of AVIRIS Data. In *Summaries of the Fourth Annual JPL Airborne Geoscience Workshop* (R. O. Green, ed.). Pasadena, CA: NASA–Jet Propulsion Laboratory, pp. 177–180.

PART THREE

# APPLICATIONS

CHAPTER FIFTEEN

# Geographic Information Systems

## 15.1. Introduction

Geographic information systems (GISs) are specialized data systems that preserve locational identities of the information they record. A digital computer provides the basis for storage, manipulation, and display of large amounts of data that have been encoded in digital form. In essence, a GIS consists of a series of overlays for a specific geographic region. These overlays may depict raw data or may show thematic information (such as soils, land use, or geology), but they must share common geographic qualities (including a common geographic coordinate system) that permit them to be merged into a single system that allows use of the varied data as an integrated unit. Although these data can be visualized as a set of superimposed images (Figure 15.1), they are in fact stored in digital form, suitable for retrieval and analysis by computer. Specialized computer programs and display equipment permit rapid analysis and display of data, with a flexibility not possible using manual methods.

Ideally, a GIS consists of several kinds of information. A detailed *planimetric base* accurately records positions of primary features on the landscape. This base forms the locational framework that permits accurate positioning and registration of the varied overlays of thematic data (such as forest types, rivers, highways, and power lines) that form the GIS. Some systems may also include a *cadastre*—a record of land parcels that depicts each unit of land, defined by its legal description of ownership. Defining and maintaining the base and the cadastre are formidable tasks, so few areas and few GISs possess suitable planimetric bases and/or cadastres. A third element consists of the varied *thematic overlays* that describe physical features (streams and rivers, soils, geology, etc.).

Sometimes the term *land information system* (LIS) describes the cadastre and the planimetric base, and the term *geographic information system* is reserved for those systems that focus on resource information (vegetation, soils, geology, etc.) and/or demographic and economic data. Further, the term *automated mapping/facilities management* (AM/FM) systems is reserved for systems devoted to mapping of utilities, transportation networks, and so on, at large scale, primarily in urban areas. In concept, there is no reason that a single system could not include these varied kinds of information, but as a practical matter, the constraints of costs, the varied expertise required to administer and maintain these systems, and the fact that each tends to be used by different kinds of organizations

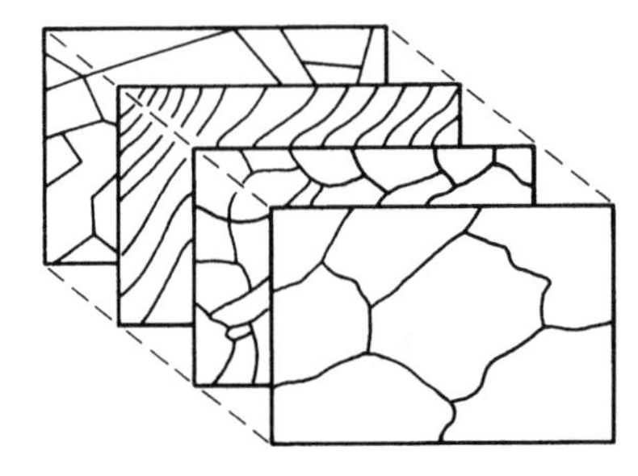

**FIGURE 15.1.** A GIS portrayed as a series of overlays.

mean that they are usually separate. However, it is still possible for separate systems within an organization to be coordinated so that information can be shared and passed from one system to another.

Superficially, a GIS may resemble *computer-aided design and display* (CADD) systems, which are designed to prepare technical graphics by computer and are effective in preparing maps and diagrams for graphic display. However, the defining characteristic of a GIS is not its graphic capability but rather its ability to conduct analytical operations based on the thematic and geographic meaning of the data. New information can be generated from the original information. Because CADD systems are graphically oriented, they have little relationships to GISs. Within a GIS, map features must always be linked to information concerning their attributes. Within a CADD, a line simply connects two points without any identification of its meaning; within a GIS, a line must always be labeled with an attribute (e.g., labeled as a highway, stream, power line, etc.).

Many of the basic concepts for GISs originated in the 1950s and 1960s, but recent innovations in computer technology have greatly increased the numbers and kinds of applications of GISs. Especially important are the decreases in computer costs and sizes and improvements in reliability and efficiency that have increased the availability of computers for businesses and governmental agencies. Such widespread availability of computer equipment has stimulated interest in applications of GISs not only among large organizations but also among small businesses and governmental units that previously would not have been able to purchase such equipment. In the United States, the increasing use of 911 emergency telephone systems, and the requirement for such systems to be based on systematic records of addresses and locations, has formed the framework for introducing GISs into local governments throughout the nation.

Geographic information systems have evolved side by side with remote sensing systems, so there are many close relationships between the two. First, remote sensing systems contribute data to GISs. Remotely sensed data can provide timely information at low cost and in a form that is compatible with the requirements of a GIS. Second, both GISs and digital remote sensing systems use similar equipment and similar computer programs for analysis and display. Therefore, the investment of funds and expertise in one field tends to form the foundation for work in the other. Third, the non-remote sensing data from a GIS can be used to assist in the analysis of remotely sensed images; such data form "ancillary," or "collateral" data, analogous to those described in Chapter 5 for manual interpretation. Thus, remote sensing and GISs have natural, mutually supporting relationships with each other. Nonetheless, the links between GISs and remote sensing are incomplete and imperfectly formed, and much more research is necessary to fully exploit the benefits of their interrelationships.

A GIS must include at least three main elements: (1) computer hardware, (2) computer programs, and (3) data. Computer hardware is much like that described in Chapter 4 and can vary in capabilities from the inexpensive microcomputers that can easily fit on an office desk to sophisticated workstations with massive storage and peripheral equipment. A computer's capabilities for storage and analysis, of course, are closely related to its size and computing capabilities, and its ability to display, print, and enter data is determined by computer hardware. For purposes of discussion in this chapter, we assume that the GIS is supported by substantial workstations and supported by peripheral devices for color display, plotting, and digitizing data.

## 15.2. Equipment for a GIS

Much of the equipment required for a GIS resembles that required for image analysis, as outlined in Chapter 9. The heart of the GIS is a digital computer. Because of the requirement to work with large data sets and complex programs, the capabilities of a GIS are greatly enhanced if the computer has a large memory and can access large volumes of mass storage. Nonetheless, many small computers, including microcomputers, can be used for the GIS if the user is willing to accept compromises in speed, convenience, and the sizes of areas that can be examined at a given time.

Because of the large amounts of data that inherently are required for a GIS, it is necessary to have access to tape drives and disk drives that permit the GIS to read information transported on computer tape or by computer networks (Chapter 4) from other computers and other GISs. A color display is often an important element of a GIS as a means of displaying several image or map patterns. Often it is important to have the capability to assign and reassign colors so the analyst can control the form of image display. A film recorder permits photography of an image displayed on the screen by means of a direct electronic link to the computer, so it is not necessary to use a hand-held camera to photograph the display screen itself.

Because a GIS often depends on data that may have been gathered by other organizations (as discussed later in this chapter), it is important for the GIS to have access to mass storage (tape units, CD-ROM drives), and to computer communications, for data transfer from other computer systems. A digitizing table (Figure 15.2) forms another method for data entry. Individual digitizing tables vary in size from perhaps 10 in. × 12 in. (*digitizing tablets*) to tables as large as 4 ft. × 8 ft. The table consists of a fine grid of thin wires or a printed circuit board encased in a dense, stable substance such as fiberglass. The wires are capable of sensing the *x–y* positions of a cursor that can be moved over the surface of the table. The finer the spacing between the wires and the more accurate their spacing, the more precise the data generated by the digitizer. The analyst tapes a map or aerial photograph to the surface of the digitizer and, after establishing a coordinate system, traces the outlines of areas or marks positions of points and lines with the cursor. As the cursor moves, the digitizer creates a digital record of its successive positions as a sequence of coordinate pairs.

Usually the digitizer is supported by a computer that allows the analyst to record data

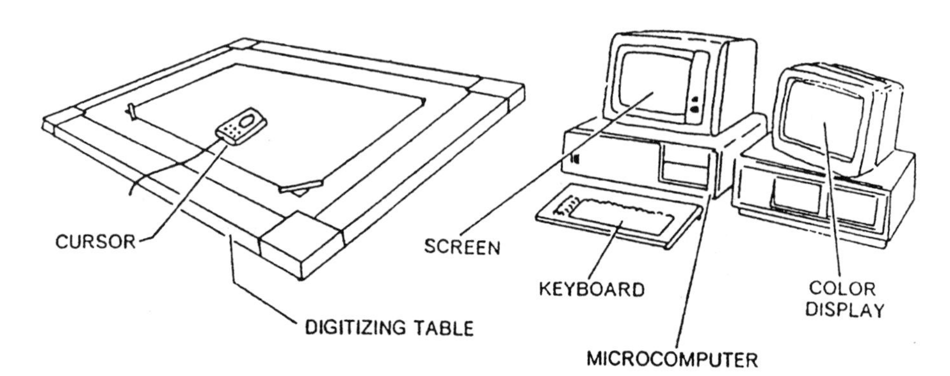


**FIGURE 15.2.** Equipment for a GIS.

with separate codes (e.g., to identify lines that record streams, highways, or power lines) and to perform geometric transformations (to correct positional errors). A display permits the analyst to see the digitized line on a computer screen as it is traced, to permit prompt identification and correction errors as they occur.

*Scanning,* another method of data entry, uses a sensor such as a CCD (Chapter 4) to systematically record the pattern represented by a map or image. Although high-quality scanners are expensive, even inexpensive scanners are often accurate enough for most purposes. The principal concern is often the process of editing scanned images to remove errors and assure logical consistency. For some kinds of data, scanning is the principal form of data entry.

## 15.3. Computer Programs

A GIS requires specialized programs tailored for manipulation of geographic data. Other kinds of data bases may have large volumes of data but do not need to retain locational information for data. Therefore, GIS software must satisfy the special needs of the analyst who needs to reference data by geographic location. Furthermore, the GIS must provide the analyst with the capability to solve the special problems that arise whenever maps or images are examined—the problems of changing coordinate systems, matching images, bringing different images into registration, and so on. A GIS must be supported by the ability to perform certain operations related to the geographic character of the data. For example, it must be capable of identifying data by location or by specified areas in order to retrieve and to display data in a map-like image. Thus, the GIS permits the analyst to display data in a map-like format so that geographic patterns and interrelationships are visible to the analyst.

Furthermore, software for a GIS must be able to perform operations that relate values at one location to those at neighboring locations. For example, to compile slope information from elevation data it is necessary to examine not only specific elevation values but also those at neighboring locations in order to calculate the magnitude and direction of the topographic gradient.

A GIS data base consists of not only a single data set but also many that together show several kinds of information for the same geographic area. Thus, a GIS may include data for topographic elevation, streams and rivers, land use, political and administrative boundaries, power lines, and other variables. This combined data set is useful only if the overlays register to one another exactly, and therefore the several kinds of data must share a common coordinate system. Because variables are likely to be derived from quite different sources, it is common for different variables to have different scales, different reference systems, and different cartographic projections. Thus, a GIS must have special programs to bring data into registration by changing scale and geometric qualities of the data. Such programs are similar to those described in Chapter 10.

Aside from problems with data acquisition, the capabilities of a GIS are determined largely by the ability it gives the analyst to ask questions concerning the geographic patterns recorded by the data. Although some queries are relatively simple ("How many acres of category *A* are present?") and can be addressed by relatively simple conventional programs, many other kinds of queries are special for the GIS. Therefore, geographic information systems require special computer programs to perform the tasks essential for the GIS (Table 15.1). *Image display* permits the analyst to present a specified data set as an image on a display screen and then to manipulate colors, scales, orientation, and other qualities as desired. An overlay capability permits the analyst to superimpose two or more data sets for display or analysis. *Visual overlay* refers to the ability to superimpose two overlays on the screen so that the two patterns can be seen together in a single image (Figure 15.3a). *Logical overlay* means that the analyst can define new variables or categories based on the matching of different overlays at each point on the map (Figure 15.3b).

The ability to overlay different images depends, of course, on the ability to manipulate geometric qualities of data sets. *Projection conversion* provides the ability to change from one map projection or geographic reference system to another. For example, geographic position in one overlay may be specified using state plane coordinates, whereas others may be based on the Universal Transverse Mercator (UTM) grid system. Unless the GIS has the capability to translate data from one reference system to another, it is not possible to match the separate overlays that form the GIS, and the GIS cannot perform one of the most basic functions—superimposing and comparing different data sets. Thus the GIS must have programs to perform *image registration* and *resampling* (Chapter 10); these capabilities permit the analyst to bring two images into registration. These capabilities are

**TABLE 15.1. Operations for GISs**

| | |
|---|---|
| Data input | Spatial interpolation |
| Display | Raster-to-vector conversion |
| Subset | Data output |
| Overlay | Data storage and retrieval |
| Projection conversion | Buffering |
| Registration/image matching | Network operations |
| Resampling | Data manipulation |
| Logical operations (Boolean) | Data reporting |
| Arithmetic operations | Statistical generation |
| Vector-to-raster conversion | Models |

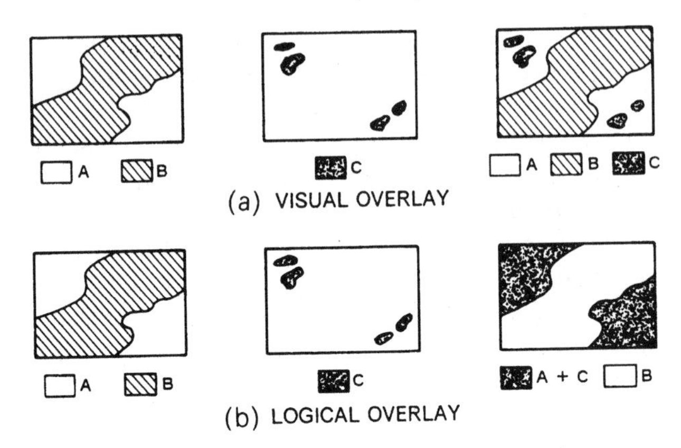


**FIGURE 15.3.** Visual and logical overlay.

among the most important of all those required for a GIS and among the most difficult to perform accurately.

Other operations permit the analyst to examine individual overlays, or combinations of individual overlays, using logical rules. For example, the analyst may wish to delineate all areas above a certain elevation or to define all areas within a certain soil category that have a certain slope or elevation. Thus, the analyst must be able to use the elevation matrix to define a slope overlay; then it is necessary to find the appropriate soil category within the soil overlay and match it to the elevation and slope overlays to find the slopes and elevations within soil classes. Such operations require a variety of general-purpose programs that permit the analyst to select regions from an image, register images, register two images to identify regions where specific regions match, calculate slopes from elevations, and so on. GIS software should also permit the analyst to define "tailor-made" geographic regions—for example, to delineate all areas within a specified distance of a particular class of stream or highway.

There are a wide range of software packages for GIS analysis, each with its own advantages and disadvantages. Even those lists too long to include here would omit many useful systems, but it is appropriate to mention a few that have been in relatively wide use for some time. At one extreme are systems designed for commercial production, with a full range of capabilities and the ability to handle very large, complex projects. These systems are expensive and require relatively long training periods before analysts become fully proficient in their use:

- **ARC/INFO** (Environmental Systems Research Institute, Inc., 380 New York Street, Redlands, CA 92373; 714-793-2853)
- **MGE** (Intergraph Corporation, 6767 Madison Pike NW, Suite 286, Huntsville, AL 35806; 800-345-4856)
- **ERDAS** (ERDAS, Inc., 2801 Buford Highway, Suite 300, Atlanta, GA 30329; 404-248-9000)

In contrast, other systems have been designed for use on PCs. These are best suited for instruction or for smaller projects that might not require a full range of analytical capabilities. These smaller systems are less expensive and easier to learn, and tend to focus on providing a selection of capabilities, rather than comprehensive detail:

- **ATLAS GIS** (Strategic Mapping, Inc., 3135 Kifer Road, Santa Clara, CA 95051; 408-970-9600)
- **IDRISI** (IDRISI Project, Clark Labs for Cartographic Technology and Geographic Analysis, Clark University, 950 Main Street, Worcester, MA 01610-1477; 508-793-7526)
- **MapInfo** (MapInfo, 1 Global View, Troy, NY 12180; 518-285-6000)

Producers of some of the more elaborate GIS packages listed previously also provide scaled-down PC versions of their systems.

## 15.4. Data and Data Entry

A GIS consists of information depicting geographic distributions of significance to those who use the GIS. Examples include topographic elevation, streams, highways, water bodies, soils, and census data representing population and economic data. Although such data are widely available as lists and tabulations, the purpose of a GIS requires that data must be coded by geographic location so that each item of information can be associated with a specific geographic location. Therefore, the analyst, for example, not only can determine the total population for a state but can display an image that shows the pattern of variation within the state and can relate the population pattern to other variables within the data base, or to population distribution at earlier dates.

A GIS retains locational identity by means of a coordinate system (Figure 15.4). The fineness of the coordinate system relative to actual ground distances determines the "res-

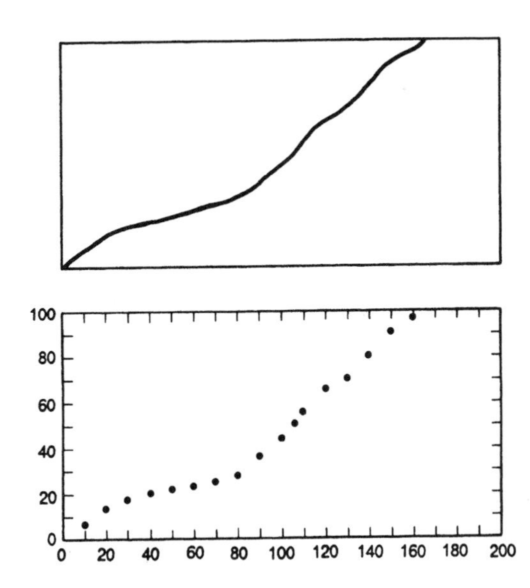


**FIGURE 15.4.** Coordinate systems.

olution" of the GIS or of the variable coded at a particular level of detail. Obviously, it would be desirable to encode locations at very fine levels of detail to provide high precision in specification of location. However, fine detail means that it is more difficult and more time-consuming to enter data for a given region, and as detail increases the amount of memory and storage required for a given area must rapidly increase. For example, doubling the number of rows and columns increases the number of locations by a factor of 4, and tripling the number of rows and columns increases the number of locations by a factor of 9. In practice, the decision to use a given level of resolution is a compromise that balances such factors as (1) the need to preserve detail and accuracy of the source documents, (2) resources available for digitization and data storage, and (3) requirements of the user.

Data for a GIS can be derived from many alternative sources. It is, of course, possible to gather data directly in the field by an original survey tailored specifically for the purposes of the GIS, although costs and time constraints are so great that rarely if ever are such surveys conducted. It is much more likely that data will be derived from a source that is already available. For example, topographic data can be derived from topographic maps, which also can provide information concerning streams, highways, and forested areas. Such information must be prepared for entry into the GIS by digitization, described next. Although data can be derived also from sources such as aerial photographs or other images, the analyst must devote extra effort to scaling the data to a specified level of detail and to correcting geometric errors inherited from the aerial images.

A third class of data is derived from digital data files generated in digital form—*archived data.* Examples include the data from the U.S. Census, available in digital form for the United States, topographic elevation data provided by the USGS and Defense Mapping Agency (DMA), land use data provided by the USGS, and remotely sensed data gathered by Landsat and other systems. Although some of these data are digitized from source documents (rather than compiled directly in digital form), they are available at modest cost in digital form, thereby providing the analyst with digital data at a small fraction of the cost and effort required for digitization locally.

## 15.5. Alternative Data Structures

Data for a GIS must be represented in a form that preserves locational identities of each unit of information so that it is possible to retrieve data by location and therefore to depict and analyze geographic patterns. Because data are frequently derived from a "conventional" (nondigital) map or image, it is necessary to convert them into digital form suitable for use by a GIS. This process, known as *geocoding,* records the pattern on each map in a form that can be accepted and manipulated by digital computers. Geocoding is often one of the most expensive and time-consuming aspects of GIS management because it requires considerable manual labor for coding and entering data and for checking and editing errors.

Two alternative GIS data structures offer contrasting advantages and disadvantages. *Raster* (or *cellular*) data structures consist of cell-like units, analogous to the pixels of an MSS scene (Chapter 6). The region of interest is subdivided into a network of such cells of uniform size and shape; each unit is then encoded with a single category or value (Fig-

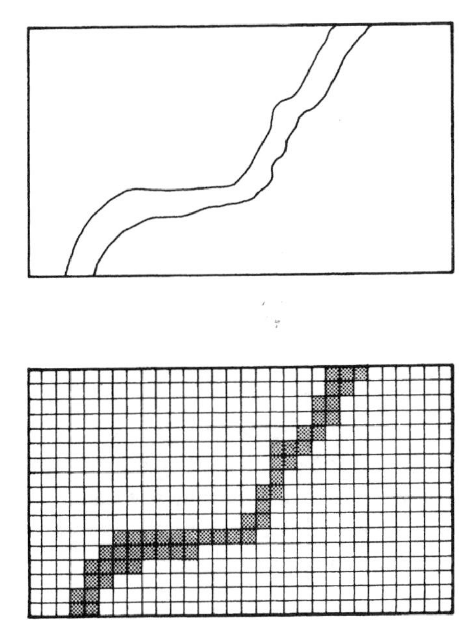

FIGURE 15.5. Raster data structure.

ure 15.5). Raster structures offer ease of data storage and manipulation and therefore permit use of relatively simple computer programs. Also, raster structures lend themselves for use with remotely sensed data because digital remotely sensed data are already presented in a raster format. Disadvantages are primarily related to losses in accuracy and detail due to the coding of each cell with a single category, even though several may be present (Figure 15.6). (This is the same kind of error caused by mixed pixels, discussed in Chapter 9.)

The alternative format is the *vector* (or *polygon*) format, which records the boundaries, or outlines, of parcels on the source document by listing the coordinates of the boundaries or coordinates of the vertices of polygons (Figure 15.7). Vector format tends to provide more efficient use of computer storage than does the raster format, as well as fin-

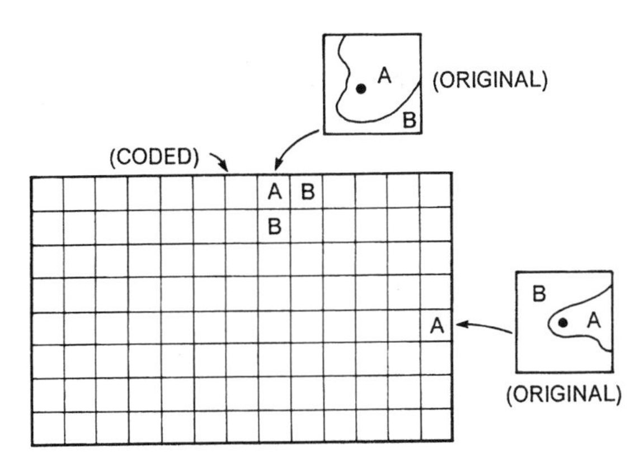


FIGURE 15.6. Errors in encoding cell data. Each cell can record only a single identifier, so cells cannot record the fine detail of the original pattern. Here cells illustrate coding by selection of the single class at the center of each cell. The two examples show how cells assigned the same label can correspond to different ground conditions.

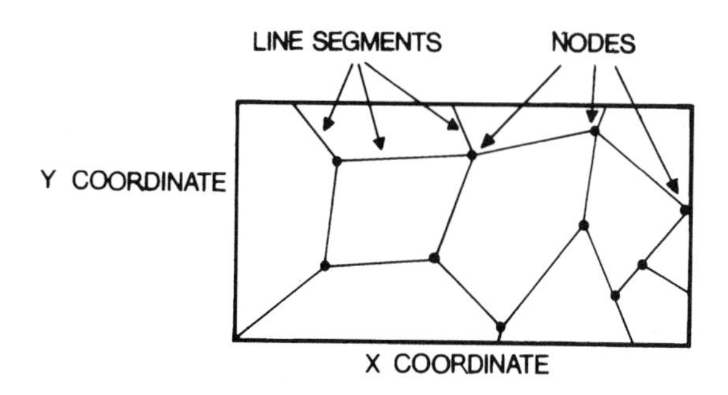


**FIGURE 15.7.** Polygon ("vector") data structure.

er detail and more accurate representation of shapes and sizes. Its main disadvantages are, in some instances, higher costs of encoding data and the greater complexity required for the computer programs that must manipulate data. Today, most GISs are designed to use vector format, although some may accommodate raster data or may use raster structures for selected analyses. Raster-to-vector conversions and vector-to-raster conversions permit mixing of the two kinds of data, although unnecessary conversions can create errors.

### *Cellular Data Structures*

Cell-based, or raster, data bases establish a uniform network of units as the locational framework for the data base. Sometimes the network of cells is completely arbitrary and has no meaning outside the context of a particular GIS. However, many GISs use preexisting geographic reference systems as the basis for locational coding of information. The UTM grid system, for example, provides a worldwide locational reference that can be applied to any level of detail that may be appropriate (Figure 15.8). In the western United States, some GISs use the U.S. Public Land Survey (Township–Range Survey System) as the basis for encoding information (Figure 15.9). Despite its faults, the township–range system has been useful in this context because it forms the basis for legal descriptions of land holdings, and as many roads, highways, and field patterns follow its boundaries, it is often visible on aerial imagery as distinct locational references. (Of course, it can be used only in that region of the United States for which it forms the legal reference for land ownership.)

Each cell in the GIS must be encoded with a single value or a single category that represents conditions at that location (Figure 15.10). For continuously varying information (e.g., topographic elevation), the single value could be the elevation at the center of the cell, or possibly the average value of elevation within the cell. However, once a choice is made, the same procedure is applied uniformly to all cells within the GIS. A cell can also encode such nominal data as a land cover class or a geologic or pedologic category. Because a given cell may include several categories, it is necessary to define a rule to consistently select which category represents the cell in the data base. The *predominant category* rule selects the single category that occupies the majority of the cell (Figure 15.10a). It is easy to understand the major problems inherent to the cell-based GIS. Variation within cells cannot be represented. Sometimes errors are insignificant, but often there are systematic errors due to the fact that some categories occur only as small parcels, or as long, narrow shapes that do not occupy large areas within cells, and therefore are systematically excluded from the GIS.

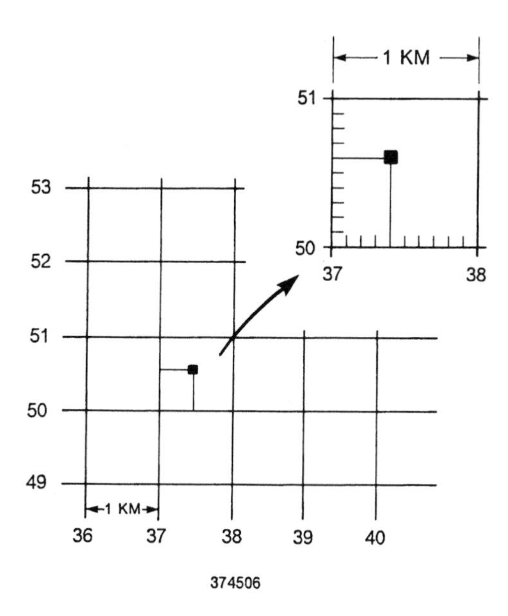


**FIGURE 15.8.** Universal Transverse Mercator reference system.

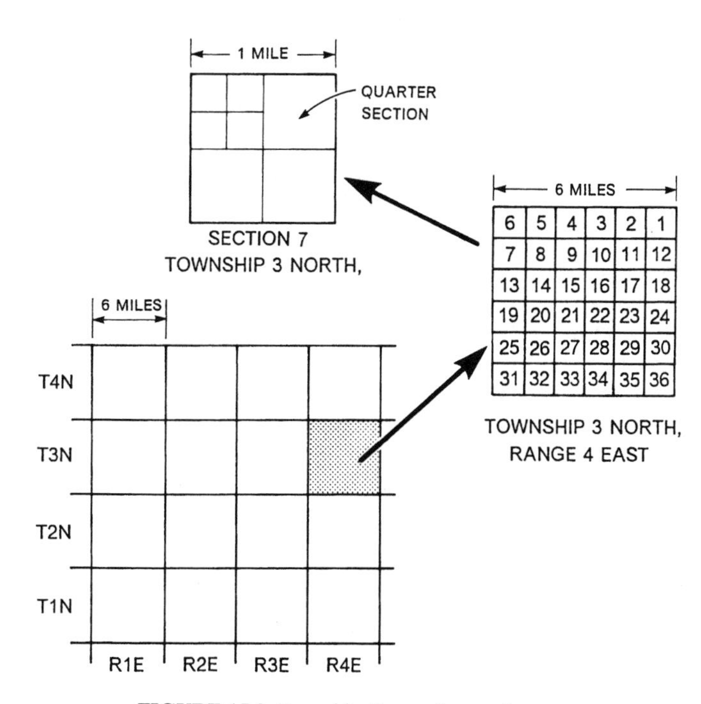


**FIGURE 15.9.** Township–Range Survey System.

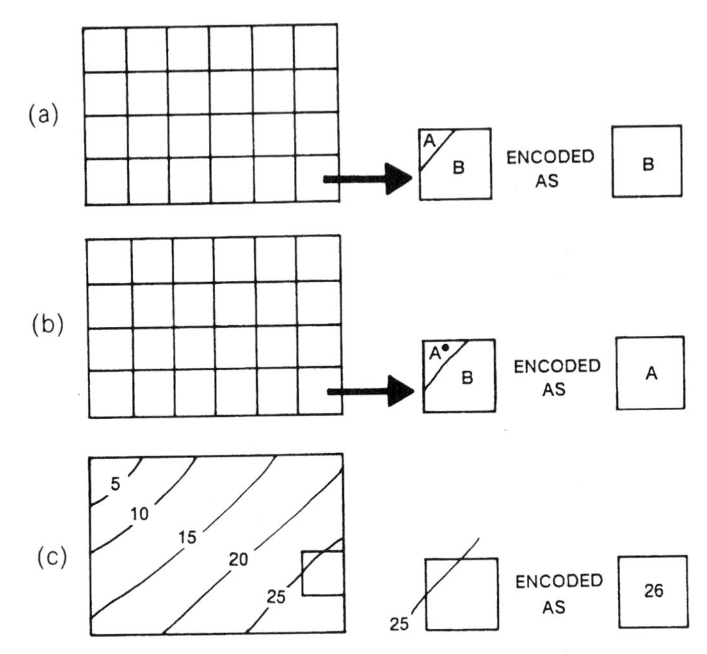


**FIGURE 15.10.** Data entry for a cell-based GIS.

Cell-based GISs therefore tend to exaggerate the importance of categories that occur in large, regular parcels, whereas small, linear, or widely dispersed units tend to be underrepresented. Although these problems can be reduced by using a smaller cell size, they become insignificant only when a cell size is very small in relation to the pattern to be studied, and the costs of using such a small cell size are often prohibitive.

Or, an alternative procedure selects the category by means of a dot in the center of each cell; the category that falls beneath the dot is entered as the category for the cell (Figure 15.10b). Or, a third procedure uses a dot randomly positioned within each cell (Figure 15.10c). (Of course, for a given data set, only a single method will be used.) Relative to the predominant category rule, these procedures improve accuracy of the data base as a whole, but the error at each cell can be quite large.

Gersmehl and Napton (1982) studied these two kinds of errors. *Inventory error* is error for the data base as a whole in reporting total areas occupied by specific categories. Inventory error is important for users of the GIS who are interested in a regional overview, who wish to learn about the character of specific regions rather than small sites within the region. Use of the dot randomly positioned within each cell reduces inventory error; for large areas, inventory errors may be quite small.

In contrast, *classification error* is the error in reporting the contents of each cell. Classification error is important for uses that require detailed information concerning specific sites. Gersmehl and Napton (1982) found that classification error is usually larger than inventory error (Figure 15.11). Clearly, classification error increases with use of the random dot method, even though the randomized selection may reduce inventory error. Both kinds of errors can be reduced if cell size is small relative to the pattern to be

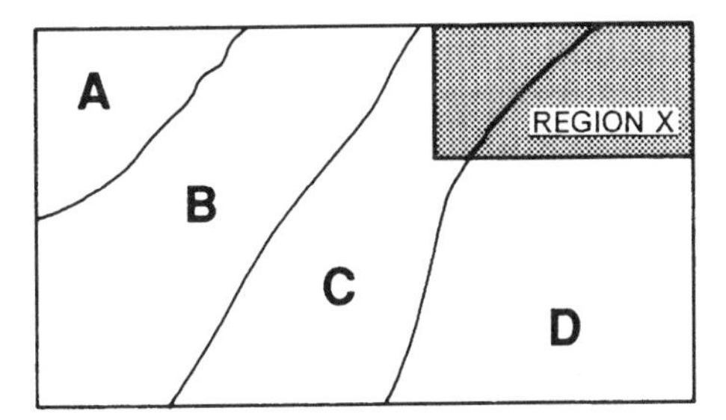


**FIGURE 15.11.** Inventory error and classification error. From Gersmehl and Napton (1982).

studied, as illustrated in Figure 15.5 and discussed in Chapter 8. However, even modest decreases in cell size greatly increase the numbers of cells; as a result, costs and practicality soon limit efforts to reduce cell size. Therefore, selection of cell size is made with great care because once a GIS is committed to a specific cell size, costs and inconvenience of changes preclude conversion to smaller cell size.

It should be noted that it is possible to interpolate data from one cell size to another (Figure 15.12). Data encoded at a 1-km grid size can be represented, for example, at a 4-km grid, as long as the analyst consistently applies a rule for deciding how to encode data for the larger cell (Figure 15.12a). However, interpolation from larger to smaller cells is a different matter. It is operationally possible to create 1-km cells for data encoded at 4-km cells; however, the change cannot improve the level of detail in the data but simply creates

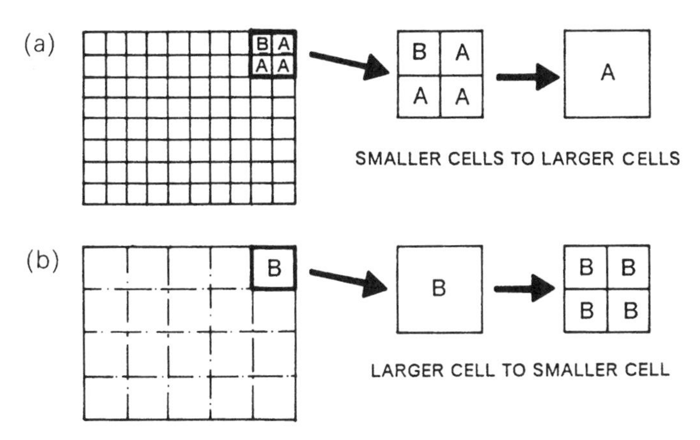


**FIGURE 15.12.** Interpolation from one cell size to another. (*a*) Smaller cell to larger cell. (*b*) Larger cell to smaller cell.

a false impression of finer detail. (There can be no method for assigning values for the smaller cells, except by replicating information of the larger cells. The visual detail has been improved, but the logical detail remains the same (Figure 15.12b).

Data for cell-based systems are entered from paper maps, aerial images, or similar documents depicting the pattern to be encoded. Often a template showing the grid system is superimposed and registered to the map. Of course, the spacing of the grid must be appropriate for the scale of the map. For example, to encode 1-km cells from a map at 1:24,000, the template should show cells 4.17 cm in size, whereas to encode from a map at 1:15,840 the template must have cells of 6.32 cm. The analyst then can read the value or category for each cell in sequence (using the specific method adopted for use), recording the data on a special form, or possibly reading them to another operator who enters them directly using a computer keyboard.

The student who is especially interested in GISs may wish to review Chapter 9, with emphasis on the material pertaining to spatial resolution. Remotely sensed data and data for GISs can differ in many important respects, so not all concepts from the discussion of image resolution apply directly, but the student can benefit from a careful comparison and contrast of the two subjects.

### *Vector Data Structures*

Numerous alternative formats exist for organization of vector data, as designed by different governmental and commercial organizations. Essentially, each format represents a different way to establish a framework of nonoverlapping spatial divisions (known as *tesselation*). A GIS must use such a framework to systematically organize the points, lines, and areas that represent geographic features. The challenge is to organize this information in such a way that it is simultaneously effective to store and retrieve it and to retain the positions and geographic meanings of the lines, points, and areas (i.e., to ensure that the computer representation of a map is not simply a network of lines but one in which the computer can distinguish between lines that represent roads, streams, and political boundaries). This task is difficult to accomplish effectively, so many strategies for organizing vector data have been designed to optimize a given quality at the expense of another.

Some of the most common vector data formats illustrate different strategies for digital encoding of the boundaries that form a geographic pattern. All depend on the use of a geographic reference system that permits definition of locations at a fine level of detail relative to the sizes of polygons. The reference system may be arbitrarily defined or tied to an established geographic reference system, such as latitude and longitude, state plane coordinates, or the UTM grid system.

*Location list data structures* (LLS) (Figure 15.13) simply encode the coordinates of the polygons that record the geographic pattern. Each polygon is encoded individually, so interior boundaries are encoded twice. Thus, although the LLS is simple and convenient to use, it makes inefficient use of computer storage and creates boundary slivers that must be removed.

The *point dictionary structure* (PDS) (Figure 15.14) records each vertex of each polygon only once. One file is used to record a list of coordinates of all vertices, with a label for each. A second file identifies each polygon in location list form using labels of

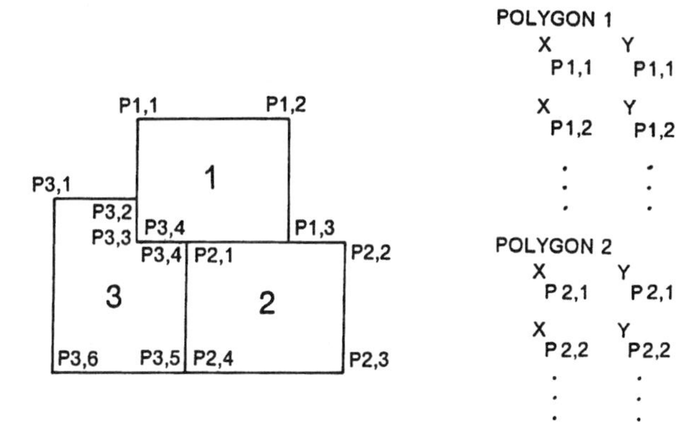


**FIGURE 15.13.** Location list file. Each polygon is treated as a separate unit; many points must be encoded several times.

points that define its outline. The PDS usually provides a great improvement in efficiency over the LLS at only a small cost in program complexity.

The *dual independent map encoding* (DIME) structure (Figure 15.15) was once one of the most widely used formats for vector data. The basic unit for the DIME format is not the polygon itself, but rather the line segments that form boundaries between polygons. The DIME format encodes a pair of coordinates for each line segment, with end points forming vertices of polygons. Each segment is labeled with the identities of polygons on either side of the line segment. The complete pattern then is formed by plotting all line segments, using a computer program that can organize line segments into a network of polygons. The DIME format has long been the primary method of encoding polygons that form the collection units for U.S. Census data but can also be used for many other kinds of vector data.

The *chain file structure* (CFS) is organized around "chains" of line segments that connect key nodes within the network of boundaries (Figure 15.16). A chain then is simply a sequence of locations that defines the boundary between two polygons. Each chain of points has an identifying label, a sequence of points and line segments that define its shape, and labels that define the identity of polygons on either side of the chain. This forms an important improvement over the DIME format, as far less coding is required during data entry.

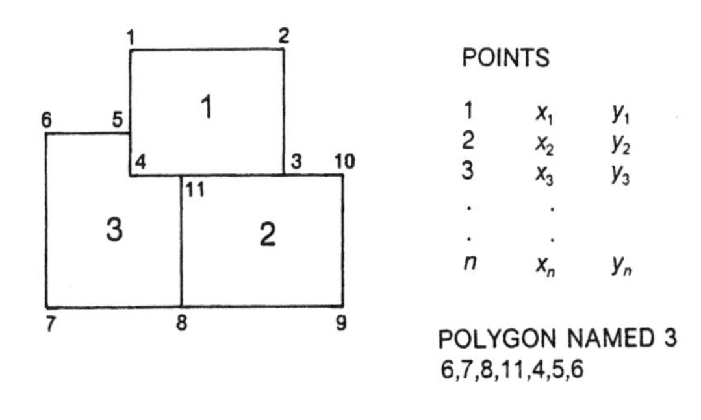


**FIGURE 15.14.** Point dictionary file. Locations of vertices are encoded independently of polygons; the "dictionary" list then defines each polygon by the points that form its perimeter.

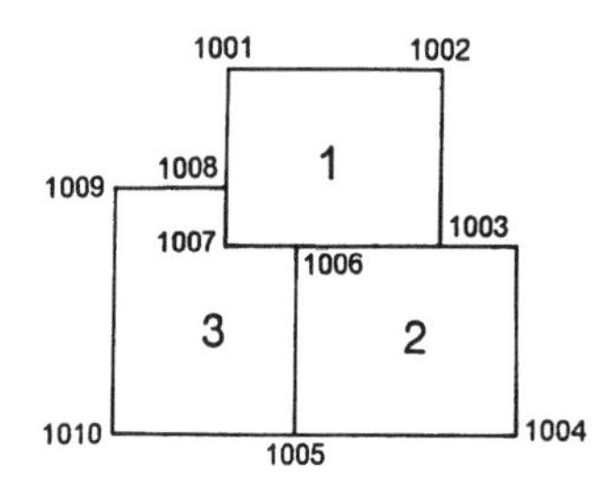


| NODES | | POLYGONS | | |
|---|---|---|---|---|
| FROM | TO | LEFT | RIGHT | COORDINATES |
| 1006 | 1005 | 3 | 2 | xy xy |
| 1007 | 1008 | 3 | 1 | xy xy |
| 1006 | 1003 | 1 | 2 | xy xy |
| 1006 | 1006 | 3 | 1 | xy xy |

**FIGURE 15.15.** DIME structure. Lines are encoded as separate entities, with polygons identified as areas on either side of each line segment.

*Digital line graphs* (DLGs) are used for digital representations of the U.S. Geological Survey's topographic map series (Figure 15.17). Each theme (streams and water bodies, topography, forest cover, etc.) is represented by a separate file with the coordinates that describe the placement of each line. A complete map can be represented only by combining all of the files.

*Topologically integrated geographic encoding and referencing* (TIGER) files, an elaboration of the chain file structure, were defined by the U.S. Bureau of the Census for the 1990 census (Broome & Meixler, 1990). A TIGER file, although treated as a single file, is in fact formed from four different file types (Figure 15.18): (1) TIGER county par-

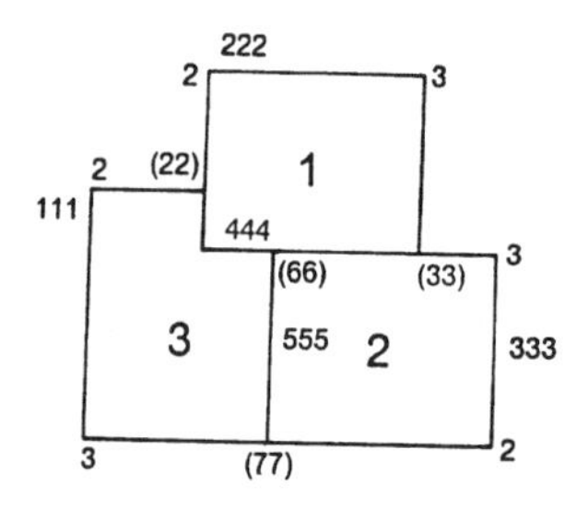


| CHAIN NAME | LENGTH | FROM | TO | LEFT | RIGHT |
|---|---|---|---|---|---|
| 111 | 3 | 22 | 77 | 0 | 3 |
| | $X_1Y_1$ . . . | | $X_3Y_3$ | | |
| 222 | 3 | 22 | 33 | 1 | 0 |
| | $X_1Y_1$ . . . | | $X_3Y_3$ | | |
| 333 | 3 | 33 | 77 | 2 | 0 |
| | $X_1Y_1$ . . . | | $X_3Y_3$ | | |

3 POLYGONS
4 NODES
5 CHAINS
1 POINT WITHIN CHAIN

**FIGURE 15.16.** Chain code file. Every line is defined by a chain, consisting of a label, a string of coordinates that form its inner path, two limiting nodes (i.e., two end points), a length (number of coordinate pairs), and polygons on either side.

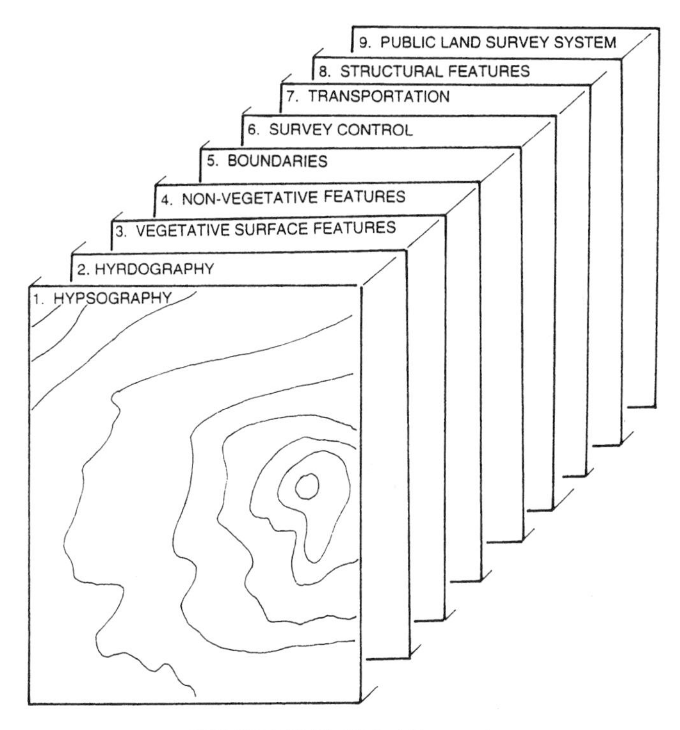


**FIGURE 15.17.** Digital line graph.

tition files, (2) GEO-CAT (*GEO*graphic *CAT*alog of Political and Statistical Areas) files, (3) the national partition file, and (4) temporary work files. The national partition file consists of boundary outlines of and identifiers for all counties, used to define count data required to represent any specific area to be mapped. The GEO-CAT files include current data and labels for political and statistical units within the system. When a specific area is to be mapped, the system refers to this file to find the correct names, populations, and so on. TIGER county partition files hold the boundaries and relationships to other units required to represent any specific region. The work files form temporary work areas to assemble data from each of the other kinds of files as a specific map is prepared.

Data for vector systems are either scanned or encoded by means of a digitizing table as described in Section 15.2. Polygons are digitized from paper or mylar copies of maps or aerial images that have been attached to the digitizing table. The analyst moves the cursor

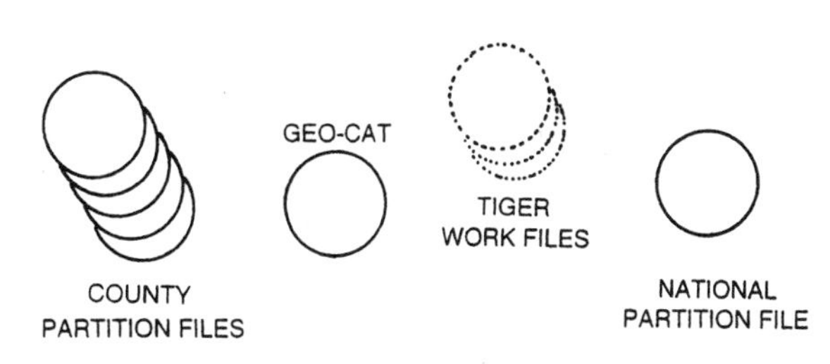


**FIGURE 15.18.** TIGER file structure.

over the document, tracing the outlines of polygons and labeling each line segment with the identities of the polygons on either side of the line. Usually the digitizer is connected to a computer, so the analyst can enter information at a keyboard and store the digitized information on a computer file. After the document has been digitized, the analyst must carefully edit the computer file that records the data because some line segments may have been omitted, misidentified, or incorrectly recorded. The analyst must identify and correct all such errors before the data can be considered completely digitized.

### *National Spatial Data Infrastructure*

The National Spatial Data Infrastructure (NSDI) (*Coordinating Geographic Data Acquisition and Access: The National Spatial Data Infrastructure*), established by an executive order signed in April 1994, is intended to form a framework for efficient exchange of geographic data between different organizations and between different computing systems. It establishes standards for geographic data and a plan for establishing a National Digital Geospatial Data Framework before the close of the century. The executive order applies only to the activities of agencies of the federal government, although governments at other levels are encouraged to participate. However, the existence of the federal effort will encourage state governments to implement similar programs, and lead to participation by private industry.

The Spatial Data Transfer Standard (SDTS) was developed as part of a broader effort to establish guidelines for information processing within the U.S. federal government. SDTS is intended to minimize problems encountered in transferring data between agencies by establishing common formats and standards. SDTS was approved by the U.S. Department of Commerce in 1992, as Federal Information Processing Standard 173. It consists of a family of standards that apply to different forms of spatial data, including Topological Vector Profiles and USGS Digital Line Graphs. Remote sensing data, as well as other raster data, are included within the Raster Profile.

## 15.6. Data for a GIS

Data for a GIS can represent virtually any topic that can be observed within the area of interest. Typical topics might include topographic elevation, land cover, soils, census information, highways, streams, power lines, and many others. Almost any variable can be useful in some context, so the decision to include specific variables often is based on the availability of the data in digital form and on issues such as accuracy and compatibility, discussed below.

*Point data* consist of observations that occur only at points, or occupy very small areas in relation to the scale of the data base (Figure 15.19). Features such as wells, for example, illustrate data that occupy a single point even at the largest levels of detail. In contrast, features such as buildings sometimes occupy significant areas, even though they may be represented as points in the data base. A polygon data base can depict point data as points and position them accurately. In contrast, a cellular data base can depict point data only at the level of detail of a single cell, so, for example, a cell can show the presence of a well within the cell but cannot show the placement of the well within the cell. Some point observa-

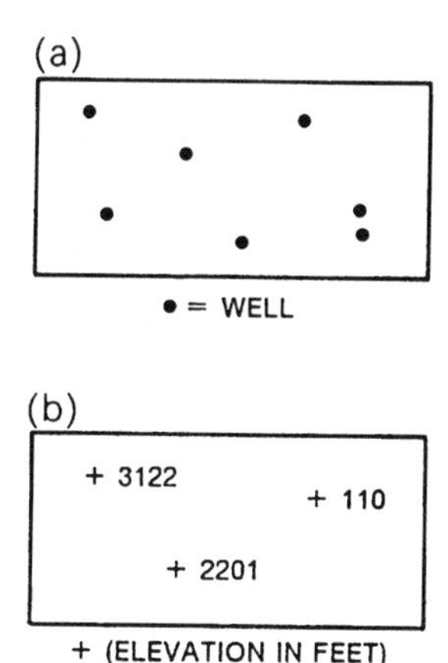


**FIGURE 15.19.** Point data. (*a*) Nominal. (*b*) Quantitative.

tions may represent such quantitative measurements as the amount of rainfall observed at a specific point or the depth of a specific geological unit below the surface.

*Line data* are, of course, exemplified by features such as highways, rivers, pipelines, or power lines (Figure 15.20). Polygon systems can show line data in fine detail, whereas cellular systems can show linear data only at the resolution of the cell. Thus, a raster data base can depict a linear feature only as a series of adjacent cells with a common code that shows the presence of a specific feature. Both raster and vector systems can depict different classes of linear features. For example, a data base can encode different classes of highways or streams that permit the analyst not only to examine the entire highway or stream network but also to select only specified classes of streams or highways for analysis.

The significance of certain forms of linear data may not be immediately appreciated. Boundaries, including political and administrative boundaries, do not necessarily form

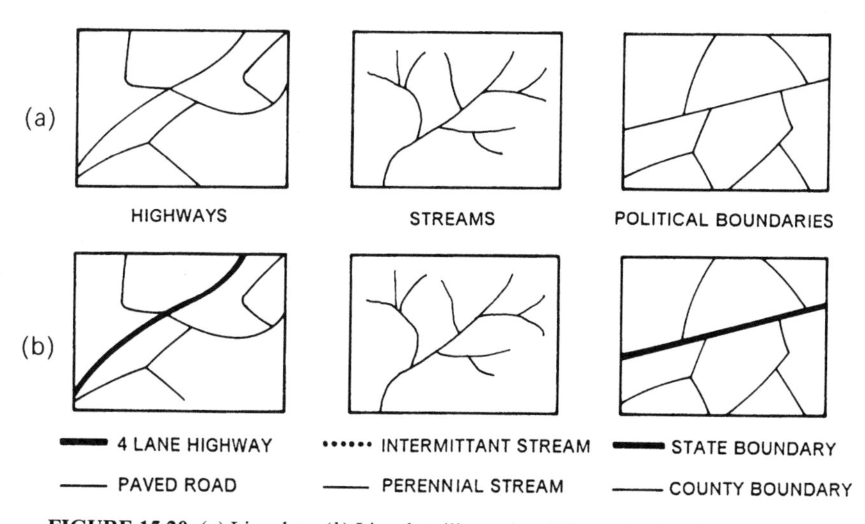


**FIGURE 15.20.** (*a*) Line data. (*b*) Line data illustrating different levels within a hierarchy.

physical features on the landscape, yet often they are extremely important in determining the usefulness of a GIS. Many of the queries and tasks that must be addressed by the GIS concern political or administrative units. Therefore, to respond to such questions as "How many acres of farmland are present within Washington County?" it is necessary to have accurately entered not only the data concerning land use and agriculture but also the information that positions the county boundary. Likewise, the boundaries of census units, drainage basins, and other features may be very important in determining the usefulness of the GIS.

Examples of *areal data* (Figure 15.21) include such distributions as soil, land cover, vegetation classes, and other patterns that occupy area at the scale of the GIS. Areal patterns often involve a nominal classification, such as land cover, in which the classes are based on distinctions in kind rather than differences in quantity. As mentioned earlier, polygon systems provide detailed delineation of the boundaries between nominal classes; raster systems inevitably involve loss of information as the data are encoded at each cell. Quantitative data that are grouped by areal units, such as census units, are depicted in finer detail by polygon structures. However, continuous data, such as topographic elevation, which represent a continuous surface by a network of equally spaced observations, are probably most directly represented by the cell-based data structure.

### *U.S. Geological Survey Digital Elevation Data*

The U.S. Geological Survey (Chapter 3) provides public access to digital elevation for the United States. Although elevation data are only one of many forms of ancillary data, they are one of the most inexpensive, widely used, and effective GIS data, so they deserve special attention here.

Digital elevation data, also referred to as *digital elevation models* (DEMs), or *digital terrain data,* are digital representations of the shape of the earth's surface. Typically digital elevation data consist of arrays of values that represent topographic elevations measured at equal intervals on the earth's surface (Figure 15.22). Such data can be considered analogous to digital remote sensing images except that each pixel represents an elevation measurement rather than a brightness value. Digital elevation data can be manipulated, classified, and analyzed in many of the same ways that digital remote sensing data can be, and can be displayed as pictorial images.

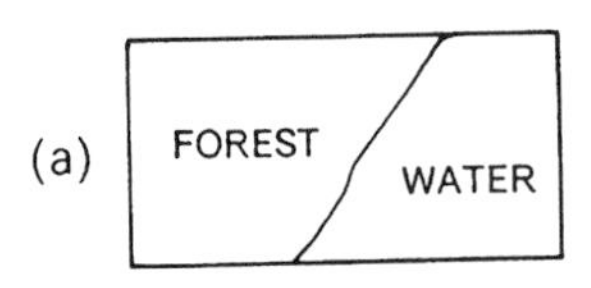


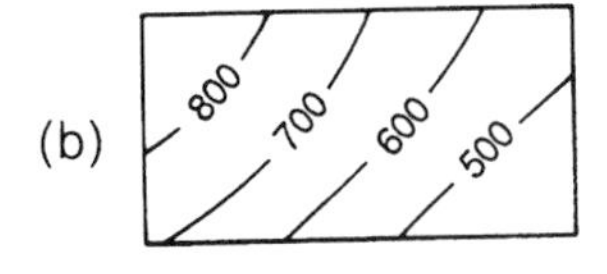


**FIGURE 15.21.** Areal data. (*a*) Nominal. (*b*) Continuous.

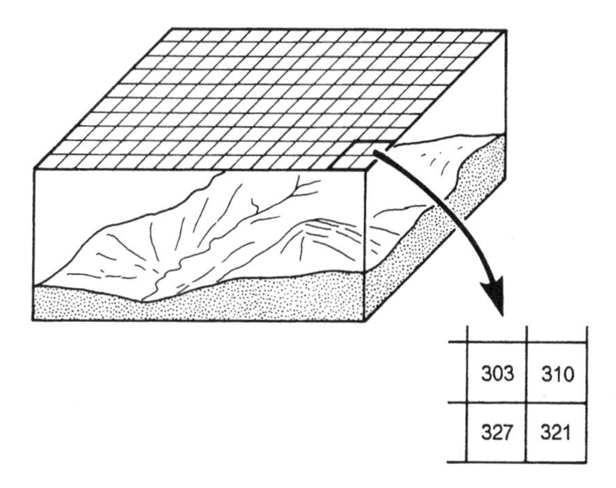


**FIGURE 15.22.** Digital elevation data.

DEMs provided by the USGS fall into three classes. Small-scale, coarse-resolution data are provided for each of the 1:250,000 map sheets that cover the United States. These data have been digitized from the map contours and resampled to provide the arrays of regularly spaced values that form the DEM. Each of the 1:250,000 map sheets covers a zone that measures 1° of latitude × 2° of longitude; the digital terrain data are usually provided in arrays that measure 1° on a side, so each 1:250,000 map sheet requires two adjacent blocks (eastern and western halves) for complete coverage. The initial version of these data, produced by the DMA, provided elevations spaced at intervals of 3 arc seconds, so the ground spacing of elevations varies with latitude. At 40° latitude, the north–south spacing is about 92 m (303 ft.) and the east–west spacing is about 71 m (233 ft.). Analysts have encountered numerous errors and defects with these data. Hutchinson (1982), for example, reported that banding of elevations, due to residual artifacts of interpolation for contours in the digitized maps, were visible in the images he produced. Therefore, alternative data, if available, may be preferred. A second set of DMA data was resampled at a uniform interval of about 60 m (200 ft.); these data appear to be of much higher quality than those of the earlier version.

For selected areas within the United States, the USGS has compiled digital elevation values directly from the stereo photographs used to compile topographic maps, rather than from the printed maps. These data are available for areas corresponding to USGS 7.5-minute quadrangles (each representing about 100 km$^2$ (60 mi.$^2$). Ground distance between elevation values is about 30 m (99 ft.). Because of the fine detail, and the direct compilation from aerial images, these data have much higher quality than either of the other DEMs. At present they are available for perhaps half of the quadrangles in the United States, and will in time become standard for USGS topographic data.

### *U.S. Census Data*

Data collected by the U.S. Census Bureau form a good illustration of archival data that differ in both form and content from the digital elevation data. The U.S. Constitution requires a census of population every 10 years. The modern census includes not only an enumeration of total population but also tabulation of varied economic and social information and compilation of a complete demographic profile of the nation's inhabitants. Publication and presentation of so much data are tasks of tremendous complexity, and to use these data it is necessary to understand their organization.

Data are aggregated at several levels including of course the nation itself as a unit but also by state, county, and minor civil division. Beginning with the 1990 census, data for the entire census are collected for census blocks, the smallest collection units, which are then aggregated into block groups and then to census tracts. Each tract may be composed of 30, 50, or even several hundred blocks. Blocks are typically defined as rectangular areas bounded by streets on all four sides, although in some areas they may be irregular in shape. Boundaries are often, but not always, defined by streets or other recognizable features such as rivers or railroads. Census units are organized hierarchically so that smaller units nest within larger ones.

Many of the nation's largest urban areas are designated *metropolitan statistical areas* (MSAs). MSAs are formed by grouping together several counties and cities in and adjacent to urban regions. Such large cities as Chicago, New York, and Los Angeles are, of course, designated as MSAs, but so are many of the nations smaller metropolitan areas, such as Roanoke, Virginia, Albany, New York, Topeka, Kansas, and Savannah, Georgia.

Data for each level—state, county, tract, MSA, and so on—are tabulated and published in printed volumes, microfiche, or CD-ROMs, or on computer tapes disseminated to libraries and other repositories open to the public.

To examine the geographic patterns recorded by census data it is necessary to assign tabulated data to their correct geographic units outlined on a map of each state, county, or city, depending on the scale of the examination. Geographic delineations of census unit boundaries are published as paper maps of differing scale and level of detail and are available in digital form on computer tapes and CD-ROMs.

Census boundaries are also given in digital format on computer tapes, which means that census boundaries, and their data, can be represented in digital form and entered into a GIS. This is important for users of census data because census data are so voluminous, and the network of tracts and blocks is so complex, that it is almost impossible to prepare maps manually even for areas of modest size. The digital representation of census boundaries is in TIGER format (pre-1990 records may be in DIME format) which uses the five-digit codes to identify polygons within each region. Even considering the advantages of a digital format, these data can present many practical problems. The TIGER file for even a small MSA might include 10,000 records, each with as many as 300 columns of data in each record.

There are some difficulties that limit or complicate uses of census data in a GIS. To preserve confidentiality of responses, census data are suppressed when the smallness of a census unit might permit one to match individual citizens with their responses to census queries, as published in census data. *Suppression* means that specified data for small census units is withheld, although the suppressed data are included as a part of totals published for larger units, where the larger numbers prevents identification of individual responses to census questions. Therefore, it may not always be possible to map income data at the finest level of detail. However, other information, such as population totals, is not suppressed and can therefore be mapped in detail.

### *Data Compatibility*

One of the major limitations to uses of GISs arises from issues related to data quality (Table 15.2), and especially from incompatibilities within and between data sets within a GIS. Compatibility is important because the entire concept of the GIS assumes logical and physical compatibility of the data within each GIS. Physical compatibility refers to the physical form of the data, which, in the instance of the computer-based GIS, is determined largely by tape and data formats such as those discussed in Chapter 4. Although incompatibilities in data formats can form serious practical problems in entering data, and in transferring data among different GISs, many such problems can be solved by special computer programs that can convert data from one format to another, or from one system to another.

Much more serious are logical incompatibilities among data within a GIS, as these are inherent to the data and cannot be easily corrected. For example, consider an ostensibly simple issue—the recording of data concerning the distinction between open and forested land for a large region—perhaps a state or a large portion of a state. Although the problem seems straightforward, the issue of compatibility raises many troublesome problems. If we start by gathering data that may have already been collected by county or local governments, we may find that they use different definitions for open and forested land, that the data have variable levels of detail and accuracy, and that they were collected at different dates. After encountering these problems, we conclude that it may be better to collect the information directly, for the specific purpose of the GIS, to be sure that the data are internally consistent. Yet even if we use aerial photography, we may find that the available photography, although complete for the area of interest, was flown at several dates, that it has several different scales, and that some areas have much higher-quality photography than others. If the area is large, we may even have difficulty acquiring Landsat coverage of the same date or season for our region. Therefore, even for the simplest of matters compatibility can be difficult to achieve.

Compatibility between different data sets can be an even more difficult problem. Because of the great costs of acquiring and digitizing data specifically for a GIS, it is often necessary to use data that were collected, or digitized, for another purpose. For example, the USGS DEMs mentioned earlier, U.S. Census Data, and the USGS Land Use and Land Cover Data (Chapter 19) are each internally consistent and available at rather low cost in a digital format. Therefore, these data, and many others of a similar nature, are often included in a GIS. Yet, use of off-the-shelf, or archival, data means that the GIS is formed, in part perhaps from data of differing levels of detail, differing levels of accuracy, differing

**TABLE 15.2. Elements of Data Quality**

| | |
|---|---|
| Age | Thematic accuracy |
| Areal coverage | Thematic detail |
| Scale | Accessibility |
| Detail | Costs |
| Format | Origins |
| Cartographic projection | Continuity with past and future data |
| Positional accuracy | Compatibility with other thematic data |

dates, differing classification systems, and differing collection units. Often these kinds of problems can greatly limit the usefulness of the data and prevent the GIS from being able to perform the function for which it was designed—that is, to permit rapid, convenient analysis and comparison of data for large regions.

## 15.7. Relationships between Remotely Sensed Data and GISs

It is easy to see some of the advantages for using remotely sensed data in a GIS. Satellite systems such as Landsat or SPOT acquire data for large areas in a short time period, thereby providing essentially uniform coverage with respect to date and level of detail. Such data are already in digital form and are provided in more or less standard formats, as outlined in Chapter 6. Furthermore, these data are available for almost all of the earth's land areas and are inexpensive relative to alternative sources. Although satellite data are not planimetrically correct, preprocessing can often bring data to acceptable levels of geometric accuracy with only modest effort. Images formed by analysis or interpretation of the raw data register to the original data. Thus remotely sensed data have the potential to address some of the difficult problems encountered in the formation of GIS.

Nonetheless, there are still numerous problems and difficulties. Data are not always available for desired dates or seasons. If a large area is to be examined, there may be problems in mosaicking data for separate dates. Accuracies of classification and analytical methods required to process the data prior to entry in the GIS may not be consistently reliable. Furthermore, the levels of detail in the satellite data may not match those from other sources.

There are several avenues for incorporating remotely sensed data into the GIS. The most satisfactory procedure depends on the specific requirements of a particular project and the kinds of equipment and financial resources that may be available.

1. Manual interpretation of aerial photographs or satellite images produces a map, or set of maps, that depicts boundaries between a set of categories (e.g., soil or land use classes). Then these boundaries are digitized to provide the digital files suitable for entry into the GIS.
2. Digital remote sensing data are analyzed or classified using automated methods to produce conventional (paper) maps and images, which are then digitized for entry into the GIS.
3. Digital remote sensing data are analyzed or classified using automated methods and then retained in digital format for entry into the GIS, using reformatting or geometric corrections as required.
4. Digital remote sensing data are entered directly in their raw form as data for the GIS (usually used only as a last resort, after other measure have proved impractical).

## 15.8. Ancillary Data

Ancillary, or collateral, data can be defined as data acquired by means other than remote sensing used to assist in the classification or analysis of remotely sensed data. For manual interpretation, ancillary data have long been useful in the identification and delineation of features on aerial images. Such uses may be an informal, implicit application of an interpreter's knowledge and experience or make explicit reference to maps, reports, and data. For digital remote sensing, uses of ancillary data are quite different. Ancillary data must be incorporated into the analysis in a structured, formalized manner that connects directly to the analysis of the remotely sensed data.

Ancillary data may consist, for example, of topographic, pedologic, or geologic data. Primary requirements are that the data be available in digital form, that the data pertain to the problem at hand, and that the data be compatible with the remotely sensed data. The conceptual basis for the use of ancillary data is that the use of additional data, collected independently of the remotely sensed data, increases the information available to separate classes and to perform other kinds of analyses. For example, in some regions vegetation patterns are closely related to topographic elevation, slope, and aspect. Therefore, the combination of elevation data with remotely sensed data can form a powerful analytical tool because the two kinds of data provide separate, mutually supporting contributions to understanding the subject of the interpretation. In other settings the same ancillary data may be less effective because of a weaker association of vegetation with topography.

*Stratification* refers to subdivision of an image into subregions that are easy to define using the ancillary data but difficult to define using only the remotely sensed data. For example, use of digital elevation data to define elevation, slope, or aspect classes provides a powerful subdivision of the image because vegetation patterns are often closely related to topography and the topographic data are independent of the spectral information in the remotely sensed image. Therefore, the ancillary data provide a means of restricting the classes that must be discriminated using the spectral data (Figure 15.23). The stratification is essentially an implementation of the layered classification strategy discussed in

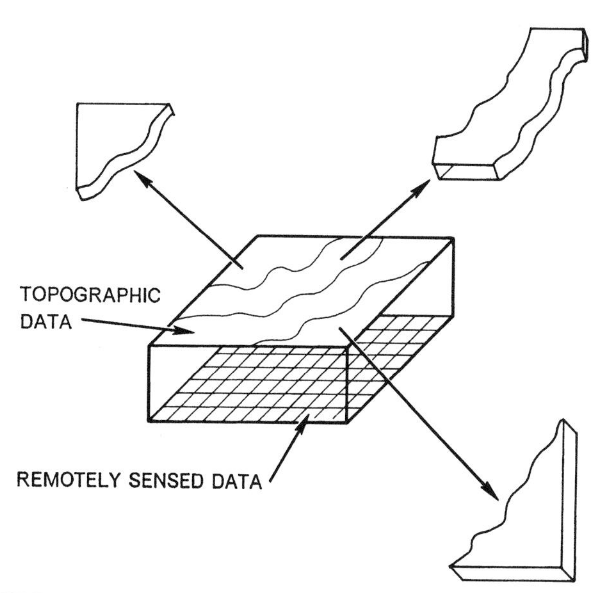


**FIGURE 15.23.** Stratification of remotely sensed data using elevation data.

Chapter 11. Some analysts have treated each strata as a completely separate image by selecting training data separately for each strata and conducting classification of each strata completely independently of one another. The several classifications are then merged for presentation as a single image for display of the final results. Hutchinson (1982) emphasized the dangers of ineffective stratification and warns of unpredictable, inconsistent effects of classification algorithms within different strata.

Ancillary data have also been used to form an additional channel. For example, digital elevation data could be incorporated with SPOT or Landsat MSS data as an additional band in the hope that additional data, independent of the remotely sensed data, will contribute to effective image classification. In general, it appears that this use of ancillary data has not been proven effective in producing substantial improvements in classification accuracy. Strahler (1980) implemented a second approach using ancillary data as a means of modifying the prior probabilities in maximum likelihood classification. Digital elevation data, as an additional band of data, were used to refine the probabilities of observing specific classes of vegetation in southern California mountains. For example, the probability of observing a specific class, such as *lodgepole pine,* varies with topography and elevation. Therefore, a given set of spectral values can have a different meaning, depending on the elevation of the pixel. This approach appears to be effective, at least in the areas studied thus far.

Finally, ancillary data can be used after the usual classifications have been completed in a process known as *postclassification sorting.* Postclassification sorting examines the confusion matrix derived from a traditional classification. Those classes that are most often confused become the candidates for postclassification sorting. Ancillary data are then used in an effort to improve discrimination between pairs of classes, thereby improving the overall accuracy by focusing on the weakest aspects of the classification. Hutchinson (1982) encountered several problems with his application of postclassification sorting but in general favors its simple, inexpensive, and convenient implementation.

Ancillary data, despite their benefits, present numerous practical and conceptual problems for the remote sensing analyst. Sometimes it may be difficult to select and acquire those data that may be most useful for a given problem. The potentially most useful data may not be available in digital form; the analyst may be forced to invest considerable effort in identifying, acquiring, and digitizing the desired information. Therefore, many if not most uses of ancillary data rely on those data that can be acquired "off the shelf" from a library of data already acquired and digitized for another purpose. Thus, there is a danger that ancillary data may be used as much because of their availability as for their appropriateness for a specific situation.

Because so much ancillary data is acquired for a purpose other than the one at hand, the problem of compatibility can be important. Ideally, ancillary data should possess compatibility with respect to scale, level of detail, accuracy, geographic reference system, and, in instances in which changes are important, date of acquisition. Because analysts are often forced to use off-the-shelf data, incompatibilities form one of the major issues in the use of ancillary data. To be useful, ancillary data must be accurately registered to the remotely sensed data. Therefore, the analyst may be required to alter the geometry, or the level of detail, in the ancillary data to bring them into registration with the remotely sensed data. Efforts to alter image geometry or to resample data change the values and therefore influence the usefulness of the data.

Sometimes ancillary data may be presented as discrete classes—for example, as specific geologic or pedologic classes. Such data are typically represented on maps as discrete parcels, with abrupt changes at the boundaries between classes. Digital remote sensor data are continuous, and therefore differences between the two kinds of data may influence the character of the results, possibly causing sharp changes where only gradual transitions should occur.

## 15.9. Modeling Spatial Processes with GISs and Remote Sensing

A logical development of the GIS concept leads from a simple description of a geographic area to modeling of specified processes. Such models are computer programs that use the information within the GIS as a basis for forecasting or estimating.

### *Example 1*

The GIS includes elevation data, and temperature data for specific climate stations. It is necessary to estimate temperatures at locations not near the climate stations. A solution to this problem is to use the elevation data to estimate temperatures at the intermediate locations, as there is a predictable relationship between elevation and average temperature. The elevation data permit derivation of slope and aspect, which together with cloud and wind data from the climate statistics permit refinement of the temperature estimates (i.e., slopes that face the equator will be warmer than these that face the poles). Thus, relatively simple knowledge of temperature variation, combined with the data of the GIS, permits estimation of temperatures at locations between the climate stations.

### *Example 2*

The GIS includes population data as well as data for topographic elevation, streams, and highways. It is necessary to study the likely effects of locating a new airport at each of several prospective sites, considering factors such as construction costs, obstructions to the flight path, access to highways, safety, exposure to aircraft noise, and similar factors.

For each site, it is possible to project the flight paths for aircraft using the existing airfield. For each flight path the analyst can determine the number of people exposed to aircraft noise and use the elevation data to determine whether obstructions are present within a prospective flight path. Likewise, the population data, together with the highway information, permit estimation of ease of access to each site. After several sites have been considered, the model can tabulate the estimates of advantages and disadvantages for each and a decision can focus only on those sites that are clearly suitable.

## 15.10. National Center for Geographic Information and Analysis

In 1988 the National Science Foundation funded the National Center for Geographic Information and Analysis (NCGIA) to conduct fundamental research on the analysis of geographic data using GISs. NCGIA consists of activities at three universities: the University of California at Santa Barbara, the State University of New York at Buffalo, and the University of Maine at Orono. NCGIA identifies research topics of significance for continued development of GIS technology and applications. Teams of scientists are then assembled to pursue these subjects, which have included design of very large spatial data bases, visualization of the quality of spatial information, formalizing cartographic knowledge, and investigating relationships between GISs and remote sensing. In addition to research programs, NCGIA is active in designing university curricula, organizing workshops and conferences, and preparing of technical papers and computer software.

## 15.11. Summary

The concept of the geographic information system, although implemented in many different forms, is the ideal that permits full utilization of the goals of remote sensing. Remote sensing achieves its maximum usefulness when images from one date, or formed from one part of the spectrum, can be integrated with those acquired at a different date or a different portion of the spectrum. Although this kind of integration can be accomplished in many ways, the GIS provides maximum flexibility in both display and analysis. The concept of the GIS has also been important in remote sensing because it has formed a focal point for scientists who work in subfields of remote sensing and diverse disciplines. By its nature, a GIS requires expertise from many fields of knowledge, so GIS research has brought scientists from diverse backgrounds together to work on common issues.

GIS research has emphasized the need for further development of capabilities of significance both to GISs and to remote sensing. Image matching, image registration, and related topics are important both for remote sensing and for geographic information systems, and both fields can benefit from improvements. Future developments in remote sensing are likely to address such topics as data compatibility, improvements in image geometry, and other subjects that may improve the usefulness of remotely sensed data for GISs.

## Review Questions

1. Design a cellular geographic information system for the county in which you live. Specify the area to be covered, the cell size, and the number and nature of the variables to be included.

2. For the information system you designed in response to question 1, calculate the number of pixels required for each overlay, and for the entire GIS as you have planned it. Identify sources for each variable you have specified (e.g., where will you get the information you desire?), and discuss problems that will occur because of differences in date, scale, accuracy, and other characteristics.

3. Consider your responses to questions 1 and 2. What kinds of products can your GIS produce? Outline the kinds of derived and interpretive variables that could be generated from the data you have selected in response to question 1.

4. Evaluate the usefulness of the following forms of archival data for your GIS:

   a. USGS digital elevation data
   b. Landsat MSS data; Landsat TM data
   c. SPOT HRV data
   d. U.S. Census data

   Identify advantages and disadvantages to the use of each in the context of your GIS.

5. Review material for Chapters 4, 8, 10, and 15 to identify topics where image registration is discussed. Prepare a list of these topics as a means of summarizing the significance of this task to remote sensing.

6. Assume you have a statewide GIS that includes digital elevation data, and census data including of course population totals. Assume also that you have the ability to superimpose boundaries of major drainage basins and political boundaries accurately. Discuss some of the ways such information could be used to address issues and problems important at the statewide level of government.

7. Prepare a list of ideas, concepts, functions, data, equipment, and other factors shared by remote sensing and GISs. Can you identify differences as well?

8. Outline steps necessary to incorporate land cover data, as derived from a digital TM image, into a GIS.

9. Outline steps required to use GIS data as ancillary data for analysis of TM data.

10. How would you decide what level of detail to use for a cell-based GIS for your state (i.e., how would you select the best cell size)? List factors you would consider.

## References

### General

Adams, Victor W. 1975. Earth Science Data in Urban and Regional Information Systems—A Review. *Geological Survey Circular 712*. Washington, DC: U.S.Geological Survey, 29 pp.

Broome, F. R., and D. B. Meixler. 1990. The TIGER Data Base Structure. *Cartography and Geographic Information Systems,* Vol. 17, pp. 39–47.

Burrough, P. A. 1986. *Principles of Geographic Information Systems for Land Resources Assessment.* New York: Oxford, 193 pp.

Clarke, Keith C. 1985. A Comparative Analysis of Polygon to Raster Interpolation Methods. *Photogrammetric Engineering and Remote Sensing,* Vol. 51, pp. 575–582.

Environmental Systems Research Institute, Inc. 1990. *Understanding GIS: The ARC-INFO Method.* Redlands, CA: Author, 411 pp.

Estes, John E. 1982. Remote Sensing and Geographic Information Systems Coming of Age in the Eighties. In *Proceedings Pecora VII Symposium* (B. F. Richason, ed.). Falls Church VA: American Society of Photogrammetry, pp. 23–40.

Laurini, Robert, and D. Thompson, 1992. *Fundamentals of Spatial Information Systems.* New York: Academic Press, 680 pp.

Maguire, D. J., M. F. Goodchild, and D. W. Rhind (eds.). 1991. *Geographical Information Systems* (2 vols.). New York: Longman, 640 pp. (Vol. I), 447 pp. (Vol. II).

Maizel, Margaret S., and R. J. Gray. 1985. *A Survey of Geographic Information Systems for Natural Resource Decision Making.* Washington, DC: American Farmland Trust, 182 pp.

Marble, Duane F., and D. J. Peuquet. 1983. Geographic Information Systems and Remote Sensing. Chapter 22 in *Manual of Remote Sensing* (R. N. Colwell, ed.). Falls Church, VA: American Society of Photogrammetry, pp. 923–958.

Marx, Robert W. 1990. The TIGER System: Yesterday, Today, and Tomorrow. *Cartography and Geographic Information Systems,* Vol. 17, pp. 89–97.

Starr, J., and J. Estes. 1990. *Geographic Information Systems.* Englewoods Cliffs, NJ: Prentice Hall, 303 pp.

Tomlin, C. D. 1990. *Geographic Information Systems and Cartographic Modeling.* Englewood Cliffs, NJ: Prentice Hall, 249 pp.

U.S. Geological Survey. 1990. *Digital Line Graphs from 1:24,000 Scale Maps.* Data User's Guide 1. Reston, VA: Author, 107 pp.

### Errors and Data Quality

Gersmehl, Philip J., and D. E. Napton. 1982. Interpretation of Resource Data: Problems of Scale and Transferability. *Papers of the Conference of the Urban and Regional Information Systems Association,* pp. 471–482.

Guptil, Stephen C., and J. L. Morrison (eds.). 1995. *Elements of Data Quality.* Tarrytown, NY: Elsevier Science, 250 pp.

Hunter, Gary J., and M. F. Goodchild. 1995. Dealing with Error in a Spatial Data Base: A Simple Case Study. *Photogrammetric Engineering and Remote Sensing,* Vol. 61, pp. 529–537.

Lunetta, R. S., R. G. Congalton, L. K. Fenstermaker, J. R. Jensen, K. C. McGwire, and L. R. Tinney. 1991. Remote Sensing and Geographic Information Systems: Error Sources and Research Issues. *Photogrammetric Engineering and Remote Sensing,* Vol. 57, pp. 677–687.

Mead, Douglas A. 1982. Assessing Quality in Geographic Information Systems. Chapter 5 in *Remote Sensing for Resource Management* (C. J. Johannsen and J. L. Sanders, eds.). Ankeny, IA: Soil Conservation Society of America, pp. 51–59.

Walsh, S. J., D. R. Lightfoort, D. R. Butler. 1987. Recognition and Assessment of Error in Geographic Information Systems. *Photogrammetric Engineering and Remote Sensing,* Vol. 53, pp. 1423–1430.

Whede, M. 1982. Grid Cell Size in Relation to Errors in Maps and Inventories Produced by Computerized Map Processing. *Photogrammetric Engineering and Remote Sensing,* Vol. 48, pp. 1289–1298.

### Combining Remotely Sensed Data with Ancillary Data

Davis, F. W., D. A. Quattrochi, M. K. Ridd, N. S–N Lam, S. J. Walsh, J. C. Michaelson, J. Franklin, D. A. Stow, C. J. Johannsen, and C. A. Johnson. 1991. Environmental Analysis Using integrated GIS and Remotely Sensed Data: Some Research Needs and Priorities. *Photogrammetric Engineering and Remote Sensing,* Vol. 57, pp. 689–697.

Ehlers, M., G. Edwards, and Y. Bédard. 1989. Integration of Remote Sensing with Geographic Information Systems: A Necessary Evolution. *Photogrammetric Engineering and Remote Sensing,* Vol. 55, pp. 1619–1627.

Ehlers, M., D. Greenlee, T. Smith, and J. Star. 1992. Integration of Remote Sensing and GIS: Data and Data Access. *Photogrammetric Engineering and Remote Sensing,* Vol. 57, pp. 99–675.

Faust, N. L., W. H. Anderson, and J. H. L. Star. 1992. Geographic Information Systems and Remote Sensing Future Computing Environment. *Photogrammetric Engineering and Remote Sensing,* Vol. 57, pp. 655–668.

Hutchinson, Charles F. 1982. Techniques for Combining Landsat and Ancillary Data for Digital Classification Improvement. *Photogrammetric Engineering and Remote Sensing,* Vol. 48, pp. 123–130.

Lunnetta, R. S., R. G. Congalton, L. K. Fenstermaker, J. R. Jensen, K. C. McGwire, and L. R. Tinney. 1991. Remote Sensing and Geographic Information System Data Integration: Error Sources and research Issues. *Photogrammetric Engineering and Remote Sensing,* Vol. 57, pp. 677–687.

Shelton, Ronald L., and J. E. Estes. 1979. Integration of Remote Sensing and Geographic Information

Systems. In *Proceedings of the 13th International Symposium on Remote Sensing of Environment.* Ann Arbor, MI: Environmental Research Institute of Michigan, pp. 463–483.

Star, J. L., J. E. Estes, and F. Davis. 1991. Improved Integration of Remote Sensing and Geographic Information Systems: A Background to NCGIA Initiative 12. *Photogrammetric Engineering and Remote Sensing,* Vol. 57, pp. 643–645.

Strahler, A. H. 1980. The Use of Prior Probabilities in Maximum Likelihood Classification of Remotely Sensed Data. *Remote Sensing of Environment,* Vol. 10, pp. 135–163.

**Spatial Data Transfer Standards**

Greenlee, David D. 1992. Developing a Raster Profile for the Spatial Data Transfer Standard. *Cartography and Geographic Information Systems,* Vol. 19, pp. 300–302.

Kidrer, David B., and Derek H. Smith. 1992. Compression of Digital Elevation Models by Huffman Coding. *Computers and Geosciences,* Vol. 18, pp. 1013–1034.

Southard, David A. 1992. Compression of Digitized Map Images. *Computers and Geosciences,* Vol. 18, pp. 1213–1253.

Stinton, David F. 1994. Geospatial Data after the Executive Order. *GeoInfo Systems,* August, p. 31.

Wortman, Kathryn. 1992. The Spatial Data Transfer Standard (FIPS 173): A Management Perspective. *Cartography and Geographic Information Systems*, Vol. 19, pp. 294–295.

CHAPTER SIXTEEN

# Plant Sciences

## 16.1. Introduction

The earth's vegetative cover is often the first surface encountered by the energy we use for remote sensing. So, for much of the earth's land area, remote sensing imagery records chiefly the character of the vegetation at the surface. We must be able to interpret these distributions to understand others, such as geologic and pedologic patterns, that are not directly recorded on the imagery.

In other situations we have a direct interest in study of the vegetation itself. Remote sensing can be useful for monitoring areas planted to specific crops, for detecting diseases and insect infestations, and for contributing to accurate production forecasts. Moreover, remote sensing imagery has been important in mapping forests, including studies of timber volume, insect infestation, and site quality.

In addition, it is becoming clear that remote sensing may provide the only practical means of mapping and monitoring changes in major ecological regions that, although not directly used for production of food or fiber, have great long-term significance for mankind. For example, the tropical forests that cover significant areas of the earth's surface have never been mapped or studied except in local regions that are unlikely to be representative of the unstudied regions. Yet these regions are of critical importance due to their role in maintaining the earth's climate (Rouse et al., 1979) and as a source of genetic diversity. Mankind's activities are rapidly destroying large areas of tropical forests, and it is only by means of remote sensing that we are ever likely to understand the nature and locations of these changes. Similar issues exist with respect to other ecological zones; remote sensing may provide a means to observe such regions at global scales and to better understand the interrelationships between the many factors that influence such patterns.

## 16.2. Manual Interpretation

### *Vegetation Classification and Mapping*

Vegetation classification can proceed along any of several alternative avenues. The most fundamental is simply to separate vegetated from nonvegetated regions or forested from open lands. Such distinctions, although ostensibly simple, can have great significance in some contexts, especially when data are aggregated over large areas or observed over long intervals. Thus, national or state governments, for example, may have an interest in knowing areas of their territories covered by forest and changes in forested land from one 10-year period to the next, even though there may be no data available regarding the different kinds of forest.

However, it is usually important to acquire information at finer levels of detail. Although individual plants can be identified on aerial photographs, seldom if ever is it practical to use the individual plant as the unit for mapping vegetation. Instead, it is more useful to define mapping units that represent groups of plants. A plant community is an aggregation of plants with mutual interrelationships among each other and with the environment. Thus, an "oak–hickory forest" is a useful designation because we know that communities are not formed by random collections of plants but consistent associations of the same groups of plants—plants that tend to prefer the same environmental conditions and to create the environments that permit certain other plants to exist nearby. A community consists of many stands—specific, individual occurrences of a given community (Figure 16.1).

Plants within communities do not occur in equal proportions; certain species tend to dominate. These species are often used to name communities (e.g., *hickory* forest), although others may be present. Dominant species may also dominate physically, forming the largest plants in a sequence of layers or *strata* that are present in virtually all commu-

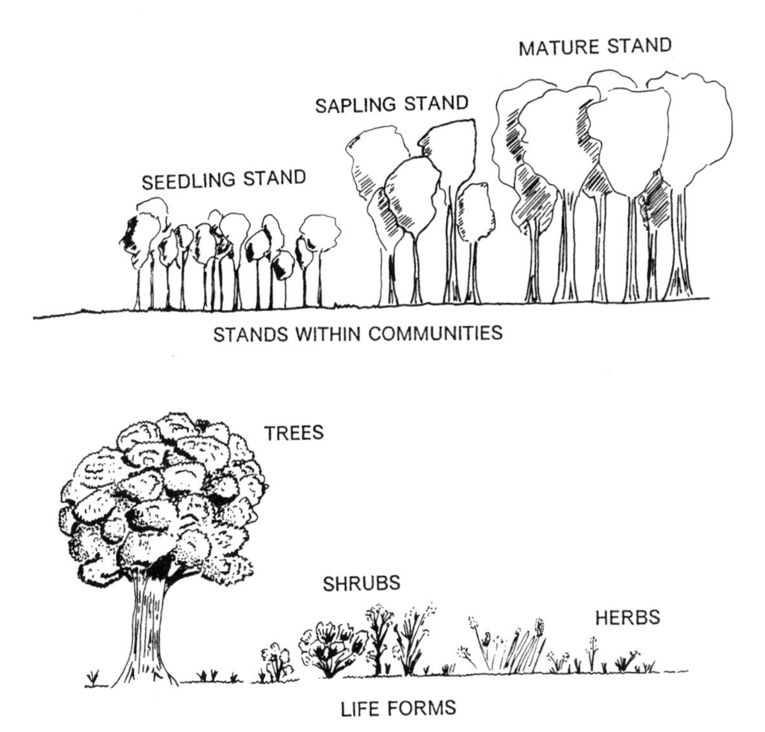


**FIGURE 16.1**. Vegetation communities, stands, vertical stratification, and different life forms.

nities. Stratification is the tendency of communities to be organized vertically, with some species forming an upper canopy, others a lower stratum, then shrubs, mosses, lichens, and so on forming other layers nearer to the ground. Even ostensibly simple vegetation communities in grasslands or arctic tundra, for example, can be shown to consist of distinct strata. Specific plants within each community tend to favor distinctive positions within each strata because they are adjusted to grow best under the conditions of light, temperature, wind, and humidity that prevail in each strata.

*Floristics* refer to the botanical classification of plants, usually based on the character of the reproductive organs, using the classification founded by Carolus Linnaeus (1707–1778). Linnaeus was a Swedish botanist who defined the binomial system of designating plants by Latin or Latinized names that specify a hierarchical nomenclature, of which the *genus* and *species* (Table 16.1) are most frequently used. The Linnaean system provides a distinctive name that places each plant in relationship to others in the taxonomy. Floristic classification reveals the genetic character and evolutionary origin of individual plants.

In contrast, *life form* or *physiognomy* describes the physical form of plants (Table 16.2). For example, common physiognomies might include "tree," "shrub," "herbaceous vegetation," and so on. Physiognomy is important because it reveals the ecological role of the plant—the nature of its relationship with the environment and other plants. Floristics and physiognomy often have little direct relationship to one another. Plants that are quite close floristically may have little similarity in their growth form, and plants that are quite similar in their ecological roles may be very different floristically. For example, the rose family (Rosaceae) includes a wide variety of trees, shrubs, and herbaceous plants that occupy diverse environments and ecological settings. A single environment, such as the alpine meadows of New England, is composed of perhaps 250 species, including such diverse families as Primulaceae (primrose), Labiatae (mint), Araliaceae (ginseng), and Umbelliferae (parsley). Thus, a purely floristic description of plants (e.g., "the rose family") seldom conveys the kind of information we would like to know about plants, their relationships with one another, and their environment.

Although it is often possible to identify specific plants and assign taxonomic designations from large-scale imagery, vegetation studies via remotely sensed images are typically based on structure and physiognomy of vegetation. For example, usually we wish to separate forest from grassland or to distinguish between various classes of forest. Although it is important to identify the dominant species for each class, our focus is usually on separation of vegetation communities based on their overall form and structure rather than on floristics alone.

**TABLE 16.1. Floristic Classification**

| Level | Example[a] |
|---|---|
| Class | Aangiospermae (broad/general) |
| Order | Sapindales |
| Family | Aceraccae |
| Genus | *Acer* |
| Species | *saccharum* (narrow/specific) |

[a]Example = sugar maple (*Acer saccharum*).

**TABLE 16.2. Classification by Physiognomy and Structure**

| | |
|---|---|
| Woody plants | Broadleaf evergreen<br>Broadleaf deciduous<br>Needleleaf evergreen<br>Leaves absent<br>Mixed<br>Semideciduous |
| Herbaceous plants | Graminoids<br>Forbs<br>Lichens and mosses |
| Special life forms | Climbers<br>Stem succulents<br>Tuft plants<br>Bamboos<br>Epiphytes |

*Note.* Data from Küchler (1967). Küchler's complete classification specifies plant height, leaf characteristics, and plant coverage.

No single approach to vegetation classification can be said to be superior to the others. At given levels of detail, and for specific purposes, each of the approaches discussed serves important functions. Floristic classification is useful when scale is large, and mapping is possible in fine detail that permits identification of specific plants. For example, analyses for forest management often require large scale both for measurement of timber volume and for identification of individual trees. Physiognomy or structure is important whenever image scale is smaller, detail is coarser, and the analyst focuses more clearly on the relationships of plants to environment. Ecological classification may be used at several scales, for analyses that require consideration of broader aspects of planning for resource policy, wildlife management, or inventory of biologic resources.

Another approach to classification considers vegetation the most easily observed component of an environmental complex including plants, soil, climate, and topography. This approach classifies regions as *ecological zones*, usually in a hierarchical system comparable to that shown in Table 16.3. Bailey (1976, 1995) defines his units as *ecore-*

**TABLE 16.3. Bailey's Ecosystem Classification**

| Level | Example |
|---|---|
| Domain | Humid temperate |
| Division | Hot continental |
| Province | Eastern deciduous forest |
| Section | Mixed mesophytic forest |
| District | |
| Land-type association | |
| Land type | |
| Land-type phase | |
| Site | |

*Note.* Data from Bailey (1978).

*gions*—"an ecoregion can best be thought of as a geographical area over which the environmental complex, produced by climate, topography, and soil is sufficiently uniform to permit development of characteristic types of ecologic associations." At the very broadest scales, ecological classification is based on long-term climate and broad-scale vegetation patterns, traditionally derived from information other than remotely sensed data. However, later sections of this chapter will show how it is now possible to use remotely sensed data to derive these classifications with much more precision and accuracy than was previously possible.

At finer levels of detail, remotely sensed imagery is essential for delineating ecoregions. The interpreter considers not only vegetation cover but also elevation, slope, aspect, and other topographic factors in defining units on the map.

### *Kinds of Aerial Imagery for Vegetation Studies*

The most important advantage of aerial imagery is the ability to conduct quick and accurate delineation of major vegetation units, providing at least preliminary identification of their nature and composition. Interpretation from aerial images does not replace ground observations as an accurate interpretation assumes that the analyst has field experience and knowledge of the area to be examined and will be able to evaluate the interpretation in the field.

Küchler (1967) recommends use of vertical photographs that provide stereoscopic coverage. Optimum choice of image scale, if the analyst has control over such variables, depends on the nature of the map and the complexity of the vegetation pattern. Detailed studies have been conducted using photography at scales as large as 1:5,000, but scales from about 1:15,000 to 1:24,000 are probably more typical for general-purpose vegetation studies. Of course, if photographs at several different scales, dates, or seasons, are available, the multiple coverage can be used to good advantage to study changes. Also, the small-scale images can be used as the basis for delineation of extents of major regions, and the greater detail of large-scale images can be used to identify specific plants and plant associations.

Panoramic photographs have been successfully used for a variety of purposes pertaining to vegetation mapping and forest management. Modern panoramic cameras permit acquisition of high-resolution imagery over very wide regions, so large areas can be surveyed quickly. Such images have extreme variations in scale and perspective so they cannot be used for measurements, but their wide areal coverage permits a rapid, inexpensive inventory.

Timing of flights may not always be under the control of the analyst but can be a critical factor in some projects. For example, mapping the understory in forested areas can be attempted only in the spring when shrubs and herbaceous vegetation have bloomed but before the forest canopy has fully emerged to obscure the smaller plants from overhead views. Because not all plants bloom at the same time, a succession of carefully timed photographic missions in the spring can sometimes record the sequence in which specific species bloom, thereby permitting mapping of detail that could not be reliably determined by a single image showing all trees in full bloom.

Infrared films and color infrared films have obvious advantages for vegetation studies of all kinds, as emphasized previously. CIR photography at high altitudes is now routinely available at modest cost for the United States through the NAPP (Chapter 3). However, custom-flown CIR photography may be considered expensive, and the usual black-and-white photographs are often satisfactory for most purposes.

### *Interpretation Procedures*

Individual plants can be identified by closely examining crown size and shape (Figure 16.2). At the edges of forested areas, the shadows of trees can form clues to identification.

At smaller scales, individual plants are not recognizable, and the interpreter must examine patterns formed by the aggregate appearance of individual stands in which individual crowns have a distinctive tone and texture. Identification may be relatively straightforward if stands are pure or composed of only a few spaces that occur in consistent proportions. If many species are present and their proportions vary, specific identification may not be possible, and broad designations such as *mixed deciduous forest* may be necessary.

At these smaller scales, an interpreter delineates separate cover types from differences in image tone, image texture, site, shadow, and so on. Delineation of cover types can be considered the interpreter's attempt to implement one of the classification strategies mentioned earlier in the context of a specific geographic region, using a specific image. Cover types may not match exactly the categories in a classification system, but they form the interpreter's best approximation of these categories for a specific set of imagery. Cover classes can be considered rather broad vegetation classes, based perhaps on predominant species present, age, and degree of crown closure. Thus, forest cover-type classes might include *aspen/mixed conifer,* and *Douglas fir,* indicating the predominant species, without precluding the residence of others. Subclasses or secondary descriptors could indicate sizes of the trees, crown closure, and presence of undergrowth. Such classifications can be prepared by manual interpretation of aerial photographs if the interpreter has field experience and knowledge of the area to be mapped. Each region has only a finite number of naturally occurring cover types, which usually prefer specific topographic sites, so the interpreter can bring several kinds of knowledge to bear upon the problem of recognition and classification. Image tone, texture, shadow, and other photographic features permit separation of major classes. Rather broad classes can be separated using small-scale photography, but finer distinctions require high-quality, large-scale, stereo photography. For

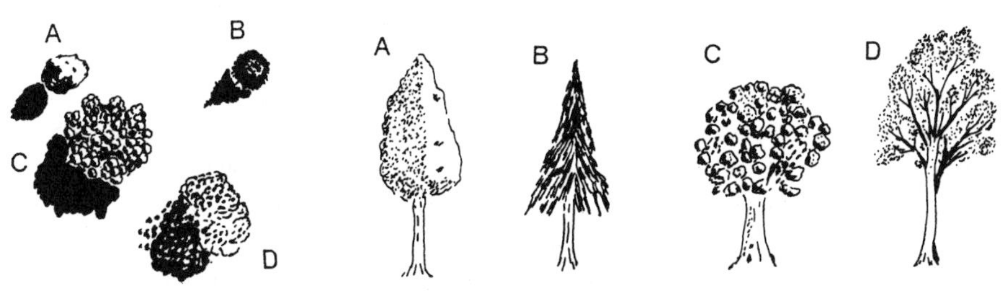


**FIGURE 16.2.** Identification of individual plants by crown size and shape

**TABLE 16.4. Examples of Cover Types**

*a.* At small scale (1:1,000,000 to 1:50,000), cover types must be defined at coarse levels of detail because it will not be possible to distinguish or to delineate classes that are uniform at the species level:
"evergreen forest"
"deciduous forest"
"chaparral"
"mixed broadleaf forest"
"mangrove forest"

*b.* At large scale (larger than about 1:20,000), cover types may be defined at fine levels of detail. Sometimes, when stands are very uniform, cover types can be mapped at the detail of individual species:
"balsam fir"
"shortleaf pine"
"aspen and white birch"
"oak–hickory forest"

example, the classes given in Table 16.4a might be appropriate for information interpreted from aerial photographs at 1:60,000, whereas classes in Table 16.4b could be interpreted only from much larger-scale photographs, perhaps at 1:15,840 to 1:24,000. At larger scales, stereoscopic coverage is important for identifying and distinguishing between classes.

Foresters desire to identify specific *stands* of timber. Stands are areas of forest with uniform species composition, age, and density that can be treated as homogeneous units by foresters. Stands form the basic unit of management, so forest managers wish to monitor their growth time to detect the effects of disease, insect damage, and fire or drought. Even when stands have been planted from seedlings by commercial foresters, aerial photography, by virtue of its map-like perspective and its wealth of environmental information provides accurate and economical information concerning the condition of the stand at a specific time.

### *Forest Photogrammetry*

If large-scale, high-quality images are available (preferably in stereo), it is possible to apply the principles of photogrammetry to the measurement of factors of significance in forestry. Tree height can be measured using lengths of shadows projected onto level, open ground (Figure 16.3), although there are numerous practical problems that make this method inconvenient for routine use (level, open ground is not often easy to find; it is necessary to find sun elevations for the latitude, day of the year, and time of day; and often shadows are too short for reliable measurements). As a result, the parallax bar and the parallax wedge (Chapter 5) are much more commonly used for routine work.

*Crown diameters* can be measured using transparent rules or graticules with tube magnifiers (Chapter 5). *Crown closure* (areal coverage of tree crowns as viewed from above) (Figure 16.4) can be estimated using a number of methods, including direct estimation and standardized templates that show examples of crown closure classes. The interpreter who uses a template compares the photo with the image on the template to select the closure class nearest to that of the area to be interpreted.

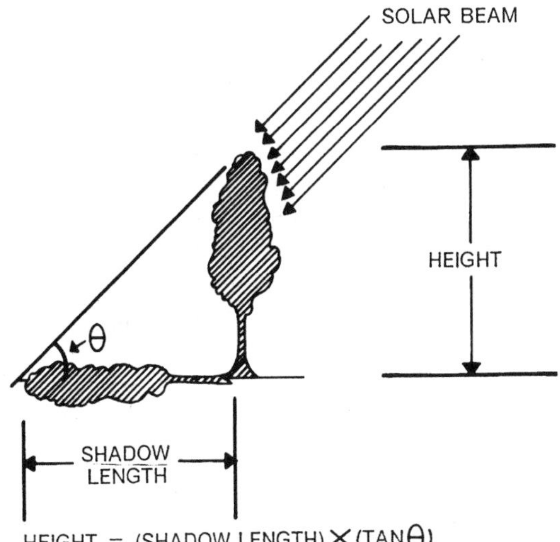


**FIGURE 16.3.** Measurement of tree height from aerial photographs

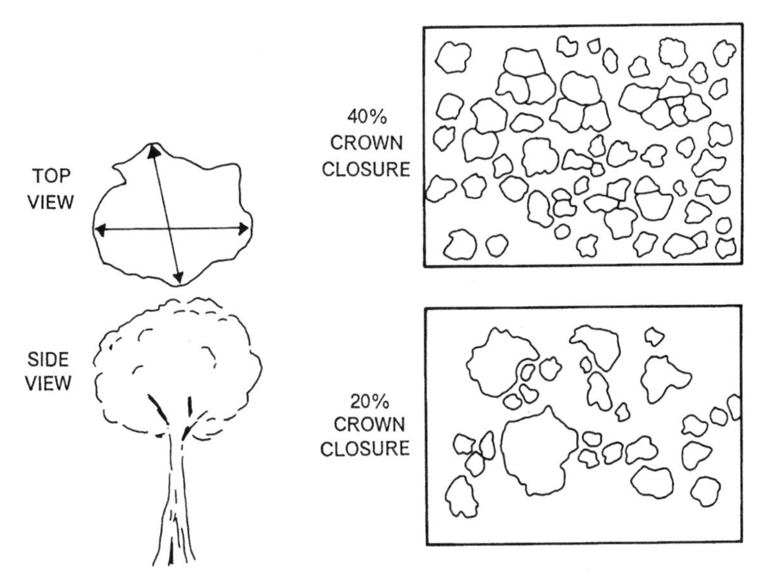


**FIGURE 16.4.** Crown diameter/crown closure.

### *Timber Volume Estimation*

Foresters are routinely interested in estimation of timber volume for a specific stand as a means of monitoring its growth over time, assessing of management practices, and determining the amount of timber to be obtained at harvest.

Volume measurement consists of estimates of board-foot, or cubic-foot volumes to be obtained from a specific tree or, more often, from a specific stand. In the field, the forester measures the diameter of the tree at breast height (dbh) and at the height of the straight section of the trunk (bole) as the two basic components to volume estimation. Measurements made for each tree are summed to give volume estimates for an entire stand.

There are several approaches to estimation of volume from measurements derived from aerial photographs; there can be no universally applicable relationship between photo measurements and timber volume as species composition, size, age, soil, and climate vary so greatly from place to place. Typically the interpreter must apply a volume table—a tabular summary of the relationship between timber volume and photo-derived measurements such as crown diameter and tree height (Table 16.5). The interpreter can make measurements directly from the photograph and enter the table to find volumes for each tree or for entire stands. New volume tables must be compiled as stand composition and the environmental setting change from place to place. Many variations on this basic strategy have been developed, but most rely on the same kinds of estimates from the photograph.

### *Crop Identification*

Crop identification by manual photo interpretation is accomplished by application of the elements of photointerpretation in the context of knowledge of the local environmental setting and the local crop calendar. In many settings crops are usually observed planted in uniform, distinct fields, a single crop to a field. In this context, mature crops can often be identified on the basis of photo tone and texture (Figure 16.5). Precise identification of specific crops may be difficult in the absence of detailed knowledge of the local setting. For example, it is usually easy to separate small grains (wheat, barley, rye) from large grain crops (such as maize or sorghum), although even experienced photo interpreters may have difficulty distinguishing crops within these classes (i.e., between wheat and barley). Therefore, careful timing of the date and season of the photographs and knowledge of the crop calendar (i.e., typical dates of planting and harvesting) are essential for deriving of maximum information concerning agricultural crops.

In some regions crops are planted in very small fields (because of terracing or the need to carefully control the timing of irrigation), or many different kinds of plants are planted together in a single field. Under such conditions crop identification may be much more difficult than in the typical mid-latitude situation in which fields are large and crops are homogeneous within fields. In other instances specific crops may be grown in fields with distinctive sizes or shapes because of the need for specialized planting, tilling, or harvesting equipment or to control movement of irrigation water. Therefore, interpreters may be able to use, along with other information, field size and shape as clues to the identity of crops.

TABLE 16.5. Example of Aerial Stand Volume Table

| Average stand height (ft.) | Crown cover (%) | | | | |
|---|---|---|---|---|---|
| | 15 | 35 | 55 | 75 | 95 |
| | 8- to 12-ft. crown diameter | | | | |
| 30 | 15 | 50 | 70 | 95 | 135 |
| 40 | 35 | 70 | 95 | 125 | 175 |
| 50 | 50 | 95 | 125 | 170 | 215 |
| 60 | 90 | 150 | 195 | 235 | 265 |
| 70 | 160 | 230 | 270 | 300 | 330 |
| 80 | 250 | 310 | 345 | 380 | 420 |
| | 13- to 17-ft. crown diameter | | | | |
| 30 | 20 | 55 | 80 | 105 | 135 |
| 40 | 40 | 80 | 100 | 145 | 195 |
| 50 | 60 | 100 | 135 | 185 | 235 |
| 60 | 100 | 160 | 205 | 245 | 285 |
| 70 | 170 | 240 | 275 | 310 | 350 |
| 80 | 260 | 320 | 360 | 390 | 430 |
| 90 | 360 | 415 | 450 | 485 | 525 |
| 100 | 450 | 515 | 555 | 595 | 635 |
| | 18- to 22-ft. crown diameter | | | | |
| 30 | 30 | 70 | 90 | 120 | 160 |
| 40 | 50 | 90 | 120 | 160 | 200 |
| 50 | 75 | 120 | 160 | 205 | 245 |
| 60 | 115 | 180 | 225 | 265 | 305 |
| 70 | 190 | 255 | 290 | 330 | 370 |
| 80 | 275 | 330 | 370 | 410 | 450 |
| 90 | 370 | 430 | 470 | 505 | 550 |
| 100 | 470 | 530 | 570 | 610 | 650 |
| 110 | 600 | 660 | 700 | 740 | 780 |

*Note.* Volume (tens of cubic feet) per acre, given average stand height, average crown diameter, and crown cover, for Rocky Mountain conifer species. Derived from field measurements in test plots. Abbreviated from a more extensive table given in Wilson (1960, pp. 480–481). Copyright 1960 by the American Society for Photogrammetry and Remote Sensing. Reproduced by permission.

Although black-and-white photography is usually suitable for crop identification if the interpreter is familiar with the local crop calendar, color emulsions may permit the interpreter to make more subtle distinctions in crop status. If color or CIR photographs are available it may be possible to interpret information concerning crop maturity as the crop approaches time for harvest. In addition, CIR photography may permit interpretation of the presence, location, and nature of insect infestations or diseases. Colwell (1956) outlined the fundamentals of using CIR photography to detect the progress of crop diseases.

Plate 14 shows a SPOT multispectral image of an agricultural landscape in eastern Oregon noted for its production of potatoes. Here the climate is arid enough to require irrigation for many crops; some irrigation is from surface water, but most fields are irrigat-

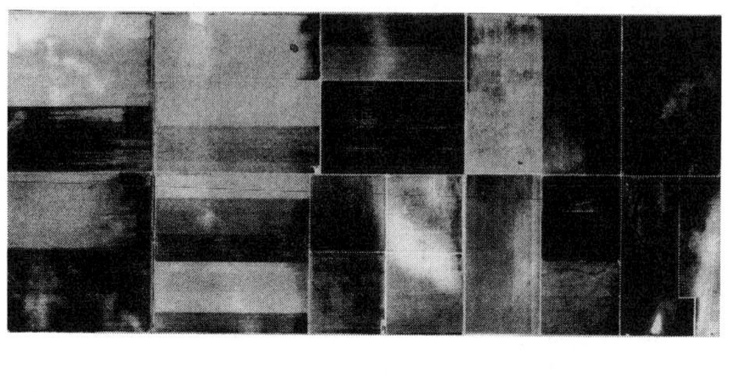


**FIGURE 16.5.** Crop identification. S, sorghum; W, wheat; C, corn; F, summer fallow.

ed by ground water using center-pivot irrigation, which creates the pattern of circular patches visible in this scene. Each circular area measures 0.5 mi. in diameter and is irrigated by a linear sprinkler system anchored at the circle's center. Careful timing of irrigation and early detection of crop stress and disease are critical for the economic success of each farm, but aerial photography and direct observation of fields on the ground are impractical due to cost and timeliness. Because of their multispectral character, broad scope, and timeliness, SPOT satellite data have been very effective in monitoring the status of potato crops, timing irrigation, and detecting effects of disease or insect infestations.

## 16.3. Structure of the Leaf

Many applications of remote sensing to vegetation patterns depend on a knowledge of spectral properties of individual leaves and plants. These properties are best understood by examining leaf structure at a rather fine level of detail.

The cross section of a typical leaf reveals its essential elements (Figure 16.6). The uppermost layer, the upper epidermis, consists of specialized cells that fit closely together without openings or gaps between cells. This upper epidermis is covered by the cuticle, a translucent, waxy layer that prevents moisture loss from the interior of the leaf. The underside of the leaf is protected by the lower epidermis, similar to the upper epidermis except that it includes openings called *stomates* (or *stomata*) that permit movement of air into the interior of the leaf. Each stomate is protected by a pair of guard cells that can open and close as necessary to facilitate or prevent movement of air to the interior of the leaf. The primary function of stomates is apparently to allow atmospheric $CO_2$ to enter the

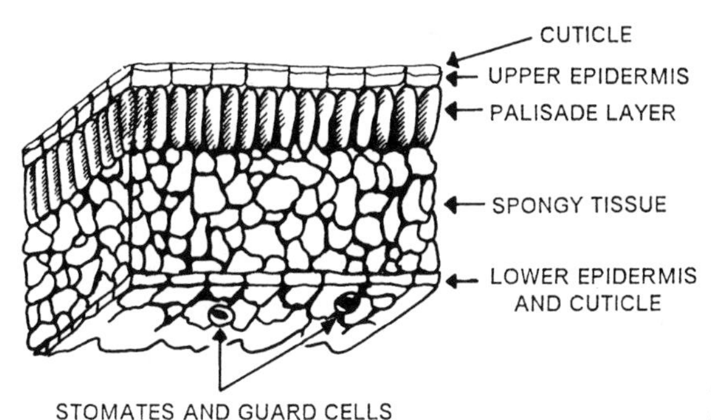


**FIGURE 16.6.** Diagram of cross section of a typical leaf.

leaf for photosynthesis. Although the guard cells and the epidermis appear to be small and inefficient, they are in fact very effective in transmitting gases from one side of the epidermis to the other. Their role in permitting $CO_2$ to enter the leaf is obviously essential for the growth of the plant, but they also play a critical role in maintaining the thermal balance of the leaf by permitting movement of moisture to and from the interior of the leaf. However, the guard cells can close to prevent excessive movement of moisture and thereby economize moisture use by the plant. Apparently the positions of stomata on the lower side of the leaf favor maximum transmission of light through the upper epidermis and minimize moisture loss when stomata are open.

On the upper side of the leaf just below the epidermis is the palisade tissue consisting of vertically elongated cells arranged in parallel, at right angles to the epidermis. Palisade cells include chloroplasts—cells composed of chlorophyll and other ("accessory") pigments active in photosynthesis, as described below. Below the palisade tissue is the spongy mesophyll tissue, which consists of irregularly shaped cells separated by interconnected openings. The surface of the mesophyll has a very large surface area; it is the site for the oxygen and carbon dioxide exchange necessary for photosynthesis and respiration. Although leaf structure is not identical for all plants, this description provides a general outline of the major elements common to most plants, especially those that are likely to be important in agricultural and forestry studies.

In the visible portion of the spectrum, chlorophyll controls much of the spectral response of the living leaf. Chlorophyll is the green pigment that is chiefly responsible for the green color of living vegetation. Chlorophyll enables the plant to absorb sunlight, thereby making photosynthesis possible; it is located in specialized lens-shaped structures, known as *chloroplasts,* found in the palisade layer. Light that passes through the upper tissues of the leaf is received by chlorophyll molecules in the palisade layer, which is apparently specialized for photosynthesis, as it contains the largest chloroplasts, in greater abundance than other portions of the plant. Carbon dioxide (as a component of the natural atmosphere) enters the leaf through stomates on the underside of the leaf, then diffuses throughout cavities within the leaf. Thus, photosynthesis creates (in a series of steps) carbohydrates from carbon dioxide and water, using the ability of chloroplasts to absorb sunlight as a source of energy. Chloroplasts include a variety of pigments, some known as *accessory pigments,* that can absorb light and pass their energy to chlorophyll. Chlorophyll occurs in two forms. The most common is chlorophyll *a,* the most important photosynthetic agent in most green plants. A second form, chlorophyll *b,* has a slightly different molecular structure; it is found in most green leaves but also in some algae and bacteria.

## 16.4. Spectral Behavior of the Living Leaf

Chlorophyll does not absorb all sunlight equally. The chlorophyll molecules preferentially absorb blue and red light for use in photosynthesis (Figure 16.7). They may absorb as much as 70% to 90% of incident light in these regions. Much less of the green light is absorbed and more is reflected, so the human observer, who can see only the visible spectrum, sees the dominant reflection of green light as the color of living vegetation (Figure 16.8).

In the near infrared spectrum, reflection of the leaf is controlled not by plant pigments but by the structure of the spongy mesophyll tissue. The cuticle and epidermis are almost completely transparent to infrared radiation, so very little infrared radiation is reflected from the outer portion of the leaf. Radiation passing through the upper epidermis is strongly scattered by mesophyll tissue and cavities within the leaf. Very little of this infrared energy is absorbed internally—most (up to 60%) is scattered upwards (which we call *reflected energy*) or downward (*transmitted energy*). Some studies suggest that palisade tissue may also be important in infrared reflectance. Thus, the internal structure of the leaf is responsible for the bright infrared reflectance of living vegetation.

At the edge of the visible spectrum as the absorption of red light by chlorophyll pigments begins to decline, reflectance rises sharply (Figure 16.8). Thus, if reflectance is considered not only in the visible but across the visible and near infrared, peak reflectance of living vegetation is not in the green but in the near infrared (Figure 16.8). This behavior explains the great utility of the near infrared spectrum for vegetation studies and, of course, facilitates separation of vegetated from nonvegetated surfaces, which are usually much darker in the near infrared (Plate 15). Furthermore, differences in reflectivities of plant species often are more pronounced in the near infrared than they are in the visible, meaning that discrimination of vegetation classes is sometimes possible using near infrared reflectance (Figure 16.9).

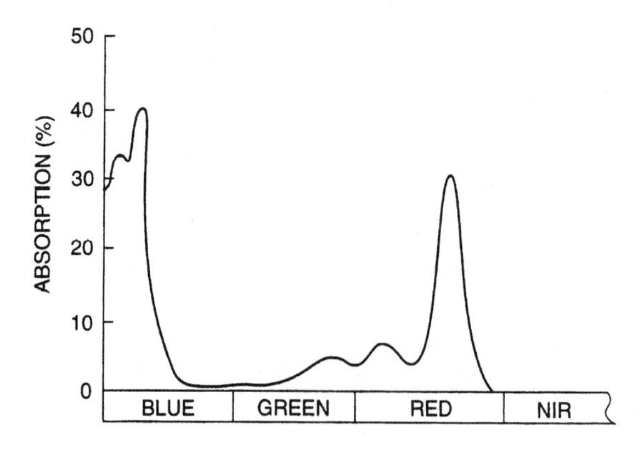


**FIGURE 16.7.** Absorption spectrum for chlorophyll.

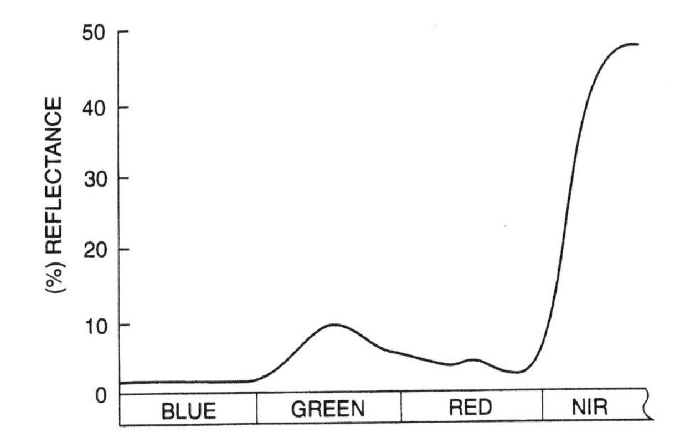


**FIGURE 16.8.** Typical spectral reflectance from a living leaf.

As a plant matures or is subjected to stress by disease, insect attack, or moisture shortage, the spectral characteristics of the leaf may change. In general, these changes apparently occur more or less simultaneously in both the visible and near infrared regions, but changes in near infrared reflectance are often more noticeable. Reflectance in the near infrared region is apparently controlled by the nature of the complex cavities within the leaf and internal reflection of infrared radiation within these cavities. Although some scientists suggest that moisture stress or natural maturity of a leaf causes these cavities to "collapse" as a plant wilts, others maintain that it is more likely that decreases in near infrared reflection are caused by deterioration of cell walls rather than physical changes in the cavities themselves. Thus, changes in the infrared reflectance can reveal changes in vegetative vigor, and infrared images have been valuable in detecting and mapping the presence, distribution, and spread of crop diseases and insect infestations (Figure 16.10). Furthermore, changes in leaf structure that accompany natural maturing of crops are sub-

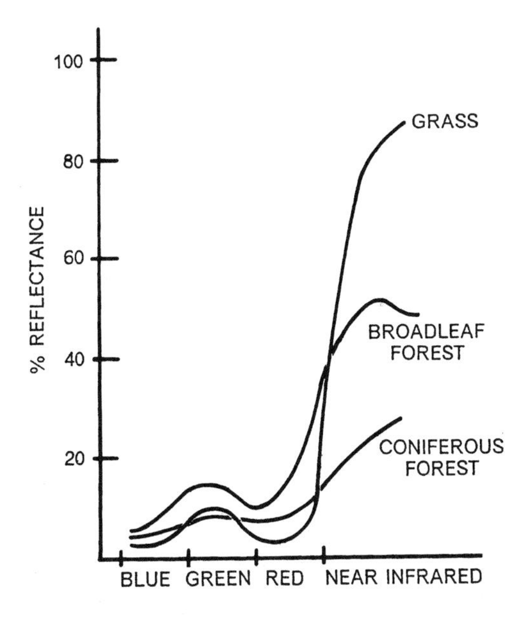


**FIGURE 16.9.** Differences between vegetation classes are often more distinct in the near infrared than in the visible.

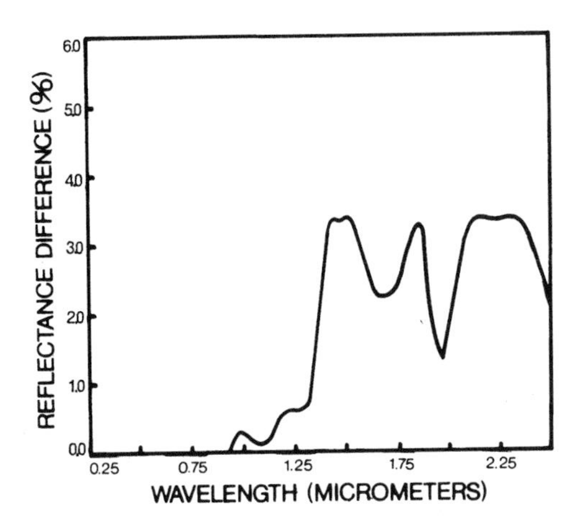


**FIGURE 16.10.** Major influences on spectral properties of the living leaf.

ject to detection with infrared imagery so that it is possible to monitor the ripening of crops as harvest time approaches. CIR film is valuable for observing such changes because of its ability to show spectral changes in both visible and near infrared regions and to provide clear images that show subtle tonal differences.

In the longer infrared wavelengths (beyond 1.3 µm) leaf water content appears to control the spectral properties of the leaf (Figures 16.11). The term *equivalent water thickness* (EWT) has been proposed to designate the thickness of a film of water that can account for the absorption spectrum of a leaf at 1.4 µm to 2.5 µm.

### *Reflection from Canopies*

Knowledge of spectral behavior of individual leaves is, of course, important for understanding the spectral characteristics of vegetation canopies but cannot itself completely

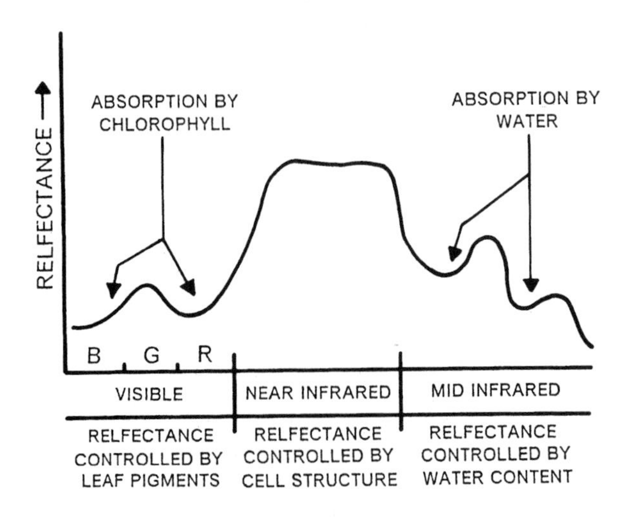


**FIGURE 16.11.** Changes in leaf water content may be pronounced in the near infrared region. This diagram illustrates *differences* in reflectance between equivalent water thicknesses of 0.018 cm and 0.014 cm; changes in reflectance are evident in the infrared region. Diagram based on simulated data in the study by Tucker (1979b, p. 10.)

explain reflectance from areas of complete vegetative cover. Vegetation canopies are composed of many separate leaves that may vary in size, orientation, shape, and coverage of the ground surface. In the field, a vegetation canopy (e.g., in a forest or a cornfield) is composed of many layers of leaves; the upper leaves form shadows that mask the lower leaves, creating an overall reflectance that is formed by a combination of leaf reflectance and shadow.

Shadowing tends to decrease canopy reflectance below the values normally observed in the laboratory for individual leaves. Knipling (1970) cited the following reported in several previously published studies:

| | Percent reflected | |
|---|---|---|
| | Visible | Near infrared |
| Single leaf | 10% | 50% |
| Canopy | 3–5% | 35% |

Thus, reflectance of a canopy is lower than values measured for individual leaves. But the relative decrease in the near infrared region is much lower than that in the visible. The brighter canopy reflection for the near infrared is attributed to the fact that plant leaves transmit near infrared radiation, perhaps as much as 50–60%. Therefore, infrared radiation is transmitted through the upper layers of the canopy, reflected in part from lower leaves, and transmitted back through the upper leaves, resulting in bright infrared reflectance (Figure 16.12). Physicists and botanists have been able to develop mathematical models for canopy reflectances by estimating optical properties of the leaves and the canopy as a whole. Because some of the transmitted infrared radiation may in fact be reflected from the soil surface below the canopy, such models have formed an important part of the research for agricultural remote sensing.

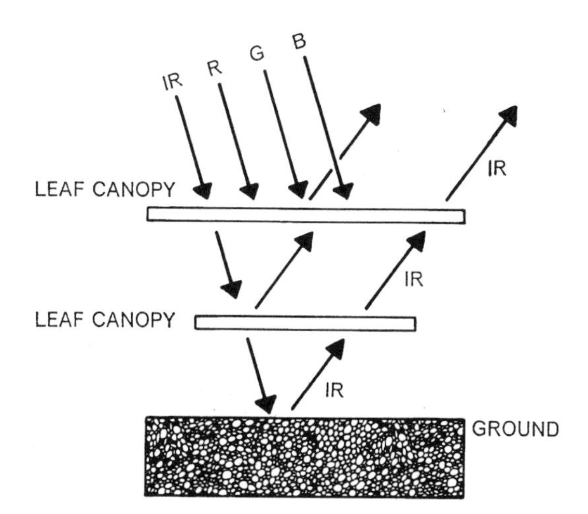


**FIGURE 16.12.** Simplified cross-sectional view of behavior of energy interacting with a vegetation canopy. In some portions of the spectrum, energy transmitted through the upper layer is available for reflection from lower layers (or from the soil surface).

### *The Red Shift*

Collins (1978) reports the results of studies that show changes in the spectral responses of crops as they approach maturity. His research used high-resolution multispectral scanner data of numerous crops at various stages of the growth cycle. Collins's research focuses on examination of the far red region of the spectrum, where chlorophyll absorption decreases and infrared reflection increases (Figure 16.13). In this zone, the spectral response of living vegetation increases sharply as wavelength increases—in the region from just below 0.7 μm to just above 0.7 μm brightness increases by about 10 times (Figure 16.8).

Collins observed that as crops approach maturity, the position of the chlorophyll absorption edge shifts toward longer wavelengths, a change he refers to as the *red shift* (Figure 16.13). The red shift is observed not only in crops but also in other plants. The magnitude of the red shift varies with crop type (it is apparently a pronounced and persistent feature in wheat).

Collins observed the red shift along the entire length of the chlorophyll absorption edge, although it was most pronounced near 0.74 μm, in the infrared region, near the shoulder of the absorption edge. He suggests that very narrow bands at about 0.745 μm and 0.780 μm would permit observation of the red shift over time and thereby provide a means of assessing differences between crops and the onset of maturity of a specific crop.

Causes of the red shift appear to be very complex and may not be completely understood. Chlorophyll *a* appears to increase in abundance as the plant matures; increased concentrations change the molecular form in a manner that adds absorption bands to the edge of the chlorophyll *a* absorption region, thereby producing the red shift. (In Chapter 17 we shall see that certain factors can alter the spectral effect of chlorophyll, thereby shifting the edge of the absorption band toward shorter wavelengths—the "blue shift" observed in geobotanical studies.)

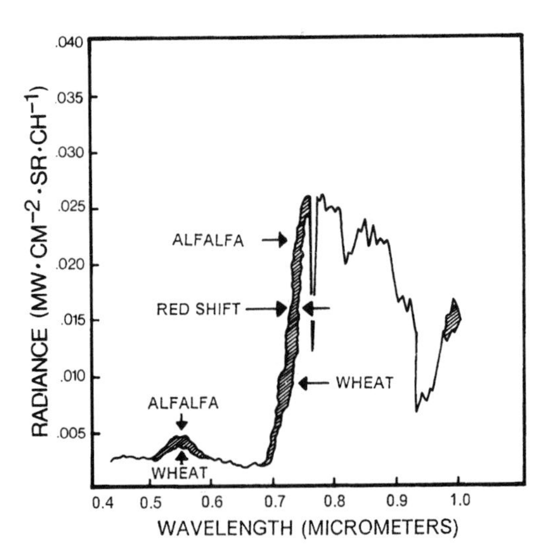


**FIGURE 16.13.** Red shift. The absorption edge of chlorphyll shifts toward longer wavelengths as plants mature. The shaded area represents the magnitude of this shift as the difference between the spectral response of headed wheat and mature alfalfa. Similar but smaller shifts have been observed in other plants. The red shift is important for distinguishing the headed stage of a crop from earlier stages of the same crop and for distinguishing headed grain crops from other green, nongrain, crops. From Collins (1978, p. 47). Copyright 1978 by the American Society for Photogrammetry and Remote Sensing. Reproduced by permission.

## 16.5. Vegetation Indices

Vegetation indices are quantitative measures, based on digital values, that attempt to measure biomass or vegetative vigor. Usually a vegetation index is formed from combinations of several spectral values that are added, divided, or multiplied in a manner designed to yield a single value that indicates the amount or vigor of vegetation within a pixel. High values of the vegetation index identify pixels covered by substantial proportions of healthy vegetation. The simplest form of vegetation index is simply a ratio between two digital values from separate spectral bands. Some band ratios have been defined by applying knowledge of the spectral behavior of living vegetation.

Band ratios are quotients between measurements of reflectance in separate portions of the spectrum. Ratios are effective in enhancing or revealing latent information when there is an inverse relationship between two spectral responses to the same biophysical phenomenon (Figure 16.14). If two features have the same spectral behavior, ratios provide little additional information (Figure 16.14a), but if they have quite different spectral responses, the ratio between the two values provides a single value that concisely expresses the contrast between the two reflectances (Figure 16.14b).

For living vegetation, the ratioing strategy can be especially effective because of the inverse relationship between vegetation brightness in the red and infrared region. That is, absorption of red light (R) by chlorophyll, and strong reflection of infrared (IR) radiation by mesophyll tissue ensures that the red and near infrared values will be quite different and that the ratio (IR/R) will be high (Figure 16.15). Nonvegetated surfaces, including open water, man-made features, bare soil, and dead or stressed vegetation, will not display this specific spectral response, and the ratios will decrease in magnitude. Thus, the IR/R ratio can provide a measure of the importance of vegetative reflectance within a given pixel.

The IR/R ratio is only one of many related measures of vegetation vigor and abundance. The green/red (G/R) ratio is based on the same concepts as used for the IR/R ratio,

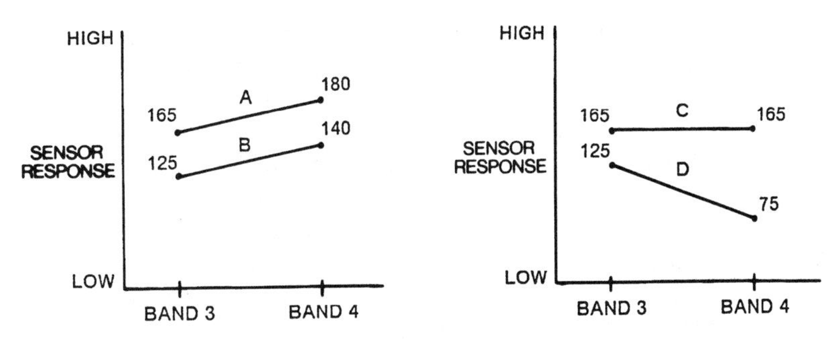


**FIGURE 16.14.** Example of band ratios. Four hypothetical materials are observed in two spectral bands. Materials *A* and *B* have similar patterns of response in bands 3 and 4. Ratios therefore reveal little about the differences between them. For *A*, the ratio of band 3 to band 4 is 0.916 (165/180); for *B*, it is 0.89 (125/140). The relationship between *C* and *D* is quite different, and the ratio clearly reveals the contrast between the two materials. For *C* it is 1.00 (165/165); for *D* it is 1.67 (125/175). From Morain (1978, p. 42).

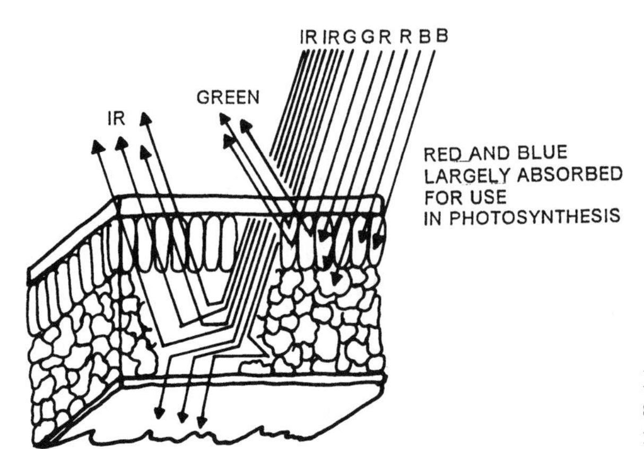


**FIGURE 16.15.** Interaction of leaf structure with visible and near infrared radiation.

although it is considered less effective. Although ratios can be applied with digital values from any remote sensing system, much of the research on this topic has been conducted using Landsat MSS data. In this context, the IR/R ratio is implemented for Landsats 4 and 5 as (MSS 4/MSS 2), although some have preferred to use MSS 3 in place of MSS 4. One of the most widely used vegetation indices is known as the normalized difference vegetation index (NDVI):

$$\text{NDVI} = \frac{\text{IR} - \text{R}}{\text{IR} + \text{R}} = \frac{\text{MSS 4} - \text{MSS 2}}{\text{MSS 4} + \text{MSS 2}} \qquad \text{(Eq. 16.1)}$$

This index in principle conveys the same kind of information as the IR/R and G/R ratios but is defined to produce desirable statistical properties in the resulting values. The studies of Tucker et al. (1979) and Perry and Lautenschlager (1984) suggest that in practice there are few differences between the many vegetation indices that have been developed.

Although such ratios have been shown to be powerful tools for studying vegetation, they must be used with care if the values are to be rigorously (rather than qualitatively) interpreted. Values of ratios and vegetation indices can be influenced by many factors external to the plant leaf, including viewing angle, soil background, and differences in row direction and spacing in the case of agricultural crops. Ratios may be sensitive to atmospheric degradation. Because atmospheric degradation typically influences some bands much more than others, atmospheric effects can greatly alter the value of the ratio from its true value (Figure 16.16). Because atmospheric path length varies with viewing angle, values calculated using off-nadir satellite data (Chapter 7) can vary according to position within the image. Clevers and Verhoef (1993) found that the relationship between leaf area index (LAI) and the vegetation index they studied was sensitive to differences in leaf orientation. Although preprocessing can sometimes address such problems, it may still be difficult to compare values of vegetation indices over time because of variation in external factors. Price (1987) and others have noted that efforts to compare ratios or indices over time, or from one sensor to another, should reduce digital values to radiances

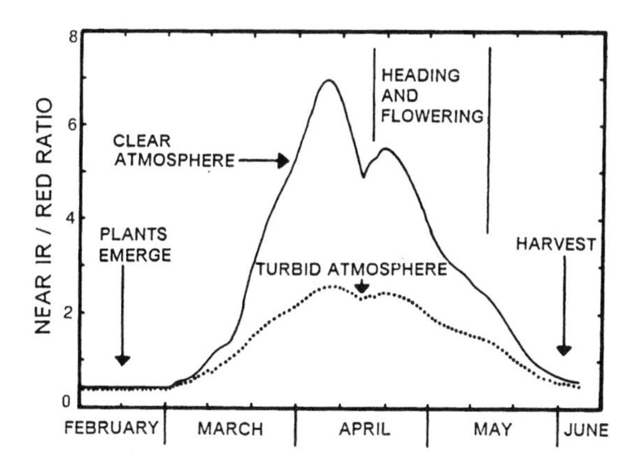


**FIGURE 16.16.** Influence of atmospheric turbidity on IR/R ratio, as estimated from simulated data. From Jackson et al. (1983, p. 195). Copyright 1983 by Elsevier Science Publishing Co. Reproduced by permission.

before calculating ratios, to account for differences in calibration of sensors (see Appendix).

## 16.6. Applications of Vegetation Indices

Vegetation indices have been employed in two separate kinds of research. Many of the first studies defining applications of vegetation indices attempted to "validate" their usefulness by establishing that values of the vegetation indices are closely related to biologic properties of plants. Typically, such studies examined test plots during an entire growing season and compared values of the vegetation indices, measured throughout the growing season, to samples of vegetation taken at the same times. The objective of such studies is ultimately to establish use of vegetation indices as a means of remote monitoring of the growth and productivity of specific crops, or of seasonal and yearly fluctuations in productivity.

Often values of the vegetation indices have been compared to in situ measurements of LAI (area of leaf surface per unit area of soil surface). LAI is an important consideration in agronomic studies because it measures leaf area exposed to the atmosphere and therefore is significant in studies of transportation and photosynthesis. Values of vegetation indices have also been compared to biomass—the weight of vegetative tissue. A number of vegetation indices appear to be closely related to LAI (at least for specific crops), but not a single vegetation index seems to be equally effective for all plants and all agricultural conditions. Results of such studies have in general confirmed the utility of the quantitative uses of vegetation indices, but details vary with specific crop considered, atmospheric conditions, and local agricultural practices.

A second category of applications uses vegetation indices as a mapping device, much more of a qualitative rather than a quantitative tool. Such applications use vegetation indices to assist in image classification, to separate vegetated from nonvegetated areas, to distinguish perhaps between different types and densities of vegetation, and to monitor seasonal variations in vegetative vigor, abundance, and distribution (Plate 15).

Yool et al. (1985) studied MSS imagery of a forested area in southern California and concluded that the ratios they examined were not superior to the original MSS values or to simple linear combinations of these values.

## 16.7. Phenology

*Phenology* is the study of the relationship between vegetative growth and environment; often phenology refers specifically to seasonal changes in vegetative growth and decline. Many phenological changes can be monitored by means of remote sensing as plants change in appearance and structure during their growth cycle. Of special significance are spectral and physiologic changes that occur as a plant matures. Each season, plants experience chemical, physical, and biologic changes, known as *senescence,* that result in progressive deterioration of leaves, stems, fruit, and flowers. In the mid-latitudes, annual plants typically lose most or all of their roots, leaves, and stems each year. Woody perennial plants typically retain some or all of their roots, woody stems, and branches but shed leaves. Evergreens, including tropical plants, experience much more elaborate phenologic cycles; individual leaves may experience senescence separately (i.e., trees do not necessarily shed all leaves simultaneously), and individual trees or branches of trees may shed leaves on cycles quite distinct from others in the same forest (Koriba, 1958).

During the onset of senescence, deterioration of cell walls in the mesophyll tissue produces a distinctive decline in infrared reflectance; an accompanying increase in visible brightness may be the result of decline in the abundance and effectiveness of chlorophyll as an absorber of visible radiation. Changes in chlorophyll produce the red shift mentioned earlier. Such changes can be observed spectrally, so remote sensing imagery can be an effective means of monitoring seasonal changes in vegetation.

The phenology of a specific plant defines its seasonal pattern of growth, flowering, senescence, and dormancy. Remotely sensed images can expand the scope of study to include overviews of vegetation communities or even of entire biomes. Dethier et al. (1973), for example, used several forms of imagery and data to observe the geographic spread of the emergence of new growth in spring, as it progressed from south to north in North America. This phenomenon has been referred to as the *green wave.* Then in late summer, the *brown wave* sweeps across the continent as plant tissues mature, dry, and are harvested.

Figure 1.1, first discussed at the beginning of Chapter 1, illustrates local phenological differences in the spread of the green wave in the early spring. The image shows a Landsat MSS band 4 (near infrared) radiation, which of course is sensitive to variations in the density, type, and vigor of living vegetation. The region represented by Figure 1.1, and shown again in somewhat different form in Figure 16.17, has uneven topography covered by a mixture of forested and open land, including large regions of cropland and pasture. When this image was acquired in mid-April, vegetation in the open land was just starting to emerge. Grasses and shrubs in these regions have bright green leaves, but leaves on the larger shrubs and trees have not yet emerged. Therefore, the pattern of white depicts the regions occupied primarily by early blooming grasses and shrubs in lower elevations. Within week or so after this image was acquired, leaves on trees began to emerge, first at the lower elevations and later at higher elevations. Thus, a second image acquired only 2

**FIGURE 16.17.** Landsat MSS band 4 image of Southwestern Virginia. Image from EROS Data Center.

weeks or so after the first would appear almost completely white due to the infrared brightness of the almost complete vegetation cover. If it were possible to observe this region on a daily basis, under cloud-free conditions, we could monitor the movement of the green wave upwards from lower to higher elevations, and from south to north, as springtime temperatures prevail over more and more of the region. In reality, of course, we can see only occasional snapshots of the movement of the green wave, as the infrequent passes of Landsat and cloud cover prevent close observation during the short period when we must watch its progress.

In an agricultural context, phenology manifests itself through the local crop calendar—the seasonal cycle of plowing, planting, emergence, growth, maturity, harvest, and fallow. Each locality and each crop has its own crop calendar, defined by the interaction of the genetic character of the crop, the local climate, and local agricultural technology and tradition.

As examples, two crop calendars significant in North America are shown in Tables 16.6 and 16.7. Winter wheat (Table 16.6) is grown chiefly in a broad zone in the semi-arid regions that border the western edge of the Great Plains, from west Texas north to western South Dakota. The crop is planted in September or October; during the cool, moist climate of mid to late autumn, plants emerge and grow to a height of a few inches. Growth

**TABLE 16.6. Crop Calendar for Winter Wheat, Western Kansas**

| Month | Stage |
|---|---|
| September | Plowing and planting |
| October, November | Emergence and early growth |
| December, January, February, March | Dormancy |
| April, May | Growth resumes |
| June, July | Heading<br>Harvest |
| August | |

*Note.* Variations in timing are caused by local differences in weather and climate.

stops when severe cold of winter arrives in December, although during winter months the soil receives critical moisture supplies from snow. The wheat crop resumes growth in spring when temperatures increase, releasing moisture and accelerating biologic processes. Wheat grows rapidly in April and May; by mid-June, the crop approaches maturity and is usually harvested in late June or early July.

In contrast, the corn crop (Table 16.7) is not planted until spring temperatures have warmed the fields and allowed excess moisture to drain, to permit farm equipment to work the soil. The crop emerges soon after planting in April or May and grows rapidly during May and June, when temperatures are warm and moisture is available. The crop has tasseled by mid-July and matures in August for harvest in late August or September.

The spectral characteristics of a given crop, observed over an entire growing season, exhibit distinctive patterns closely related to the local crop calendar. The opportunities for regular coverage provided by earth observation satellites such as Landsat mean that specific areas can be observed over time and that knowledge of the crop calendar can be applied in a manner that is not always practical if there is imagery for only a single date. For example, the winter wheat crop is characterized by bare fields in September and October and by a mixture of soil and newly emerged vegetation in mid to late autumn. During spring and early summer, plant cover increases, completely masking the soil by late May early June. Then, as the crop matures, spectral evidence of senescence records the approach of harvest time in late June or early July. In contrast, the corn crop is not planted

**TABLE 16.7. Crop Calendar for Corn, Central Midwestern United States**

| Month | Stage |
|---|---|
| April, May | Plowing and planting |
| May, June | Emergence and early growth |
| July | Tasseling |
| August, September | Maturity and harvest |

until spring, does not attain complete coverage until June, and is not mature until August. Thus, each crop displays a characteristic signature over time, which permits tracking of the progress of the crop (the objective of the Large Area Crop Inventory Experiment; see pp. 480–481). Or, knowledge of the local crop calendar permits selection of a single date that will provide maximum contrast between two crops. In the instance of discrimination between corn and wheat, for example, selection of a date in late spring should show complete vegetative cover in wheatfields, while the cornfields show only newly emerged plants against a background of bare soil.

## 16.8. Advanced High-Resolution Radiometer

Vegetation indices and knowledge of phenology can also be applied in a much different context, using data acquired by a meteorological satellite not yet discussed here. The advanced high-resolution radiometer (AVHRR) is a multispectral radiometer carried by a series of meteorological satellites operated by NOAA in near-polar, sun-synchronous orbits. They can acquire imagery over a swath width of approximately 2,800 km, providing global coverage on a daily basis. Although AVHRR was designed primarily for meteorological studies, it has been successfully used to monitor vegetation patterns over broad geographic regions. Areal coverage is extensive enough that entire biomes, or major ecological zones, can be directly observed and monitored in ways that were not previously feasible.

The satellite makes about 14 passes in a 24-hour period, collecting data for each 2,800-km swath twice daily, at 12-hour intervals. Resolution at the nadir is about 1.1 km, but an onboard computer can generalize data to 4-km resolution, before data are transmitted to ground stations, to permit broader geographic coverage for a given volume of data. Data are recorded at 10 bits. The wide angular view means that areas recorded near the edges of images suffer from severe geometric and angular effects. As a result, AVHRR data selected from the regions near the nadir provide the most accurate information. Although designed initially for much narrow purposes, Tucker et al. (1984) report that 800 to 900 km of the 2,800-km swath are usable. (The unusable data are from the edges of the image, where angular effects of perspective and atmospheric path length present problems.)

Details of spectral coverage varies with the specific mission, but in general AVHRR sensors have been designed to attain meteorological objectives, including discrimination of clouds, snow, ice, land, and open water. Nonetheless, AVHRR has been employed on an ad hoc basis for land resource studies. AVHRR collects five channels of data at a spatial resolution of 1.1 km in the visible, near infrared, and thermal infrared (Table 16.8). One channel in the visible, one in the near infrared, and three in the thermal infrared region. Channel 1 was originally proposed as 0.55–0.90 µm, but after use on the prototype was redefined to 0.58 to 0.68 µm to improve separation of snow-free and snow-covered land areas. For scientists this change provided the benefit of a channel positioned in the red region that permits calculation of vegetation indices.

The visible and near infrared data can be ratioed to form a "greenness index" or vegetation index, similar to those discussed above, known as the normalized difference vegetation index (NDVI):

**TABLE 16.8. Spectral Channels for AVHRR**

| Channel | Spectral limits | Region |
|---|---|---|
| 1 | 0.58–0.68 μm | Visible |
| 2 | 0.72–1.10 μm | Near infrared |
| 3 | 3.55–3.93 μm | Thermal infrared |
| 4 | 10.30–11.30 μm | Thermal infrared |
| 5 | 11.5–12.5 μm | Thermal infrared |

*Note.* Spectral definitions differ between different AVHRR missions.

$$\text{NDVI} = \frac{\text{Channel 2} - \text{Channel 1}}{\text{Channel 2} + \text{Channel 1}} \qquad \text{(Eq. 16.2)}$$

Although the spectral bands differ from those mentioned earlier, the meaning of the ratio is the same—high values reveal pixels dominated by high proportions of green biomass. The resolution is much coarser than that of Landsat MSS and TM data, but the areal coverage is much broader, and the opportunity for repeat coverage is much greater. The frequent repeat coverage permits gathering of data for areas that may be obscured by clouds on one date but not the next. Therefore, over time, cloud-free coverage of continental or subcontinental areas can be acquired. Such images do not, of course, show the detail of MSS or TM data, but they do give a broad geographic perspective that is not portrayed by other images.

### *Compiled AVHRR Data*

Full-resolution data, at about 1.1 km at nadir, can be acquired during an given day from a restricted region of the earth, usually only when the satellite is within line of sight of a receiving station. This process produces *local area coverage* (LAC). LAC archives have been compiled for selected regions of the earth.

Eidenshink (1992) describes the AVHRR data set for the United States derived from 1990 AVHRR data, since repeated for other years. These composites are designed to record phenological variations throughout a growing season. Full-resolution AVHRR scenes were selected to form a set of 19 biweekly composites from March to December. Compositing over the 2-week period permits selection of a least one cloud-free date for each pixel. A special processing program was designed to select data based on viewing geometry, solar illumination, sensor calibration, and cloud cover and then to prepare geometrically registered NDVI composites. The process produces composite images composed of the original AVHRR bands, NDVI data, data describing the compositing process, and a statistical summary.

*Global area coverage* (GAC) is compiled by onboard sampling and processing, transmitted to ground stations on a daily basis. A GAC pixel is formed by averaging the first four pixels a row, skipping the fifth pixel, averaging the next four pixels, skipping the next, continuing to the end of the line. The algorithm skips the next two lines, then processes the fourth line in the same manner as it did the first line, and so on until the entire scene is processed. These data provide spatial resolution of about 4 km.

*Global vegetation index* (GVI) data are created from the daily AVHRR data by examining the differences between the red and near infrared channels; the highest values for a 7-day interval are then used to select the dates for which to calculate an NDVI value, which represents a pixel for the entire 7-day period. The preliminary selection using the red–near infrared difference is designed both to select cloud-free pixels (if possible) for the 7-day interval and also to minimize atmospheric effects. NDVI data are processed in such a manner that they represent about a 15-km resolution. These data have been compiled since 1982, although algorithms and formats have varied considerable during this interval (Townshend, 1994).

### *Some AVHRR Examples*

These broad-scale images provide an opportunity to observe major ecological zones and seasonal changes in a manner that was not previously possible. Figures 16.18 and 16.19 show such images for two dates for the Northern Hemisphere. Darker areas show high values for the vegetative index (a reversal of the usual convention of representing high values as bright tones). The summer scene shows the continent blanketed with a broad

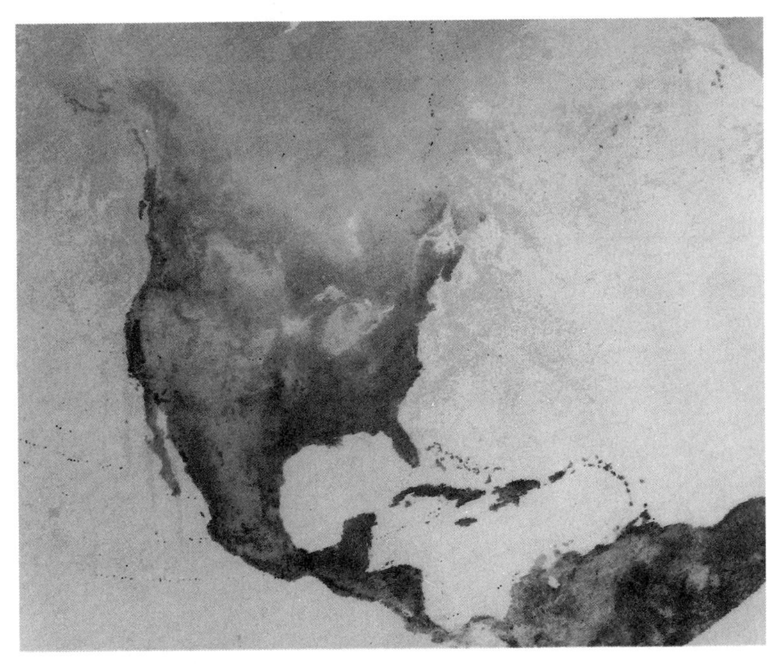

**FIGURE 16.18.** AVHRR image for North America (March 1982). This image shows the values of normalized difference calculated from red and near infrared bands of the AVHRR (Table 16.8). Contrary to the usual conventions, this image shows the highest values of the NDVI as dark tones; the lowest values are sparse green vegetation. This image (March 1982) shows most of the continent to be covered by dormant vegetation, and the far north by clouds, snow, and ice. Areas along the Gulf Coast, the central valley of California, and tropical forests of Mexico and Central America and the Caribbean Islands are covered by green vegetation. The western deserts are shown to be sparsely vegetated. Image courtesy of National Space Science Data Center.

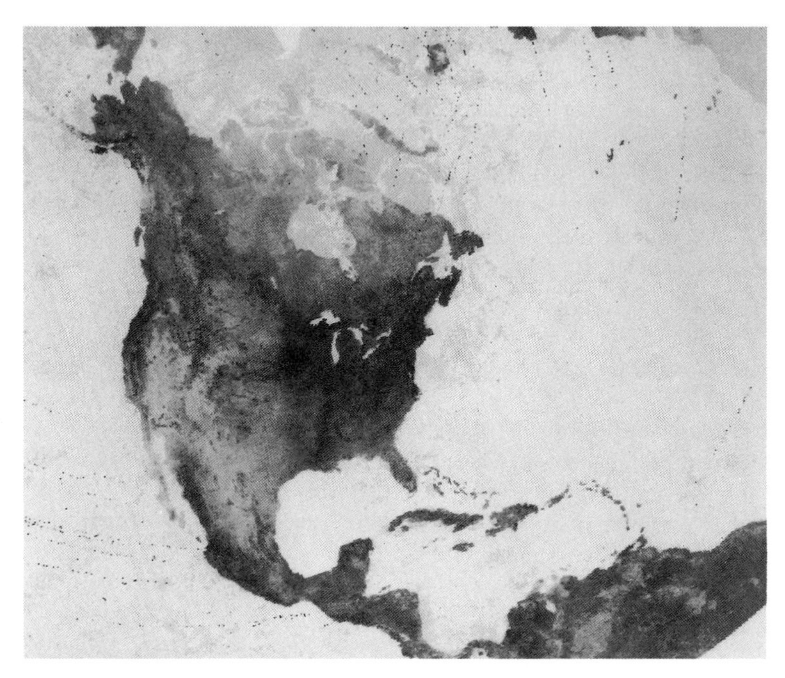

**FIGURE 16.19.** AVHRR Image for North America (August 1982). Green vegetation has emerged through most of the humid climates of the continent. Image courtesy of National Space Science Data Center.

cover of vegetation. The lighter tones in the Great Basin and southwestern United States reveal sparsely vegetated arid regions and the tundra zone north of the arctic tree line. The March scene shows (in addition to snow, ice, and cloud cover in northern areas) much lower biomass, except for localized patches in tropical central America, southern California, and the Gulf Coast of the United States, where winter temperature and moisture permit year-round vegetative growth.

Such products are produced on a weekly basis by NOAA. They cannot be used to study specific sites, but they provide an excellent overview of ecological conditions and zones and permit identification of regions worthy of more detailed study with high-resolution data. Tucker et al. (1984) used AVHRR data to examine crop phenology and agricultural practices in the Nile delta. They collected AVHRR data for 15 dates between May and October 1981, excluding data with severe angular perspective, poor atmospheric conditions, and insufficient geometric control. For each date, they calculated a normalized vegetation index for each pixel. Because clouds were not always distinct in channels 1 and 2, they used a thermal band to detect, then mask, cloud-covered areas. (That is, they did not calculate ratios for pixels influenced by clouds, but simply depicted the cloud-covered regions as black pixels.)

Although these data were much too coarse to attempt mapping of individual crops and fields, their results form an interesting overview of the crop cycle in this region of the earth. For each of the 15 dates, they prepared a map of the vegetation index. Individually, the maps show pattern of vegetation and agriculture; in sequence, they reveal seasonal patterns in irrigation, crop growth, maturity, and harvest.

AVHRR data have been archived to form a global data base and have been processed to provide a variety of products depicting seasonal changes in vegetative cover over areas of continental size. Some of these data are described in Chapter 20, and Internet access to some of them has been outlined previously in Chapter 4.

## 16.9. Separating Soil Reflectance from Vegetation Reflectance

Digital data collected by satellites, especially at the coarse resolution of the Landsat MSS, mix vegetation reflectance with that of the soil surface. Some mixing of soil and vegetation reflectances occurs because of mixed pixels (Chapter 8). In agricultural scenes, however, reflections from individual plants, or individual rows of plants, are closely intermingled with the bare soil between plants and between rows of plants, so that reflectances are mixed even at the finest resolutions. Of course, this mixing is especially important after plants have emerged, as large proportions of soil are exposed to the sensor. But even after leaves have fully emerged, soil can still contribute to reflectance because of penetration of some wavelengths through the vegetation canopy. Mixing of soil and vegetation reflectances can therefore be a serious barrier to implementation of the concepts presented earlier in this chapter. Hutchinson (1982) found that vegetation information is difficult to extract when vegetation cover is less than 30%. In arid and semi-arid regions, vegetative cover may be this low, but even in agricultural settings, plant cover is low when crops are emerging early in the growing season.

If many spectral observations are made of surfaces of bare soil, the values of red and near infrared brightnesses resemble those in Figure 16.20. The soil brightnesses tend to fall on a straight line—as a soil becomes brighter in the near infrared, so does it tend to get brighter in the red. Dry soils tend to be bright in both spectral regions and appear

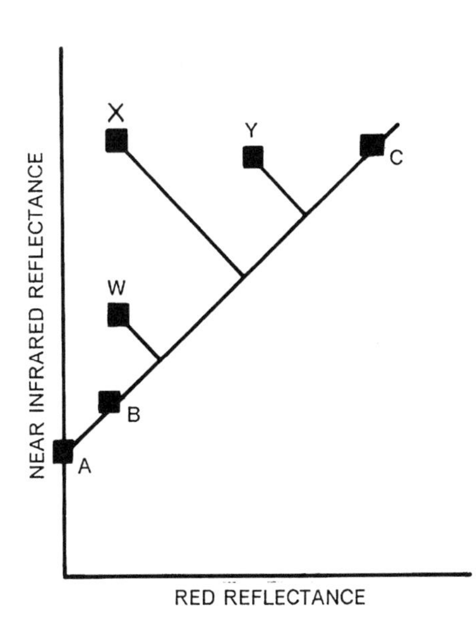


**FIGURE 16.20.** Perpendicular vegetation index. Based on Richardson and Wiegand (1977, p. 1547). Copyright 1977 by the American Society for Photogrammetry and Remote Sensing. Reproduced by permission.

at the high end of the line (C); wet soils tend to be dark, and are positioned at the low end (B).

Richardson and Wiegand (1977) defined this relationship, known as the *soil brightness line,* and recognized that spectral response of living vegetation will always have a consistent relationship to the line. Soils typically have high or modest response in the red and infrared regions, whereas living vegetation must display low values in the red (due to the absorption spectra of chlorophyll) and high values in the near infrared (due to IR brightness of mesophyll tissue). Thus, points representing "pure" vegetation response will be positioned in the upper left of Figure 16.20, where values on the red axis are low and those on the IR axis are high. Furthermore, Richardson and Wiegand defined an index to portray the relative magnitudes of soil background and vegetative cover to a given spectral response. Thus, point $X$ typifies a "pure" vegetation pixel, with a spectral response determined by vegetation alone, with no spectral contribution from soil. In contrast, point $Y$ typifies a response from a partially vegetated pixel—it is brighter in the red and darker in the near infrared than is $X$. Richardson and Wiegand quantified this difference by defining the perpendicular vegetation index (PVI) as a measure of the distance of a pixel (in spectral data space) from the soil brightness line. The PVI is simply a Euclidean distance measure similar to those discussed in Chapter 11:

$$\text{PVI} = \sqrt{(S_{\text{R}} - V_{\text{R}})^2 + (S_{\text{IR}} - V_{\text{IR}})^2} \qquad \text{(Eq. 16.4)}$$

where $S$ is the soil reflectance, $V$ is the vegetation reflectance, R represents red radiation, and IR represents near infrared radiation.

In practice, the analyst must identify pixels known to be composed of bare soil to identify the local soil brightness line and pixels known to be fully covered by vegetation to identify the local value for full vegetative cover (point $X$ in Figure 16.20). Then, intermediate values of PVI indicate the contributions of soil and vegetation to the spectral response.

Baret et al. (1993), investigated the soil brightness line from both experimental and theoretical perspectives. They concluded that the theoretical basis for the soil brightness line is sound, although effects of variations in soil moisture and surface roughness are not well understood. Their experimental data revealed that it is not feasible to define a single, universally applicable soil brightness line as local variations in soil types lead to spectral variations. However, these variations were found to be minor in the red and infrared regions and the use of a single soil brightness line is a reasonable approximation, especially in the context of analysis of course-resolution satellite data.

## 16.10. Tasseled Cap Transformation

The "tasseled cap" transformation (Kauth & Thomas, 1976) is a linear transformation of Landsat MSS data that projects soil and vegetation information into a single plane in multispectral data space—a plane in which the major spectral components (axes of maximum variance) of an agricultural scene are displayed in two dimensions. Although defined initially for MSS data, subsequent research (Crist & Cicone, 1984) has extended the concept to the six nonthermal bands of the TM. The transformation can be visualized as a rotation of a solid multidimensional figure (representing all spectral bands) in a manner that permits the analyst to view the major spectral components of an agricultural scene as a two-dimensional figure.

The transformation consists of linear combinations of the four MSS bands to produce a set of four new variables:

$$TC1 = +0.433 \text{ MSS } 4 + 0.632 \text{ MSS } 5 + 0.586 \text{ MSS } 6 + 0.264 \text{ MSS } 7 \qquad \text{(Eq. 16.4)}$$

$$TC2 = -0.290 \text{ MSS } 4 - 0.562 \text{ MSS } 5 + 0.600 \text{ MSS } 6 + 0.491 \text{ MSS } 7 \qquad \text{(Eq. 16.5)}$$

$$TC3 = -0.829 \text{ MSS } 4 + 0.522 \text{ MSS } 5 - 0.039 \text{ MSS } 6 + 0.194 \text{ MSS } 7 \qquad \text{(Eq. 16.6)}$$

$$TC4 = +0.223 \text{ MSS } 4 + 0.012 \text{ MSS } 5 - 0.543 \text{ MSS } 6 + 0.810 \text{ MSS } 7 \qquad \text{(Eq. 16.7)}$$

These coefficients from Kauth and Thomas (1976) apply to the Landsat 2 MSS. They are calculated by means of an iterative procedure that can be applied to as many bands as may be available (Jackson et al., 1983). Although these coefficients can be considered universally applicable, detailed studies should probably use locally defined coefficients. Here the new bands are designated as, for example, "TC1," for "Tasseled Cap Band 1." Although these four new bands do not match directly to observable spectral bands, they do carry specific information concerning agricultural scenes.

Kauth and Thomas interpret TC1 as *brightness,* a weighted sum of all four bands. TC2 is designated as *greenness,* a band that conveys information concerning the abundance and vigor of living vegetation, derived from the contrast between visible and near infrared bands. TC3 depicts *yellowness,* derived from the contrast between red and green bands; it is an axis oriented at right angles to the other two. Finally, TC4 is referred to as *nonesuch* because it cannot be clearly matched to observable landscape features and is likely to carry system noise and atmospheric information.

The first two bands (TC1 and TC2, brightness and greenness) usually convey almost all the information present in an agricultural scene—often 95% or more. Therefore, the essential components of an agricultural landscape are conveyed by a two-dimensional diagram, using TC1 and TC2 (Figure 16.21), that is in many respects similar to that defined by Richardson and Wiegand (Figure 16.20).

Over the interval of an entire growing season, TC1 and TC2 values for a specific field follow a stereotyped trajectory (Figure 16.22). Initially, the spectral response of a field is dominated by soil, as the field is plowed, disced, and planted. The field has a position near the soil brightness line. As the crop emerges and grows, it simultaneously increases in greenness and decreases in soil brightness, as the green canopy covers more and more of the soil surface. Then, as senescence, maturity, and harvest occur, the field decreases in greenness and increases in soil brightness to return the field near its original position on the diagram.

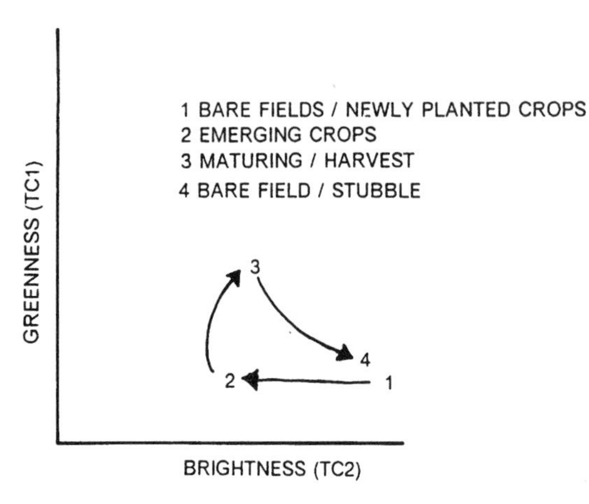


**FIGURE 16.21.** Seasonal variation of a field in data space defined by the greenness and brightness axes. From Crist, *Symposium on Machine Processing of Remotely Sensed Data 1983 with Special Emphasis on Natural Resources Evaluation*. Copyright 1983 by Purdue Research Foundation, West Lafayette, Indiana. Reproduced by permission.

A plot of data for an entire growing season, or for an image that shows many fields at various points in the crop cycle, has a distinctive shape similar to that of Figure 16.22; the resemblance of this shape to a tasseled ("Santa Claus"-type) cap provides the basis for the name that Kauth and Thomas gave to their technique. The analogy holds as the data are examined in other TC dimensions (Figure 16.23). Crist and Cicone (1984) show that the intermediate values are chiefly vegetated pixels in which plants are extending their canopy cover as they mature. Pixels representing senescent plant cover retain high greenness values until the canopy cover is so low that bare soil is again exposed.

Unlike transformations of the type discussed at the beginning of Chapter 10, in which the transformation applies only to a specific scene, the TC is consistent from one scene to another, permitting use of a single transformation to monitor crops throughout a growing season. In this context, the analyst should realize that TC transformations may be sensitive to atmospheric turbidity, angle of illumination, and other factors not related to vegetation status. These other factors become especially important when changes over time are studied, as it is then necessary to use data that vary greatly with respect to illumination, atmospheric clarity, and other factors.

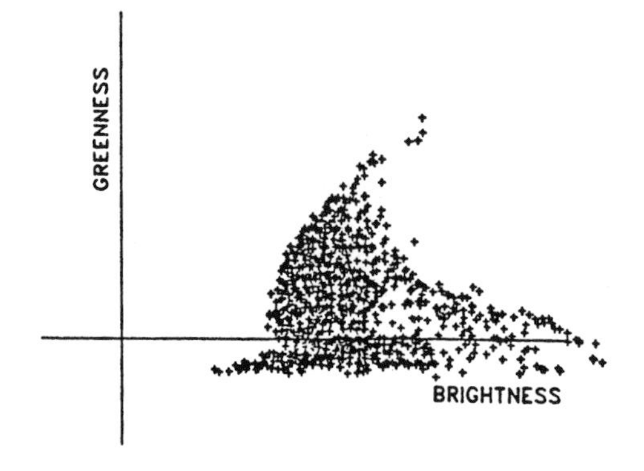


**FIGURE 16.22.** Tasseled cap. Spectral values from an agricultural scene plotted on TC1 and TC2 axes. Compare with Figures 16.21 and 16.23. These data are simulated TM data for agricultural regions in the midwestern United States. From Crist and Cicone (1984, p. 347). Copyright 1984 by the American Society for Photogrammetry and Remote Sensing. Reproduced by permission.

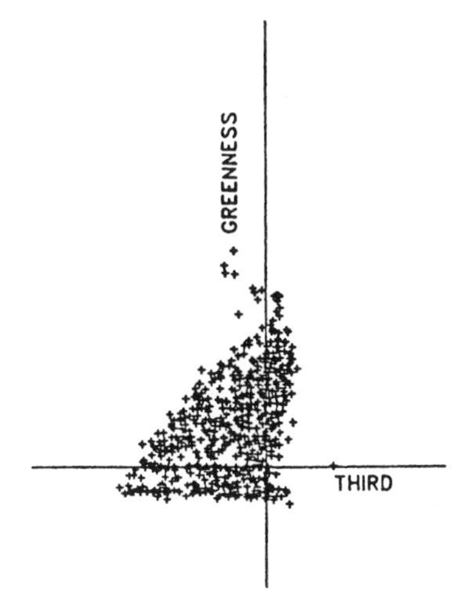


**FIGURE 16.23.** Tasseled cap viewed from another angle. These are the same data depicted in Figure 16.22 but viewed from another angle. The vertical axis is the same as Figure 13.22 (greenness) but the horizontal axis depicts another dimension, chiefly variations in soil moisture content. From Crist and Cicone (1984, p. 347). Copyright 1984 by the American Society for Photogrammetry and Remote Sensing. Reproduced by permission.

## 16.11. Summary

For a large proportion of the earth, vegetation cover forms the surface observed by remote sensing instruments. In some instances we have a direct interest in examining this vegetative blanket, to map the patterns of different forests, rangelands, and agricultural production. In other instances, we must use vegetation as a means of understanding those patterns that may lie hidden beneath the plant over. In either instance, it is essential that we be able to observe and understand the information conveyed by the vegetated surface.

Given the obvious significance of agriculture and forestry, and the immense difficulties in monitoring use of these resources, the techniques outlined in this chapter are likely to form one of the most significant contributions of remote sensing to the well-being of mankind. The ability to examine vegetation patterns using the vegetation indices described here, combined with the synoptic view and repetitive coverage of satellite sensors, provides an opportunity to survey agricultural patterns in a manner that was not possible even a few years ago. If the information gathered by remote sensing systems can be integrated into the decision making and transportation/distribution systems so important for agricultural production, there is a prospect for improvements in the effectiveness of food production.

Chapter 17 demonstrates how some of the information introduced here has significance in geological remote sensing. Full pursuit of extraction of geological information from an image requires intimate knowledge of how the spectral behavior of living plants responds to variations in the geologic substratum—a good example of how the fabric of remote sensing is so closely woven that it is not possible to isolate individual components as separate units.

## Review Questions

1. Summarize differences between classification of vegetation from ground observations and classification from aerial images. Consider such factors as (a) basis of the classification

and (b) the units classified. Identify distinctions for which remote sensing is especially well suited, and those for which it is not likely to be useful.

2. Remote sensing to monitor crop development is much more difficult than it might appear initially. List some of the practical problems you might encounter as you plan an experiment to use satellite data to study the development of the winter wheat crop in western Kansas (or another crop region specified by your instructor).
3. Return to Tables 11.2 and 11.3 and to the category labels tabulated in the review questions to Chapter 11. Use at least two of the vegetation indices discussed in this chapter (as specified by your instructor) to assess the forested, pasture, and crop pixels. Examine differences between classes and between dates. Do some indices seem more effective than others?
4. To apply some of the knowledge presented in this chapter (e.g., the red shift) it is necessary to have data with very fine spectral, radiometric, and spatial detail. From your knowledge of remote sensing, discuss how this requirement presents difficulties for operational applications.
5. List some of the reasons why an understanding of the spectral behavior of an individual plant leaf is not itself sufficient to conduct remote sensing of vegetation patterns.
6. List some reasons why the multispectral satellite images might be especially well suited for observation of vegetation patterns. Or, if you prefer, list some reasons why it is *not* quite so useful for vegetation studies. Briefly compare the MSS, TM, and SPOT with respect to utility for vegetation studies.
7. Write a short description of a design for a multispectral sensor tailored specifically for recording information about living vegetation and vegetation patterns, disregarding all other applications. Suggest optimum timing for a satellite to carry the sensor in a sun-synchronous orbit.
8. Describe some of the ways that image classification (Chapter 11) might be useful in the study of vegetation patterns. Identify also some of the limitations of such methods in the study of vegetation.
9. Study of vegetative patterns by remote sensing is greatly complicated by the effects of soil background. Summarize contributions of soil background to spectral responses of a variety of natural and man-made landscapes, including cultivated crops, dense forest, orchards, and pasture. How might effects differ as season changes?
10. How do changes in sun angle and sun azimuth (due to differences in season and time of day) influence the way in which vegetative patterns are recorded on remotely sensed images?

## References

### General

Badhwar, G. D. 1980. Crop Emergence Date Determination from Spectral Data. *Photogrammetric Engineering and Remote Sensing,* Vol. 46, pp. 369–377.

Badhwar, G. D., J. G. Carnes, and W. W. Austin. 1982. Use of Landsat-Derived Temporal Profiles for Corn–Soybean Feature Extraction and Classification. *Remote Sensing of Environment,* Vol. 12, pp. 57–79.

Bailey, Robert G. 1976. *Ecoregions of the United States* (Map at 1:7,500,000). Ogden, UT: U.S. Forest Service.

Bailey, Robert G. 1978. *Description of Ecoregions of the United States*. Ogden, UT: U.S. Forest Service, 77 pp.

Bailey, Robert G. 1995. *Description of the Ecoregions of the United States* (with Map at 1:7,500,000). 2nd ed. Ogden, UT: U.S. Forest Service, 108 pp.

Baret, S. Jacquemond, and J. F. Hanocq. 1993. The Soil Line Concept in Remote Sensing. *Remote Sensing Reviews,* Vol. 7, pp. 65–82.

Brisco, B., and R. J. Brown. 1995. Multidate SAR/TM Synergism for Crop Classification in Western Canada. *Photogrammetric Engineering and Remote Sensing,* Vol. 61, pp. 1009–1014.

Collins, W. 1978. Remote Sensing of Crop Type and Maturity. *Photogrammetric Engineering and Remote Sensing,* Vol. 44, pp. 43–55.

Colwell, R. N. 1956. Determining the Prevelance of Certain Cereal Crop Diseases by Means of Aerial Photography. *Hilgardia,* Vol. 26, No. 5, pp. 223–286.

Condit, H. R. 1970. The Spectral Reflectance of American Soils. *Photogrammetric Engineering,* Vol. 36, pp. 955–966.

Eidenshink, J. C. 1992. The 1990 Conterminous U.S. AVHRR Data Set. *Photogrammetric Engineering and Remote Sensing,* Vol. 58, pp. 809–813.

Gausman, H. W. 1974. Leaf Reflectance of Near-Infrared. *Photogrammetric Engineering,* Vol. 40, pp. 183–191.

Howard, J. A. 1970. *Aerial Photo-Ecology* New York: Elsevier, 325 pp.

Jackson, R. D., P. J. Pinter, S. B. Idso, and R. J. Reginato. 1979. Wheat Spectral Reflectance: Interactions Between Crop Configuration Sun Elevation and Azimuth Angle. *Applied Optics,* Vol. 18, pp. 3730–3732.

Jasinski, M. F., and P. S. Eagleson. 1990. Estimation of Subpixel Vegetation Cover Using Red–Infrared Scattergrams. *IEEE Transactions on Geoscience and Remote Sensing,* Vol. 28, pp. 253–267.

Jensen, John R. 1978. Digital Land Cover Mapping Using Layered Classification Logic and Physical Composition Attributes. *The American Cartographer,* Vol. 5, pp. 121–132.

Jensen, John R. 1983. Biophysical Remote Sensing. *Annals of the Association of American Geographers,* Vol. 73, pp. 111–132.

Kanemasu, E. T. 1974. Seasonal Canopy Reflectance Patterns of Wheat Sorghum and Soybean. *Remote Sensing of Environment,* Vol. 3, pp. 43–57.

Kimes, D. S., W. W. Newcomb, R. F. Nelson, and J. B. Schutt. 1986. Directional Reflectance Distributions of a Hardwood and Pine Forest Canopy. *IEEE Transactions on Geoscience and Remote Sensing,* Vol. GE-24, pp. 281–293.

Kimes, D. S., J. A. Smith and K. J. Ranson. 1980. Vegetation Reflectance Measurements as a Function of Solar Zenith Angle. *Photogrammetric Engineering and Remote Sensing,* Vol. 46, pp. 1563–1573.

Knipling, Edward B. 1970. Physical and Physiological Basis for the Reflectance of Visible and Near Infrared Radiation from Vegetation. *Remote Sensing of Environment,* Vol. 1, pp. 155–159.

Küchler, A. W. 1967. *Vegetation Mapping*. New York: Ronald Press, 472 pp.

McClain, E. Paul. 1980. Environmental Satellites. Entry in *McGraw-Hill Encyclopedia of Environmental Science*. New York: McGraw-Hill, pp. 15–30.

Meyers, Victor I. 1975. Crops and Soils. Chapter 22 in *Manual of Remote Sensing* (R. Reeves, ed.). Falls Church, VA: American Society of Photogrammetry, pp. 1715–1816.

Ray, Peter Martin. 1963. *The Living Plant.* New York: Holt, Rinehart & Winston, 127 pp.

Richardson, A. J., and C. L. Wiegand. 1977. Distinguishing Vegetation from Soil Background. *Photogrammetric Engineering and Remote Sensing,* Vol. 43, pp. 1541–1552.

Rouse, J. W., R. H. Haas, J. A. Schell, D. W. Deering, and J. C. Sagan, C., O. B. Toon, and J. B. Pollock. 1979. Anthropogenic Changes and the Earth's Climate. *Science,* Vol. 206, pp. 1363–1368.

Townsend, J. R. G. 1994. Global Data Sets for Land Applications from the Advanced Very High Resolution Radiometer: An Introduction. *International Journal of Remote Sensing,* Vol. 15, pp. 3319–3332.

Tucker, C. J., J. A. Gatlin, and S. R. Scheider. 1984. Monitoring Vegetation of the Nile Delta NOAA-6 and NOAA-7 AVHRR Data. *Photogrammetric Engineering and Remote Sensing,* Vol. 50, pp. 53–61.

Tucker, Compton J., H. H. Elgin, and J. E. McMurtrey. 1979. *Relationship of Red and Photographic Infrared Spectral Radiances to Alfalfa Biomass, Forage Water Content, Percentage Canopy Cover, and Severity of Drought Stress.* NASA Technical Memorandum 80272. Greenbelt, MD: Goddard Space Flight Center, 13 pp.

Tucker, C. J., J. R. G. Townshend, and T. E. Goff. 1985. African Land-Cover Classification Using Satellite Data. *Science,* Vol. 227, pp. 369–374.

Wilson, Richard C. 1960. Photo Interpretation in Forestry. Chapter 7 in *Manual of Photographic Interpretation* (R. N. Colwell, ed.). Falls Church, VA: American Society of Photogrammetry, pp. 457–520.

Yool, S. R., D. W. Eckhardt, Jeffery L. Star, T. L. Becking, and J. E. Estes. 1985. Image Processing for Surveying Natural Vegetation: Possible Effects on Classification Accuracy. In *Technical Papers, American Society for Photogrammetry and Remote Sensing.* Falls Church, VA: American Society for Photogrammetry and Remote Sensing, pp. 595–603.

### Phenology

Dethier, B. E., M. D. Ashley, B. Blair, and R. J. Hopp. 1973. Phenology Satellite Experiment. In *Symposium on Significant Results Obtained from ERTS-1,* Vol. 1. Washington, DC: NASA, pp. 157–165.

Harlan, C. 1974. *Monitoring the Vernal Advancement and Retrogradation (Greenwave Effect) of Natural Vegetation.* Type III Final Report. Greenbelt, MD: NASA Goddard Space Flight Center, 371 pp.

Loveland, T. R., J. W. Merchant, J. F. Brown, D. O. Ohlen, B. C. Reed, P. Olson, and J. Hutchinson. 1995. Seasonal Land-Cover Regions of the United States. *Annals of the Association of American Geographers,* Vol. 85, pp. 339–355.

Koriba, Kwan. 1958. On the Periodicity of Tree-Growth in the Tropics, With Special Reference to the Mode of Branching, the Leaf Fall, and the Formation of the Resting Bud. *Gardens' Bulletin, Singapore,* Series 3, Vol. 17, pp. 11–81.

Rouse, J. W., R. H. Haas, J. A. Schell, and D. W. Deering. 1973. Monitoring Vegetation Systems in the Great Plains with ERTS. In *Third ERTS Symposium.* NASA Special Publication SP-351I, pp. 309–317.

### Vegetation Indices

Clevers, J. G. P. W., and W. Verhoef. 1993. LAI Estimation by Means of the WDVI: A Sensitivity Analysis with a Combined PROSPECT-SAIL Model. *Remote Sensing Reviews,* Vol. 7, pp. 43–64.

Cohen, Warren B. 1991. Response of Vegetation Indices to Changes in Three Measures of Leaf Water Stress. *Photogrammetric Engineering and Remote Sensing,* Vol. 57, pp. 195–202.

Crist, E. P. 1983. The Thematic Mapper Tasseled Cap—A Preliminary Formulation. In *Proceedings, Machine Processing of Remotely Sensed Data Symposium.* West Lafayette IN: Laboratory for the Applications of Remote Sensing, pp. 357–364.

Crist, E. P., and R. C. Cicone. 1984. Application of the Tasseled Cap Concept to Simulated Thematic Mapper Data. *Photogrammetric Engineering and Remote Sensing,* Vol. 50, pp. 343–352.

Curran, P. 1980. Multispectral Remote Sensing of Vegetation Amount. *Progress in Physical Geography,* Vol. 4, pp. 315–341.

Jackson, R. D., P. N. Slater, and P. J. Pinter. 1983. Discrimination of Growth and Water in Stress in Wheat by Various Vegetation Indices through Clear and Turbid Atmospheres. *Remote Sensing of Environment,* Vol. 13, pp. 187–208.

Kauth, R. J. and G. S. Thomas. 1976. The Tasseled Cap—A Graphic Description of the Spectral–Temporal Development of Agricultural Crops as Seen by Landsat. In *LARS: Proceedings of the Symposium on Machine Processing of Remotely Sensed Data.* West Lafayette, IN: Purdue University, pp. 4B-41–4B-51.

Morain, Stanley A. 1978. *A Primer on Image Processing Techniques.* Albuquerque: University of New Mexico Technology Applications Center, TAC TR #78-009, 54 pp.

Perry, Charles R., and L. F. Lautenschlager. 1984. Functional Equivalence of Spectral Vegetation Indices. *Remote Sensing of Environment,* Vol. 14, pp. 169–182.

Price, J. C. 1987. Calibration of Satellite Radiometers and Comparison of Vegetation Indicies. *Remote Sensing of Environment,* Vol. 18, pp. 35–48.

Singh, S. M., and R. J. Saull. 1988. The Effect of Atmospheric Correction on Interpretation of Multitemporal AVHRR-Derived Vegetation Index Dynamics. *Remote Sensing of Environment,* Vol. 25, pp. 37–51.

Then Kabil, P. S., A. D. Ward, J. G. Lyon, and C. J. Merry. 1994. Thematic Mapper Vegetation Indices for Determining Soybean and Corn Crop Parameters. *Photogrammetric Engineering and Remote Sensing,* Vol. 60, pp. 437–442.

Tucker, Compton J. 1979a. Red and Photographic Infrared Linear Combinations for Monitoring Vegetation. *Remote Sensing of Environment,* Vol. 8, pp. 127–150.

Tucker, Compton J. 1979b. *Remote Sensing of Leaf Water Content in the Near Infrared.* NASA Technical Memorandum 80291. Greenbelt, MD: Goddard Space Flight Center, 17 pp.

## THE CORN BLIGHT WATCH

In 1970 the southern corn leaf blight (SCLB) spread through most of the corn crop in the central and eastern United States; it is estimated that the SCLB reduced the yield of the 1970 crop by about 15%. SCLB blight is a fungus that propagates by windblown spores that rapidly infect susceptible varieties of corn, which then formed about 85% of the crop. During warm, moist weather, the SCLB attacks rapidly, infecting lower and then upper leaves of plants; severe infection reduces both quality and quantity of grain that may be harvested.

Agricultural scientists expected that spores of SCLB fungus would survive the winter to damage the 1971 corn crop. Because of the significance of the corn crop not only to individual farmers but to consumers of corn and corn products in the United and abroad, persistence of SCLB in the central United States could have had severe impact upon food production.

Planning for an experimental monitoring program began late in the summer of 1970. Then, in April 1971, the U.S. Department of Agriculture (USDA) initiated a program to employ remote sensing to assess the spread of infection during the 1971 growing season. The plan required several states to cooperate with federal agencies in the use of remotely sensed images to study the SCLB; this project was known as the "Corn Blight Watch Experiment." The Corn Blight Watch was designed not only to detect spread of SCLB in 1971 but to evaluate usefulness of remotely sensed images and to develop techniques to be used in similar situations that might arise in the future.

Sample segments measuring 1 mi. × 8 mi. were selected within the seven midwestern states selected for the experiment (Missouri, Indiana, Iowa, Illinois, Minnesota, Ohio, and Nebraska). Within each segment, USDA statisticians selected 6 to 10 typical corn fields, for a total of about 1,806 fields. These were covered by about 30 north–south flight lines. In addition, a region in western Indiana was studied intensively by photos taken along eight overlapping flight lines.

NASA aircraft acquired high-altitude photography using several film–filter combinations; this photography covered the sample segments throughout the seven-state region, while coverage of the intensive study area in Indiana was acquired by a multispectral scanner operated by the University of Michigan. Each segment was photographed several times. During the first phase (15–30 April) black-and-white photography was used to make a land use map of the study areas. Then during the second phase (10–30 May) interpretations were made of soil conditions. Finally, during the third phase (14 June–1 October) interpreters used CIR imagery and the multispectral data to detect and map the spread of SCLB. At the same time, a team of specialists from the USDA and state agricultural agencies directly examined the study areas, collecting leaf samples when SCLB was detected. Photographs were acquired every two weeks and delivered to the Laboratory for Applications of Remote Sensing (LARS) (see Chapter 9), where they were examined by photointerpretation teams. (In all, more than 18,000 frames were acquired and interpreted.) On the CIR images, healthy crops appear as a bright red color; infected crops appear as pink, brown, and then gray as the infection increases in severity. Experienced photointerpreters were able to classify infested areas into several categories. In addition, the digital data from the multispectral scanner were analyzed using image classification techniques (Chapter 11) to derive the severity classes.

SCLB was reported in 600 counties by mid-July 1971. Then, in late July, several weeks of cool, dry weather throughout the Midwest halted the spread and arrested the progress of infection in areas already reporting SCLB. In addition, damage was also reduced by increased use of blight-resistant varieties. Interpreters achieved high accuracy in identification of corn from the other crops present within this region and produced reliable estimates of severity of damage from SCLB.

## LARGE AREA CROP INVENTORY EXPERIMENT

Advance information concerning food production is important for forecasting areas that may experience food shortages and for providing timely, accurate information necessary to stabilize fluctuations commodity markets, which are especially sensitive to uncertainties, or wide fluctuations in supply or demand. In 1972, and again in 1974, commodity markets in the United States were unprepared for large purchases by the Soviet Union, which required foreign grain to supplement its small harvests. The U.S. markets lacked timely knowledge of the Soviet Union's need for grain and therefore priced its products low. Soviet agents were able to make large purchases at low prices, leaving the market's regular customers to pay higher prices demanded for the remaining supplies of grain. Public concern in the aftermath of these events created an extra incentive for devising improved forecasts of grain production in the major grain exporting nations of the world.

Worldwide crop forecasts are the responsibility of the USDA, and international agencies such as the United Nations Food and Agricultural Organization (FAO). Such reports are often based on reports provided by each individual country, which vary greatly in accuracy and reliability or may not be released until after the harvest is complete. The classic survey methods in the United States depend heavily on reports sent by individual farmers, and on ground sampling by agricultural specialists sent to specific sites to observe the progress of the crop throughout the growing season.

In 1974, the U.S. government attempted to improve forecasting capabilities by applying remote sensing to crop forecasting in a project known as the Large Area Crop Inventory Experiment (LACIE). LACIE was applied to forecasts of a single crop—wheat. Wheat is one of the most important grains in international commerce; is subject to fluctuations in supply due to inherent variability of the climates in which it is usually grown. Furthermore, wheat is often grown in rather large fields, easily detected by remote sensing imagery. LACIE was to provide both of the elements necessary for reliable forecasts: (1) estimates of area planted to wheat and (2) yield per unit area based on knowledge of the varieties planted, local agricultural practices, and weather conditions throughout the growing season.

LACIE was conducted by a team of three agencies of the U.S. government: (1) NASA, which would derive estimates of cropped areas based on analysis of Landsat MSS data; (2) NOAA, which would analyze data from meteorological satellites to provide information concerning weather and climate; and (3) USDA, which would derive its estimates of total production from the data provided by NOAA and NASA. Estimates were to be made for the United States, Canada, and the Soviet Union. Pilot studies were conducted in China, Australia, India, Brazil, and Argentina. LACIE was conducted over 3 crop years. The 1974–1975 winter wheat crop was studied in the central United States. Then during the next year the project was expanded to include Canada and the Soviet Union. Finally, the 1977 wheat crop was forecast using new methods developed from the earlier studies, and evaluation of the procedure continued in the U.S. test areas.

LACIE sample segments consisted of areas measuring 5 by 6 nautical miles on a side. Sample segments were located randomly within sampling strata defined by political or census units (in the United States these were counties; in the Soviet Unions, oblasts; and in Canada, census subdivisions) (see Chapter 12 for descriptions of sampling strategies). Area planted to wheat was estimated from examination of Landsat MSS data for each sample segment; these values were then aggregated to provide a value for a larger area consisting of many counties, called a zone. Each zone was defined so that it would match to the areal weather data provided by NOAA.

A special procedure was devised to interpret the extent and the growth stage of the wheat crop within each sample segment, based in part on experience acquired during the Corn Blight Watch (pp. 479–480). Interpreters had access to collateral information consisting of knowledge of the local crop calendar, weather information, historical data, and forecasts of crop progress but did not have access to ground observations. The entire segment (117 × 192 pixels in size) was classified by unsupervised classification (Chapter 11), using a systematic random sample of pixels within segments to select initial centers for clusters. At the same time, an experienced image interpreter examined false color composites of the same segment, using a grid to locate 100 randomly selected pixels for classification as either small grains (potentially wheat) or other (not wheat). Some of these interpreter-identified pixels were used to label the classes produced by unsupervised classification, so that every pixel was labeled either small grain or other. These clusters were then used as training data for a maximum likelihood classification of the pixels within the segment. (Swain, 1984, feels this last step was not really necessary, as the initial unsupervised classification was sufficiently accurate in his view.) The remaining interpreter-identified pixels were used to evaluate the performance of the classification and to adjust the estimate if it appeared that the computer classification under- or overestimated the area planted to small grains.

At this point, the procedure had produced estimates of the area within each segment that had been planted to small grains; another step was necessary to estimate how much of this total was actually wheat. Nonimage information was then used to estimate the proportion of wheat within the total of small grain crops for the region examined. (Because the classification produced only two categories, "small grains," and "other," the "small grains" area had to be adjusted to estimate the area planted in wheat.) These values estimated the area planted to wheat, which was then used as input to the broader forecasting model as outlined earlier.

CHAPTER SEVENTEEN

# Earth Sciences

## 17.1. Introduction

This chapter addresses applications of remote sensing in the earth sciences, which are loosely defined here to include geology, geomorphology, and soil science. Despite their many differences, these disciplines share a common focus on the earth's shape and structure and on the nature of the soils and sediments at its surface. Applications of remote sensing in the earth sciences are especially difficult. Often the subjects of investigation are geologic structures, soil horizons, and other features entirely or partially hidden beneath the earth's surface. Seldom can remotely sensed images directly record such information—instead, we must search for indirect evidence that may be visible at the surface. Thus, interpretations of geologic, pedologic, and geomorphic information are often based on inference rather than direct sensing of the qualities to be studied. Furthermore, interpretations of geoscience information are frequently based on rather small differences in tone, texture, and spectral response. Even direct examination on the ground of many geologic and pedologic materials is subject to error and controversy, so it should be no surprise that applications of remote sensing in these fields can be equally difficult. Finally, we must often depend on composite signatures formed as individual, "pure" signatures of various types of soil, rock, and vegetation are mixed together into a single, combined spectral response that may be unlike any of its components (Chapter 9). Field observations are made by necessity at points such as outcrops, soil profiles, or boreholes. Because remote sensing instruments record radiation from ground *areas* rather than points, it may be difficult to relate remotely sensed data to the ground observations with which we traditionally have experience and confidence. Such problems present both practical and conceptual difficulties for applications of remote sensing in the earth sciences.

Remote sensing can, however, provide opportunities not normally available to geoscientists. The ability to observe reflectance and emittance over a range of wavelengths opens the door to the study of subjects that would not otherwise be possible. Also, the synoptic view of satellite images gives a broad-scale perspective of patterns not discernible within the confines of the large-scale, close-up view of ground observation. A single satellite image can record an integrated signature formed of many kinds of information, including shadow, soil, rock, and vegetation. Although these mixtures may present problems, they also portray subtle differences in terrain that cannot be easily derived from other sources. While use of remotely sensed images cannot replace direct ground observation or data derived from field and laboratory studies, they can form valuable supplements to more traditional methods and sometimes provide information and a perspective not otherwise available.

## 17.2. Photogeology

Geologists study many aspects of the earth's surface in an effort to understand the earth's structure, to guide the search for minerals and fuels, and to understand geologic hazards. Remote sensing contributes to several dimensions of the geological sciences by providing information concerning lithology, structure, and vegetation patterns. Lithology refers to fundamental physical and chemical properties of rocks, including, for example, the gross distinctions between sedimentary, igneous, and metamorphic rocks. Structure defines the kinds of deformation experienced by rocks, including folding, fracturing, and faulting. Geobotanical studies focus on relationships between plant cover at the earth's surface and the lithology of underlying rocks.

*Photogeology* is the derivation of geologic information from interpretation of aerial photographs. Photogeology originated early in the development of air photo interpretation. Many of its basic techniques were developed in the 1920s and then refined and applied into the 1950s and 1960s, when they seemed to approach the limits of their capabilities and were assimilated into newly developing research in geologic remote sensing. Today photogeology is routinely applied to good effect, although most research and innovation are likely to occur in the broader context of geologic remote sensing, as aerial photographs are now only one of many forms of aerial images routinely available to the geologist.

Many of the techniques of photogeology are the direct application of the principles of image interpretation (Chapter 4) to geological problems. Image texture, size, shape, tone, shadow, and so on, have special significance in the realm of the geologist's view of the terrain. Likewise, the principles of photogrammetry (Chapter 5) have special applications in the context of photogeology for calculation of thicknesses of beds and determination of strike and dip from aerial photography. Such measurements permit derivation of structural information from aerial images.

Arid and semi-arid regions, where vegetative cover may be sparse, may be especially favorable for lithologic interpretation of aerial photographs. Bedding in sedimentary rocks, especially those that are exposed at hillsides or by folding or faulting, is one of the strongest clues to lithologic composition. More resistant beds stand out in relief, while less resistant beds are lowered by differential erosion. Differences in image tone of specific beds may reveal differences in lithology. Image tone of strata is often directly influenced by grain size of the rocks, although the relationship between grain size and image tone may vary so much that it is best not to attempt to establish general rules for interpretation. Nonetheless, guidelines can be defined for specific localities. Examination of drainage density, drainage pattern, and vegetation patterns may also provide clues to lithology, even when beds are not directly exposed. Several of these points are illustrated in Figure 17.1, which shows an aerial photograph of a landscape in Texas.

Igneous rocks are usually more difficult to interpret. Extrusive igneous rocks may be recognizable by distinctive landforms, such as cones and flows, especially if they are

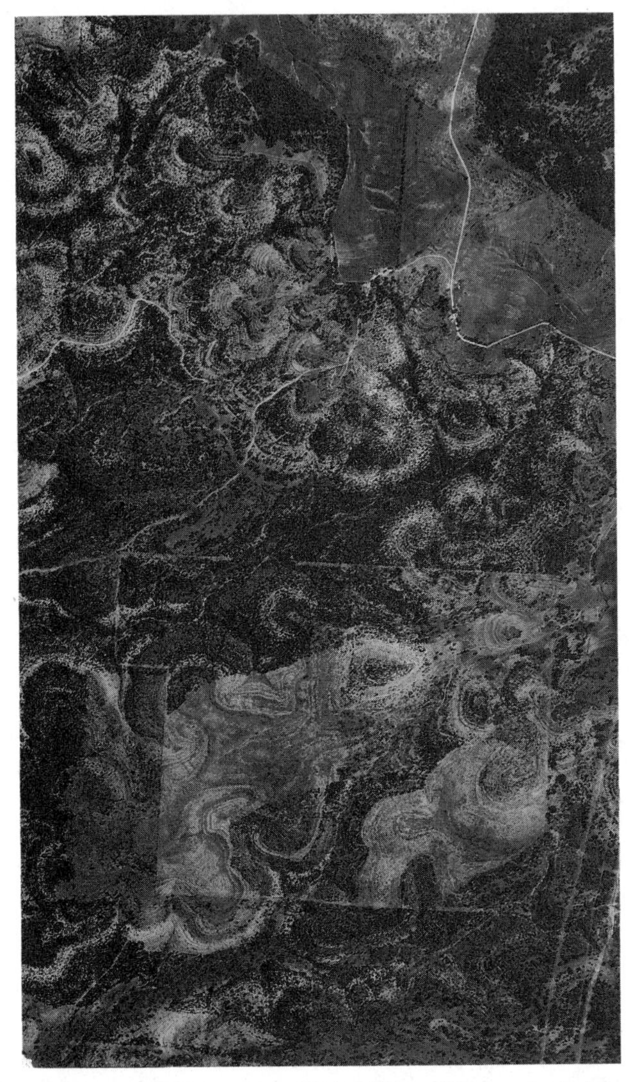

**FIGURE 17.1.** Aerial photograph of Hays County, Texas, illustrating several principles of photogeology. Interbedded shales and sandstones are exposed here in horizontal strata. Lithologic differences are visible as differences in tone, topographic expression (visible in stereo), and vegetative pattern and density. Note also the complex interplay between natural vegetation patterns and those that are clearly related to mankind's activities, chiefly differences in intensity of grazing from one side of a fenceline to the other. U.S. Department of Agriculture photograph.

young, and by differences in image tone or by drainage patterns and vegetation distributions. Intrusive igneous rocks can sometimes be easily recognized by their distinctive image tones and by topographic expression through differential erosion, but often their close integration with surrounding sedimentary strata means that they are more subtly expressed on aerial photographs than are extrusive rocks and therefore can be interpreted with less confidence.

Information concerning metamorphic rocks can also be interpreted from aerial photographs, but typically only with great difficulty because of complex local structures induced by metamorphism. Metamorphism may also reduce the differences in resistance to erosion that are so important in inferring lithology.

Manual interpretation for lithologic information typically subdivides an image into distinctive photomorphic regions that show the combined influence of soil color, vegetation, land use, drainage, and shadowing. In the second step, the geologist interprets each region to identify specific lithologies, or combinations of lithologies, within each region, using field observations to guide and verify interpretations. The analyst uses areas of known geology as a means of identifying analogous regions elsewhere on the image. Thus, the image provides a map-like perspective and an integrated representation of many different factors that permits separation of regions of differing geology; however, identification of the geological significance of each region usually requires application of other knowledge not visible on the image, generally derived from field observations.

In some situations, geologic information can be derived using the techniques of digital classification described in Chapter 11. Areas of known geologic identity can form training areas for the identification of analogous areas elsewhere on the image. In the context of Landsat data, such techniques can be effective only when the composite signatures formed by such elements as soil color, vegetation, shadowing, and other factors form distinctive spectra that are clearly associated with geological patterns. Although such situations exist, they cannot be considered the typical conditions for geological remote sensing, so other techniques are usually required. In some situations, simple density slicing or band ratioing may permit visual recognition of distinctive geologic units on remotely sensed images.

## 17.3. Lineaments

*Lineament* is the name given by geologists to lines or edges, of presumed geologic origin, visible on remotely sensed images. Such features have also been referred to as *linears*, or *lineations,* although O'Leary et al. (1976) establish *lineament* as the preferred term. Use of the term *lineament* in a geologic context dates to 1904 and apparently has even earlier analogs in other languages. These early uses, prior to availability of aerial images, applied to such specific geologic or geomorphic features as topographic features (ridgelines, drainage systems, or coastlines), lithologic contacts, or zones of fracture.

As early as the 1930s, photogeologists studied fracture patterns visible on aerial photographs as a means of inferring geologic structure. These photo features apparently corresponded rather closely to faults and fractures defined in the field. More recently, in the context of geologic remote sensing, the term has assumed a broader meaning. Any linear feature visible on an aerial image can be referred to as a *lineament.* A problem arises because it is difficult to establish a clear link between the features on images and corresponding features, if any, on the ground.

The strength of this link depends in part on the nature of the imagery. In the instance of interpretations from aerial photographs and mosaics, scales are usually large and ample detail is visible. Linear features visible on such images are likely to match to those that can be confirmed in the field. However, more extensive and more subtle features cannot be easily detected on aerial photography. Each photograph shows only a small area, and if mosaics are formed to show larger regions, variations in illumination and shadowing (Figure 3.29) obscure more subtle lineaments that might extend over many photographs. Furthermore, in the era prior to routine availability of the CIR films now used for high-altitude photography, high-altitude photographs (which *can* show large areas under uniform illumination) were not of good visual quality due to effects of atmospheric scatter.

The advent of nonphotographic sensors and the availability of the broad-scale view of satellite images changed this situation because it then became possible to view large areas with images of good visual clarity. Early radar images depicted large areas illuminated from a single direction, at rather low illumination angles—conditions that tended to increase topographic shadowing in a manner that enhanced the visibility of linear features (Figure 17.2). Locations of these features did not always match previously known faults or fractures, so the term *lineament* was evoked to avoid an explicit statement of a geological origin. Later, similar features were again detected on Landsat MSS imagery (Figures 17.3, 17.4), and the detection of lineaments and the interpretation of their meaning have attracted considerable attention and controversy. Because of the broad scale and coarse resolution of such images, the lineaments they show are much more subtly expressed and much more difficult to relate to previously known geologic features than were faults and fractures observed previously on aerial photographs.

Lineaments are, of course, "real" features if we consider them simply to be linear

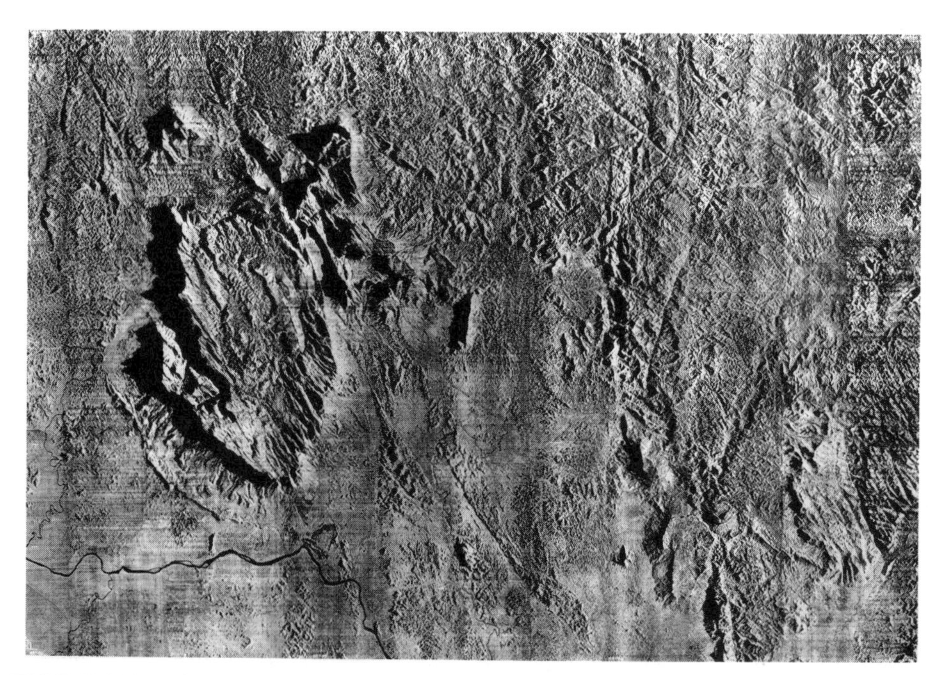

**FIGURE 17.2.** Radar mosaic of southern Venezuela. Here topography and drainage are clearly visible, as are the linear features in the upper right that illustrate the kinds of features important in lineament analysis. Image courtesy of Goodyear Aeospace Corporation.

**FIGURE 17.3.** Band 4 of Landsat MSS image of northwestern Arizona. Date: 15 January 1973. Sun elevation: 27°. Sun azimuth: 148°. Scene ID: 1176-17382-7. Image from U.S. Geological Survey, EROS Data Center.

features detected on aerial imagery (e.g., Figures 17.1, 17.2, 17.3, and 17.4 show a variety of such features). The controversy arises when we attempt to judge the geologic significance (if any) of these features. There are sound reasons for assigning a geologic meaning to some lineaments, even if they do not always correspond to clearly observable physical features. In the simplest instances, a dip–slip fault may leave a subtle topographic feature that is visible as a linear shadow on aerial photography when illuminated obliquely from the elevated side (Figure 17.4). Of course, if it is oriented parallel to the direction of illumination, it may be indistinct, or invisible. Not all faults are expressed topographically, but the fault plane may offer preferred avenues of movement for moisture and for the growth of plant roots (Figure 17.5). Therefore, the trace of the fault may be revealed by vegetation patterns, even though it has no obvious topographic expression. Faults of any form can, of course, alter drainage in a manner that creates linear drainage segments, which are then visible on aerial images. Such features, when clearly expressed, may have structural origins, and it may be possible to verify their existence through field observations. However, more subtle features may be observable only as broad-scale features, not

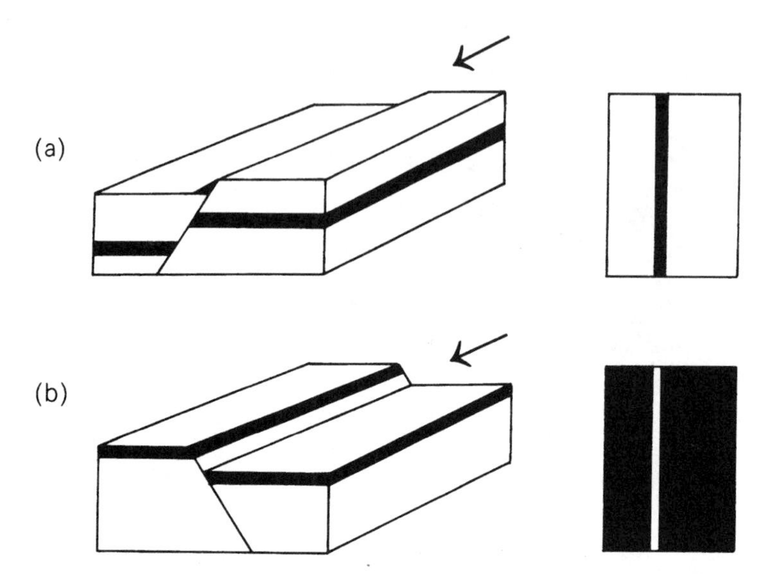


**FIGURE 17.4.** Schematic illustration of a dip–slip fault. (*a*) Illumination from the elevated side produces a strip-like shadow on the image. (*b*) Illumination from the downthrust side may produce a bright strip as the sun is reflected from the linear facet that now faces the solar beam.

easily verifiable at a given point. Such lineaments may be genuine structural features but not easily confirmable as such.

Linear features that are clearly of structural origin are significant because they indicate zones of fracturing and faulting. It is often assumed that regions of intersection of lineaments of differing orientation are of special significance, as in theory they might identify zones of mineralization, stratigraphic traps, regions of abundant groundwater, or zones of structural compression. For this reason, much is often made of the orientation and of the angles of intersection of lineaments.

Other lineaments may not have structural origins. It is conceivable that some may be artifacts of the imaging system or of preprocessing algorithms. In Figure 17.1, note the numerous man-made linear features, and in Figure 17.3 note the linear features caused by

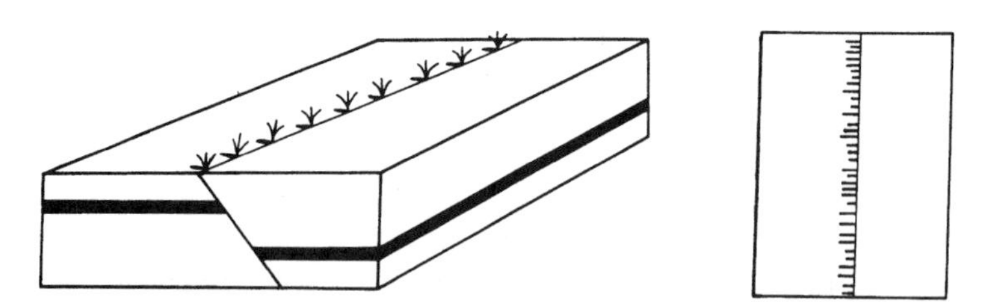

**FIGURE 17.5.** Hypothetical situation in which a fault with little or no surface expression forms a preferred zone of movement for moisture and for the growth of plant roots. Such a zone may create a favorable habitat for certain species, or speed the growth of more prevalent plants located near the fault. The linear arrangement of such plants and their shadows may form a linear feature on an image.

a jet contrail and its shadow, visible as the parallel white and dark streaks across the upper image. Some linear features may be purely surface features, such as wind-blown sediments, that do not reflect subsurface structure (Figure 17.6). Cultural patterns, including land cover boundaries, edges of landownership parcels, and political borders can all have linear form and can be aligned in a manner that creates the linear features observed on aerial images (Figure 17.1). Because these features may not be readily distinguished from those of geologic origin, and because of the inconsistency of individual delineations of lineaments from the same image, lineament analysis has been regarded with skepticism by a significant proportion of the geologic community (Wise, 1982).

Lineaments have been identified by manual image interpretation (Figure 17.7) and by computer analysis. For manual interpretation, transparencies or prints can be examined using normal equipment and procedures. If multispectral images are available, lineaments may be more distinct on particular bands. For example, interpretations using Landsat MSS images may be easiest using the infrared bands (frequently band 4 is used) because shadows are sharp and vegetation patterns are distinct. Prior to interpretation, images can be enhanced by application of contrast stretch (Chapter 5) to improve visibility of lineaments.

Abrams et al. (1983) studied lineaments interpreted from a Landsat MSS color composite of a region in southwestern Arizona. Their analysis focused upon the Helvetia–Rosemont area, known to be rich in porphyry copper. They attempted to determine whether their techniques would locate previously discovered mineralized zones. Paleozoic limestones, quartzites, shales, and dolomites cover pre-Cambrian granites and shists in the study area. The Paleozoic deposits themselves were altered by Mesozoic uplift and erosion and then covered by further deposition. The Laramide orogeny during the late Cretaceous produced intrusion of granitic rocks with thrust faulting along a predominant northeast–southwest trend. This tectonic activity, together with later Paleocene intrusions, caused mineralization along some of the faults. Tectonic activity continued into the late Tertiary, producing extensive faulting and a complex pattern of lineaments.

Abrams et al. examined orientations of lineaments they interpreted from the Landsat image and plotted their results in the form of a rose diagram (Figure 17.8). Each wedge in the circular pattern represents the numbers of lineaments oriented along specific compass azimuths. Because each lineament has two azimuths (e.g., a "north–south" line is oriented

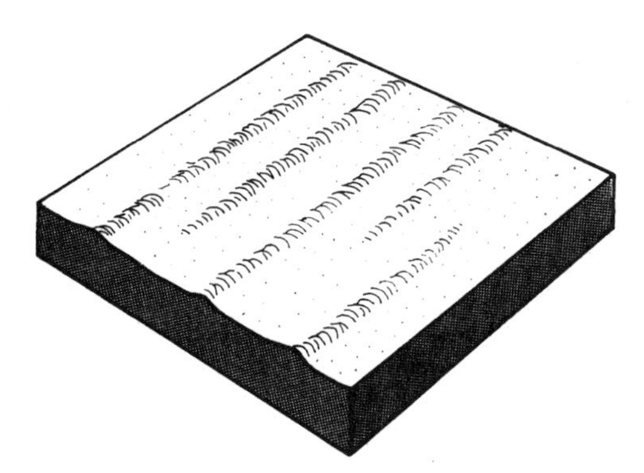

**FIGURE 17.6.** Example of lineaments that do not reflect geologic structure. Wind-blown sand has been aligned in linear ridges that indicate the nature of the surface sediments only.

**FIGURE 17.7.** Example of manual interpretation of lineaments. From an MSS image of northwestern Arizona (shown in Figure 17.3). From Vaughn (1983, p. 4).

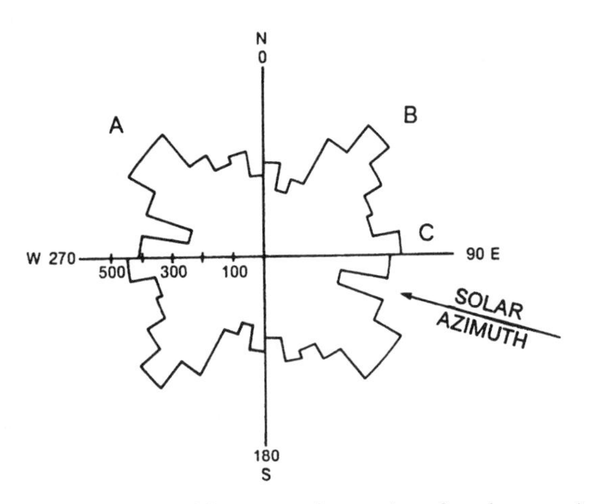


**FIGURE 17.8.** Strike-frequency diagram of lineaments for a region of southeastern Arizona. Features, as discussed in the text, include: A, predominant northwest–southeast orientation; B, predominant northest–southwest orientation; and C, low frequency along the axis that parallels the solar beam at the time of image acquisition. Based on Abrams et al. (1983, p. 593). Reproduced by permission.

as much to the south as it is to the north), the diagram is symmetric, and some investigators prefer to show only half of the diagram.

Several features are noteworthy. First, the low numbers of lineaments oriented at (approximately) 110 (and 290), roughly east-southeast–west-northwest, correspond to the axis that parallels the orientation of the solar beam at the time that the Landsat data were acquired—the lineaments that might have this orientation will not be easily detected because shadowing is minimized. The northeast–southwest trend (*B* in Figure 17.8) is said to represent the predominant trend of pre-Cambrian faulting that is observed throughout Arizona. Superimposed over this pattern is the northwest–southeast trend (*A*) arising from faulting in the Mesozoic rocks mentioned above. Finally, the east–west trend (*C*) is interpreted as the result of the Laramide faulting known to be associated with the mineralization that produced the copper deposits present in this region. Abrams and his colleagues counted the frequencies of the lineaments in each cell of a 10-km grid superimposed over the image and concluded that the highest frequencies, which they interpreted to form favorable locations for intrusion of magmas and mineralization, correspond to zones of known mineral deposits. This kind of study does not form a conclusive test of the effectiveness of lineament analysis because it is conducted with foreknowledge of the nature and location of the resources that are to be detected, but it does illustrate the kinds of techniques that can be applied in the study of lineaments in less well known regions.

### *Automated Analysis of Lineaments*

Because the human visual system tends to generalize, manual interpretations may identify continuous, linear features that are later found to be formed of separate segments of unrelated origin. Lineaments that are found to correspond to geologic features often are extensions of known fault systems rather than completely new systems. Results of manual interpretations vary greatly from interpreter to interpreter (Podwysocki et al., 1975)—a fact that has detracted from the credibility of lineament analysis. As a result, other research has attempted to automate the interpretation of lineaments. Because detection of edges and lines has long been an important task in the fields of image analysis and pattern recognition, automatic detection of lineaments would seem to be a direct application of an existing methodology.

The first application of digital analysis in the search for lineaments was simply to apply the basic procedures for image enhancement discussed in Chapter 5 as a means of preparing the image for visual interpretation. In the enhanced images, lineaments are shown with greater clarity and distinctness. Berlin et al. (1976) and Chavez et al. (1976) applied spatial filtering and both linear and sinusoidal contrast stretch to enhance color Landsat MSS images of the desert regions of southwestern Jordan. Visual interpretation of these images revealed the presence of lineaments that the authors judge to correspond to locations of previously undiscovered faults.

Later research takes digital lineament analysis a step further by using enhanced (or otherwise preprocessed) images as input for an algorithm that replaces the manual interpretation by automatically searching for linear features. Moore et al. (1983) examined Landsat MSS band 4 by first using a low-pass filter to remove the high-frequency component of the image. Such a filter excludes the short-range image variation that would tend to be noisy, presumably leaving the broader-scale variations that would be more likely to portray useful information concerning lineaments. Next they applied a series of directional filters; each filter enhanced features that tended to be linearly oriented in one of eight compass directions. This step produced a set of eight images, each showing the linear features (if any) that tended to be oriented in a specific direction. Then they again applied a smoothing procedure to remove undesirable high-frequency variability and rescaled the brightness values to produce a visually sharper image. The final image shows only the linear features detected by this procedure; it can be superimposed on the original image to provide reference to known geographical features.

Vaughn (1983) applied a different procedure that relied on an analogous strategy. He applied filters to isolate the low-frequency component, then sampled every fourth row and column, applied a median filter to remove noise while retaining edges, and requantized. He then applied his own algorithm to detect the linear features present within the preprocessed image.

## 17.4. Geobotany

Geologic materials and processes can directly influence the character and amount of nutrients available for plant growth. Weathering of rocks and minerals influences the abundance and nature of clay minerals present in the soil cover, thereby influencing soil fertility and plant growth. Variations in the density, vigor, and kinds of plants may then be recorded on remotely sensed images (Figure 17.9).

Geologic processes may concentrate trace elements in specific strata or regions. Geologic weathering may release trace elements in a form that can be absorbed by vegetation. Plants that absorb these trace elements at higher than normal levels may display abnormal spectral signatures, thereby signaling the existence and location of the anomalous concentration of elements.

The study of geobotany depends on knowledge of how geologic materials release elements into the nutrient pool, how these elements are absorbed by the soil and concentrated in plant tissues, and how they can alter the spectral signatures of plants. Because specific kinds of plants may thrive under certain unusual conditions of soil fertility, detailed knowledge of the spectral characteristics of such plants can be very valuable. Geobotanical studies are especially valuable in heavily vegetated regions where soil and rock are not exposed to direct view of the sensor, but they may also be useful in sparsely vegetated regions where it may be possible to identify individual plants, or where vegetation patterns may be especially sensitive to subtle environmental variations (Figure 17.1).

The practice of geobotanical reconnaissance is restricted by several factors. First, geobotanical studies may depend on observations of subtle distinctions in vegetation vigor and pattern, so successful application may require data of very fine spatial, spectral, and radiometric resolution—resolution much finer than that of Landsat or SPOT data, for example. Thus, some of the concepts of geobotanical exploration cannot be routinely applied with imagery and data most commonly available. Hyperspectral remote sensing (Chapter 14) provides the necessary capabilities for developing geobotanical knowledge. This work, now conducted mainly in a research context, may provide the basis for development of sensors and analytical techniques applicable on an operational basis. Second,

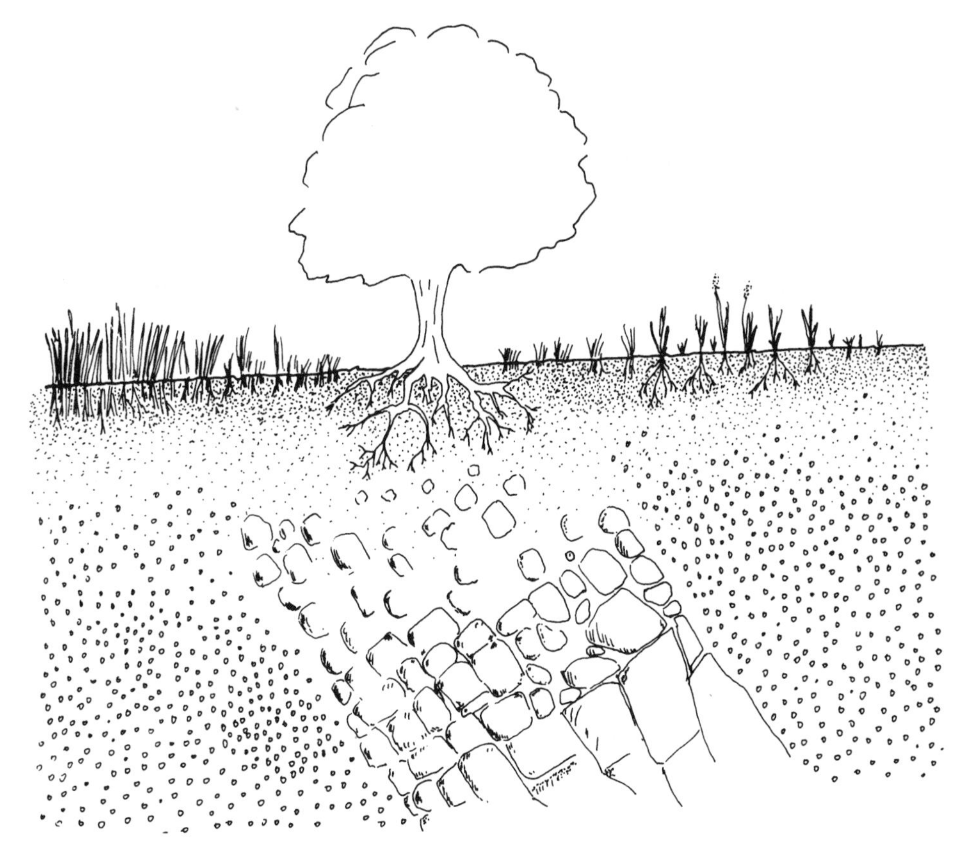

**FIGURE 17.9**. General model for geobotanical studies. Abundance, vigor, and patterns of plants are related to the geologic substratum. If geologists can isolate these geologic factors from others that influence vegetation patterns, botanical information can be used (with other data) to contribute to geological knowledge.

geology is only one of the many factors that influence plant growth. Site, aspect, and disturbance (both by mankind and by nature) influence vegetation distributions in complex patterns. Effects of these several causes cannot always be clearly separated.

Timing of image acquisition is critical. Some geobotanical influences may be detectable only at specific seasons or may be observable as, for example, advances in the timing of otherwise normal seasonal changes in vegetation coloring. The issue of timing may be especially important because the remote sensing analyst may not often have control over the timing of image acquisition or may have imagery only for a single date, whereas several may be necessary to observe the critical changes. Finally, many geobotanical anomalies may have a distinctively local character—specific indicators may have meaning only for restricted regions, so intimate knowledge of regional vegetation and geology may be necessary to fully exploit knowledge of geobotany.

Geobotanical knowledge can be considered first in the context of the individual plant and its response to the geologic substratum. Growth of most plants is sensitive not only to the availability of the major nutrients (P, K, N) but also to the micronutrients—those elements (including B, Mg, S, and Ca) that are required in very small amounts. It is well established that growth and health of plants depend on the availability of elements present in low concentrations, and that sensitivity may be high in some plant species. In contrast, other elements are known to have toxic effects, even at low concentrations, if present in the soil in soluble form. Heavy metals, including Ni, Cu, Cr, Co, and Pb may be present at sufficient levels to reveal their presence through stunted plant growth or by the localized absence of specific species that are especially sensitive to such elements. From such evidence, it may be possible to use the concentrations of these elements indirectly to reveal the locations of specific geologic formations or zones of mineralization worthy of further investigation.

Hydrocarbons may significantly influence plant growth if they are present in the root zone. Hydrocarbon gases may migrate from subsurface locations; at the surface, concentrations may be locally sufficient to influence plant growth. Hydrocarbons may be present in greater concentrations at the surface as petroleum seeps or coal seams. The presence of hydrocarbons in the soil may favor plant growth by increasing soil organic matter but may also alter soil atmosphere and soil biology in a manner that is toxic to some plants.

Specific geobotanical indicators include the following, which, of course, may have varied geologic interpretation depending on the local setting. Variations in biomass, either significantly higher or lower than expected, may indicate geobotanical anomalies. Or, the coloring of vegetation may be significant. The term *chlorosis* refers to a general yellow discoloration of leaves due to a deficiency in the abundance of chlorophyll. Chlorosis can be caused by many of the geologic factors mentioned above, but it is not uniquely of geologic origin so its interpretation requires knowledge of the local geologic and biologic setting.

Observation of these indicators may be facilitated by use of the vegetation indices and ratios mentioned in Chapter 13. Manual interpretation of CIR film often provides especially good information concerning vigor and distribution of vegetation. Of special significance in the context of geobotany is the *blue shift* in the chlorophyll absorption spectra. The characteristic spectral response for living vegetation was discussed in Chapter 16. In the visible, peak reflectance is in the green region; the absorption spectra of chlorophyll decreases reflectance in both the blue and the red regions. However, in the near infrared, reflectance increases markedly (Figure 17.10). Collins et al. (1983) and others have observed that geochemical stress is most easily observed in the spectral region from about 0.55 to 0.75 µm (which includes portions of the green, red, and infrared regions). Most notably, the position and slope of the line that portrays the increase in reflectance at the edge of the visible region, although constant for healthy green plants, is especially sensitive to geochemical influences. As geochemical stress occurs, the position of this line shifts toward shorter wavelengths (i.e., toward the blue end of the spectrum; hence the term *blue shift*) and its slope becomes steeper (Figure 17.11).

It must be emphasized that, by everyday standards, the change is very subtle (0.007 to 0.010 µm) and has been observed only in data recorded by high-resolution sensors processed to filter out background reflectance. Such data are not acquired by remote sensing instruments available for routine use, although increased spectral, spatial, and radiometric resolution provided by hyperspectral remote sensing (Chapter 14) opens up opportunities to exploit these kinds of knowledge.

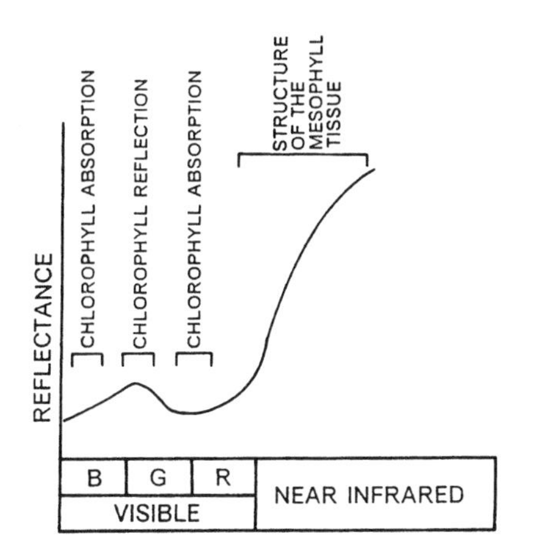


**FIGURE 17.10.** General pattern of spectral reflection from living vegetation. Reflection in the visible spectrum is controlled by leaf pigments, chiefly chlorophyll, which reflect in the green and absorb in the blue and red regions. The far edge of the chlorophyll absorption region (where the reflectance rises sharply in the near infrared) is of special interest in geobotanical studies. The high reflectances in the near infrared are controlled largely by cell structure within the leaf, and at even longer wavelengths the water content is the dominant factor.

Nonetheless, the blue shift has been observed in all plant species studied thus far. Collins et al. (1983) report that they believe this effect to be "universal for most green plants" (p. 739), so it is not confined to a few species. The cause of the blue shift is not clearly understood at present. Unusual concentrations of heavy metals in the soil are absorbed by plants and apparently tend to be transported toward the actively growing portions of the plant, including the leaves (Figure 17.9). Concentrations of such metals can cause chlorosis, often observed by more conventional methods. The relationship between heavy metals and the blue shift is clear, but work by Chang and Collins (1983) indicates that the presence of heavy metals in the plant tissues does not alter the chlorophyll itself—the metallic ions apparently do not enter the structure of the chlorophyll. The heavy metals appear to retard the development of chlorophyll, thereby influencing the abundance, but not the quality, of the chlorophyll in the plant tissue.

Chang and Collins (1983) conducted laboratory and greenhouse experiments to confirm the results of spectral measurements in the field, and to study the effects of different metals and their concentrations. Their experiments used oxides, sulfates, sulfides, carbonates, and chlorides of Cu, Zn, Fe, Ni, Mn, Mo, V. Se, Ni, Cu, and Zn produced stress at

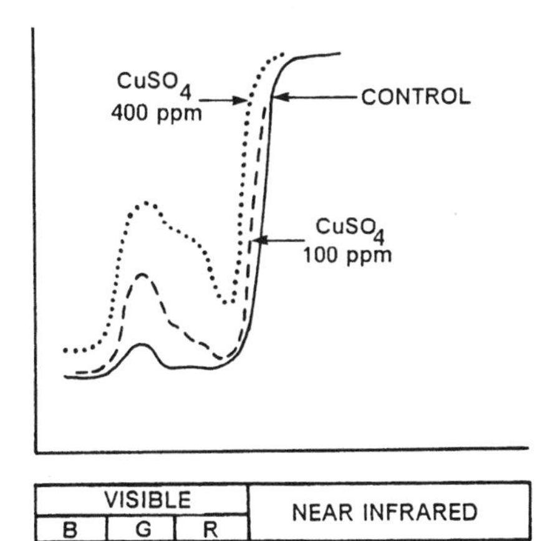


**FIGURE 17.11.** Blue shift in the edge of the chlorophyll absorption band. As heavy mineral concentration in plant tissue increases, the edge of the absorption band shifts toward shorter wavelengths (i.e., toward the blue end of the spectrum). The solid line shows the spectrum for a control plant. The dotted and dashed lines show how the magnitude of the shift increases as concentration of the heavy minerals increases. From Chang and Collins (1983, p. 727). Reproduced by permisssion.

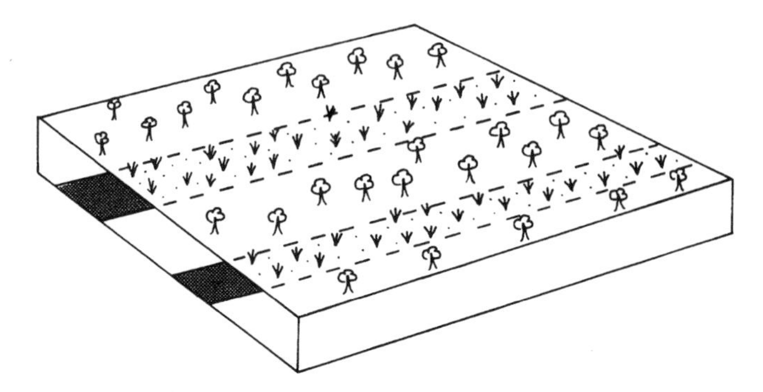

**FIGURE 17.12.** Model for regional geobotany. This diagram depicts a situation in which the geologic differences between two stratigraphic units are not directly observable from surface color or relief, yet differences in soil and drainage lead to noticeable differences in vegetative cover of the two units. See Figure 17.1 for an actual example.

concentrations as low as 100 ppm. Fe and Pb appeared to have beneficial effects on plant growth; Mo and V produced little effect at the concentrations studied. In combination, some elements counteract each other, and some vary in effect as concentrations change.

*Regional geobotany* focuses on the study of vegetation patterns as indicators of lithologic variations. Specific plant species may have local significance as indicator species, as they tend to avoid or favor certain lithologic units. In other instances, variations in abundance or vigor may signal the occurrence of certain lithologies (Figure 17.12).

## 17.5. Direct Multispectral Observation of Rocks and Minerals

A second broad strategy to remote sensing of lithology depends on accurate observation of spectra from areas of soil and rock exposed to observation. Color and, by extension, the spectral characteristics of rocks and minerals can be distinctive identifying features in the direct examination of geologic samples. Geologists use color for identification of samples in the field, and in the laboratory sensitive instruments are used to measure spectral properties across a range of wavelengths in the ultraviolet, visible, and infrared regions. Spectral emittance and reflectance of rocks and minerals are important properties that are often closely related to their physical and chemical properties. In the laboratory, spectral characteristics can be observed in sufficient detail that they can sometimes form diagnostic tests for the presence of specific minerals. In addition, field-portable radiometers (Chapter 12) permit in situ observation of rocks and minerals in nonvisible portions of the spectrum. As a result, geologists have considerable experience in observing spectral properties of geologic materials and have developed an extensive body of knowledge concerning the spectral properties of rocks and minerals.

The application of this knowledge in the context of remote sensing can be very difficult. In the laboratory, spectra can be observed without the contributions of atmospheric attenuation, vegetation, or shadowing to the observed spectra. In the laboratory setting, scientists can hold secondary properties (such as moisture content) constant from one sample to the next or can compensate for their effects. In remotely sensed data, the effects of such variables often cannot be assessed. The usual remote sensing instruments do not have the fine spectral and radiometric resolutions required to make many of the subtle distinctions this strategy requires. Furthermore, the relatively coarse spatial resolutions of many remote sensing systems means that the analyst must consider composite signatures formed by the interaction of many landscape variables rather than the pure signatures that can be observed in the laboratory. Whereas photogeology distinguishes between units without attempting to make fine lithologic distinctions, much of the research in multispectral remote sensing in geology is devoted to more precise identification of specific minerals thought both to be spectrally distinctive and especially valuable in mineral exploration.

Locations of zones of hydrothermal alteration may be revealed by the presence of limonite at the surface. *Limonite* refers broadly to minerals bearing oxides and oxyhydrites of ferric iron, including goethite and hematite; such minerals tend to exhibit broad absorption bands in the near infrared, visible, and ultraviolet regions. Typically, their spectra decline below 0.5 μm, producing a decrease in the ultraviolet not normally observed in other minerals (Figure 17.13). A broad, shallow absorption region from 0.85 to 0.95 μm is observed in the near infrared. The presence of limonite, either as a primary mineral or secondarily as the product of geologic weathering, may identify the location of a zone of hydrothermal alteration, thereby suggesting the possibility of mineralized zones. However, the identification of limonite is not definitive evidence of hydrothermal activity because limonite can be present without hydrothermal alteration and hydrothermal activity may occur without the presence of limonite. Thus the observation of these spectra, or any reflectance spectra, is not uniquely specific, and the analyst must always consider such evidence in the context of other information and knowledge of the local geologic setting.

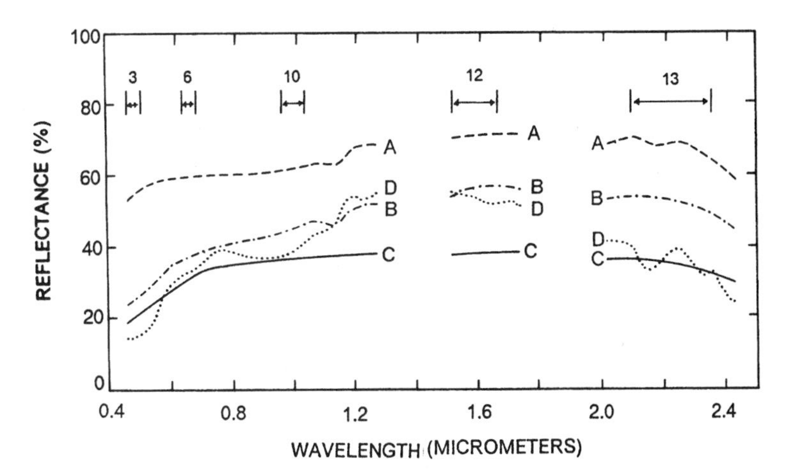


**FIGURE 17.13.** Spectra of some natural geologic surfaces: A, White Joe Lott Tuff member of Mt. Belknap volcanics; B, limonitic Joe Lott Tuff member of Mt. Belknap volcanics; C, tan soil on rhyolite; D, orange altered latite. Data were collected in the field near the surface using portable spectroradiometers. Limonitic minerals (B, C, D) show low reflectances at wavelengths below 0.5 μm, relative to nonlimonitic materials (A). From Goetz et al. (1975), as given by Williams (1983, p. 1748).

Clay minerals often cause a decrease in reflectance beyond 1.6 μm; this behavior is common to enough minerals and occurs over a spectral region that is broad enough that it has been used to locate zones of hydrothermal alteration by means of the surface materials rich in clay minerals. At finer spectral resolution, clay minerals and carbonates (characterized by AlOH and MgOH structures) display spectra with a narrow but distinctive absorption band from 2.1 to 2.4 μm (Figure 17.13).

Emittance in the spectral region from 8 to 12 μm (in the mid infrared) permits the identification of some silicate minerals and the separation of silicate minerals from nonsilicate minerals. Silicate minerals typically exhibit emittance minima in this region, whereas nonsilicate minerals have minima at longer wavelengths.

## 17.6. Photoclinometry

*Photoclinometry*, loosely defined, is the derivation of topographic information from image brightness. If an irregular surface of uniform reflectance is illuminated at an angle, variations in image brightness carry information concerning the slopes of individual facets on the ground, and therefore the full image, composed of many such pixels, depicts the shape of the earth's terrain.

A diffuse reflector (Chapter 2) will have a brightness that is predictable from the angle of illumination (Figure 17.14). It is intuitively obvious that as the angle of illumination changes, so also will the brightness of the surface. The relationship between brightness and angle of illumination can be expressed more precisely using *Lambert's cosine law* (Chapter 2):

$$G(x, y) = I \cos \theta(x, y) \qquad \text{(Eq. 17.1)}$$

**FIGURE 17.14.** Brightnesses of surfaces depend on the orientation of the surface in relation to the direction of illumination. If these surfaces all have the same surface material and behave in accordance with Lambert's cosine law, their brightnesses are related to the angles at which they are oriented.

where $G$ is the image gray tone value, $(x,y)$ are the image pixel row and column coordinates, $I$ is the intensity of the solar illumination, and $\theta$ is the angle of illumination, measured from the zenith.

In the context of remote sensing, the surface and the source of illumination have a fixed geometric relationship, but the scene itself is composed of many individual facets, each with specific orientation with respect to the solar beam. That is, the solar beam illuminates the earth at a fixed angle at any given instant, but topographic irregularities cause the image to be formed of many varied brightnesses. Thus, for a uniform surface type of irregular topography, variations in $G$ portray variations in surface slope and orientation. Here, as a simplification, we will refer to such an image as the *topographic image* ($T$).

Although few surfaces on the earth meet the assumption of uniform reflectance, some extraterrestrial surfaces display only small variations in reflectivity, primarily because of the absence of vegetative cover and differences in color caused by weathering of geologic materials. Photogeologists who have studied the moon's surface and that of the planets have formalized the methods necessary to derive topographic information from images. Photoclinometry encompasses elements of remote sensing, photogrammetry, photometry, and radiometry. Although both the theory and the practice of photoclinometry must still deal with many unsolved issues, photoclinometry has important applications in remote sensing.

In the context of terrestrial remote sensing, photoclinometry must address problems not present in the extraterrestrial case. The varied surface materials that form the earth's surface (open water, living vegetation, bare soil, etc.) contribute to the brightnesses on the image, confusing the purely topographic interpretation of image brightness (Figure 17.15). (Even though the moon's surface is not completely uniform, its spectral variations are small compared to those of the earth.) Therefore, equation 17.1 must be modified to include different reflectances ($R$) of varied surfaces:

$$G = RT \qquad \text{(Eq. 17.2)}$$

where $R$ represents the reflectivities of different surface materials that contribute to different brightnesses on the image. Thus, brightnesses caused by different reflectivities are mixed together with brightness caused by topography, so it is necessary to separate the two if topographic information is to be derived from the image.

Separation of these two kinds of information is accomplished by means of band ratios, a subject briefly introduced in Chapter 13 and one of fundamental significance in many aspects of remote sensing. In a ratio of two bands, the gray values in the separate bands for a specific pixel are paired, then divided. That is, we take brightness values for the same pixel and divide, using the quotient as a new value for that pixel. The set of all ratio values for a specific image and pair of bands forms a *ratio image*. (Because the numeric value of the ratio is often a decimal fraction, the quotient is usually multiplied by a constant to scale the values over an appropriate range of integers.)

At each pixel, information pertaining to topography (i.e., cos $\theta$, designated here as $T$), reflectance of the surface material ($R$), and diffuse light ($D$) are combined in the observed gray value. The diffuse light produces another element in the relationship.

$$G = RT + D \qquad \text{(Eq. 17.3)}$$

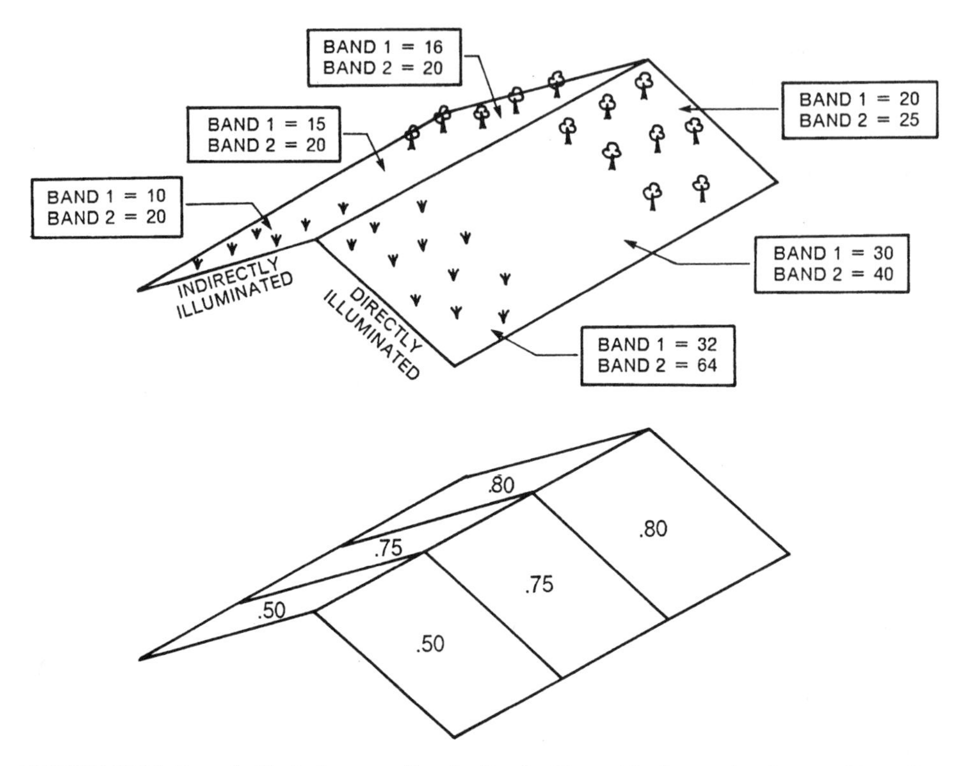


**FIGURE 17.15.** Example illustrating use of band ratios. In this contrived example a topographic surface illuminated from the right side of the illustration consists of three contrasting surface materials, each partially illuminated by the solar beam, and on the reverse slope illuminated indirectly by diffuse light. The raw digital values (top) convey the mixed information of surface reflectance and shadowing. Because of shadowing, separate surface materials cannot be clearly separated into their distinctive classes. In the bottom illustration, ratios of the two bands isolate the effects of the different surface materials (removing the topographic effects) to permit recognition of distinct spectral classes. Based on Taranick (1978, p. 28).

Reflectance and the effects of diffuse light differ greatly from one band to the next, so the value of $G$ at any given pixel varies from one band to the next. For diffuse light, the difference due to separate spectral bands can be approximated by a proportional relationship such that diffuse light in band 1 is estimated by the value in band 2 multiplied by a constant value $c$:

$$D_{\text{band 1}} = c\, D_{\text{band 2}} \qquad \text{(Eq. 17.4)}$$

Although this relationship (and its companion, $T_1 = cT_2$) is not exactly correct, it often can serve as a satisfactory approximation. Note that we can make no such approximation concerning the differences in reflectance between the two bands, so the $R$ components of the relationship are not influenced by this approximation. Thus, a ratio between two separate bands can be expressed as:

$$\frac{G_{\text{band 1}}}{G_{\text{band 2}}} = \frac{R_1(T_1 + D_1)}{R_2(T_2 + D_2)} \qquad \text{(Eq. 17.5)}$$

Because of the relationship between $I_1$ and $I_2$, and $D_1$ and $D_2$, the relationship now becomes

$$\frac{G_{\text{band 1}}}{G_{\text{band 2}}} = \frac{R_1 c(T_2 + D_2)}{R_2(T_2 + D_2)} \quad \text{(Eq. 17.6)}$$

The terms for $T$ and $D$ in the numerator and denominator are now equivalent, and cancel out, leaving the reflectance terms as the only components of the ratio image:

$$\frac{G_{\text{band 1}}}{G_{\text{band 2}}} = \frac{R_1}{R_2} \quad \text{(Eq. 17.7)}$$

This ratio image ($R_1/R_2$) has several useful properties. First, the relationship holds both for shadowed pixels and for directly illuminated pixels. Therefore, a ratio image shows pure reflectance information without effects of topography. This result permits geologists to examine spectral properties of surfaces without the confusing effects of mixed brightnesses of topography and material reflectances. In the raw image, a difference in pixel brightness can be caused by a difference in slope, by shadowing, or by differences in the color of the surface material. In a ratio image the geologist knows that differences in brightness are caused by differences in reflectance only.

This result is illustrated by the example in Figure 17.15. In the upper diagram the brightness of the scene is caused by combined differences in topography and surface reflectance, and the analyst cannot resolve differences between the two effects. In the lower diagram the ratios clearly separate the separate categories of surface reflectance, even though each is present on two different topographic surfaces.

In the context of photoclinometry, ratio images have a special application. An unsupervised classification of the several ratio images derived from a multispectral image permits identification of regions on the earth's surface that have uniform spectral behavior. Remember that the ratio images convey only spectral, not topographic, information; note also that the classification is performed on the ratio images, not on the original digital values. The classification of the ratio values produces classes with uniform spectral properties regardless of slope. If the *original* pixel values for each class are examined, it can be determined that some are light (the directly illuminated pixels) and others are dark (the shadowed pixels). Therefore, all pixels in the scene, regardless of their spectral class, can be designated either *shadow* or *directly illuminated* (Figure 17.16a).

Because brightnesses of shadowed pixels are determined only by indirect illumination, the values in the shadowed regions provide estimates of the diffuse light present throughout the image. So these values can be subtracted from the entire image to remove the atmospheric influences that remain after preprocessing. Then each pixel is assigned a value determined by the mean value of its spectral cluster (as found by using the ratios, as mentioned above) to produce estimates of the spectral brightness of the terrain. This image is formed from a mosaic of patches, each representing a distinctive class of ground cover. It is not necessary to "identify" such classes as "forest," "water," "pasture," and so on—it is sufficient simply to know that they are spectrally homogeneous, distinctive groups of pixels. This image can be called the reflectance image (*Rf*). *Rf* differs from $R$ only in that the estimated reflectance $R$ at each pixel has been replaced by the mean value for its spectral class. Because it is formed by averaging together reflectances from many pixels at varied topographic positions, the *Rf* image approximates the reflectance ($R$) of

**FIGURE 17.16.** (*a*) Shadow image. (*b*) Topographic image. Brightnesses within the topographic image are determined solely by topographic slope; brightnesses caused by different surface materials have been removed..

each class as if it were observed on level topography, thereby defining an estimate of $R$ that does not include the contribution of $T$.

The objective of photoclinometry is the isolation of the purely topographic image. Because the steps described in the preceding paragraphs produce a pure reflectance image, it is possible to separate the original image value into its two components, thereby deriving the topographic content. The final topographic image ($Tp$) is found simply by dividing values in the dehazed image by the corresponding values in the reflectance image ($Rf$). This step can be visualized as

$$Tp = \frac{TR}{Rf} \qquad \text{(Eq. 17.8)}$$

Because $Rf$ estimates $R$, essentially the same term is present in both numerator and denominator and the reflectance component cancels out, leaving only the topographic information (Figure 17.16b). In this final estimate of the topographic image, brightnesses are due solely to differences in topographic slope, and it provides clear information concerning the shape of the earth's surface.

## 17.7. Band Ratios

Band ratios have further significance in geologic remote sensing. The previous sections illustrate that they convey purely spectral information, so they remove effects of shadowing, which otherwise are mixed with the spectral information necessary to make lithologic discriminations. Furthermore, ratios have other advantages (at least under certain circumstances) because they may minimize differences in brightness between lithologic units (i.e., ratios tend to emphasize color information, deemphasizing absolute brightness) and may facilitate comparisons of data collected on different dates, which will differ in solar angle.

Goetz et al. (1975) examined rock reflectances in Utah using a portable spectrometer capable of collecting high-resolution spectra for a variety of natural surfaces. They found several ratios to be useful in discriminating between lithologic units. Because their bands are based on high-resolution data that do not match familiar subdivisions of the spectrum, their ratios must be defined using their own designations, given in Figure 17.13. For example, a ratio between their bands 6 and 10 (roughly, red/near infrared) provided a good distinction between bare rock and vegetation-covered areas. Their 12/3 ratio (mid IR/blue) was successful in distinguishing between limonitic and nonlimonitic rocks. The spectrum for limonite decreases sharply in channel 3 (blue) because of the presence of ferric iron, which absorbs strongly in this region, whereas nonlimonitic rocks tend to be brighter in this region. Thus, the 12/3 ratio tends to accentuate differences between limonitic and nonlimonitic rocks. Hydrothermally altered rocks containing clay minerals display absorption bands in channel 13, so the 12/13 ratio tended to identify hydrothermally altered regions.

Plate 16 shows a ratio image of the Cuprite region, a mining area in southwestern Nevada near the California border. This image depicts a sparsely vegetated region of known geology and proven mineral resources about 4.5 km × 7.0 km (about 2.8 mi. × 4.3 mi.) in size. The geology here is composed essentially of tufts and basalts within an area otherwise composed chiefly of Cambrian limestones and clastic sediments.

This image was produced for a study proposed by Vincent et al. (1984), in which data from two airborne multispectral scanners were obtained to evaluate geological uses of 12 different spectral regions. Plate 16 is formed from a selection of these data. In this image, the intensity of each of the three primaries is determined by the value of the ratio between two spectral channels. As the value of the ratio varies from pixel to pixel, so does the intensity of the color at each pixel. Because the image is formed from three primaries, and each primary represents data from two bands, the image is derived from a total of six different bands.

Full description of the selection of spectral channels, and their interpretation of the image, requires the extended description given by Vincent et al. (1984). However, a concise summary can be presented as follows:

1. The red band in the image is determined by the ratio (0.63–0.69 μm) : (0.52–0.60 μm); greater intensity of red indicates high levels of ferric oxide.
2. The green band is a function of the ratio (1.55–1.75 μm) : (2.08–2.35 μm); the intensity of green indicates areas where there is high clay content in the surface soil.
3. The blue band is controlled by the value of the ratio (8.20–8.54 μm) : (8.60–8.95 μm); the presence of high levels of silica at the soil surface.

Thus, a yellow color (both red and green) indicates the presence of both ferric oxide and clay, an indication of hydrothermal alteration, and a possible presence of a zone of mineralization. The blue region near the center of the image delineates a region of silification (high silica); it is surrounded by a zone of hydrothermal alternation, marked by yellow regions surrounding the blue center.

This example illustrates the power of using ratios between carefully selected spectral regions as aids for geologic investigation. However, it is important to identify aspects of this study that differ from those that might be feasible on a routine basis. Although these data have 30-m resolution, comparable to those of satellite sensors, they were collected at aircraft altitudes so are much less affected by atmospheric effects than would be the case for comparable satellite data. Also, this region has a sparsely vegetated even topography so it favors isolation of geologic information that otherwise can be mixed with vegetation, shadowing, and topographic effects, as noted in the discussion of composite signatures in Chapter 14.

The use of ratio images carries certain risks, as mentioned in Chapter 16. In a geologic context, it is important that data be free of atmospheric effects, or that such effects be removed by preprocessing. The significance of the atmosphere can be appreciated by reexamining the earlier discussion, where the diffuse light terms in the two band values were estimated to have predictable values that would cancel each other out. If severe atmospheric effects are present, they will differ from one band to the other and the value of the ratio will no longer portray only spectral properties of the ground surface but will have values greatly altered by the varied atmospheric contributions to the separate bands.

## 17.8. Soil and Landscape Mapping

Soil scientists study the mantle of soil at the earth's surface in an effort to understand soil formation and to map patterns of soil variation. Soil is a complex mixture of inorganic material weathered from the geologic substratum and mixed with decayed organic matter from tissues of plants and animals. Typically, a "soil" consists of three layers (known as *soil horizons*) of varying thickness and composition (Figure 17.17). Nearest the surface is the A horizon, usually dark in color, and rich in decayed organic matter. The A horizon is sometimes known popularly as *topsoil*—the kind of soil one would like to have for a lawn or garden. In this layer, plant roots are abundant, as are micro-organisms, insects, and other animals. Below the A horizon is the second layer—the B horizon—usually lighter in color and more compact, where plant roots and biologic activity are less abundant. Sometimes the B horizon is known as the *subsoil*—the kind of hard, infertile soil one would prefer *not* to have at the surface of a lawn or garden. At even greater depth is the C horizon, the deepest pedologic horizon, which consists of weathered geologic material, decayed or fractured into material that is softer and looser than the unweathered geologic strata below. This is the material that usually forms the raw material for the A and B horizons. Finally, below the C horizon is the R horizon, consisting of unaltered bedrock.

The exact nature of the horizons at a given place is determined by the interaction between local climatic, topographic, geologic, and biologic elements as they act over time. Because varied combinations of climate, vegetation, and topography produce different soils from similar geologic materials, soil science is distinct from, although closely related to, geology and geomorphology.

The landscape is covered by a mosaic of patches of different kinds of soil, each distinct from its neighbors with respect to character and thickness (Figure 17.18). Soil surveyors outline on maps those areas covered by specific kinds of soil as a means of showing the variation of soils on the landscape. Each symbol represents a specific kind of soil, or, when the pattern is very complex, sometimes two or three kinds of soil that occur in an intimate pattern (Figure 17.19). Soil maps portray distributions of pedologic units, which together

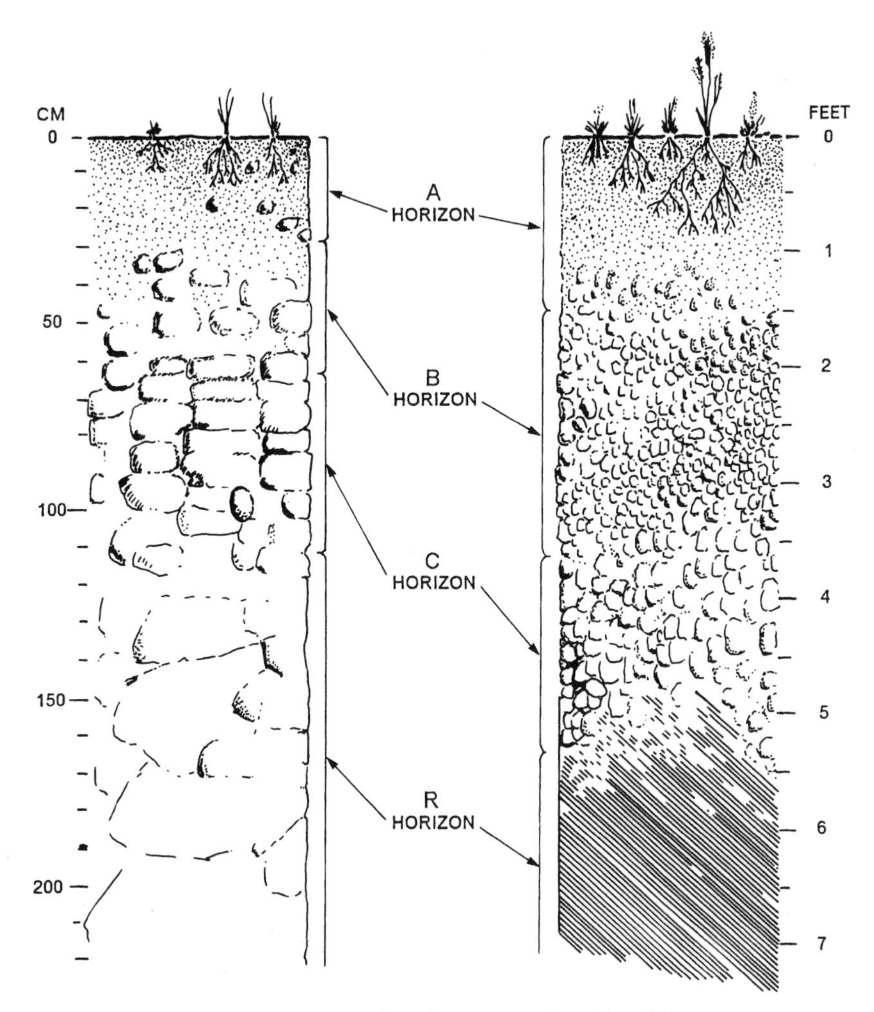


**FIGURE 17.17.** Sketches of two contrasting soil profiles.

with other maps and data convey valuable information concerning topography, geology, geomorphology, hydrology, and other landscape elements. In the hands of a knowledgeable reader, they convey a comprehensive picture of the physical landscape. As a result, they can be considered among the most practical of all forms of landscape maps and are used by farmers, planners, and others who must judge the best locations for specific agricultural activities, community facilities, or construction of buildings and highways.

Soil mapping is conducted routinely by national soil surveys (or their equivalents) in most countries, and by their regional counterparts at lower administrative levels. Details of mapping technique vary from one organization to another, but the main outlines of the procedure are common to most soil survey organizations. A soil map is formed by subdividing the landscape into a mosaic of discrete parcels. Each parcel is assigned to a class of soil—a *mapping unit*—that is characterized by specific kinds of soil horizons. Each mapping unit is represented on the map by a specific symbol—usually a color or an alphanu-

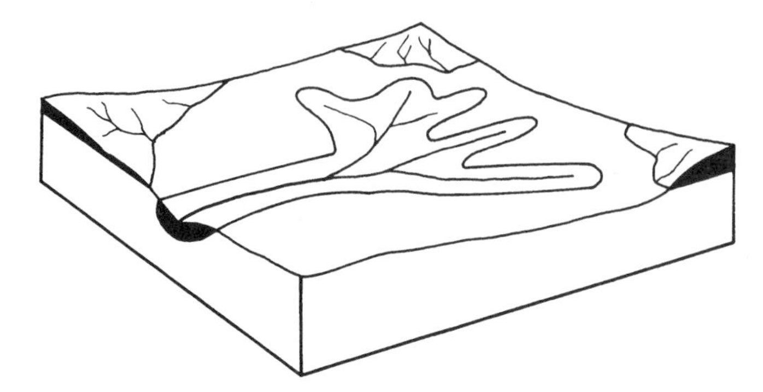

**FIGURE 17.18.** The soil landscape.

meric designation. In theory, mapping units are homogeneous with respect to pedologic properties, although experienced users of soil maps acknowledge the existence of internal variation as well as the presence of foreign inclusions. Mapping units are usually defined with links to a broader system of soil classification defined by the soil survey organization, so that mapping units are consistently defined across, for example, county and state borders. For each region, a mapping unit can be evaluated with respect to the kind of agriculture that is important locally, so the map can serve as a guide to select the best uses for

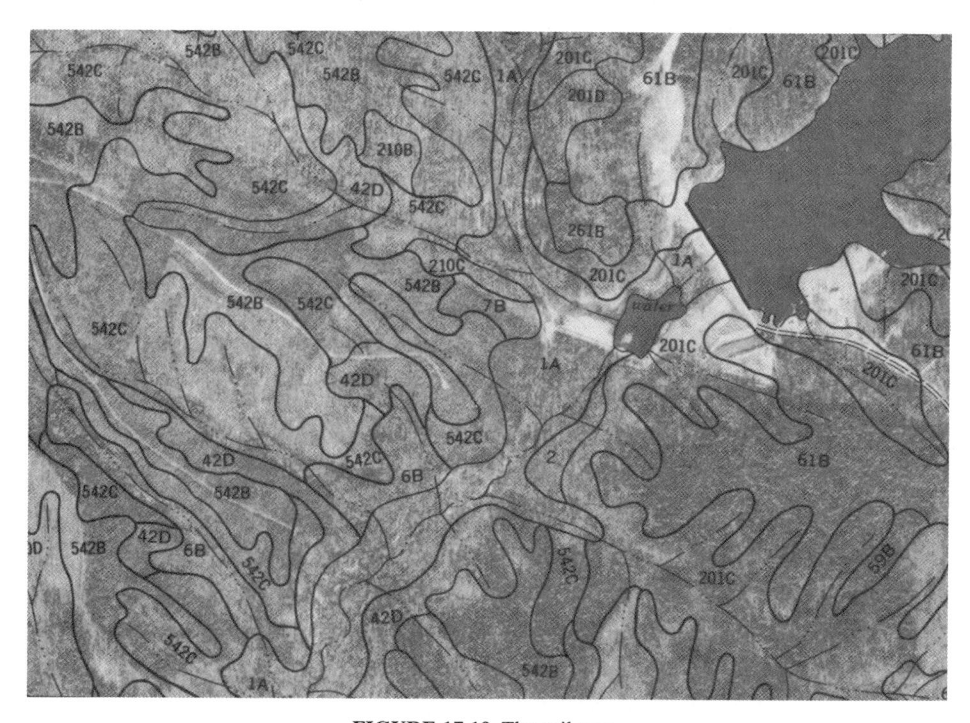


**FIGURE 17.19.** The soil map.

each soil. The following paragraphs outline an example of the soil survey process, although details may vary from one organization to the next.

### *Compilation of a Soil Map*

The first step in the actual preparation of the draft map is the *mapping* process. The term *mapping* in this context means specifically to designate the delineation of landscape units on an aerial photograph or map base, rather than in the broader sense that refers to the entire set of activities necessary to generate the map. In the mapping step, the soil scientist examines aerial photographs, often using a stereoscope, to define the major boundaries between classes of soil. The soil scientist has already learned much about the region's soils through field observations, so he or she has a knowledge of the kinds of soils present and their approximate locations. The soil scientist can use this field knowledge in interpretation of the photograph, using breaks in slope, boundaries between vegetation classes, and drainage pattern to define boundaries between soil mapping units. Seldom if ever can the soil classes actually be identified from the photographs—identification must be based on field observations—but in the mapping step, the scientist can subdivide the landscape into a mosaic of soil parcels that are then each treated as independent units in the later steps.

Completion of the mapping step requires considerable field experience within the region to be mapped, as the surveyor must acquire knowledge of the numbers and kinds of soils present within a region, their properties and occurrence, and their uses and limitations. As a result, completion of this step requires a much longer time than that actually devoted to marking the aerial photographs if we include the time required to learn about the local landscape. Morphologic descriptions of each mapping unit are made in the field, and mapping units are sampled for later analysis.

The second step is *characterization.* Samples are collected from each prospective mapping unit, then subjected to laboratory analysis for physical, chemical, and mineralogical properties.

These measurements form the basis for *classification*, the third step. In the United States, classification is the implementation of the classification criteria specified by *Soil Taxonomy*, the official classification system of the Soil Conservation Service, U.S. Department of Agriculture. In other countries, classification is conducted by applying the classification criteria established by each national soil survey organization. *Correlation*, possibly the most difficult step, matches mapping units within the mapped region to those in adjacent regions and to those in ecologically analogous areas. Whereas the other steps in a soil survey may be conducted largely on a local basis, correlation requires the participation of experienced scientists from adjacent regions, and those from national or international levels in the organization, to provide the broader experience and perspective often required for successful correlation.

The final step is *interpretation,* in which each mapping unit is evaluated with respect to prospective agricultural and engineering uses. The interpretation step provides the user map with information concerning the likely suitability of each mapping unit to the land uses most common within the region.

Within this rather broad framework there are, of course, many variations in details of technique and in overall philosophy and strategy. The result is a map that shows the pattern of soils within a region and a report that describes the kind of soil that is encountered within each mapping unit. Each mapping unit is evaluated with regard to the kinds of uses that might be possible for the region, so that the map serves as a guide to wise use of soil resources.

### *Interpretation by Proxy*

Remote sensing can be applied to soil studies using either of two approaches: (1) *interpretation by proxy,* or (2) *direct sensing of soil properties.* A *proxy*, or *surrogate*, is a substitute for something else (Chapter 5). Proxies are important in soil studies because remotely sensed images cannot show the feature that we wish to study—the soil profile, which is mainly below the ground surface, which is itself often hidden from view by vegetation. As a result, identification of soil units on aerial imagery often must be accomplished indirectly, by means of proxies such as topography, vegetation, and drainage patterns that serve as clues for the interpretation of soils (Figure 17.20).

This is the traditional approach used in the *mapping* step described above, developed over the years for the derivation of soil information from aerial photographs. In essence, the focus is on the remote sensing image as a base on which to plot the boundaries between mapping units, using evidence on the image to guide correct placement of the boundaries. However, the actual identification of the mapping unit—information concerning the soil profile—is derived from other sources, mainly direct observations in the field.

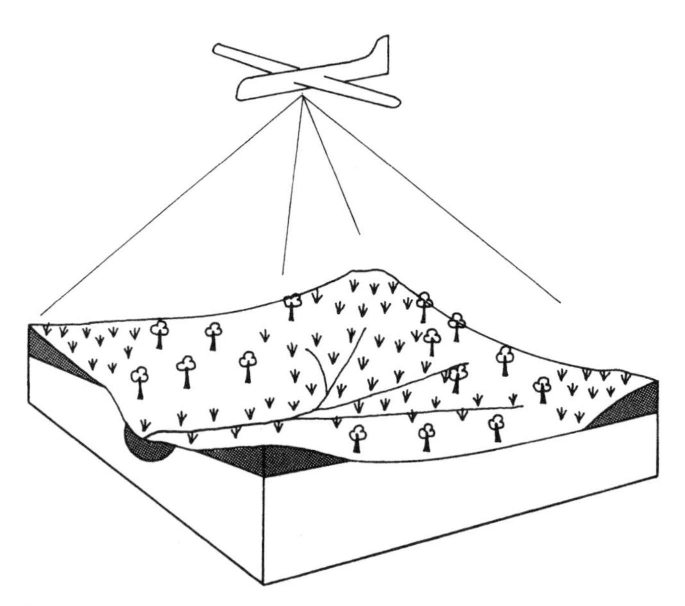

**FIGURE 17.20.** Interpretation by proxy. The analyst uses characteristics such as drainage, topography, and vegetative cover to estimate the outlines of the major soil units, applying also a local knowledge of the geographic setting.

Although interpretation by proxy is clearly a subjective procedure, it requires systematic and disciplined examination of images. Interpreters perceive image tone and texture; recognize terrain features by their sizes, shapes, and locations; and employ stereoscopic interpretation to assess slope and relative elevations. These visible features permit the interpreter to estimate the soil pattern, which of course is not itself visible on the image. Whenever interpretation by proxy is applied, it is necessary to establish the link between the proxy and the mapped distribution using nonimage information. Therefore, the soil scientist must make extensive field observations to establish the links between the patterns visible on the images and the soil profiles hidden beneath the surface. Once a link is established, it can be applied with great effectiveness to speed the otherwise very slow pace of soil survey. However, the link that connects the image pattern to the mapped pattern is valid only for a restricted geographic region (typically 1,000 $km^2$ or so in size) so the interpreters must always check and refine their interpretation procedures based upon field observations.

The interpreter works by drawing lines that separate different mapping units, usually using clues such as breaks in slope, topographic position, vegetation patterns, image tone (if the soil surface is directly visible), drainage lines, and other features, to identify positions of boundaries (Figure 17.19). In the United States, current maps are published on planimetrically correct orthophotomaps (Chapter 2). Therefore, such maps can be used as the basis for measurements of distance or area. (Earlier maps had been published on uncorrected photographic images that often exhibited severe positional errors.) Other procedures sometimes call for the boundary information to be transferred to a map base, then recompiled as a separate map.

Sometimes interpreters prepare separate overlays of drainage, slope, vegetation, land use, surficial materials, and other topics and then superimpose the overlays so that the coincidence of the several patterns reveals the outlines of the soil pattern. This is the method of *interpretive procedures*, or *factor overlays* (Chapter 5), used by some agencies as a standard procedure.

### *Direct Sensing of Soil Properties*

The alternative strategy for interpretation by inference is to directly sense soil properties. In theory, direct observation of the soil surface would release interpreters from dependence on proxies and possibly permit quantitative measurements of soil properties (Figure 17.21). Soil scientists could use remotely sensed images and data not only to delineate boundaries between mapping units but also to identify some of the properties that characterize the mapping units, thereby reducing the requirement for detailed fieldwork. Such a capability could be especially important in remote regions where mapping is very difficult, or in the preparation of reconnaissance surveys that might guide the course of more detailed soil surveys. Or, direct sensing of highly variable properties, such as moisture content at the soil surface, might be especially valuable in contexts unrelated to the mapping effort (e.g., daily or weekly monitoring of changes in soil moisture for agricultural or hydrologic purposes). Such data could contribute to more accurate assessment of soil conditions in times of drought or in forecasting floods.

For most practical purposes, spectral characteristics of unvegetated soils are deter-

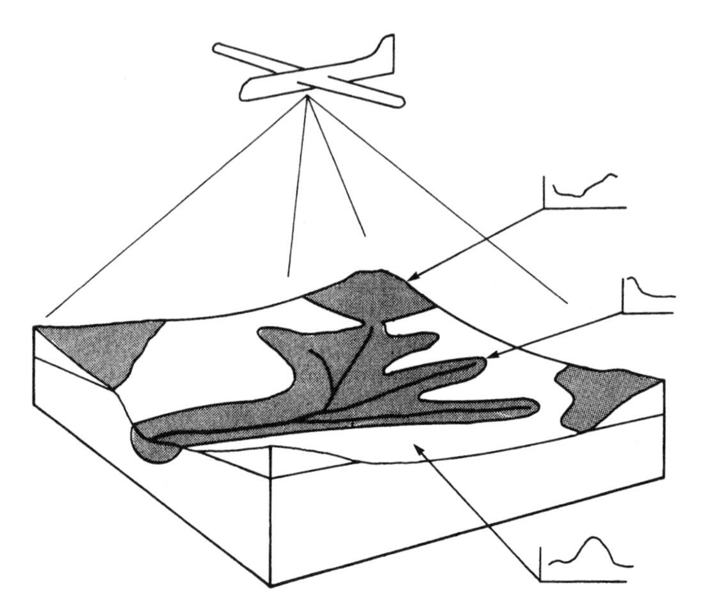

**FIGURE 17.21.** Direct multispectral sensing of soils. The sensor directly observes the soil surface, or a partially vegetated soil surface, to derive information concerning soil properties and soil patterns. Soil properties are studied by analysis of their spectral characteristics.

mined by soil characteristics at or very close to the soil surface. In the visible and near infrared, radiation penetrates only 1 mm to 2 mm in fine-textured materials, and 1 cm to 2 cm in coarse sands (Lee, 1978). Penetration of longer wavelengths depends on frequency of the energy and moisture content of the soil. Passive microwave sensors may be sensitive to moisture conditions at depths of 10 cm or more. Active microwave energy (at long wavelengths) is thought to penetrate much deeper (approximately 1 m) under favorable conditions, but few believe that it would be feasible to routinely use such radiation to interpret subsurface soil properties (Myers, 1975). The determination of characteristics of soils at or very near the soil surface limits the application of direct sensing of soil properties, as many key differences between soils may reside in subsurface horizons beneath spectrally similar surface materials and therefore beyond the reach of direct sensing of soil properties. However, these subsurface differences may be susceptible to indirect interpretation using proxies as described above.

Table 17.1 lists some of the soil properties that have been studied using their spectral properties. It should be noted that this problem is complex, and that in each specific instance it is necessary to consider all the issues discussed in the following paragraphs. As a broad generalization, soils observed in the field display spectra with peak reflectances in or near the red portion of the spectrum, and it is in this region that the greatest spectral contrasts are observed. Condit (1970) studied spectral properties of 160 soil samples from the United States and found that the spectral properties could be characterized by measurements at only five wavelengths: 0.45 μm (in the blue), 0.54 μm (in the green), 0.64 μm (in the red), 0.74 μm (near infrared), and 0.86 μm (also near infrared).

TABLE 17.1. Soil Properties Subject to Direct Remote Sensing

| Property | References |
|---|---|
| Mineral content | Hovis (1966) |
| Texture | Bowers and Hanks (1965)<br>Gerbermann and Neher (1979) |
| Soil moisture | Bowers and Hanks (1965)<br>Planet (1970)<br>Estes et al. (1977) |
| Organic matter | Obukhov and Orlov (1964)<br>Cipra et al. (1980) |
| Iron oxide | Obukhov and Orlov (1964) |
| Albedo | Ångström (1925) |
| Temperature | Myers and Heilman (1969) |
| Soil structure | Janza (1975) |

*Note.* Only a few of many references can be cited here. See Myers and Allen (1968) and Janza (1975) for comprehensive bibliographies.

Some of the diverse research concerning spectral properties of soils can be summarized by noting the nature of the materials studied and the contributions of the studies to the effort to measure soil properties by remote sensing.

1. ARTIFICIAL MATERIALS

Zwermann and Andrews (1940) studied ceramic powders to determine the influence of particle size on reflectivity. By using identical materials ground to different sizes, the effect of particle diameter on reflectance could be isolated. Fine-textured particles tend to have brighter surfaces than coarse particles due to the smoother surface, with less shadowing by particles at the surface. In nature, fine-textured soils are often associated with high moisture or organic matter content, both of which cause lower reflectivities and therefore counteract the effect of small size alone.

2. ARTIFICIAL MIXTURES OF NATURAL SOIL COMPONENTS

Individual components of natural soils can be separated, then reconstituted in desired proportions to create the desired variation in the property to be studied. Gerbermann and Neher (1979) used this strategy to examine the spectral consequences of varying the proportions of sand and clay, holding other factors constant.

3. LABORATORY ANALYSES OF NATURAL SOILS

Bowers and Hanks (1965) examined soil samples under laboratory conditions, then used their laboratory spectra to assist in the classification of the same soils as observed in the field using an airborne multispectral scanner. Stoner et al. (1980) observed differences between spectra of in situ soils and those observed in the laboratory.

4. IN SITU SOIL

Cipra et al. (1980) examined unvegetated soil surfaces in the field and then compared their results with corresponding spectra acquired by the Landsat MSS. Examination of several soil properties revealed that color was the property most closely related to spectral variations in the MSS data.

These results emphasize the dichotomy between component and composite signatures. Research devoted to (1) and (2) above attempts to identify component signatures by isolating effects of specific soil properties on observed spectra. Thus, Mathews (1972) specifically excluded variations in organic matter content in a portion of his study. Bowers and Hanks (1965) isolated the influence of soil moisture. Such research addresses the question, "Is it possible to measure the magnitude of soil property *X*, given spectral information from several soils?"

In contrast, naturally occurring soils possess *composite* signatures, formed by contributions from numerous individual properties interacting in a complex fashion. Thus, studies of naturally occurring soils ([3] and [4] above) define de facto signatures for mapping units, without attempting precise measurement of the contributions of specific properties. Such research addresses the question, "Is it possible to distinguish between soil *A* and soil *B?*" A question of this kind is a return to the same kinds of questions that mapping by proxies addresses because it leads to the establishment, then recognition, of classes of soil rather than measurement of the levels of specific soil properties.

Further obstacles to the application of direct sensing of soil properties can be enumerated briefly. Masking of the soil by vegetation is an obvious problem, and one that has unexpected complications. First, vegetative cover is not necessarily opaque in the nonvisible portion of the spectrum, so the soil can contribute to the spectral properties in ways that may not be immediately obvious. Second, reflection from partially vegetated surfaces can be very complex. Siegal and Goetz (1977) found that the presence of vegetative cover has the greatest influence on signatures of dark soil surfaces, which "may be altered beyond recognition with only ten percent green vegetative cover" (p. 161). Signatures of soil surfaces can vary greatly as angles of illumination and observation vary, and in the case of partially vegetated surfaces, angular effects can be especially severe. Furthermore, in agricultural landscapes, the effects of cultural practices can be significant. Stoner and Horvath (1971) report that "tillage practices appear to alter greatly the spectral radiance of surface soil. When areas of identical or similar surface practices can be isolated and classified separately, a useful map of certain soil properties can be produced" (p. 2110).

## 17.9. Integrated Terrain Units

Another approach—definition of *integrated terrain units*—directly contrasts with the methods described thus far. Direct multispectral recognition of rocks or soils depends on isolation of spectral information pertaining to a general class of data and defining specific categories within the broader class. For example, the analyst must separate soil spectra from those of other features and then define spectra for various classes of soil, recognizing that a single class of soil may have different spectra due to differences in moisture, slope, or other factors. This process can present formidable difficulties because of the complexity of the natural landscape and the requirement for fine spatial and spectral resolution. In contrast, the method of integrated terrain units recognizes the complex spectral mixtures that are formed as spectral responses from vegetation, soil, rock, shadow, moisture, and drainage are combined, especially at the coarse resolution of satellite sensors. Thus, for example, the method does not attempt to identify the specific spectral response for a class of soil but rather a response for the ecological setting of which the soil forms one of several elements.

The method of integrated terrain units is based on the concept that the varied and complex assemblages of soil, terrain, vegetation, and so on, form distinctive spectral responses that can be recognized and mapped. This general approach has long been applied, under many different names, to the interpretation of resource information from aerial photographs and other forms of imagery. The best known, and oldest, of these methods is *land system mapping.*

Land system mapping subdivides a region into sets of recurring landscape elements based on comprehensive examination of distributions of soils, vegetation, hydrology, and physiography. The formalization of land system mapping procedures dates from the period just after World War II, although some of the fundamental concepts were defined earlier. Today a family of similar systems is in use; all systems are based on similar principles and methods, although the terminology and details vary considerably. One of oldest, best known, and most formally defined systems is that developed by the Australian CISRO Division of Land Research and Regional Survey (Christian, 1959); variations have been developed and applied throughout the world.

The method is based on a hierarchical subdivision of landscapes. *Land systems* are recurring contiguous associations of landforms, soils, and vegetation, composed of component *land units.* The basic units of the system (which are assigned varied definitions and names in alternative versions of this basic strategy) are areas of uniform lithology with relatively uniform soil and drainage. A characteristic feature of all versions of the land system method is a hierarchical spatial organization, so that subcategories are nested within the broader categories defined at higher levels. Designation of separate levels within the hierarchy differs among alternative versions. For example, Thomas (1969) defines the following sequence: site (the smallest units), facet, unit landform, landform complex, landform system, and landform region. Wendt et al. (1975) define a system using landtypes (the smallest regions), landtype associations, landtype phases, subsections, sections, and provinces. Although many versions imply that both land systems and their component elements recur in widely separated geographic regions (although under analogous ecological circumstances), closer examination of the method reveals that land systems are most effective when defined to be essentially local units.

Implicit, if not explicit, in most applications is the use of aerial photography as a primary tool for land system mapping, together with direct observation in the field. Aerial photography provides the broad overview and the map-like perspective favoring convenient definition and delineation of land systems and portrays the complex spatial patterns of topography, vegetation, and drainage in an integrated form that is compatible with the assumptions, methods, and objectives of this approach to terrain analysis. "There would be no point in defining any terrain class if its chances of being recognized from the air photographs and background information were small" (Webster & Beckett, 1970).

An assumption of the procedure is that easily identified landscape features (such as vegetation and physiography) form surrogates for more subtle soil features not as easily defined from analysis of aerial photographs. Although the procedure has been applied at a range of scales from 1:1,000,000 to 1:25,000, most applications are probably at fairly broad scales, for reconnaissance surveys, or mapping of rather simple landscape divisions under circumstances in which the ultimate use will be rather extensive.

The method, and its many variations, have been criticized for their subjectivity and the variability of the results obtained by different analysts. Compared to most large-scale soil surveys, land system mapping presents a rather rough subdivision of the landscape, as the mapping units display much greater internal variability and are not as carefully defined and correlated as one would expect in more intensive surveys. As a result, the procedure may be best suited for application in reconnaissance mapping, where broad-scale, low-resolution mapping is required as the basis for planning more detailed surveys.

The method of integrated terrain units has also been applied to analysis of digital data. At the coarse resolution of satellite sensors, individual spectral responses from many landscape components are combined to form composite responses. In this context, the usual methods of image classification for mapping of individual classes of soil, vegetation, or geology may be extraordinarily difficult. Therefore, many analysts have applied the integrated terrain unit strategy to digital classification in an effort to define image classes in a more realistic manner. In the ideal, we might prefer to define pure categories that each show only a single thematic class. But, given the complexity of the natural landscape, composite categories are often well suited to representation of the gradations and mixtures that characterize many environments (Robinove, 1981; Green, 1986).

## 17.10. Summary

Aerial photography and remote sensing have long been applied to problems in the earth sciences. Geology, topographic mapping, and related topics formed one of the earliest routine applications of aerial photography, and today remote sensing, used with other techniques, continues to form one of the most important tools for geologic mapping, exploration, and research.

Research for geologic remote sensing spans the full range of subjects within the field of remote sensing, including use of additional data in the thermal, near infrared, visible, and microwave regions of the spectrum. More than most other subjects, efforts such as geobotanical research are especially interesting because they bring knowledge of so many different disciplines together to bear on a single problem. In addition, geologic investigations have formed important facets of research and applications programs for Landsat, SPOT, RADARSAT, ERS-1, and other systems. Geologic studies are said to form the major economic component of the practice of remote sensing. Certainly they form one of the most important elements in both theoretical and practical advances in remote sensing.

## Review Questions

1. Why is the timing of overpasses of earth observation satellites such as Landsat or SPOT likely to be of special significance for geologic remote sensing?
2. Explain why the ability to monitor the presence of moisture (both open water bodies as well as soil moisture at the ground surface) might be of special significance in geologic, geomorphic, and pedologic studies.
3. In the past there has been discussion of conflicts between earth scientists and plant scien-

tists concerning the design of spectral sensitivity of sensors for the MSS and TM, based upon the notion that the two fields have quite different requirements for spectral information. Based upon information in this chapter, explain why such a distinct separation of information requirements may not be sensible.

4. Why should Landsat imagery have been such an important innovation in studies of geologic structure?
5. Summarize the significance of lineaments in geologic studies, and the significance of remote sensing to the study of lineaments.
6. Compare relative advantages and disadvantages of photogeology and geologic remote sensing, as outlined in this chapter.
7. Some might consider lineament analysis, application of geobotany, and other elements of geologic remote sensing to be modern, state-of-the-art techniques. If so, why should photogeology, which is certainly not as technologically sophisticated, be so widely practiced today?
8. Full application of the principles and techniques of geologic remote sensing requires much finer spectral, radiometric, and spatial resolution than most operational sensors have at present. What difficulties can you envision in attempts to use such sensors for routine geologic studies?
9. Refer back to the first paragraphs of Section 11.1; compare and contrast the concept of image classification with that of integrated terrain units (Section 17.9). Refer also to Section 5.10. Write two or three paragraphs that summarize the major differences between the two strategies.
10. In what ways might radar imagery (Chapter 7) be especially useful for studies of geology and other earth sciences?
11. Write a short description of a design for a multispectral remote sensing system tailored specifically for recording geologic information. Disregard all other applications. Suggest how you would select the optimum spatial and radiometric resolution and the most useful spectral regions.

## References

### General

Abrams, M. J., D. Brown, L. Lepley, and R. Sadowski. 1983. Remote Sensing for Porphyry Copper Deposits in Southern Arizona. *Economic Geology,* Vol. 78, pp. 591–604.

Chang, S., and W. Collins. 1983. Confirmation of the Airborne Biogeophysical Mineral Exploration Technique Using Laboratory Methods. *Economic Geology,* Vol. 78, pp. 723–726.

Clarke, John I. 1966. Morphometry from Maps. Chapter in *Essays in Geomorphology* (G. H. Dury, ed.). New York: American Elsevier, pp. 235–274.

Cole, Monica M. 1982. Integrated Use of Remote Sensing Imagery in Mineral Exploration. *Geological Society of America Transactions,* Vol. 85, Part 1, pp. 13–28.

Collins, W., G. Raines, F. Canney, and R. Ashley. 1983. Airborne Biogeophysical Mapping of Hidden Mineral Deposits. *Economic Geology,* Vol. 78, pp. 737–749.

Goetz, A. H. 1989. Spectral Remote Sensing in Geology. Chapter 12 in *Theory and Applications of Optical Remote Sensing* (G. Assar, ed.). New York: Wiley, pp. 491–547.

Goetz, Alexander F. H., G. Vane, J. E. Solomon, B. N. Rock. 1985. Imaging Spectrometry for Earth Remote Sensing. *Science,* Vol. 228, pp. 1147–1153.

Goetz, A. F. H., B. N. Rock, and L. C. Rowan. 1983. Remote Sensing for Exploration: An Overview. *Economic Geology,* Vol. 78, pp. 573–590.

Goetz, A. F. H. 1984. High Spectral Resolution Remote Sensing of the Land. In *Remote Sensing* (P. N. Slater, ed.). SPIE Proc. 475. Bellingham, WA: SPIE, pp. 56–68.

Goetz, A. F. H., F. C. Billingsley, A. R. Gillespie, M. J. Abrahms, R. L. Squires, E. N. Shoemaker, I. Lucchita, and D. P. Elston. 1975. *Applications of ERIS Images and Image Processing to Regional Geologic Problems and Geologic Mapping in Northern Arizona.* Pasadena, CA: California Institute of Technology, Jet Propulsion Laboratory Technical Report 32-1597, 188 pp.

Gupta, Ravi P. 1991. *Remote Sensing Geology.* New York: Springer-Verlag, 356 pp.

Kiefer, Ralph W. 1967. Terrain Analysis for Metropolitan Fringe Area Planning. *Journal of the Urban Planning and Development Division, Proceedings of the American Society of Civil Engineers,* pp. 119–139.

King, R. B. 1970. A Parametric Approach to Land System Classification. *Geoderma,* Vol. 4, pp. 37–46.

Meyer, B. S., D. B. Anderson, R. H. Bohning, and D. G. Fratianne. 1973 *Introduction to Plant Physiology.* New York: Van Nostrand, 565 pp.

Mitchell, Colin W. 1973 *Terrain Evaluation.* London: Longman. 221 pp.

Myers, V. I. (ed.). 1975. Crops and Soils. Chapter 22 in *Manual of Remote Sensing* (R. G. Reeves, ed.). Bethesda, MD: American Society for Photogrammetry and Remote Sensing, pp. 1715–1813.

Peel, R. F., L. F. Curtis, and E. C. Barrett. 1977. *Remote Sensing of the Terrestrial Environment.* London: Butterworths, 275 pp.

Rast, Michael, J. S. Hook, C. D. Elvidge, and R. E. Alley. 1991. An Evaluation of Techniques for the Extraction of Mineral Absorption Features from High Spectral Resolution Remote Sensing Data. *Photogrammetric Engineering and Remote Sensing,* Vol. 57, pp. 1303–1309.

Rowan, L. C., A. F. H. Goetz, and R. P. Ashley. 1977. Discrimination of Hydrothermally Altered and Unaltered Rocks in Visible and Near Infrared Multispectral Images. *Geophysics,* Vol. 42, pp. 522–535.

Settle, Mark. 1982. Use of Remote Sensing Techniques for Geologic Mapping. In *Proceedings, Harvard Computer Graphics Week.* Cambridge, MA: Harvard University Graduate School of Design, pp. 1–12.

Schneider, Stanley R., D. F. McGinnis, and John A. Pritchard. 1979. Use of Satellite Infrared Data for Geomorphology Studies. *Remote Sensing of Environment,* Vol. 8, pp. 313–330.

Siegal, Barry S., and A. R. Gillespie (eds.). 1980. *Remote Sensing in Geology.* New York: Wiley, 702 pp.

Taranick, James V. 1978. *Characteristics of the Landsat Multispectral Data System.* U. S. Geological Survey Open File Report 78–187, 76 pp.

Vincent, R. K., P. K. Pleitner, and M. L. Wilson. 1984. Integration of Airborne Thematic Mapper and Thermal Infrared Multispectral Scanner Data for Lithologic and Hydrothermal Alteration Mapping. *Proceedings, International Symposium on Remote Sensing, Third Thematic Conference, Remote Sensing for Exploration Geology.* Ann Arbor: Environmental Research Institute of Michigan, pp. 219–226.

Watson, K. 1975. Geologic Applications of Thermal Infrared Imagery. *Proceedings of the IEEE,* Vol. 63, pp. 128–137.

Williams, R. S. 1983. Geological Applications. Chapter 31 in *Manual of Remote Sensing* (2nd edition) (R. N. Colwell, ed.). Bethesda, MD: American Society of Photogrammetry and Remote Sensing, pp. 1667–1953.

**Photogeology**

Lueder, D. R. 1959. *Aerial Photographic Interpretation: Principles and Applications.* New York: McGraw-Hill, 462 pp.

Miller, Victor C. 1961. *Photogeology.* New York: McGraw-Hill, 248 pp.

Ray, Richard G. 1960. *Aerial Photographs in Geologic Interpretation and Mapping.* U.S. Geological Survey Professional Paper 373. Washington, DC: U.S. Government Printing Office, 230 pp.

**Lineaments**

Berlin, G. L., P. S. Chavez, T. E. Grow, and L. A. Soderblom. 1976. Preliminary Geologic Analysis of Southwest Jordan from Computer Enhanced Landsat I Image Data. In *Proceedings, American Society of Photogrammetry.* Bethesda, MD: American Society of Photogrammetry, pp. 543–563.

Chavez, P. S., G. L. Berlin, and A. V. Costa. 1976. Computer Processing of Landsat MSS Digital Data for Linear Enhancements. *Proceedings of the Second Annual William T. Pecora Memorial Symposium,* pp. 235–250.

Clark, C. D., and C. Watson. 1994. Spatial Analysis of Lineaments. *Computers and Geosciences,* Vol. 20, pp. 1237–1258.

Ehrich, Roger W. 1977. Detection of Global Edges in Textured Images. *IEEE Transactions on Computers,* Vol. C-26, pp. 589–603.

Ehrich, R. W. 1979. Detection of Global Lines and Edges in Heavily Textured Images. *Proceedings, Second International Symposium on Basement Tectonics,* Newark, DE, pp. 508–513.

Moore, Gerald K., and F. A. Waltz. 1983. Objective Procedures for Lineament Enhancement and Extraction. *Photogrammetric Engineering and Remote Sensing,* Vol. 49, pp. 641–647.

O'Leary, J. D. Friedman, and H. A. Pohn. 1976. Lineament, Linear, Lineation: Some Proposed New Standards for Old Terms. *Geological Society of America Bulletin,* Vol. 87, pp. 1463–1469.

Podwysocki, M. H., J. G. Moik, and W. D. Shoup. 1975. Quantification of Geologic Lineaments by Manual and Machine Processing Techniques. *Proceedings, NASA Earth Resources Survey Symposium,* Houston, TX, pp. 885–903.

Raghavan, V., K. Wadatsumi, and S. Masummoto. 1994. SMILES: A FORTRAN-77 Program for Sequential Machine Interpreted Lineament Extraction Using Digital Images. *Computers and Geosciences,* Vol. 20, pp. 121–159.

Vanderbrug, G. J. 1976. Line Detection in Satellite Imagery. *IEEE Transactions on Geoscience Electronics,* Vol. GE-14, pp. 37–44.

Vaughan, Russell W. 1983. *A Topographic Approach for Lineament Recognition in Satellite Images.* Unpublished M.S. thesis. Blacksburg, VA: Virginia Polytechnic Institute, Department of Computer Science, 70 pp.

Wheeler, Russell L., and Donald U. Wise. 1983. Linesmanship and the Practice of Linear Geo-Art: Discussion and Reply. *Geological Society of America Bulletin,* Vol. 94, pp. 1377–1379.

Wise, Donald U. 1982. Linesmanship and the Practice of Linear Geo-Art. *Geological Society of America Bulletin,* Vol. 93, pp. 886–888.

Yamaguchi, Y. 1985. Image-Scale and Look-Direction Effects on the Detectability of Lineamants in Radar Images. *Remote Sensing of Environment,* Vol. 17, pp. 117–127.

### Integrated Terrain Units

Christian, C. S. 1959. The Eco-Complex and Its Importance for Agricultural Assessment: Chapter 36, Biogeography and Ecology in Australia. *Monographiae Biologicae,* Vol. 8, pp. 587–605.

Green, G. M. 1986. Use of SIR-A and Landsat MSS Data in Mapping Shrub and Intershrub Vegetation at Koonamore, South Australia. *Photogrammetric Engineering and Remote Sensing,* Vol. 52, pp. 659–670.

Robinove, C. J. 1981. The Logic of Multispectral Classification and Mapping of Land. *Remote Sensing of Environment,* Vol. 11, pp. 231–244.

Wendt, G. E., R. A. Thompson, and K. N. Larson. 1975. *Land Systems Inventory: Boise National Forest, Idaho, A Basic Inventory for Planning and Management.* Ogden, UT: U.S. Forest Service Intermountain Region, 49 pp.

### Soils and Geomorphology

Agbu, Patrick A., and E. Nizeyimana. 1991. Comparisons between Spectral Mapping Units Derived from SPOT Image Texture and Field Soil Map Units. *Photogrammetric Engineering and Remote Sensing,* Vol. 57, pp. 397–405.

Bleeker, P., and J. G. Speight. 1978. Soil–Landform Relationships at Two Localities in Papua New Guinea. *Geoderma,* Vol. 21, pp. 183–198.

Goudie, Andrew, et al. 1981. *Geomorphological Techniques.* London: Allen & Unwin, 395 pp.

Post, D. F., E. H. Horvath, W. M. Lucas, S. A. White, M. J. Ehasz, and A. K. Batchily. 1994. Relations Between Soil Color and Landsat Reflectance on Semiarid Rangelands. *Soil Science Society of America Journal,* Vol. 58, pp. 1809–1816.

Thornbury, William D. 1967. *Regional Geomorphology of the United States.* New York: Wiley, 609 pp.

Thomas, M. F. 1969. Geomorphology and Land Classification in Tropical Africa. Chapter in *Environment and Land Use in Africa* (M. F. Thomas and G. W. Whittington, eds.). London: Methuen, pp. 103–105.

Townshend, J. R. G. (ed.). 1981. *Terrain Analysis and Remote Sensing.* London: Allen & Unwin, 232 pp.

Webster, R., and P. H. T. Beckett. 1970. Terrain Classification and Evaluation by Air Photography: A Review of Recent Work at Oxford. *Photogrammetria,* Vol. 26, pp. 51–75.

Weismiller, R. A., and S. A. Kaminsky. 1978. An Overview of Remote Sensing as Related to Soil Survey Research. *Journal of Soil and Water Conservation,* Vol. 33, pp. 287–289.

Wright, J. S., T. C. Vogel, A. R. Pearson, and J. A. Messmore. 1981. *Terrain Analysis Procedural Guide for Soil.* ETL-0254. Ft. Belvoir, VA: U.S. Army Corps of Engineers, Geographic Services Laboratory, Engineering Topographic Laboratories, 89 pp.

### Direct Sensing of Soil Properties

Ångström, A. 1925. The Albedo of Various Surfaces of Ground. *Geografiska Annaler,* Vol. 7, pp. 323–342.

Bowers, S. A., and R. J. Hanks. 1965. Reflection of Radiant Energy from Soils. *Soil Science,* Vol. 100, pp. 130–138.

Cipra, J. E., D. P. Franzmeir, M. E. Bauer, and R. K. Boyd. 1980. Comparison of Multispectral Measurements from Some Nonvegetated Soils Using Landsat Digital Data and a spectroradiometer. *Soil Science Society of America Journal,* Vol. 44, pp. 80–84.

Condit, H. R. 1970. The Spectral Reflectance of American Soils. *Photogrammetric Engineering,* Vol. 36, pp. 955–966.

Estes, J. E., M. R. Mel, and J. D. Hooper. 1977. Measuring Soil Moisture with an Airborne Imaging Passive Microwave Radiometer. *Photogrammetric Engineering and Remote Sensing,* Vol. 43, pp. 1273–1281.

Gerbermann, A. H., and D. D. Neher. 1979. Reflectance of Varying Mixtures of a Clay Soil and Sand. *Photogrammetric Engineering and Remote Sensing,* Vol. 45, pp. 1145–1151.

Hovis, W. A. 1966. Infrared Spectral Reflectance of Some Common Minerals. *Applied Optics,* Vol. 5, pp. 245–248.

Janza, F. J. 1975. Interaction Mechanisms. Chapter 4 in *Manual of Remote Sensing,* 1st ed. (R. G. Reeves, ed.). Falls Church, VA: American Society of Photogrammetry, pp. 75–179.

Lee, R. 1978. *Forest Microclimatology.* New York: Columbia University Press, 276 pp.

Mathews, H. L. 1972. *Application of Multispectral Remote Sensing and Spectral Reflectance Patterns to Soil Survey Research.* Unpublished Ph.D. dissertation. University Park, PA: Pennsylvania State University, 110 pp.

Myers, V. I., and W. I. Allen. 1968. Electro-optical Remote Sensing Methods as Nondestructive Testing and Measuring Techniques in Agriculture. *Applied Optics,* Vol. 7, pp. 1819–1838.

Myers, V. I., and M. D. Heilman. 1969. Thermal Infrared for Soil Temperature Studies. *Photogrammetric Engineering,* Vol. 35, pp. 1024–1032.

Obukhov, A. I., and D. S. Orlov. 1964. Spectral Reflectivity of the Major Soil Groups and Possibility of Using Diffuse Reflection in Soil Investigations. *Soviet Soil Science,* pp. 174–184.

Planet, W. G. 1970. Some Comments on Reflectance Measurements of Wet Soils. *Remote Sensing of Environment,* Vol. 1, pp. 127–129.

Siegal, B. S., and A. F. H. Goetz. 1977. Effect of Vegetation on Rock and Soil Type Discrimination. *Photogrammetric Engineering and Remote Sensing,* Vol. 43, pp. 191–196.

Stoner, E. R., and E. H. Horvath. 1971. The Effect of Cultural Practices on Multispectral Response from Surface Soil. In *Proceedings of the Seventh International Symposium on Remote Sensing of Environment.* Ann Arbor: University of Michigan, pp. 2109–2110.

Stoner, E. R., M. F. Baumgardner, R. A. Weismiller, L. L. Biehl, and B. F. Robinson. 1980. Extension of Laboratory-Measured Soil Spectra to Field Conditions. *Soil Science of America Journal,* Vol. 44, pp. 572–574.

Zwermann, C. H., and A. I. Andrews. 1940. Relation of Particle Size and Characteristic of Light Reflected from Porcelain Enamel Surfaces. *Journal of the American Ceramic Society,* Vol. 23, pp. 93–102.

CHAPTER EIGHTEEN

# Hydrospheric Sciences

## 18.1. Introduction

Open water covers about 74% of our planet's surface. Its largest areas are, of course, the oceans, which account for about 95% of the surface area of open water, but freshwater lakes and rivers (about 0.4%) have a significance for mankind that exceeds their small percentages (Table 18.1). In addition, soil and rock near the earth's surface hold significant quantities of freshwater (but only about 0.01% by volume of the earth's total water), as do the ice and snow of polar regions. Moisture in the form of "permanent" ice (about 5% of the earth's surface) is largely beyond the reach of mankind, although the seasonal accumulation and melt of snowpack in temperate mountains is an important source of moisture for some agricultural regions in otherwise arid zones (such as the Great Basin of North America). Hydrologists and meteorologists monitor water as it occurs in all of these forms, as it changes from liquid to vapor, condenses to rain and snow, and moves on and under the earth's surface (Figure 18.1). In addition, studies of sea ice, movement of pollutants, and ocean currents are other important areas that attract the attention of the many scientists who study hydrology, oceanography, and related subjects.

Most traditional means of monitoring the earth's water depend largely on measurements made at specific points or collections of samples from discrete locations. Oceans, lakes, and rivers can be studied by samples collected at the surface, or by special devices that fill a collection bottle with water from a specified depth. Groundwater can be studied by collection of samples from wells or boreholes. Samples can be subjected to chemical and physical tests to measure levels of pollutants, to detect bacteria and other biological phenomena, and to examine oxygen content, sediment content, and many other qualities of the water.

Such measurements, or samples, of course, provide information about discrete points within the water body, whereas the analyst usually is interested in examining entire water bodies or regions within water bodies. Although measurements can be collected at several locations to build up a record of place-to-place variation within the water body, such efforts usually can form only a piecemeal approach to studying the complex and dynamic characteristics of water bodies that are of interest to hydrologists.

**TABLE 18.1. Water on Earth**

| | Percentage | |
|---|---|---|
| | By surface area[a] | By volume[b] |
| Oceans | 94.90 | 97.1 |
| Rivers and lakes | 0.40 | 0.02 |
| Groundwater | — | 0.60 |
| Permanent ice cap | 4.69 | 2.20 |
| Earth's atmosphere | — | 0.001 |

*Note.* Calculated from values given by Nace (1967).
[a]Percentage by area of earth's total water surface.
[b]Percentage by volume of earth's water.

Therefore, remote sensing provides a valuable perspective concerning broad-scale, dynamic patterns that can be difficult to examine in detail using only point measurements. Careful coordination and placement of surface samples permits establishment of the relationship between the sample data and those collected by the remote sensor. Remotely sensed data can be especially valuable in studying phenomena over large areas, and satellite sensors provide the opportunity for regular observation of even very remote regions. Although remotely sensed images seldom replace the usual sources of information concerning water resources, they can provide valuable supplements to field data by revealing broad-scale patterns not recognizable at the surface, recording changes over time, and providing data for inaccessible regions.

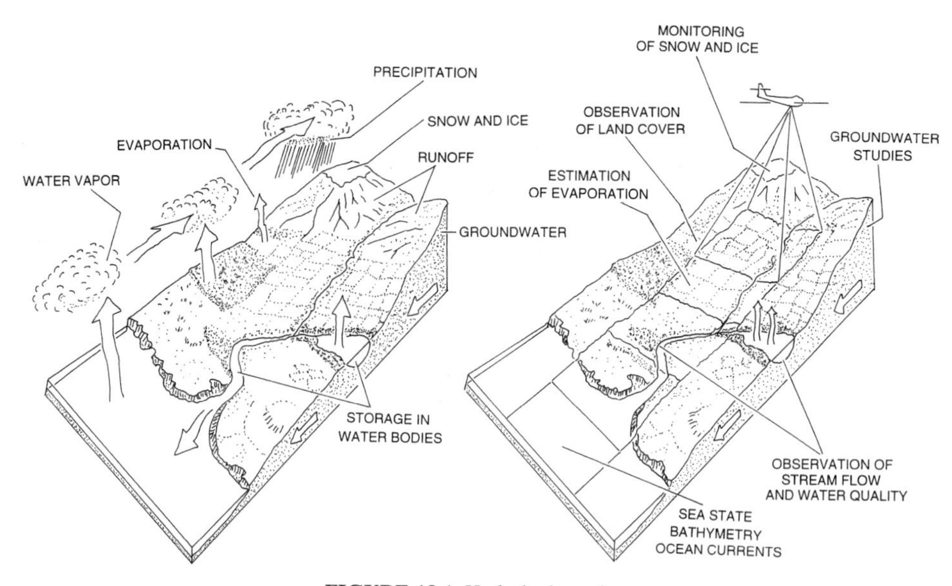


**FIGURE 18.1.** Hydrologic cycle.

## 18.2. Spectral Characteristics of Water Bodies

Spectral qualities of water bodies are determined by the interaction of several factors, including the radiation incident to the water surface, optical properties of the water, roughness of the surface, angles of observation and illumination, and, in some instances, reflection of light from the bottom (Figure 18.2). As incident light strikes the water surface, some is reflected back to the atmosphere; this reflected radiation carries little information about the water itself, although it may convey information about roughness of the surface and, and therefore, about wind and waves. Instead, the spectral properties (i.e., "color") of a water body are determined largely by energy that is scattered and reflected within the water body itself, known as *volume reflection* because it occurs over a range of depths rather than at the surface. Some of this energy is directed back toward the surface, where it again passes into the atmosphere, and then to the sensor (Figure 18.2). This light, sometimes known as *underlight*, is the primary source of the color of a water body.

The light that enters a water body is influenced by (1) absorption and scattering by pure water, and (2) scattering, reflection, and diffraction by particles that may be suspended in the water. For pure water, some of the same principles described previously for atmospheric scattering apply. Scattering by particles that are small relative to wavelength (Rayleigh scattering) causes shorter wavelengths to be scattered the most. Thus, for deep water bodies, we expect (in the absence of impurities) water to be blue or blue-green in color (Figure 18.3). Maximum transmittance of light by clear water occurs in the range 0.44 to 0.54 μm, with peak transmittance at 0.48 μm. Because the color of water is determined by volume scattering, rather than surface reflection, spectral properties of water

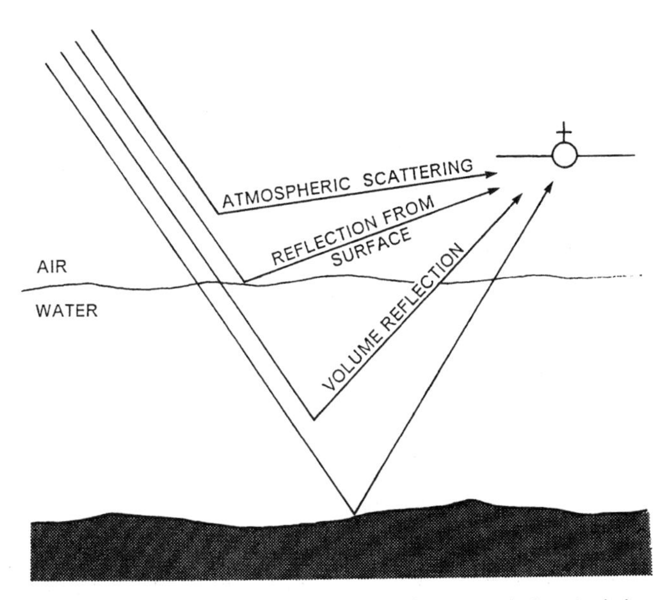


**FIGURE 18.2.** Diagram illustrating major factors influencing spectral characteristics of a water body. Modified from Alföldi (1982, p. 318).

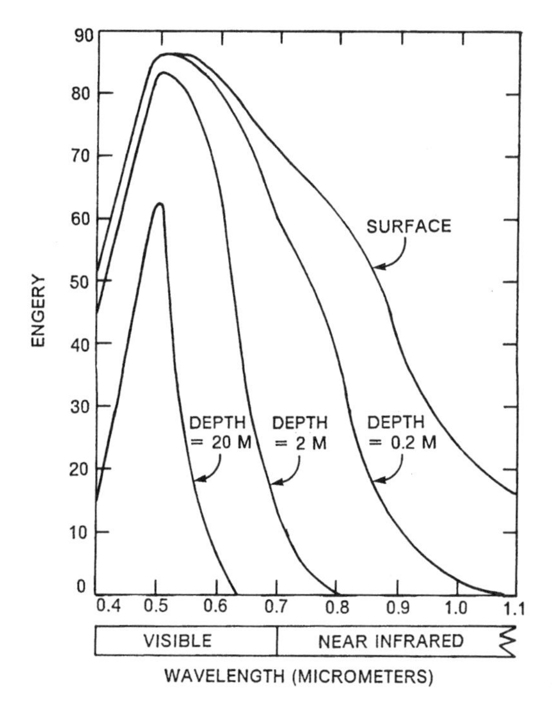


**FIGURE 18.3.** Light penetration within a clear water body. From Moore (1978, p. 458). Reproduced by permission of the Environmental Research Institute of Michigan.

bodies (unlike those of land features) are determined by transmittance rather than surface characteristics alone. In the blue region, light penetration is not at its optimum, but at slightly longer wavelengths; in the blue-green region, penetration is greater, and at these wavelengths the opportunity for recording features on the bottom of the water body are greatest. At longer wavelengths, in the red region, absorption of sunlight is much greater, and only shallow features can be detected. Finally, in the near infrared region, absorption is so great that only land–water distinctions can be made.

As impurities are added to a water body, its spectral properties change. Sediments are introduced both from natural sources and by mankind's activities. Such sediments consist of fine-textured silts and clays eroded from stream banks, or by water running off disturbed land, that are fine enough to be carried in suspension by moving water. As moving water erodes the land surface, or the shoreline, it carries small particles as suspended sediment; faster-flowing streams can erode and carry more, and larger, particles than can slower moving streams. As a stream enters a lake or ocean, the decrease in velocity causes coarser materials to be deposited. But even slow-moving rivers and currents can carry large amounts of such fine-textured sediments, as clays and silts, which can also be found in calm water bodies. Sediment-laden water is referred to as *turbid water;* scientists can measure turbidity by sampling the water body or using devices that estimate turbidity by the transparency of the water. One such device is the *Secchi disk,* a white disk of specified diameter that can be lowered on a line from the side of a small boat. Because turbidity decreases the transparency of the water body, the depth at which the disk is no longer visible can be related to sediment content. Nephometric turbidity units (NTU) are measured by the intensity of light that passes through a water sample. A special instrument uses a light

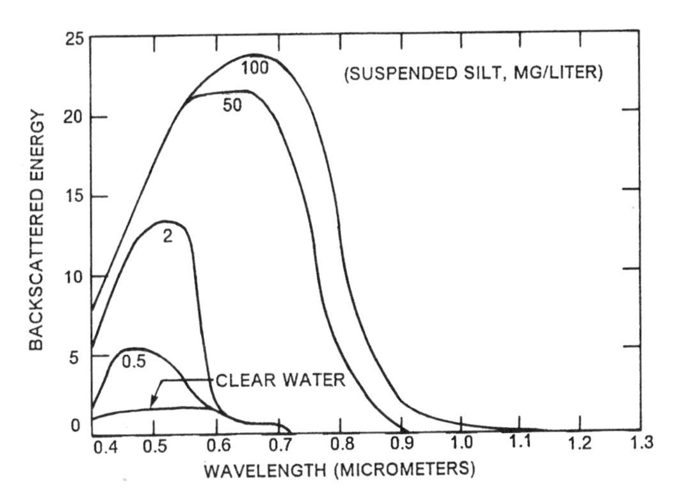


**FIGURE 18.4.** Effects of turbidity upon spectral properties of water. From Moore (1978, p. 460). Reproduced by permission of the Environmental Research Institute of Michigan.

beam and a sensor to detect differences in light intensity. Water of high turbidity decreases the intensity of the light in a manner that can be related to sediment content.

Thus, as sediment concentration increases, spectral properties of a water body change. First, its overall brightness in the visible region increases, so the water body ceases to act as a "dark" object but becomes more and more of a "bright" object as sediment content increases. Second, as sediment concentrations increase, the wavelength of peak reflectance shifts from a maximum in the blue region toward the green. The presence of larger particles means that the wavelength of maximum scattering shifts toward the blue-green and green regions (Figure 18.4). Therefore, as sediment content increases there tends to be a simultaneous increase in brightness and a shift in peak reflectance toward longer wavelengths, and the peak itself becomes broader, so that at high levels of turbidity, the color becomes a less precise indicator of sediment content. As sediment content approaches very high levels, the color of the water begins to approach that of the sediment itself (Plate 3). Plate 4 illustrates the turbidity patterns that can be recorded using CIR films. The green reflectance of turbid water appears as light blue on the processed CIR film (Chapter 2), which provides a sharp contrast in color with the dark blue or black of clear water bodies.

## 18.3. Spectral Changes as Water Depth Increases

Figure 18.3 shows spectral characteristics of sunlight as it penetrates a clear water body. Near the surface, the overall shape of the curve resembles the spectrum of solar radiation (Figure 2.5), but the water body increasingly influences the spectral composition of the light as depth increases. At a depth of 20 m little or no infrared radiation is present, as the water body is an effective absorber of these longer wavelengths. At this depth, only blue-green wavelengths remain—these wavelengths are therefore available for scattering back to the surface, from the water itself, and from the bottom of the water body.

The attenuation coefficient ($k$) describes the rate at which light becomes dimmer as depth increases. If $E_0$ is the brightness at the surface, the brightness at depth $z$ ($E_z$) is given by

$$E_z = E_0 e^{-kz} \qquad \text{(Eq. 18.1)}$$

In hydrologic studies, the influence of the atmosphere can be especially important. The atmosphere, of course, alters the spectral properties of incident radiation and also influences the characteristics of the reflected signal. Although these influences are also present in remote sensing of land surfaces, they assume special significance in hydrologic studies, in part because such studies often depend on subtle spectral differences (easily lost in atmospheric haze) and also perhaps because much of the hydrologic information is carried by the short wavelengths that are most easily scattered by the atmosphere.

Water bodies are typically dark, so the analyst must work with a rather restricted range of brightnesses relative to those available for study of land surfaces (Figure 18.5). As a result, analysts who specialize in remote sensing of water bodies must devote special attention to the radiometric qualities of the remotely sensed data. Typically, data to be an-

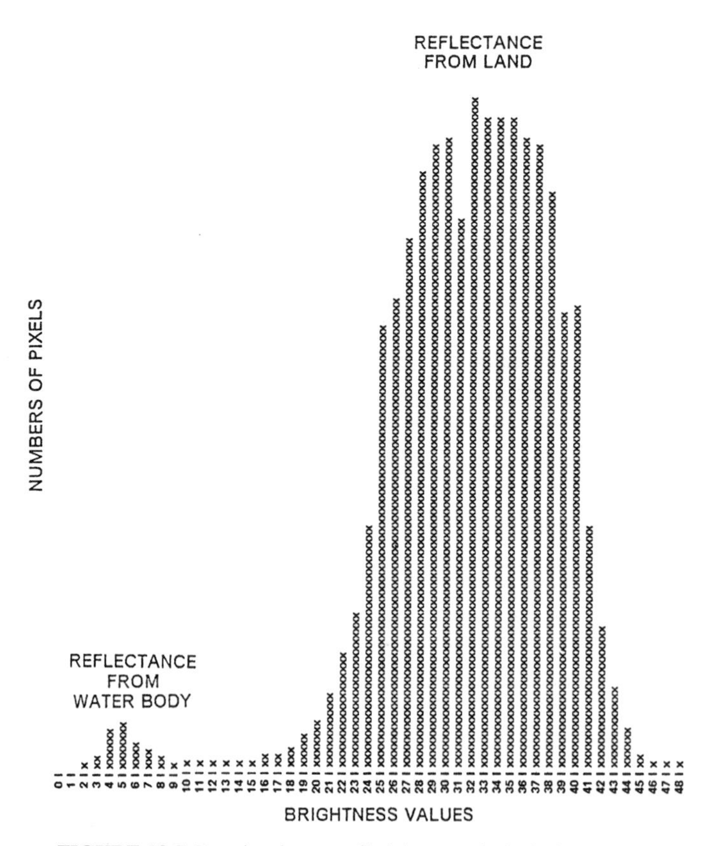


**FIGURE 18.5**. Restricted range of brightnesses in hydrologic setting.

alyzed for hydrologic information are examined carefully to assess their quality, the effects of the atmosphere, and sun angle. Geometric preprocessing is used with caution to avoid unnecessary alteration of radiometric qualities of the data. In some instances, analysts calculate average brightness over blocks of contiguous pixels to reduce transient, noisy effects of clouds and whitecaps. Or, sometimes several scenes of the same area, acquired at different times, can be used to isolate permanent features (such as shallows and shoals) from temporary features (including waves and atmospheric effects). Also, it is often advantageous to estimate original radiometric brightnesses from digital values, as a means of accurately assessing differences in color and brightness.

## 18.4. Location and Extent of Water Bodies

Remote sensing provides a straightforward means to map the extent of water bodies, to inventory area occupied by open water, and to monitor changes in water bodies over time. Although such tasks may seem simple, there are numerous situations in which simple determination of the land–water boundary can be very important. The study of coastal erosion requires accurate determination of shoreline position and configuration at several dates. Likewise, comparisons of shoreline position before and after flooding permits measurement of areas flooded, as well as determination of locations of flooded areas. Such information can be difficult to acquire by conventional means.

Determination of the land–water body is usually easiest in the near infrared region, where land, especially if vegetated, is bright and open water dark. Usually it is possible to determine a sharp contact between the two (Figure 18.6). With Landsat 1 to 3 MSS data, the contrast is usually clear on bands 6 or 7. With photography, black-and-white infrared film is usually satisfactory. CIR film is suitable if the water is free of sediment. However, be-

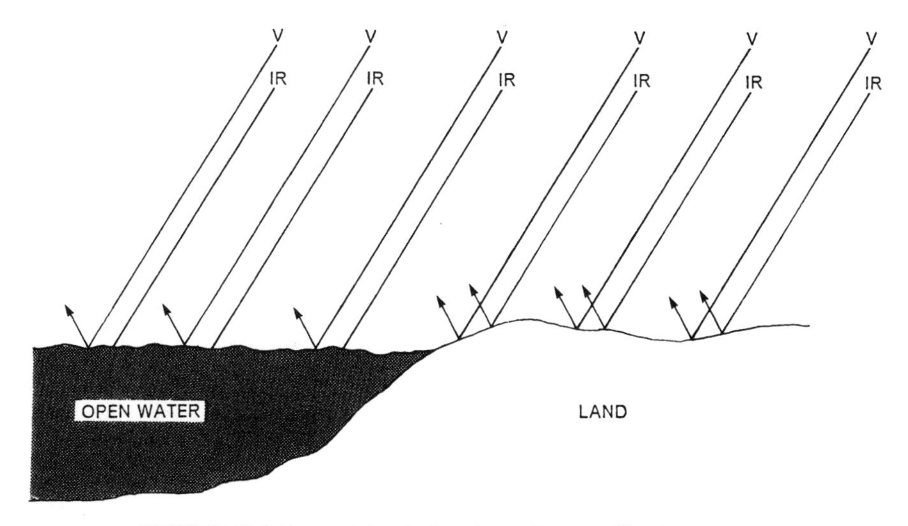


**FIGURE 18.6.** Determining the location and extent of land–water contact.

cause CIR film is sensitive to green light, turbid water will appear bright, thereby preventing clear discrimination of the land–water boundary.

In the early 1970s Landsat MSS data were used to inventory impounded water throughout the United States. The failure of earth dams at Rapid City, South Dakota, in June 1972, and Buffalo Creek, West Virginia, in February 1972 produced catastrophic loss of life and property. In both instances poorly constructed earth dams failed after periods of unusually heavy rainfall, which both increased the volume of impounded water and weakened the earthen dams. In August 1972, the U.S. Congress passed Public Law 92-367 requiring federal inspection of all dams impounding water bodies exceeding specified sizes. Because there existed no comprehensive record of such water bodies, NASA conducted a study to determine the feasibility of using Landsat MSS data as the means of locating water bodies impounded by dams requiring inspection under Public Law 92-367. NASA's method required selection of GCPs (Chapter 10) to register the data to accurate maps. A density slice technique separated water bodies from land areas; a printed output showing water pixels could then be matched to maps to locate impoundments not already delineated on the maps.

Philipson and Hafker (1981) studied application of Landsat MSS data to delineation of flooded areas. They concluded that simple visual interpretation of MSS band 4 images was as accurate as manual interpretation of multiband composite images and as accurate as digital analysis of MSS band 4 and combinations of both bands 2 and 4.

They experienced problems in applying the digital approach because of difficulties in registering the flooded and unflooded scenes. In addition, turbidity associated with flooding increased band 4 brightness to the extent that there was insufficient contrast between the two dates to unambiguously separate flooded and unflooded areas. Nonetheless, their approach is worth consideration. For their digital interpretation they defined a ratio that compares the brightness values in band 4 on the two dates:

$$\frac{\text{(MSS 1, unflooded)} - \text{(MSS 4, flooded or unflooded)}}{\text{(MSS 1, unflooded)}} \qquad \text{(Eq. 18.2)}$$

Because open, clear water is very dark in band 4, the ratio approaches 1 for flooded areas and 0 for unflooded areas. For example, if dry land has digital values near 35 in both bands, for flooded areas the ratio is $(35 - 0)/35 = 1$, and for unflooded areas, $(35 - 35)/35 = 0$. Although Philipson and Hafker found that when water turbiditity was high the procedure was not always successful, there are other procedures that can identify turbid water and separate such areas from unflooded land.

## 18.5. Roughness of the Water Surface

Figure 18.7 shows spectra for calm and wave-roughened surfaces in the visible and the near infrared. Wave-roughened surfaces are brighter than are smoother surfaces. Calm, smooth water surfaces direct only volume-reflected radiation to the sensor, but rough, wavy water surfaces direct a portion of the solar beam directly to the sensor. As a result, wavy surfaces are much brighter, especially in the visible portion of the spectrum. McKim et al. (1984) describe a procedure that uses polarizing filters to separate the surface brightness from that of the water body itself, using high-resolution data.

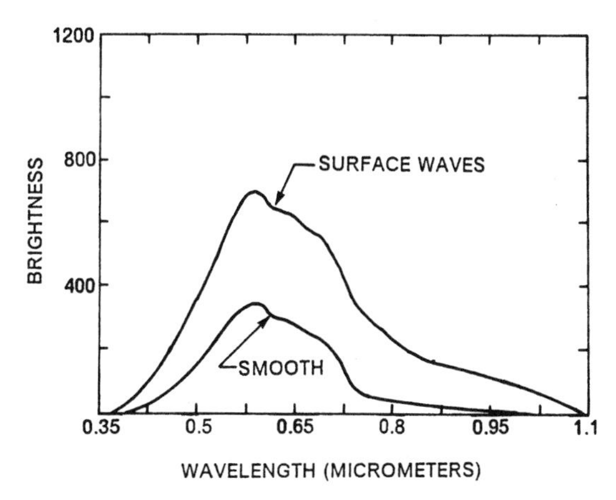


**FIGURE 18.7.** Spectra of calm and wind-roughened water surfaces (from laboratory experiments). Based on McKim et al. (1984, p. 358). Copyright 1984 by the American Society for Photogrammetry and Remote Sensing. Reproduced by permission.

The most intensive studies to study ocean roughness have used active microwave sensors (Chapter 6). *Sea state* refers to the roughness of the ocean surface, which is determined by wind speed and direction, as they interact locally with currents and tides. Sea state is an important oceanographic and meteorological quality because if studied over large areas and over time, it permits inference of wind speed and direction—valuable information for research and forecasts. Radars provide data concerning ocean roughness; the backscattering coefficient increases as wave height increases. Radars therefore provide a means of indirectly observing sea state over large areas and a basis for inferring wind speed and direction at the water's surface. If the imaging radar is carried by a satellite, as is the case with Seasat, ERS, RADARSAT, and the Shuttle Imaging Radar, the analyst has the opportunity to observe sea state over very large areas at regular intervals. These observations permit oceanographic studies that were impractical or very difficult using conventional data. Furthermore, timely information regarding sea state has obvious benefits for marine navigation and, can, in principle, form the basis for inferences of wind speed and direction that can contribute to meteorological data. Direct observation of sea state is relatively straightforward, but because ocean areas are so large and conditions so rapidly changeable, the usual observations from ships in transit are much too sparse to provide reliable data.

Rigorous study of sea state by radar started soon after World War II and continued during the 1950s and 1960s, culminating with the radars carried by Seasat and the Space Shuttle. Although many experiments have been conducted using a variety of microwave instruments, one of the most important broad-scale sea state experiments was based on the Seasat A project, which used several instruments to monitor the earth's oceans. Seasat, ERS, RADARSAT, and SIR (Chapter 6) provided an opportunity to observe large ocean areas on a repetitive basis.

Calm ocean surfaces, with waves that are small relative to the radar wavelength, act as specular reflectors and appear as dark regions on the image, as energy is reflected away from the antenna. As wind speed increases, the ocean surface becomes rougher and acts more like a diffuse reflector causing bright regions on the imagery (Figures 18.8 and 18.9). Because radar wavelength and system geometry are known, the received signal can form the basis for estimates of wave height and velocity.

In a test in the Gulf of Alaska, scientists found agreement between data derived from the Seasat SAR and those observed at the ocean surface. Wave direction, height, and wavelength were all estimated from the Seasat SAR imagery. Waves were most accurately observed when their direction of travel was within 25° of the look direction. Other studies have shown the SAR data to be useful for studying other kinds of ocean swells and currents.

## 18.6. Bathymetry

Information concerning water depth and configuration of the ocean floor is one of the most basic forms of hydrographic data. Bathymetry is especially important near coastlines, in harbors, and near shoals and banks, where shallow water can present hazards to navigation and where changes can occur rapidly as sedimentation, erosion, and scouring of channels alters underwater topography.

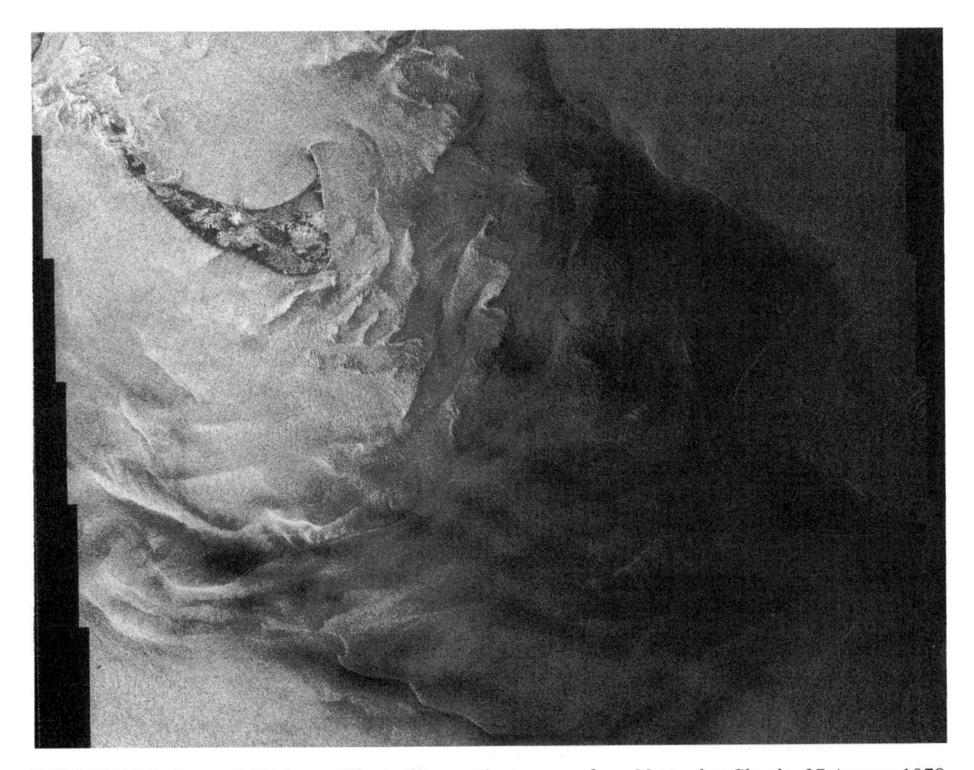

**FIGURE 18.8.** Seasat SAR image illustrating rough ocean surface: Nantucket Shoals, 27 August 1978. Image courtesy of Jet Propulsion Laboratory, Pasadena, California.

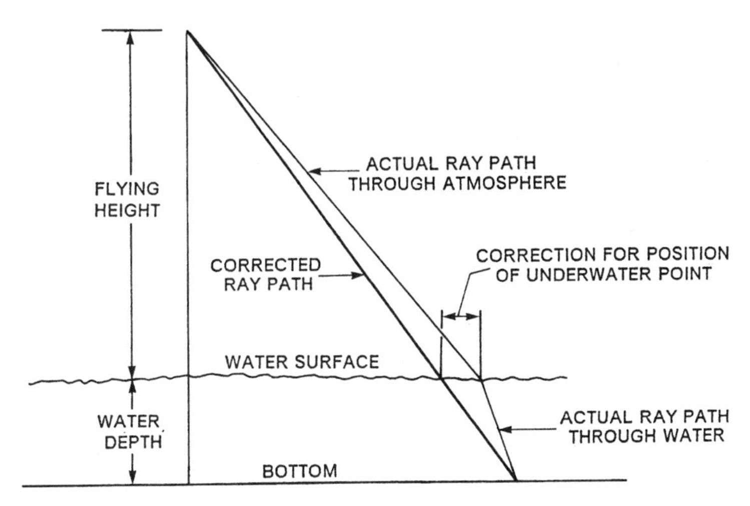


**FIGURE 18.9.** Bathymetry by photogrammetry.

Water depth can be measured by instruments carried on vessels, especially by acoustic (sonar) instruments that measure depths directly below the vessel. Ideally, bathymetric maps should be compiled from a more or less uniform network of depth measurements, rather than from a limited set of data from traverses. Although modern side-scanning sonar can yield a continuous surface of depth information, aerial imagery forms an important means of mapping subsurface topography in shallow water.

Photogrammetry can be applied to bathymetric measurements if high-quality, large-scale, photography is available and if the water is clear. Filters have been used to separate radiation in the spectral region 0.44 μm to 0.54 μm, where solar energy is most easily transmitted by clear water. From Figure 18.3 it can be seen that in principle, sunlight can penetrate to depths of about 20 m. Aerial photographs record information only from relatively shallow depths, although depths as deep as 16 m have been mapped using aerial photographs.

Special problems in applications of photogrammetry to bathymetry include estimation of differences in refraction between air and water (Figure 18.9), and the difficulty in acquiring a good set of underwater control points. Furthermore, mapping is difficult in regions far removed from the shoreline, because of the difficulty in extending control across zones of deep water. Nonetheless, photogrammetric methods have been successfully used for mapping zones of shallow water.

Water depth can also be estimated by observing refraction of waves as they break in shallow water. As a wave approaches the shore, it begins to break at a depth that is determined by its velocity, wavelength, and period. If velocities and wavelengths can be measured on successive overlapping photographs, it is possible to calculate an estimate of depth. The method works well only when the water body is large, and if the bottom drops slowly with increasing distance from shore. The wave refraction method is not as accurate as other procedures, but it does have the advantage of applicability in turbid water, when other methods cannot be applied.

### *Multispectral Bathymetry*

Depth of penetration of solar energy varies with wavelength. Longer wavelengths are absorbed most strongly; even modest depths—perhaps 2 m or so—of clear water will absorb all infrared radiation so that longer wavelengths convey no information concerning depth of water or the nature of the bottom. Visible radiation penetrates to greater depths but is still influenced by wavelength-dependent factors. For clear water, maximum penetration occurs in the blue–green region, which is approximated by Landsat MSS 1, although TM 1 is likely to be more effective for bathymetric studies (Chapter 5). Polcyn and Lyzenga (1979) report that for their study region in the Bahamas under optimum conditions, maximum penetration for MSS 1 is as deep as 25 m, and about 6 m for MSS 2. (For consistency in designation, Landsats 4 and 5 MSS numbering is used here.)

Intensity of radiation decreases exponentially with depth (i.e., brightness decreases as depth increases). Thus MSS 1 (or TM 1; both are positioned at or near the spectral region of maximum penetration) should be useful for estimating depth. Dark values suggest that the bottom is deep, beyond the 20-m range represented in Table 18.2. Bright values suggest that the bottom is near the surface (Figure 18.10). Many depth extraction algorithms use a logarithmic transformation of MSS 1 brightness as a means of incorporating our knowledge that brightness decreases exponentially with depth.

Polcyn and Lyzenga (1979) defined a depth-finding algorithm:

$$L_{sen} = L_{atm} + T_{atm}(L_{surf} + L_{wat}) \quad \text{(Eq. 18.3)}$$

where $L_{atm}$ is the radiance contributed by the atmosphere, $T_{atm}$ is atmospheric transmission, $L_{surf}$ is radiance from the water surface, and $L_{wat}$ is radiance from beneath the water surface. $L_{wat}$ for a specific band $i$ is usually designated as $L_i$, and has been estimated as

$$L_i = L_{w(i)} + (L_{b(i)}\ L_{w(i)})\ e^{-2k_i z} \quad \text{(Eq. 18.4)}$$

where $i$ is a specific spectral band, $L_{w(i)}$ is the deep-water radiance for band $i$, $L_{b(i)}$ is the bottom reflectance for band $i$, and $z$ is water depth.

$L_w$ is defined as the energy reflected from a water body so deep that incident solar radiation is absorbed, and the energy observed by the sensor is so dim that it cannot be distinguished from sensor and environmental noise. Values for $L_w$ vary depending on sensor sensitivity, spectral band, and sensor resolution; but for a given sensor, $L_w$ is influenced also by atmospheric clarity, solar angle, and other factors. Therefore, for each spe-

**TABLE 18.2. Logarithmic Transformation of Brightnesses**

| $x$ | ln $(x)$ | Brightness | Depth |
|---|---|---|---|
| 2 | 0.69 | Dim | Deep |
| 6 | 1.79 | | |
| 10 | 2.30 | | |
| 14 | 2.64 | | |
| 18 | 2.89 | | |
| 22 | 3.09 | | |
| 26 | 3.26 | Bright | Shallow |

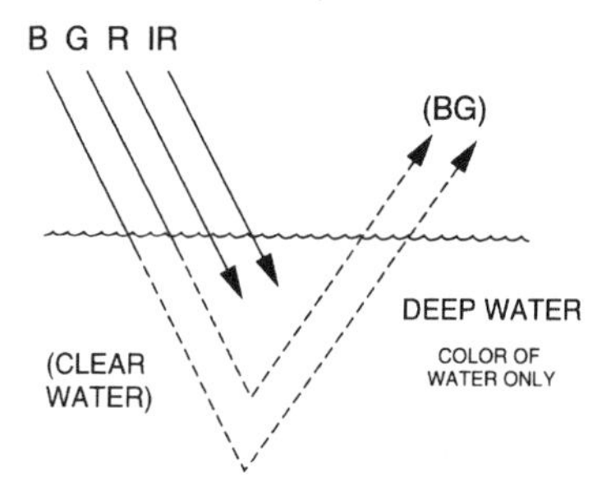


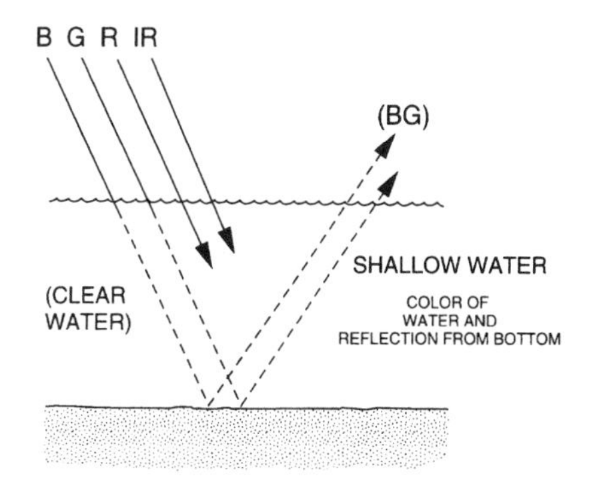


**FIGURE 18.10.** Multispectral bathymetry. Shallow water bodies filter radiation reflected from the bottom; the balance between spectral bands observed at the sensor reveals the approximate depth of the water body. See Figure 18.11.

cific study $L_w$ must be estimated, usually by selecting a set of pixels known to correspond to depths well beyond the range of penetration of solar energy.

Jupp et al. (1985) have summarized published values to produce the following estimates for maximum penetration of solar radiation within clear, calm, ocean water under a clear sky:

- MSS band 1: 15–20 m (50–60 ft.)
- MSS band 2: 4–5 m (13–17 ft.)
- MSS band 3: 0.5 m (1.5 ft.)
- MSS band 4: 0

As expected from inspection of Figure 18.4, penetration is greatest within the blue-green region of MSS band 1 and least in MSS band 4, where solar radiation is totally absorbed. It is possible to use this information to designate each MSS pixel as a member of a specific depth class.

If we find the numeric difference between the observed brightness in band $i$ ($L_i$) and the deep-water signal for the same band ($L_{w(i)}$), then the quantity ($L_i - L_{w(i)}$) measures the degree to which a given pixel is brighter than the dark pixels of open water.

If the value for band 1 exceeds the band 1 deep-water limit, but all other pixels are at deep-water values for their respective bands, then it is known that water depth is within the 15- to 20-m range (Figure 18.11). If both band 1 and band 2 are brighter than their respective values for $L_w$, but bands 3 and 4 are at their $L_w$ values, then water depth is within the 4- to 5-m range suggested above. If a pixel displays values in bands 1, 2, and 3 that all exceed their respective thresholds for $L_w$, then water depth is likely to fall within the 0.5-

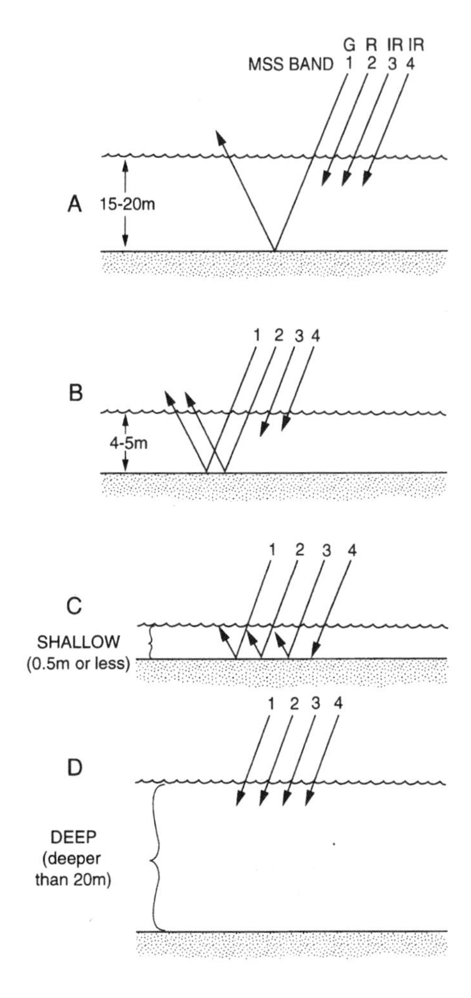


**FIGURE 18.11.** Multispectral bathymetry using Landsat MSS data. The relative brightnesses of the several spectral bands indicate depth classes, as discussed in the text.

m range. Finally, if all four bands are at their threshold values, the pixel corresponds to a depth beyond the 20-m limit. Because band 4 values for water pixels should always be very dim, bright values in this band suggest that water areas may have been delineated incorrectly, marsh vegetation rises above the water surface, or other factors have altered the spectral properties of the water.

This procedure assigns pixels to depth classes but does not provide quantitative estimates of depth, which, of course, would be more useful. Because of the exponential effect of the attenuation coefficient (equation 18.1) brightnesses of water bodies are not linearly related to depth. That is, we cannot observe a given brightness in band *i*, then directly match that brightness to a specific depth. Lyzenga (1981) has attempted to solve this problem by devising a relationship that creates a new variable, $x_i$, for each band *i*; $x_i$ is linearly related to water depth and therefore can be used to estimate depth within a given depth zone:

$$x_i = \ln (L_i - L_{\mathrm{w}(i)}) \qquad \text{(Eq. 18.5)}$$

Thus, $x_i$ is defined as the natural logarithm of the difference between observed brightness in land $i$ ($L_i$) and the deep-water reflectance ($L_{w(i)}$) in the same band. Table 18.2 shows that the logarithmic transformation enhances the significance of dim values from deep water. In Table 18.2, $x$ represents a hypothetical set of digital brightnesses observed by a sensor; ln($x$) does not itself measure depth but is linearly related to water depth so that the investigator can define the local relationship between $x_i$ and water depth.

Some methods have relied on estimates of $k$, which varies with wavelength and depends on local effects that influence the clarity of the water. Polcyn and Lyzenga (1979) used $k = 0.075\ m^{-1}$ for MSS 1 and $k = 0.325\ m^{-1}$ for MSS 2 for their studies of water depths in the Bahamas. They were able to use this algorithm to achieve accuracies of 10% at depths of about 20 m (70 ft.). Under ideal conditions, they measured depths as deep as 40 m (130 ft.).

These procedures seem to be effective provided that the water body is clear (otherwise turbidity contributes to brightness) and that the bottom reflectance is uniform. If attenuation of the water, and bottom reflectivity are known, accuracies as high as 2.5% have been achieved (although typical accuracies are lower). If the bottom reflectance is not uniform, differences in brightness will be caused not only by differences in depth but also by differences in reflectivity from the bottom materials. Because such differences are commonly present, due to contrasting reflectances of different sediments and vegetation, it is often necessary to apply procedures to adjust for differing reflectances from the bottom of the water body. The effectiveness of this procedure varies with sun angle (due to variations in intensity of illumination) and is most effective when data of high radiometric resolution are available.

In some instances ratios of two bands may remove differences in bottom reflectivity. Lyzenga (1979) defines an index that separates different reflectivities using different spectral bands. His index, calculated for each pixel, can be used to classify pixels into different bottom reflectivity classes, which can then each be examined, as classes, to estimate water depth.

## 18.7. Landsat Chromaticity Diagram

Colorimetry—the precise measurement of color—is of significance in several aspects of remote sensing but is especially significant in applications of remote sensing to studies of water quality. Even casual observation reveals that water bodies differ significantly in color, yet it can be very difficult to accurately measure these color differences and to relate color to the variables we wish to study.

Specification of color is not as self-evident as it might seem, and scientists have established several alternative systems for describing color. Most establish a three-dimensional color space in which three primaries (red, green, and blue) form three orthogonal axes. Unique specification of a color requires identification of a location within this color space, giving hue (dominant wavelength), saturation (purity), and value (brightness) of a given color (Chapter 2). These axes define *tristimulus space,* in which any color can be represented as a mixture of the three primaries—that is, as a location within the coordinate system represented in Figure 18.12. Such a system defines a color solid, which, in principle, permits specification of any color. For many purposes, it is convenient to define a two-dimensional plane within the color solid by setting:

$$R = \frac{r}{r+g+b} \qquad G = \frac{g}{r+g+b} \qquad B = \frac{b}{r+g+b} \qquad \text{(Eq. 18.6)}$$

The values *R, G,* and *B* define locations within a two-dimensional surface that shows variations in hue and saturation but does not portray differences in brightness. In effect, *R, G,* and *B* have been standardized by dividing by total brightness in each primary without conveying information concerning absolute brightness. Thus colors represented by this convention vary only with respect to hue and saturation because brightness information has been removed by standardization. For example, *R* specifies *percent red,* but does not convey information concerning whether it is a percentage of a very bright color or only a dark color.

In 1931, the *Commission Internationale de l'Eclairage* (CIE) (International Commission for Illumination) met in England to define international standards for specifying color. The CIE system defines a tristimulus space as follows:

$$x = \frac{X}{X+Y+Z} \qquad y = \frac{Y}{X+Y+Z} \qquad z = \frac{Z}{X+Y+Z} \qquad \text{(Eq. 18.7)}$$

where $x$, $y$, and $z$ correspond approximately to the red, green, and blue primaries, and $X$, $Y$, and $Z$ are the CIE chromaticity coordinates. Also, $x + y + z = 1$.

The three CIE primaries are "artificial" because, as a matter of convenience, they are defined to facilitate mathematical manipulation and do not correspond to physically real colors. Because $x,y$, and $z$ sum to 1.0, only two are necessary to uniquely specify a given color, as the third can be determined from the other two. Therefore the CIE chromaticity diagram has only two axes, using the $x$ and $y$ coordinates to locate a specific color in the diagram.

Figure 18.12 shows the CIE chromaticity diagram. The vertical axis defines the $y$ (green) axis, and the horizontal axis defines the $x$ (red) primary, each scaled from 0 to 1.0 (to portray the proportion of total brightness in each band). At coordinates $x = 0.333$ and $y = 0.333$ (and, of course, $z$ must equal 0.333), all three primaries have equal brightness and the color is pure white. Colors to which the human visual system is sensitive are inside the horseshoe shaped outline in Figure 18.12. The white colors near the center are "without color" as there is no dominant color; purity (*saturation*) increases toward the edges of the horseshoe where colors are "fully saturated." The line at the base of the horseshoe, known as the *purple line*, locates *nonspectral* colors. (Nonspectral colors are those that do not appear in a spectrum of colors because they are formed by mixtures of colors from opposite ends of the spectrum.) Spectral colors are positioned in sequence around the curved edge of the horseshoe.

Because each of the $x$, $y$, and $z$ primaries has been defined by dividing by the total brightness to yield the percentage brightness in each primary, they convey hue information without brightness information. That is, colors have been standardized to remove brightness, leaving only hue and saturation information. Because $x$, $y$, and $z$ sum to 1.0, only two are necessary to specify a specific hue, as the third can be determined from the other two. Therefore, the CIE color diagram uses only two axes, usually the $x$ and $y$ primaries, as a means of specifying color. The CIE diagram is useful for precisely specifying

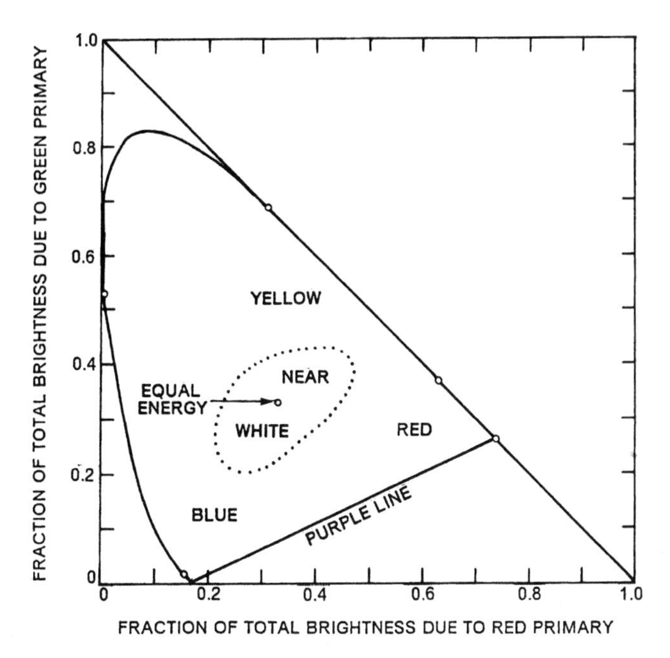


**FIGURE 18.12.** CIE chromaticity diagram.

color and for transferring color specifications from one context to another. Note, once again, that the CIE diagram conveys differences in color without portraying differences in brightnesses of colors—that is, it shows only hue and saturation.

Alföldi and Munday (1977) have applied the chromaticity concept to analysis of Landsat data. They defined a *Landsat chromaticity diagram*, analogous to the CIE chromaticity diagram, except that it defines a color space using MSS bands 1, 2, and 3 instead of the red, green, and blue primaries used for the CIE diagram. Thus, infrared data from band 3 are substituted for the missing primary. Because MSS digital counts do not accurately portray absolute brightnesses, it is necessary to use the appropriate calibration data (Table 18.3, Chapter 4, and Appendix) to estimate the brightnesses in each band. (Here for consistency Landsats 4 and 5 designations are used to refer to data from Landsat 1.)

For example, digital data from the Landsat 1 MSS require the following transformation:

$$\frac{\text{MSS 1 digital value}}{127} \times 24.8 = x \qquad \text{(Eq. 18.8)}$$

$$\frac{\text{MSS 2 digital value}}{127} \times 20.0 = y \qquad \text{(Eq. 18.9)}$$

$$\frac{\text{MSS 3 digital value}}{127} \times 17.6 = z \qquad \text{(Eq. 18.10)}$$

where 127 represents the maximum digital value possible for the sensor, and the appropriate constants are taken from Table 18.3. $x$, $y$, and $z$ now represent coordinates for the Landsat chromaticity diagram.

**TABLE 18.3. Landsat MSS Calibration Data**

| | | Band | | | |
|---|---|---|---|---|---|
| | | 1 | 2 | 3 | 4 |
| Landsat 1 | Min. | 0.00 | 0.00 | 0.00 | 0.00 |
| | Max. | 24.8 | 20.0 | 17.6 | 15.3 |
| Landsat 2[a] | Min. | 1.0 | 0.7 | 0.7 | 0.5 |
| | Max. | 21.0 | 15.6 | 14.0 | 13.8 |
| Landsat 2[b] | Min. | 0.8 | 0.6 | 0.6 | 0.4 |
| | Max. | 26.3 | 17.6 | 15.2 | 13.0 |
| Landsat 3[c] | Min. | 0.4 | 0.3 | 0.3 | 0.1 |
| | Max. | 22.0 | 17.5 | 14.5 | 14.7 |
| Landsat 3[d] | Min. | 0.4 | 0.3 | 0.3 | 0.1 |
| | Max. | 25.9 | 17.9 | 14.9 | 12.8 |

*Note.* Values in $mW \cdot cm^{-2} \cdot sr^{-1} \cdot \mu m^{-1}$. From Markham and Barker (1986). MSS bands 1, 2, 3, and 4 were formerly designated as 4, 5, 6, and 7.

[a]For data processed before 16 July 1975.
[b]For data processed after 16 July 1975.
[c]For data processed before 1 June 1978.
[d]For data processed after 1 June 1978.

The Landsat chromaticity coordinates can be used to assess the turbidity of water bodies imaged by the MSS. Figure 18.13 shows the Landsat chromaticity diagram, which is identical to the CIE chromaticity diagram except for the use of MSS bands 4, 5, and 6 as substitutes for the CIE primaries. "E" signifies the *equal radiance point*, where radiances in MSS bands 4, 5 and 6 are equal. (Note that the labels in Figure 18.12 do not apply to the Landsat chromaticity diagram, as it does not use the CIE primaries.) The curved line extending to the right of the equal radiance point is the experimentally defined locus that defines the positions of water pixels as their radiances are plotted on the diagram. Pixels with positions near the right end of this line represent water bodies with clear water. As chromaticity coordinates shift closer to the other end of the line, near the equal radiance point, the spectral properties of the pixels change, indicating increases in turbidity as the pixel position approaches the left-hand end of the line. This line was defined experimentally by Alföldi and Munday for their study area, using sediment samples collected at the time the imagery was acquired. (For detailed studies, it is important to define a specific locus for each individual study region, but for illustrative purposes it is sufficient to use their line as a general-purpose locus.)

Thus, in Figure 18.14, the position of pixel A reveals a clear water body, and pixel B represents a turbid water body. As pixels known to represent water bodies drift away from the line (pixel C is an example), so do we suspect that atmospheric haze or thin clouds have altered their radiances. Although our inspection of the chromaticity diagram is essentially a qualitative analysis of pixel positions on the chromaticity diagram, Alföldi and Munday have conducted more quantitative studies, using observed values of suspended sediment to calibrate the technique.

Table 18.4 lists Landsat MSS data for the same water body observed at several dates. The first two pixels are worked through as examples, but the others are left for the student to work as problems and to plot on Figure 18.15. In practice, of course, these calculations

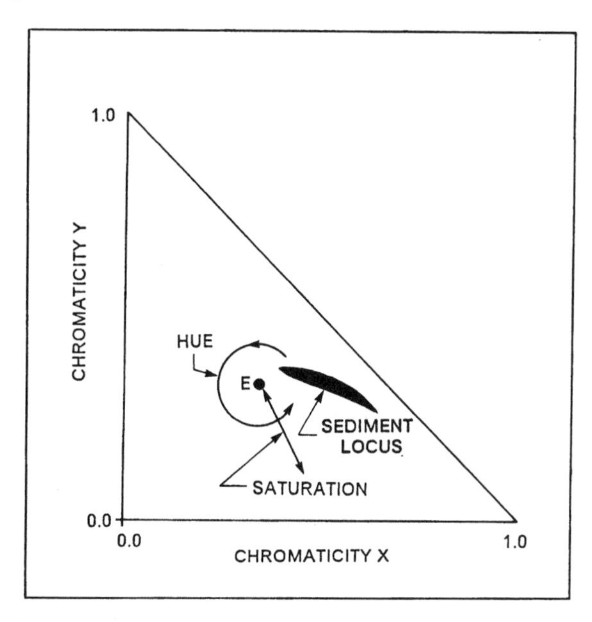


**FIGURE 18.13.** Landsat MSS chromaticity space. The sediment locus is experimentally derived and may vary slightly from region to region, or over time within a given region. The radial dimension (outward from E) presents spectral purity information, and radial shifts in the position of the locus are attributed to atmospheric scattering, thin clouds, air pollution, or whitecaps. Diagram based on Alföldi and Munday (1977).

are easily accomplished by computer. Also, it is often convenient to use groups of contiguous pixels as a means of reducing noisy variability that might be caused by system noise, whitecaps, and other spectral anomalies.

## 18.8. Drainage Basin Hydrology

A less obvious component of water resources concerns the management of land areas to promote orderly retention and flow of moisture over the land surface, and then within the stream system, so that flooding and erosion are minimized and an even, steady flow of moisture is maximized. The nature and distribution of land use and land cover within drainage basins influences runoff from the land surface and eventually also influences the flow of water within stream channels. Although there is no set relationship between these elements that can be said to form a universally "normal" balance, each climatic and ecological situation tends to be characterized by an established pattern of rainfall, evapotranspiration, and streamflow. Extensive changes to natural land cover patterns can disrupt this local balance and cause accelerated soil erosion, flooding, and other undesirable consequences. Hydrologists attempt to study such processes by constructing hydrologic models—mathematical simulations that use measurements of local rainfall, together with information concerning patterns of land use, soil, topography, and drainage, to predict the nature of the streamflow that might be caused by storms of a given magnitude. Such mod-

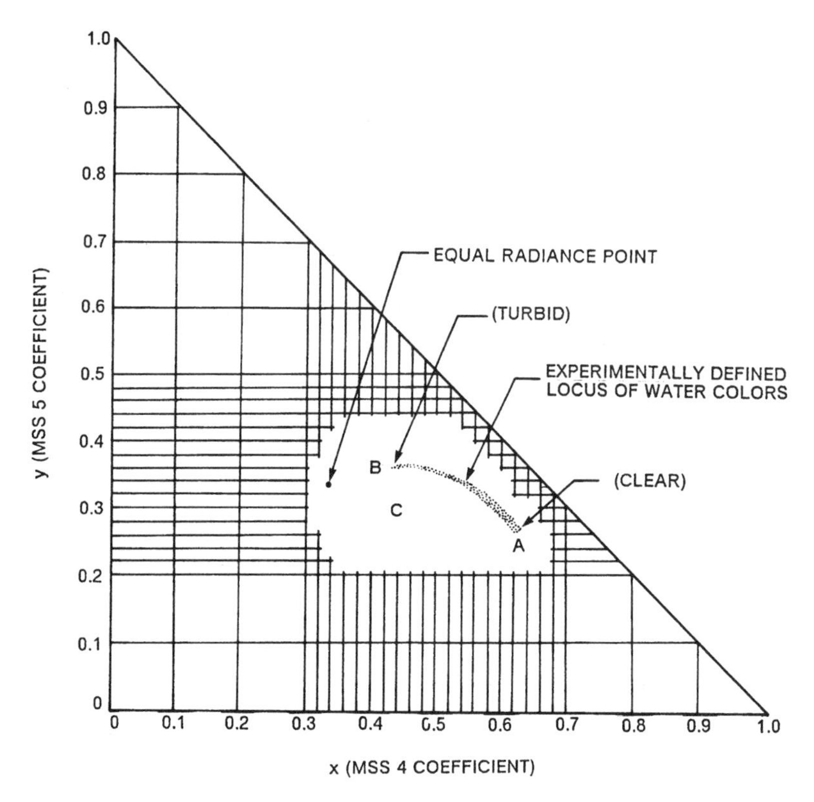


**FIGURE 18.14.** Landsat MSS chromaticity diagram. From Alföldi and Munday (1977, p. 336). Reproduced by permission of the Canadian Aeronautics and Space Institute.

els provide a means of assessing effects of proposed changes in land use (e.g., new construction or harvesting of timber) or in channel configuration because they can estimate the effects of such changes before they actually occur. Often such models can be evaluated and refined using data from watersheds that have very accurate records of rainfall and streamflow.

Hydrologic models require large amounts of detailed and accurate data, so many of the costs involved in applying a given model to a specific region arise from the effort required to collect accurate information concerning soil and land cover patterns—two of the important variables required for such models. Some of the information concerning land use can be derived from remotely sensed data. Manual interpretation of land use and land cover from aerial photographs (Chapter 16) is a standard procedure, and, of course, aerial photographs are a source of much of the information regarding topography and soils. Many studies have evaluated the use of image classification of digital remote sensing data (Chapter 11) as a source of land cover information for such models. Advantages include reduced costs and the ability to quickly prepare new information as changes occur.

Jackson et al. (1977) applied this kind of approach in a study of a small watershed in Northern Virginia. Their study used both manual interpretations of aerial photography and computer classifications of Landsat MSS data and was able to compare results based on

TABLE 18.4. Data for Chromaticity Diagram

| Date | MSS digital counts | Radiance | Sum | Chromaticity coordinates |
|---|---|---|---|---|
| May[a] | 36 33 24 07 | 7.02 5.20 3.33 | 15.55 | 0.451 0.334 |
| May[a] | 36 36 27 08 | 7.02 5.67 3.74 | 16.43 | 0.427 0.345 |
| Feb. | 30 29 19 03 | | | |
| Feb. | 29 32 17 03 | | | |
| Feb. | 21 16 22 12 | | | |
| Apr. | 28 21 14 04 | | | |
| Apr. | 28 21 14 05 | | | |
| Apr. | 31 22 15 03 | | | |
| May | 40 44 24 05 | | | |
| May | 40 44 25 06 | | | |
| May | 42 39 22 05 | | | |
| June | 37 28 26 08 | | | |
| June | 37 30 31 11 | | | |
| June | 37 30 29 12 | | | |
| Sept. | 26 25 16 03 | | | |
| Sept. | 28 25 16 03 | | | |
| Sept. | 27 25 16 02 | | | |
| Nov. | 18 09 05 01 | | | |
| Nov. | 16 11 06 02 | | | |
| Nov. | 18 11 07 01 | | | |

[a]As an illustration, positions of these two pixels are plotted on Figure 18.15. The student should use the remaining data as practice for application of the chromaticity technique.

MSS data with those of conventional estimates. Their results indicated that the use of Landsat data was much cheaper than the manual method and tended to produce results of comparable reliability.

## 18.9. Evapotranspiration

*Evapotranspiration* is defined as combined evaporation (moisture lost to the atmosphere from soil and open water) and transpiration (moisture lost to the atmosphere by plants). Evapotranspiration measures the total moisture lost to the atmosphere from the land surface, and as such it (together with other variables) forms an important variable in understanding the local operation of the hydrologic cycle.

However, accurate measurement of evapotranspiration is difficult and reliable data are usually available only for well-equipped meteorological stations and agricultural re-

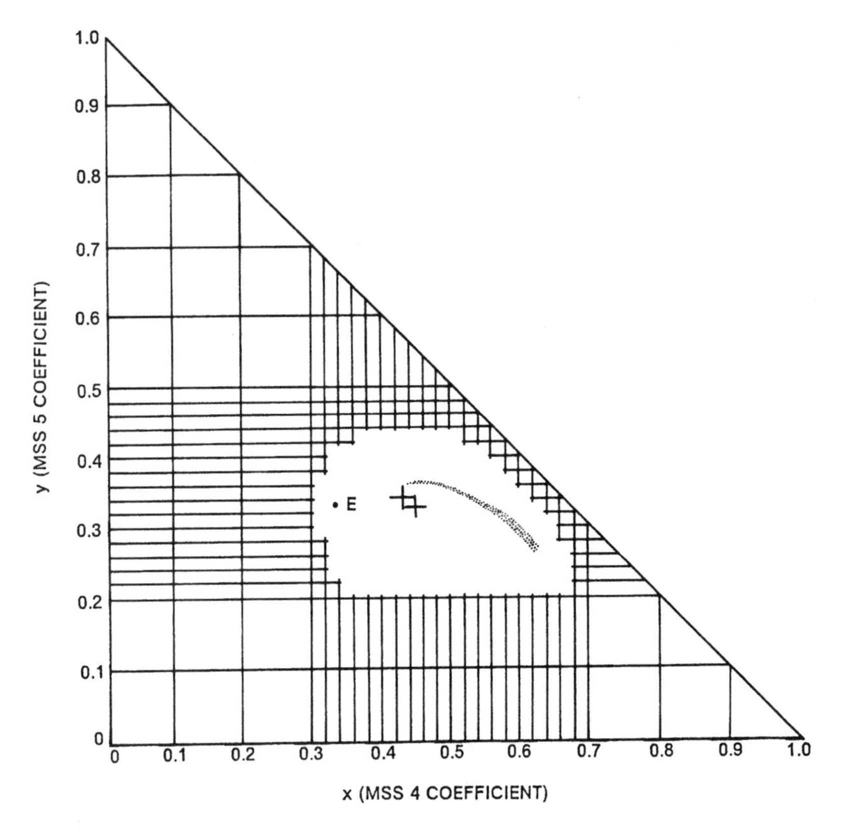


**FIGURE 18.15.** Data from Table 18.4 plotted in Landsat chromaticity space.

search facilities that are regularly attended by a trained staff. The vast majority of the earth's land area is not near such installations, so reliable measurements of evapotranspiration are not available. Evapotranspiration can, of course, be estimated using measurements of air temperature, wind speed, plant cover, cloudiness, and other variables, but such estimates often lack the detail and accuracy required for good environmental studies and also require data that are themselves difficult to obtain.

Remotely sensed data can form the basis for "spatially referenced" estimates of evapotranspiration—that is, estimates that apply to specific areas of small areal extent (Jackson, 1985). In contrast, estimates based on meteorological measurements can usually provide only a single estimate for a large region in the vicinity of a climate station. In effect, spatially referenced estimates provide detailed, independent estimates of evapotranspiration for each pixel.

Direct sensing of the soil surface using remote sensing instruments (especially passive) can provide estimates of soil moisture, or calculation of such estimates can be based on digital elevation data (described in detail previously in Chapter 17) that represent topographic elevation as a matrix of values, similar in concept to digital remote sensing images. For the elevation data, each pixel value represents the topographic elevation at the center of each pixel rather than the brightness of the energy reflected from the land at that pixel. From such data, a computer program can calculate the topographic slope at each pixel and also determine the aspect (the compass direction that the slope faces).

With the elevation data in hand, it is possible to overlay Landsat data and data from meteorological satellites, such as AVHRR (Chapters 16 and 20), so that all three forms of data register to one another. A classification of the Landsat data (as described in Chapter 11) can provide a map of the major vegetation classes (forest, pasture, bare ground, etc.) present. And the data gathered by meteorological satellites can be used to estimate temperature and total solar radiation received during a 24-hour interval. These vary depending on the slope and aspect of each pixel, as derived from the elevation data. From these separate forms of data it is then possible to estimate evapotranspiration for each pixel. With independent estimates of evapotranspiration for each pixel within a study area, the analyst can then prepare much more detailed hydrologic data. Jackson (1985) describes the estimation of evapotranspiration from remotely sensed data and observes that some difficulties still remain. Atmospheric degradation of the remotely sensed data can cause errors in the estimates, and the usual procedures cannot consider effects of factors such as wind speed or surface roughness. Also, estimates apply only to a specific time (when the remotely sensed data were acquired), although usually the requirement is to estimate evapotranspiration over intervals of a day, several days, or even longer. Therefore it is necessary to extend the estimates from a specific time to a longer interval—a process likely to introduce error. Nonetheless, it should be emphasized that further development of such procedures is one of the most interesting and practical applications of remote sensing, as such procedures can make major contributions to studies of hydrology, climatology, and agriculture in many of the earth's remote, inaccessible regions.

## 18.10. Manual Interpretation

The following examples illustrate how techniques of manual image interpretation (Chapter 4), applied with a knowledge of films and filters (Chapter 3), and of the spectral behavior of water bodies as outlined earlier in this chapter, can be used to derive information concerning water bodies and coastal features.

Figures 18.16 and 18.17 show the mouth of a tributary at the east bank of the Patuxent River, a tidal estuary in Maryland, south of Washington, D.C. The two images show the same scene as recorded by different film/filter combinations. The panchromatic image (Figure 18.16) shows the land area near the river as a pattern of cropped agricultural fields bordered by hedgerows and groves of trees. Here the banks of the Patuxent are fairly steep, and the land–water contact is well defined, but near the center of Figure 18.16 it is possible to see a sparsely vegetated patch of land that is periodically exposed and submerged as river and tidal influences change the water level of the river. The tributary carries a high sediment load, visible as the darker gray color that contrasts with the bright tone of the main channel.

In Figure 18.17 the region is depicted by black-and-white infrared film. Here the land–water distinction is especially clear as there is a sharp contrast between the dark of the open water and the bright infrared reflectance of the vegetated land areas. Some of the variations in vegetative cover are visible, but variations in tone within the water bodies (previously visible in Figure 18.16) are no longer visible.

**FIGURE 18.16.** Panchromatic photograph of the same area shown in Figure 18.17. Image courtesy of NASA.

We have in general a clearer view of the turbidity patterns that were at least partially visible in Figure 18.16. The Patuxent River itself carries a modest load of sediment, but the tributary has much higher levels, possibly due to runoff from the cultivated land that occupies most of its drainage basin. As it discharges into the Patuxent, the current carries the turbid water downstream in a plume that parallels the shoreline. Within the wetlands at the mouth of the tributary, the black-and-white infrared film portrays details of the differences in kinds and densities of wetland vegetation that are not clearly visible on the other image.

Plate 17 shows the mouth of the Chesapeake Bay, as photographed 8 June 1991 from the Space Shuttle. Land areas appear in silhouette due to underexposure to record hydro-

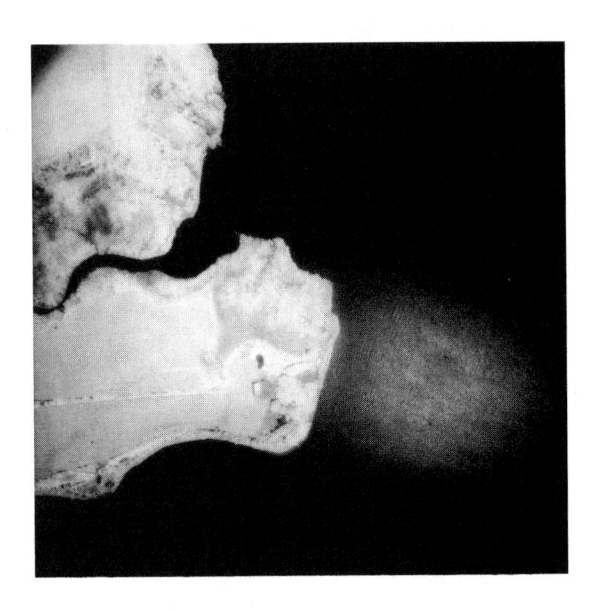

**FIGURE 18.17.** Black-and-white infrared photograph of the same area depicted in Figure 18.16. Image courtesy of NASA.

graphic features. As a result, the land–water interface is very sharply delineated. The Chesapeake Bay occupies the left-hand portion of the image; Norfolk, Virginia, and associated naval facilities occupy the lower left corner, but are not visible due to the underexposure. The Chesapeake Bay Bridge–Tunnel is visible as the dark line at the mouth of the bay. Just to the right of the bridge–tunnel, a curved front separates the waters of the bay from those of the Atlantic. The coastal waters of the Virginia coast are visually distinct from those of the bay, due to currents, surface oils, temperature, and wind and wave patterns. Near this front, wakes of several ships are visible.

Figures 18.18 through 18.21 show this same region from a different perspective. Thermal images of the Atlantic coast of the eastern shore of Virginia were acquired just after noon on 23 July 1972, using an infrared scanner. The setting is depicted in its broader context in Plate 17 and Figures 18.18 and 18.19; the coastline is protected by a series of barrier islands formed by sediment transported from the north and periodically reshaped by cur-

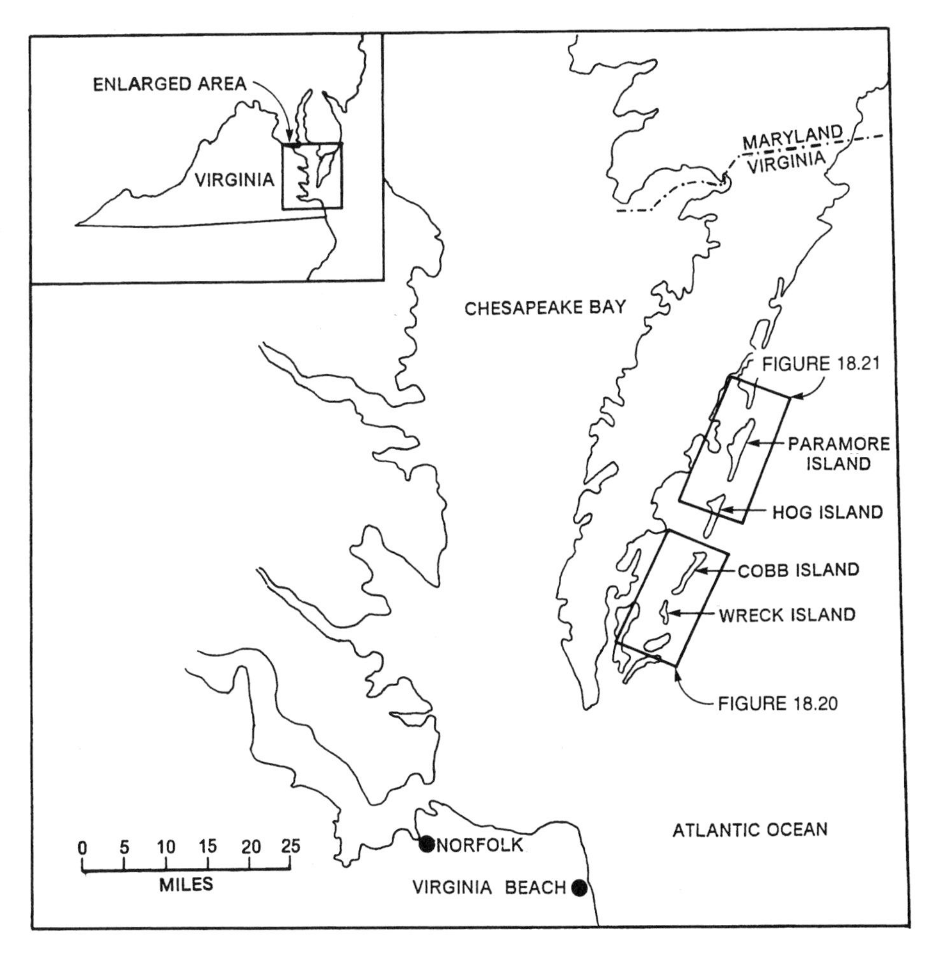


**FIGURE 18.18.** Eastern shore of Virginia: overview showing areas imaged in Figures 18.20 and 18.21.

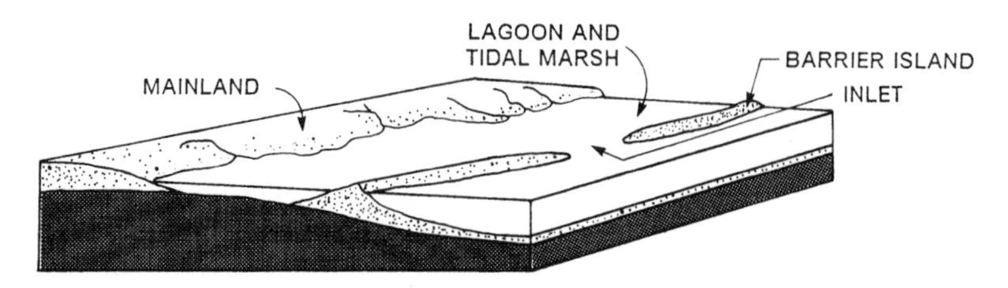


**FIGURE 18.19.** Eastern shore of Virginia: diagram of barrier islands and tidal lagoons.

rents and storms. The islands themselves are long, low strips of sand shaped by the action of both water and wind. At the edge of the ocean, a long beach slopes toward the water; the lowest sections of the beach are influenced every day by the effects of tides and currents. The upper, higher sections are affected only by the highest tides and strongest storms, so these regions may experience major changes only once or twice a year, or even less often. Inland from the upper beach ridges is wind-blown sand formed into dunes that are generally above the reach of waves, although very strong storms may alter these regions. The dunes are reshaped by wind, but their general configuration is often stabilized by grasses and small shrubs that cover much of this zone. Inland from the dunes, elevations decrease, and water again assumes the dominant influence in shaping the ecosystems.

Water from the tidal marshes between the islands and the mainland rises and falls

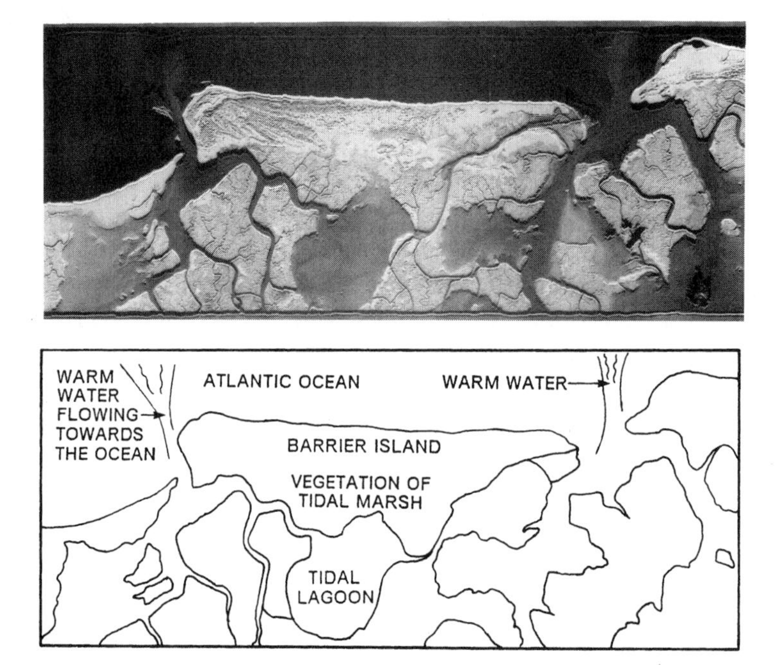


**FIGURE 18.20.** Thermal infrared image of eastern shore of Virginia, 23 July 1972, approximately 12:50 P.M.: Cobb Island and Wreck Island. Image courtesy of NASA.

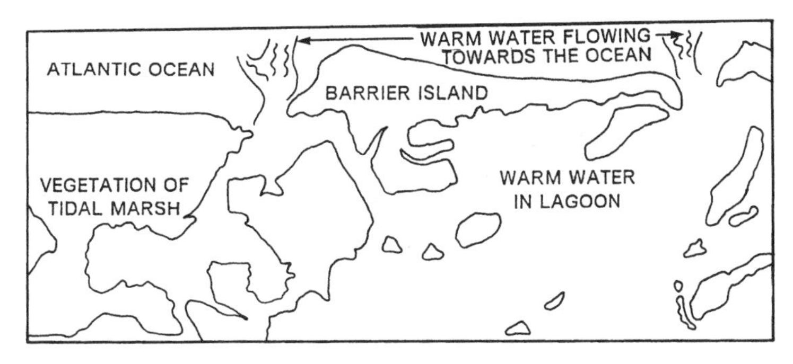


**FIGURE 18.21.** Thermal infrared image of eastern shore of Virginia, 23 July 1972, approximately 12:50 P.M.: Paramore Island. Image courtesy of NASA.

with the tides but is without the strong waves that characterize the seaward side of the islands. These tidal flats are covered with vegetation adapted to the brackish water, experience tidal fluctuations, and have poorly drained soils.

Many of these features are clearly visible on the infrared image of this region (Figure 18.20). The bright white strip along the seaward side of the island contrasts sharply with the cool, dark of the open ocean. This bright strip is the hot surface of the open sandy beach, which has received direct energy of the solar beam for several hours and now, at midday, is very hot. Inland from the beach the dunes are visible as a slightly darker region on the image with darker areas caused by shadowing. The topographic structure of the beach ridges is clearly visible in several regions. In the tidal marshes, the image appearance is controlled largely by vegetative cover, which gives a clear delineation of the edges of the open channel not usually visible on other images. The open water within the tidal lagoons has a lighter tone than that of the open ocean, indicating warmer temperatures. Here the shallow water and the restricted circulation have enabled the water in these areas to absorb the solar radiation it has been receiving now for several hours. Nonetheless, this water is still cooler than water flowing from the land surface; it is possible to see the bright (warm) plumes of water from the streams that flow into the tidal marshes from the mainland. Finally, tidal currents are clearly visible as the bright (warm) stream of water passing between the barrier islands to enter the darker (cooler) water of the open ocean. The tide is now flowing out (ebbing); compare some of the patterns visible on this image with those of a hydrographic chart of the same region.

Plate 18 provides another example of the use of remotely sensed data to study currents and tidal flow. Here the images show a portion of the Belgian port of Zeebrugge as observed using 12 multispectral channels encompassing portions of the visible, near infrared, and thermal regions. Image data have been processed to provide correct geometry and radiometry and, with the use of on-site observations, analyzed to reveal sediment content within the upper 1 m of the water column. Reds and yellows indicate high sediment content; blues and greens, low sediment content. The left image shows conditions at low tide; the right image, conditions at high tide. From such images it is possible to estimate the broad-scale patterns of sediment transport and deposition, and to plan efficient dredging and construction activities.

## 18.11. Summary

Hydrologic studies in general cover a broad range of subject matter, from the movement of currents in bodies of open water to the evaporation of moisture from a soil surface. Such studies can be very difficult, especially if the goal is to examine the changes in hydrologic variables as they occur over time and space, because the usual methods of surface observation gather data at isolated points or specific times. The great advantages of remote sensing in this context are the synoptic view of the aerial perspective and the opportunity to examine dynamic patterns at frequent intervals.

Yet remote sensing encounters difficulties when applied to hydrologic studies. Many of the standard sensors and analytical techniques have been developed for study of land areas and are not easily applied to the special problems of studying water bodies. Analyses often depend on detection of rather subtle differences in color, which are easily lost by effects of atmospheric degradation of the remotely sensed energy. Currents and other dynamic features may change rapidly, requiring frequent observation to record the characteristics of significance to the analyst. Important hydrologic variables, such as ground water, are not usually directly visible, and others, such as evapotranspiration, are not at all visible but must be estimated through other quantities that may themselves be difficult to observe. These problems and others mean that further development of hydrologic remote sensing is likely to be one of the most challenging research areas.

## Review Questions

1. List qualities of water bodies that present difficulties for those who study them using only surface observations collected from a ship. Identify those difficulties that are at least partially alleviated by use of some form of remote sensing.

2. Review Chapter 7 to refresh your memory on the qualities of Seasat, and Chapter 6 to find corresponding information for Landsat. List differences between the two systems (orbit, frequency of repeat observation, spectral bands, and so on), identifying those qualities that are particularly well or especially poorly suited for hydrographic applications.

3. It is probably best to compile bathymetric information using directly observed surface data because of their significance for navigation. Yet there are some special situations in which

use of remotely sensed data may be especially advantageous; can you identify at least two such circumstances?

4. Can you think of important applications of accurate delineation of edges of water bodies (i.e., simple separation of land vs. open water)? Be sure to consider observations over time as well as use of images from a single data.

5. Contrast SPOT (Chapter 6) and Seasat (Chapter 7) with respect to their usefulness for oceanographic studies. (Chapter 7 discusses only some of the Seasat sensors, but it provides enough information to conduct a partial comparison.)

6. Plot the data listed in Table 18.4 on the chromaticity diagram (Figures 18.14 and 18.15) and assess both the degree of turbidity (high, moderate, low) and the atmospheric clarity for each entry.

7. Write a short description of a design for a multispectral remote sensing system tailored specifically for collecting hydrologic information and accurate location of the edges of water bodies. Disregard all other applications. Suggest some of the factors that might be considered in choosing the optimum time of day for using such a system.

8. Outline the dilemma faced by scientists who wish to use preprocessing in the application of MSS or TM data for bathymetric information. Describe the reasons why a scientist would very much like to use preprocessing in some instances, as well as the counterbalancing reasons why he or she would prefer to avoid preprocessing.

9. Outline some of the ways that the methods of image classification (Chapter 11) might be useful for hydrologic studies. Outline also problems and difficulties that limit the usefulness of these methods for studies of water bodies.

## References

### General

Alföldi, T. T. 1982. Remote Sensing for Water Quality Monitoring. Chapter 27 in *Remote Sensing for Resource Management* (C. J. Johannsen and J. L. Sanders, eds.). Ankeny, IA: Soil Conservation Society of America, pp. 317–328.

Bhargava, D. S., and D. W. Mariam. 1991. Effects of Suspended Particle Size and Concentration on Reflectance Measurements. *Photogrammetric Engineering and Remote Sensing,* Vol. 57, pp. 519–529.

Curran, P. J., and E. M. M. Novo. 1988. The Relationship between Suspended Sediment Concentration and Remotely Sensed Spectral Radiance: A Review. *Journal of Coastal Research,* Vol. 4, pp. 351–368.

Fishes, L. T., Frank Scarpace, and Richard Thomson. 1979. Multidate Landsat Lake Quality Monitoring Program. *Photogrammetric Engineering and Remote Sensing,* Vol. 45, pp. 623–633.

Jackson, Ray D. 1985. Evaluating Evapotranspiration at Local and Regional Scales. *Proceedings of the IEEE,* Vol. 73, pp. 1086–1096.

Jackson, T. J., R. M. Ragan, and W. N. Fitch. 1977. Test of Landsat-Based Urban Hydrologic Modeling. *Journal of Water Resources Planning and Management Division, American Society of Civil Engineers,* Vol. 103, No. WRI, pp. 141–158.

Khorram, S. 1980. Water Quality Mapping from Landsat Digital Data. *International Journal of Remote Sensing,* Vol. 2, pp. 143–153.

Klemas, V., R. Sicna, W. Treasure, and M. Otley. 1973. Applicability of ERTS-1 Imagery to the Study of Suspended Sediment and Aquatic Forms. In *Symposium on Significant Results Obtained from Earth Resources Technology Satellite-1*. Greenbelt, MD: Goddard Space Flight Center, pp. 1275–1290.

Liedtke, T., A. Roberts, and J. Luternauer. 1995. Practical Remote Sensing of Suspended Sediment Concentration. *Photogrammetric Engineering and Remote Sensing,* Vol. 61, pp. 167–175.

McKim, H. L., C. J. Merry, and R. W. Layman. 1984. Water Quality Monitoring Using an Airborne Spectroradiometer. *Photogrammetric Engineering and Remote Sensing,* Vol. 50, pp. 353–360.

Moore, G. K. 1978. Satellite Surveillance of Physical Water Quality Characteristics. In *Proceedings of the Twelfth International Symposium on Remote Sensing of Environment*. Ann Arbor, MI: Environmental Research Institute of Michigan, pp. 445–462.

Nace, R. L. 1967. *Are We Running Out of Water?* U.S. Geological Survey Circular 586. Washington, DC: U.S. Government Printing Office, 7 pp.

Philipson, Warren R., and W. R. Hafker. 1981. Manual versus Digital Landsat Analysis for Delineating River Flooding. *Photogrammetric Engineering and Remote Sensing,* Vol. 47, pp. 1351–1356.

Schwab, D. J., G. A. Leshkevich, and G. C. Muhr. 1992. Satellite Measurements of Surface Water Temperatures in the Great Lakes. *Journal of Great Lakes Research,* Vol. 18, pp. 247–258.

Strandberg, C. 1966. Water Quality Analysis. *Photogrammetric Engineering,* Vol. 32, pp. 234–249.

Stumpf, R. P. 1992. Remote Sensing of Water Clarity and Suspended Sediment in Coastal Waters. *Proceedings of the First Thematic Conference on Remote Sensing for Marine and Coastal Environments. Proceedings of the International Society for Optical Engineering,* Vol. 1930, pp. 14363–1437.

### Chromaticity Analysis

Alföldi, T. T., and J. C. Munday. 1977. Progress Toward a Landsat Water Quality Monitoring System. In *Proceedings, Fourth Canadian Symposium on Remote Sensing.* Quebec, Canada: Canadian Remote Sensing Society of the Canadian Aeronautics and Space Institute, pp. 325–340.

Markham, B. L., and J. L. Barker. 1986. Landsat MSS and TM Post-calibration Dynamic Ranges, Exoatmospheric Reflectance, and At-Satellite Temperatures. *Landsat Technical Notes,* No. 1, pp. 3–8.

Munday, J. C. Jr., and T. T Alföldi. 1975. Chromaticity Changes from Isoluminous Techniques Used to Enhance Multispectral Remote Sensing Data. *Remote Sensing of Environment,* Vol. 4, pp. 221–236.

Munday, J. C., Jr., T. T. Alföldi, and C. L. Amos. 1979. Bay of Funday Verification of a System for Multidate Landsat Measurement of Suspended Sediment. In *Satellite Hydrology* (M. Deutsch, D. R. Wiesner, and A. Rango, eds.). Minneapolis, MN: American Water Resource Association, pp. 622–640.

Nelson, Ross. 1985. Reducing Landsat MSS Scene Variability. *Photogrammetric Engineering and Remote Sensing,* Vol. 51, pp. 583–593.

Stimson, A. 1974. *Photometry and Radiometry for Engineers*. New York: Wiley, 446 pp.

### Bathymetric Mapping

Hallada, W. A. 1984. Mapping Bathymetry with Landsat 4 Thematic Mapper: Preliminary Findings. In *Proceedings, Ninth Canadian Symposium on Remote Sensing.* Quebec, Canada: Canadian Aeronautics and Space Institute, pp. 629–643.

Jupp, D. L. B., K. K. Mayo, D. A. Kucker, D. Van R. Classen, R. A. Kenchington, and P. R. Guerin. 1985. Remote Sensing for Planning and Managing the Great Barrier Reef of Australia. *Photogrammetria,* Vol. 40, pp. 21–42.

Lyzenga, David R. 1979. Shallow-Water Reflectance Modeling with Applications to Remote Sensing of the Ocean Floor. In *Proceedings, Thirteenth International Symposium on Remote Sensing of Environment.* Ann Arbor, MI: Environmental Research Institute of Michigan, pp. 583–602.

Lyzenga, David R. 1981. Remote Sensing of Bottom Reflectance and Water Attenuating Parameters in Shallow Water Using Aircraft and Landsat Data. *Journal of Remote Sensing,* Vol. 2, pp. 71–82.

Lyzenga, David R., R. A. Shuchman, and R. A. Arnone. 1979. Evaluation of an Algorithm for Mapping Bottom Features Under a Variable Depth of Water. In *Proceedings, Thirteenth International Sympo-*

*sium on Remote Sensing of Environment.* Ann Arbor, MI: Environmental Research Institute of Michigan, pp. 1767–1780.

Polcyn, F. C., and D. R. Lyzenga. 1979. Landsat Bathymetric Mapping by Multispectral Processing. In *Proceedings, Thirteenth International Symposium on Remote Sensing of Environment.* Ann Arbor, MI: Environmental Research Institute of Michigan, pp. 1269–1276.

Satzman, B. (ed.). 1985. *Satellite Oceanic Remote Sensing. Advances in Geophysics,* Vol. 27. New York: Academic Press, 511 pp.

Tanis, F. J., and W. A. Hallada. 1984. Evaluation of Landsat Thematic Mapper Data for Shallow Water Bathymetry. In *Proceedings, Eighteenth International Symposium on Remote Sensing of Environment.* Ann Arbor, MI: Environmental Research Institute of Michigan, pp. 629–643.

CHAPTER NINETEEN

# Land Use and Land Cover

## 19.1. Introduction

*Land use* can be defined as the use of land by humans, usually with emphasis on the functional role of land in economic activities. Land use forms an abstraction, not always directly observable even by the closest inspection. We cannot see the actual use of a parcel of land but only the physical artifacts of that use. Sometimes the implications of the artifacts are quite clear—a steel mill, for example, can be clearly associated with specific economic activities and land use categories. In contrast, a large extent of forested land may display little if any physical evidence of its varied uses, which might include production of timber, supply of water for distant urban areas, and space for recreation. In addition, some land areas may be characterized by contrasting activities (belonging perhaps to separate classes of land use) at separate seasons of the year. For example, some farmland might be used alternatively as cropland and as pasture at different times in the agricultural calendar.

In contrast, *land cover*, in its narrowest sense, often designates only the vegetation, either natural or man-made, on the earth's surface (Figure 19.1). In a much broader sense, land cover designates the visible evidence of land use to include both vegetative and nonvegetative features. In this meaning, dense forest, plowed land, urban structures, and paved parking lots all constitute land cover. Whereas land use is abstract, land cover is concrete and therefore is subject to direct observation.

Another distinction is that land cover lacks the emphasis on economic function that is essential to the concept of land use. Hydrologists can focus solely on land cover because of their concern with only the physical components of the landscape that pertain to the movement of moisture. But a traffic engineer must consider land use a component of a traffic flow model and must address the economic function of each parcel of land as a contributor of automobile traffic to the region's highways. Usually the distinction between land use and land cover becomes more important as the scale of a study becomes larger and detail becomes finer.

Land use patterns reflect the character of a society's interaction with its physical environment, a fact that becomes obvious when it is possible to see different economic and social systems occupying the same or similar environments. Plate 19 shows a Landsat

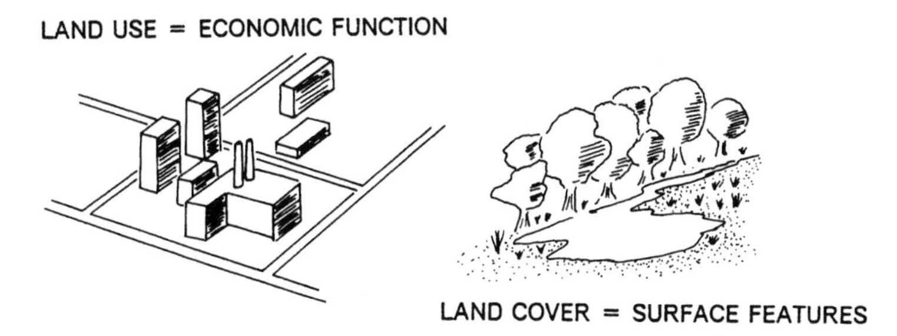


**FIGURE 19.1.** Land use and land cover.

TM quarter scene depicting Santa Rosa del Palmar, Bolivia (northwest of Santa Bruise), acquired in July 1992. Here four distinct land use patterns are visible, each reflecting its distinctive relationship with the landscape it occupies. In the southeastern portion of the scene, the land use pattern is dominated by broad-scale mechanized agriculture practiced by a Mennonite community. The upper right (northeastern) region of the image is a mountainous area occupied by a diminishing population of Indians who practice a form of slash-and-burn agriculture, visible as dispersed patches of light green. The Bolivian government has encouraged broad-scale clearing of forest for agriculture, visible throughout the northwestern and central portions of the image. In the upper left region of the image, these clearings take the form of light green spots aligned northwest to southeast; at the center of each patch of cleared land is a central facility providing colonists with fertilizers, pesticides, and staples. In the southwestern corner of the image, the complex field pattern reflects an established agricultural landscape occupied by Japanese immigrants.

## 19.2. Significance of Land Use and Land Cover Information

Almost all governmental units have a continuing requirement to form and implement laws and policies that directly or indirectly involve existing or future land use. If we exclude the possibility of annexation, each community has only a fixed amount of land area to allocate among the varied economic and social activities required to support its citizens. There is increasing recognition that sensible uses of finite, or possibly shrinking, resources requires comprehensive planning of community activities to coordinate the amount and placement of private and public facilities with the amounts and locations of human resources. Uncoordinated development can lead to inefficient and undesirable environmental, social, and economic conditions. As existing development and increasing population limit options for use of land resources, development may be attracted to sites of marginal suitability or cause displacement of activities from their optimum sites. For example, residential or commercial development on steep slopes, on flood-prone areas, or on high-quality agricultural land may inflict long-term damage to the community. In recognition of these problems, many states have legal requirements for local jurisdictions to prepare comprehensive plans outlining the kinds of land use patterns to be encouraged or discouraged in specific sites and defining favored locations for specific uses.

Land use information forms an important part of decisions made at the state level. For example, traffic flow models used to plan highway development at the state level require land use data as input to estimate traffic generated by neighborhoods supplying traffic to specific highways. State legislatures must address issues regarding allocation of land to alternative uses, either in specific geographic regions (e.g., a decision to establish a state park or scenic reserve) or through general policies tailored for specific statewide goals (e.g., laws to assist in preserving farmland from urbanization). In either context, the availability of accurate information regarding existing uses of the state's land is an important element in making sound decisions.

At national levels, land use information is an important element in forming policies regarding economic, demographic, and environmental issues. In the United States, such policies might pertain to determining the location, extent, and character of surface mining; losses of agricultural land to urbanization; national parks and defense installations; or storage and disposal of hazardous wastes, to mention only a few of today's many controversial issues pertaining to land use.

International requirements for land use data also focus on many of today's major concerns considered at their broadest possible scales. For example, major changes in land use within the world's major biomes (most notably tropical forests, but also elsewhere) may have generated as yet unknown effects on global biochemical cycles (Chapter 20) and on the global energy balance. Other issues that require worldwide perspective include changes in global patterns of agricultural and forest lands, settlement patterns within zones of uncertain and variable climate, and efforts to control environmentally questionable agricultural practices.

Examination of these issues requires collection of data from many diverse sources—data that are compatible with respect to scale, detail, accuracy, and categorization. For example, examination of land use within the tropical forests, considered worldwide, requires data from many diverse areas, separated geographically and with contrasting political and administrative traditions. Existing data, gathered independently by each nation, would probably be of only minimal utility in a serious effort to examine these most important issues from a global perspective.

At the broad scale of a statewide perspective, the character of the required information differs greatly from that needed at local scales. For statewide land use and land cover information to be effective it must be collected for the entire state at comparable levels of detail and accuracy and at comparable dates. If compatible, hierarchical classifications have been used, it is perhaps theoretically possible to form a statewide data set from information collected at local levels (assuming the unlikely precondition that all component jurisdictions collected information). In practice, however, it is unlikely that such an effort could be effective due to the innumerable practical problems in coordination and administration. As a result, discussion of statewide land use data must assume the availability of a staff administered at the state level to ensure compatibility and quality of information.

At local levels, it may be practical to gather data by direct observation or to combine direct observation with the use of remote sensing imagery. At the state level it would seem to be impractical to rely on any method except aerial photography or other remote sensing imagery.

### *Significance for Science and Research*

Land use information is, of course, also of great significance in scientific and scholarly research. National and regional land use patterns reflect the character of the interaction between man and environment and the influence of distance and resources based on mankind's basic economic activities. As a result, geographers, economists, and others have long regarded knowledge of regional land use patterns as a fundamental element in their studies of economic systems. Land use patterns are also recognized as influential elements in hydrological and meteorological processes. The importance of land use in the theories developed by von Thünen, Lösch, and others in the fields of regional science, economics, and geography demonstrates evidence of the fundamental importance of land use in both theoretical and applied research.

## 19.3. Applications of Remote Sensing

Remotely sensed images lend themselves to accurate land cover and land use mapping in part because land cover information can be interpreted more or less directly from evidence visible on aerial images. Relatively little inference is required in most situations.

Land use maps are routinely prepared at a wide variety of scales, typically ranging from 1:12,500 and larger to 1:250,000 and smaller. At one end of this spectrum (the larger-scale maps) remotely sensed imagery may itself contribute relatively little information

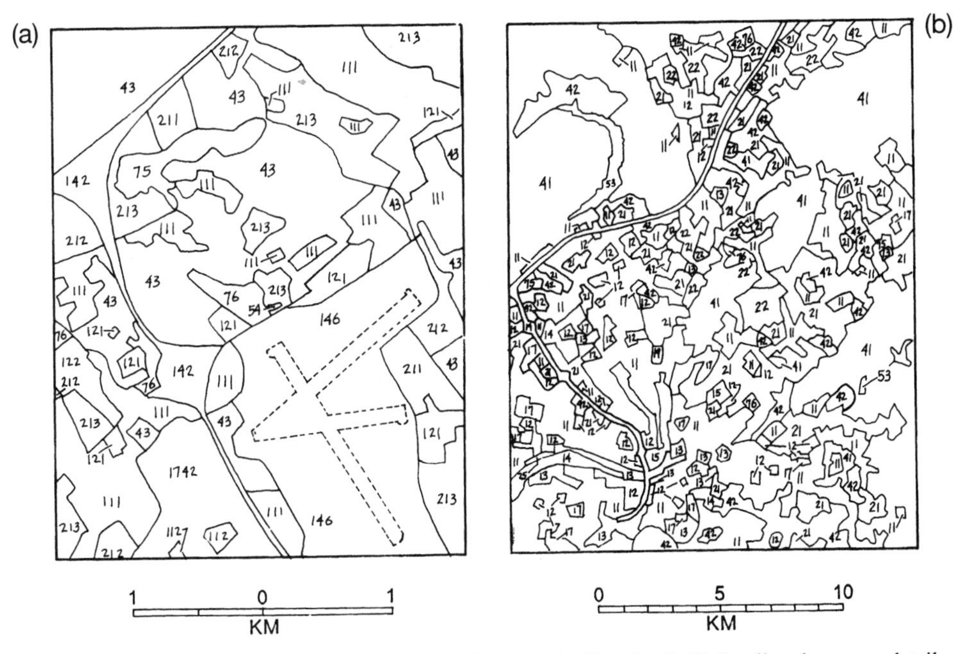


**FIGURE 19.2.** Land use and land cover maps. (*a*) Large scale, fine detail. (*b*) Small scale, coarse detail.

to the survey; its main role may be to form a highly detailed base for recording data gathered by other means (Figure 19.2). At such large scales the land use map may actually form a kind of reference map, having little cartographic generalization. Such products are often used at the lowest levels of local governments, perhaps mainly urban areas, that have requirements for such detailed information and the financial resources to acquire it.

As the scale of the survey becomes broader (i.e., as map scale becomes smaller), the contribution of the image to the informational content of the map becomes greater, although even at the smallest scales there must always be some contribution from collateral information. Differing scales and levels of detail serve different purposes and different users. For the regional planner, the loss of resolution and detail at smaller scales may actually form an advantage in the sense that the integration and a simplification of information occur that must be examined. For the medium- and small-scale land use surveys that are most often compiled by use of remotely sensed images, the product is a thematic map that depicts the predominant land cover within relatively homogeneous areas, delineated subject to limitations of scale, resolution, generalization, and other constraints accepted (or at least recognized) by both compilers and users of thematic maps.

## 19.4. Land Use Classification

Preparation of a land use map from aerial imagery is in essence a process of segmenting the image into a mosaic of parcels, with each parcel assigned to a land use class. The novice interpreter is often tempted to devise land use classes based on categories easily recognized and delineated from examination of the imagery. For example, *suburban land* may seem to be a reasonable land use category because of ease of identification and correspondence with informal land use classes developed from personal experience. In practice, of course, the most useful categories are those that match the informational needs of the map user. Typically, categories such as *suburban land* are unsatisfactory for the user who requires division of land into classes such as *residential, commercial,* and *industrial land.* In addition, it is important to use classification systems that are compatible with others that have been used in the past or are used for neighboring jurisdictions or by higher levels of government. Compatibility is important if maps and data are to be compared over time (to determine changes), aggregated over counties or states, or compared with other forms of data. Therefore, although land use classification is not in itself a task of great difficulty, it requires considerable knowledge and care to prepare a satisfactory land classification scheme and to apply it to a specific landscape.

Because of the unpredictable interplay between image detail, classification detail, and map scale, the interpreter must find a balance between the precision of the classification system and the sizes of the parcels that can be interpreted and then portrayed legibly on a map. For example, detailed categories for cemeteries is of no practical value if all cemeteries in the mapped region occur in parcels of land too small to be legibly represented at the scale of the final map, or if the image has such coarse resolution that cemeteries cannot be reliably interpreted. As a result, the image interpreter must prepare a classification system that is simultaneously compatible with the needs of the map user and consistent with image detail and map scale.

Although many land use classification systems have been devised, it is probably best to describe the system that is widely used in the United States (and elsewhere as well); other systems can be used if necessary once the student has gained some experience in the practice of land use survey. In the early 1970s, the U.S. Geological Survey (USGS) began a program for mapping of the land use and land cover of the United States using aerial photography. The land use and land cover classification system (Table 19.1) devised for

**TABLE 19.1. 1976 USGS Land Use and Land Cover Classification**

| Level I | | Level II | |
|---|---|---|---|
| 1 | Urban or built-up land | 11 | Residential |
| | | 12 | Commercial and services |
| | | 13 | Industrial |
| | | 14 | Transportation, communications, and utilities |
| | | 15 | Industrial and commercial complexes |
| | | 16 | Mixed urban or built-up land |
| | | 17 | Other urban or built-up land |
| 2 | Agricultural land | 21 | Croplands and pasture |
| | | 22 | Orchards, groves, vineyards, nurseries, and ornamental horticultural areas |
| | | 23 | Confined feeding operations |
| | | 24 | Other agricultural land |
| 3 | Rangeland | 31 | Herbaceous rangeland |
| | | 32 | Shrub and brush rangeland |
| | | 33 | Mixed rangeland |
| 4 | Forest land | 41 | Deciduous forest land |
| | | 42 | Evergreen forest land |
| | | 43 | Mixed forest land |
| 5 | Water | 51 | Streams and canals |
| | | 52 | Lakes |
| | | 53 | Reservoirs |
| | | 54 | Bays and estuaries |
| 6 | Wetland | 61 | Forested wetland |
| | | 62 | Nonforested wetland |
| 7 | Barren land | 71 | Dry salt flats |
| | | 72 | Beaches |
| | | 73 | Sandy areas other than beaches |
| | | 74 | Bare exposed rock |
| | | 75 | Strip mines, quarries, and gravel pits |
| | | 76 | Transitional areas |
| | | 77 | Mixed barren land |
| 8 | Tundra | 81 | Shrub and brush tundra |
| | | 82 | Herbaceous tundra |
| | | 83 | Bare ground tundra |
| | | 84 | Wet tundra |
| | | 85 | Mixed tundra |
| 9 | Perennial snow or ice | 91 | Perennial snowfields |
| | | 92 | Glaciers |

*Note.* From Anderson et al. (1976, p. 8).

the USGS program has become one of the most widely used classification systems for land use maps prepared by interpretation of remotely sensed images.

The USGS system has many useful features. First, it is prepared specifically for use with remotely sensed imagery. Its categories are appropriate for information interpreted from aerial images, and it has a hierarchical structure that lends itself for use with images of differing scales and resolutions. Level I, for example, is tailored for use with broad-scale, coarse-resolution imagery (Landsat imagery or high-altitude aerial photography). Levels II and III are composed of more detailed classes that can be interpreted from large-scale, fine-resolution images. Although the USGS system specifies the level I and II categories, the level III categories must be defined by the analyst to meet the specific requirements of a particular study and a specific region. As the analyst defines level III categories, it is important that the level I and II categories be used as a framework for the more detailed level III classes.

## 19.5. Mapping Land Use by Manual Interpretation

Use of aerial imagery for mapping land cover requires application of skills not normally required in the simple, intuitive examination of images for recognition of individual objects. Individual land cover categories are formed from collections of diverse objects, features, and structures that are often not individually resolved on the image; the interpreter's task is not so much one of identifying separate objects as it is the accurate delineation of regions of relatively uniform composition and appearance. The interpreter of land use information must, then, generalize to define the areal units that compose the subject of the interpretation. The goal should be to perform this mental generalization in a consistent, logical manner and to describe the procedures accurately in a written account of the interpretation process.

### *The Image Overlay and the Final Map*

The interpretation process (Figure 19.3) begins with the assembling of imagery, collateral information, equipment, and materials required to conduct the interpretation. Imagery should be inspected to note defects, holidays in coverage, and areas obscured by clouds. If there is no index to the coverage of separate frames or images, the interpreter should prepare his own index to permit convenient identification of specific images covering a given area. If there are gaps (*holidays*) in the coverage, the interpreter should initiate the process of acquiring additional imagery to provide the best possible information for voids in the primary coverage.

The interpretation process consists of two related but distinct steps. The first consists of constructing an appropriate classification system. Although many different approaches to classification of land use can be applied, the system proposed by Anderson et al. (1976) provides one of the best and the most widely applicable outlines for structuring the classification. Their system (Table 19.1) provides the more general categories, but the interpreter must become involved in the design of the more detailed categories at levels II and III. The definition and design of these detailed categories must be accomplished in coor-

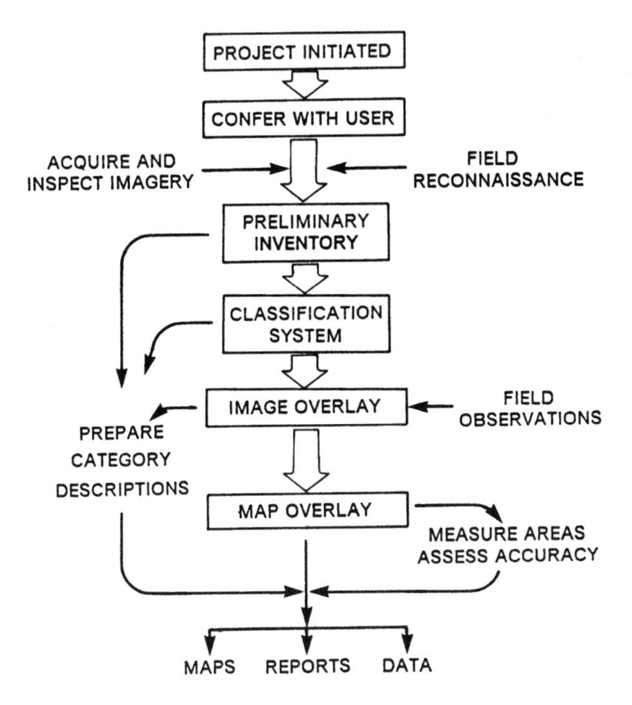


**FIGURE 19.3.** Interpretation process.

dination with those who will ultimately use the final maps and data. If it is determined that the users require information or detail not feasible with the use of aerial imagery, the project must turn to other imagery, or to collateral information, to supply the needed information. If collateral information is to be used, the interpreter should carefully investigate the alternatives to ensure maximum accuracy and maximum compatibility with the aerial imagery.

When a tentative outline of the classification is available, the interpreter should then carefully inspect the imagery to compile a list of the categories present, estimate the sizes of parcels likely to result from the application of the classification, and in general anticipate problems in identifying or delineating categories. If several interpreters are to work on the same project, all should participate in this process to ensure uniformity of perspective and gain the benefit of independent contributions. A revised list of categories then forms the basis of further discussions with the user before a final classification is accepted for the project.

The second step applies the classification system to the imagery. This step is accomplished by marking the boundaries between categories as they occur on the imagery, following consistently the guidelines outlined below. Each separate parcel is outlined, then identified with a symbol (usually one to three numerals) corresponding to taxa in the classification system (Figure 19.4). Important considerations include application of standard size for smallest parcels to be shown, as described below, and the principles of consistency, clarity, and legibility. If several interpreters are to work on separate portions of the same project, special attention must be devoted to coordination of their efforts to ensure

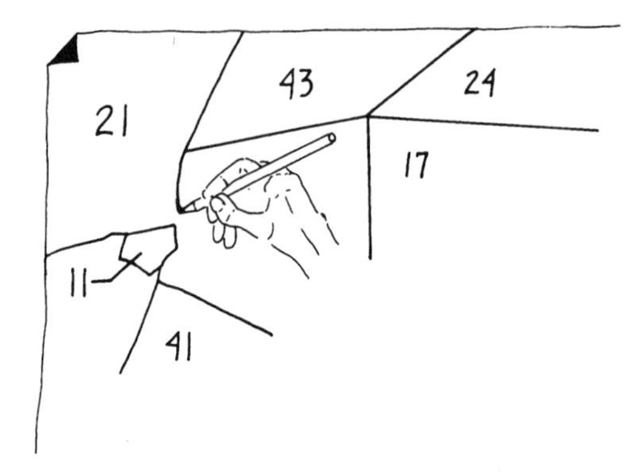


**FIGURE 19.4.** Delineating and labeling land use parcels. Boundaries completely enclose parcels; each is labeled. Leader lines label small parcels.

uniform application of the guidelines outlined below. Individual interpreters share responsibility for coordination with those working on adjacent areas to be sure that parcel boundaries and labels match at edges of sheets (Figure 19.5).

The results of the interpretation are recorded on a translucent overlay that registers to the image; this image overlay records the boundaries between parcels and identifying symbols (Figure 19.6). As the interpreter prepares the image overlay, he or she begins to compile the classification table (described below) that forms a summary of category definitions and identifying characteristics. When the image overlay is complete, it will be necessary to generate a map overlay that portrays the land cover/land use information on an accurate planimetric base. The map overlay is required because the image overlay inherits from the image the positional errors inherent to all remotely sensed imagery, so measurements of distance or area made directly from the image overlay will be in error.

Another important consideration is recognition that the map cannot stand by itself; a

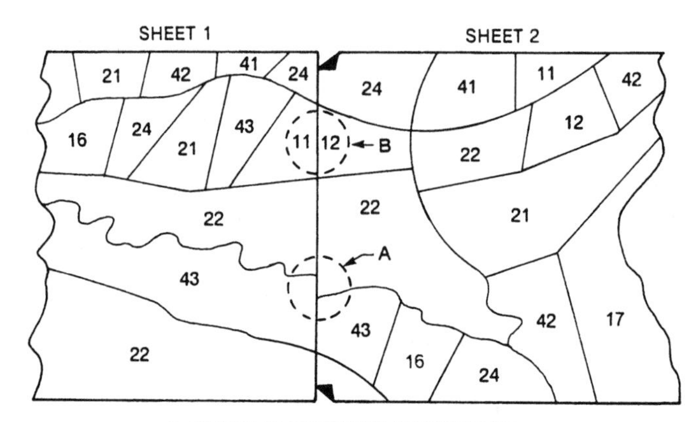


**FIGURE 19.5.** Matching of boundaries and categories at edges of parcels.

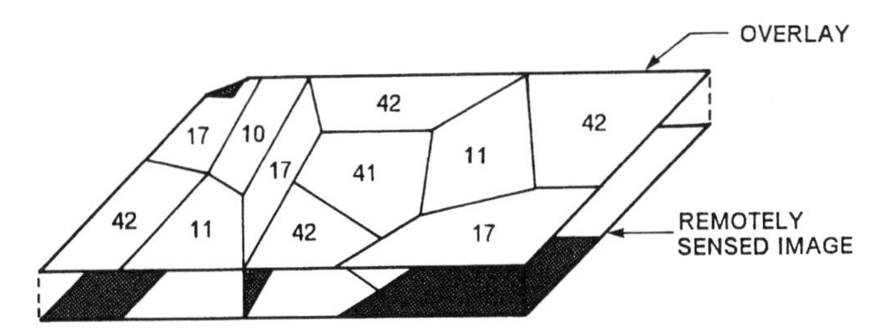


**FIGURE 19.6.** The image overlay.

written report must accompany each map so that the map reader has a summary of the interpretation process and a clear definition and description of each mapping unit. Without the written report the reader cannot understand the classification system or how it has been applied to the imagery. Analysts may well devote as much time to recording and describing the map and its preparation as they do to preparing the map itself.

## *Preparation of the Image Overlay*

The following paragraphs outline some of the essential considerations in preparing the image overlay (Figure 19.6) and recording information to appear in the accompanying report. Some of these considerations follow from an obvious interest in visual and logical clarity; others are simply conventions established by long application of rather arbitrary decisions. For the novice, it is probably best to follow the guidelines as outlined below; later, as additional experience is acquired, variations can be devised as required. But always the guiding principles should be attainment of visual and logical clarity and explicit description of the interpretation process.

1. Identification of land cover parcels is based on the elements of image interpretation as discussed previously. Sometimes identification may require a focus on identities of specific objects or facilities, but usually the primary task is one of consistent delineation of uniform parcels that match the classification system. The greater the uniformity of actual land use within areas represented on the map under a single symbol, the greater the usefulness of the map to the user. As outlined below (items 9 and 11), it may sometimes be necessary to violate this principle, especially at small scales.

2. The classification system must dovetail with categories accepted as useful by the map user. Maps and reports that organize information in a manner inappropriate for users' requirements are of little practical use; the image analyst has responsibility for ensuring that the final product is consistent with users' needs.

3. Sometimes classifications are proposed that tailor categories with respect to ease of recognition on specific forms of imagery. For the user-oriented maps discussed here, such classification systems would not be suitable because they organize informational content of the map around the image rather than on users' requirements. The analyst should remember that users' requirements may be dictated by legal definitions, by policy of a governing body, or by requirements for compatibility with data collected previously

or by a neighboring jurisdiction and therefore are not easily redefined to meet desires of the image interpreter.

4. As a result, the interpreter should not define categories solely from appearance on the imagery. For example, categories such as *suburban land* and *strip development* are probably poor categories for most land use maps because they are based largely on ease of recognition on the image rather than correspondence with the land use categories of interest to planners and geographers. Thus, *suburban land* should be redefined and mapped, as appropriate, into residential and commercial land use categories accepted in the user's definitions. (Note that any category, if properly defined, and tailored to users' needs, may have its merits; as a rule, however, it is best not to devise unconventional categories for the convenience of the interpreter.) In summary, land use categories used in making the map should have a clear meaning to those who will use the map.

5. The beginning interpreter is well advised to use the system proposed by Anderson et al. (1976) as a framework for organizing a land use classification. The Anderson system (Table 19.1) provides a flexible framework for developing classifications at several levels of detail. Its hierarchical structure provides a capability to permit consistent, compatible classifications using imagery at varied scales and resolutions and the ability for convenient cartographic and statistical generalization. The USGS publication that describes their classification provides a clear summary of the major issues of significance in the development of land use classification from aerial imagery. The Anderson system need not be perceived as the ultimate classification (e.g., Drake, 1977), but for most purposes it seems to form a good standard to be accepted unless there is good reason to use another system. Nunnally and Witmer (1970) have observed that one of the problems with land use classification prior to the proposal of the USGS system was the incompatibility of the many systems then in use.

6. Use of collateral material may be necessary if the interpreter is not intimately familiar with the region or if unusual categories are encountered. Nonimage collateral information might include topographic maps, existing land cover maps (at scales or dates differing from the one in preparation), or tabulated economic statistics. Additional imagery at large scale can be used to resolve uncertainties emerging from analysis of small-scale, coarse-resolution imagery.

7. The image overlay records, in manuscript form, the boundaries between land cover parcels. Each parcel is completely enclosed by a boundary and is labeled with a symbol keyed to the category descriptions in the classification system (Figure 19.4). As a general rule, the image overlay shows only those features that occupy areas at the scale of the final map; usually point or linear features are not mapped. Thus, a highway would not normally be shown unless the publication scale permits legible delineation of both sides of the highway right-of-way. For example, at a small scale, even four-lane expressways will be represented (if at all) as lines; at a somewhat larger scale, the cloverleaf interchanges are large enough to be mapped (Figure 19.7). Finally, a large-scale land cover map might be able to show the entire highway as an areal feature. This same rule should be applied to such other linear features as streams and rivers, railways, and power lines. Selected point or linear features may, of course, be useful as landmarks or locational references on the map, but they are not classified as areas.

8. The principle of consistent composition at each category means that the map user can be confident that the map presents a consistent representation of variation present on

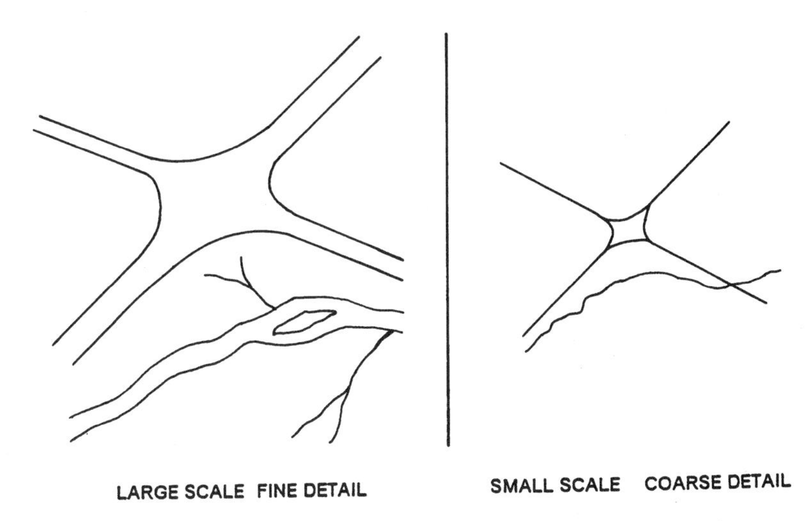


**FIGURE 19.7.** Representation of linear and areal features at different scales.

the landscape. Each parcel will by necessity encompass areas of categories other than that named by the parcel label; these inclusions are permissible but must be clearly described in the category descriptions and must be consistent throughout the map. The issue of consistency is especially important when several interpreters work on the same project. Individual interpreters must coordinate their work with those working on neighboring areas, and supervisors must check to be sure that detail is uniform throughout the mapped area.

9. The entire area devoted to a specific use is delineated on the overlay (Figure 19.8). Thus, the delineation of an airfield normally includes not only the runway but also the hanger, passenger terminals, parking areas, access roads, and in general all features inside the limits of the perimeter fence (i.e., the outline of the parcel encompasses areas occupied by all these features, even though they are not shown individually on the map). In a similar manner, the delineation of an interstate highway includes not only the two paved roadways but also the median strip and the right-of-way.

10. The issue of multiple use is discussed in detail by Anderson et al. (1976). In brief, the problem is caused by the practice of assigning parcels to single categories even though we know that there may in fact be several uses. A forested area may simultaneously serve as a source of timber and a recreational area for hunters and hikers. In general, the interpreter must make a decision, apply it consistently throughout the image, and clearly document the procedure in the written report.

11. The interpreter must select an appropriate minimum size for the smallest parcels to be represented on the final map (Figure 19.9). The interpreter may be able to identify on the image parcels much too small to be legibly represented on the map; therefore, it is necessary to select (as outlined below) a minimum size for the smallest parcels to be shown on the map. If several interpreters are working on the same project, all must apply the same minimum parcel size to assure that the variations in map detail reflect actual variations in parcel size on the ground.

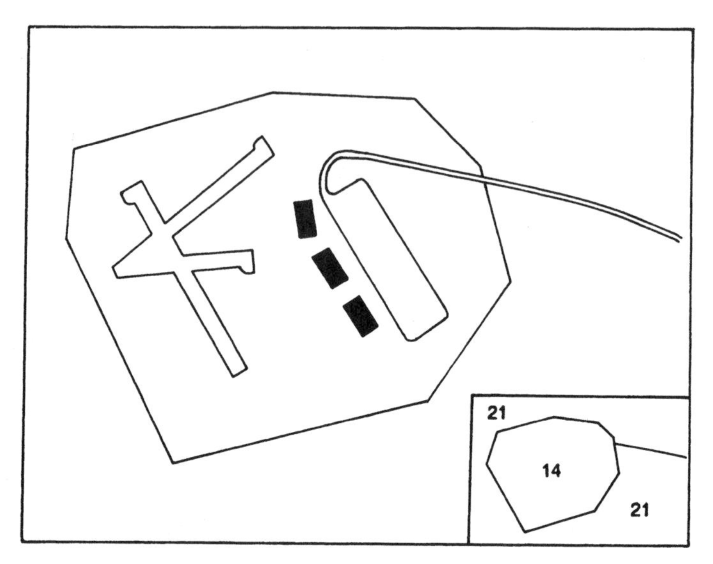


**FIGURE 19.8.** Delineation of the entire area devoted to a given use.

12. In determining an appropriate minimum size for land cover parcels, it is important to remember that it is the minimum size of parcels on the final map overlay that are of interest, as it is the legibility of this map (rather than the image overlay) that is of interest to the user. Because the map overlay may be presented at a different scale than that of the image overlay, the interpreter must extrapolate the minimum parcel size from the scale of the final map to the working scale of the image. Because it is difficult to perform this extrapolation mentally, it is sometimes useful to prepare a rough template of the approximate correct size to aid the interpreter in maintaining the correct level of spatial detail on the image overlay.

13. Anderson et al. (1976) recommend that parcels on the final map be no smaller than 0.1 in (2.54 mm) on a side (about 6.5 mm$^2$ in area). Loelkes (1977) presents values for minimum sizes of parcels to be represented on USGS land use maps. Parcels of urban land, water, and certain other specified categories are to have minimum sizes of 4 ha (10

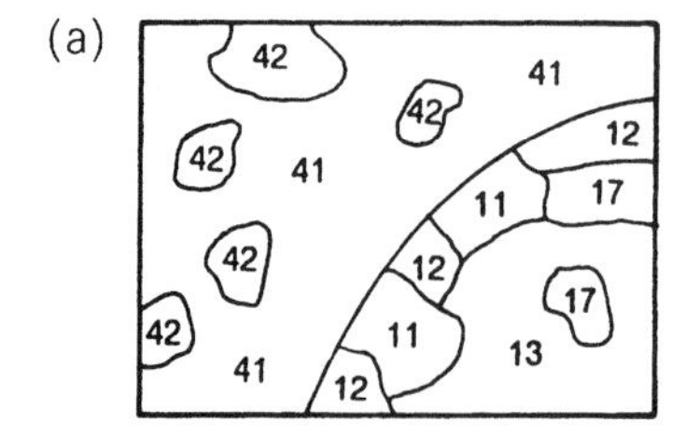


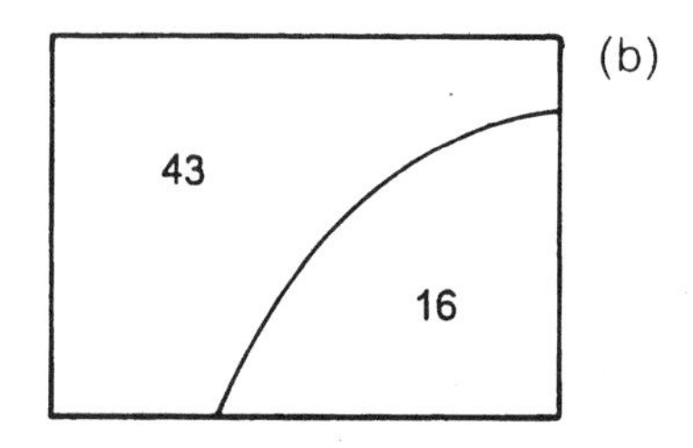


**FIGURE 19.9.** Minimum parcel size. (*a*) Small minimum parcel size. (*b*) Large minimum parcel size.

acres). At 1:100,000, such parcels would occupy about 4 $mm^2$ on the map; at 1:250,000 they would require about 0.64 $mm^2$. For all other categories, the minimum size should be 16 ha (40 acres). At 1:100,000, these parcels occupy about 16 $mm^2$, or about 2.6 $mm^2$ at 1:250,000. These guidelines apply for rather compact parcels; Loelkes (1977) proposes additional guidelines governing minimum widths for long, narrow delineations. Although these sizes may be appropriate for the USGS land use maps, they seem too small for routine use by interpreters using less sophisticated equipment and techniques. It should be obvious, however, that any single value proposed as a minimum size for mapped parcels should be interpreted as appropriate for the occasional presence of small parcels; if the interpreter encounters an extremely complex pattern of extremely small units, an attempt to represent them all at the sizes mentioned above would produce an illegible map. In such a situation, some form of generalization is clearly required to produce a visually and logically clear map at publication scale.

14. Usually the label of each category identifies the predominant category present within each parcel. At small mapping scales especially, there may be inclusions of other categories; the mapping effort should aspire to define categories that include relatively consistent mixtures, identities, and proportions of such inclusions, and to accurately describe their presence within each category.

15. Sometimes the identification or correct placement of boundaries can be a problem, especially if the interpreter can discern a wide transition zone between categories. Usually the interpreter can place the mapped boundary at the center of the transition zone, then describe the situation in the written report that accompanies the map. Sometimes wide transition zones occur consistently within the mapped area; if so, it may be appropriate to define a separate category: (e.g., "419. Transitional Zone between Evergreen and Deciduous Forest Land").

16. Mosaics of contrasting categories can present problems if the individual parcels are too small to be represented legibly at the scale of the final map (Figure 19.10). In these situations, it may be appropriate to create a category tailored to a description of the situation (e.g., "215. Mosaic of Cropland and Pasture"). The written category description then specifies the sizes and shapes of the parcels and presents an estimate of the percentages of the areal extent of each member of the mosaic. These composite mapping units are often a necessary departure from the principle of uniform composition of mapping units outlined above, but there is ample precedent for their use (Christian, 1959; Robinove, 1981).

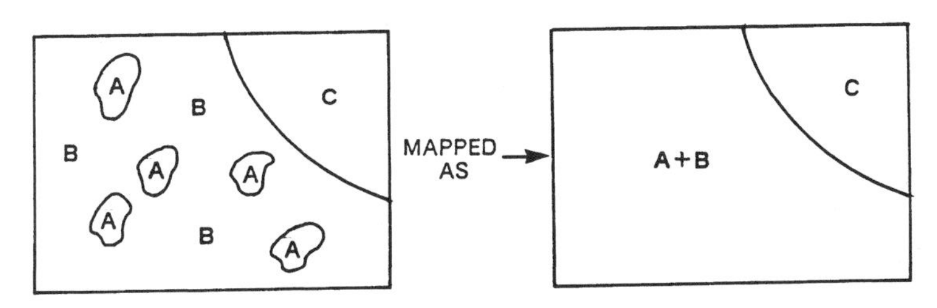


**FIGURE 19.10.** Mosaics.

### *Field Observations*

Even the most thorough, accurate interpretations require use of field observations as confirmation of manuscript maps and as a means of resolving uncertainties in the interpretation process. Ideally, field observations should be acquired on at least three occasions during preparation of the land use map: (1) before the interpretation of imagery begins, during preparation of the classification system, as a means of familiarizing interpreters with the region and its major land uses; (2) as the image overlay nears completion, to verify uncertain interpretations, and to confirm consistency of the interpretation; and (3) when the preliminary draft of the map overlay is complete, to detect and resolve any final problems before the final inked copy of the map is prepared. These three excursions to the field serve different purposes, so they may vary in respect to duration, route, and intensity.

It is probably not practical to specify a universally applicable rule for determining the effort to be devoted to field observations. General guidelines can be suggested as follows. The analyst should be sure that field observations are made throughout the study area, and that all land use classes are observed in the field. That is, observations should not be confined to specific regions of the study area or to certain classes of land use. Thus, the analyst must make an effort to systematically observe representative areas of each land use class and each sector of the study region. These rather subjective guidelines must, of course, be applied as appropriate to specific situations. If the field observations are to be used for a quantitative role in the analysis, it may be necessary to apply a more rigorous approach to gathering field data by using the sampling patterns and principles outlined in Chapter 12.

Although details of field excursions will depend greatly on individual preferences and local circumstances, the following observations seem generally applicable:

1. Imagery should be taken to the field, together with manuscript overlays and supporting notes and maps. Provision must be made for annotating images (on overlays) or correcting manuscript overlays in the field. This usually means that maps and overlays should be temporarily clipped to a hardboard or cardboard surface small enough to carry in the field and to use in a vehicle but large enough to present a sizable portion of the map for navigation and annotation.
2. The route should be planned carefully to select an efficient itinerary that covers all essential areas. If timing is critical, it is important to allow time for unexpected delays. It may be wise to assign priorities to specific portions of the route so the most important areas can be visited first.
3. Notes, photographs, and sketches should be made in a systematic manner that ensures their usefulness later in the lab. If machine (Xerox) copies of maps and aerial photographs are made beforehand, they can be used as a medium for making notes and recording locations of photographs and notes. These images will probably not record all detail visible on the original, but they permit the analyst to make annotations and notes without damaging the original.
4. If several interpreters are participating, it is usually best to work in teams, with division of labor such that each has responsibility for specific tasks. If interpreters have been assigned specific geographic areas, each is responsible for planning the itinerary for that area.

5. If the project has not already initiated contact with local officials and personnel from local extension services, planning agencies, and other offices, these organizations should be solicited for advice and information.

6. At the completion of the trip, a systematic effort must be made to organize all material and information and to be sure that field notes are clearly transcribed into a more formal format while their meaning is still clear to all involved. Notes and annotated maps must be dated and labeled with names of the persons who made the observations.

### *Checking and Editing*

Errors are inevitable. As a result, checking and editing are essential steps in the preparation of a map, equal in significance to the more immediately obvious steps of acquiring imagery and preparation of the image overlay. The search for errors is continuous throughout the preparation of the map but should be focused at specific stages. Each parcel on the image overlay should be compared to the original image to confirm its identity and correct boundary placement and to adhere to the minimum parcel size. On the final draft of the map overlay, a check is made to confirm the presence and legibility of all boundary segments, label parcels with their correct symbols, and register land use boundaries to detail on the base map. For large projects with several map sheets, a check must be made to ascertain that boundaries match at the edges of sheets and that parcel identities match across sheet edges.

There is little benefit to be gained from an attempt to list all of the errors that can occur in the preparation of a land use map from aerial imagery. But it may be useful to propose two principles that can be applied to detect and reduce errors of all kinds. First, checks for errors should be made throughout the preparation of the map; the earlier errors are detected, the easier they are to correct. However, if the check for errors is not focused at specific stages in the production of the map, the search for errors becomes so diffuse that it loses meaning. Therefore, specific steps in the production process (perhaps at the completion of the image overlay, completion of the map overlay, etc.) should be designated as opportunities for checks of the work completed thus far—a hurdle that must be passed before the next step begins. Second, errors are easier to control if specific individuals or groups assume responsibility for specific portions of the project. Checking and editing should then be clearly separated from the preparation process by designating individuals to check the work of others or having individuals check their own work at a time and place different from those of original preparation. These steps assist in promoting a critical attitude during the process of checking work for errors.

### *The Map Overlay*

The map overlay is formed by plotting boundaries from the image overlay onto an accurate planimetric base (Figure 19.11). Parcel boundaries on the image overlay include positional errors inherent to all remotely sensed images. (Sources of such errors include relief displacement [aerial photography], skew [images generated by electro-optical scanners], radar layover, and radar foreshortening [SLAR imagery].) Therefore, the image overlay

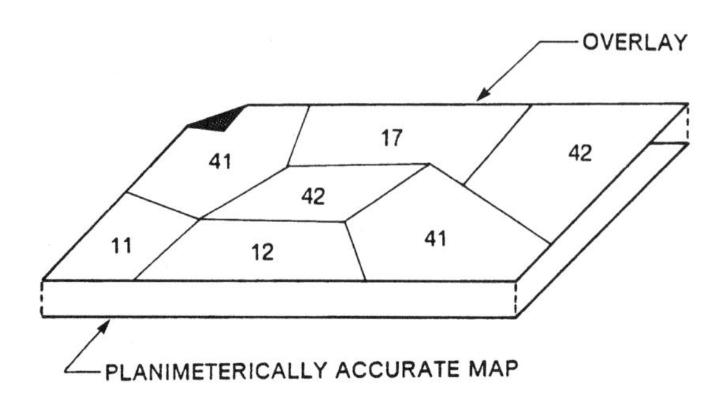


**FIGURE 19.11.** The map overlay registers to an accurate map.

cannot be used as the basis for accurate measurements of distance or area and as a result must be considered a preliminary document. The map overlay registers to an accurate map so that corresponding detail (when present) on the two maps match exactly; it forms the basis for the final land use map, suitable for use by the planners and administrators who require the land use data. Even in instances when geometric errors have been minimized by preprocessing of the remotely sensed data, it is often desirable to plot land use boundaries on a map base familiar to those who will use the map.

Changes in image scale and geometry can be made in a number of ways, including use of reflecting projectors or image transfer instruments such as those mentioned in Chapter 5. Often preparation of the map overlay requires not only changes in map scale but also changes in map geometry necessitated by geometric properties of the remotely sensed images. Image transfer instruments are useful for this purpose because of their ability to permit convenient changes in image geometry—a process that can be accomplished only with difficulty using the usual reflecting projectors.

Regardless of the procedure to be used, the process of matching the two images is easiest when there is ample detail common to both the image and the map. As a result, the image overlay should be prepared to show locations of drainage, topographic, and transportation features, for example, visible both on the map and on the remotely sensed image, even if such features do not always correspond to land use boundaries to be shown on the final map. Such features provide the common detail necessary to permit convenient and accurate registration of the two maps.

The map overlay shows land use boundaries and symbols plus other features (such as major highways, rivers, and place names) that may be useful to the reader in orientation and interpretation of the land use map. The completed map should include a bar scale, legend, title, north arrow, coverage diagram, and other information required for accurate interpretation.

### *The Report*

Few if any maps stand alone, to be accepted by readers at face value, without supporting information. The notion of "supporting information" in this context should be interpreted broadly to include not only formally presented written material that accompanies the map but also the wider realm of knowledge that the reader uses to examine and evaluate a map.

Some maps include written explanations of mapping technique and mapping unit characteristics, either as text on the map itself or in documents that accompany the map. This kind of information forms explicit information formally presented to the map reader.

Although many maps are without supporting information in such explicit form, the map reader often has the benefit of implicit, informal knowledge derived from experience with similar maps. For example, the reader of a USGS topographic quadrangle is presented with very little explicit information concerning mapping technique if the map itself is considered the only source of information. However, the reader usually has access to substantial implicit information acquired in the examination of other similar maps and through knowledge of cartographic conventions. As a result, the reader can employ in an interpretation a wide range of knowledge not obvious from inspection of the map itself, pertaining to, for example, symbolization, accuracy, cartographic conventions, and degree of generalization.

The usual land use maps, despite many superficial similarities to one another, are characterized by notable diversity with respect to purpose, categorization, detail, accuracy, and symbolization. Therefore, the reader must depend largely on explicit information, formally presented in written documents that accompany the map. As a result, careful preparation of supporting documentation assumes an importance possibly equal in significance to that of preparation of the map itself.

Specifically, the reader requires knowledge regarding (1) the regional setting, (2) methods and materials used to prepare the map, (3) definitions of mapping units, and (4) the summarized results. Emphasis devoted to each of these topics may vary in relation to requirements of the organization that will use the study, purposes of the study, and the experience of the intended audience. For example, the description of the regional setting may be abbreviated if it is known that the users of the map are already familiar with the area to be mapped. As a general rule, however, each item should be discussed to provide a complete document that can stand by itself.

### *Regional Setting*

The regional description outlines the geographic setting of the mapped region with emphasis on those factors most likely to influence the development of land use patterns (Figure 19.12). Unless the area is unusually large and diverse, a brief narrative of a few paragraphs should be sufficient. The reader who requires a detailed description should be directed to longer and more elaborate documents that focus on specific aspects of the region. The purpose of this section is simply to set the stage for subsequent description and analysis of land use patterns. These patterns can be best understood in relation to the physical and economic context of the region. The physical setting is described in brief outlines of climate, topography, soils, drainage, and natural vegetation. The economic setting is best described in terms of key elements of the industrial, commercial, and agricultural life of the region. In most instances this section should also include a brief description of the regional transportation system, with emphasis of links to other regions. The brevity of the regional description precludes completeness; it should, however, sketch the main features of the regional economic pattern, with emphasis on interplay of physical and cultural elements that determine the broad features of regional land use patterns.

REGIONAL SETTING

This region consists largely of mountainous topography, with valley bottoms at elevations of 150 to 300 meters (500 to 1,000 feet), plateaus at 520 to 600 meters (1,700 to 2,000 feet), and some peaks as high as 1000 meters (3,500 feet). Valleys are narrow, with steep sides and even floors. Plateau topography forms rather broad, level, surfaces, with gentle slopes.

Climate is characterized by short, cool summers, and long, severe winters. Average annual temperature is -3 to 10 degrees Celsius (near 27 to 50 degrees Fahrenheit). Annual precipitation is between 760 and 1000 millimeters (30 to 40 inches). Maximum precipitation falls during the summer months. Much of the winter precipitation falls as snow. Ample precipitation and cool temperatures mean that moisture is plentiful; perennial streams are abundant, and moisture deficits are not common.

Typical soils are moderately deep, medium-textured, and stony. In valleys slopes are gentle, and soils tend to be well-drained, although poorly-drained soils are important locally. On the uplands and valley sides slopes are steeper, soils are shallower, and stones are more abundant.

In valleys, most of the land is in farms, although most farms include sections of mixed hardwood forest. Pasture, hay, small grains, and potatoes are among the most important crops. Dairy cattle and poultry are important, and some land is devoted to orchards. Abandoned or fallow land is covered with grass and shrubs. Valley sides and uplands are primarily in forest.

**FIGURE 19.12.** The report: regional setting.

### *Methods and Materials*

This section is also very brief, consisting of only a few concise paragraphs that describe the imagery and interpretation techniques used in preparing the map and supporting data and documents (Figure 19.13). Imagery should be described in detail with respect to scale, resolution, date, quality, format, coverage, and source. If it is necessary to use several missions to obtain complete coverage of the study area (due perhaps to holidays, clouds, or partial coverage by the primary imagery), a coverage diagram should be prepared to depict respective coverage of each form of imagery (Figure 19.14). The character and sources of collateral information are described. The character and timing of field observations are described. The interpreter also provides an account of the interpretation procedure, with mention of any special equipment used.

### *Mapping Unit Descriptions*

The descriptions of mapping units form the most important part of the written report; they describe each and every category used in making the map, as they have been defined for this specific report and have been applied to this specific image (Baker et al., 1979). Information should be presented concisely but clearly and in sufficient detail to be of use to

METHODS AND MATERIALS

Interpretation was made using black and white aerial photography at a scale of about 1;20,000. Photography was flown in May of 1987; this imagery is of good quality, free of cloud cover and other defects. Interpretation of urban land use was sometimes aided by the use of larger scale photography at 1:12,000, dated April 1985, that provided additional detail.

Paper prints were examined using magnification and stereoscopes; land cover parcels were then delineated on translucent overlays that registered to the photographs. Information was later transferred, with the aid of a Zoom Transfer Scope, to a second overlay that registers to USGS 7.5 minute quadrangles at 1:24,000. Land cover parcels are identified by three-digit symbols that match to a classification based upon the land cover classification system defined by Anderson et al (1976).

**Figure 19.13.** The report: methods and materials.

the reader in interpreting the map. Each category is described by specifying four separate elements (Table 19.2). First, the name and symbol are presented exactly as they are used in the map legend and on the map itself. Every category used on the map appears in the written report.

Second, category definitions give precise, clear definitions. The reader may understand the general, conceptual definition of, for example, *urban land,* but cannot be expected to know the specific operational interpretation applied using specific imagery of a specific geographic region. Often the interpreter may be required to make very subtle or arbitrary distinctions in applying the classification to a specific image, and these distinctions may vary at differing scales and resolutions, and with differing forms of aerial imagery. The reader is entitled to an explanation of the exact procedure used for making the map; without the benefit of the information presented by the interpreter, the map reader has no means of reconstructing the operational meaning of the categories on the map. Descriptions should usually be concise; if elaborate descriptions are required, they probably should be presented in an appendix.

Third, the "ground features" section presents an inventory of the primary objects and

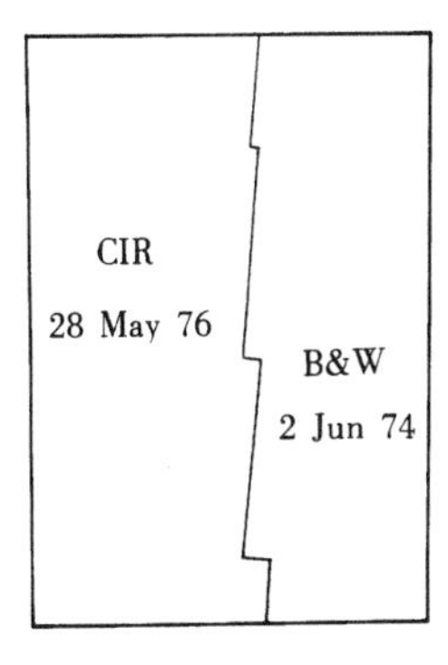


**FIGURE 19.14.** Coverage diagram.

**TABLE 19.2. Examples of Descriptions of Land Cover Categories Defined from Remote Sensing Imagery**

| Symbol | Name | Definition | Ground features | Image appearance |
|---|---|---|---|---|
| 111 | Single-unit residential | Land occupied primarily by detached dwellings and associated structures | Individual homes<br>Lawns<br>Streets<br>Trees | Tree crowns usually dominant features (medium dark tone, coarse texture)<br>Regular street pattern<br>Driveways, sidewalks, lawns visible<br>Rooftops visible |
| 112 | Multiple-unit residential | Dwelling units designed for occupation by several families | Apartment buildings<br>Grassed areas<br>Parking areas | Large buildings, usually rectangular, arranged in clusters<br>Rooftops, parking lots visible<br>Trees usually absent or sparse |
| 211 | Cropland | Land used for harvested cash crops | Plowed farmland<br>Land planted in crops<br>Fencelines, hedgerows<br>Farm roads | Fields frequently have straight, or even sides<br>Fine, even texture<br>Photo tone usually light (very dark for plowed land)<br>Boundaries sharp<br>Contour plowing, strip cropping frequently found<br>Field size usually small or moderate |
| 212 | Pasture | Land used primarily for grazing livestock or for hay | Open grassland<br>Occasional isolated shrubs, trees<br>Fencelines, hedgerows<br>Farm roads | Field frequently large<br>Irregular shape, indistinct boundaries<br>Medium photo tones<br>Texture<br>Mottled appearance |

*Note.* From Campbell (1983, p. 64). Copyright 1983 by the Association of American Geographers. Reproduced by permission.

features that occur within each mapping unit. This section serves several purposes. It permits the reader to acquire a very precise understanding of the way that the interpreter has applied the classification to the image; in effect, this section reveals the interpreter's operational definition of each category. It also maintains the interpreter's discipline in defining mapping units; the interpreter who has difficulty in preparing concise inventories for each category discovers in effect that the categories have not been carefully defined or consistently applied to the image. The interpreter discovers earlier rather than later that any efforts include errors.

Finally, the most important function of this section is to present an accounting, for each category, of the presence of foreign inclusions. For example, the limitations of mapping scale may require that unmapped parcels of forest be included within areas designated on the map as cropland. If so, the mapping unit description for cropland should specify the presence, identity, proportion, and (if possible) the pattern of occurrence of the unmapped inclusions (e.g., "Includes small isolated patches of deciduous forest too small to be mapped up to a total of about 15% of the area mapped as cropland; size and frequency of these areas decreases toward the southern edge of the mapped region"). The image appearance section describes each category as it appears on the image, using the "elements of image interpretation" as a framework for description. Table 19.2 lists a suggested vocabulary, with examples, as a means of describing the image appearance of land use categories, as they appear on black-and-white aerial photography. Variations can be devised to suit other forms of imagery and a range of scales and resolutions.

The image appearance section does not attempt to describe the ground appearance of the category but the appearance of the category as it is represented on the imagery used for the study. It does not attempt to describe the image appearance at seasons other than the one in question. In brief, it does not form a general, universally applicable description but merely an account of the facts that apply to the specific interpretation at hand.

The mapping unit descriptions can be presented in either of two formats. A brief narrative section, such as that used by Baker et al. (1979) presents each mapping unit description in a few concise sentences, organized to present all the information outlined above (Figure 19.15). Or, it may be appropriate to present the same information in a table (the *classification table*) organized as illustrated in Table 19.3). The classification table serves two complementary purposes. First, it provides explicit information for the map reader regarding definitions and compositions of categories as they have been defined for this specific study. Second, compilation of the classification table forms a means for the

MAPPING UNIT DESCRIPTIONS

**2. AGRICULTURAL LAND**

**21. Cropland and pasture**

Parcels of varying shapes and sizes often bordered by roads and highways. This category typically appears as a mosaic of parcels of varied tones and textures that occupy valley bottoms in rural regions.

**211. Cropland**

Areas in cropland are often characterized by straight edges or gently rounded curves, aligned in groups along rural roads. Distinctions between fields of different crops can be made on the basis of image tone and texture, and sometimes from field size and shape. Textures are usually smooth to medium; tones are usually light to medium. Textures and tones within fields tend to be uniform. Farm buildings and isolated groves of trees are sometimes recognizable.

**212. Pasture and meadow (includes Pasture/Forest regrowth)**

Larger parcels of pasture usually border cropland. Typically, tones vary from light to medium gray, forming a mottled appearance. Textures are smooth to medium. This category is usually positioned on moderate slopes. Often one side of a parcel will border a large tract of forest, while another side borders cropland.

**FIGURE 19.15.** The report: mapping unit descriptions.

**TABLE 19.3. Suggested Terminology for Written Descriptions of Land Use Categories Defined from Aerial Imagery**

| Element of image interpretation | Some suggested qualitative descriptors |
|---|---|
| Size | "Small," "medium," "large"; also, is size "uniform" or "varied"? |
| Shape | "Compact," "regular," "elongate," "square," "irregular," "rectangular" |
| Tone | "Light," "medium," "dark," "very light," "very dark" |
| Texture | "Coarse," "medium," "fine"; also, "even," "uneven," "mottled," "uniform" |
| Association | State whether there exists a consistent spatial association with other categories. What is the character of the boundaries with neighboring categories? |
| Shadow | Can you determine whether shadow contributes to the appearance of a category? Consider not only objects, but *areas* as well. For example, scattered bushes in a pasture may be too small to be identified (except perhaps as tiny dots), but often their *shadows* are large enough to contribute to the photo appearance of the pasture. (Result: a speckled appearance.) Or, often the edge of a forest at a pasture may be sharp enough to cause a shadow to fall in the pasture. (Result: an "enhancement" of the boundary on one side, but not on others.) |
| Site | In some instances, topographic position (site) may be an important means of describing the distinctive characteristics of features or categories. |
| Pattern | Specify whether objects within a specific category are arranged in a distinctive manner. An obvious example: an orchard. Possibly the arrangement of greens and fairways in a golf course could be considered to be a distinctive pattern. |

*Note.* Often a precise and accurate description of a manual interpretation of remote sensing image is rendered difficult by an unfamiliarity with appropriate descriptive terms. Here a variety of qualitative descriptive terms are listed as suggestions for your interpretations. These terms are in a sense imprecise, and must apply only to specific images, but they do offer a means of specifying image characteristics of land cover categories. Develop your own modifications as you gain experience. From Campbell (1983, p. 65). 

interpreter to evaluate the logic of the interpretation process. If an interpreter attempts to map ill-defined categories, or categories that cannot be clearly separated on the basis of image appearance, the problems quickly become evident in the preparation of the mapping unit descriptions.

## Summary

The report concludes by summarizing the results of the inventory. Here the interpreter can describe problems encountered during the preparation of the study. If an evaluation of the map's accuracy has been conducted (Chapter 13), results are reported here.

For most studies, however, the main portion of this section is devoted to the summary of areas occupied by each of the categories on the map. Our interest here is essentially with inventories of existing land use, so this summary forms a description of land use patterns as observed at the time the imagery was acquired. Usually an evaluation or interpretation of the appropriateness of the observed patterns is not appropriate in this context, although the results of the inventory may form the starting point for a separate study that does assess the relationship of existing patterns to ideal patterns.

| SYMBOL | NAME | AREA (HA) |
|---|---|---|
| 111 | SINGLE FAMILY RESIDENTIAL | 445.2 |
| 146 | AIRPORT | 207.6 |
| 142 | FOUR LANE LIMITED ACCESS HIGHWAY | 150.0 |
| 212 | PASTURE | 149.9 |
| 761 | ROAD CUTS | 138.6 |
| 163 | MIXED URBAN AND RESIDENTIAL | 137.5 |
| 122 | PRIMARY AND SECONDARY EDUCATION | 125.9 |
| 171 | CEMETERIES | 121.1 |
| 211 | CROPLAND | 118.3 |
| 161 | MIXED COMMERCIAL/RESIDENTIAL | 115.3 |
| 213 | PASTURE (OVERGROWN) | 108.7 |
| 112 | MULTIPLE UNIT RESIDENTIAL | 103.9 |
| 762 | BARREN AREAS | 101.3 |
| 173 | RECREATIONAL AREAS | 100.3 |
| | TOTAL | 2,123.6 |

**FIGURE 19.16.** Summary of areas occupied by each class.

A brief narrative may be appropriate, but the heart of the summary is a tabulation of the areas occupied by each land use category (Figure 19.16). The area of each parcel on the final map is measured using a planimeter, electronic digitizer, or other means as appropriate. (Note that accurate measurement of areas depends completely on careful preparation of the map overlay, and that measurements made from the image overlay will be in error.) Areas of separate parcels are summarized by category to yield a single total for each category. The final tabulation of areas shows each mapping unit by name and symbol, with its total area for the mapped region (reported in acres, square miles, or hectares as appropriate), and its areal percentage of the total mapped area. Detailed categories are collapsed into broader categories, so the listing reports all possible levels of detail. If the mapped area has been subdivided into political or census units, areas of each category are reported by each subdivision as well.

## 19.6. Land Cover Mapping by Image Classification

Land cover can be mapped by applying image classification techniques discussed in Chapter 11 to digital remote sensing images. In principle, the process is straightforward; in practice, many of the most significant factors are concealed among apparently routine considerations.

1. *Selection of images.* Success of classification for land cover analysis depends on the astute selection of images with respect to season and date. Therefore, the earlier dis-

cussion (Section 6.10) of the design and interpretation of searches of image archives, although ostensibly mundane in nature, assumes vital significance for the success of a project. What season will provide the optimum contrasts between the classes to be mapped? Possibly two or more dates might be required to separate all of the classes of significance.

2. *Preprocessing.* Accurate registration of images and correction for atmospheric and system errors (Chapter 10) are required preliminary steps for successful classification. Subsetting of the region to be examined requires careful thought, as discussed in Section 10.2.

3. *Selection of classification algorithm.* The previous discussion in Chapter 11 reviewed many of the classification algorithms available for land cover analysis. Selection of the classification procedure should also be made on the basis of local experience. AMOEBA, for example, tends to be accurate in landscapes dominated by large, homogeneous patches, such as the agricultural landscapes of the midwestern United States, and less satisfactory in landscapes composed of many smaller, heterogeneous parcels, such as those found in mountainous regions. Local experience and expertise are likely to be more reliable guides for selection of classification procedures than are universal declaration statements about their performance. Even when there is comparative information on the classification effectiveness, it is difficult to anticipate the balance between effects of the choice of classifier, selection of image data, characteristics of the landscape, and other factors.

4. *Selection of training data.* Accurate selection of training data is one process that is universally significant for image classification, as emphasized earlier in Chapter 11. Training data for each class must be carefully examined to be sure they are represented by an appropriate selection of spectral subclasses, to account for variations in spectral appearance due to shadowing, composition, and so on. Many individual laboratories and image analysis software packages have applied unsupervised classification in various forms to define homogeneous regions from which to select training fields for supervised classification (e.g., Chuvieco & Congalton, 1988). Another approach to the same question is an algorithm that permits the analyst to select a pixel, or group of pixels, that then forms the focal point for a region that grows outward until a sharp discontinuity is encountered. This process identifies a region of homogeneous pixels from which the analyst may select training fields for that class.

5. *Assignment of spectral classes to informational classes.* Because of the many subclasses that must be defined to accurately map an area by digital classification, a key process is the aggregation of spectral classes and their assignment to informational classes. For example, accurate classification of the informational class *deciduous forest* may require several spectral subclasses, such as *north-facing forest, south-facing forest, shadowed forest,* and the like. When the classification is complete, these subclasses should be assigned a common symbol to represent a single informational class.

6. *Display and symbolization.* The wide range of colors that can be presented on color displays, and the flexibility in their assignment, provide unprecedented opportunities for effective display of land cover information. Although unconventional choices of colors can sometimes be effective, it is probably sensible to seek some consistency in symbolization of land cover information to permits users to quickly grasp the meaning of a specific map or image without detailed examination of the legend. Therefore, the color symbols recommended by Anderson et al. (1976) (Table 19.4) may be useful guides. Another strat-

**TABLE 19.4. Color Symbolization of Land Cover Classes**

| Anderson/USGS level I categories[a] | |
|---|---|
| Urban or built-up land | Red |
| Agricultural land | Light brown |
| Rangeland | Light orange |
| Forested land | Green |
| Water | Dark blue |
| Wetland | Light blue |
| Barren land | Gray |
| Tundra | Green–gray |
| Perennial snow or ice | White |
| **USGS 7.5-minute quadrangles[b]** | |
| Urban or built-up land | Pink |
| Nonforested land | White |
| Forested land | Green |
| Water and wetland | Blue |
| Exposed soil | Brown |
| Principal highways | Red |
| Structures, secondary highways | Black |

[a]From Anderson et al. (1976); recommended for work to be published.

[b]Offered here as an alternative, for experimentation, or for student projects.

egy for assignment of colors to classes is to mimic the colors used for USGS 7.5-minute quadrangles (Table 19.4).

Within such general strategies, it is usually effective to assign related colors to related classes to symbolize level II and level III categories. For example, subclasses of agricultural land can be represented in shades of brown (using Anderson et al.'s strategy as a starting point), water and wetlands in shades of blue, subclasses of forest in shades of green, and so forth.

## 19.7. Mapping Land Use Change

### *Significance*

Land use patterns change over time in response to economic, social, and environmental forces. The practical significance of such changes is obvious. For planners and administrators they reveal the areas that require the greatest attention if communities are to develop in a harmonious and orderly manner. From a conceptual perspective, study of land use changes permits identification of long-term trends in time and space and the formation of policy in anticipation of the problems that accompany changes in land use (Estes & Senger, 1972; Anderson, 1977; Jensen & Toll, 1982).

Because a given map can show only a single image of the many that form the evolving pattern of land use in a region, any land use map begins to decrease in accuracy from the time it is completed. Users who are familiar with the mapped region will accumulate an informal knowledge of changes that have occurred after the map was prepared—a mental map of the changes. For systematic study of changes, however, it is necessary to prepare maps that formally document changes in land use between two specific dates. Aerial imagery provides the unique capability to reconstruct previous land use patterns using archived images and thereby form the basis for a study of former patterns even though no map was prepared at the time.

### *Manual Compilation of Land Use Change Maps*

Although in theory preparation of change maps is very simple, a number of practical problems are encountered in practice. Preparation of a change map requires comparison of two separate land use maps prepared from imagery acquired at two dates; areas that experience changes and land use are noted (usually by superimposition of the two maps), then recorded on a third map (Figure 19.17). This third map shows only the changes, which can then be tabulated by area and category to reveal the extent and placement of land use changes. If imagery for several dates is available, a series of change maps can record the evolution of land use patterns over time and possibly reveal long-term patterns of change, rates of change in specific areas, and intermediate steps in the development of land use patterns.

Although this procedure is essentially simple and straightforward, a number of practical problems must be anticipated. First, the two maps must share a common base before they can be registered to one another. Even if both maps already use a common base, the interpreter must work carefully, as minor differences in placement of boundaries can create differences that do not form evidence of land use change but are merely artifacts of the interpretation process. Furthermore, the two maps must be prepared at a consistent level of detail so that the change map records true changes rather than differences in interpretation technique. Finally, the two maps must use the same classification system or it will not be possible to compare the two maps. For these reasons, it is important that the same interpreters work on all phases of a change map, or that interpreters be supervised by a single individual with overall knowledge of the project.

The image interpreter may find that preparation of a change map requires examina-

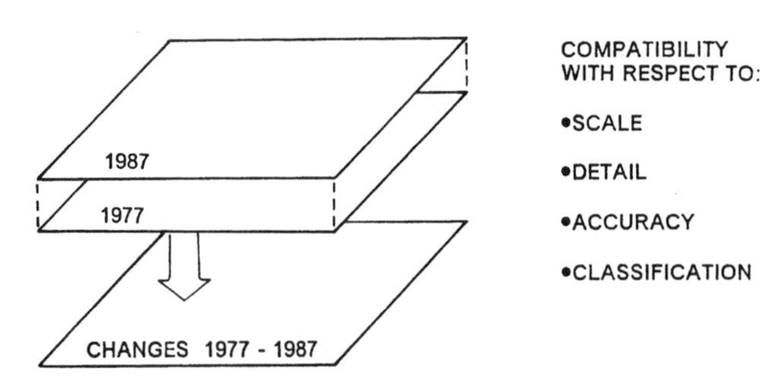


**FIGURE 19.17.** Manual compilation of land use change.

tion of imagery at varying scales, resolutions, and qualities. Each interpreter must assess the imagery in relation to map scale and classification detail to determine the level of detail most suitable for the change map. As a general guideline, it seems sensible to recommend that all change maps be prepared at a level of detail consistent with that obtained from interpretation of the lowest-quality, coarsest-resolution imagery that will be used in the project. Otherwise, the interpreter will be faced with the problem of comparing two maps that differ greatly in detail and accuracy.

These considerations lead to the general observation that preparation of change maps requires continuity in technique and in personnel and close coordination throughout the mapping process. Source maps must be compatible with respect to classification, spatial detail, and map base. As a result, it would seem unwise to use maps prepared by different individuals or organizations as the basis for change maps.

### *Photographic Compilation of Spectral Changes*

Eyton (1983) is one of several who have studied methods for combining film images of several Landsat scenes of the same area acquired at different times into a single composite image. Three film positives showing the same geographic area at different dates are used to form a single false-color composite in which colors depict changes in reflectivity. The black-and-white film positives are used to make separate color separations so that the image for each date forms one of the three separate emulsions for a color reversal image, similar to those described in Chapter 3. On the composite, black denotes objects that are dark on all images; white identifies objects that are bright on all three dates. Objects that experience big spectral change exhibit bright colors on the composite; those with small changes are represented as more subtle, neutral colors.

Although this method is limited somewhat by the difficulty in registering the separate images, it forms a relatively quick, inexpensive method of identifying areas that have experienced spectral change. Convenience in preparation of change maps means that it is possible to explore a wide variety of topics. For example, seasonal change maps display changes that occur between contrasting seasons to show differences in extent of water bodies, vegetation cover, and so on. Anniversary change maps illustrate changes that have occurred from one year to the next.

### *Digital Compilation of Land Use Change Maps*

The simple logic and mechanical nature of change compilation procedures lend themselves to implementation in the digital domain. Jensen (1996) enumerates change detection algorithms and notes that not all land cover changes are identical. For example, suburban land cover typically progresses through several stages—from initial clearing of land through construction and landscaping—any or all of which may be present on a given image. Because each stage has distinctive spectral characteristics, a given area may appear in different form on each image, and the analyst may encounter different levels of difficulty in detection and classification at each stage.

The issue of compatibility retains its significance in the digital domain. Images must register and have comparable levels of spatial, spectral, and radiometric resolution. Unless the objective is to compare images from one season to another, the two images typically represent the same season, but different years. Jensen (1981) lists a selection of preprocessing operations effective in preparing data for use in change detection comparisons. In any specific situation, the analyst must apply an intimate knowledge of the area to be studied, then apply a selection of methods tailored to the specifics of the study area. It seems unlikely that there is any single procedure that can be equally effective in all situations.

Jensen's (1996) list of digital change detection procedures includes the following:

1. *Image algebra* applies arithmetic operations to pixels in each image, then forms the change image from the resulting values. *Image differencing,* for example, simply subtracts one digital image from another digital image of the same area acquired at a different date. After registration, the two images are compared pixel by pixel to generate a third image composed of numerical differences between paired pixels from the two images. Values at or near zero identify pixels that have similar spectral values and therefore presumably have experienced no change between the two dates. This procedure is typically applied to a single band of a multispectral data set; usually a constant value is added to eliminate negative values. The analyst must select (sometimes by trial and error) a threshold level to separate those pixels that have experienced change from those that have not changed land cover but may exhibit small spectral variations caused by other factors. Jensen (1981, 1982) reports that image differencing is among the most accurate change detection algorithms, but it is not equally effective in detecting all forms of land use change. *Image ratios* (Chapters 16 and 17), another form of image algebra, can be important in change detection because they may assist in standardizing for variations in illumination, thereby isolating spectral changes caused by land cover changes.

2. *Postclassification comparison* requires two or more independent classifications of each individual scene, using comparable classification strategies. The two classifications are then compared, pixel by pixel, to generate an image that shows pixels placed in different classes on the two scenes. Successful application of this method requires accurate classifications of both scenes, so that differences between the two scenes portray true differences in land use rather than differences in classification accuracy. In urban and suburban landscapes, the high percentages of mixed pixels (at Landsat MSS resolution) have tended to decrease classification accuracy and therefore to generate inflated estimates of the numbers of pixels that have experienced change. Because postclassification comparison permits compilation of a matrix of *from–to changes,* it provides more useful results than some of the other methods (e.g., image algebra, which simply identifies pixels that have changed, without specifying the classes involved).

3. *Multidate composites* are formed by assembling all image bands from two or more dates together into a single data set. The composite data set can then be examined using methods described in earlier chapters, including principal components analysis (Chapter 10) and image classification (Chapter 11). Thus, for two dates of Landsat MSS data, the data would consist of eight bands of spectral values. A classification of the composite identifies not only the usual land use categories but also classes composed of pixels that have experienced change from one date to the next. Due to the large size of the data set and its unwieldy character, this approach is usually inefficient and has been reported to be less accurate.

4. *Spectral change vector analysis* examines each pixel's position in multispectral data space. If the corresponding pixels in images from two dates occupy similar positions in multispectral data space, then the ground area represented by the pixel has not changed much during the interval between the two dates. If the two pixels occupy different positions, then the ground area has experienced changes. Usually the procedure is applied by preparing a pair of images representing the magnitude of the multispectral changes and the detection of the change; these images can then be inspected by the analyst to set thresholds separating substantive changes from incidental changes caused by atmospheric effects, shadowing, or other ephemeral causes.

5. Use of a *binary change mask* requires classification of the image from the first date. Image algebra is then applied to original image data for both dates to generate an image of changes, as described in (1) above. The image is used to prepare a binary mask, representing only changed and unchanged pixels. The binary change mask is then superimposed over the image data for the second date, and the classification of this image examines only those pixels that have been identified as changed. This method is often effective, provided that the binary change mask is accurate.

6. *On-screen digitization* requires specialized image analysis software that permits the analyst both to view images side by side on the same screen and to outline changed areas manually using on-screen digitization based on visual interpretation. This method has been used primarily, if not exclusively, with large-scale digital images.

7. *Change detection by image display* is essentially a digital version of the photographic change detection methods mentioned above. Corresponding bands from different dates are used as separate overlays in a red–green–blue color display (Chapter 4). Pixels that have experienced change appear in distinctive colors (depending on the assignment of images to colors), thereby flagging those pixels that have changed during the interval between the two dates. This technique does not reveal the from–to change information provided by some other techniques.

Jensen (1996) provides additional detail, examples, and discussion of other methods.

## 19.8. Broad-Scale Land Cover Studies

The availability of multispectral AVHRR data (and data from similar meteorological satellites) on a regular basis has provided the capability to directly compile broad-scale land cover maps and data. In this context, *broad scale* refers to images that represent entire continents, or even entire hemispheres, by data collected over a short period of time, perhaps about 10 days to 2 weeks. Previously, data for such large regions could be acquired only by generalizing more detailed information—a task that was difficult and inaccurate because of the incompleteness of coverage and the inconsistencies of the many detailed maps required to prepare small-scale maps. The finer-resolution data from the Landsat and SPOT systems provide information of local and regional interest but are not really suitable for compilation of data at continental scales because of the effects of cloud cover, differences in sun angle, and other factors that prevent convenient comparisons and mosaicking of many scenes into a single data set representing a large region.

AVHRR data, described in Chapter 16, provide essentially uniform coverage of entire continents over relatively short periods. Accumulation of data over a period of 1 week to 10 days usually permits each pixel to be observed at least once under cloud-free conditions. Although the scan angle varies greatly, data are acquired at such frequent intervals that it is often possible to select coverage of the region of interest from the central section of each scene to reduce the effects of the extreme perspective at the edges of each scene.

Tucker et al. (1985) examined AVHRR data for Africa, using images acquired over a 19-month interval. They examined changes in the vegetation index (the "normalized difference" described in Chapter 16) at intervals of 21 days; their results clearly illustrate the climatic and ecological differences between the major biomes, as observed using the vegetation index, and seasonal variations in the vegetation index. They conducted a second analysis using eight of the 21-day segments selected from the 19-month interval mentioned above. They calculated principal components (Chapter 10) as a means of condensing the many variables into a concise, yet potent, data set that describes both seasonal and place-to-place variations in the vegetation index. Their land cover map (based on the first three principal components) is an extraordinary representation of key environmental conditions over an entire continent. It shows what are clearly major climate and ecological zones, as defined by conventional criteria, although they have been derived from data that are completely independent of the usual climatic data. The map has a uniformity, a level of detail, and (apparently) an accuracy not obtainable using conventional vegetation and climatic analysis. Although the authors have not fully evaluated their map by systematic comparison with field observations, it seems clear that products of this sort will provide an opportunity to examine broad ecological zones at continental scales and, furthermore, to examine seasonal and year-to-year changes in such zones.

More recently, AVHRR data have been used to compile other kinds of broad-scale land cover maps, to be discussed in Chapter 20.

## 19.9. Summary

Study of land use and land cover reveals the overall pattern of mankind's occupation of the earth's surface and the geographic organization of its activities. At broad scales the land cover map provides a delineation of the broad patterns of climate and vegetation that form the environmental context for our activities. At local and regional scales, knowledge of land use and land cover forms a basic dimension of resources available to any political unit; both the citizens and the leaders of any community must understand the land resources available to them and the constraints that limit uses of land and environmental resources.

Although the formal study of land use and land cover dates from the early 1800s, systematic mapping at large scale was not attempted until the 1920s, and aerial photography and remote sensing were not routinely applied until the 1960s. Thus, effective land use mapping is a relatively recent capability, and we have yet to fully assemble and evaluate all the data that are available and to develop the techniques for acquiring and interpreting imagery.

Without the aerial images acquired by remote sensing, there can be no really practical method of observing the pattern of land cover or of monitoring changes. Systems such as MSS, TM, SPOT, and AVHRR have provided a capability for observing land cover at broad scales and at intervals that previously were not practical. Images from such systems have not only provided vital information but also presented data at new scales that have changed the intellectual perspective with which we consider the environment by recording broad-scale patterns and relationships that otherwise could not be accurately perceived or analyzed.

## Review Questions

1. Using aerial photographs and other information provided by your instructor, design level III categories (compatible with the USGS classification) for a nearby region.
2. Review Chapter 9 and then Chapter 8 to refresh your memory of resolutions of satellite sensors and the effects of mixed pixels. A typical city block is said to be about 300 ft. × 800 ft. in size. Make rough assessments of the effectiveness of the MSS, TM, and SPOT systems for depicting land use and land cover within urban regions. List factors other than the sizes of objects that would be important in making such assessments.
3. Outline some of the difficulties that would be encountered in compiling land use and land cover maps if aerial photography and remotely sensed data were not available.
4. Compare relative advantages and disadvantages of alternative kinds of imagery, including photography, thermal imagery, or radar imagery (at comparable scales and resolutions), for compiling land use and land cover maps. List advantages and some problems that might be encountered in using all three kinds of images in combination.
5. Review Section 19.7, and then prepare a diagram or flow chart that illustrates manual preparation of a land use change map.
6. The following issues all require, directly or indirectly, use of accurate land use and land cover information. For each, identify, in a few sentences or in short paragraphs, the role of accurate land use maps and data.
   a. Solid waste disposal
   b. Selecting a location for a new electrical power plant.
   c. Establishing boundaries of a state park or wildlife preserve
   d. Zoning decisions in a suburban region near a large city
   e. Abandoned toxic waste dumps
7. Section 19.8 is closely related to material in Chapters 16 (Section 16.8) and 20 (Sections 20.3 and 20.4). Review these sections and prepare a summary of links between the topics, as well as some differences.
8. Some scientists have advocated development of a classification system with categories based upon the appearance of features on specific kinds of remotely sensed images. In contrast, the approach used by Anderson et al. (Table 19.1) is based upon the idea that remotely sensed data should be categorized using classes that remain the same for all forms of remotely sensed images, and that match those used by planners. Compare advantages and disadvantages of both strategies, considering both the ease of application to the imagery as well as the ease of use by those who must actually apply the information.

9. About 18% of urban land is devoted to streets. About 20% of the land area of large cities is said to be undeveloped. Assess the ability of remotely sensed images to contribute to assessing the amount and pattern of these two kinds of land use; consider TM and SPOT data as well as aerial photography at 1:10,000.

## References

### General

Anderson, James R., E. E. Hardy, J. T. Roach, and R. E. Witmer. 1976. *A Land Use and Land Cover Classification for Use with Remote Sensor Data.* U.S. Geological Survey Professional Paper 964. Washington, DC: U.S. Government Printing Office, 28 pp.

Baker, Robert D., J. E. deSteiger, D. E. Grant, and M. J. Newton. 1979. Land-Use/Land Cover Mapping From Aerial Photographs. *Photogrammetric Engineering and Remote Sensing,* Vol. 45, pp. 661–668.

Campbell, J. B. 1983. *Mapping the Land: Aerial Imagery for Land Use Information.* Resource Publications in Geography. Washington, DC: Association of American Geographers, 96 pp.

Campbell, James B. 1996 (in press). Land Use Inventory. Chapter 11 in *Photographic Interpretation* (W. R. Philipson, ed.). Bethesda, MD: American Society for Photogrammetry and Remote Sensing.

Christian, C. S. 1959. The Eco-Complex and Its Importance for Agricultural Assessment. *Monographic Biologicae,* Vol. 8, pp. 587–605.

Chuvieco, Emilio, and R. G. Congalton. 1988. Using Cluster Analysis to Improve the Selection of Training Statistics in Classifying Remotely Sensed Data. *Photogrammetric Engineering and Remote Sensing,* Vol. 54, pp. 1275–1281.

Cicone, R. C., and M. D. Metzler. 1984. Comparison of Landsat MSS, Nimbus-7 CZCS, and NOAA-7 AVHRR Feature for Land-Cover Analysis. *Remote Sensing of Environment,* Vol. 15, pp. 257–265.

Cowardin, L. M., V. Carter, F. C. Golet, and E. T. LaRoe. 1979. *Classification of Wetlands and Deepwater Habitats of the United States.* Washington, DC: Fish and Wildlife Service, 103 pp.

Drake, B. 1977. Necessity to Adapt Land Use and Land Cover Classification Systems to Readily Accept Radar Data. In *Proceedings of the Eleventh International Symposium on Remote Sensing of Environment.* Ann Arbor: University of Michigan, pp. 993–1000.

Jensen, John R. 1978. Digital Land Cover Mapping Using Layered Classification Logic and Physical Composition Attributes. *The American Cartographer,* Vol. 5, pp. 121–132.

Jensen, John R., and D. L. Toll. 1982. Detecting Residential Land-Use Development at the Urban Fringe. *Photogrammetric Engineering and Remote Sensing,* Vol. 48, pp. 629–643.

Loelkes, G. L. 1977. *Specifications for Land Cover and Associated Maps.* U.S. Geological Survey Open File Report 77-555. Reston, VA: USGS, 51 pp.

Loveland, Thomas R., J. W. Merchant, D. O. Ohlen, and J. F. Brown. 1991. Development of a Land-Cover Characteristics Database for the Coterminous U.S. *Photogrammetric Engineering and Remote Sensing,* Vol. 57, pp. 1453–1463.

Loveland, Thomas R., J. W. Merchant, J. F. Brown, D. O. Ohlen, B. C. Reed, P. Olson, and J. Hutchinson. 1995. Seasonal Land-Cover Regions of the United States. *Annals of the Association of American Geographers,* Vol. 85, pp. 339–355.

Nunnally, N. R., and R. E. Witmer. 1970. Remote Sensing for Land Use Studies. *Photogrammetric Engineering,* Vol. 36, pp. 449–453.

Philipson, W. R. (ed.). 1996 (in press). *Photographic Interpretation.* Bethesda, MD: American Society for Photogrammetry and Remote Sensing.

Robinove, C. 1981. The Logic of Multispectral Classification and Mapping of Land. *Remote Sensing of Environment,* Vol. 11, pp. 231–244.

### Change Detection

Estes, J. E., and L. W. Senger. 1974. *Remote Sensing: Techniques for Environmental Analysis.* Santa Barbara, CA: Hamilton, 340 pp.

Eyton, J. Ronald. 1983. Landsat Multispectral Color Composites. *Photogrammetric Engineering and Remote Sensing,* Vol. 49, pp. 231–235.
Jensen, John R. 1979. Spectral and Textural Features to Classify Elusive Land Cover at the Urban Fringe. *Professional Geographer,* Vol. 31, pp. 400–409.
Jensen, John R. 1981. Urban Change Detection Mapping Using Landsat Digital Data. *The American Cartographer,* Vol. 8, pp. 127–147.
Jensen, John R. 1982. Detecting Residential Land-Use Development at the Urban Fringe. *Photogrammetric Engineering and Remote Sensing,* Vol. 48, pp. 629–643.
Jensen, John R. 1996. *Introductory Digital Image Processing: A Remote Sensing Perspective.* Upper Saddle River, NJ: Prentice Hall, 316 pp.

## AVHRR for Land Cover Studies

Defries, R. S., and J. R. G. Townsshend. 1994. NDVI-Derived Land Cover Classifications at a Global Scale. *International Journal of Remote Sensing,* Vol. 15, pp. 3567–3586.
Tucker, C. J., J. R. G. Townshend, and T. E. Goff. 1985. African Land Cover Classification Using Satellite Data. *Science,* Vol. 227, pp. 369–375.

CHAPTER TWENTY

# Global Remote Sensing

## 20.1. Introduction

The history of remote sensing is one of increasing scope of vision, with respect to both space and time (Figure 20.1). Before the availability of aerial photography, human observation of the landscape was limited in scope to the area visible from a high hill or building. As aerial photography became available, it became possible to view larger areas over longer intervals; later, as the routine availability of infrared emulsions permitted high-altitude aerial photography, larger areas could be represented within a single frame. Later, the synoptic views of satellite images increased our geographic scope even further. Finally, by merging images from several dates to remove effects of cloud cover, it was possible to compile images representing continents, hemispheres, and the entire globe (Plate 20). Such images have made it possible to directly examine broad-scale environmental patterns and their changes over years, decades, and even longer intervals.

## 20.2. Biogeochemical Cycles

Such capabilities open new opportunities for understanding basic processes that underlie the earth's climatic and biologic patterns. The elements and nutrients that form the basis for life on earth reside differentially in biologic, geologic, and atmospheric components of the earth. Atmospheric and geologic processes cycle these nutrients and chemicals through the different components of the biosphere for different regions of the earth. The earth's atmosphere and oceans move at rates measured in hours, minutes, or days. At the other extreme, minerals of the earth's crust can be considered mobile only when viewed over intervals of many millions of years. Understanding the earth's environment requires examination of movements and interchanges between its components. For example, minerals weather from geologic material, enter the soil, are absorbed by plant roots, are incorporated into plant tissues as plants grow, are released into groundwater or surface runoff as plants die and decompose, then again assume geologic form as sedimentological and lithological processes concentrate and mineralize materials. Such are the biogeochemical processes that govern the movement of carbon, sulfur, nitrogen, oxygen, and other elements. Although such processes are well understood in general, we do not have a good un-

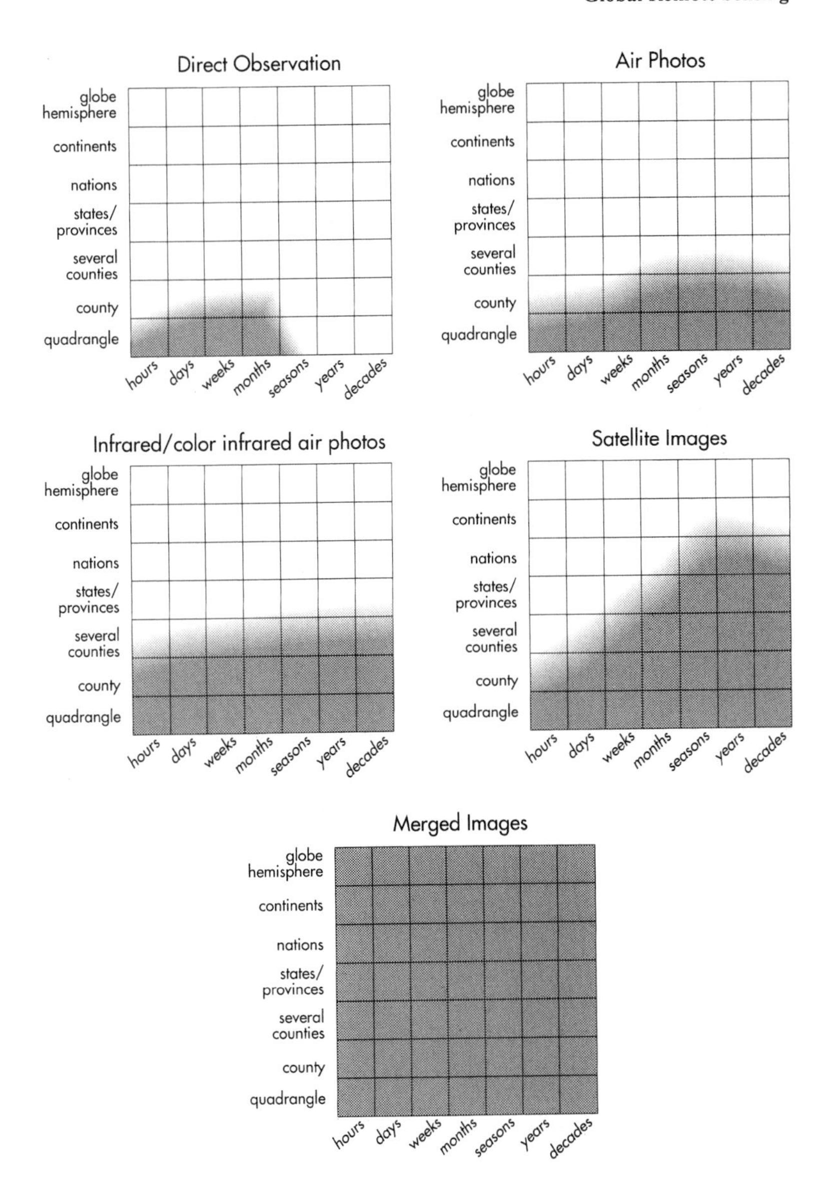


**FIGURE 20.1.** Increasing scope of human vision of the earth. The horizontal axis in each graph represents time (in an arbitrary scale of increasing intervals, from hours and days at one extreme to years and decades at the other). The vertical axis represents space (in arbitrary units, from quadrangle-sized areas at one extreme to continents and hemispheres at the other). From MacEachren et al. (1992). Copyright 1992 by Rutgers University Press. Reproduced by permission.

derstanding of specific details—amounts, rates of movement, how rates might change from one ecosystem to another, and how they might change in response to disturbances caused either by natural events or by the actions of humans.

The following sections outline major components of the most important biogeochemical cycles; these are not intended to provide complete descriptions but to illustrate the kinds of issues that require global perspectives.

### *Hydrologic Cycle*

The hydrologic cycle is one of the best known and most central of the many processes that cycle materials through the earth's biosphere (Figure 20.2). Because of the central role of moisture in biologic, climatic, and geologic processes, the hydrologic cycle (discussed earlier in Chapter 18) has obvious significance for all processes that cycle materials through the earth's environment. Hydrologic processes at all scales are intimately connected with land cover (see Chapters 16 and 19), so the broad-scale land cover information provided by AVHRR data form a central component of efforts to examine the hydrologic cycle at continental and global scales. Further, as mentioned in Chapter 6, CORONA imagery provides, at least in some regions, a record of land cover extending into the early 1960s, which offers one of the longest broad-scale image records of land cover available.

### *Carbon Cycle*

Carbon, in its varied forms, is one of the essential components of living organisms and is of key interest in any consideration of global cycling within the biosphere (Figure 20.3).

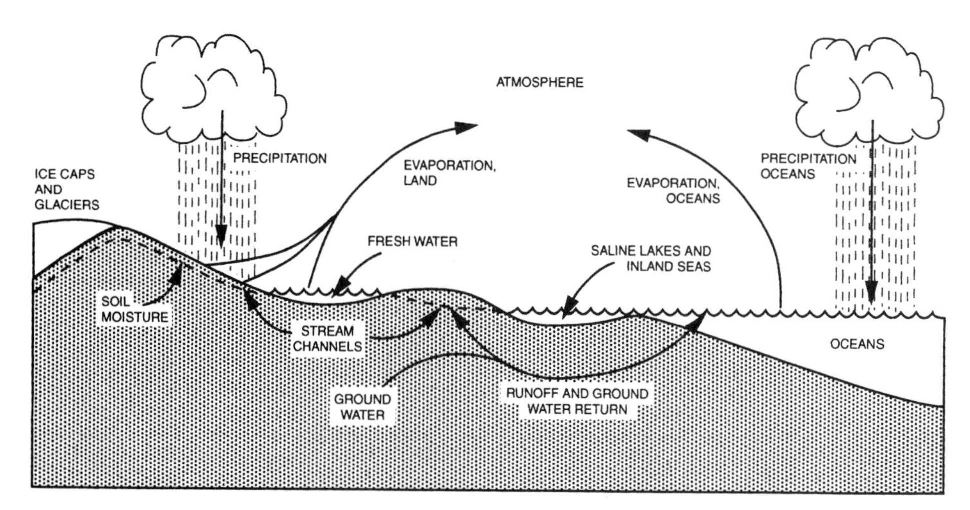


**FIGURE 20.2.** Hydrologic cycle. Adapted from Strahler and Strahler (1973). Copyright 1973 by A. N. Strahler. Reproduced by permission.

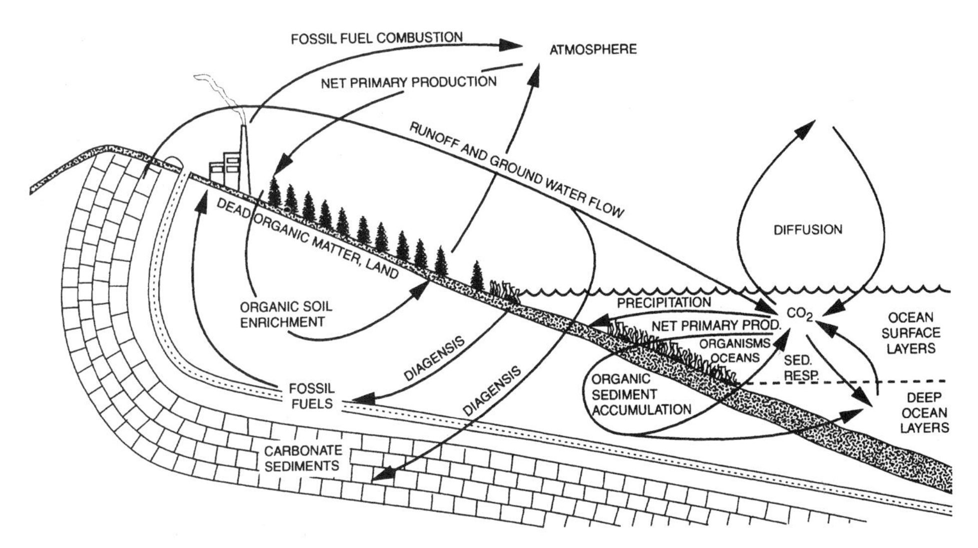


**FIGURE 20.3.** Carbon cycle. Adapted from Strahler and Strahler (1973). Copyright 1973 by A. N. Strahler. Reproduced by permission.

Carbon is available for living organisms as $CO_2$ and as organic carbon (plant and animal tissues), sometimes known as *fixed carbon* because it is much less mobile than $CO_2$. Photosynthesis reduces $CO_2$ to organic carbon; within organisms, respiration can convert organic carbon to its gaseous form.

Chapter 2 described the significance of atmospheric $CO_2$ in the earth's energy balance; although atmosphere holds only a small proportion of the earth's carbon, atmospheric $CO_2$ is very effective in absorbing long-wave radiation emitted by the earth. Larger amounts of carbon are stored in oceans.

The largest reservoir of carbon is carbonate rocks (such as limestone). Very slowly, carbon from the atmosphere, oceans, and living organisms enters this reservoir as sedimentary rocks are formed. Slowly, carbon is released through the weathering of these rocks. An exception, of course, is carbon residing as fossil fuels, which is rapidly released as it is burned for heat, to generate electrical power, and to propel automobiles and aircraft. Estimation of the amount of carbon released by these activities and its effects on the earth's climate is a high priority for those who hope to gauge effects of human activities upon earth's climates.

Major components in the carbon cycle are illustrated in Figure 20.3. Among the key processes is the incorporation of atmospheric $CO_2$, into the structure of living organisms by photosynthesis. The decay of organic matter releases $CO_2$ into the atmosphere. $CO_2$ is also released by burning of fossil fuels.

Remote sensing instruments assist scientists in understanding the carbon cycle by estimating the areas covered by plants, identifying the kinds of plants, and estimating the period for which they are photosynthetically active. Therefore, the use of broad-scale remote sensing data and vegetation indices contributes to efforts to estimate carbon incorporated into biomass, rates of carbon fixation, rates of release as plant tissues decompose, and the impact of human disruption of established vegetation patterns.

### *Nitrogen Cycle*

Nitrogen occurs in large amounts on earth and is of great significance in the food chain because it occurs as a principal element in the atmosphere and in tissues of plants and animals (Figure 20.4). It occurs in several forms, both organic and inorganic, and participates in important biologic processes. Large amounts of the earth's nitrogen are considered inactive as they are held in sedimentary and crustal rocks and do not participate in the short-term cycling of nitrogen. Within the active pool, the largest amount of nitrogen resides in the atmosphere, as one of the major constituents of the atmosphere. Atmospheric nitrogen, however, is not available for biologic use.

Nitrogen becomes available for biologic use through fixation—the chemical reaction that combines with hydrogen to create ammonia ($NH_3$). Although nitrogen fixation can occur within the atmosphere (through lightning strikes), in the burning of fossil fuels, and industrially (in the manufacture of artificial fertilizers), the most significant source by far is biologically based. Biologic fixation occurs by the action of living plants to change atmospheric nitrogen ($N_2$) to fixed form ($NH_3$), which can be used by plants and animals. Microorganisms, both free-living and those symbiotic with specific plants (legumes), are the only animals that can accomplish this process. Amino acids from plant or animal tissue can be broken down by the process of ammonification, which releases energy, $H_2O$, and $CO_2$. Ammonification is accomplished by decomposing microorganisms.

By this process, nitrogen enters the food chain and the tissues of plants and animals. Denitrification, the process by which nitrites ($NO^-{}_2$) or nitrates ($NO^-{}_3$) are converted to molecular nitrogen or nitrous oxide, is accomplished by soil bacteria under anaerobic con-

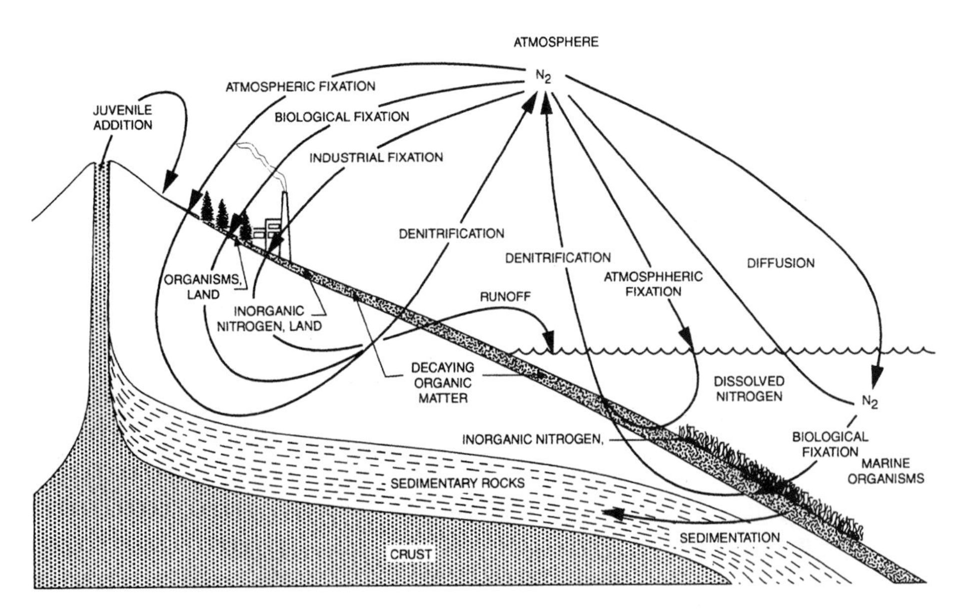


**FIGURE 20.4.** Nitrogen cycle. Adapted from Strahler and Strahler (1973). Copyright 1973 by A. N. Strahler. Reproduced by permission.

ditions. This process in which ammonia is oxidized to nitrate occurs in two steps. The first, completed by the bacteria *Nitrosommonas*, yields nitrate ions and water. The second step, accomplished by the bacteria *Nitrobacter*, consumes nitrite ions to yield nitrate.

Industrial fixation goes through the same process, applying energy under high pressures to transform atmospheric nitrogen to ammonia, which can be used in other industrial processes to manufacture fertilizers and explosives.

One of the largest reservoirs of active nitrogen is decaying organic matter; remote sensing assists in understanding the nitrogen cycle by providing accurate estimates of areas in which organic matter accumulates for decay (e.g., swamps and marshes).

### *Sulfur Cycle*

In comparison with the other biogeochemical cycles discussed here, the sulfur cycle (Figure 20.5) is somewhat less significant with respect to total amounts in active circulation and with respect to its role in the food chain. But it has great significance if we examine ways in which humans influence the sulfur cycle. Sulfur resembles nitrogen in the sense that it occurs in several forms within the environment and participates in several organic and inorganic reactions that alter its forms and behavior. By far the largest amount of sulfur is in geologic form, in sedimentary deposits, removed from short-term participation in the sulfur cycle. Weathering releases small amounts of geologic sulfur to the atmosphere. Of greater interest, however, is the release of sulfur into the atmosphere by removing and burning fossil fuels.

Oceans form the main reservoir for sulfur in active circulation. Only small amounts reside in the atmosphere. On land, some is incorporated in the tissues of plants and ani-

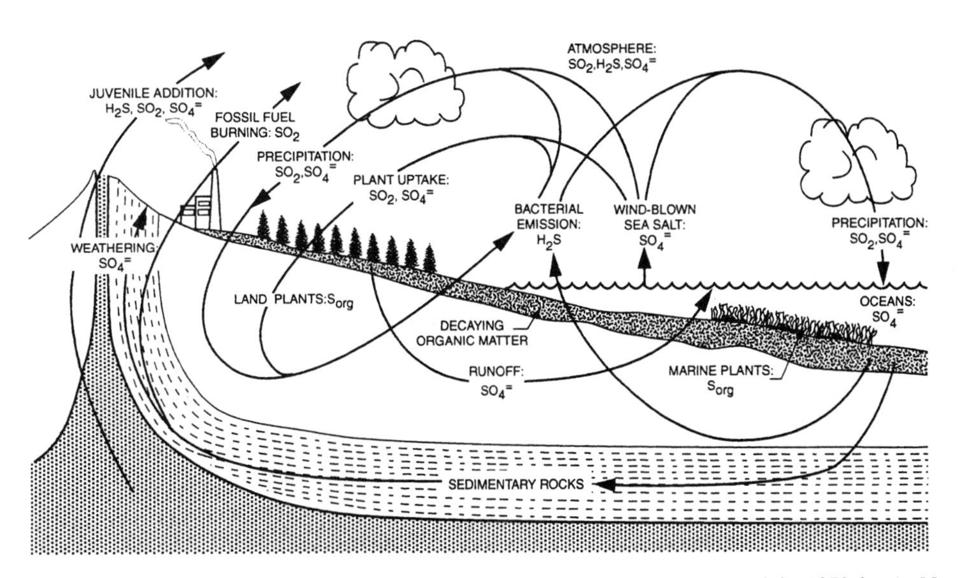


**FIGURE 20.5.** Sulfur cycle. Adapted from Strahler and Strahler (1973). Copyright 1973 by A. N. Strahler. Reproduced by permission.

mals and some is retained as decaying organic matter (e.g., in swamps and marshes). Sulfur enters the atmosphere from four sources: emissions which form bacteria as organic matter decays; combustion of fossil fuels; airborne dust, as sea spray evaporates; and volcanic sources.

Accelerated release of geologic sulfur by burning of fossil fuels has contributed to reactions within the atmosphere that have increased the acidity of rainfall near industrialized regions of northern latitudes, which in turn is believed to have decreased the pH of lakes and rivers and soils in some regions. Such changes in soils and water acidity, if sustained, would have profound impact on wildlife, vegetation patterns, and soil fertility. Because such impacts have already been observed, governments have instituted efforts to reduce sulfur emissions and to encourage industries to switch to low-sulfur fuels.

### *Significance of Remote Sensing*

Remote sensing is the primary means by which we can observe the dynamic character of the earth's biosphere. Although basic outlines of biogeochemical cycles are well-known, specific quantities, rates of movement, and regional variation are not well understood, nor is the true impact of human activities on their functioning. Such detail is necessary for scientists to assess the significance of human influences on biogeochemical processes and their significance in relation to natural variations. Therefore, scientists seek not only to understand the general structure of such cycles but also to develop a more precise knowledge of quantities and rates of movement within these cycles. Preparation of such estimates is very difficult as the figures commonly given are necessarily rough estimates and conventional data are not adequate. For example, the existing network of climate stations consists of a series of point observations at locations not representative of specific environments. This network can provide only the crudest estimates of temperature patterns and is clearly inadequate for answering the critical questions concerning the possible warming of the earth's atmosphere during recent history.

Mather and Sdasyuk (1991, pp. 96–113) review existing sources of environmental data suitable for assessing broad-scale environmental trends, identifying the many defects in the present sources of data. They, like many others, conclude that broad-scale remote sensing provides the only practical means to acquire data of sufficient scope for addressing these issues.

## 20.3. Advanced High-Resolution Radiometer

Previously introduced in Chapter 16, AVHRR is a multispectral radiometer carried by a series of meteorological satellites operated by NOAA in near-polar, sun-synchronous orbits. It can acquire imagery over a swath width of approximately 2,400 km. AVHRR acquires global coverage on a daily basis; the frequent global coverage is possible because of the wide-angular field of view, but areas recorded near the edges of images suffer from severe geometric and angular effects. As a result, AVHRR data selected from the regions near the nadir provide the most accurate information. Although designed initially for much narrow purposes, AVHRR is the first sensor to provide the basis for collection of worldwide data sets that permit assessment of environmental issues.

AVHRR pixels are about 1.1 km on each side at nadir, with fine radiometric resolution at 10 bits. Details of spectral coverage vary with specific mission, but in general AVHRR sensors have recorded radiation in the visible and infrared regions of the spectrum. AVHRR spectral regions have been selected to attain meteorologic objectives, including spectral data to distinguish clouds, snow, ice, and open water. Nonetheless, AVHRR has been employed on an ad hoc basis for land resource studies. The AVHRR sensor includes one channel in the visible spectrum, one in the near infrared, and three in the thermal infrared. It is possible to use the data to calculate vegetation indices of the type described in Chapter 16.

From these AVHRR data, the global vegetation index (GVI) data set has been prepared for the period 1982–1985. GVI data are presented using an arbitrary grid system, with cells that vary in size with latitude, from about 13 km at the equator to 26 km at the poles. Data collected within a 7-day period are examined for the brightest ("greenest") NDVI value in each cell for the reporting interval. These data form the AVHRR composites described in Chapter 16 and referred to indirectly in Chapter 4. GVI images remove effects of clouds, except for instances in which clouds are present in all seven images.

Such data provided one of the first tools that permitted study of the earth's major biomes. Previously, the areas and extents of major regions of forest, savanna, tundra, deserts and the like could be estimated roughly from climate and vegetation patterns as observed at ground level. Although satellite data show that these earlier approximations were quite accurate, they permit closer studies of detail of the patterns and examination of seasonal variations and long-term trends with accuracy and detail that were not previously possible. Specifically, AVHRR permits closer examination of effects of droughts and other broad-scale climatic changes, as well as changes related to human activities.

### *Pathfinder Data*

In 1990, NASA and NOAA began a program to compile *Pathfinder data sets,* designed to provide data for analysis of long-term environmental processes at continental and global scales. Pathfinder data are compiled from systems placed in service prior to the Earth Observing System, to provide experience with global scale data systems and to extend the period of analysis to as early a date as possible. These objectives require the use of systems with accurate calibration and consistent qualities to permit analysis over time and from one sensor system to another. The first Pathfinder data sets have been prepared from AVHRR data, other meteorological satellite data, and selected Landsat data. Pathfinder data are sampled over time (in part to select cloud-free data and to reduce other atmospheric effects) and over space (to create consistent levels of detail and data sets of manageable sizes) and processed to permit convenient use by a wide variety of users. Separate Pathfinder initiatives are under way to compile data describing ocean surface temperature, atmospheric conditions, and land surfaces.

Pathfinder AVHRR land data are global in scope but sampled over time and space. The *daily data set* consists of AVHRR terrestrial data, mapped to an equal area projection on an eight km grid. Pixels near the edges of the AVHRR swath are excluded. The daily data can be used directly for analysis or as source data for composite data sets. The *composite data set* is derived from the daily data set; the composting process filters out effects of clouds and selects data for each date from the scene with the highest NDVI. The *climate data set* is compiled to a resolution of 1° × 1° resolution, for 8- to 11-day composite periods. The result forms a set of cloud-free, averaged, NDVI data. Thirty -six such images are produced each year to form data for examination by scientists studying broad-scale climatic and biosphere changes. *Browse images* are prepared from the daily data set and the composite data set to provide a reduced data set to assist in data selection and evaluation. Browse images are not designed for analysis.

Following is a list of Internet addresses that will yield all sorts of valuable information.

| | |
|---|---|
| AVHRR Pathfinder Program: | http://xtreme.gsfc.nasa.gov/pathfinder/ |
| Pathfinder Data Sets: | http://pathfinder.arc.nasa.gov/ |
| AVHRR Pathfinder Home Page: | http://xtreme.gsfc.nasa.gov/ |
| AVHRR 10-Day Composite: | http://edcwww.cr.usgs.gov/landaac/comp10d.html/ |
| AVHRR Oceans Pathfinder: | http://podaac-www.jpl.nasa.gov/sst/ |

## 20.4. Earth Observing System

Although existing satellite data such as AVHRR and those described in previous chapters might seem at first consideration to provide a means to study global environmental problems, in fact these images provide piecemeal rather than systematic, long-term coverage. Likewise, much of the existing archive of imagery records arbitrary regions of the spectrum (rather than regions targeted on those required for the specific tasks) and is not coordinated with related data collection efforts. Attacking these broad-scale, long-term data collection requirements requires instruments tailored for specific tasks at hand. This is the context in which NASA, in the late 1980s and early 1990s, developed a program to acquire the environmental data required to address questions posed by concerns over global environmental change.

*Mission to Plant Earth* (MTPE) is a long-term effort to observe the earth's environments and how they change over time as an essential component to developing an understanding of the nature, rates, and consequences of broad-scale environmental change. MTPE's early activities (in the 1990s) focused on the design of numerous satellites that require specialized orbits and instruments.

The second, long-term, component of MTPE is an integrated system of earth-orbiting satellites designed to provide a continuous stream of data for a period of 15 years or more. The Earth Observing System (EOS) consists of several satellites carrying numerous sensors designed to monitor environmental variables on a worldwide basis.

EOS efforts have been coordinated among agencies of the U.S. government and internationally through agreements with other nations. The design of EOS has changed in response to scientific, operational, and budgetary considerations and will no doubt continue to change, so the following sections will outline some of the instruments likely to be of interest to readers of this text rather than attempt an exhaustive description of the complete system.

## 20.5. EOS Instruments

EOS is designed to collect data to provide a comprehensive overview of the dynamic components of the earth's atmosphere and land and water surfaces (Figure 20.6). To accomplish this objective, the EOS plan has included 30 or more instruments designed to monitor physical and biologic components of the biosphere. However, as budget priorities have changed, so has the program design evolved and no doubt will continue to evolve. Therefore, individual sensors have been redesigned or dropped from the plan. This chapter outlines only those instruments proposed for EOS that are most closely related to those already discussed in this volume. The reader may find that further program changes will alter even the brief list presented here.

### *Moderate-Resolution Imaging System*

The Moderate-Resolution Imaging System (MODIS; Figure 20.7) is designed to acquire imagery that will permit compilation of global data sets at frequent intervals (Running et al., 1994). It uses a cross-track scanning mirror, CCD detectors in 36 spectral bands, covering spectral regions from the visible to the thermal infrared. Spatial resolutions will vary from 500 m to 1 km; the satellite provides coverage every 2 days. MODIS instruments are designed to extend the kinds of data collected by AVHRR, TM, and some of the meteorologic satellites.

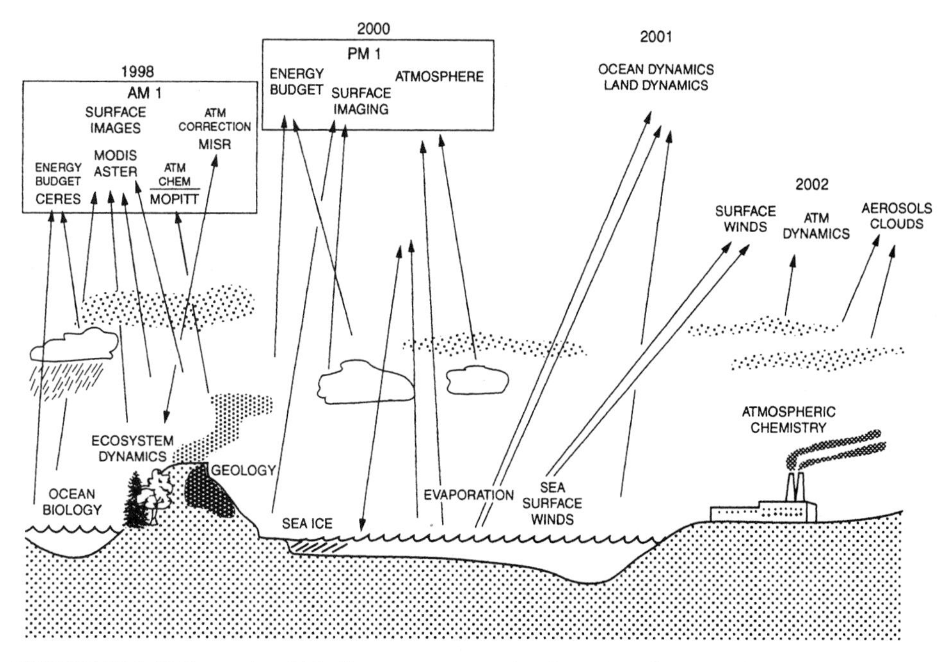


**FIGURE 20.6.** EOS overview. This diagram represents the planned configuration for AM-1, planned for 1998 and, in a more tentative manner, later EOS missions. Based on NASA diagram given in Assrar and Dokken (1993).

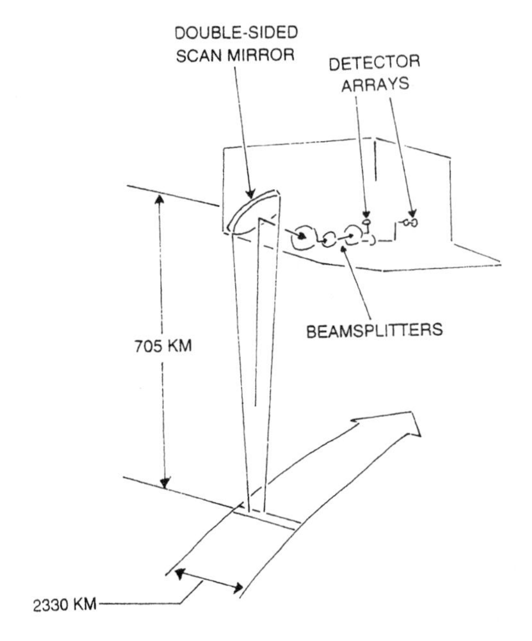


**FIGURE 20.7.** Schematic sketch of MODIS. Based on Running et al. (1994).

MODIS is designed for monitoring the earth's land areas; in essence, it is an instrument similar to the AVHRR but tailored to monitoring land resources (rather than the meteorologic mission of the AVHRR). *Moderate resolution* refers to its proposed 500-m pixels, which may seem coarse relative to SPOT or Landsat but are fine relative to the 1-km pixels of AVHRR and achieve the required global coverage to meet EOS objectives.

MODIS data will permit analysts to monitor biologic and physical processes at the earth's land surfaces and water bodies and collect data describing cloud cover and atmospheric qualities. MODIS will collect data describing range and forest fires, snow cover, chlorophyll concentrations, surface temperature, and other variables.

### *Advanced Spaceborne Thermal Emission and Reflection Radiometer*

The Advanced Spaceborne Thermal Emission and Reflection Radiometer (ASTER) is an imaging radiometer designed to collect data in 14 spectral channels (Figure 20.8). Three channels in the visible and near infrared provide 15-m resolution in the region 0.5 to 0.9 μm; six channels in the shortwave infrared provide 30-m resolution in the interval 1.6 to 2.5 μm, and five channels in the thermal infrared provide 90-m resolution in the interval 8 to 12 μm. Imagery can be acquired in a 60-km swath centered on a pointable field of view ±8.5° in the mid infrared and thermal infrared and ±2.4° in the visible and near infrared. The instrument acquires a stereo capability through use of an aft-pointing telescope that acquires imagery in channel 3. ASTER can provide 16-day repeat coverage in all 14 bands and 5-day coverage in the visible and near infrared channels.

The instrument is designed to provide data to describe land use, cloud coverage and

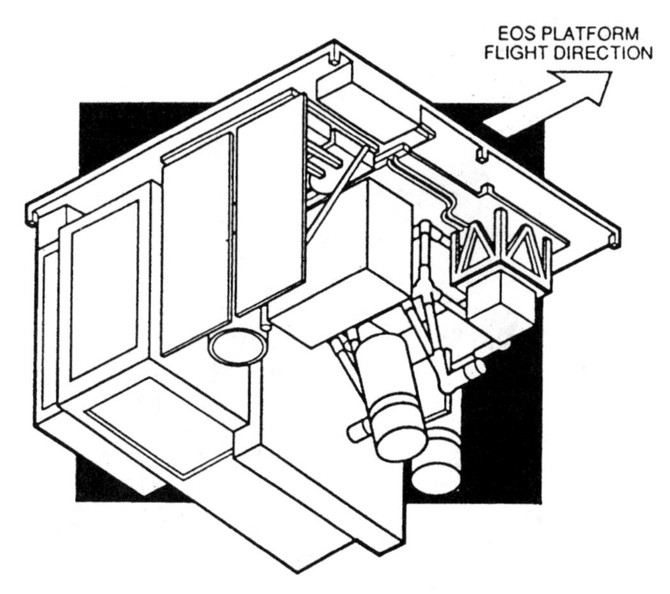


**FIGURE 20.8.** Schematic sketch of ASTER. NASA diagram: Assrar and Dokken (1993).

kind and heights of clouds, land and ocean temperature, topographic data, and information on glacial activity and snow cover.

### *Multi-Angle Imaging Spectroradiometer*

The Multi-Angle Imaging Spectroradiometer (MISR) employs nine CCDs to observe the earth's surface from nine angles. One looks to the nadir and four each look to the fore and aft along the ground track of the satellite in each of four spectral regions, centered on 0.443 μm, 0.555 μm, 0.67 μm, and 0.865 μm. These channels have resolutions of 240 m, 480 m, 960 m, and 1.92 km, respectively.

MISR is designed to observe the atmosphere at different angles, to observe features through differing path lengths and angles of observation. Data from MISR will permit the study of atmospheric properties, including effects of pollutants, and the derivation of information to correct data from other sensors for atmospheric effects. The instrument will also contribute to the development of models for the optical behavior of surfaces (such as forest canopies) by developing accurate BRDFs. The configuration of the sensors provide the ability to acquire stereo imagery of both ground surfaces and clouds.

### *Multifrequency Imaging Microwave Radiometer*

The Multifrequency Imaging Microwave Radiometer (MIMR; Figure 20.9) is a passive microwave radiometer designed to acquire data at six frequencies (6.8, 10.65, 18.7, 23.8, 36.5, and 90 GHz) at dual polarization. It observes a 1,400-km swath at resolutions vary-

**FIGURE 20.9.** Schematic sketch of MISR. NASA diagram: Assrar and Dokken (1993).

ing from 60 to 5 km. It is designed to observe surface temperature, sea ice, sea surface, atmospheric water vapor, and soil moisture patterns.

## 20.6. EOS Bus

EOS sensors are to be placed on a series of satellites designed to provide standard support functions, including communications, thermal control, power supply, and related functions. The standardized components are known as the *EOS bus* (Figure 20.10). One satellite, EOS-AM, planned for launch in 1998, will carry several of the sensors mentioned above, plus others, in an orbit designed for morning observation (local sun time), which optimizes observation of terrestrial surfaces by minimizing land areas obscured by cloud cover. A second satellite, EOS-PM, planned for launch in the year 2000, observes the earth later in the day, designed for optimum observation of ocean and atmospheric features as well as some terrestrial phenomena. EOS-PM will carry some of the same sensors (e.g., MODIS) that will permit features to be imaged by the same instrument at different times of day, providing the ability to examine diurnal changes such as those suggested in Chapter 8.

## 20.7. EOS Data and Information System

The EOS Data and Information System (EOSDIS) was established to handle the unprecedented volume of data to be acquired by the system of satellite sensors. EOS requires a

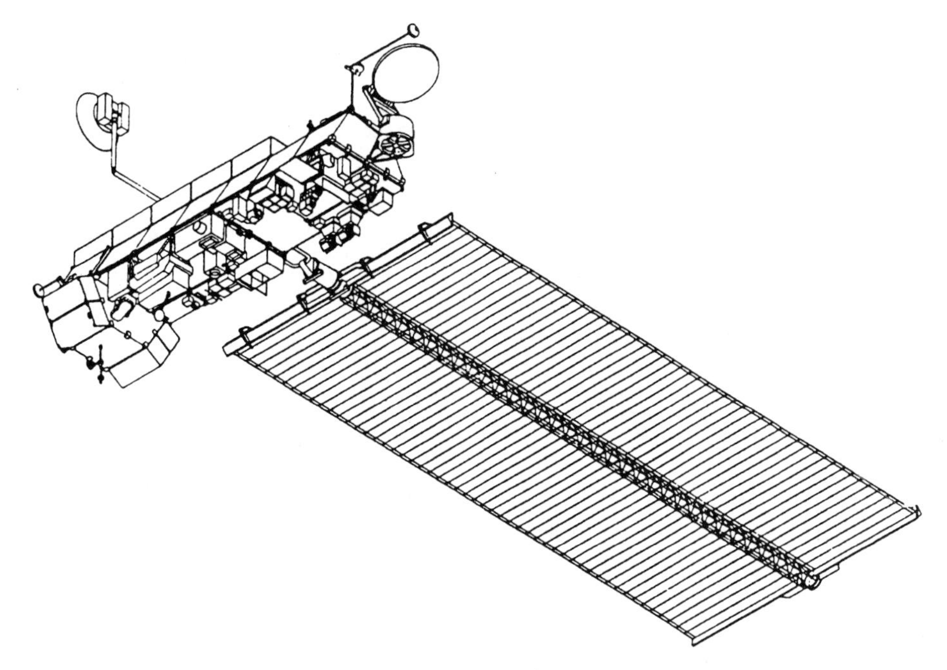

**FIGURE 20.10.** EOS bus, as planned for AM-1. NASA diagram.

system for acquiring and processing data and for distributing data to scientists for interpretation and analysis. Key components within EOSDIS are the eight distributed active archive centers (DAACs), established to process and distribute data. Each DAAC specializes in specific kinds of data (e.g., "land processes," "sea ice and polar processes," or "hydrology"), with the objective of providing sustained access to the data for a broad community of users (Figure 20.11 and Table 20.1). Components of EOSDIS are connected electronically, so users can quickly access archives, indices, and data from their own institutions, as outlined in Chapter 4.

## 20.8. Long-Term Environmental Research Sites

In this context, information provided by field data assumes a new meaning. The usual field observations, collected to match specific images on a one-time basis, at isolated times and places, have less significance when the objectives extend over large areas and long intervals. To support the search for long-term, broad-scale patterns of change, field data must exhibit continuity over time and be positioned to represent the ecosystems at continental scales. To address these needs, a network of long-term ecological research sites has been established within many of the earth's major ecological zones. In each site, long-term studies are under way to document ecological behavior and changes over time and to understand relationships between natural changes and those that might be caused or accelerated by human activities.

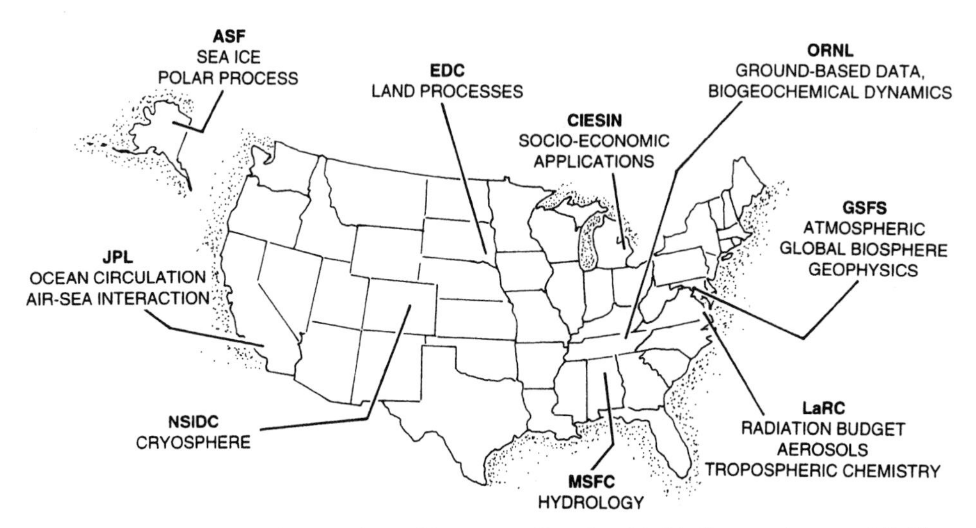


**FIGURE 20.11.** EOS distributed active archive centers. Based on NASA diagram: Assrar and Dokken (1993).

**TABLE 20.1. EOS Distributed Active Archive Centers**

| | |
|---|---|
| **GSFC**<br>Goddard Space Flight Center<br>Greenbelt, MD | Upper atmosphere, atmospheric dynamics, global biosphere and geophysics |
| **LaRC**<br>Langley Research Center<br>Langley, VA | Radiation budget, aerosols, tropospheric chemistry |
| **EDC**<br>EROS Data Center<br>Sioux Falls, SD | Land process data |
| **UAF**<br>University of Alaska<br>Fairbanks, AK | Sea ice and polar process data |
| **CU**<br>University of Colorado<br>Boulder, CO | Cryosphere (snow, glaciers, ice) |
| **JPL**<br>Jet Propulsion Laboratory<br>Pasadena, CA | Oceanic circulation and air–sea interactions |
| **MSFC**<br>Marshall Space Flight Center<br>Huntsville, AL | Hydrology |
| **ORNL**<br>Oak Ridge National Laboratory<br>Oak Ridge, TN | Biogeochemical dynamics |
| **CIESIN**<br>Consortium for International<br>Earth Science Information Network<br>Saginaw, MI | Socioeconomic data |

*Note.* From Assrar and Dokken (1993).

Research under way at some sites records local patterns of nutrient cycling as a component in understanding changes in the global cycles. Some long-term ecological research sites (LTERs) (Table 20.2) were established previously as reserves or research sites, so in some instances, some may have 30 or 40 years of records of climate and ecological conditions. Figure 20.12 and Table 20.2 show a network of LTER stations supported in part by the U.S. National Science Foundation. These sites represent many of the principal ecological zones of our hemisphere. Similar efforts are under way throughout the world to establish a network of international sites representing a broader range of biomes.

## 20.9. Global Land Information System

Global Land Information System (GLIS), developed by the U.S. Geological Survey (EROS Data Center, GLIS User Assistance, Sioux Falls, SD 57198; 1-800-252-4547; GLIS@GLIS.CR.USGS.GOV), is an interactive computer system designed to permit scientists to identify information pertaining to the earth's land surfaces. GLIS consists of metadata (Chapter 6) describing dates, coverages, cloud cover, quality ratings, and other characteristics of broad-scale data sets of the earth's land areas. Included are summaries of land cover, soil, geologic, cultural, topographic, and remotely sensed data. Users can connect to GLIS computers remotely to review data for their specific regions. GLIS permits examination of suitability of data for specific projects and inspection of sample data projects, and provides a vehicle for placing orders to acquire data.

GLIS descriptions are organized at three levels of detail. *Directories* provide summary information about data sets to permit scientists to search for the availability of specific data. For example, users could search directories by using key words to identify specific topics, by providing coordinates to search for coverage of specific areas, or by date to ascertain availability of data for specific years or seasons. *User guides* present detailed in-

**TABLE 20.2. Long-Term Ecological Research Sites**

1. Arctic tundra: Arctic Lakes and Tundra, Brooks Range, AK
2. Taiga: Bonanza Creek Natural History Area, Fairbanks, AK
3. Temperate coniferous forest: H. J. Andrews Experimental Forest, Blue River, OR
4. Eastern deciduous forest and tall-grass prairie: Cedar Creek Natural History Area, Minneapolis, MN
5. Northern temperate lakes: North Temperate Lakes, Madison WI
6. Row–crop agriculture: Kellogg Biological Station, Hickory Corners, MI
7. Hot desert: Jordana Experimental Range, Las Cruces, NM
8. Alpine tundra: Niwot Ridge/Green Lakes Valley, Boulder, CO
9. Eastern deciduous forest: Hubbard Brook Experimental Forest, West Thornton, NH
10. Eastern deciduous forest: Harvard Forest, Petersham, MA
11. Coastal barrier island: Virginia Coastal Reserve, Oyster, VA
12. Eastern deciduous forest, Coweeta Hydrological Laboratory, Otto, NC
13. Tall-grass prairie: Konza Prairie, Manhattan, KS
14. Short-grass prairie: Central Plains Experimental Range, Nunn, CO
15. Tropical rain forest: Luquillo Experimental Forest, San Juan, PR
16. Arid mountains/hot desert/cold desert: Sevilleta National Wildlife Refuge, Albuquerque, NM
17. Polar marine: Palmer Station, Antarctica
18. Antarctic dry valleys: McMurdo Dry Valleys, Antarctica

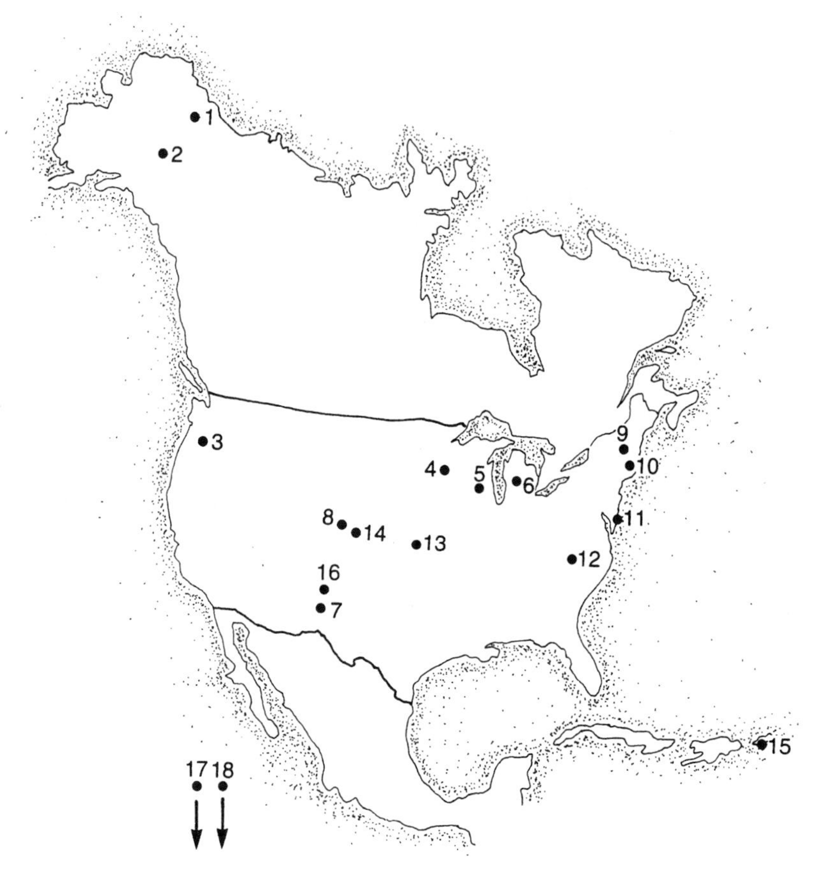


**FIGURE 20.12.** Long-term ecological research sites (see Table 20.2).

formation about the technical characteristics of specific sensors or imagery. *Inventories* describe specific data sets collected by given sensor systems (e.g., Landsat MSS or AVHRR).

## 20.10. Summary

The instruments and programs outlined in this chapter are designed to gather data that will provide the geographic coverage and consistency required to compile global data sets, and to examine fundamental questions concerning place-to-place variations in climate, vegetation patterns, and biophysical processes, and their changes over time. Although existing data have permitted humans to define the general framework for understanding these phenomena, it is only by collection of consistent data over large areas, over long intervals of time, that we will be able to develop an intimate understanding that will permit us to assess effects of human actions in the context of inherent geographic and historical variations.

## Review Questions

1. Enumerate some of the factors important in preventing, until recently, the practice of global remote sensing.
2. How might the practice of global remote sensing differ from the practice of other, more conventional, remote sensing?
3. Discuss problems in collecting and evaluating field data to support global remote sensing.
4. How might global remote sensing assist in assessing issues of global climate change?
5. In what ways might hyperspectral data assist in global change studies?
6. Global remote sensing developed initially from applications of data collected by meteorologic satellites for purposes other than their intended function. Can you identify problems (either practical or conceptual) that arise from these origins?
7. What are some of the special practical and theoretical problems inherent in the practice of global remote sensing?
8. Enumerate some of the special problems or research questions that can be addressed by global remote sensing, but not by more conventional remote sensing.
9. The question of continuity of data (i.e., maintaining a continuous and consistent data archive) is a critical issue for global change research. Discuss some of the problems this presents. For example, what is a sensible response when new, improved technology becomes available? Is it better to improve instruments or to maintain continuity of the older instruments?
10. Discuss some of the problems you might expect to encounter in maintaining an archive for global change data and in distribution of data to scientific analysts throughout the world.

## References


Assrar, Ghassem, and D. J. Dokken. 1993. *EOS Reference Handbook.* Washington, DC: NASA, 140 pp.

Crowley, Thomas J. and G. R. North. 1991. *Paleoclimatolgy.* New York: Oxford University Press, 339 pp.

Danson, F. Mark, and S. E. Plummer. 1995. *Advances in Environmental Remote Sensing.* New York: Wiley, 184 pp.

Defries, R. S., and J. R. G. Townshend. 1994. NDVI-Derived Land Cover Classifications at Global Scale. *International Journal of Remote Sensing,* Vol. 15, pp. 3567–3586.

Delwiche, C. C. 1970. The Nitrogen Cycle. *Scientific American,* Vol. 223, pp. 136–146.

Eidenshink, Jeffrey C. 1992. The 1990 Conterminous AVHRR Data Set. *Photogrammetric Engineering and Remote Sensing,* Vol. 58, pp. 809–813.

EOS Science Steering Committee. (n.d.). *From Pattern to Process: The Strategy of the Earth Observing System.* EOS Science Steering Committee Report, Vol. III. Washington, DC: NASA, 140 pp.

Foody, Giles, and P. Curran. 1994. *Environmental Remote Sensing From Global to Regional Scales.* New York: Wiley, 238 pp.

Gabrynowicz, J. I. 1995. The Global Land 1-km AVHRR Project: An Emerging Model for Earth Observations Institutions. *Photogrammetric Engineering and Remote Sensing,* Vol. 61, pp. 153–160.

Gallo, K. P., and J. C. Eidenshink. 1988. Differences in Visible and Near-IR Responses, and Derived Vegetation Indices, for the NOAA-9 and NOAA-10 AVHRRs: A Case Study. *Photogrammetric Engineering and Remote Sensing,* Vol. 54, pp. 485–490.

Goward, Samuel N., D. Dye, A. Kerber, and V. Kalb. 1987. Comparison of North and South American Biomes from AVHRR Observations. *Geocarto International,* Vol. 1, pp. 27–39.

Goward, Samuel N., C. J. Tucker, and D. Dye. 1985. North American Vegetation Patterns Observed with the NOAA-7 Advanced Very High Resolution Radiometer. *Vegetatio,* Vol. 64, pp. 3–14.

Harries, J. E., D. T. Llewellyn-Jones, C. T. Mutlow, M. J. Murray, I. J. Barton, and A. J. Prata. 1995. The ASTER Programme: Instruments, Data and Science. Chapter 1 in *TERRA 2: Understanding the Terrestrial Environment: Remote Sensing Data Systems and Networks* (P. M. Mather, ed.). New York: Wiley, pp. 19–28.

Hastings, David A., and W. J. Emery. 1992. The Advanced Very High Resolution Radiometer (AVHRR): A Brief Reference Guide. *Photogrammetric Engineering and Remote Sensing,* Vol. 58, pp. 1183–1188.

Iverson, L. R., E. A. Cook, and R. L. Graham. 1989. A Technique for Extrapolating and Validating Forest Cover across Large Regions: Calibrating AVHRR Data with TM Data. *International Journal of Remote Sensing,* Vol. 10, pp. 1085–1812.

Justice, C. O., and J. R. Townshend. 1994. Data Sets for Global Remote Sensing: Lessons Learnt. *International Journal of Remote Sensing,* Vol. 17, pp. 3621–3639.

Justice, C. O., J. R. G. Townshend, B. N. Holben, and C. J. Tucker. 1985. Analysis of the Phenology of Global Vegetation Using Meteorological Satellite Data. *International Journal of Remote Sensing,* Vol. 6, pp. 1271–1318.

Loveland, T. R., J. W. Merchant, J. F. Brown, D. O. Ohlen, B. K. Reed, P. Olson, and J. Hutchinson. 1995. Seasonal Land Cover Regions of the United States. *Annals of the Association of American Geographers,* Vol. 85, pp. 339–335 (with map at 1:7,500,000).

MacEachren, A. M., et al. 1992. Visualization. Chapter 6 in *Geography's Inner Worlds* (R. F. Abler, M. G. Marcus, and J. M. Olson, eds.). New Brunswick, NJ: Rutgers University Press, pp. 99–137.

Maiden, M. F., and S. Grieco. 1994. NASA's Pathfinder Data Set Programme: Land Surface Parameters. *International Journal of Remote Sensing,* Vol. 15, pp. 3333–3345.

Mather, John R., and G. V. Sdasyuk. 1991. *Global Change: Geographical Approaches*. Tucson: University of Arizona Press, 289 pp.

Mather, P. M. (ed.). 1995. *TERRA 2: Understanding the Terrestrial Environment: Remote Sensing Data Systems and Networks*. New York: Wiley, 236 pp.

Mounsey, Helen (ed.), and R. Tomlinson (gen. ed.). 1988. *Building Databases for Global Science.* Philadelphia: Taylor & Francis, 419 pp.

Roller, Norman E. G., and J. E. Colwell. 1986. Coarse-Resolution Satellite Data for Ecological Surveys. *BioScience,* Vol. 36, pp. 468–475.

Running, S. W. et al. 1994. Terrestrial Remote Sensing Science and Algorithms Planned for EOS/MODIS. *International Journal of Remote Sensing,* Vol. 15, pp. 3587–3620.

Skekielda, Karl-Heinz. 1988. *Satellite Monitoring of the Earth.* New York: Wiley, 326 pp.

Steinwand, D. R., J. R. Hutchinson, and J. P. Snyder. 1995. Map Projections for Global and Continental Data Sets and an Analysis of Pixel Distortion Caused by Reprojection. *Photogrammetric Engineering and Remote Sensing,* Vol. 61, pp. 1487–1497.

Strahler, A. N., and A. H. Strahler. 1973. *Environmental Geoscience: Interaction between Natural Systems and Man.* New York: Wiley, 511 pp.

Townshend, J. R. G. 1994. Global Data Sets for Land Applications From the Advanced Very High Resolution Radiometer: An Introduction. *International Journal of Remote Sensing,* Vol. 15, pp. 3319–3332.

Tucker, C. J., W. W. Newcomb, and H. E. Dregne. 1994. AVHRR Data Sets for Determination of Desert Spatial Extent. *International Journal of Remote Sensing,* Vol. 15, pp. 3547–3565.

# Calculation of Radiances

## A.1. Introduction

Analysts often need to calculate radiances from the digital values provided as image data, usually because specific analyses require radiances or because it is necessary to compare data from one scene to another. As an instructional aid, this Appendix provides data for several kinds of digital satellite data described in Chapter 6. Chavez (1988) describes the procedure for converting Landsat digital numbers to radiances, and (together with Robinove, 1982; Nelson, 1985; Price, 1987) gives calibration values for each of the Landsat sensors. Calibration data for other commonly used sensors are typically given in user handbooks or descriptive records that accompany digital data. These values, if available, should supersede values given here.

## A.2. Landsat MSS and TM Data

For Landsat MSS data, the calculation requires knowledge of the sensor's minimum brightness ($L_{min}$), maximum brightness ($L_{max}$), maximum digital number, and range of brightnesses ($L_{max} - L_{min}$). These are required to calculate radiances:

$$\text{Radiance} = \frac{DN}{D_{max}} (L_{max} - L_{min}) + L_{min} \qquad \text{(Eq. A.1)}$$

where $L$ represents radiance (milliwatts per square centimeter per steradian per micrometer), the subscripts min and max designate, respectively, the threshold and saturation values, $DN$ designates a specific digital number from the scene in a given band, and $D_{max}$ designates the maximum digital number for a given sensor. Users should check header records to confirm minimum and maximum values for specific images; typically, $D_{min}$ and $D_{max}$ are 0 and 127 for MSS data, with the exception of band 4 MSS data acquired by Landsats 1–3 before 1 February 1979 and band 4 data acquired by Landsat 4 MSS processed before 22 October 1982, which

**TABLE A.1. Sensor Calibration Data for Landsat 1 MSS**

| Sensor and band | Landsat | | |
|---|---|---|---|
| | Min. | Max. | Range |
| MSS 1 | 0.00 | 24.8 | 24.8 |
| MSS 2 | 0.00 | 20.0 | 20.0 |
| MSS 3 | 0.00 | 17.6 | 17.6 |
| MSS 4 | 0.00 | 15.3 | 15.3 |

*Note.* Values record milliwatts per square centimeter per steradian per micrometer. From Markham and Barker (1986). Dates here signify processing dates.

**TABLE A.2. Sensor Calibration Data for Landsat 2 MSS**

| Sensor and band | Before 16 Jul. 1975 | | | After 16 Jul. 1975 | | |
|---|---|---|---|---|---|---|
| | Min. | Max. | Range | Min. | Max. | Range |
| MSS 1 | 1.0 | 21.0 | 20.0 | 0.8 | 26.3 | 25.5 |
| MSS 2 | 0.7 | 15.6 | 14.9 | 0.6 | 17.6 | 17.0 |
| MSS 3 | 0.7 | 14.0 | 13.3 | 0.6 | 15.2 | 14.6 |
| MSS 4 | 0.5 | 13.8 | 13.3 | 0.4 | 13.0 | 12.6 |

*Note.* Values record milliwatts per square centimeter per steradian per micrometer. From Markham and Barker (1986). Dates here signify processing dates.

**TABLE A.3. Sensor Calibration Data for Landsat 3 MSS**

| Sensor and band | Before 16 Jul. 1978 | | | After 16 Jul. 1978 | | |
|---|---|---|---|---|---|---|
| | Min. | Max. | Range | Min. | Max. | Range |
| MSS 1 | 0.4 | 22.0 | 21.6 | 0.4 | 25.9 | 25.5 |
| MSS 2 | 0.3 | 17.5 | 17.2 | 0.3 | 17.9 | 17.6 |
| MSS 3 | 0.3 | 14.5 | 14.2 | 0.3 | 14.9 | 14.6 |
| MSS 4 | 0.1 | 14.7 | 14.6 | 0.1 | 12.8 | 12.7 |

*Note.* Values record milliwatts per square centimeter per steradian per micrometer. From Markham and Barker (1986). Dates here signify processing dates.

**TABLE A.4. Sensor Calibration Data for Landsat 4 MSS**

| Sensor and band | Before 26 Aug. 1982 | | | 26 Aug. 1982–31 Mar. 1983 | | | After 1 Apr. 1983 | | |
|---|---|---|---|---|---|---|---|---|---|
| | Min. | Max. | Range | Min. | Max. | Range | Min. | Max. | Range |
| MSS 1 | 0.2 | 25.0 | 24.8 | 0.2 | 23.0 | 22.8 | 0.4 | 23.8 | 23.4 |
| MSS 2 | 0.4 | 18.0 | 17.6 | 0.4 | 18.0 | 17.6 | 0.4 | 16.4 | 16.0 |
| MSS 3 | 0.4 | 15.0 | 14.6 | 0.4 | 13.0 | 12.6 | 0.5 | 14.2 | 13.7 |
| MSS 4 | 0.3 | 13.3 | 13.0 | 0.3 | 13.3 | 13.0 | 0.4 | 11.6 | 11.2 |

*Note.* Values record milliwatts per square centimeter per steradian per micrometer. From Markham and Barker (1986). Dates here signify acquisition dates.

TABLE A.5. Sensor Calibration Data for Landsat 5 MSS

| Sensor and band | Before 6 Apr. 1984 | | | 6 Apr. 1984–8 Nov. 1984 | | | After 9 Nov. 1984 | | |
|---|---|---|---|---|---|---|---|---|---|
| | Min. | Max. | Range | Min. | Max. | Range | Min. | Max. | Range |
| MSS 1 | 0.4 | 24.0 | 23.6 | 0.3 | 26.8 | 26.5 | 0.3 | 26.8 | 26.5 |
| MSS 2 | 0.3 | 17.0 | 16.7 | 0.3 | 17.9 | 17.6 | 0.3 | 17.9 | 17.6 |
| MSS 3 | 0.4 | 15.0 | 14.6 | 0.4 | 15.9 | 15.5 | 0.5 | 14.8 | 14.3 |
| MSS 4 | 0.2 | 12.7 | 12.5 | 0.3 | 12.3 | 12.0 | 0.3 | 12.3 | 12.0 |

*Note.* Values record milliwatts per square centimeter per steradian per micrometer. From Markham and Barker (1986). Dates here signify acquisition dates.

both used values of 0 and 63. Note that the bands here designated as MSS 1 (green), MSS 2 (red), MSS 3 (near infrared), and MSS 4 (near infrared) were previously designated as MSS 4, MSS 5, MSS 6, and MSS 7, respectively. TM data range between 0 and 255.

Values for $L_{max}$ and $L_{min}$ for MSS and TM data are tabulated in Tables A.1–A.6, as given by Markham and Barker (1986). Users should always check header records that accompany each digital scene to verify conditions under which the data were acquired and processed, and be alert for announcements that might supersede listed values. Markham and Barker (1986) and other references provide further details.

## A.3. SPOT XS Data

For SPOT data, the analyst requires values for coefficients α and β, to be used in the relationship

$$R_i = \alpha \, DN_i + \beta_i \qquad \text{(Eq. A.2)}$$

where the subscript $i$ designates separate spectral channels. Table A.7 gives initial values for SPOT 1 as reported by Price (1987), which have since been revised periodically during the satellite's service and reported in data header records.

TABLE A.6. Sensor Calibration Data for TM

| Sensor and band | Min. | Max. | Range |
|---|---|---|---|
| TM 1 | –0.15 | 15.21 | 15.36 |
| TM 2 | –0.28 | 29.68 | 29.96 |
| TM 3 | –0.12 | 20.43 | 20.55 |
| TM 4 | –0.15 | 20.62 | 20.77 |
| TM 5 | –0.037 | 2.719 | 2.756 |
| TM 6 | 0.1238 | 1.5600 | 1.4362 |
| TM 7 | –0.015 | 1.438 | 1.453 |

*Note.* Values record milliwatts per square centimeter per steradian per micrometer. From Markham and Barker (1986), for data processed after 15 January 1989. See also Chavez (1989).

**TABLE A.7. Sensor Calibration Data for SPOT 1 Sensors**

| | HRV 1 | | HRV 2 | |
|---|---|---|---|---|
| Sensor and band | α | β | α | β |
| HRV XS1 | 1.08 | 0.00 | 1.06 | 0.00 |
| HRV XS2 | 1.14 | 0.00 | 1.01 | 0.00 |
| HRV XS3 | 1.06 | 0.00 | 1.02 | 0.00 |

*Note.* Values record milliwatts per square centimeter per steradian. From Price (1987).

## References

Chavez, P. S. 1975. Atmospheric, Solar, and MTF Corrections for ERTS Digital Imagery. *Proceedings of the American Society of Photogrammetry,* Bethesda, MD, p. 69.

Chavez, P. S. 1988. An Improved Dark-Object Subtraction Technique for Atmospheric Scattering Correction of Multispectral Data. *Remote Sensing of Environment,* Vol. 24, pp. 450–479.

Chavez, P. S. 1989. Radiometric Calibration of Landsat Thematic Mapper Multispectral Images. *Photogrammetric Engineering and Remote Sensing,* Vol. 55, pp. 1285–1294.

Courtois, M., and G. Weill. 1985. The SPOT Satellite System: Monitoring Earth's Ocean, Land and Atmosphere from Space. Chapter in *Progress in Astronautics and Aeronautics,* Vol. 97 (Abraham Schnapf, ed.). New York: American Institute of Aeronautics and Astronautics, pp. 493–523.

Markham, B. L., and J. L. Barker. 1986. Landsat MSS and TM Post-calibration Dynamic Ranges, Exoatmospheric Reflectance, and At-Satellite Temperatures. *Landsat Technical Notes*, No. 1, pp. 3–8.

Nelson, R. 1985. Sensor Induced Temporal Variability of Landsat MSS Data. *Remote Sensing of Environment,* Vol. 18, pp. 35–48.

Nelson, R. 1985. Reducing MSS Scene Variability. *Photogrammetric Engineering and Remote Sensing,* Vol. 54, pp. 583–593.

Price, J. C. 1987. Calibration of Satellite Radiometers and the Comparison of Vegetation Indices. *Remote Sensing of Environment,* Vol. 21, pp. 15–27.

Robinove, C. J. 1982. Computation of Physical Values from Digital Landsat Digital Data. *Photogrammetric Engineering and Remote Sensing,* Vol. 48, pp. 781–784.

# Index

**B**

**D**



**E**

## N



## O

## Q



## R

## T

## W



## X



## Y